The Compact NASA Atlas of the Solar System

The Compact NASA Atlas of the Solar System

Ronald Greeley
Arizona State University

Raymond Batson
formerly United States Geological Survey, Flagstaff

CAMBRIDGE
UNIVERSITY PRESS

PUBLISHED BY THE PRESS SYNDICATE OF THE UNIVERSITY OF CAMBRIDGE
The Pitt Building, Trumpington Street, Cambridge, United Kingdom

CAMBRIDGE UNIVERSITY PRESS
The Edinburgh Building, Cambridge CB2 2RU, UK
40 West 20th Street, New York, NY 10011-4211, USA
10 Stamford Road, Oakleigh, VIC 3166, Australia
Ruiz de Alarcón 13, 28014 Madrid, Spain
Dock House, The Waterfront, Cape Town 8001, South Africa
http://www.cambridge.org

Derived from The NASA Atlas of the Solar System

First published 1997

First published 2001

Printed in Belgium at 

Typeface TimesNewRoman *System* QuarkXPress®

A catalogue record for this book is available from the British Library

ISBN 0 521 80633X hardback

Table of Contents

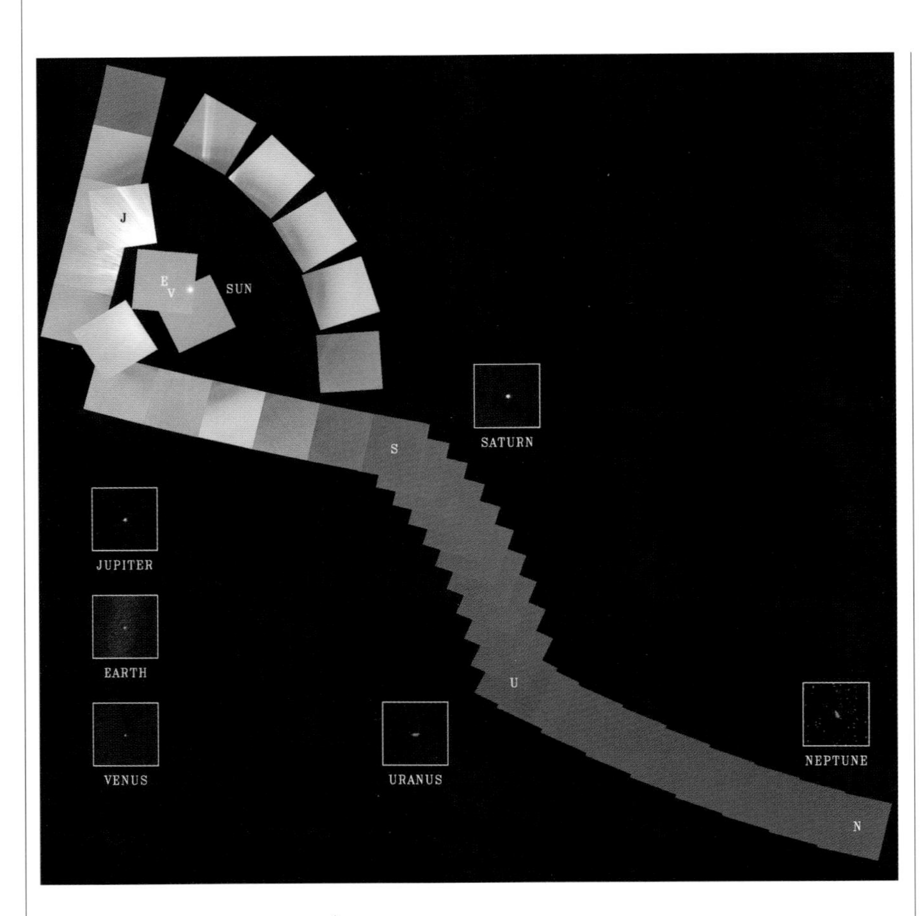

After a journey of more than 12 years through the Solar System, *Voyager 1* pointed its camera system back toward the Sun and took a series of pictures forming this first-ever portrait of the Solar System seen from the outside; 60 individual images were used to compile this composite mosaic (JPL photograph P-36087).

Credits

National Aeronautics and Space Administration
- Editing
  - Elsie Diven Weigel
  - Howard S. Golden
- Resource Administrators
  - Lennard A. Fisk, Ph.D.
  - Wesley T. Huntress, Jr., Ph.D.
- Scientific and Technical Monitor
  - Joseph M. Boyce
  - Ted H. Maxwell
- Contracting Officer
  - John E. Moore
- Printing Officers
  - Paul Willis
  - Anthony Armstrong

U.S. Geological Survey
- Project Coordinator
  - Haig F. Morgan
  - assisted by Christopher E. Isbell, D. Anthony Manley, and Robert M. Sucharski
- Geologic Map Editors
  - Richard C. Kozak
  - Nancy R. Isbell
- Nomenclature and Gazetteer Editors
  - Jennifer S. Blue
  - Derrick D. Hirsch
  - Joel F. Russell
  - Mary E. Strobell
- Airbrush Cartographers
  - Eric L. Blinn
  - Patricia M. Bridges
  - Susan L. Davis
  - Patricia Hagerty Gray
  - Jay L. Inge
  - Barbara J. Peacock
- Illustrations
  - Ralph O. Aeschliman, Jr.
- Photomechanical and Cartographic Support
  - Hugh F. Thomas
  - Roger D. Carroll

Geologic Map Contributors
- Alexander T. Basilevsky (Venus)
- Kelley Bender (Callisto)
- David Crown (Io)
- Stephen Croft (Uranian satellites and Triton)
- Ronald Greeley (Io and Callisto)
- Richard Kozak (Europa and Venus)
- Roger L. Larson (Earth)
- Baerbel Lucchitta (Ganymede)
- Jeffrey M. Moore (Rhea and Tethys)
- Quinn R. Passey (Enceladus)
- Jeffrey Plescia (Dione)
- Gerald Schaber (Venus)
- Paul Spudis (Mercury)
- Kenneth L. Tanaka (Mars)
- Don Wilhelms (Moon)

Scientific Reviewers
- Raymond Arvidson, Ph.D., Washington University
- Michael Belton, Ph.D., National Optical Astronomy Observatories
- Geoffrey Briggs, Ph.D., NASA Ames Research Center
- Michael Carr, Ph.D., U.S. Geological Survey
- Stephen Croft, Ph.D., University of Arizona
- E. Julius Dasch, Ph.D., NASA Headquarters Education Division
- Bruce Jakosky, Ph.D., University of Colorado
- Michael Malin, Ph.D., Malin Space Science Systems
- Ted Maxwell, Ph.D., Smithsonian Air and Space Museum
- Jeffrey Moore, Ph.D., NASA Ames Research Center
- David Morrison, Ph.D., NASA Ames Research Center
- Gerald Schaber, Ph.D., U.S. Geological Survey
- Paul Spudis, Ph.D., Lunar and Planetary Institute
- Robert Strom, Ph.D., University of Arizona

Arizona State University
- Photographic Support
  - Tim Askins
  - Daniel Ball
  - Stephen Meszaros
  - Josh Oliver
- Computerized Image Processing Support
  - Larry Bolef
  - Edisanter Lo
- Word Processing
  - Shana Blixt
  - Byrnece Erwin
  - Maureen Geringer
  - Stephanie Holaday
  - Susan Oliver
- Index Preparation
  - Cynthia R. Greeley
- Administrative Support
  - Evelyn Anderson
  - Maa-Ling Chen
  - Kevin Gorman
  - Carmen LaBelle
  - Robert Lange
- Regional Planetary Image Facility
  - Charles Hewett

Graphic Design and Production
- Designer
  - Kevin R. Osborn
- Project Coordinator
  - Charles O. Hyman
- Map Nomenclature Production
  - Philip Zimmermann

Illustration Contributors
- Raymond Arvidson
- Rita Beebe
- Don Campbell
- B. R. Frieden
- Howard S. Friedman
- Donald Gault
- Matt Golombek
- James D. Griggs
- Robert Haight
- Torrence Johnson
- Mark Lemmon
- Paul Lowman
- Alfred McEwen
- William McKinnon
- Davis Meltzer
- Stephen Meszaros
- Donald Michels
- Jeffrey Moore
- David Morrison
- David Muench
- Gerhard Neukum
- Mark Parmentier
- Carle Pieters
- Harold Reitsema
- Roger Ressmeyer
- Stephen J. Reynolds
- Nathan Seaver
- Mark Schoeberl
- Peter Schultz
- David Smith
- Peter Smith
- Jurie Van der Woude
- Robert Wallace
- Don Wilhelms
- David Williams
- Veronica Zabala
- Maria Zuber

Diagrams (unless noted in caption)

Foreword

Maps have always figured prominently in the exploration of new lands. Some of these have been conceived as works of art, and many maps and charts have acquired that character through the devotion exercised in their creation. Not surprisingly, the progress of civilization is reflected to some degree in the creation and publication of maps and charts of ever increasing scope and accuracy.

One of the most spectacular products of our everwidening exploration of space has been the maps that have resulted from the deep space missions of the United States and the former Soviet Union. Over the last three decades the creative talents of thousands of engineers, scientists and support personnel have allowed us to land on, orbit or make close encounters with all of the planets in the solar system, and many of their moons, save Pluto.

Early in this golden age of planetary exploration, it was recognized that maps would be critical for planning purposes and for providing a framework for reporting scientific results. Consequently, in the last quarter century, thousands of planetary maps have been produced to record scientific findings and exploration results.

The U.S. Geological Survey has been in a close collaboration with NASA since the beginning, and the cartographic products in this atlas were all created by the Astrogeology branch of that Agency. Mapping planets poses special problems not faced in making maps of Earth. What, for example, does one use as "sea level" for reference to elevations when only Earth has oceans? In addition, nearly all planetary maps must be made without benefit of ground control. These and other issues have been resolved, but only after many years of perfecting the techniques required for making planetary maps.

This Atlas is a reflection of the growing maturity of planetary science and of the techniques of planetary cartography. For the first time, it draws together systematic maps. Photographs, and overviews of the planets and major satellites, representing more than a quarter century of Solar System exploration.

Wesley T. Huntress, Jr
Associate Administrator
for Space Science
National Aeronautics
and Space Administration

Preface

Even though only the first tentative steps have been taken beyond the boundaries of planet Earth, in the past three decades the preliminary exploration of the Solar System by automated spacecraft has become a reality. Space probes have been sent throughout the Solar System to all major planets except Pluto and have returned a tremendous wealth of information. From the data collected it has been possible to piece together the first maps of the solid-surface planets and most of their satellites. The maps presented in this atlas are based only on this available data. Where data were not collected, maps have not been projected. More than a thousand maps and charts have been prepared and published for the Moon, Mars, and other planetary objects. Many of these maps are now rare, out of print, or otherwise difficult to obtain. Consequently, the goal of this atlas is to provide a set of maps of uniform format and consistent scales, organized by planetary system for all of the objects seen thus far.

It should be noted that the atlas is not intended to be a textbook on the Solar System; such an effort could easily expand to encyclopedic proportions. Rather, text and illustrations are provided to give the reader sufficient background to place the maps in an overall context of the Solar System and to gain some insight into the nature of the mapped planetary bodies. Although we have aimed for consistency throughout the atlas, this goal could not always be achieved because the level of knowledge about the mapped bodies is quite variable.

Producing the atlas has been a tremendous effort that has involved the contributions of a great many people. First, Geoffrey A. Briggs (NASA Ames Research Center) and Barbara Beatty (formerly, University of Arizona) are acknowledged as conceiving the atlas. Lennard A. Fisk and Wesley T. Huntress, Jr. (both of NASA) are thanked for providing support for all phases of atlas production; we appreciate the perseverance of Joseph M. Boyce and Howard S. Golden (both of NASA) needed to bring the atlas to fruition.

We acknowledge with gratitude Haig F. Morgan (U.S. Geological Survey) for preparing the maps in formats suitable to the atlas; Jurie Van der Woude (Jet Propulsion Laboratory) for providing photographs; Stephen Meszaros (Bureau of Land Management); Ralph O. Aeschliman, Jr. (U.S. Geological Survey) and Daniel Ball (Arizona State University) for aiding in preparation of the illustrations and photographs; Richard Kozak and Nancy Isbell (U.S. Geological Survey) for synthesizing the geologic maps; Jennifer S. Blue, Derrick D. Hirsch, Joel Russell, Mary E. Strobell (U.S. Geological Survey) and the late Harold Masursky for the gazetteer and nomenclature; Raymond Arvidson (Washington University), Michael Belton (National Optical Astronomy Observatories), Michael Carr (U.S. Geological Survey), Stephen Croft (University of Arizona), E. Julius Dasch (NASA Headquarters Education Division), Bruce Jakosky (University of Colorado), Jim Klemaszewski (Arizona State University), Michael Malin (Malin Space Science Systems), Ted Maxwell (Smithsonian Air and Space Museum), Jeffrey Moore (NASA Ames Research Center), David Morrison (NASA Ames Research Center), Gerald Schaber (U.S. Geological Survey), Paul Spudis (Lunar and Planetary Institute), Robert Strom (University of Arizona), and David Williams (Arizona State University), for reviewing sections of the text; Ted Maxwell (Smithsonian Air and Space Museum) and Geoffrey Briggs (NASA Ames Research Center) for reviewing the entire text; Elsie DivenWeigel and Cynthia Greeley for general editing; and Susan Oliver, Maureen Geringer, Shana Blixt and Byrnece Erwin for word processing. Kevin R. Osborn and Charles O. Hyman are thanked for their superb book design and for assisting in the production stages.

Whether you are a planetary scientist or a person interested in Solar System exploration on a more casual basis, we hope that you find this atlas useful.

Ronald Greeley and Raymond Batson

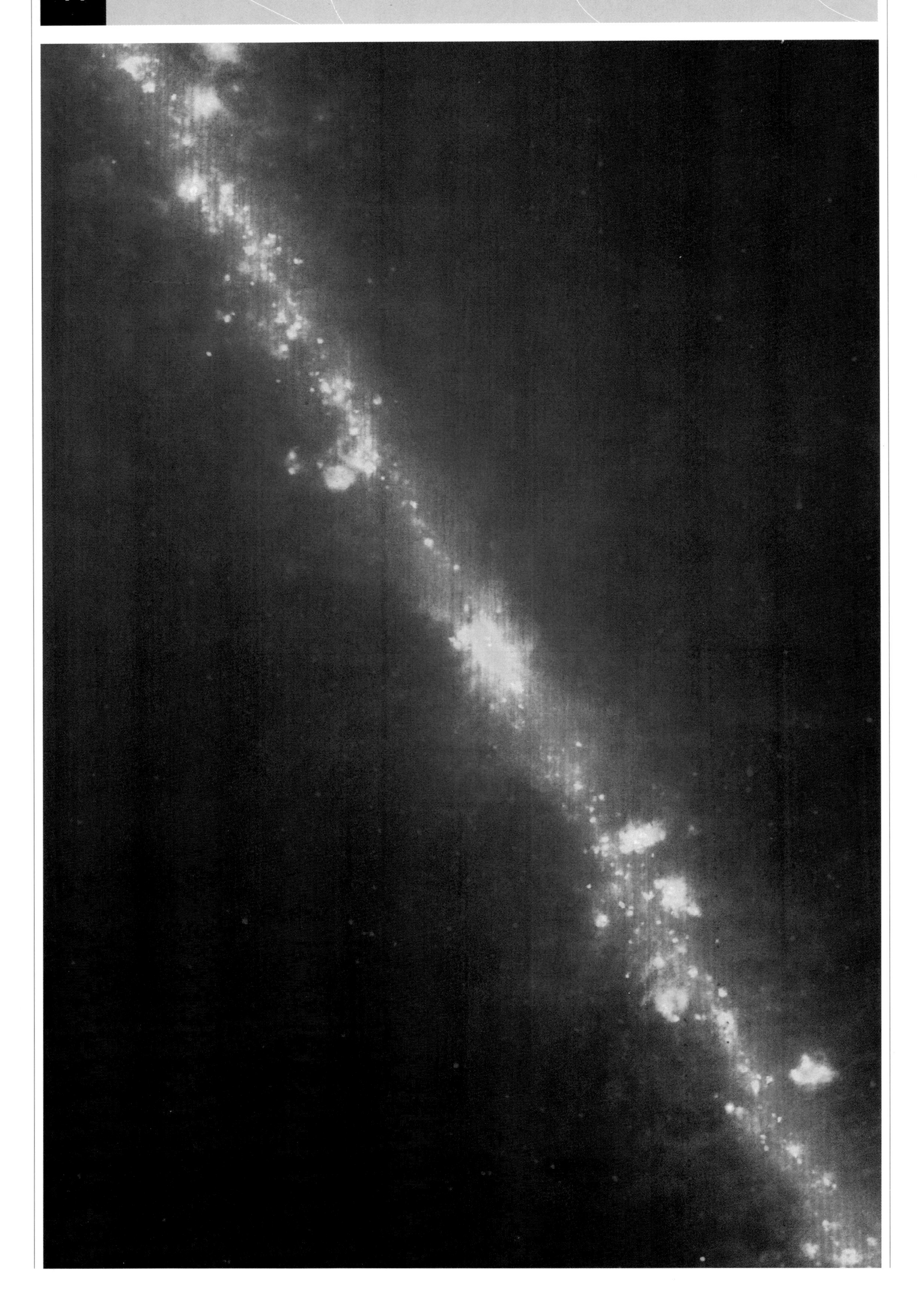

Introduction

The first maps were probably made by early hunters long before the dawn of civilization. Most likely, the maps were rough sketches drawn in the dirt with a stick. Through time, maps became progressively more important for establishing political boundaries, developing resources, conducting military operations, and charting the exploration of new areas. Today, maps in one form or another are used by nearly all inhabitants of Earth. Map-making plays such an important role in a nation's well-being that most countries devote substantial budgets to producing maps.

At the time of Columbus's voyages of exploration to the Western Hemisphere 500 years ago, the position of the continents and their shapes were largely unknown. In some respects, the positions of the stars and other objects in the nighttime sky were better known than surface features on Earth. From observations of the heavens, it had long been recognized that some celestial objects follow unusual paths. The Greeks applied the term planet – meaning wanderer – to such objects to distinguish them from stars. Despite the recognition that planets were different, the Earth-centered view of the universe was common in Columbus's time and prevailed for many decades.

Mapping the surfaces of Solar System objects began in the astronomical community. Beginning with Galileo's observations in the seventeenth century, charts of the Moon became more sophisticated with improvements in the telescope. Observations were also made of the other objects in the Solar System, and sketch maps were produced for Mars, Venus, and Mercury. But because of limitations in telescopes, these maps were simple and were often better portrayals of the imagination of the observer than of features actually existing on the planets.

Sputnik 1 initiated the Space Age in 1957, and the race to the Moon was on. Maps of the Moon were urgently needed for lunar exploration by the United States and the Soviet Union. Some of the most detailed maps using Earth-based telescopic observations were produced by the United States Air Force's Aeronautical Chart and Information Center and the Army Map Service in preparation for automated and eventual manned landings on the Moon. Although these maps were excellent for general planning, it was clear that they were inadequate for selecting landing sites and making the detailed engineering and scientific studies that would accompany such landings. Consequently, both the United States and the Soviet Union undertook a series of unmanned space probes to the Moon primarily to obtain high quality photographs and other data to enable the preparation of detailed maps. At this stage of exploration, cartographers began to develop the methods that would become the standards for portraying planetary surfaces.

Geologists E. M. Shoemaker and H. G. Stevens using field methods to recover positions from Major John Wesley Powell's historic survey of the Colorado River and the Grand Canyon (U.S. Geological Survey photograph).

Facing page: Image of the center of our galaxy, the Milky Way, taken by the Infrared Astronomical Telescope. Yellow and green areas along the band are clouds of interstellar gas and dust that have been heated by nearby stars. The warmest material is shown as blue, while the coldest material is shown as red (NASA photograph P-26295).

Planetary versus Terrestrial Maps

Making maps of extraterrestrial planets and satellites – termed planetary cartography – poses special problems not faced in terrestrial mapmaking. On Earth, for example, mountain heights and other elevations are referenced to sea level, but Earth is the only planet that has oceans. What does one use for the reference, or datum, on other planets? In addition, maps of Earth are tied to precise control points that have been surveyed on the ground. This precision is not feasible with current data for extraterrestrial objects. Because of these and other differences, special techniques have been developed for planetary cartography.

Planetary maps consist almost entirely of representations of landforms, whereas maps of Earth are often dominated by political boundaries and cultural features (such as roads). The mapping of Earth as a globe has involved surveying and mapping small areas and assembling the resulting maps into montages covering larger areas. Until the Space Age, this was the only avenue available to Earth-bound explorers because they could not get far enough above the planet for a global view. Planetary explorers, on the other hand, have the benefit of an initial global perspective and have progressed through regional to local vantage points only after many complex space-flights. As of 2001, only a few extraterrestrial localities, such as the *Apollo* landing sites on the Moon and the landing sites on Mars, have been mapped in detail.

One of the antennas at Goldstone, California that is part of the Deep Space Network, used to receive data from spacecraft (JPL photograph 332-10415).

The primary data for making maps of planetary surfaces are photographs and electronically produced images taken from spacecraft. The first stages in the exploration of a planet typically involve flyby spaceflights. On these one-way journeys, a spacecraft takes pictures and records other data as it passes a planet or planetary system. The *Voyager* missions to Jupiter, Saturn, Uranus, and Neptune involved flybys in which careful planning allowed extensive data to be returned from these planets and their satellites and ring systems.

Because of the importance of pictures in planetary exploration and mapping, most spacecraft in flyby missions have carried instruments capable of obtaining high-resolution images. These instruments use telescopic lenses that produce images suitable for planimetric mapping, but the images are marginal for deriving elevations for topographic mapping. Moreover, most flyby missions observe only part of a planet and cannot provide global coverage. Only in the later stages of exploration are spacecraft put into orbits from which images can be gathered systematically for global or near-global coverage. Although orbiting spacecraft are far better platforms for obtaining map data than flyby missions, the orbits typically are elliptical and result in images that have inconsistent resolution and illumination. These and other complexities require improvisations, compromises, and techniques not normally required by makers of terrestrial maps.

Before mapping can begin, control points must be defined. On Earth, these points are typically bench marks that have been located precisely by surveying on the ground. On extraterrestrial surfaces, control points consist of well-defined latitudes and longitudes for distinctive surface features seen on images. The point locations are compiled by combining astronomical measurements of the positions and orientations of the planets, spacecraft position, camera orientation angles, and geometric analyses of spacecraft images. Synthesis of these measurements, termed photogrammetry, is the final step in producing accurate control points and map coordinates for planetary surface features. On the Earth, photogrammetry uses photographs taken with wide-angle cameras to make precise measurements of features on the ground. Conventional cameras designed for photogrammetric work are too heavy to be used in planetary exploration and require film that must be returned to Earth for analysis. Although similar, but modified, cameras were taken to the Moon on some United States and Soviet missions, most planetary exploration involves electronic cameras that do not use film. Early electronic cameras on spacecraft were vidicon tube systems (similar to television cameras) that are rather primitive compared with the solid-state electronic cameras used today, such as the *Mars Global Surveyor* camera.

Mapping the planets with electronic cameras has many advantages. Because film is not required, there is no need for the spacecraft to return to Earth; thus, the cost, complexity, and time needed for receipt of the data are greatly reduced. Electronic images are returned to Earth at the speed of light. For example, images from Saturn were received 90 minutes after they were taken, whereas at least 2 years would have been needed to recover them by the return of the spacecraft to Earth. In addition, electronic cameras can record surface brightness values more precisely and over a much wider range than film. Hence, subtle shadings can be discriminated that cannot be recorded on film. Finally, the number of pictures taken from a spacecraft is limited only by available power rather than by the supply of film, and it is not unusual for electronic cameras on spacecraft to function for many years and to return tens of thousands of pictures.

Once the appropriate spacecraft data are obtained, making planetary maps involves computer processing of electronic images, assembling the images into photomosaics, and, in some cases, generating charts of the planetary surface using an airbrush technique. Because no single version of a map can serve all purposes, maps are commonly produced in several versions to emphasize landforms, surface markings, albedo, or other planetary characteristics.

Map of the Moon made in 1680 by the Italian astronomer, Gian Domenico Cassini.

Image Processing

An electronic picture, more commonly called a digital image, is a checkerboard array of picture elements, or pixels. The gray level of each pixel is given a density number, or DN, which typically ranges from 0 (black) to 255 (white), yielding 256 shades of gray, including black and white. Digital image processing involves mathematical manipulations to obtain the best possible pictures from unprocessed data. These manipulations were originally developed by space scientists at the NASA Jet Propulsion Laboratory in California to enhance images of the Moon taken in 1967 by the *Ranger* spacecraft. It was not long before the techniques were refined to enhance medical analysis of electronic X-Ray and other forms of electronic imaging, for study of images of the Earth, until the present time when digital image processing and computer aided design virtually dominate the graphic arts.

Photomosaics

Large surface areas seldom can be mapped precisely with single images because global images do not show enough detail. Large areas are mapped by mosaicking many pictures into a collage, or photomosaic. Mosaicking requires sophisticated image processing to correct for the curvature of the planet or satellite being mapped, the complexities of orbits, the illumination from the Sun, and the viewing geometry.

Photomosaics differ from true maps because photographs record complicated light and dark patterns related to both lighting and the colors on a surface. In addition, photographs are distorted by the camera system and project three-dimensional terrain, such as mountains, onto two-dimensional surfaces. Nevertheless, mosaics of photographs nearly always serve as the starting point for more sophisticated mapping of planetary surfaces.

Digital Maps

The word map calls to mind a piece of paper with various symbols and notations indicating

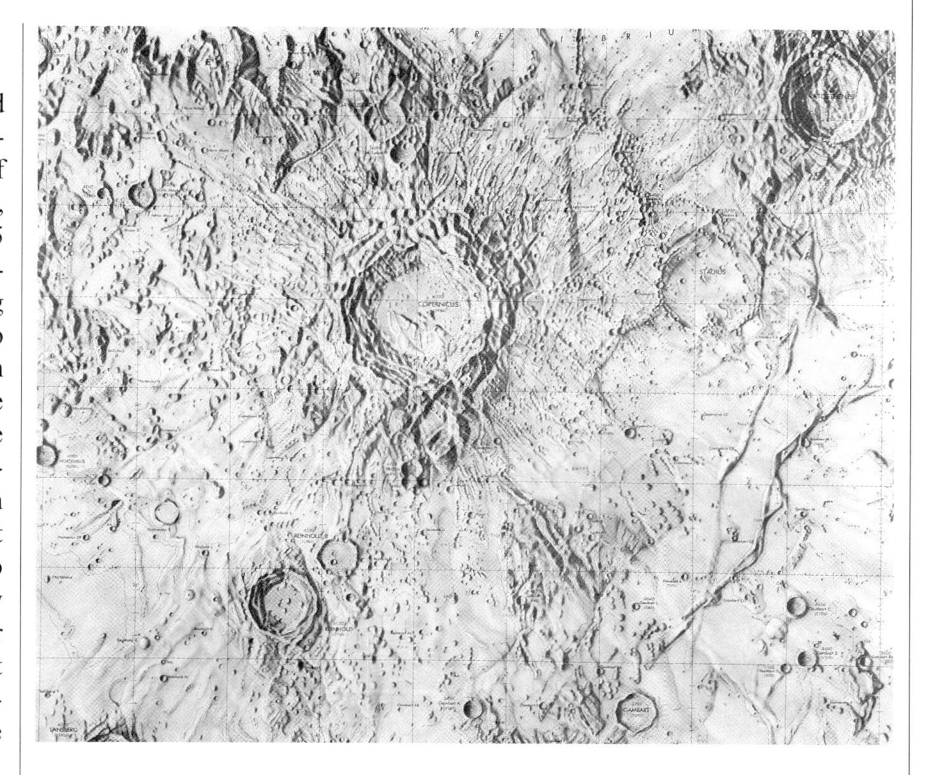

U.S. Air Force, Aeronautical Chart and Information Center map of the Copernicus region of the Moon, published in 1964.

features such as mountains and roads. This traditional kind of map is designed around the limitations of the human eye, which can perceive only about 16 shades of gray along with the visible part (red to violet) of the color spectrum. In addition to distinguishing 256 shades of gray, some spacecraft imaging systems can discriminate more than 200 separate colors. Digital maps produced from these imaging systems can preserve all of this information in computer storage, so that customized maps can be made for specific purposes. Digital maps contain topographic, geophysical, geochemical, and geologic data, as well as photographic mosaics in a form that allows easy comparisons.

Airbrush Maps

Cartographers have long experimented with ways to show three-dimensional landforms, such as mountains, on paper. Although mosaics provide photographic views of planetary surfaces, generally they do not give a clear impression of the three-dimensional terrain. This is because individual pictures in the mosaic were acquired at different times of day, producing different shadow patterns.

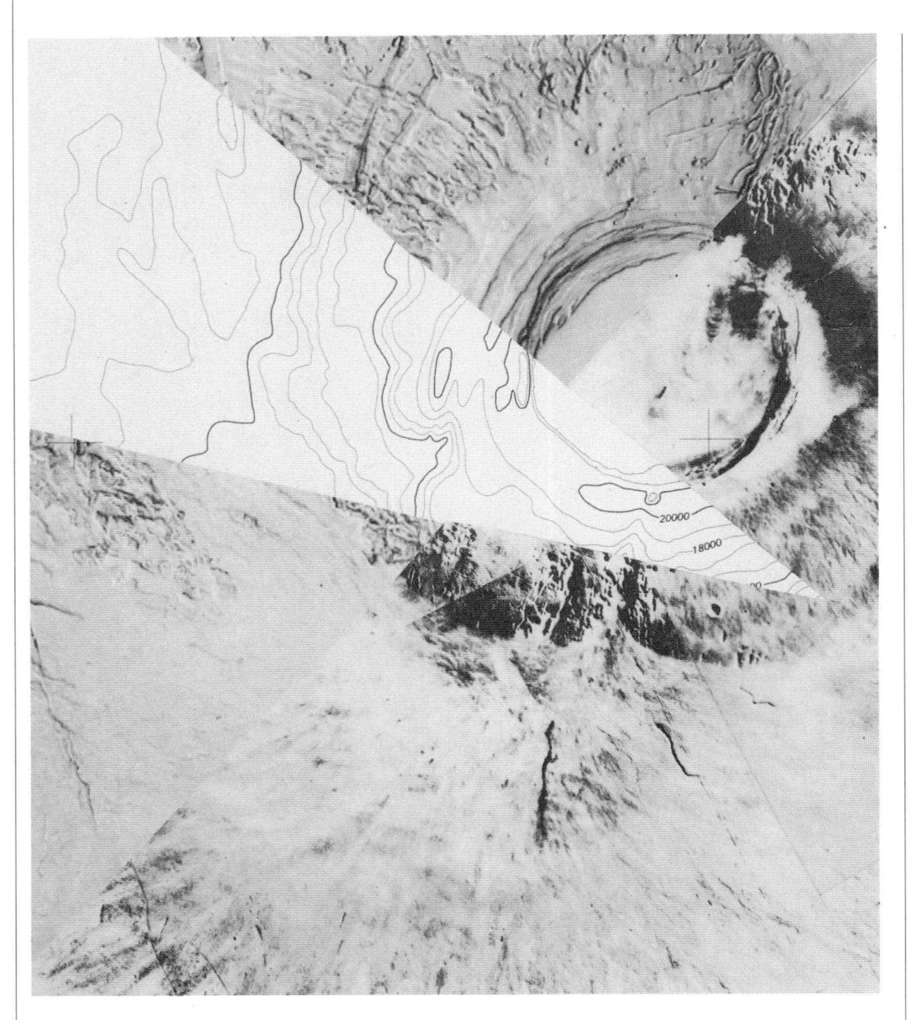

The Arsia Mons caldera (125 kilometers in diameter) on Mars and some of its surroundings on three kinds of maps: airbrushed shaded relief (northwest half), photomosaic (southeast half), and a contour map (interval = 1 km) of parts of both areas (U.S. Geological Survey)

In addition, terrain shading is often masked by coloration and mottling by surface deposits such as dust. To solve these problems, many planetary maps were commonly produced as shaded relief ("hill-shaded") maps. Specially trained cartographers produce shaded relief maps by making a composite painting of the surface using a tiny spray gun, or airbrush. The airbrush technique requires artistic talent, skill in the use of the airbrush, and, most importantly, an ability to visualize landforms by examining many different photographs. After developing a mental image based on this examination, airbrush cartographers can produce that image on the map so that it is visible to others.

Airbrush mapping was originally devised to extract information from rather limited data on the Moon. Prior to lunar spaceflights, airbrushed maps of the Moon were derived from photographs and direct observation through telescopes. Telescopic observations allowed cartographers to see more detail than could be photographed and to observe features that appeared for only an instant under the best of viewing conditions.

After the lunar mapping program, airbrush cartography was applied to Mars, Mercury, Venus, and the outer-planet satellites, although direct telescopic observations are no longer used for making maps. Airbrush cartography has resulted in maps that display a level of detail that cannot be produced from either photographs or digital cartography alone.

Topographic Maps

Of the various methods of portraying terrain, the most precise are topographic maps that show contour lines connecting points of equal elevation. To make topographic maps of Earth, stereoscopic aerial photographs are taken from aircraft with distortion-free, wide-angle, film cameras. The photographs are taken at uniform separation; each image overlaps its predecessor by about 60 percent. When the overlapping images are viewed together stereoscopically, a strong three-dimensional effect is perceived. Stereoplotting instruments use this effect to produce an optical, three-dimensional image model. An operator viewing the image model is able to move an index mark within the model to trace contour lines directly onto maps or into computer-based files.

Vertical measurement accurate to 1/10,000 of the aircraft altitude is common in making topographic maps of Earth from stereoscopic photographs. For example, pictures taken from an aircraft flying at an altitude of 10,000 meters can be used to measure the relative heights of features such as hills to within one meter accuracy. Although this technology is available to planetary scientists, the cameras used for Earth cannot be carried easily on planetary spacecraft. By necessity, the technique has been modified extensively for use with electronic spacecraft cameras. Although the results are not as accurate as those of conventional systems used on Earth, useful topo-

Diagram of a global digital image map. The top layer in this illustration is a geologic map of Mars, the middle layer is a color-coded map of elevations, and the bottom layer is an image map (U.S. Geological Survey).

graphic maps have already been produced for most of Mars.

In addition to stereophotography, other techniques can be used to make topographic maps. For example, some spacecraft instruments can provide direct measurements of distances between orbiting spacecraft and the surface of the planet. By combining these measurements with precise tracking of the spacecraft, a three-dimensional model of the surface can be derived to produce topographic maps. Topographic maps have been made for Venus using radar-ranging from the United States *Pioneer-Venus* and *Magellan* spacecraft and the Soviet *Venera 15/16* spacecraft. The technique was also used with a laser-ranging device on the *Mars Global Surveyor* mission to Mars.

Mapping Small Satellites and Asteroids

Maps of spheroidal planets and satellites are familiar and well understood. As projections of spheres onto sheets of paper, they have mathematical properties that enable areas and distances to be measured. However, most observed moons smaller than 100 kilometers across, such as the martian moons, are not spherical but are very irregular, "potato-shaped" objects. The same is true of the asteroids that have been seen by spacecraft. Presumably, the shapes of these moons and asteroids are typical of countless other small bodies that exist in the Solar System but have yet to be seen. Map projections commonly used for spherical objects are difficult to use on irregular-shaped bodies. Rather, analyses

Artist's rendition of the launch of the *Magellan* spacecraft from *Shuttle* in May 1989. *Magellan* spent 15 months in cruise, arriving at Venus in August 1990 to begin radar mapping (Photograph courtesy of Martin Marietta Corporation).

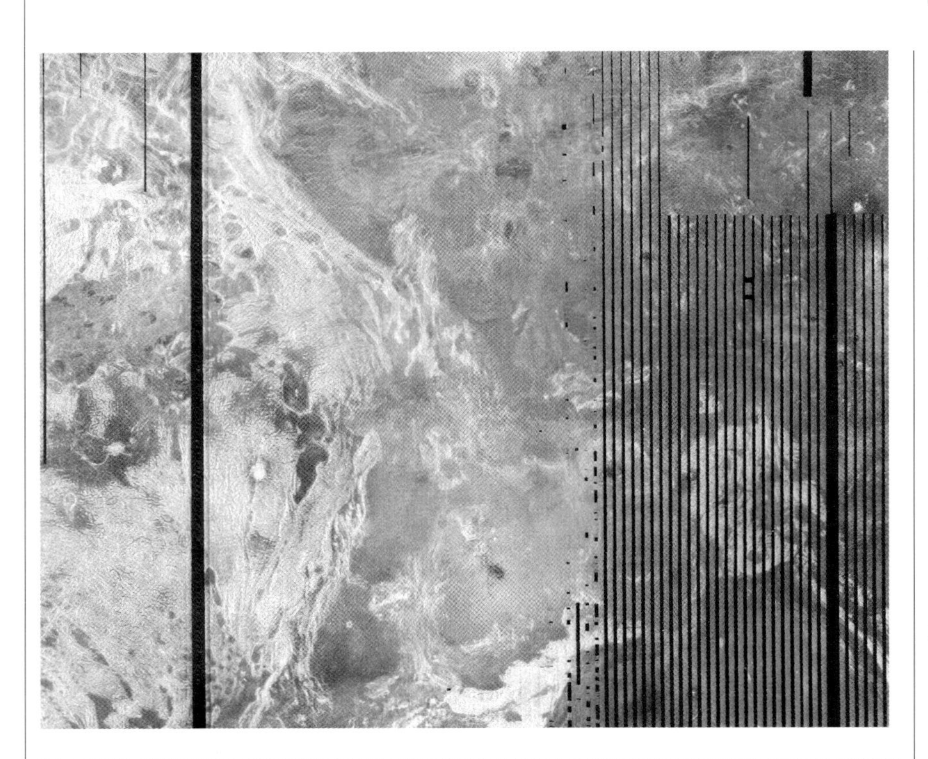

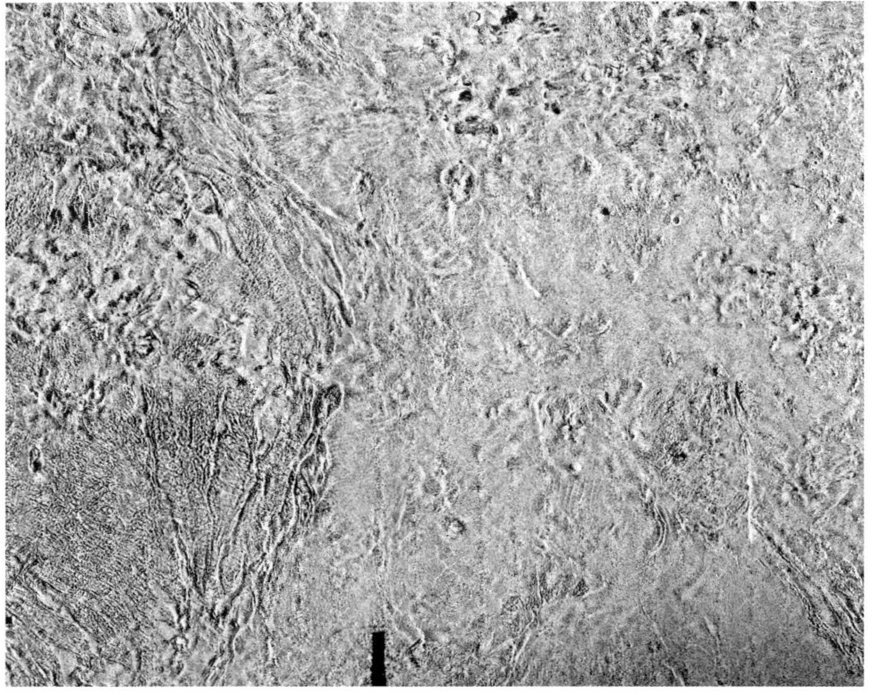

Above and facing page: Maps of part of the Tellus Tessera area of Venus. All have the same scale and cover the same area.

Top: Mosaic of radar images acquired by *Magellan* with black stripes showing the width (20 to 25 kilometers) of the image strips. Mosaic images emphasize radar-reflection intensity and cover an area 2,400 kilometers by 3,100 kilometers, nearly the size of the conterminous United States.

Bottom: Mosaic of radar images acquired by *Venera 15/16* with lower resolution than those of *Magellan* but providing complementary information about surface properties.

of their dimensions and volumes and study of the spatial distributions of materials on their surfaces are best done with computerized digital maps.

Digital maps of irregular objects involve combining a sinusoidal equal-area array of radii (lengths from the geometric center of the body to the surface) with spacecraft images. The maps are made by a process similar to that of projecting pictures onto a physical model of a satellite or asteroid, resulting in a collection of image intensities, latitudes, longitudes, and radii. The initial model upon which the images are projected is derived from photogrammetric measurements of surface features that serve as control points. Once the digital map has been made, it can be processed by computers to make image maps such as those in this atlas.

Names on Planets and Satellites

Geographic place names are critical for communication, and people have been naming features for as many years as language has existed. Astronomers began to name features on the Moon and Mars soon after the invention of the telescope. Several competing schemes of nomenclature were developed, and by the early 1900s, many of the conspicuous features on the near side of the Moon were known by three or more different names. In an attempt to standardize lunar names, Great Britain's Royal Astronomical Society and the International Association of Academies appointed a committee in 1907 to collate the existing nomenclatures. When the International Astronomical Union was formed in 1919, a subcommittee was appointed to deal with martian and lunar nomenclature.

With Space Age exploration of the Moon and planets, the compelling need for an unambiguous system of names and procedures for applying them became obvious. Competition was intense between organizations and nations for use of their explorers' prerogative to name their discoveries. However, rather than allow the system to descend into chaos, scientists of many nations followed the procedures of the International Astronomical

Union, and that body remains responsible for naming planetary features.

The United States *Mariner 6* and *7* missions to Mars returned photographs of previously unknown landforms and details of the surface markings observed by astronomers. The "canali" noted by astronomer G. V. Schiaparelli in the late nineteenth century were not seen on Mars. As the focus of scientific attention shifted to newly discovered martian landscapes, it became clear that a new system of nomenclature would be required. The International Astronomical Union decided to combine the albedo names used by classical astronomers for martian features with Latin terms that had been used to describe lunar topographic features.

As the basis for the new nomenclature, the International Astronomical Union selected a Mars map by E. M. Antoniadi, which was published in 1929 and was based on Schiaparelli's map made in the late 1800s. Thus, when a place name was needed for a newly discovered feature on Mars, the name of the nearest albedo feature on the Antoniadi map was combined with a descriptive term. For example, Nix Olympica ("Snows of Olympus") was the name applied to a distinctive feature seen telescopically. When spacecraft images revealed the feature to be a plume-cloud associated with an enormous mountain, the name Olympus Mons (the Latin for mountain) was applied to the mountain.

The International Astronomical Union devised similar systems for naming features on other bodies. Basic rules include exclusion of names of political, religious, and living people. Themes have been defined for many planets and satellites. For example, most features on Venus have names of feminine derivation. Some categories of features also carry themes; martian valleys are given names that mean "Mars" in various languages.

Early in planetary exploration, several years were needed to obtain official names for newly discovered features. To expedite the process, suitable official names are now collected in "name banks," to be applied to as-yet-undiscovered features on specified planets as new information becomes available and as new maps are produced.

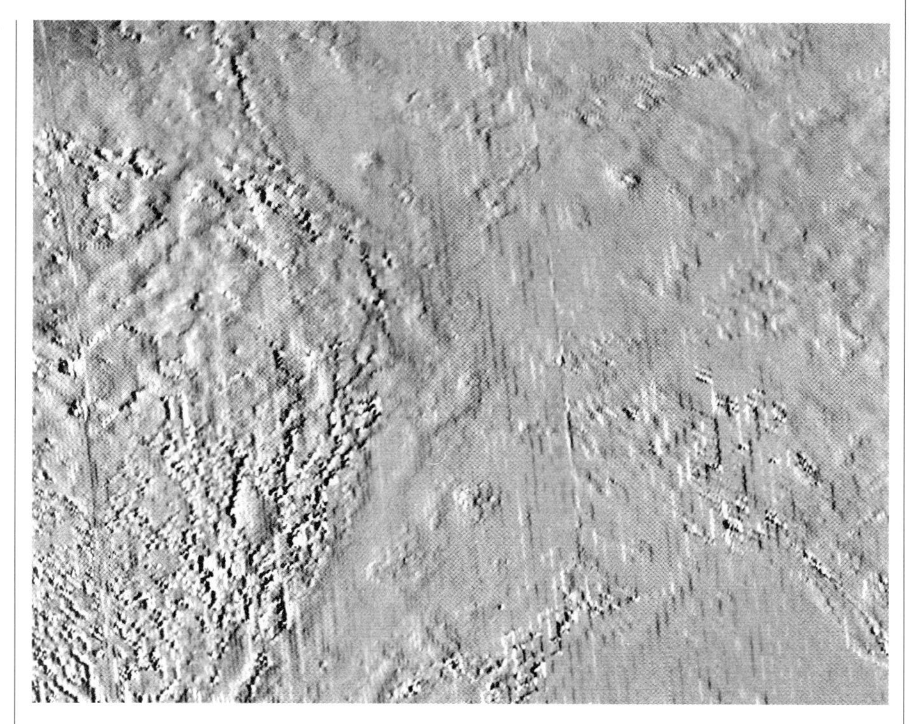

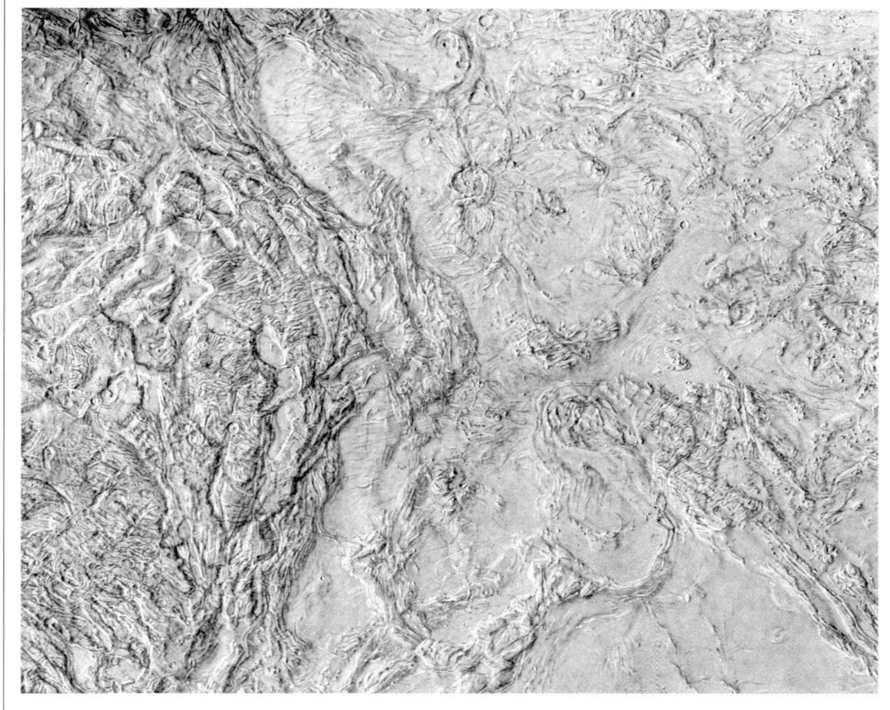

Top: Shaded relief image made from *Magellan* radar altimetry soundings showing the relationships between surface form and reflectivity.

Bottom: Hand-drawn shaded relief map by a cartographer using airbrush and data interpretation (U.S. Geological Survey).

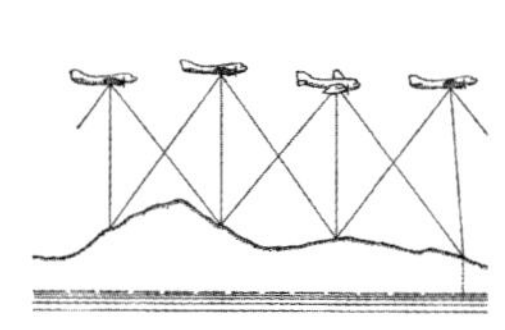

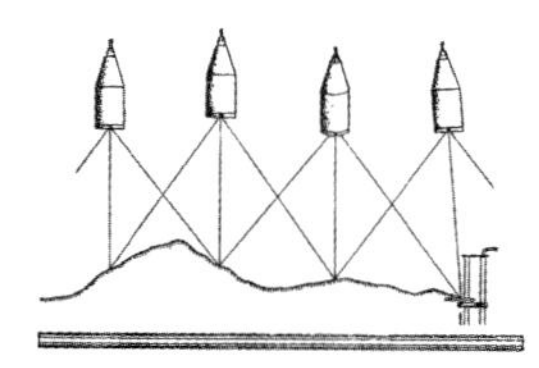

Geometric parallelism between aerialphotography and multiplex reconstruction.

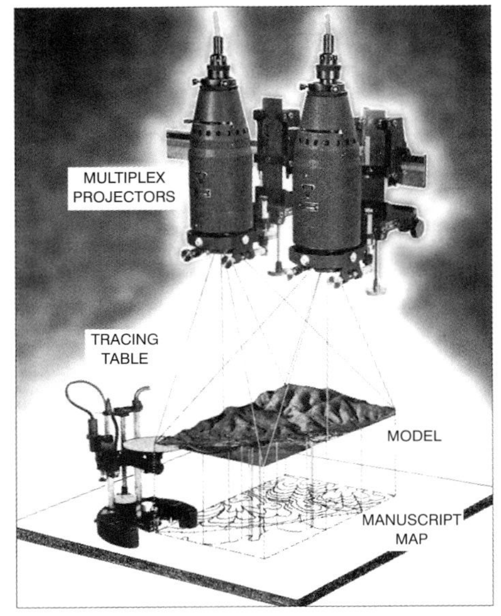

Diagramatic sketch of multiplex model.

Making a map with a 1930s stereoplotter. A miniature, three-dimensional optical model can be created and measured precisely by projecting aerial photographs from the relative positions occupied by the camera when the pictures were taken (©American Society of Photogrammetry and Remote Sensing).

The ancient art and science of mapmaking has evolved rapidly in the past two decades. The mapping of Earth and other objects in the Solar System has benefitted from computer processing and manipulation of digital data. The unique and non-Earthlike aspects of many planets and satellites and the relative paucity of mapping data have required the development of special innovative techniques for mapping the planets.

The Solar System Atlas: A Preview

More than a thousand maps have been made for planets and satellites other than the Earth. Only Pluto and its satellite, Charon, and Saturn's moon, Titan, were unmapped as of 2001. Pluto remained unmapped because no spacecraft had been sent there; Titan's surface was and is obscured by clouds.

Maps include mosaics of photographs, relief maps, topographic maps, and digital image mosaics, as described earlier in this chapter. Maps in the atlas were adapted from these materials and redesigned to produce a uniform set for the Solar System. However, completely uniform treatment could not always be achieved. For example, many maps of Venus are being made, using *Magellan* data. To show the Venus maps at the same scale as those of Mars would result in many more pages than could be reasonably included in the atlas.

Dividing planets for presentation on atlas pages presents problems not encountered in making an atlas of the Earth. For example, maps of Earth are commonly centered on the continents so that only oceans are interrupted by the page edges. Oceans do not exist on other planets, and global segments must be designed according to some systematic scheme that divides the planet arbitrarily. The maps used in the atlas overlap from one page to the next.

Map Content

All maps are snapshots in time – their subject matter changes with time. Earth mappers constantly revise their maps because of political changes and expanding human development. Weather maps have useful lifetimes measured in hours, whereas maps of solid surfaces should be valid for thousands or even millions of years depending on geologic activities. Exceptions abound, however. For example, the surface of Jupiter's moon, Io, is disrupted daily by volcanic activity, and the Io seen by the *Voyager* spacecraft in 1979 is substantially different from the Io mapped by the *Galileo* spacecraft in the 1990s.

Solid surfaces are not observed on the giant gas planets of Jupiter, Saturn, Uranus, and Neptune. Photographs of these bodies show only atmospheric features and are essentially weather maps. Examples of these photographs are included in the atlas as illustrations but not in map formats.

Bright colors are rare on the planets and satellites, and shades of gray dominate. Consequently, some of the atlas maps are printed in subtle colors to enhance landforms and to make place names legible. Tints of different colors are used in these cases, but the colors

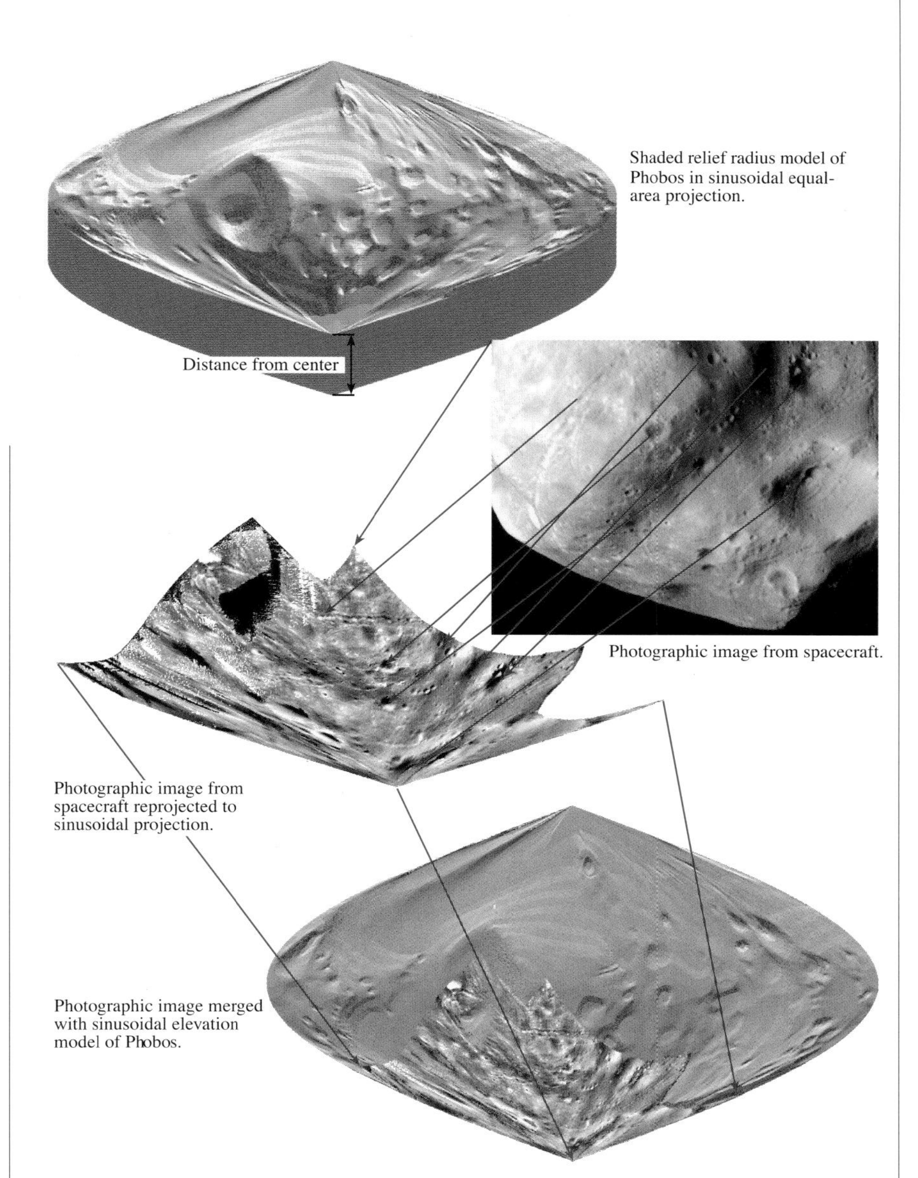

Mapping small, irregular satellites and asteroids. A rough model of the shape of the object is sculptured and images from a spacecraft are projected onto it to produce a three-dimensional image mosaic.

are not intended to depict, even approximately, the actual colors of the planets. Color-enhanced (and exaggerated) photographic maps of some bodies, such as Mars and Io, are included to augment the primary maps, but their true colors are less vivid.

Geologic Maps

The need for geologic maps of the Earth arose during the industrial revolution in the early nineteenth century. To exploit mineral resources and develop transportation systems, the distribution of rocks and minerals must be known. Geologic maps were derived to provide this information systematically, as well as to provide information on geologic structures, such as folded and faulted rocks. The principal units on geologic maps, termed formations, are shown in different colors and represent rocks that formed generally during a specific interval of geologic time in a given locality. For example, a cooled lava flow from Mount St. Helens could be considered a formation for mapping purposes. More typically, multiple eruptions occur from volcanoes – leading to a series of lava flows – and the entire sequence could also be considered a formation. In planetary geologic mapping, individual formations are often difficult to identify because of insufficient information. Consequently, it is more common to map large areas of relatively homogeneous rocks as the same unit. For example, a region consisting of volcanic rocks erupted from many sources, but representing the same span of geologic time, might be mapped as a single unit.

Geologic maps of Earth are normally made from field observations, study of rock samples, and use of aerial photographs. Lacking the opportunity to do field work or to have access to rock samples, planetary scientists must rely on remote sensing data to make photogeologic maps of the planets and satellites. For example, many planetary surface materials have distinctive colors and, as a consequence, spectral mapping – precise color determinations – provides clues to surface compositions. Spectral mapping is also accomplished using telescopes to observe planets and satellites. These data allow rock units to be defined and placed in a relative time sequence. In some respects, these maps are comparable to the preliminary maps made by oil and mineral exploration geologists in preparation for field work. Although both planetary and terrestrial maps involve interpretations, they are necessary for planning future work and for compiling other data.

Photogeologic maps have been prepared for nearly all explored planets and satellites. These maps range in complexity from highly detailed documentations of the *Apollo* landing sites on the Moon to simple sketch maps of outer planet satellites. Generalized geologic maps are provided in the atlas within each chapter for the planet and planet systems. The level of detail and the accuracy of the maps, however, depend on available information, which is not uniform for the Solar System.

Projections and Scales

Consistent map formats were designed to enhance intuitive impressions of similarities, differences, sizes, and distributions of landforms. No map printed on flat paper can por-

The launch of the *Viking 1* spacecraft to Mars. The launch was made by a *Titan Centaur 3* booster from Cape Canaveral, Florida, August 20, 1975, with the spacecraft arriving at Mars the following June 1976 (NASA photograph P-35344).

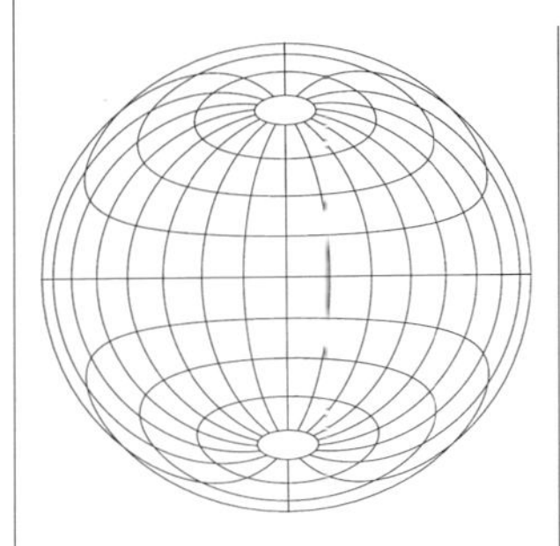

Lambert azimuthal equal-area projection. This projection can show an entire globe, but distortion of shapes increases away from center point and only the central part of the projection is included on most maps.

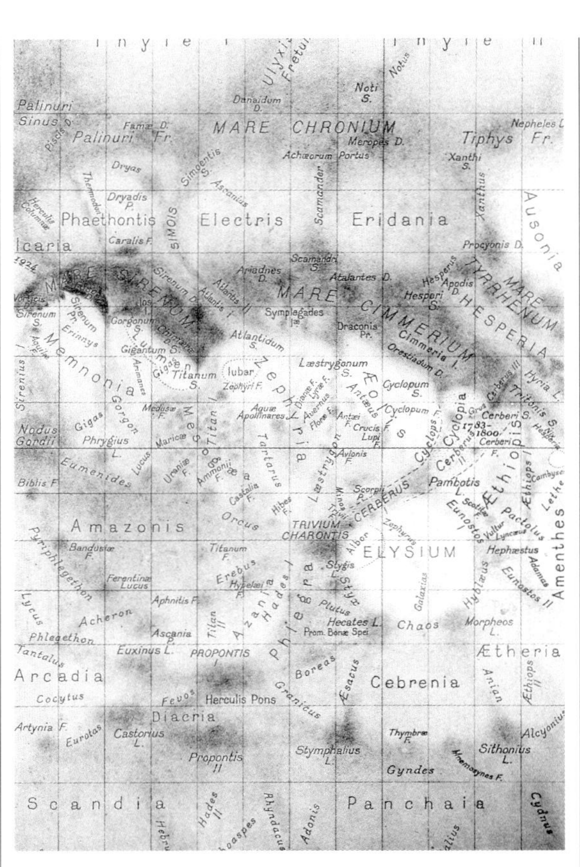

Part of Antoniadi's map of Mars showing patterns and colorations made from telescopic observations in the late nineteenth century. North is at the bottom of his map, because images seen through the telescopes of the time – like some telescopes today – were inverted.

tray spherical surfaces of planets without distortion. Numerous map projections – the mathematical bases for making maps – have been designed. For example, a "conformal" projection (such as the Mercator or stereographic) accurately portrays the shapes of small features. Equal-area projections, however, allow areas to be measured easily and accurately. For example, a coin placed on an equal-area map will cover the same number of square kilometers no matter where the coin is placed, whereas the measurement of distances is complex regardless of projection. Lambert azimuthal equal-area projection and the same map scale (1:10,000,000, where 1 millimeter represents 10 kilometers) are used for all of the maps in the atlas except for the very large or very small objects. A scale bar in the corner of each atlas map is only approximate, whereas the area-scales are exact. Each map has its own projection "center" to reduce the distortions in scale; however, distortion increases with distance from the center of the maps.

Gazetteer

A gazetteer of named features on all the planets and satellites (abbreviated for Earth) is included in the atlas, but many names in the listing do not appear on the atlas maps. Names identify the dominant features in the atlas in order to produce uniform and uncluttered maps that do not obscure surface detail.

The systems of classical names for bright and dark areas and various schemes of double- and triple-letter designations of craters that were devised for the Moon and Mars are rarely used in modern scientific literature and are not included in the gazetteer.

Glossary

This publication limits the use of specialized terms, but many of these terms and jargon are commonly used in space science. Although many of the terms are adapted from terrestrial geosciences and cartography, they may be unfamiliar to the general reader. Consequently, the atlas includes a glossary of the terms that are common in space science.

The first tentative steps in the exploration of the Solar System have been taken. Automated spacecraft have visited every major planet and satellite except Pluto and have returned mapping data ranging from preliminary glimpses of their surfaces to near-global observations. The methods for analyzing the quantity and complexity of this information have been developed and refined so that the general evolution of planetary surfaces can be formulated. The next chapter outlines the major bodies that make up the Solar System. In the chapters that follow, each planet or planetary system is portrayed by maps, charts, and spacecraft images that convey both the similarity and the unique aspects of the Solar System.

Solar System

Following the launch of *Sputnik 1*, an armada of spacecraft was sent to the Moon and throughout the Solar System in the 1960s and 1970s. The initial focus was on the Moon (see Appendix A) with United States manned and Soviet unmanned landings. These landings culminated between 1969 and 1972 with the delivery of lunar samples to Earth. Except for meteorites, these are the only samples from other planetary objects available for study on Earth.

With this series of lunar missions concluded, attention shifted to the exploration of the terrestrial planets with missions to Mercury, Venus, and Mars. For a variety of reasons, the United States concentrated on martian exploration, while the Soviets carried out a series of missions to Venus. In the late 1980s, these roles reversed somewhat, as the United States launched the *Magellan* mission on its journey to Venus, and the Soviets sent two ill-fated *Phobos* spacecraft to Mars. Only the United States *Mariner 10* spacecraft has visited Mercury.

Exploration of the outer Solar System began with the Pioneer spacecraft to Jupiter and Saturn in the late 1970s, followed by two *Voyager* spacecraft. *Voyagers* returned information on all the outer planet systems except Pluto (as yet unvisited by any spacecraft) and provided the primary data for mapping the planets and satellites of the outer Solar System. The launch of the *Galileo* spacecraft to Jupiter in 1989 carried with it the goal of intensive study of the largest of all planets and its bizarre family of satellites. On its way to Jupiter, the *Galileo* camera returned the first-ever close-up views of asteroids (Gaspra and Ida). This feat was followed by the *NEAR-Shoemaker* spacecraft which returned data for asteroids Mathilde and Eros, landing on the surface of Eros in early 2001.

The return of Comet Halley to the inner Solar System in 1986 provided the opportunity to send space probes to observe this famous object, an important component of the Solar System. The Soviets, Europeans, and Japanese all sent space probes to fly by Comet Halley, each probe carrying a variety of instruments. Results from these missions provided the first close observations of the nature

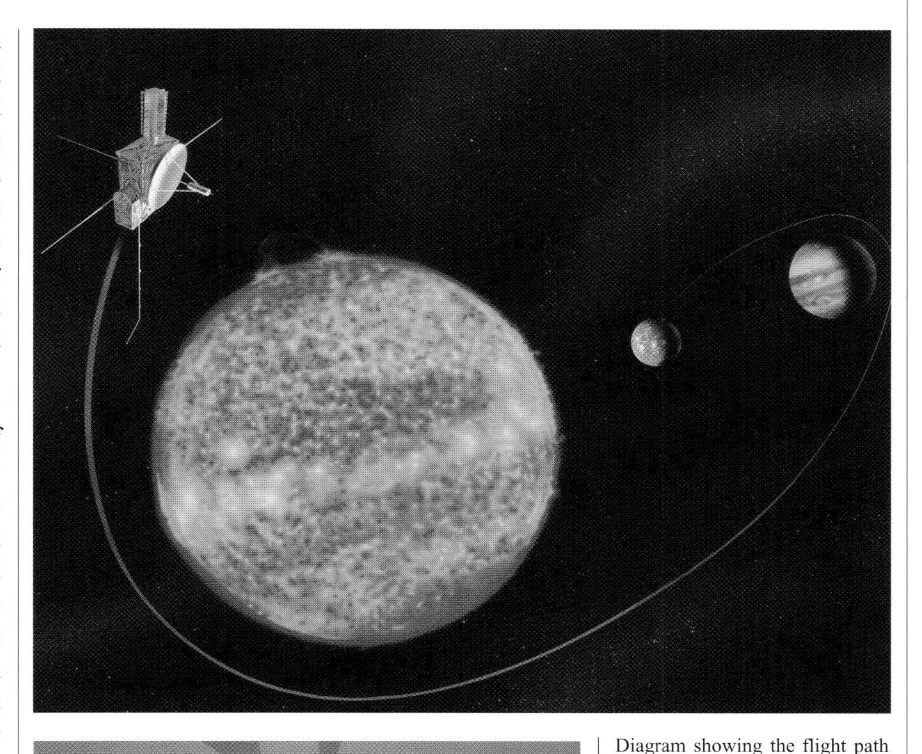

Diagram showing the flight path of the *Ulysses* spacecraft after its launch in October 1990. Instruments on *Ulysses* were designed to study magnetic fields in space, particles streaming from the Sun, X-rays and energetic particles from solar flares, cosmic dust, cosmic rays, and radio sources from the Sun (NASA P-35344).

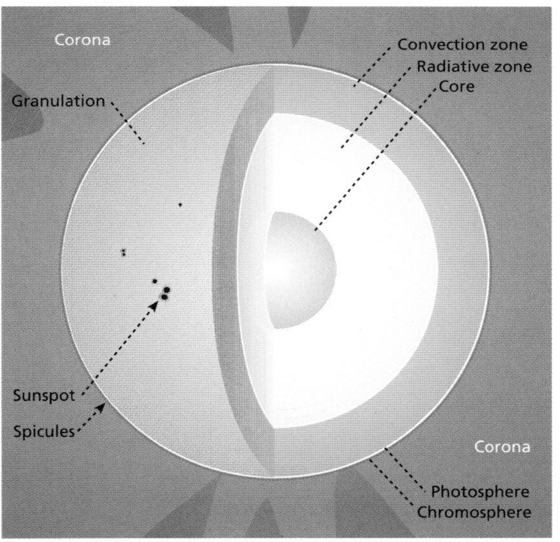

Diagram showing the principal features of the Sun.

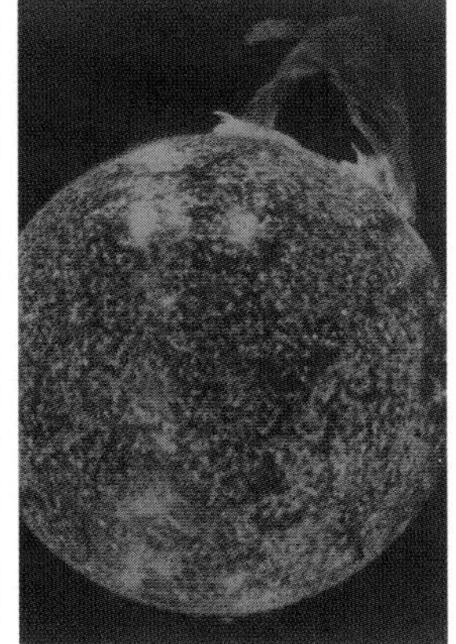

Photograph of the Sun taken by the *Skylab* astronauts in 1973 shows several flares (yellow areas) and an eruptive prominence extending more than 400,000 kilometers above the surface (NASA *Skylab* photograph).

Facing page: multiple exposure over the telescope facilities on Mauna Kea, Hawaii, showing phases of the solar eclipse on July 11, 1991 (Copyright © 1992 by ROGER RESSMEYER/©Corbis).

Spacecraft image montage of the Solar System, showing small, rocky inner planets, also called the terrestrial planets (Mercury, Venus, Earth, and Mars), and the Jovian planets (Jupiter, Saturn, Uranus, and Neptune) composed primarily of volatile elements; the Jovian planets and Pluto comprise the outer planets. (U.S. Geological Survey).

PLUTO
NEPTUNE
URANUS
SATURN

 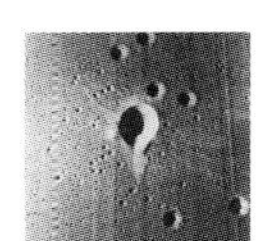 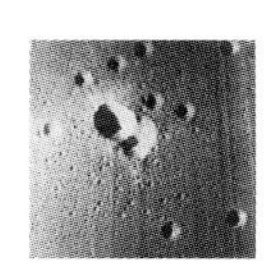 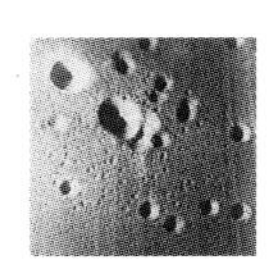 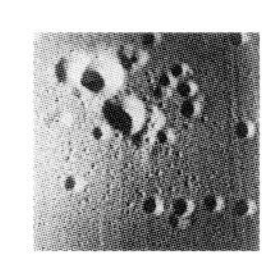 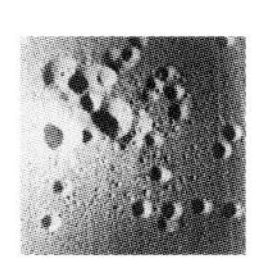   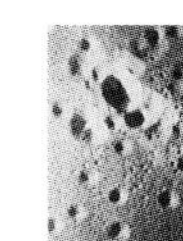 

of comets. For a variety of reasons linked to cost, the United States did not launch a spacecraft to Comet Halley.

From Earth-based and spacecraft observations, we now have insight into the general history and expanded knowledge of the major

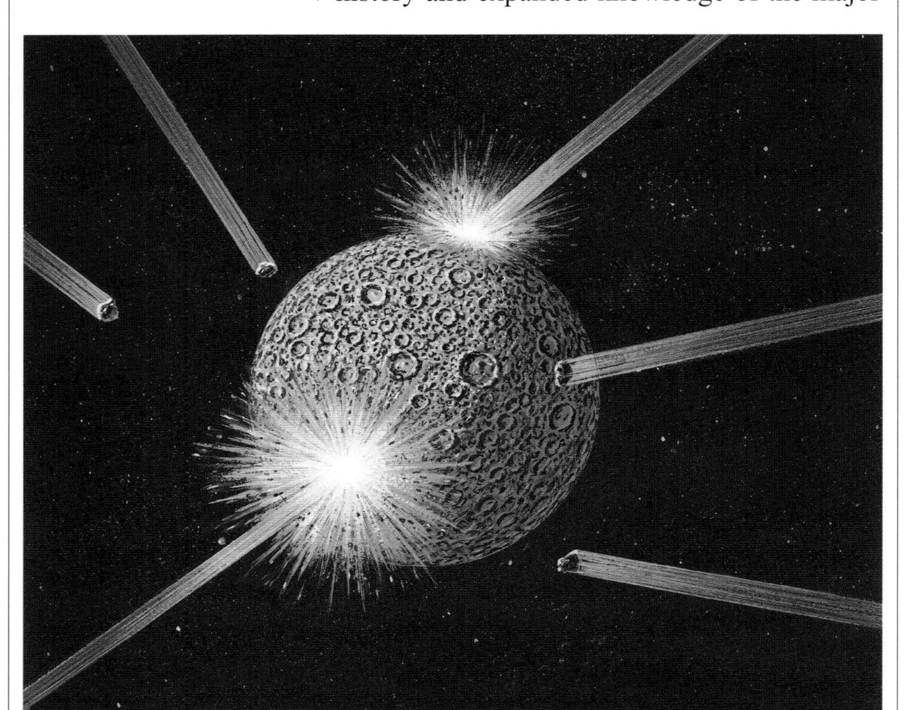

Diagram showing planetary formation by the accretion of small bodies such as planetesimals. Heat generated by the collisions was sufficient to melt the accreted "protoplanet." With time and cooling, the material separated into a core (composed of the densest materials), a mantle (which behaves as a plastic), and an outer rigid lithosphere, which may also incorporate a crust. This diagram is appropriate for Earth, but probably is also applicable to the other terrestrial planets (Illustration by Howard S. Friedman).

parts of the Solar System. The Solar System originated some 4.6 billion years ago through the accretion of gases and dust grains. Today the Solar System consists of a remarkable array of objects, from the enormous nuclear engine of the Sun to tiny dust and ice grains. Although this atlas focuses on bodies that have solid surfaces, such as Mars and the Moon, other objects are also discussed briefly. Four groups of objects constitute the Solar System: the Sun, the terrestrial planets (including Earth's satellite), the giant planets and their satellites, and the small bodies.

The Sun

Most of the mass of the Solar System is contained within the Sun. It is composed primarily of the elements hydrogen and helium with tiny amounts of other elements, including carbon, nitrogen, oxygen, silicon, and iron. The Sun is more than 1,390,000 kilometers across. By comparison, Jupiter – the largest planet – is only about one-tenth the diameter of the Sun. At the center of the Sun, temperatures are thought to reach nearly 15,000,000°C. Energy from the Sun comes mainly from the thermonuclear fusion of hydrogen to form helium. Even at the incredible rate of conversion of some 5 million tons per second, the Sun will continue to shine for at least another 4 billion years. The Sun is the only self-luminous object in the Solar System – we observe all of the other bodies by sunlight that they reflect.

The Terrestrial Planets

The terrestrial planets consist of Mercury, Venus, Earth, Mars, and the Earth's Moon. Although by strict definition Earth's Moon is not a planet, it is usually included with a discussion of the terrestrial planets because of its large size, its rock composition, and the similarity of its early geologic history to the other terrestrial planets.

The terrestrial planets, also called the inner planets because of their place in the Solar System, show many common attributes. All are relatively high in density, suggesting that, like Earth, they are composed of elements such as silicon, iron, and nickel. Models of the origin and early evolution of the Solar System show that these dense elements would have been retained close to the Sun, while lighter elements, such as hydrogen and helium, would have escaped to the outer reaches of the Solar System.

The individual planets evolved by the collection of smaller bodies – planetesimals – through accretion. As these "protoplanets" grew in size, their increasing mass attracted still more bodies, eventually incorporating most stray objects in their vicinity. The process continues today, although on a much smaller scale, as evidenced by the daily infall of meteorites on Earth. It is estimated that over 10,000 tons of such material are added to Earth every year. Although this is an enormous amount, it is insignificant when compared to the total mass of the Earth – it represents only 0.00000000000000017 percent of our planet. Moreover, the amount pales in comparison to the infall of planete-

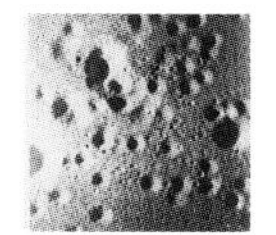
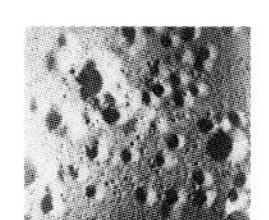
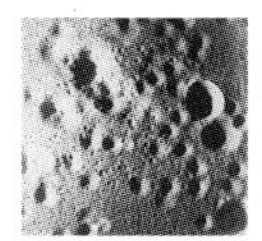

simals and other bodies on the terrestrial planets in the early stages of planetary formation. In fact, so much material was swept up by Earth that the heat generated by the collisions probably resulted in complete melting of the Earth. During this time, most of the heavier elements – such as iron and nickel – sank to the interior to form Earth's core and the lighter elements rose toward the surface to form the crust. This process of separation of the elements is termed differentiation. The crust is part of the lithosphere, or the outer, rigid part of the planet. The mantle forms the plasticlike zone between the core and the lithosphere.

Within the first half-billion years of Solar System history, much of the debris had been swept up by the planets. Gradually, the terrestrial planets, including Earth, began to cool and form crusts as elements combined and crystallized as rocks and minerals. For the most part, these elements include silicon, oxygen, iron, magnesium, sodium, calcium, potassium, and aluminum in various combinations that collectively make up the silicate minerals. Silicates are the basic building blocks of most rocks composing the crusts of Earth and the Moon, and they are thought to make up most of Mercury, Venus, and Mars as well. The presence of large impact craters on all of the terrestrial planets shows that their crusts had cooled and solidified before all of the debris was incorporated into other celestial bodies.

Venus, Earth, and Mars all have significant atmospheres – gases gravitationally bound to the planets in the outermost layer of these differentiated bodies. Probably some gases were accumulated by each protoplanet from the original solar nebula, but it is thought that these original atmospheres have been lost to space. Secondary atmospheres were outgassed from the hot interiors of these planets and subsequently changed by chemical interactions with the surface. Mercury and the Moon are too small to retain anything but the most tenuous atmospheres (measurable only by sensitive instruments). Earth's atmosphere may be termed a tertiary atmosphere – one greatly modified by the living organisms that populate our planet. Also, one of the gases

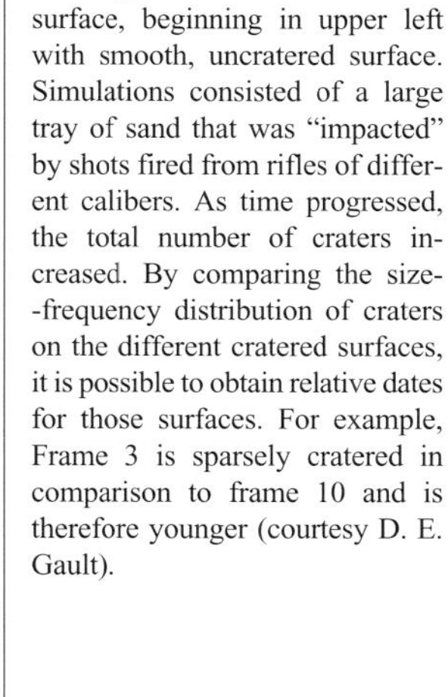

Above: Sequence of photographs from laboratory simulations showing evolution of a cratered surface, beginning in upper left with smooth, uncratered surface. Simulations consisted of a large tray of sand that was "impacted" by shots fired from rifles of different calibers. As time progressed, the total number of craters increased. By comparing the size-frequency distribution of craters on the different cratered surfaces, it is possible to obtain relative dates for those surfaces. For example, Frame 3 is sparsely cratered in comparison to frame 10 and is therefore younger (courtesy D. E. Gault).

Impact cratering involves the collision of planetary objects. These diagrams show the impact of an asteroid (traveling at a typical speed of 30 kilometers per second) on Earth and the near-instantaneous release of energy. The first stage (top) sends an intense shock wave through the crust; in the second stage (middle), the shock wave is reflected back toward the surface, and the fractured rock, called ejecta, is thrown outward to form the crater. In the final stage (bottom) the ejecta is emplaced around the crater; clots of rock in the ejecta may cause the formation of secondary craters. Note that the rocks in the crater rim are overturned in relation to their normal sequence (Illustrations by Howard S. Friedman after D. E. Gault).

Tectonism involves the deformation of planetary surfaces, as may happen by compression, in which rocks are folded into anticlines and synclines or faulted (reverse faults); tension on the crust may lead to normal faults and the formation of fault valleys, termed grabens, or fault-block mountains called horsts. Joints are relatively simple fractures or cracks.

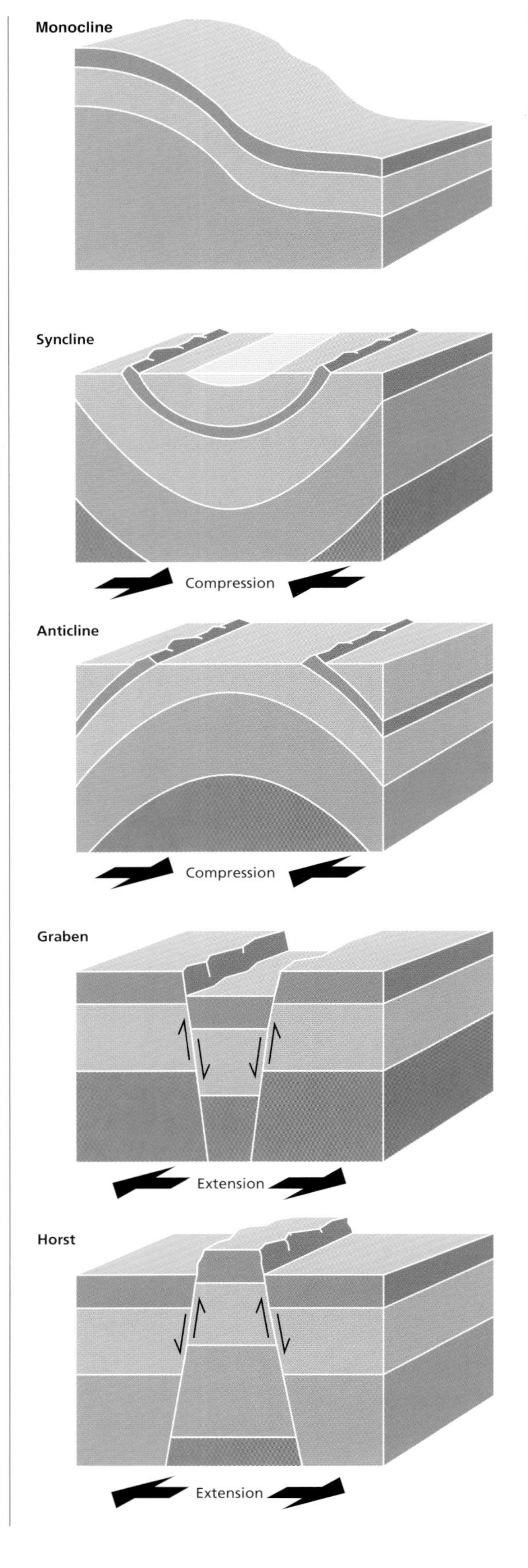

exhaled from the Earth's interior – water vapor – has condensed to form a liquid and contribute to the development of our oceans, lakes, and rivers.

With the formation of a crust, the geologic evolution for the planets was initiated, and each developed along different paths. One of the principal puzzles in the exploration of the Solar System is to determine what these paths have been and the reasons for their differences.

The Giant Planets

Jupiter, Saturn, Uranus, and Neptune are referred to as giant planets. All are gaseous objects and lack solid surfaces like those of Earth or Mars. Unlike the terrestrial planets, they are enormous and contain most of the mass in the Solar System outside the Sun.

The early history of the giant planets is in some ways similar to that of the terrestrial planets. They, too, formed by the accretion of smaller solid bodies to form a nucleus of sufficient size to capture gravitationally the lighter elements that had escaped from the hot inner Solar System to the outer frigid parts of the Solar System. As this process continued, the giant planets grew to consist mostly of hydrogen, helium and other gases, plus much smaller amounts of other elements. As such, each of the giant planets resembles the Sun in composition, but not even the largest, Jupiter, was destined to grow to a size sufficient to initiate nuclear fusion reactions in the deep interior. The great internal temperatures of the giant planets are the result of their original gravitational accretion together with the decay of radioactive elements.

The giant planets resemble the Sun in one other important way – each grew and evolved to have a family of smaller bodies in orbit about them so that each resembles the Solar System in miniature. These moons are typically composed of ices as well as rocky materials. One – Saturn's moon, Titan – even has a thick atmosphere of its own.

Photographs of the giant planets show only the upper layers of their atmospheres as a "snapshot" in time. Consequently, maps are not commonly made for the giant gas planets,

Photograph taken by *Lunar Orbiter V* of Schröter's Valley in the northern hemisphere of the Moon. Area shown is about 45 kilometers wide (*Lunar Orbiter V* photograph M202).

although names are given to dominant cloud features as shown in the atlas. Of interest for mapping purposes are the satellites of the giant planets. Collectively, these satellites represent a myriad of objects of different sizes, compositions, and geologic histories. Some, such as the Jupiter moons Ganymede and Callisto, are about the size of the planet Mercury. Others, such as the saturnian satellite Enceladus and the uranian satellite Miranda, have experienced tremendous crustal upheavals and deformation. Still others appear to have remained relatively unaltered since their initial formation. At least two, the Jupiter satellite Io and Neptune's Triton, are currently volcanically active – in fact, Io may be the most geologically active object in the Solar System.

The outer planet satellites provide an opportunity to compare a diverse set of bodies with the terrestrial planets. Many of the geologic processes that operate on terrestrial planets are also seen on outer planet satellites. But because of their different compositions (mostly ices plus some silicates) and extremely cold environment, they also display features representing processes unique to the outer Solar System.

Small Bodies and Pluto

Asteroids, comets, and the smaller planetary satellites are often called small bodies. However, some are not small; the largest asteroid, Ceres, is more than 1,000 kilometers across. Most asteroids occur in the zone between the orbits of the planets Mars and Jupiter. Whether asteroids are planetesimals that either never accreted to form a planet or are parts of one or more former objects, or whether they represent the remnants of comets or other objects, remains a subject of current study.

Comets are probably primordial material left over from the early stages of the development of the Solar System. Often described as "dirty snowballs," comets appear to be composed of solid dust grains embedded in water ice. Telescopic observations of comets allow their point of origin to be calculated and it has been concluded that most comets reside in the "Oort cloud," and the "Kuiper belt", beyond the orbit of Pluto. Having spent most of their lives in this "deep freeze," comets may represent material changed relatively little from the time of Solar System origin.

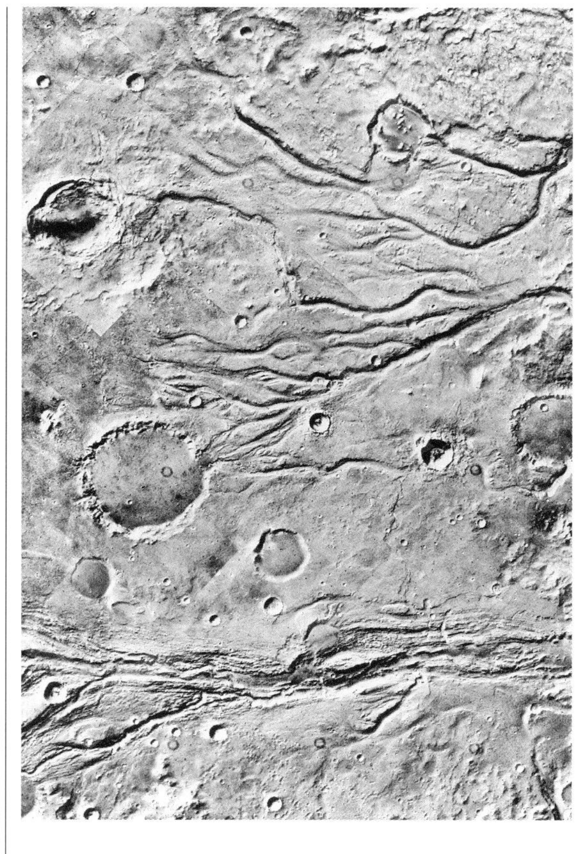

View of the martian surface taken by the *Viking Orbiter* spacecraft in 1978 showing an ancient river bed carved by flowing water. Area shown is 150 kilometers wide (*Viking Orbiter* mosaic 211-5190).

Of the estimated 10 to 1,000 billion comets, only a few have orbits that swing through the inner Solar System. But these, such as Comet Halley, engender great excitement. As discussed in the last chapter, comets may afford an opportunity to study a part of the Solar System record that is unattainable by any other means.

Pluto does not fit into the subdivision of planets as well as terrestrial and gas giant objects. Little is known of its properties; almost nothing is known of its surface. It is included in the last chapter with "small bodies" because it may, in fact, share more characteristics with comets than with any of the other planets.

Planetary Processes

In many respects, it is fortunate that our own planet, Earth, forms the basis for understanding the processes that shape planetary sur-

Caldera formation
Explosions
Lava plateau
Cinder cone
Cinder cones
Shield volcano
Eroded laccolith
Ash cone with lava dome
Volcanic neck with dikes
Fissure
Composite volcano
Maar
Conduit
Diatreme
Laccolith
Sill
Dike
Stock
magma

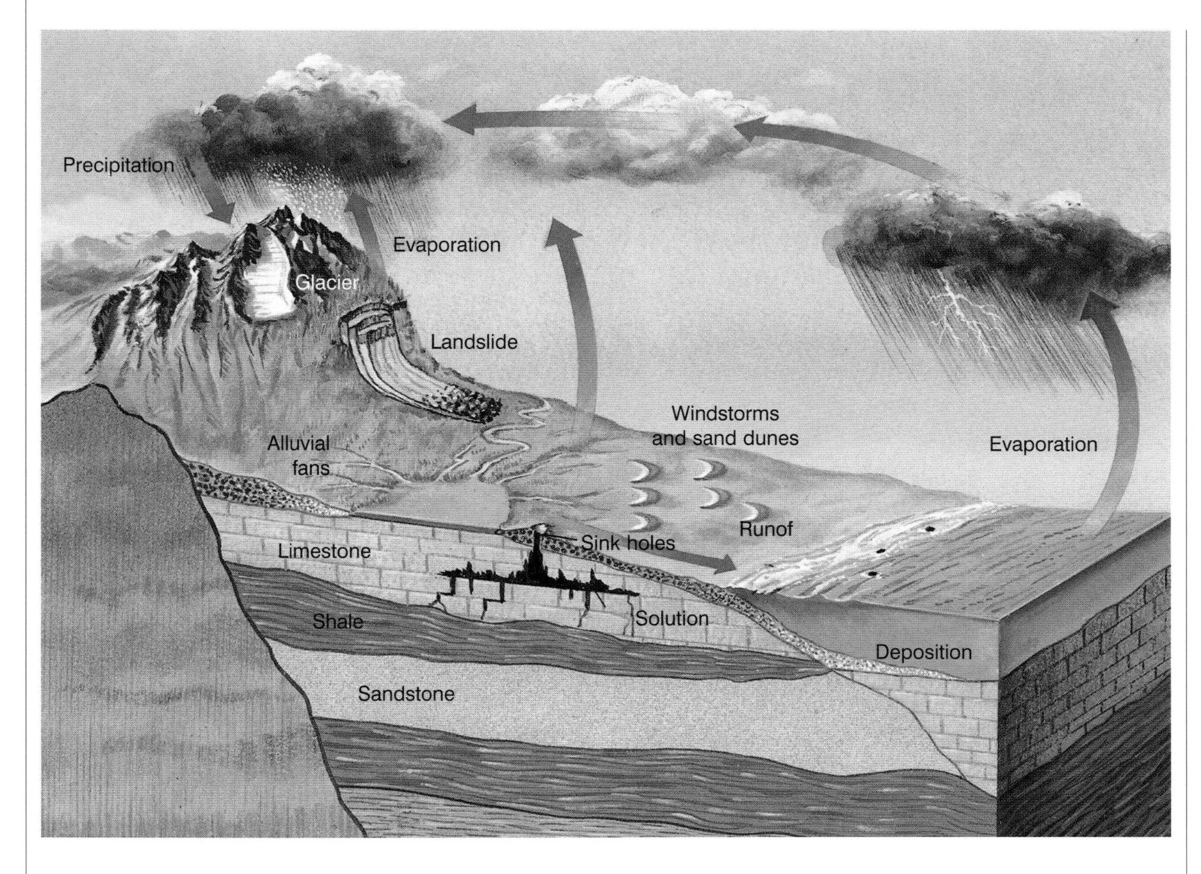

Gradation involves the weathering, erosion, transportation, and deposition of rocks by wind, water, or gravity (Illustration by Howard S. Friedman).

faces. Of the terrestrial planets, it is by far the most complex and diverse. Consequently, there is substantial knowledge to draw upon as other planets and satellites are explored. Of course, one of the exciting aspects of exploration is the search for non-Earthly features and processes. There have been many notable discoveries, especially in the frigid worlds of the outer Solar System.

Earth's surface is the result of many complex events that take place on time scales ranging from seconds to centuries. These events can be classified into four processes. Processes that originate below the surface in response to heat from Earth's interior include igneous activity (which can lead to eruptions of molten rock called magma and the formation of volcanoes) and tectonic processes (which disrupt the crust, such as by earthquakes). Processes which originate above the surface include the impact of objects on Earth and gradation (which is the erosion, removal, and deposition of material by wind, water, or gravity).

Most planets and satellites with solid surfaces show evidence for the activity of all or most of these four processes. The relative importance of the processes depends on the local environment and geologic history of the planetary object of concern. For example, today Earth is the only planet where liquid water can remain on the surface for any length of time. All other planets are either too hot or too cold to support liquid water. Consequently, processes involving running water – such as erosion by rivers – can occur only on our planet. Geologic evidence shows ancient river beds on Mars that were carved by running water, meaning that sometime in the past the climate on Mars must have been more similar to that of the Earth (warmer and wetter) and demonstrating the importance of the surface record in deciphering the past.

While Earth has been important for understanding planetary processes, particularly for the terrestrial planets, the study of other planets has also provided insight for understanding the Earth. Prior to the Space Age, impact cratering was little appreciated as a geologic process because on Earth impact craters are rapidly degraded and lost from the record.

Facing page: igneous activity involves the formation of molten rock (magma) and its intrusion within the crust to form features such as dikes and sills, or its eruption on the surface to form volcanoes, ash, and lava flows. Eruptions may be relatively quiet, forming vast sheets of lava as from fissure eruptions, or they may be violently explosive (Illustration by Howard S. Friedman, after U.S. Geological Survey).

Facing page: diagram showing the history of the Earth since the origin of the Solar System 4.6 billion years ago to the present time. Humans have been on Earth only a very short time. Most of the early record of the Earth (prior to about 1 billion years ago) has been lost through crustal "recycling."

Solar System exploration has shown impact craters to be ubiquitous and has caused geologists to reexamine Earth's record for impact events. Nearly 150 craters or remnants of craters of impact origin have been found on Earth, and each year one or more craters are discovered and added to the inventory. In fact, some of the great mysteries of Earth may be solved when impact cratering is considered. For example, there is substantial evidence to support the idea that a major impact occurred on Earth at about the time of the extinction of the dinosaurs some 70 million years ago. Although no one believes that the death of the dinosaurs was caused by direct impact, their demise – and the demise of a great many other organisms at the same time – may have occurred as a consequence of events triggered by the impact. The theory is that a large impact ejected millions of tons of rock and soil into the atmosphere and blocked solar energy for many years. The cooler temperatures and attendant loss of vegetation and reduction of the food chain then may have caused the extinction not only of the dinosaurs but of many other organisms.

Geologic Time

It is difficult to visualize the enormous distances involved in the Solar System. Similarly, geologic time is hard to comprehend because it is so long in comparison with our basic time reference – a human lifetime. A geologic time scale has been derived for the Earth, beginning with planetary accretion. Determining the ages of rocks is accomplished primarily by using radiogenic "clocks" based on the principle that certain unstable radioactive elements contained in some rocks decay or convert to more stable elements at a known rate. By knowing this rate and by measuring the amounts of unstable and stable elements in a rock sample, it is possible to determine the age of the rock. Of course, radioactive elements of the right type must be available for measurement in the sample and, unfortunately, not all samples contain these elements. Consequently, only some rock types can be dated. Typically, radiogenic dating provides the age of a rock in years from its first formation.

Geologic time is also often given in relative terms rather than specific time segments. Rocks are classified as being older, younger, or the same age as other rocks. On Earth, combinations of radiogenic and relative ages are commonly used to describe the geology of a given area.

How is geologic time assessed on other planets and satellites? Practical limitations require that rock samples be analyzed in laboratories on Earth to determine radiometric ages because automated dating systems for robotic spacecraft have not yet been developed. Consequently, dates have been obtained only for rock samples from the Moon and for meteorites. Using a combination of dates from samples and relative ages from lunar surfaces, a geologic time scale has been derived for the Moon (see Earth-Moon chapter).

Because no direct samples are available from other planets, only relative ages can be assigned for all of the other planets and satellites. The principle of superposition was developed for assigning ages to rocks on Earth, in which rock units on top of an undisturbed sequence of rocks are younger. This principle is also applied to planet and satellite surfaces. Another method for establishing the relative ages of surfaces is based on the presence of impact craters. Old surfaces have been exposed to the impact environment longer than young surfaces and, statistically, they should preserve more impact craters. By counting the numbers of craters superposed on planetary surfaces, their age relative to other surfaces can be determined. This concept was developed for the Moon and was verified when rock samples were returned to Earth for analysis and radiogenic dating. By this method, geologic time scales have been derived for Mercury, Venus, and Mars.

Geologic time: 4600 million years

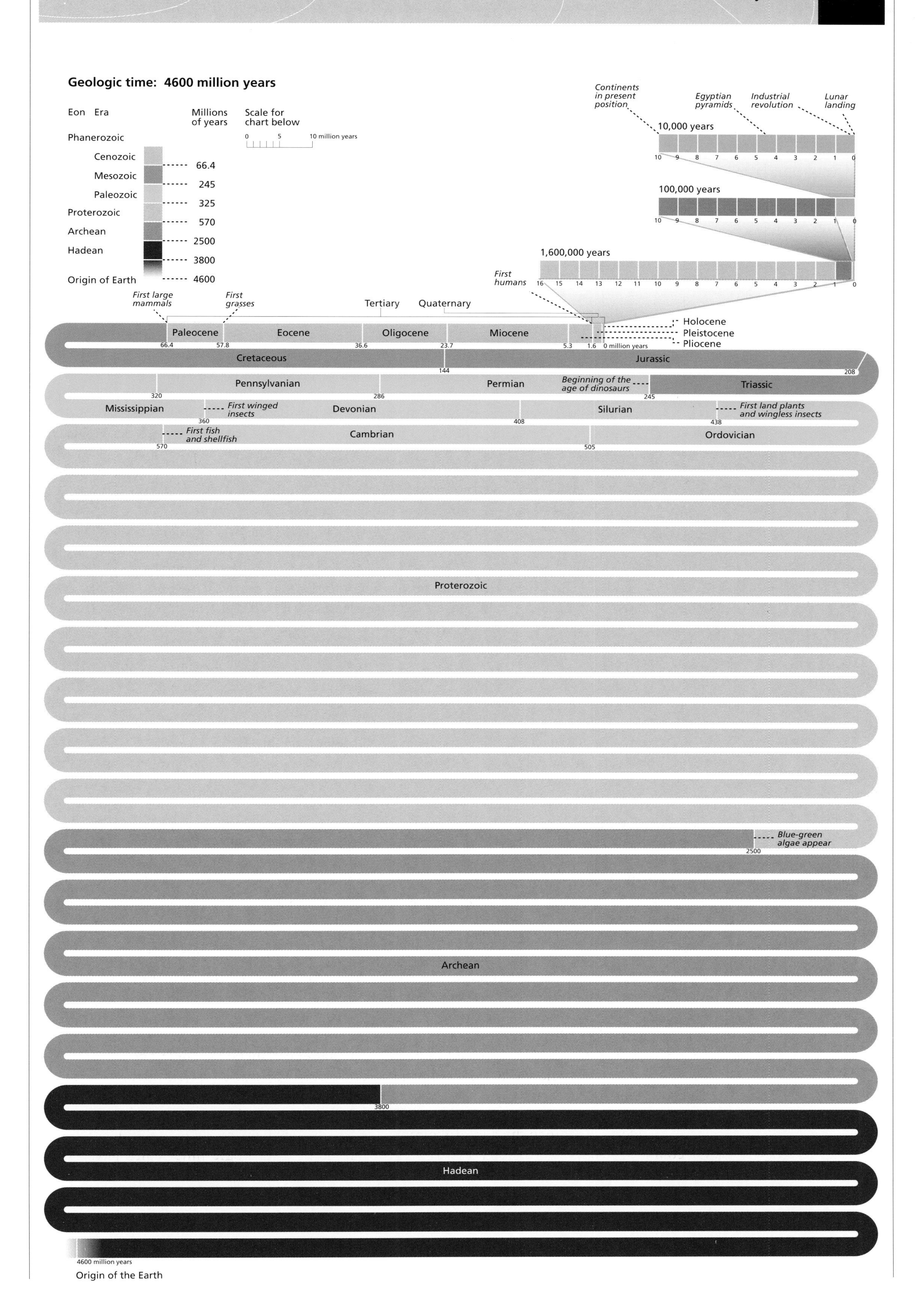

Eon Era
Millions of years
Scale for chart below
0
5
10 million years
Phanerozoic
Cenozoic 66.4
Mesozoic 245
Paleozoic 325
Proterozoic 570
Archean 2500
Hadean 3800
Origin of Earth 4600
Continents in present position
Egyptian pyramids
Industrial revolution
Lunar landing
10,000 years
100,000 years
1,600,000 years
First humans
First large mammals
First grasses
Tertiary
Quaternary
Paleocene
Eocene
Oligocene
Miocene
Holocene
Pleistocene
Pliocene
0 million years
Cretaceous
Jurassic
Pennsylvanian
Permian
Beginning of the age of dinosaurs
Triassic
Mississippian
First winged insects
Devonian
Silurian
First land plants and wingless insects
First fish and shellfish
Cambrian
Ordovician
Proterozoic
Blue-green algae appear
Archean
Hadean
4600 million years
Origin of the Earth

Mercury

The planet Mercury, named after the Roman Messenger of the Gods, was undoubtedly a familiar celestial object to primitive peoples. Although the first recorded mention of Mercury was in 265 B.C. by the Greek astronomer, Timocharis, there are indications that the Egyptians were aware of Mercury and discovered that the planet orbits the Sun. Because Mercury is the innermost planet, it can be observed from Earth only with great difficulty and only during the twilight hours. Consequently, even the best Earth-based telescopes provide poor data for the planet. The first serious observations of Mercury were made in the early 1800s by J. H. Schröter and K. L. Harding. They noted light and dark patterns on the surface and proposed (erroneously) the existence of mountains reaching 20 kilometers above the surface.

The first maps of Mercury were made in the late 1800s, first by Giovanni Schiaparelli and later by Percival Lowell, both famous for their observations and maps of Mars. These and subsequent Mercury watchers assigned a variety of names to real and perceived features on the planet. Although most of the names bear little resemblance to features actually existing on Mercury, the names used by E. M. Antoniadi have been adopted where practical on modern maps.

Observations prior to the exploration of Mercury by spacecraft revealed three primary facts: (1) Mercury has a very high density compared to the other terrestrial planets, (2) its surface has physical properties similar to Earth's Moon, and (3) the rotation period of Mercury is in 3:2 resonance with its orbital period, meaning that it spins on its axis exactly 3 times during 2 orbits around the Sun. Knowledge of the surface features on Mercury and clues to its geologic history came only with data returned from the highly successful *Mariner 10* mission. After a journey of nearly 5 months from Earth, *Mariner 10* first flew past Mercury on March 29, 1974, and returned the first close-up pictures of the planet's surface and other data to Earth. Engineers at NASA's Jet Propulsion Laboratory were able to take advantage of the orbital geometry of Mercury and the trajectory of the spacecraft to permit a second and a third flyby of Mercury.

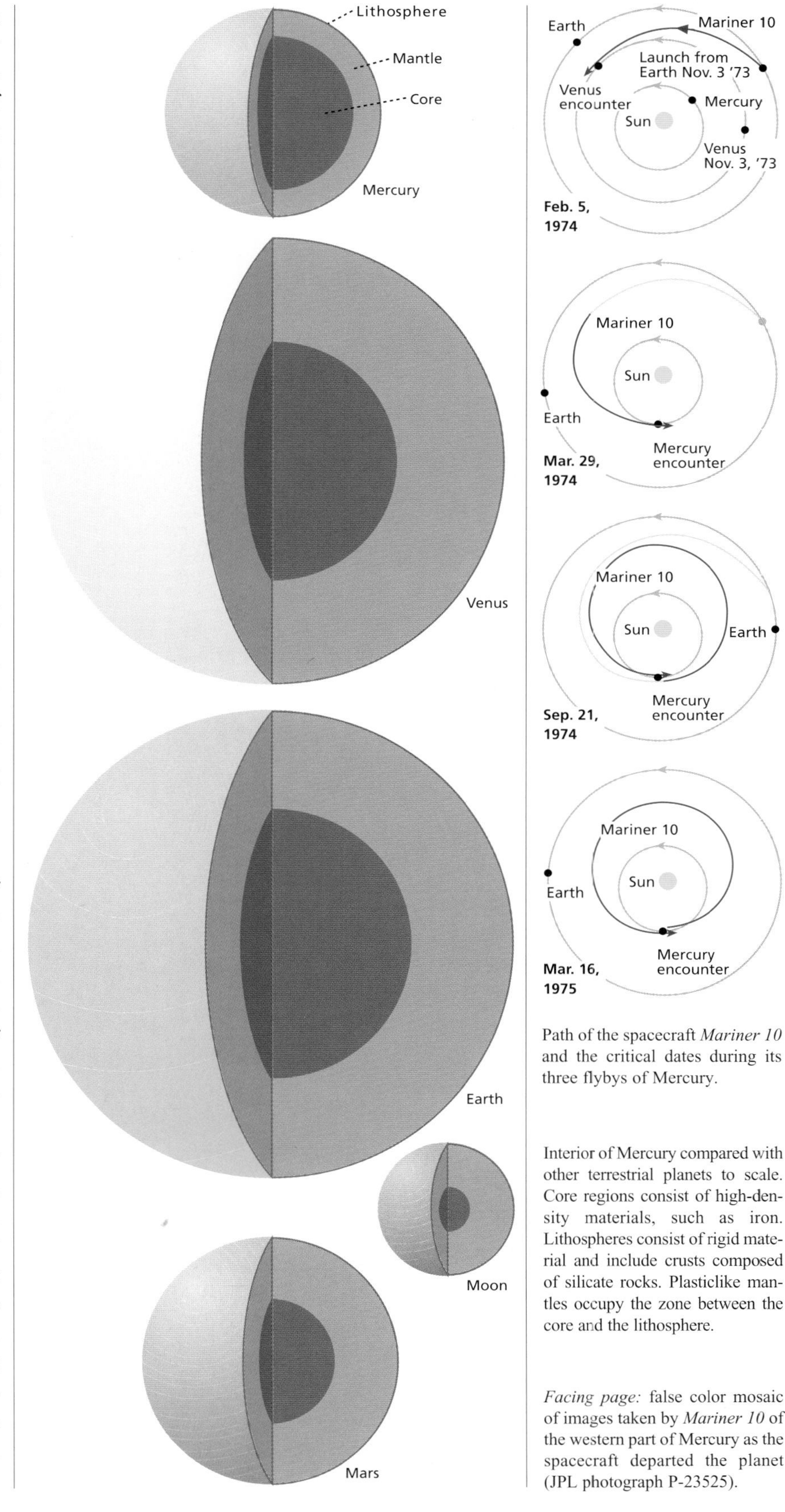

Path of the spacecraft *Mariner 10* and the critical dates during its three flybys of Mercury.

Interior of Mercury compared with other terrestrial planets to scale. Core regions consist of high-density materials, such as iron. Lithospheres consist of rigid material and include crusts composed of silicate rocks. Plasticlike mantles occupy the zone between the core and the lithosphere.

Facing page: false color mosaic of images taken by *Mariner 10* of the western part of Mercury as the spacecraft departed the planet (JPL photograph P-23525).

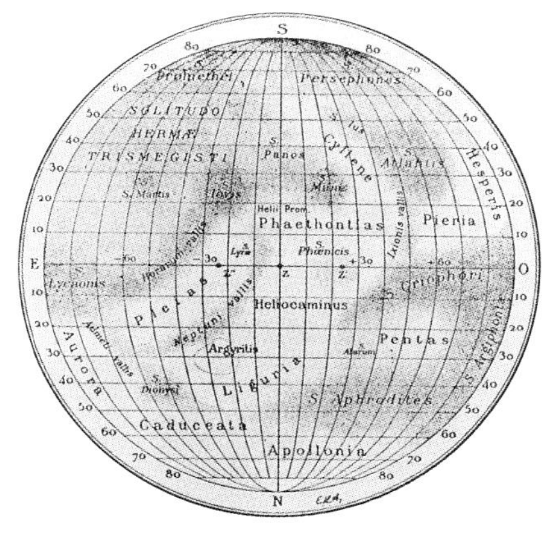

Antoniadi's map of Mercury published in 1934 portrays several features retained on modern maps.

Mosaic of images taken by *Mariner 10* of the eastern part of Mercury approaching the planet (JPL photograph P-14470).

In all, *Mariner 10* yielded more than 2,700 useful photographs that show about half the surface of Mercury at resolutions of 100 meters to 4 kilometers. This coverage and resolution is similar to that achieved for the Moon by Earth telescopes. Consequently, our knowledge of the surface of Mercury is still quite incomplete.

Magnetometer data from the *Mariner 10* mission revealed the presence of a magnetic field around Mercury, a notable discovery with important implications for the physical state of the planetary interior. The density of Mercury is the greatest of all the planets and implies a high abundance of iron – 65 to 70 percent by weight. Mercury's high density and its observed magnetic field suggest that the planet has a large iron core that is at least partially molten even today.

There is no evidence from *Mariner 10* data or other considerations to suggest that Mercury has ever had an atmosphere of any significance. This means that surface erosion resulting from wind and water have not occurred on Mercury and that the modification of its surface is the result primarily of impact cratering, tectonic processes, and probably volcanism.

Geology

In many respects, Mercury outwardly resembles Earth's satellite, the Moon. It displays a surface pockmarked with craters and locally blanketed with various plains-forming deposits. Craters formed by impact range in size from the limit of resolution (about 100 meters) to more than 1,300 kilometers across. Undoubtedly, even smaller craters exist – in the absence of an atmosphere, even the tiniest of meteoroids would pepper the surface of Mercury just as they do on the Moon.

When pictures from *Mariner 10* were first analyzed, investigators thought that there was a lack of large, multiringed, impact basins on Mercury compared to the Moon. However, later study showed this conclusion to be erroneous. The larger, older craters are simply difficult to recognize, having been nearly obliterated as a result mainly of blanketing by material that forms a terrain called intercrater plains. These

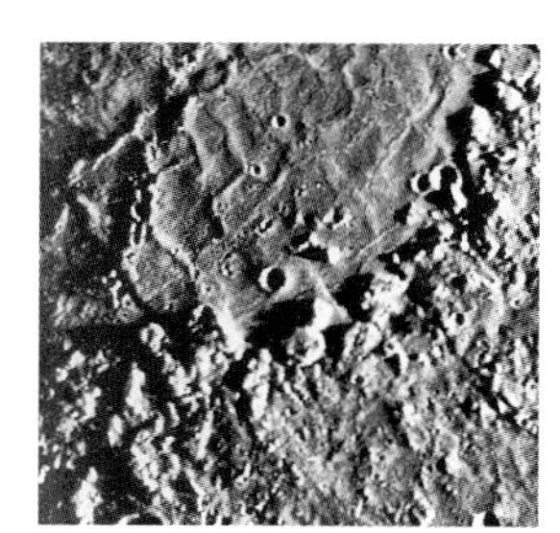

Mariner 10 image of part of the Caloris basin showing the rugged mountains 1 to 2 kilometers high that define the basin ring (*Mariner 10* photograph, FDS 229).

plains are gently rolling areas that are thought to be either volcanic in origin (perhaps lava flows) or to represent a type of impact ejecta deposit. In either case, intercrater plains bury or partly blanket many of the larger impact basins on Mercury.

Study of impact craters of all sizes on Mercury shows some important differences in comparison to impact craters on the Moon. On the Moon, impacts excavate large fragments that strike the surface, forming secondary craters. Secondary craters occur closer to the rim of the primary craters on Mercury than on the Moon. This is a result of the higher gravitational acceleration on Mercury as compared with the Moon.

Tectonic features of several types are seen on Mercury. One type, lobate scarps, consists of cliffs that cut across the plains and craters. From their geometry, many lobate scarps appear to be thrust faults, which are the result of crustal compression. Such compression may have taken place through a global shrinkage of Mercury's surface in its early history when the planet may have been cooling and contracting, perhaps aided by the gradual slowing of the planet's rotation. Another type of tectonic feature resembles mare ridges observed on the Moon. The ridges on Mercury appear, as on the Moon, to result from local compression of the crust.

The most prominent feature on the half of Mercury seen from *Mariner 10* is the Caloris basin, an enormous impact scar. The Caloris basin consists of a ring of mountain blocks about 1,300 kilometers in diameter. The mountain blocks are some 30 to 50 kilometers long and stand 1 to 2 kilometers high. Interspersed among the mountains are surfaces that are peppered with small craters. A hilly and lineated terrain, found opposite to the center of the Caloris basin, represents the impact that created the basin. The floor of the basin, which shows few craters, has been fractured, displaying giant cracks. Although the origin of the Caloris floor deposits is controversial, they are probably either rocks that cooled from melts derived from magmas "tapped" by the impact that formed the basin, or they formed from crust melted by heat from the impact.

Mosaic of *Mariner 10* images of part of the 1,300-kilometer-diameter Caloris basin, showing radial texture (arrows) or ejecta deposits and terrain modified during impact. The interior of the basin includes fissured and ridged terrain thought to consist of volcanic lavas emplaced following the impact.

View of Discovery Rupes, centered at 52° N, 38° W. This scarp is more than 500 kilometers long and stands about 2 kilometers high, making it one of the largest scarps on Mercury (*Mariner 10* photomosaic H-11).

Oblique view of Mercury showing impact craters and large areas of intercrater plains (*Mariner 10* photograph 27328).

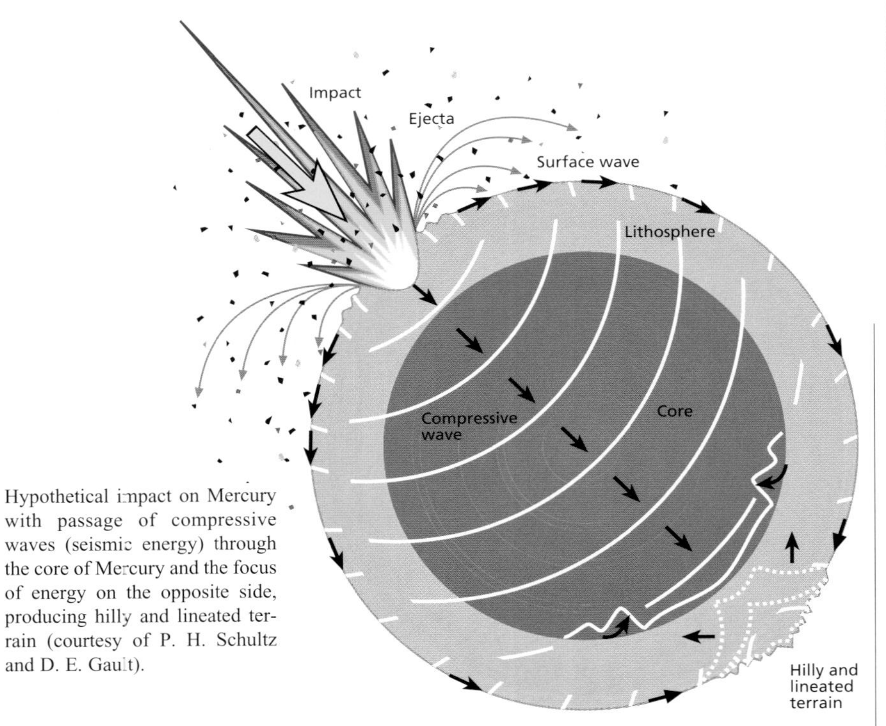

Hypothetical impact on Mercury with passage of compressive waves (seismic energy) through the core of Mercury and the focus of energy on the opposite side, producing hilly and lineated terrain (courtesy of P. H. Schultz and D. E. Gault).

High resolution image of hilly and lineated terrain in the area of Mercury antipodal to the Caloris Basin. Area shown is about 120 kilometers across (*Mariner 10* photograph 27463).

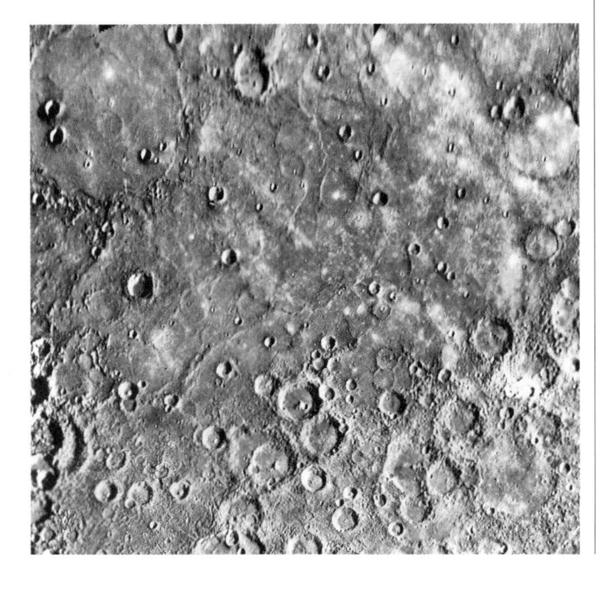

Photomosaic of the north polar region showing dark smooth plains that flood and bury older, heavily cratered terrain. These plains may be lava flows, although definitive evidence for a volcanic origin is lacking (U.S. Geological Survey *Mariner 10* photomosaic H-1).

Geologic History

Mercury's high density and analysis of geologic maps of Mercury provide the means to reconstruct its geologic history. Given its high density and close proximity to the Sun, Mercury is probably composed mostly of iron and silicate materials. Most scientists think that Mercury went through an early molten phase in which the lighter elements, such as aluminum, rose to the surface, while iron and the other heavy elements concentrated in the center to form a massive core. With subsequent cooling, a crust would have developed and thickened. Impact craters and basins seen today reflect the final stages of planetary accumulation. These features, in turn, were partly buried by intercrater plains material that may represent early volcanic processes. This phase of mercurian history is the pre-Tolstojan Period.

The next geologic phases are marked by two giant impact events. The Tolstojan Period is represented by the formation of the Tolstoj basin, an impact structure about 500 kilometers in diameter, found in the southern hemisphere. The formation of the Caloris impact basin is the most prominent event in the history of Mercury and marks the beginning of the Calorian Period. Ejecta and related surface features from the Caloris basin form the most important geologic markers on the surface of Mercury. An impact of the size that formed this basin inevitably generated tremendous seismic waves, which were felt over the entire surface of Mercury. Seismic energy may even have been focused on the side directly opposite the impact where the surface is observed to be disrupted to form hilly and lineated terrain. Smooth plains seen in many places on Mercury, including the floor of Caloris and other basins, may represent eruptions of vast sheets of lavas that covered many areas of the planet.

The two youngest periods in the geologic history of Mercury are named the Mansurian and Kuiperian Periods. Only impact cratering appears to have occurred during these two youngest periods.

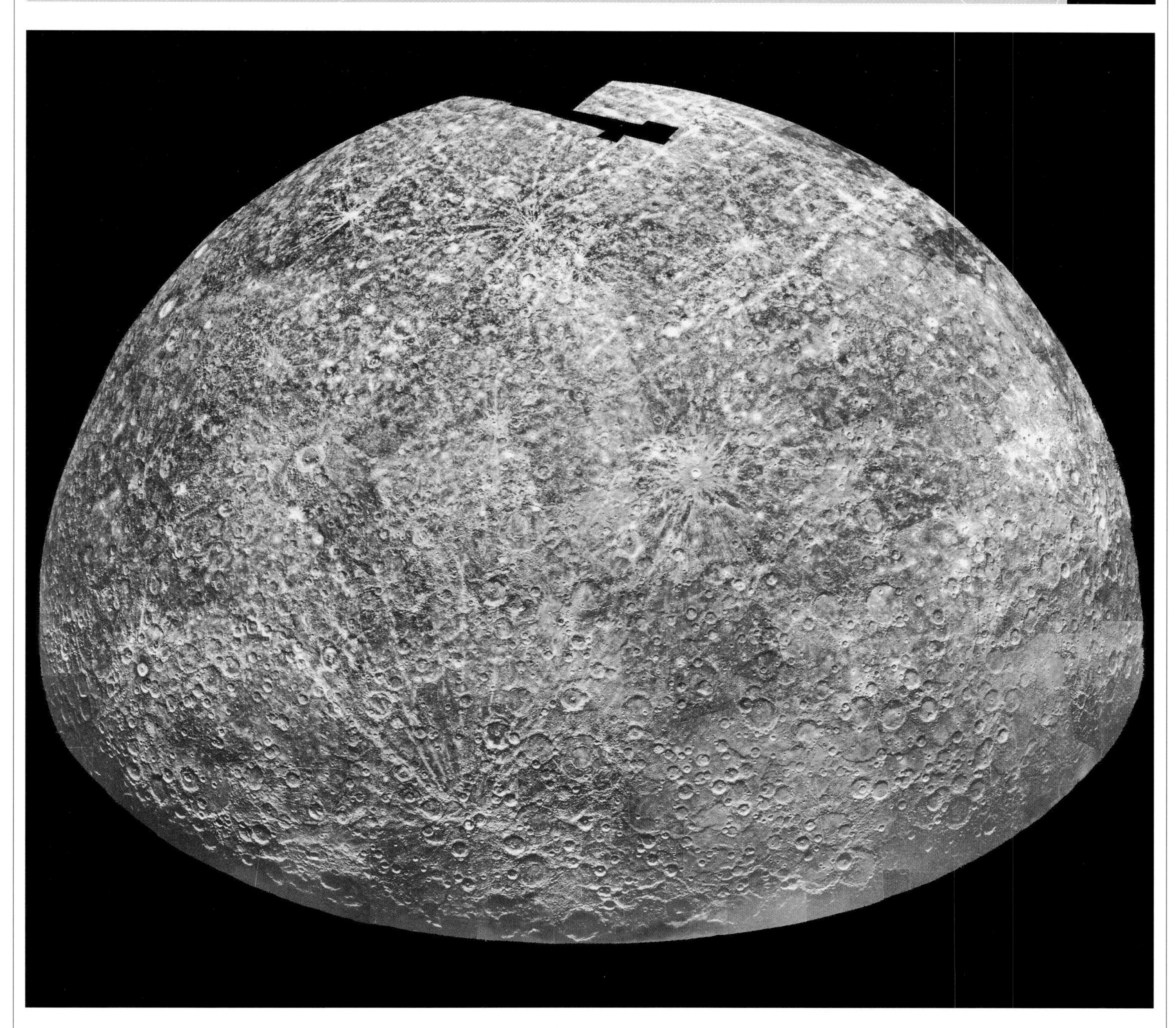

Mosaic of images taken by *Mariner* 10 of the southern region of Mercury (courtesy of U.S. Geological Survey).

For many years spacecraft missions to Mercury more ambitious than *Mariner 10* seemed unfeasible because of the planet's proximity to the Sun and the peculiarities of its orbital geometry. Placing spacecraft into orbit around the planet or making automated landings seemed far too difficult to achieve. Recent studies, however, have shown that many of these difficulties can be overcome. Current NASA plans include a spacecraft to orbit around Mercury, collecting high-resolution photographs and other remote sensing data. Later, an automated spacecraft may land on its surface to collect and analyze samples.

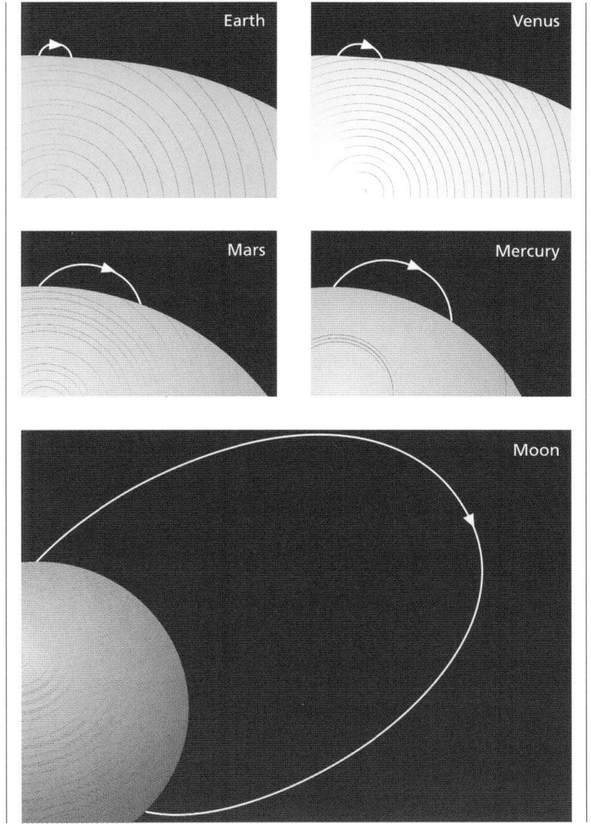

Comparisons of the distance ejecta would travel under a given set of conditions on the terrestrial planets as a function of gravity. The high gravitational acceleration of Earth and Venus cause ejecta to fall close to the crater rim, in contrast to the Moon. Mercury, although smaller than Mars, has similar mass (because of its high density), thus permitting the ejecta to travel only a little farther on its surface than it would on Mars (courtesy of P. H. Schultz).

Mercury 100° Hemisphere Geologic Map

1 cm²:
Center lat. 0°
Center lon. 100°

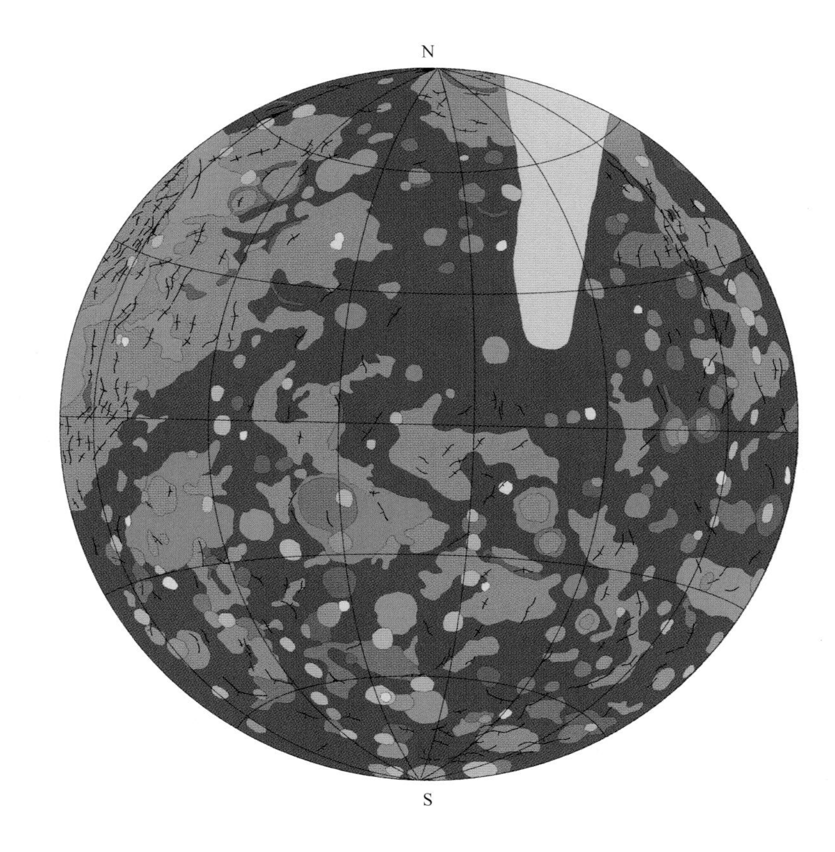

Description of Map Units

Impact Materials

Crater Materials

- Kuiperian System (K)
- Mansurian System (M)
- Calorian System (C)
- Tolstojan System (T)
- Pre-Tolstojan System (pT)

Basin Materials

- Calorian System (C)
- Tolstojan System (T)
- Pre-Tolstojan System (pT)
- Pre-Tolstojan System (pT) basin debris

Plains Materials

- Calorian System (C)
- Calorian (C) and Tolstojan (T) Systems

Unmapped

- Data resolution insufficient

Structure Symbols

- Lobate fault scarp
- Crack or graben
- Mare ridge

Mercury 100° Hemisphere Reference Map

1 cm²:
Center lat. 0°
Center lon. 100°

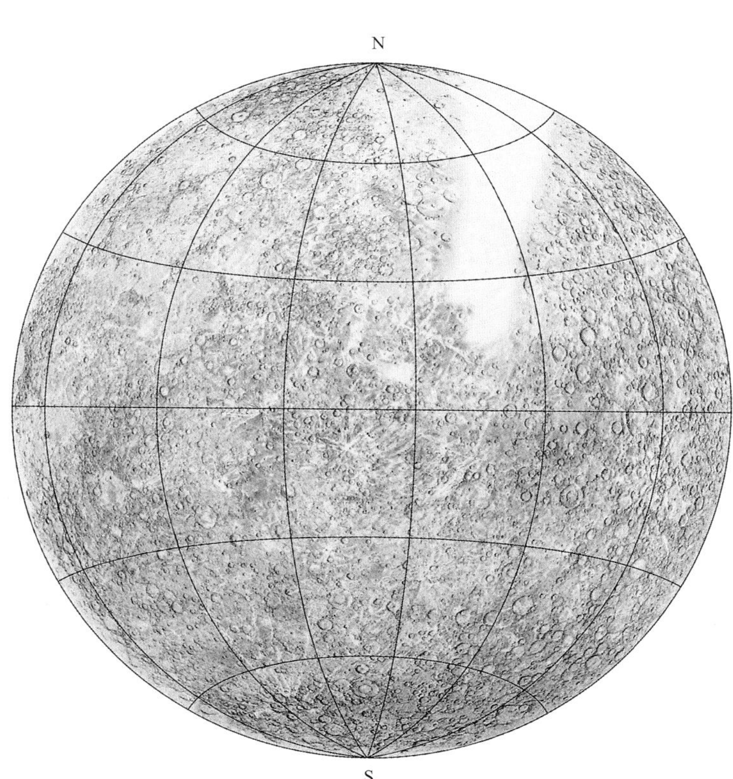

Correlation of Map Units

Youngest at top

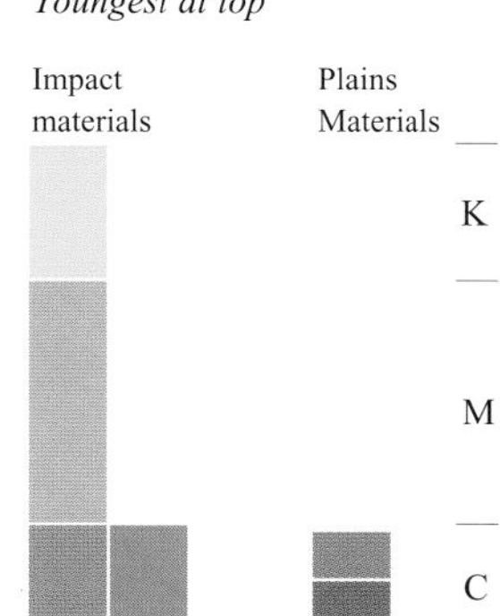

Opposite hemisphere not mapped: no available data

Mercury Airbrush Map Index

Center lat. 0°
Center lon. 100°

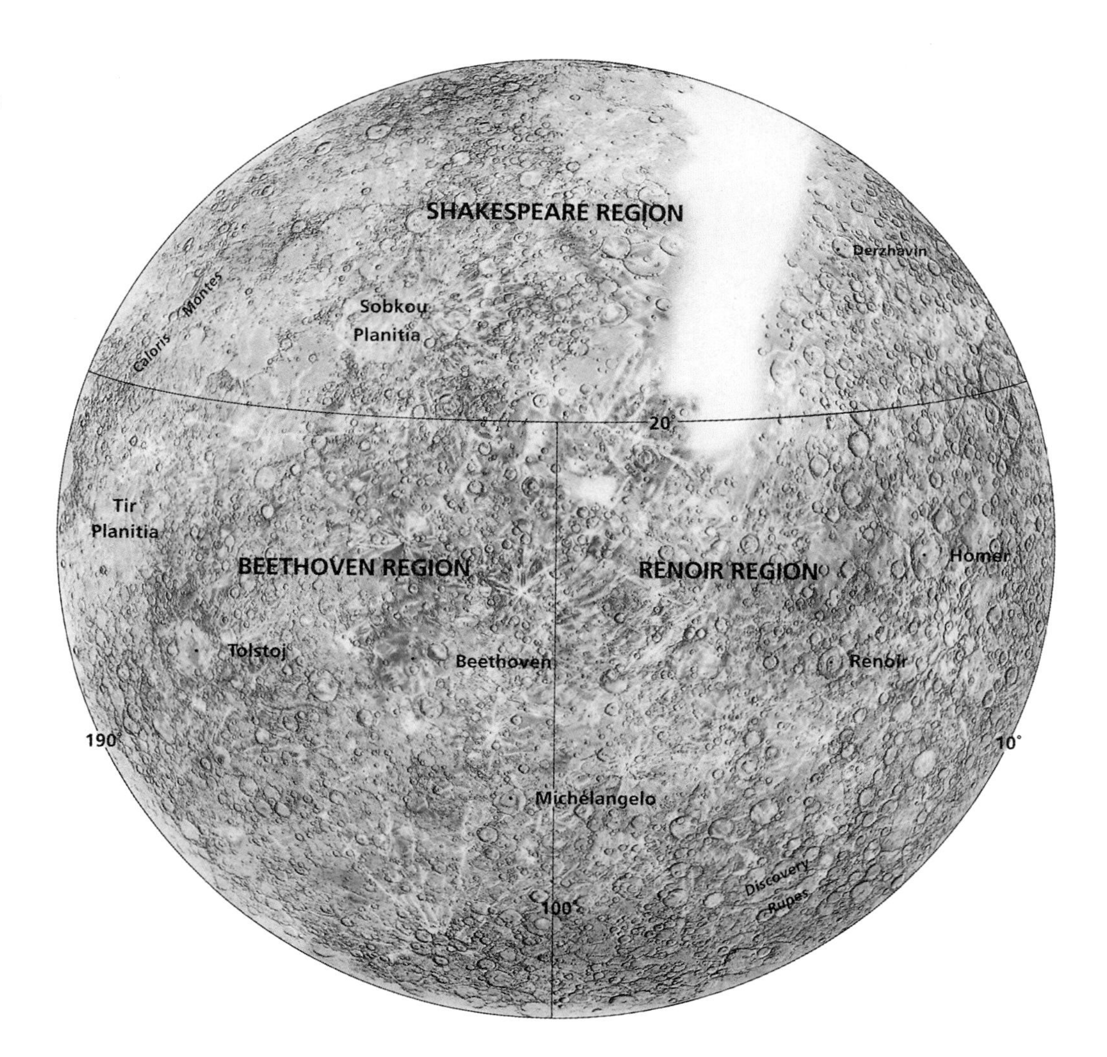

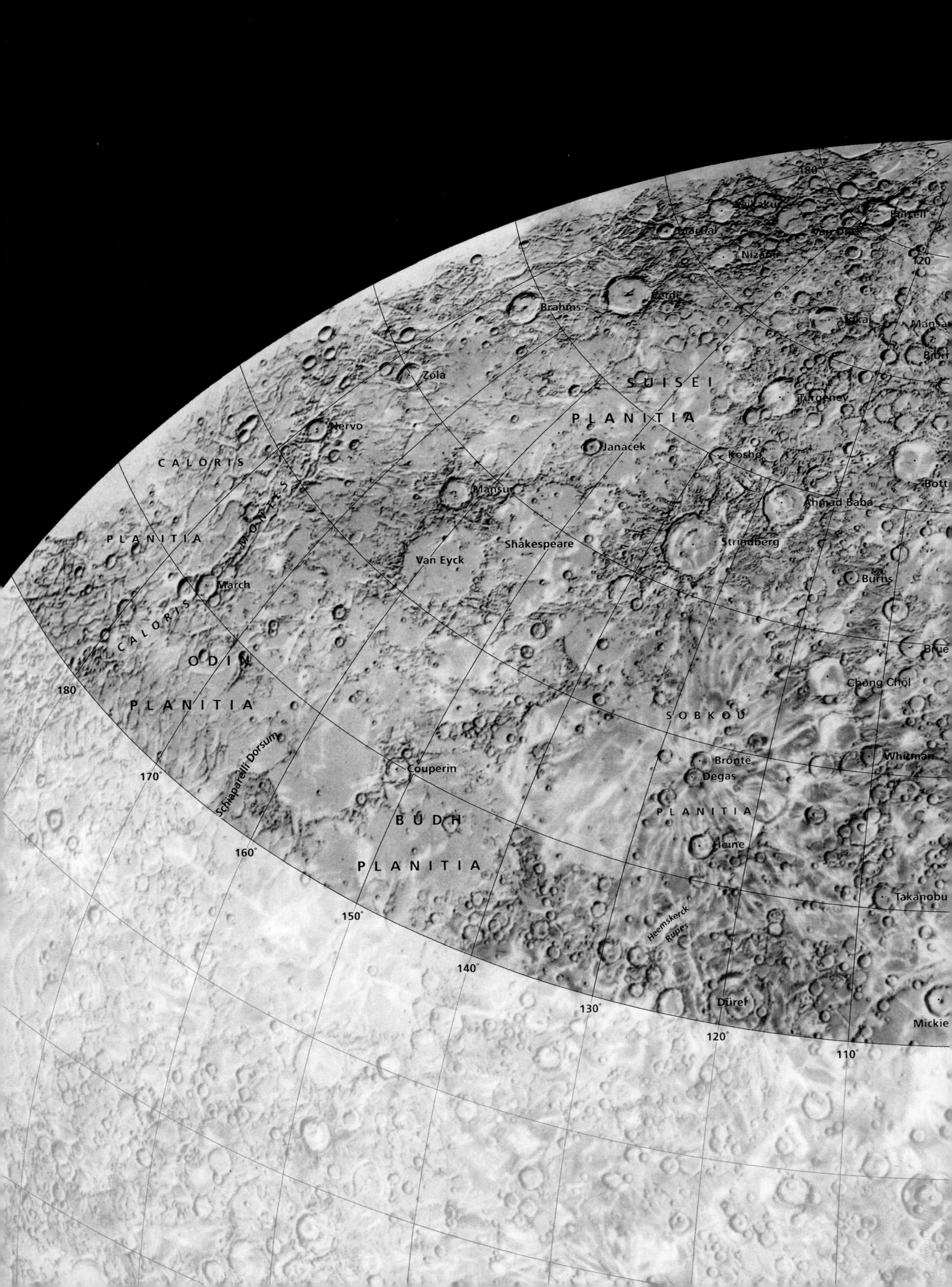
180°
Purcell
120°
Brahms
Verdi
Zola
SUISEI
PLANITIA
Turgenev
Nervo
Janaček
CALORIS
PLANITIA
MONTES
Mansur
Ahmad Baba
Shakespeare
Strindberg
Van Eyck
March
CALORIS
Burns
ODIN
PLANITIA
180°
Chong Chol
SOBKOU
PLANITIA
Brontë
Degas
Whitman
Couperin
Schiaparelli Dorsum
170°
BUDH
PLANITIA
Heine
160°
150°
Takanobu
Heemskerck Rupes
140°
130°
Dürer
120°
110°

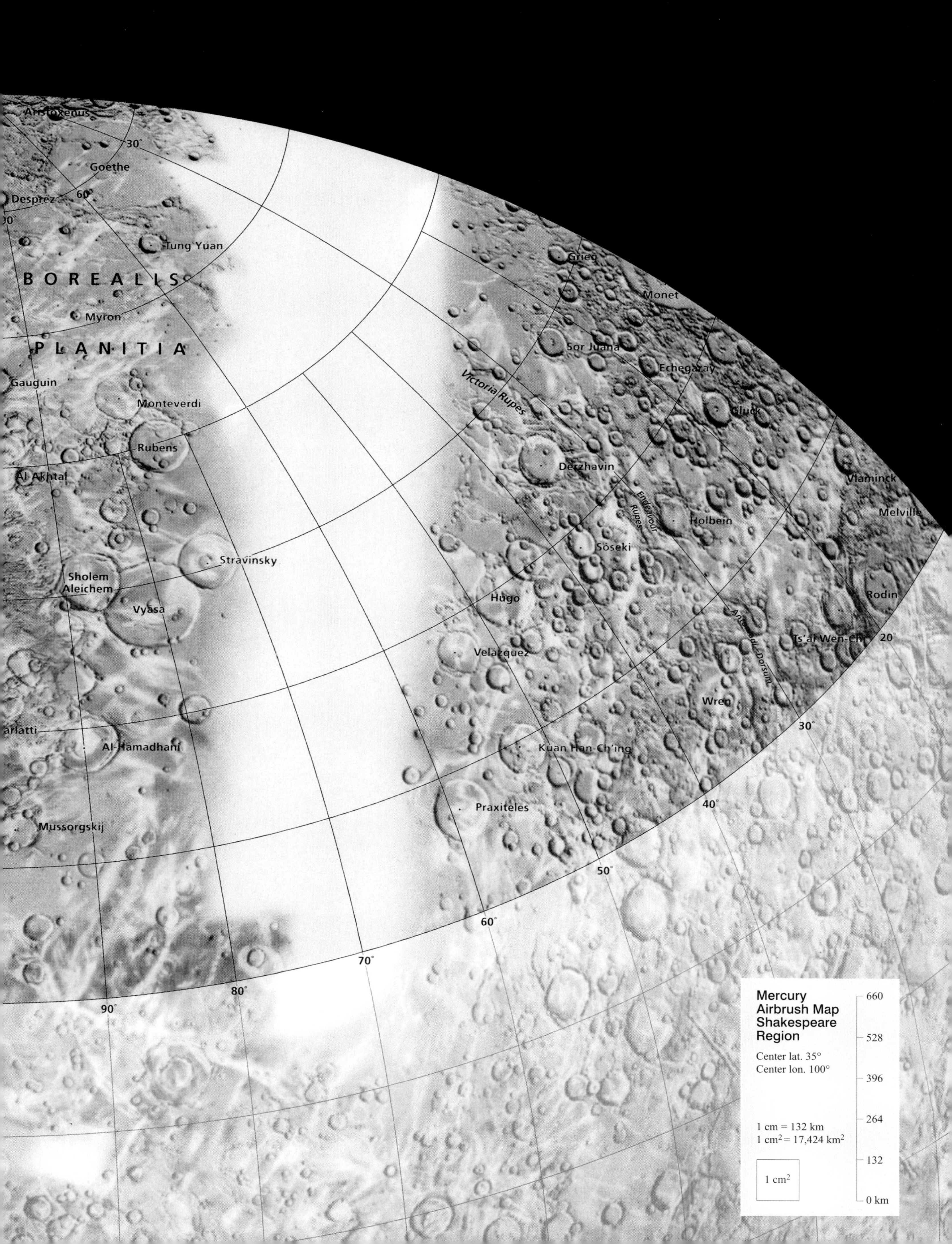
Aristoxenus
30°
Goethe
Desprez
60°
90°
Tung Yüan
BOREALIS
Myron
PLANITIA
Gauguin
Monteverdi
Rubens
Al-Akhtal
Stravinsky
Sholem Aleichem
Vyāsa
arlatti
Al-Hamadhani
Mussorgskij
Grieg
Monet
Sor Juana
Echegaray
Victoria Rupes
Gluck
Derzhavin
Endeavour Rupes
Vlaminck
Melville
Holbein
Sōseki
Hugo
Antoniadi Dorsum
Rodin
Ts'ai Wen-Chi
20°
Velázquez
Wren
30°
Kuan Han-Ch'ing
Praxiteles
40°
50°
60°
70°
80°
90°
Mercury Airbrush Map Shakespeare Region
Center lat. 35°
Center lon. 100°
1 cm = 132 km
1 cm² = 17,424 km²
1 cm²
660
528
396
264
132
0 km

70°
80°
90°
100°
Vivaldi
Tansen
Polygnotus
Rajnis
Boethius
Thakur
Machaut
Snorri
Simeiz Vallis
Flaubert
Repin
Futabatei
Sullivan
Raphael
Haydn
Matisse
Unkei
Lessing
Rude
Chekhov
Carducci
Copley
Astrolabe Rupes
Smetana
Ghiberti
Sūr Dās
Coleridge
Riemenschneider
Fram Rupes
Ma Chih-Yuan
Adventure Rupes
Holberg
Sei
Spitteler
Okyo
Li Ch'ing-Chao
90°
100°
120°

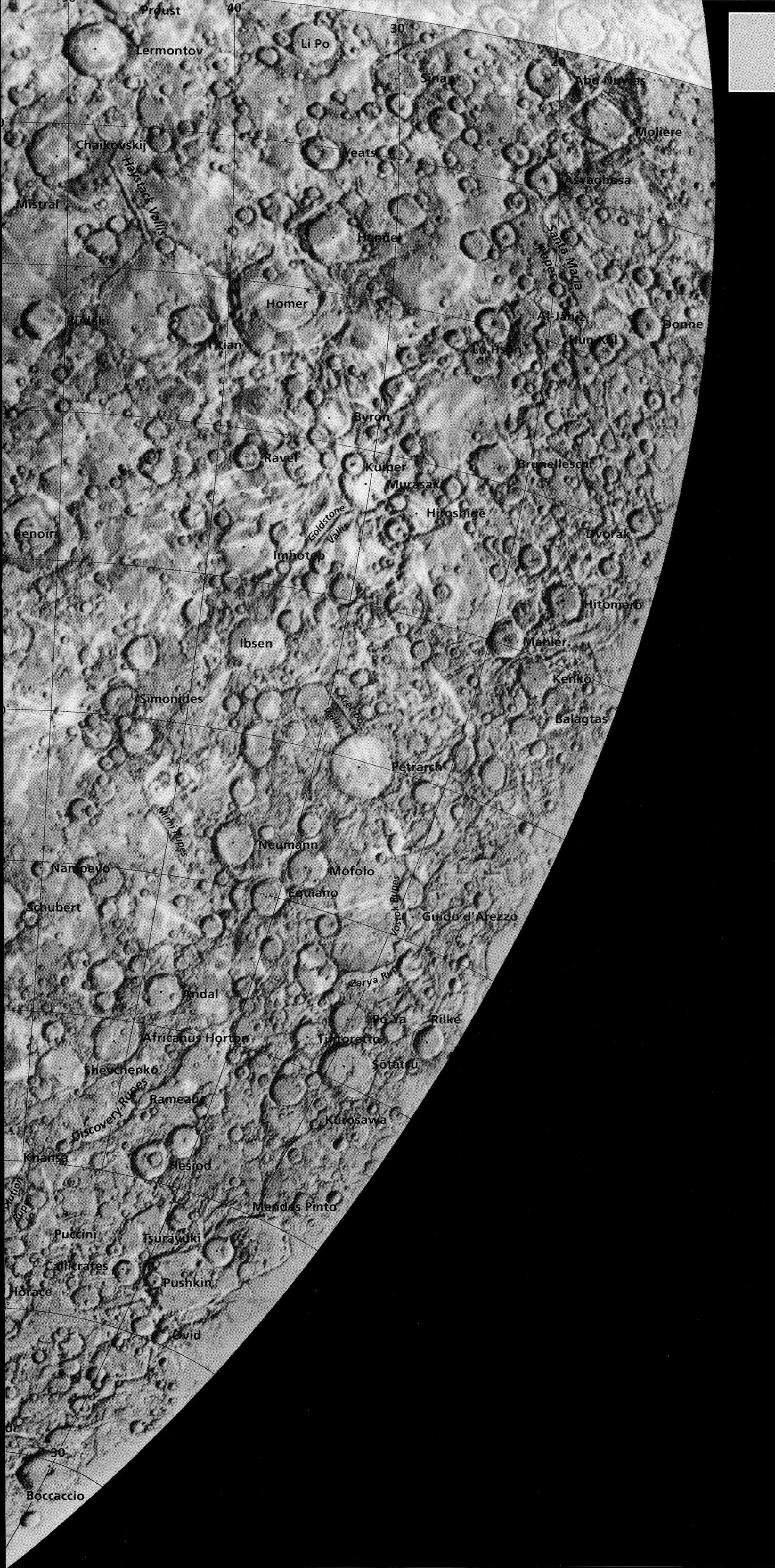

Mercury Airbrush Map Renoir Region

Center lat. –35°
Center lon. 55°

1 cm = 132 km
1 cm^2 = 17,424 km^2

1 cm^2

660
528
396
264
132
0 km

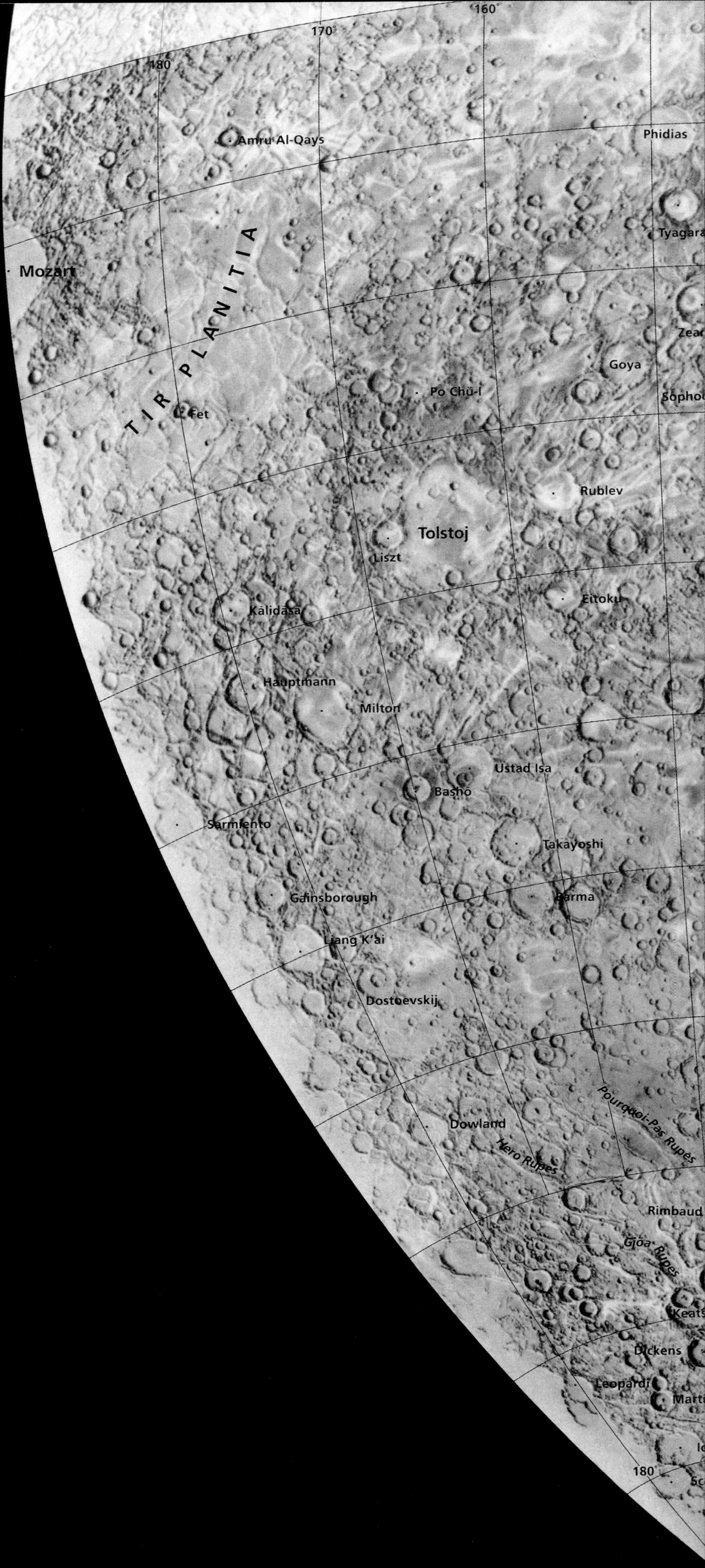

160°
170°
180°
Amru Al-Qays
Phidias
Tyagara
Mozart
TIR PLANITIA
Goya
Po Chü-I
Fet
Rublev
Tolstoj
Liszt
Eitoku
Kālidāsa
Hauptmann
Milton
Ustad Isa
Bashō
Sarmiento
Takayoshi
Gainsborough
Sarma
Liang K'ai
Dostoevskij
Pourquoi-Pas Rupes
Dowland
Hero Rupes
Rimbaud
Gjöa Rupes
Dickens
Leopardi
180°

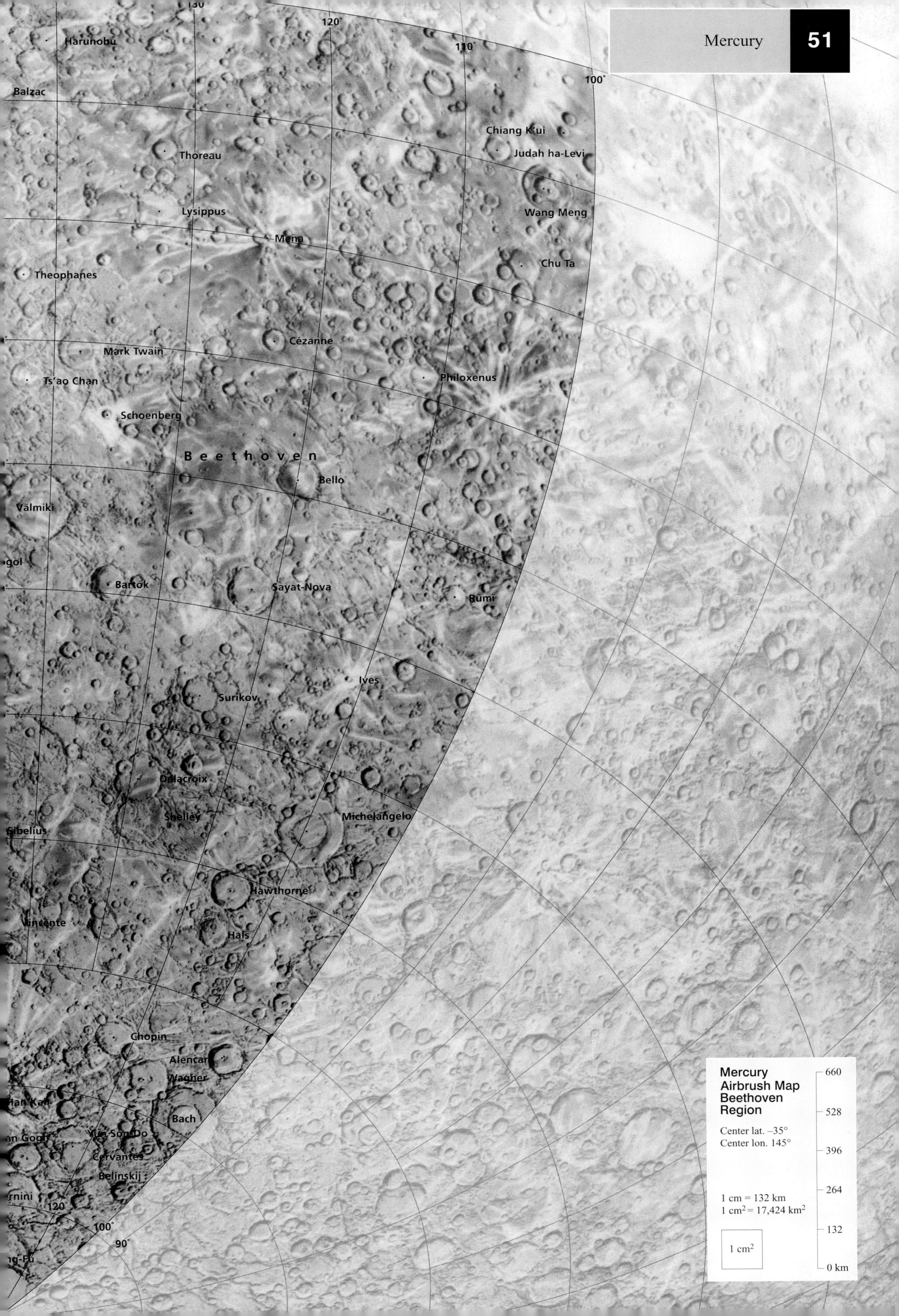
Harunobu
Balzac
Thoreau
Lysippus
Mena
Theophanes
Chiang K'ui
Judah ha-Levi
Wang Meng
Chu Ta
Cézanne
Mark Twain
Ts'ao Chan
Philoxenus
Schoenberg
Beethoven
Bello
Vālmiki
Bartók
Sayat-Nova
Rūmi
Ives
Surikov
Delacroix
Shelley
Sibelius
Michelangelo
Hawthorne
Vincente
Hals
Chopin
Alencar
Wagner
Bach
Cervantes
Belinskij
130°
120°
110°
100°
90°
Mercury
Airbrush Map
Beethoven
Region
Center lat. –35°
Center lon. 145°
1 cm = 132 km
1 cm² = 17,424 km²
1 cm²
660
528
396
264
132
0 km

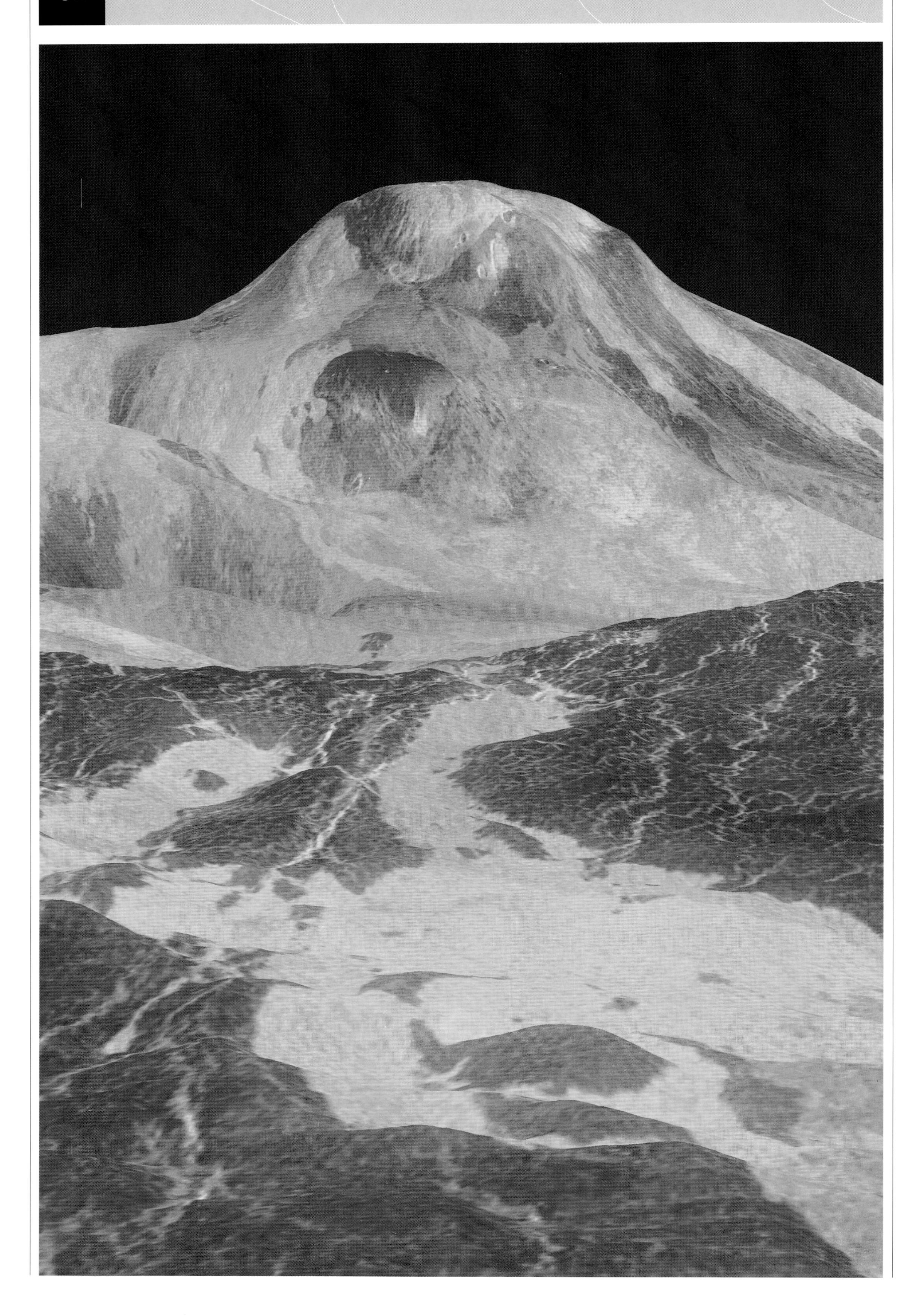

Venus

Even to the casual observer, Venus is distinctive. After the Sun and the Moon, it is the brightest object in the sky and has been called both the "evening star" because of its prominence after sunset and the "morning star" because of its appearance at dawn. Unlike most planets, Venus is in retrograde rotation, spinning in the direction opposite to its orbit around the Sun. The rate of rotation is so slow, however, that the length of one day on Venus (from noon to noon) is equal to more than 115 Earth days.

Like Mars, Venus has also stimulated fertile imaginations and has frequently been the subject of science fiction. Ideas of its surface environment included steamy swamps teaming with life and dry, wind-swept deserts. Both Earth-based and spacecraft observations show that Venus's surface is inhospitable, at least from the perspective of terrestrial life. The predominantly carbon dioxide atmosphere of Venus is so dense that it exerts a pressure of 90 bars on the surface, or 90 times that on Earth – comparable to being underwater on the seafloor at a depth of more than 900 meters. Droplets of sulfuric acid and other compounds in the atmosphere form thick, swirling clouds that hide the surface from view. The clouds form a layer some 30 kilometers thick 60 kilometers above the surface.

The thick venusian atmosphere creates a thermal blanket above the surface and causes temperatures to rise to more than 450°C (lead would be molten on Venus) in most places on the planet. This "greenhouse effect," or heat-trapping by the carbon dioxide atmosphere, was recognized as a dominant process for Venus long before the effect was appreciated for Earth.

Amazingly, machines can survive on the surface of Venus. Soviet robotic probes have landed and survived nearly two hours before the crushing pressure and searing heat silenced the radio transmitters. The Soviets successfully landed 12 *Venera* spacecraft and tracked balloon probes high in the atmosphere. United States entry probe missions have been carried out on a much smaller scale. Nevertheless, the measurements made simultaneously by four *Pioneer* entry probes in 1978 have provided essential data on the physical and chemical properties of the venusian atmosphere to com-

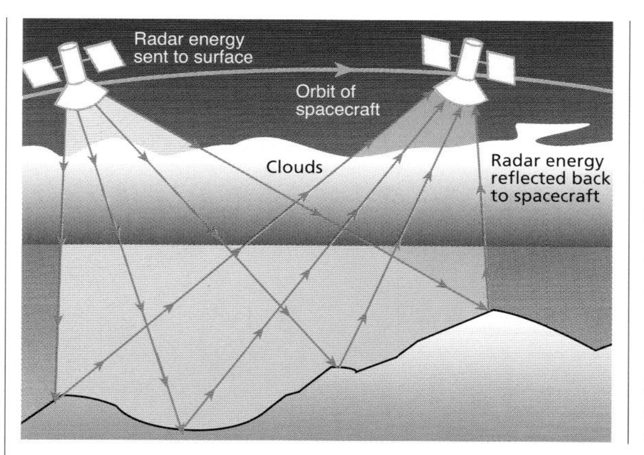

Pulses of radar energy are sent from a spacecraft to the surface of Venus where some of the energy is reflected back to the spacecraft depending on terrain shape, surface roughness, and other factors. Computers process the complex signals to synthesize radar images of the surface, such as those obtained by the *Magellan* spacecraft.

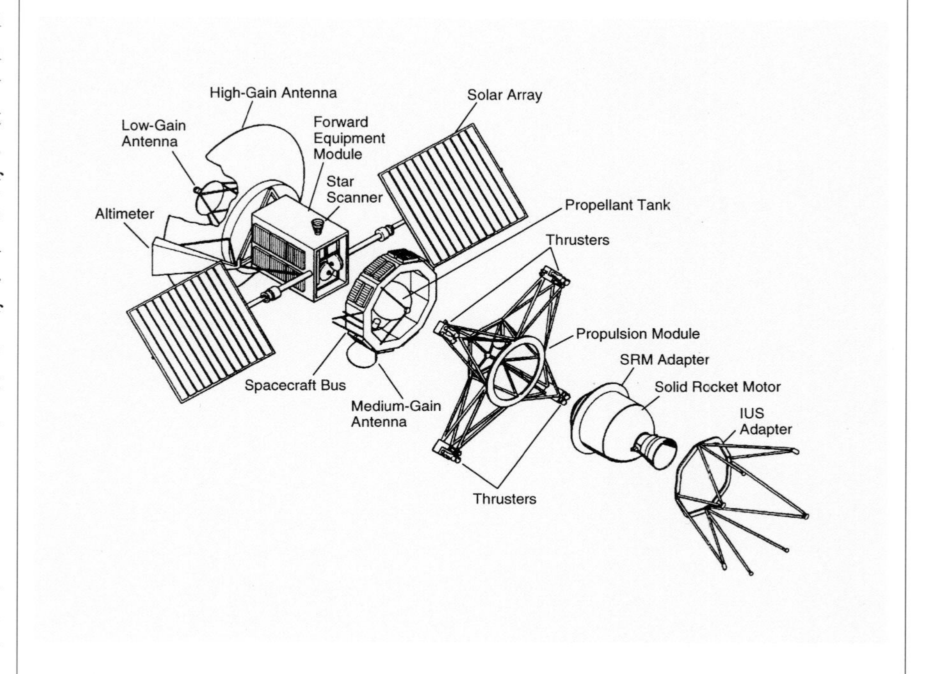

Diagram showing the principal components of the *Magellan* spacecraft.

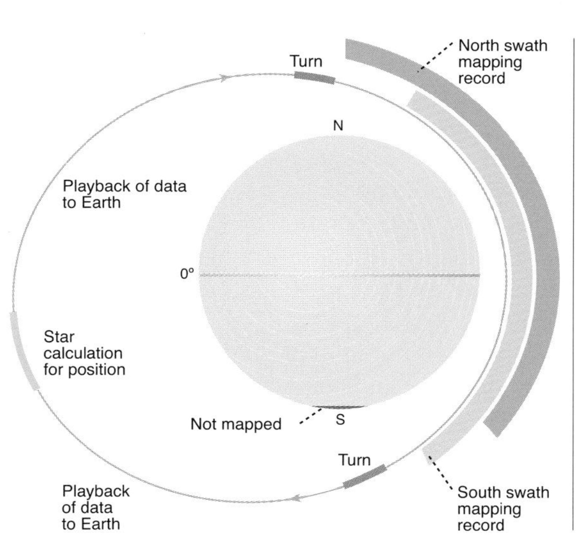

Diagram showing typical mapping sequence by the *Magellan* spacecraft in its orbit around Venus.

Facing page: Magellan radar image of Maat Mons, an 8-kilometer high volcano located at 0.9° N, 194.5° E. Visible in the foreground are geologically young lava flows (bright areas) and tectonically fractured terrain. False color has been added and the vertical scale exaggerated to enhance detail (JPL photograph P-39326).

Facing page: image taken by the *Pioneer Venus Orbiter* in 1979 showing thick clouds which hide the surface of Venus from view by conventional cameras. Color has been added to enhance detail in this computer-generated image. (JPL photograph P-25316).

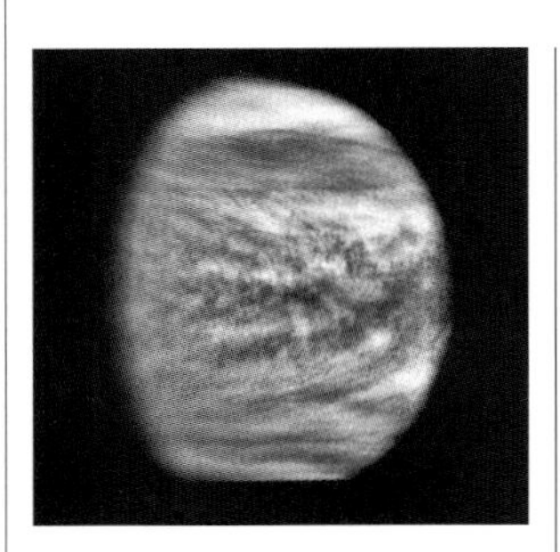

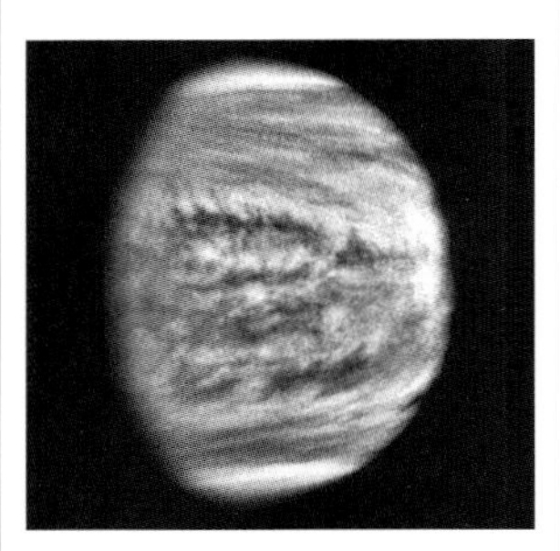

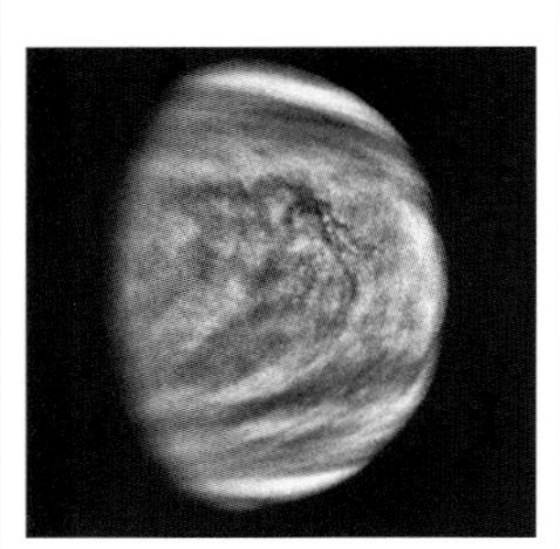

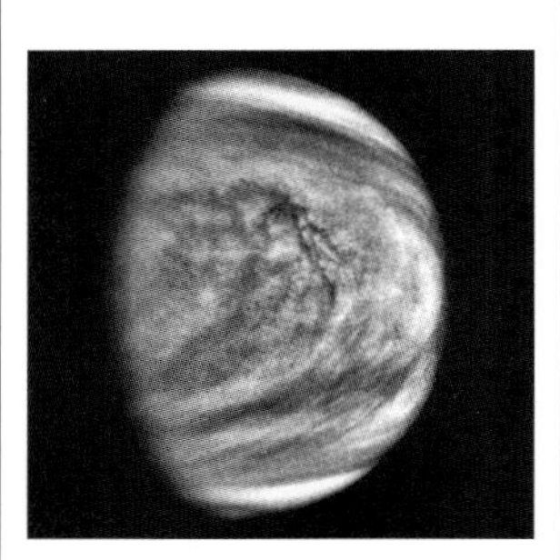

Four views of Venus taken by the *Galileo* spacecraft. Top 2 pictures were taken about 4 and 5 days before closest approach, bottom 2 pictures were taken 6 days after closest approach, 2 hours apart. Images have been processed to show detail in the clouds and to enhance the bright polar regions (JPL photograph P-37215).

plement the *Venera* results. Pictures taken by the *Venera* spacecraft of the surface, along with measurements of the atmospheric pressure, surface temperatures, wind speeds, chemical compositions of the rocks, and other data, confirm the view of a blistering, hostile world devoid of liquid water and of life.

From orbit, only radar can penetrate the thick, acid-laden clouds to reach the surface. In 1978, the United States *Pioneer-Venus* spacecraft carried a radar system in orbit and provided the first global assessment of the topography of Venus. The data showed that Venus has huge mountains, valleys, and vast plains. Because it lacks an ocean and a reference "sea level" for elevations, the zero reference used for topography is the average radius from the center of the planet, or 6,051 kilometers.

During the 1980s, Earth-based radar telescopes and the Soviet radar-mapping spacecraft, *Veneras 15* and *16*, returned pictures of the surface that provided clues to the geologic diversity of Venus. It was not until the successful operation of the United States *Magellan* spacecraft, beginning in 1990, however, that the incredible complexity of Venus was revealed.

Geology

Because Venus is nearly the same size and density as Earth, and because the planets are neighbors in the Solar System, the two are often considered "sister" planets. Theories of the interior configuration of Venus, based on its global properties and geophysical measurements made from spacecraft, suggest that Venus has a lithosphere 20 to 40 kilometers thick overlying an upper mantle and a core that may be about the same size as Earth's core. Earth and Venus might share geologic processes such as volcanism and tectonism. Consequently, one of the key goals for the exploration of Venus was to search for evidence of Earthlike plate tectonics. For example, does Venus show evidence of crust that has spread apart or been subjected to heat rising from the interior and forming hot spots? *Magellan* results show no organized arrangements of such features, other than possible hot spots.

The best evidence for possible hot spots includes huge circular features, termed coronae, seen on the surface. Some of the coronae are more than 1,000 kilometers wide. Coronae consist of discontinuous ridges and grooves arranged in concentric patterns, suggesting deformation of the crust by upwelling from the venusian mantle. However, models of the interior suggest that the upper mantle of Venus may be rather stiff, inhibiting convection. Alternative explanations propose that coronae may be deformed volcanoes or are structures related to impact processes.

Venus has its own unique set of tectonic features. For example, some areas are called tesserae terrain from the Greek word (tessara) referring to mosaics. Tesserae terrain consists of crust fractured into blocks from several to 20 kilometers across and provides evidence of extensive tectonic deformation. In other places, belts of ridges and grooves hundreds of kilometers long may reflect large-scale tectonic deformation not apparently related to Earthlike plate motion.

Volcanic features are abundant on Venus. Some of the large mountains, such as Theia Mons and Rhea Mons, are shield volcanoes several hundred kilometers across which are composed of accumulations of lava and are 4 kilometers high. Both volcanoes are associated with an enormous rift valley that can be traced more than 3,000 kilometers, a scale comparable to the rift valley of east Africa. Other areas, such as Ishtar Terra, are punctuated by huge volcanic craters, or calderas. In addition, fields of hundreds of small volcanic domes, cones, and shields attest to local eruptions. Radar images returned by the *Magellan* spacecraft show that the vast expanses of plains on Venus are covered by sheets of lava erupted from fissures and calderas and often carried hundreds or even thousands of kilometers by rivers of lava.

As a consequence of deep underground magma chambers erupting their lavas on the surface, some parts of the crust collapsed, leaving irregular depressions and pit craters. One such area is in Lakshmi Planum in the northern hemisphere. This area displays linear depressions and collapse pits typical of a complex volcanic region.

Venus also has at least one type of volcano not seen elsewhere in the Solar System. These volcanoes resemble giant pancakes and consist

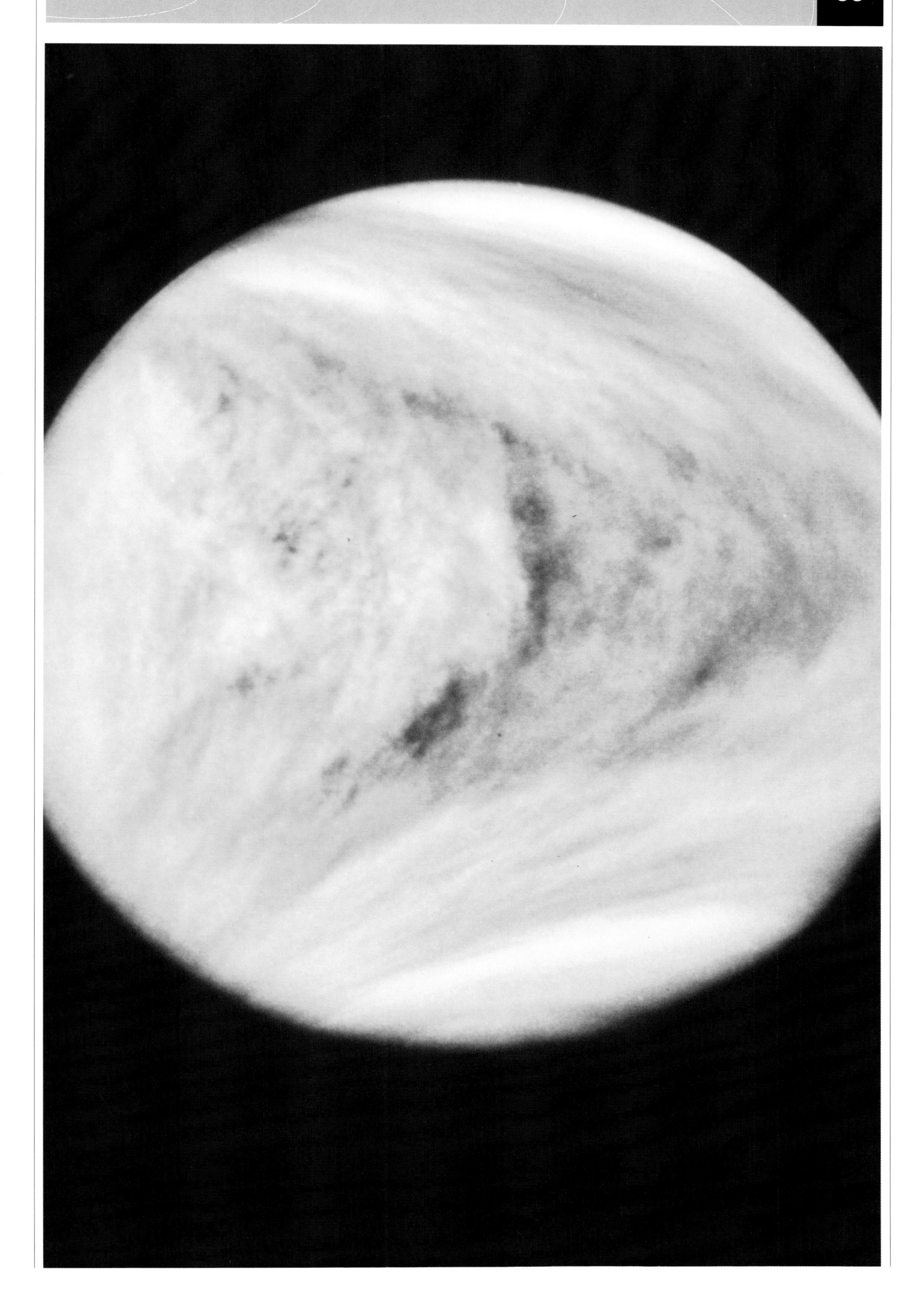

of circular domes or flat-topped mounds of lava flows presumed to have been rather viscous and pastelike at the time of their eruption. They are as large as 25 kilometers across and 1 kilometer high and may represent a style of eruption or a type of magma not previously encountered in planetary exploration.

Artist's conception of the radar mapper, *Magellan*, in orbit around Venus. For 37 minutes of each orbit, the Synthetic Aperture Radar (SAR) imaged a ~24-kilometer wide swath and obtained altimetry and radiometry data. As the spacecraft moved to the high point of its orbit, the data were transmitted to Earth (NASA P-31315).

Does Venus currently experience active volcanism? Observations of large variations of sulfur dioxide (a common product of active volcanoes on Earth) in the atmosphere lead many planetary scientists to think so. Some relatively fresh lava flows visible on *Magellan* images could also support the notion of geologically-recent volcanism. But, as yet, no direct evidence of active volcanoes, such as flowing lava, has been found.

The Soviet *Venera* lander spacecraft provided clues to the composition of the lava flows and other rocks on Venus. Most of the landings occurred on the flanks of Beta Regio, a prominent volcanic region astride the equator. Composition measurements made by instruments on the landers show that the rocks are similar to basalts found on the sea floor of Earth. However, a kind of basalt, rare on Earth, with a high percentage of potassium and a basalt rich in sulfur were also identified. Two other Soviet probes, each from the *Vega* mission, landed in Rusalka Planitia on the northern flanks of Aphrodite Terra and also returned measurements typical for basaltic rocks. Thus, as has been found on Earth, Earth's Moon, and Mars, eruptions of basalt and basaltlike lava flows were probably common on Venus.

In addition to volcanism and tectonic deformation, the surface of Venus has been altered

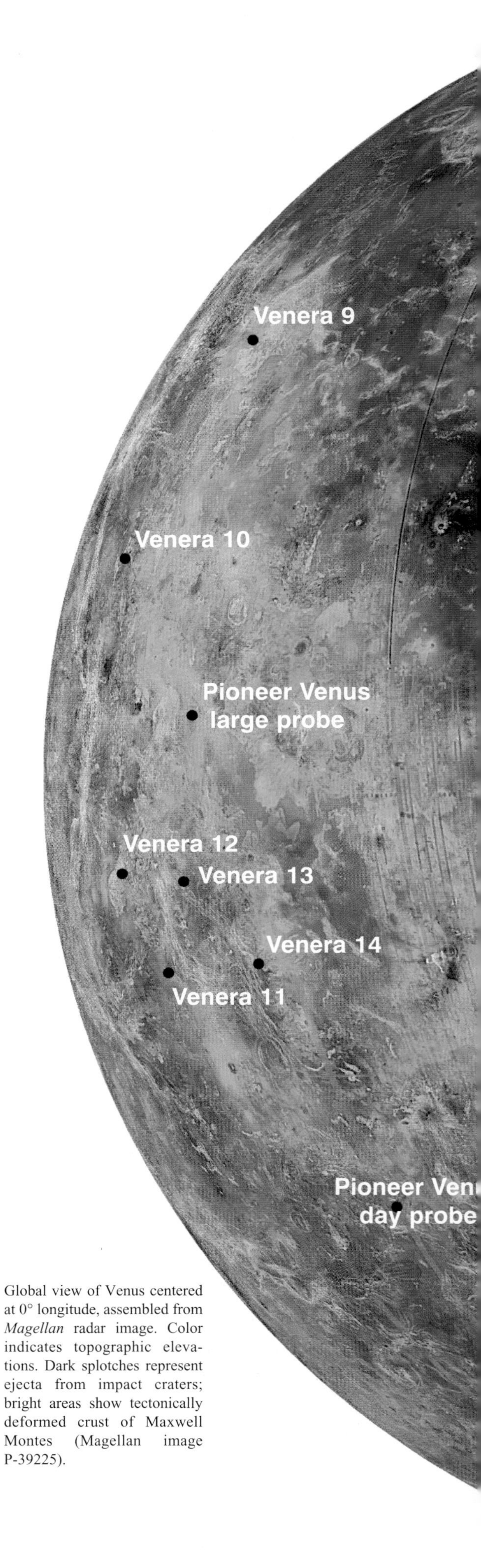

Global view of Venus centered at 0° longitude, assembled from *Magellan* radar image. Color indicates topographic elevations. Dark splotches represent ejecta from impact craters; bright areas show tectonically deformed crust of Maxwell Montes (Magellan image P-39225).

Pioneer Venus north probe
Venera 4
Venera 5
Venera 6
Venera 7
Venera 8
Pioneer Venus night probe
Planetary Radius (km)
Below
6050
6052
6054
6056
6058
and higher

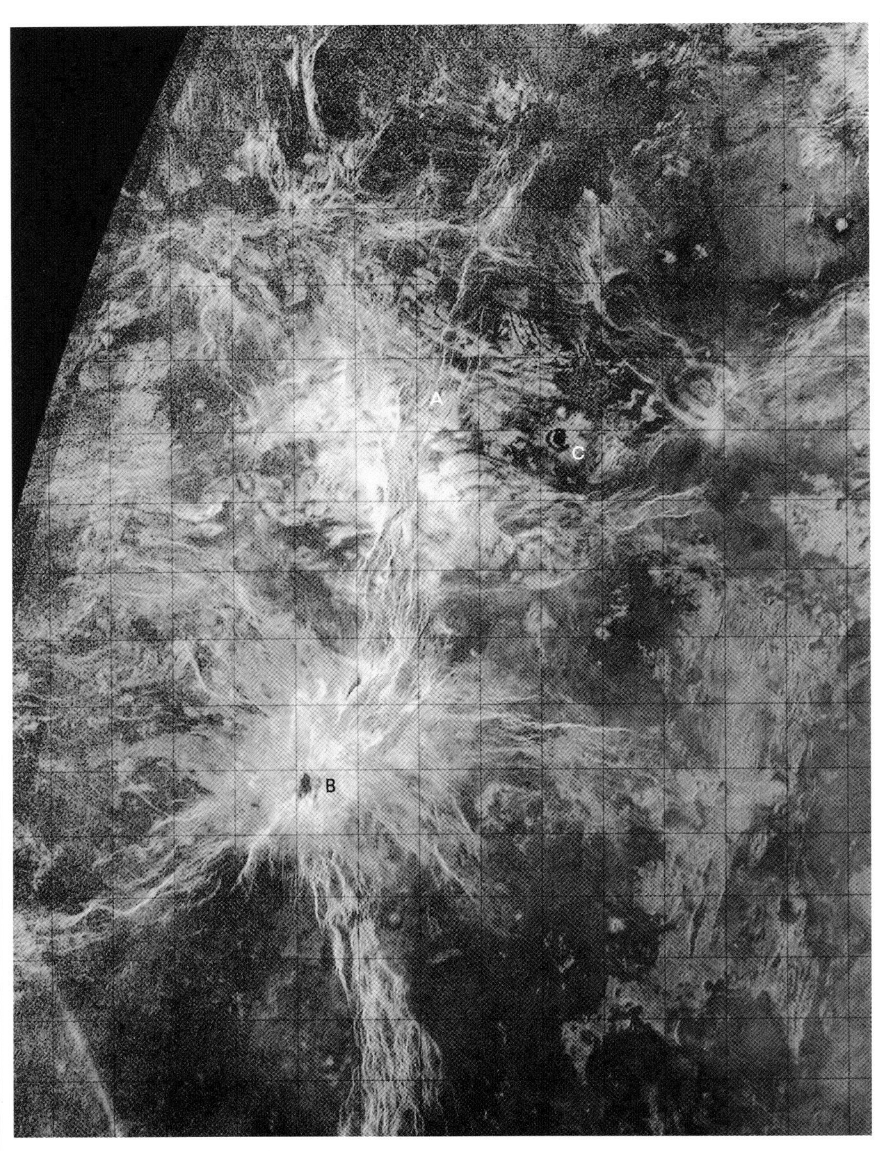

Image of Beta Regio in the northern hemisphere of Venus obtained by the Arecibo radar telescope at Puerto Rico. Volcanoes Rhea Mons (A) and Theia Mons (B) are connected by a rift valley, Devana Chasma, whose fractures and walls show as radar-reflective (bright) linear features. Also visible are an impact crater (C) and its radar reflective ejecta (*Magellan* MRPS 32722; image courtesy of D. B. Campbell).

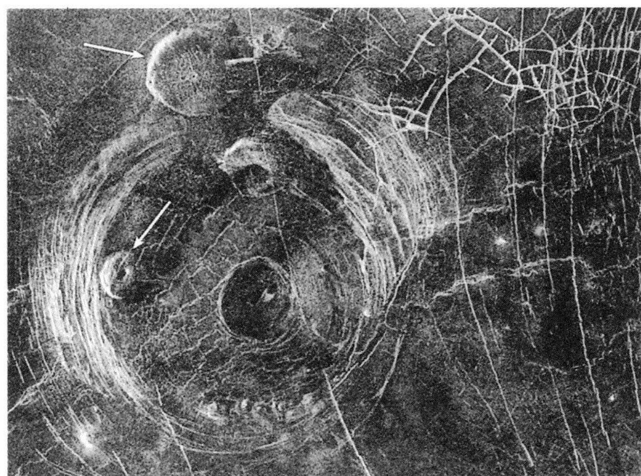

Radar image of a corona (300 kilometers across) centered at 59° S, 164° E, south of Aphrodite Terra and flat-topped, dome-shaped volcanoes (arrows) (*Magellan* image P-38340).

by erosion and impact cratering. However, the number of impact craters found on Venus is low when compared to Mercury and Earth's Moon, suggesting that much of the venusian surface is geologically young. Most venusian impact craters show fields of rugged ejecta, sharp crater rims, and other signs of little erosion. The floors of many of the craters also appear to be partly filled with lava or with rock that was melted during the impact.

One of the many surprises revealed by the *Magellan* mission was the unusual flowlike features associated with some impact craters. The crater outflow features include wide channels, as long as 150 kilometers, that apparently originated from the ejecta of large impact craters. The channels have sinuous walls a hundred or more meters high, show "islandlike" features on their floors, and empty imperceptibly onto the surrounding plains. It is debatable whether the crater outflow features resulted from melted rock generated as part of the impact process, whether they represent volcanism "triggered" by the impact, or whether they resulted from some as yet unknown process. Although the features superficially resemble water channels, this is not possible because of the high surface temperatures throughout most of Venus's history.

If water has not modified the surface of Venus, does erosion occur? Photographs taken by *Venera* and *Magellan* suggest that some weathering and erosion does take place. The dense, hot, acid-rich atmosphere is very corrosive, so rocks are sure to be altered chemically. Sediments and other fine-grained material, at least partly weathered from rocks, are seen in all four of the landing sites where the *Venera* spacecraft took pictures. Only the amount of sediments differs from one site to another.

Although only sluggish winds of 0.5 to 2 meters per second have been measured on the surface of Venus, they are sufficient to move sand grains in the high-density atmosphere. Radar signals suggest that most of the venusian surface is bedrock, only about one-fourth being covered with porous or loose materials. Consequently, agents of abrasion, such as sand grains, may not have eroded the surface. Nonetheless, *Magellan* images do show evidence of wind in some regions. Fan-shaped,

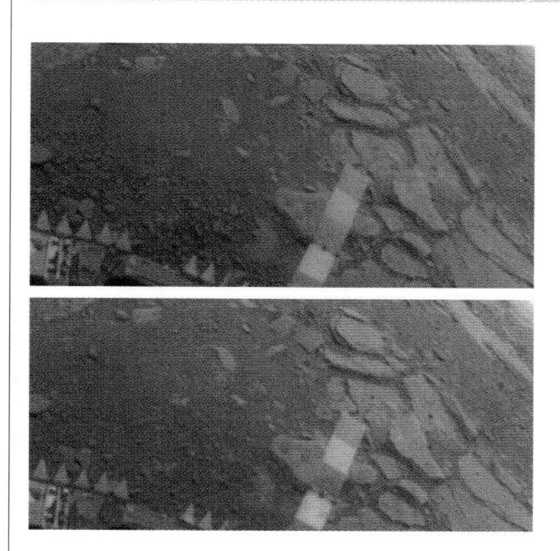

Color panorama of the surface of Venus taken by the Soviet *Venera 13* lander. The color properties of the rocks suggest the presence of iron-rich minerals that have been oxidized, indicating that chemical weathering has taken place on Venus. The lower part of the pictures shows the spacecraft and a color scale bar used to adjust the pictures' color. The top view is as seen from the spacecraft; the bottom view shows the same scene as it would appear in "white" light without the interference of the atmosphere (courtesy of C. Pieters).

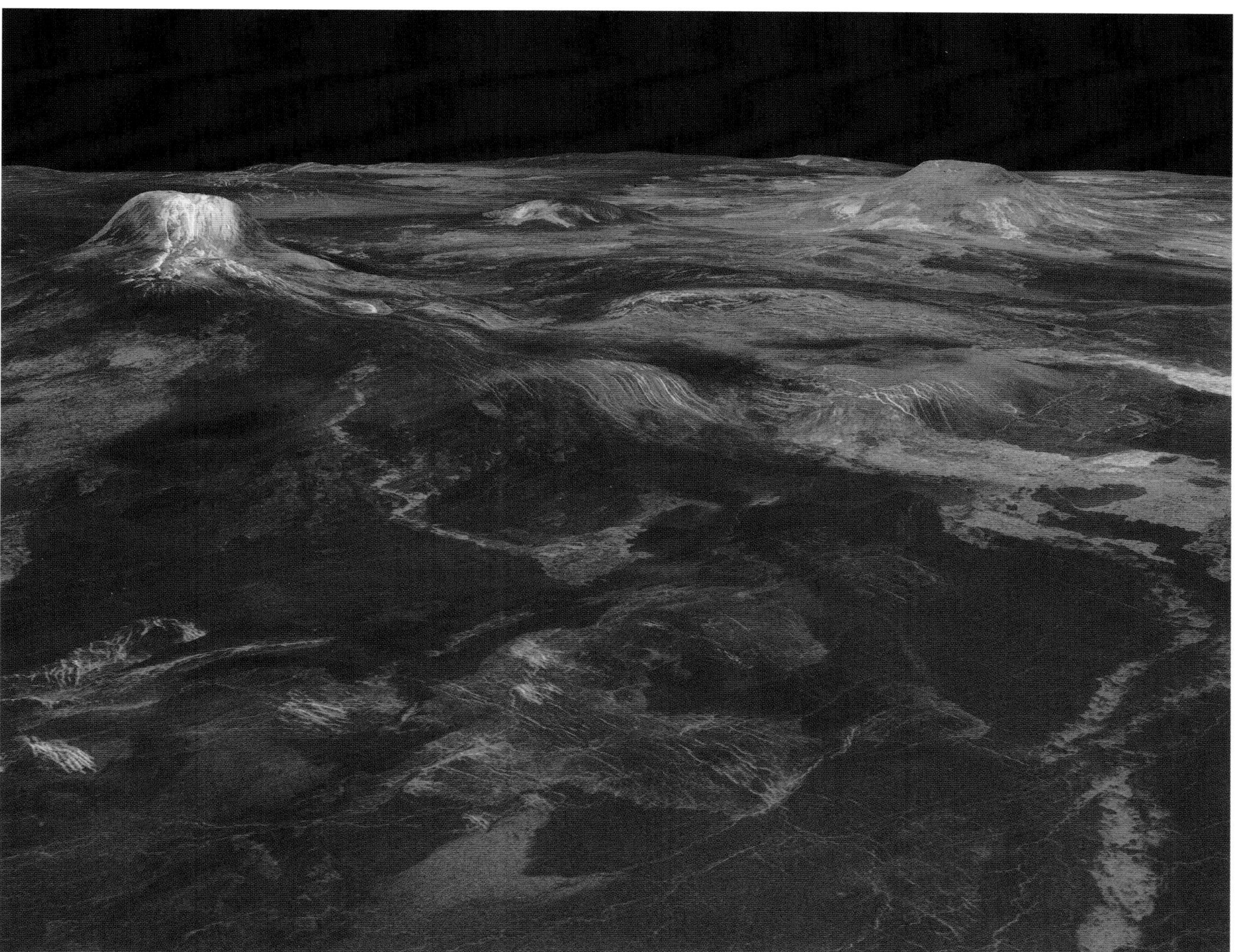

Left: the eastern part of Alpha Regio at 30° S, 11.8° E, viewed from the *Magellan* spacecraft, showing three dome-like volcanoes. These and similar features average some 25 kilometers across. False color has been added and the vertical scale exaggerated to enhance detail (JPL photograph P-38870).

Above: radar view across western Eistla Regio showing two prominent volcanoes, Gula Mons on the left and Sif Mons on the right, and various lava flows. Color has been added to enhance detail, and the height of the features has been exaggerated about 20 times. Gula Mons rises about 3 kilometers above the surrounding plains, while Sif Mons is 2 kilometers high and about 300 kilometers across (*Magellan* image P-38724).

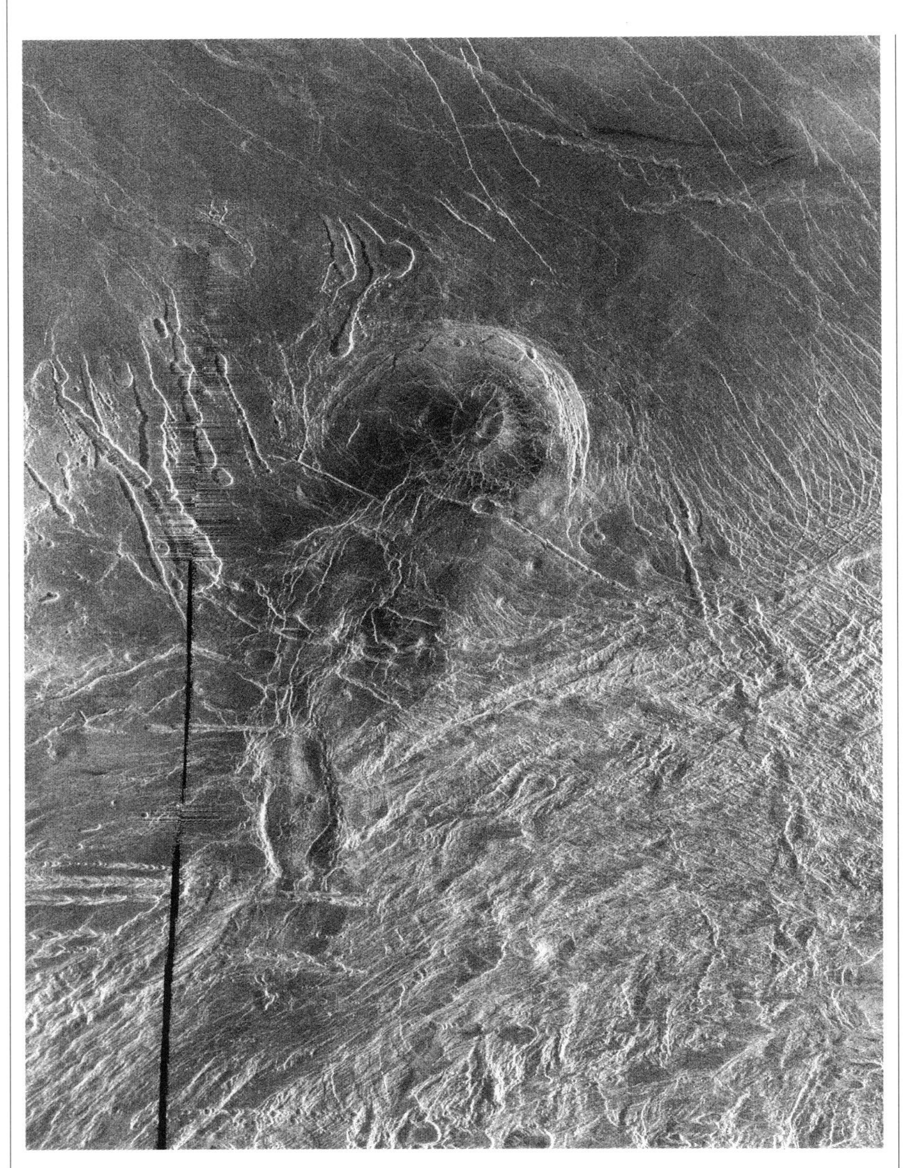

Mosaic of radar images of Lakshmi Planum, the smooth area in the north (top), and Clotho Tessera to the south. Irregular depressions and pits formed by collapse of the crust as magma and lava drained from subsurface chambers. Area shown is centered at 61° N, 341° E and is about 200 kilometers wide (*Magellan* image P-37139).

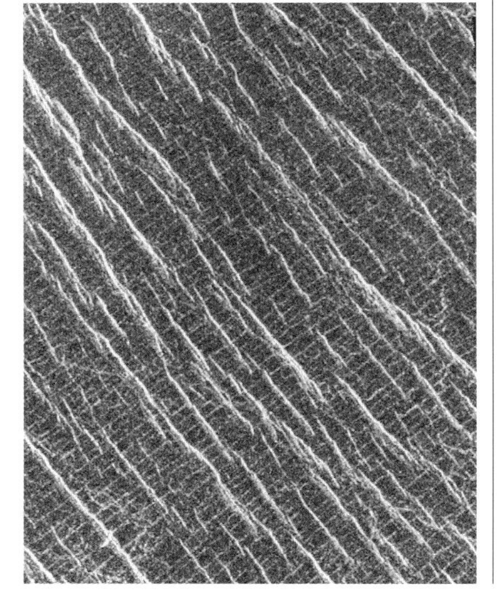

Radar image of the Guinevere region, showing deformation of the crust by northwest-southeast fractures breaking the terrain into rectangular blocks about 1 by 2 kilometers. Area shown is about 36 kilometers by 49 kilometers (*Magellan* image P-36699).

bright (radar-reflective) features up to 20 kilometers long are found in association with small hills. Most of the surface around the hills is only moderately radar reflective, indicating the presence of loose particles. The bright zones probably denote locations where winds streaming around the hills have stripped the loose material from the surface, exposing bedrock that is radar reflective. These wind streaks are found in several places on Venus and serve as local "wind vanes," recording the prevailing wind direction at the time of their formation.

In addition to wind streaks, several areas on Venus resemble sand dunes on Earth. These features are a few hundred meters wide and occur in fields nearly 100 kilometers wide. One such field is found in association with ejecta deposits of a large impact crater, named Aglaonice. Planetary scientists suggest that dune fields can form on Venus wherever sufficiently strong winds and a supply of small grains occur. Both conditions are apparently met at some craters where impact ejecta deposits provide a ready source of sand.

Geologic History

Geologic histories for planets and satellites are derived primarily from the ages and distributions of rocks seen on their surfaces, coupled with knowledge or inferences about their interior characteristics. Unfortunately, the interpretation of information for Venus has only begun with the return of *Magellan* images. Only the northern part of the planet has been systematically mapped (using *Venera 15/16* results) and no seismometer measurements (the primary source for information about planetary interiors) have been obtained. Consequently, there is only speculation on the geologic history of Venus.

Like the other terrestrial planets, Venus was probably assembled from small bodies about 4.6 billion years ago. Sufficient heat was generated by collisions to melt the proto-planet completely. With time and cooling, a crust, mantle, and core developed from the melt. Because Venus and Earth are comparable in size and density (and, hence, probably similar in compositions), the interior configurations of both planets are probably similar.

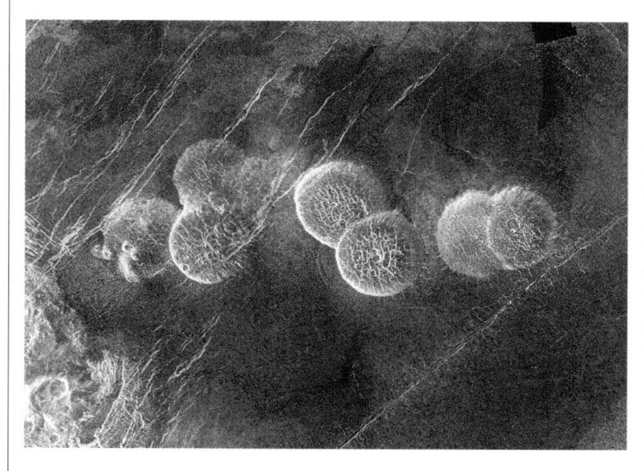

Magellan image of the eastern edge of Alpha Regio showing volcanic domes, each about 25 kilometers across. The domes appear to be composed of viscous lavas that cracked into concentric and radial fracture patterns as they cooled. Area shown is about 160 kilometers wide (*Magellan* image P-37125).

The lack of abundant impact craters on Venus suggests that, like Earth, most of the crust that formed early in its history has been destroyed, either by burial by lava flows, tectonic over-turning of the crust, or by erosion and burial of the surface by sediments. Of these three possible explanations, over-turning of the crust one or more times in the past seems to be most likely. Surface erosion and burial by sediments are least likely because erosional processes do not appear to be vigorous on Venus. Destruction of the impact craters by Earthlike plate tectonics is possible, but evidence of such processes has not been found, at least not in a style similar to the Earth's.

The processes that dominate the part of Venus's history accessible to study are volcanism and tectonism. The formation of both large and small volcanoes and the emplacement of vast sheets of lava over hundreds of millions of years on Venus were accompanied by deformation of the crust. Both of these processes show heat loss from the interior of the planet. It is possible – perhaps even likely – that these processes are still active. Are other processes active? Sluggish winds do sweep the surface with sufficient energy to move loose sand and dust, but the amounts of material available for transport by the wind may be rather limited. Impact cratering can also occur just as it does on other planets, but because of the dense atmosphere on Venus, only the larger objects will reach the surface. For example, most iron meteors less than about 30 to 100 meters in diameter will be consumed by heating, vaporization, and break up as they pass through the dense venusian atmosphere. Consequently, the smallest craters found on Venus are about 1.5 to 2 kilometers across.

Perhaps the most puzzling aspect of Venus's history is the evolution of its atmosphere. One might have expected the atmospheres of Earth and Venus to be similar – given their other similarities – and perhaps they were early in their histories. Yet, atmospheric evolution on the two planets took drastically different paths at some point in their evolution, leading to Earth's mild, nitrogen-oxygen atmosphere and Venus's dense, carbon dioxide atmosphere. There are many possible explanations for these differences. For example, like

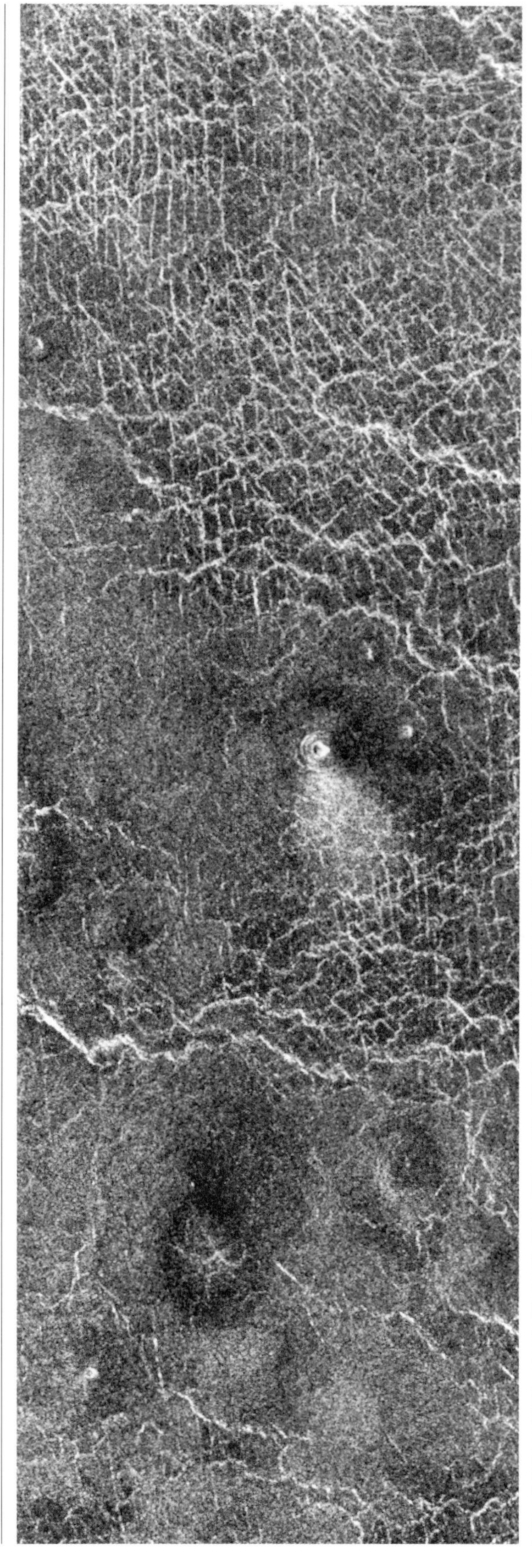

Magellan radar image of the area northeast of Ushas Mons, Venus, showing bright wind streaks, extending downwind from small volcanic cones. The streaks are about 10 kilometers long and may indicate the prevailing wind direction at the time of their formation (*Magellan* image P-36698).

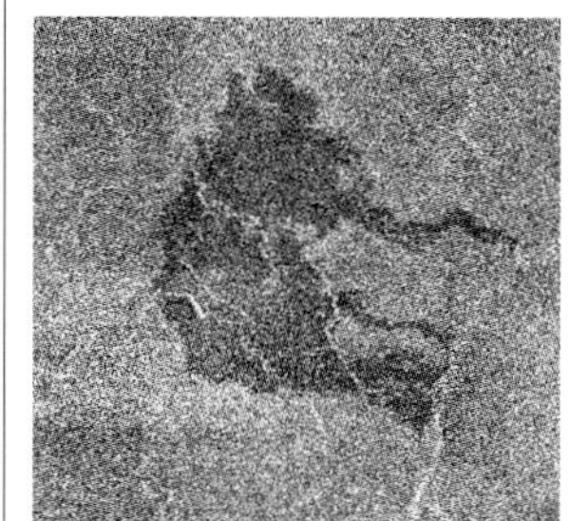

Magellan radar picture showing dark volcanic lava flow erupted from fissures (bright features to the right), with flow toward the left over older lava plains. Area shown is located at 5° N, 132° E and covers about 45 kilometers by 45 kilometers (*Magellan* image P-36704).

Facing page: oblique *Magellan* radar view of 3 impact craters in Lavinia Planitia: Saskia crater (lower center), about 37 kilometers across; Danilova, a 48-kilometer diameter crater (right); and Aglaonice crater (left). Clearly visible are ejecta deposits surrounding the crater rims, central-peak complexes in the middle of the craters, and the dark crater floors (*Magellan* image P-39146).

Venus, Earth also evolved a large quantity of carbon dioxide. Most of the carbon dioxide is in limestone and other rocky deposits, formed primarily by precipitation from sea water. If all of the carbon dioxide presently contained in rocks on Earth were released, Earth as a planet might also be enveloped in a thick carbon dioxide atmosphere. Moreover, it is likely that surface temperatures would rise through the greenhouse effect, just as they are thought to have risen on Venus.

Thus, part of the difference between Earth and Venus is linked to the development of oceans on Earth. Based on geochemical evidence, some scientists have suggested that Venus, too, may have evolved sufficient water to have had oceans – oceans that would have long since boiled. One of the puzzles is whether large quantities of water were present early in its history. Exploration of Venus is still in its infancy. Until a global, uniform view is obtained, knowledge of the geologic history of Venus remains fragmentary.

More data have been returned for Venus from the *Magellan* mission than from all previous missions. Analyses of these data will continue well into the future and are already yielding an enormous bounty. When coupled with knowledge derived from previous missions, *Magellan* data will undoubtedly help solve many current venusian puzzles.

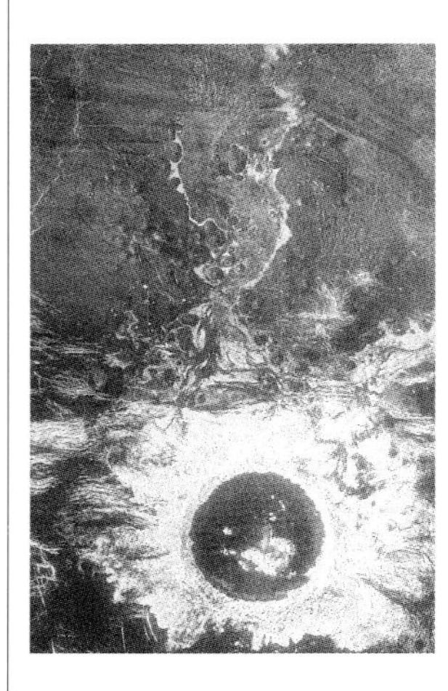

Radar picture showing channellike feature 150 kilometers long which appears to be associated with ejecta from the Aglaonice impact crater in the Lavinia region of Venus near 27° S, 339° E. Area shown is 160 kilometers by 250 kilometers (part of *Magellan* image P-36711).

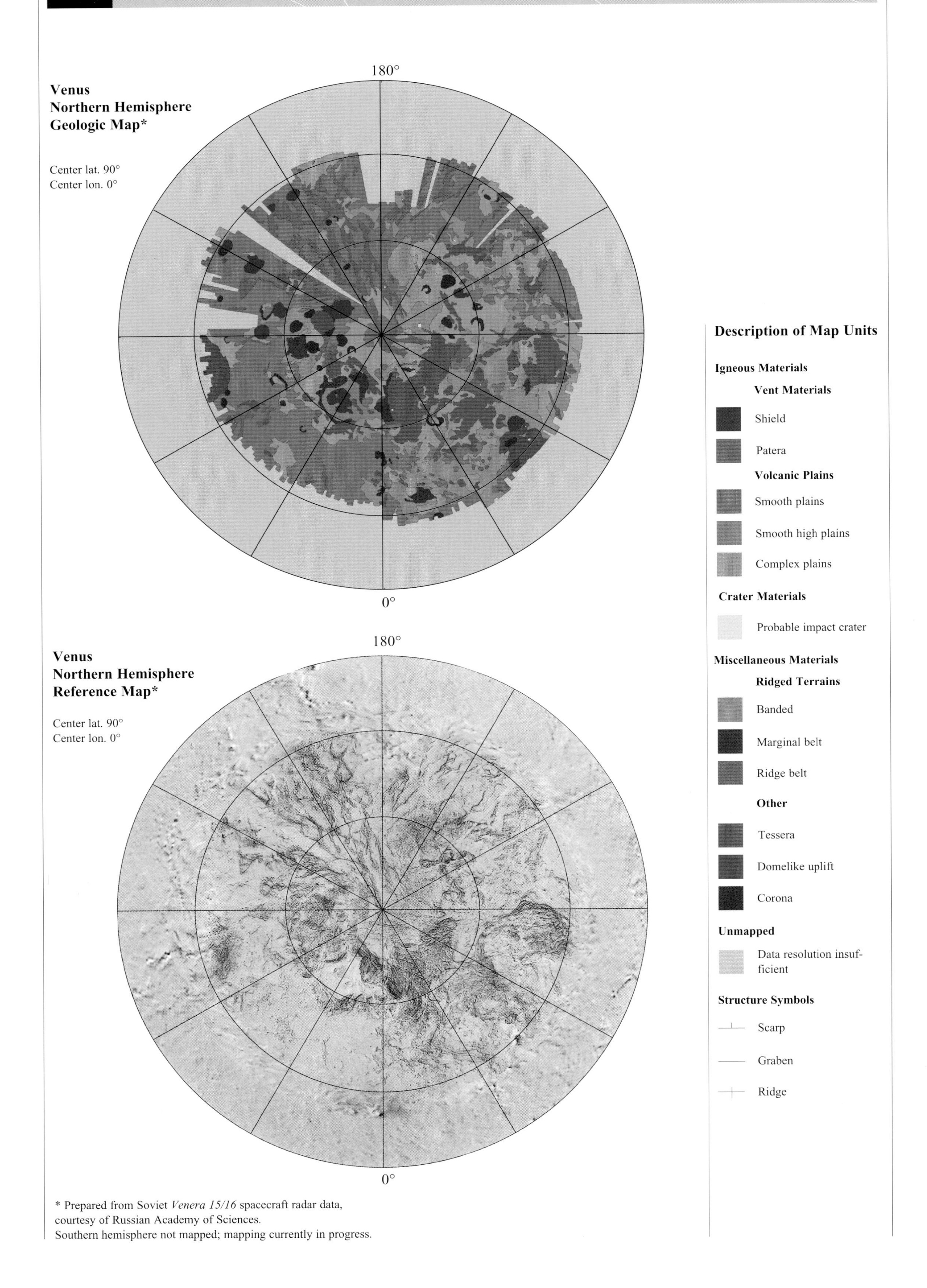

* Prepared from Soviet *Venera 15/16* spacecraft radar data, courtesy of Russian Academy of Sciences.
Southern hemisphere not mapped; mapping currently in progress.

Venus
Index Map

Center lat. –30°
Center lon. 180°

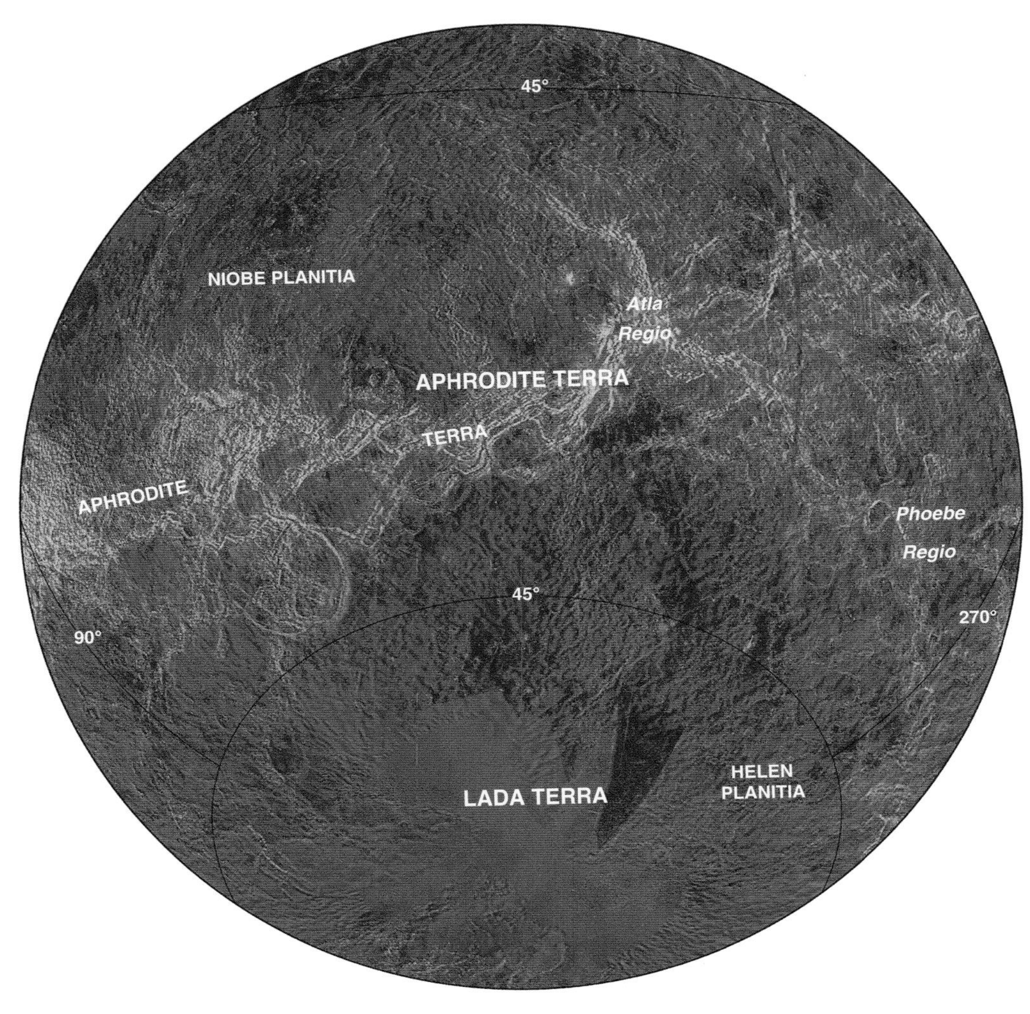

Venus
Index Map

Center lat. 30°
Center lon. 0°

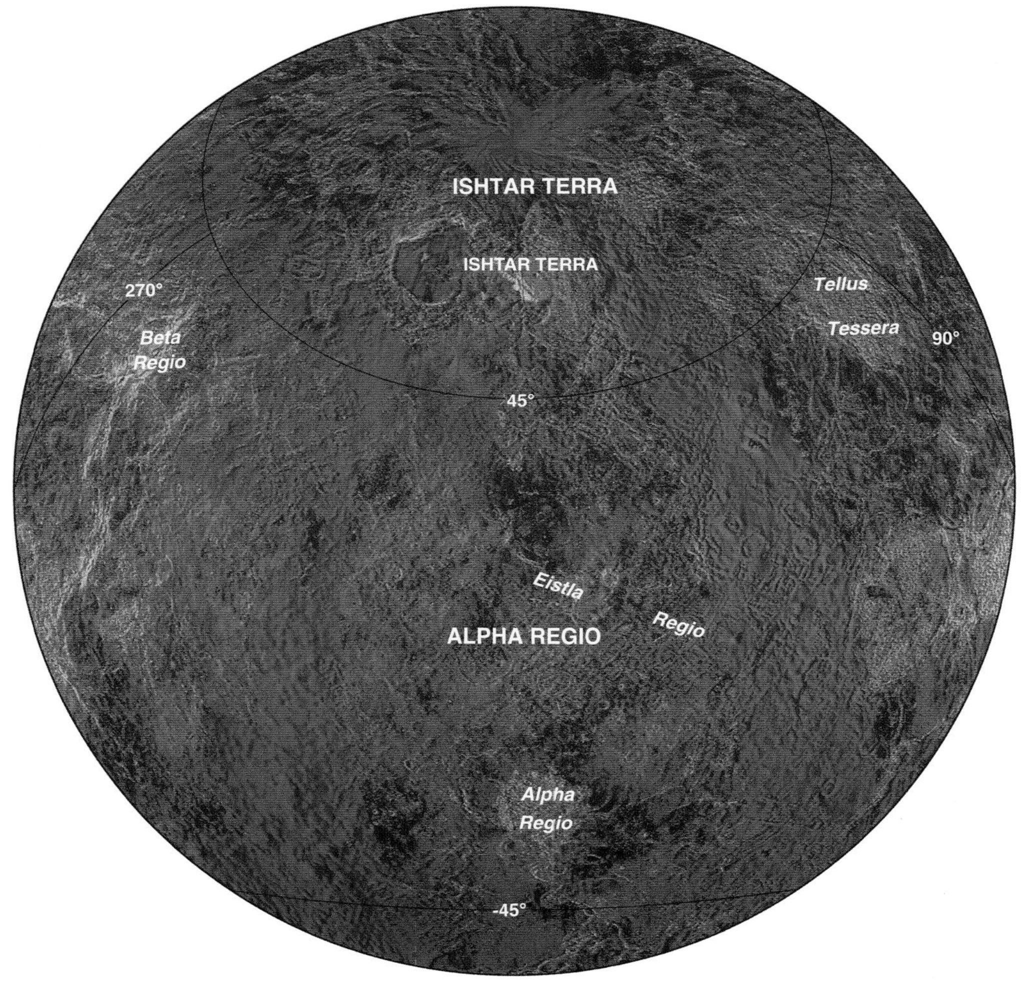

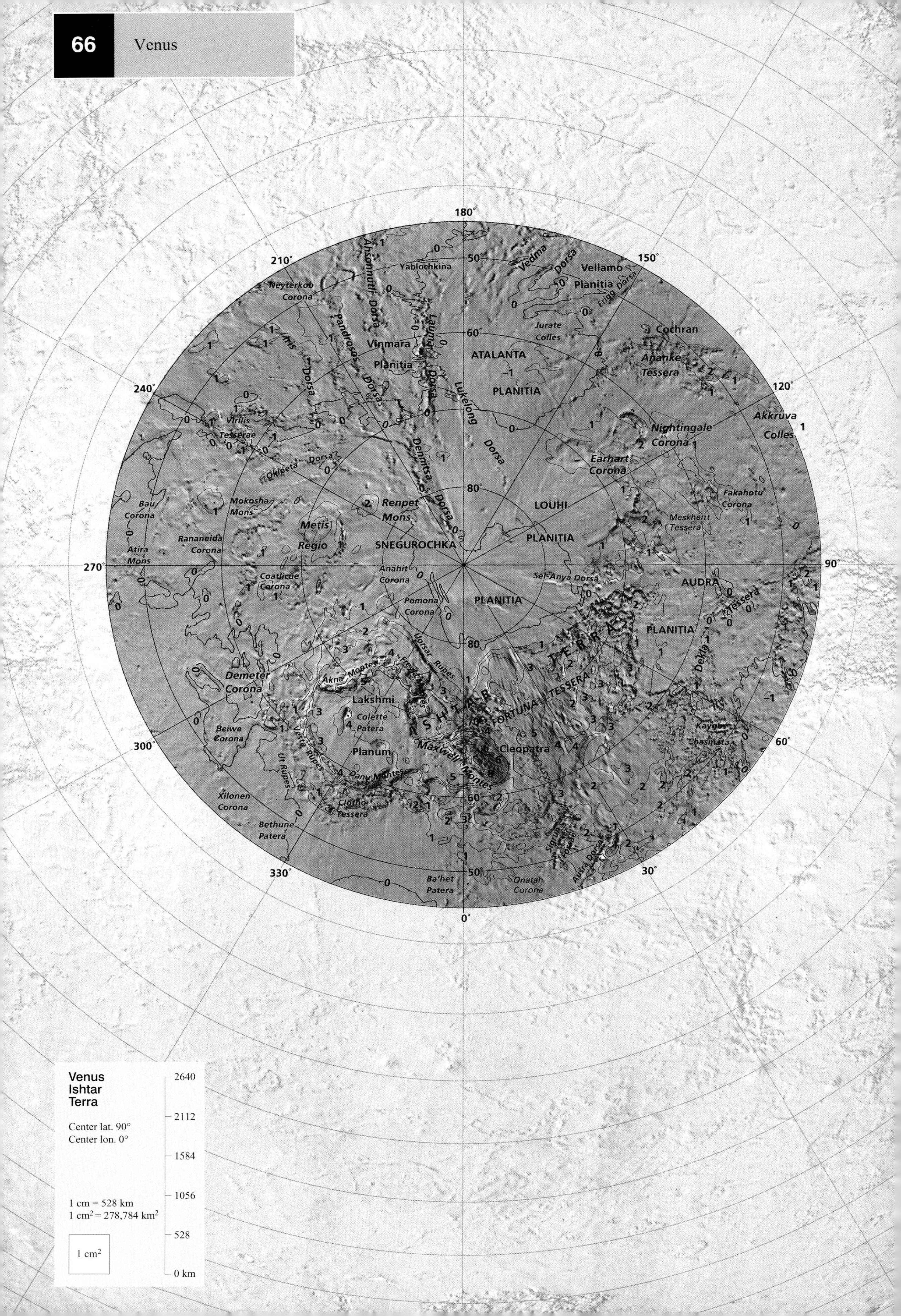

180°
210°
150°
240°
120°
270°
90°
300°
60°
330°
30°
0°
Yablochkina
Neyterkob Corona
Ahsonnutli Dorsa
Vedma Dorsa
Vellamo Planitia
Frigg Dorsa
Jurate Colles
Cochran
Pandrosos Dorsa
Laüma Dorsa
Vinmara Planitia
ATALANTA PLANITIA
Ananke Tessera
Iris Dorsa
Lukelong Dorsa
Akkruva Colles
Virilis Tesserae
Nightingale Corona
Earhart Corona
Okipeta Dorsa
Dennitsa Dorsa
Fakahotu Corona
Bau Corona
Mokosha Mons
Renpet Mons
LOUHI PLANITIA
Meskhent Tessera
Metis Regio
Rananeida Corona
SNEGUROCHKA PLANITIA
Atira Mons
Anahit Corona
Sel-Anya Dorsa
AUDRA PLANITIA
Coatlicue Corona
Pomona Corona
Tessera
Dekla Tessera
Demeter Corona
Akna Montes
Freyja Montes
Uorsar Rupes
ISHTAR TERRA
FORTUNA TESSERA
Lakshmi Planum
Colette Patera
Beiwe Corona
Vesta Rupes
Maxwell Montes
Cleopatra
Kaygus Chasmata
Ut Rupes
Danu Montes
Xilonen Corona
Clotho Tessera
Bethune Patera
Sigrun Fossae
Ausra Dorsa
Ba'het Patera
Onatah Corona
Venus
Ishtar
Terra
Center lat. 90°
Center lon. 0°
1 cm = 528 km
1 cm² = 278,784 km²
1 cm²
2640
2112
1584
1056
528
0 km

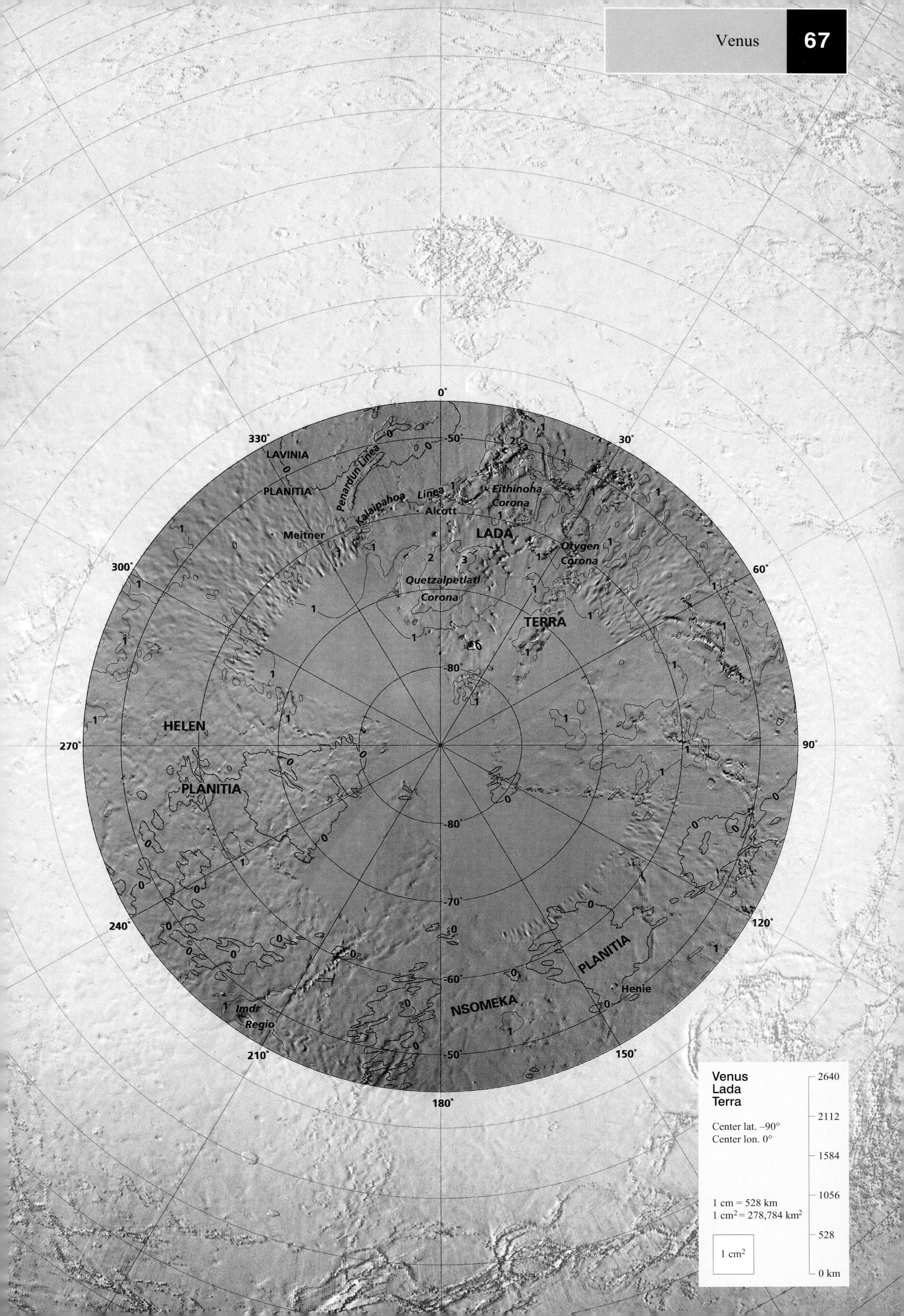
0°
30°
60°
90°
120°
150°
180°
210°
240°
270°
300°
330°
-50°
-60°
-70°
-80°
LAVINIA
PLANITIA
Penardun Linea
Kalaipahoa Linea
Alcott
Meitner
Eithinoha
Corona
LADA
TERRA
Otygen
Corona
Quetzalpetlatl
Corona
HELEN
PLANITIA
NSOMEKA
PLANITIA
Henie
Imdr
Regio
Venus
Lada
Terra
Center lat. –90°
Center lon. 0°
1 cm = 528 km
1 cm² = 278,784 km²
1 cm²
2640
2112
1584
1056
528
0 km

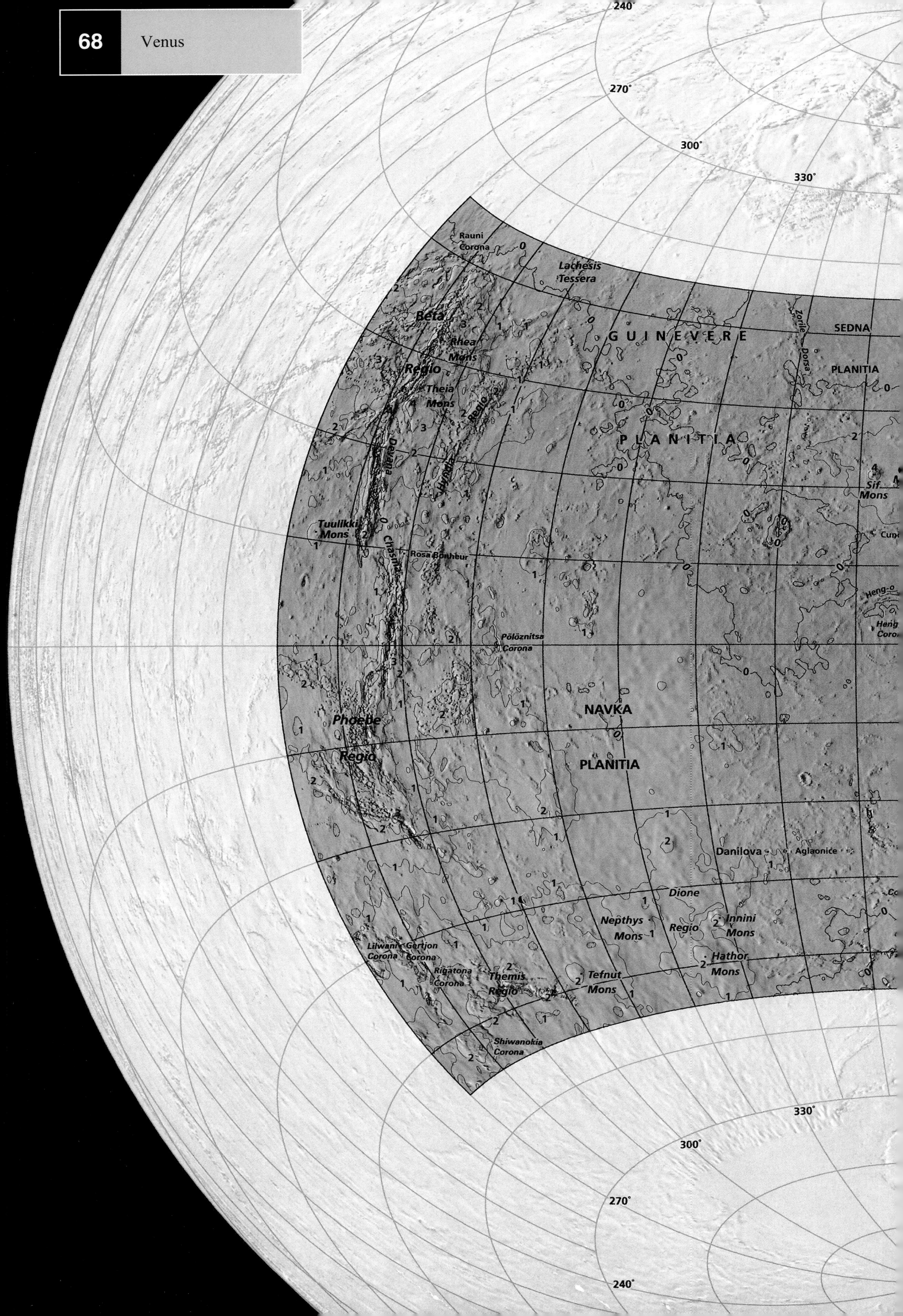
240°
270°
300°
330°
Rauni Corona
Lachesis Tessera
Beta Regio
Rhea Mons
Theia Mons
Rhea Regio
Devana
Hyndla
Tuulikki Mons
Chasma
Rosa Bonheur
GUINEVERE PLANITIA
Zorile Dorsa
SEDNA PLANITIA
Sif Mons
Heng-o Corona
Pölöznitsa Corona
NAVKA PLANITIA
Phoebe Regio
Danilova
Aglaonice
Dione Regio
Nepthys Mons
Innini Mons
Hathor Mons
Tefnut Mons
Lilwani Corona
Gertjon Corona
Rigatona Corona
Themis Regio
Shiwanokia Corona
330°
300°
270°
240°

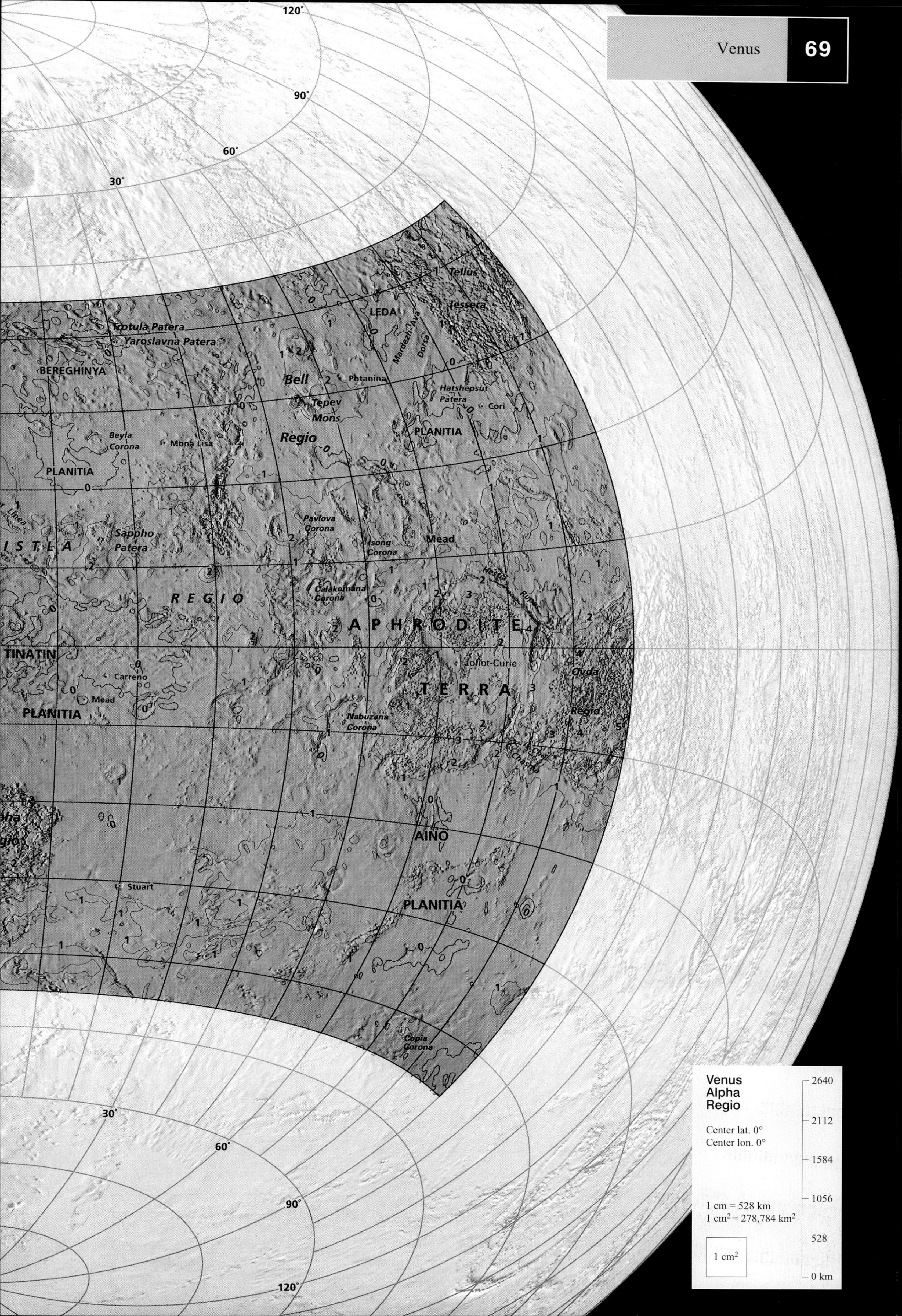

120°
90°
60°
30°
Tellus
Tessera
LEDA
Mardezh-Ava
Dorsa
Trotula Patera
Yaroslavna Patera
BEREGHINYA
Bell
Potanina
Tepev
Mons
Regio
Hatshepsut
Patera
Cori
PLANITIA
Beyla
Corona
Mona Lisa
PLANITIA
Linea
Pavlova
Corona
Sappho
Patera
ISTLA
Isong
Corona
Mead
Hestia
Rupes
REGIO
Calakomana
Corona
APHRODITE
TINATIN
Joliot-Curie
Carreno
Ovda
Mead
TERRA
Regio
PLANITIA
Nabuzana
Corona
Chasma
AINO
Stuart
PLANITIA
Copia
Corona
30°
60°
90°
120°
Venus
Alpha
Regio
Center lat. 0°
Center lon. 0°
1 cm = 528 km
1 cm² = 278,784 km²
1 cm²
2640
2112
1584
1056
528
0 km

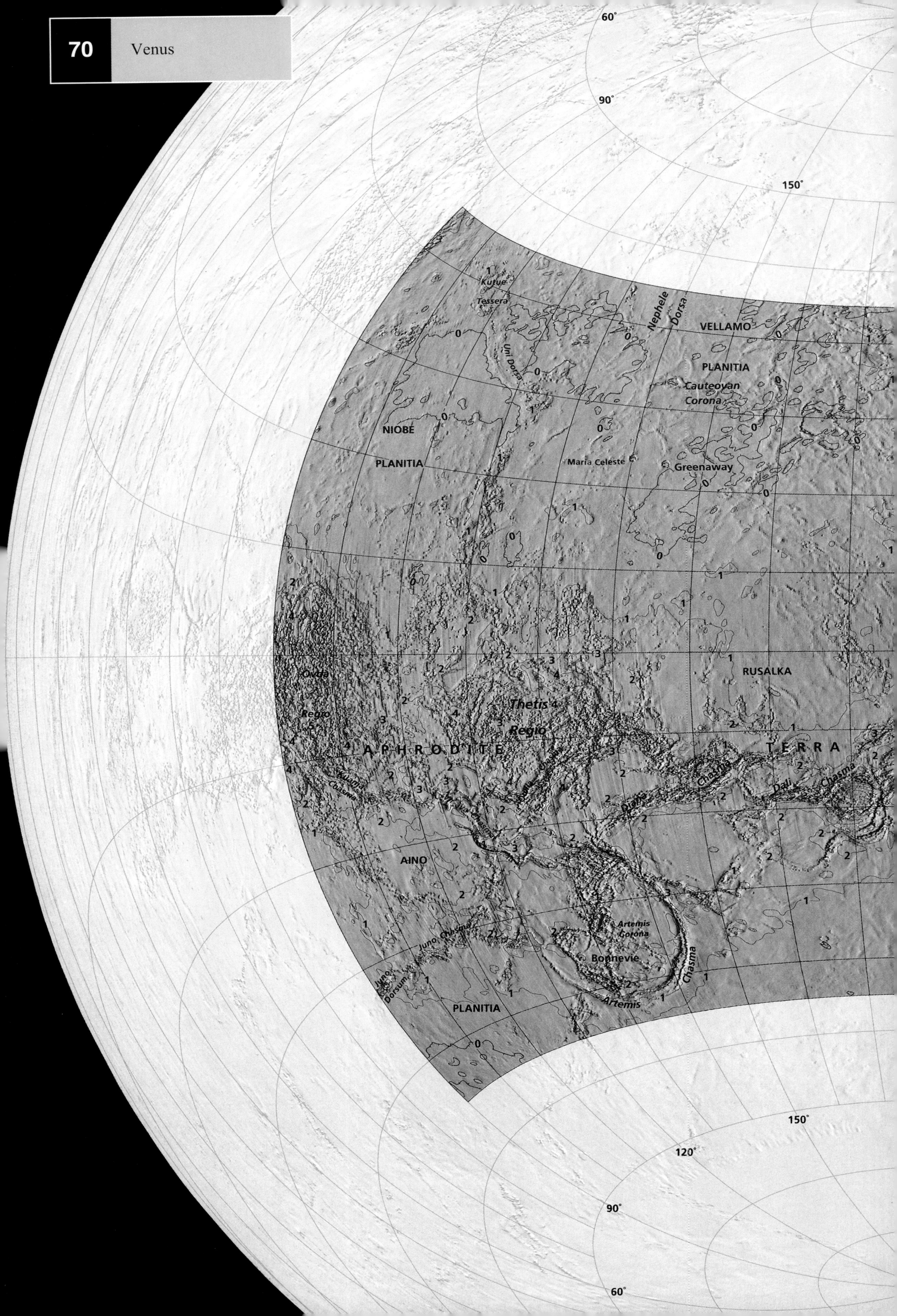
60°
90°
150°
Kutue
Tessera
Uni Dorsa
Nephele
Dorsa
VELLAMO
PLANITIA
Cauteovan
Corona
NIOBE
PLANITIA
Maria Celeste
Greenaway
Ovda
Regio
Thetis
Regio
RUSALKA
A P H R O D I T E
T E R R A
Kuanja
Chasma
Diana
Chasma
Dali
Chasma
AINO
Juno Chasma
Juno
Dorsum
Artemis
Corona
Bonnevie
Artemis
Chasma
PLANITIA
150°
120°
90°
60°

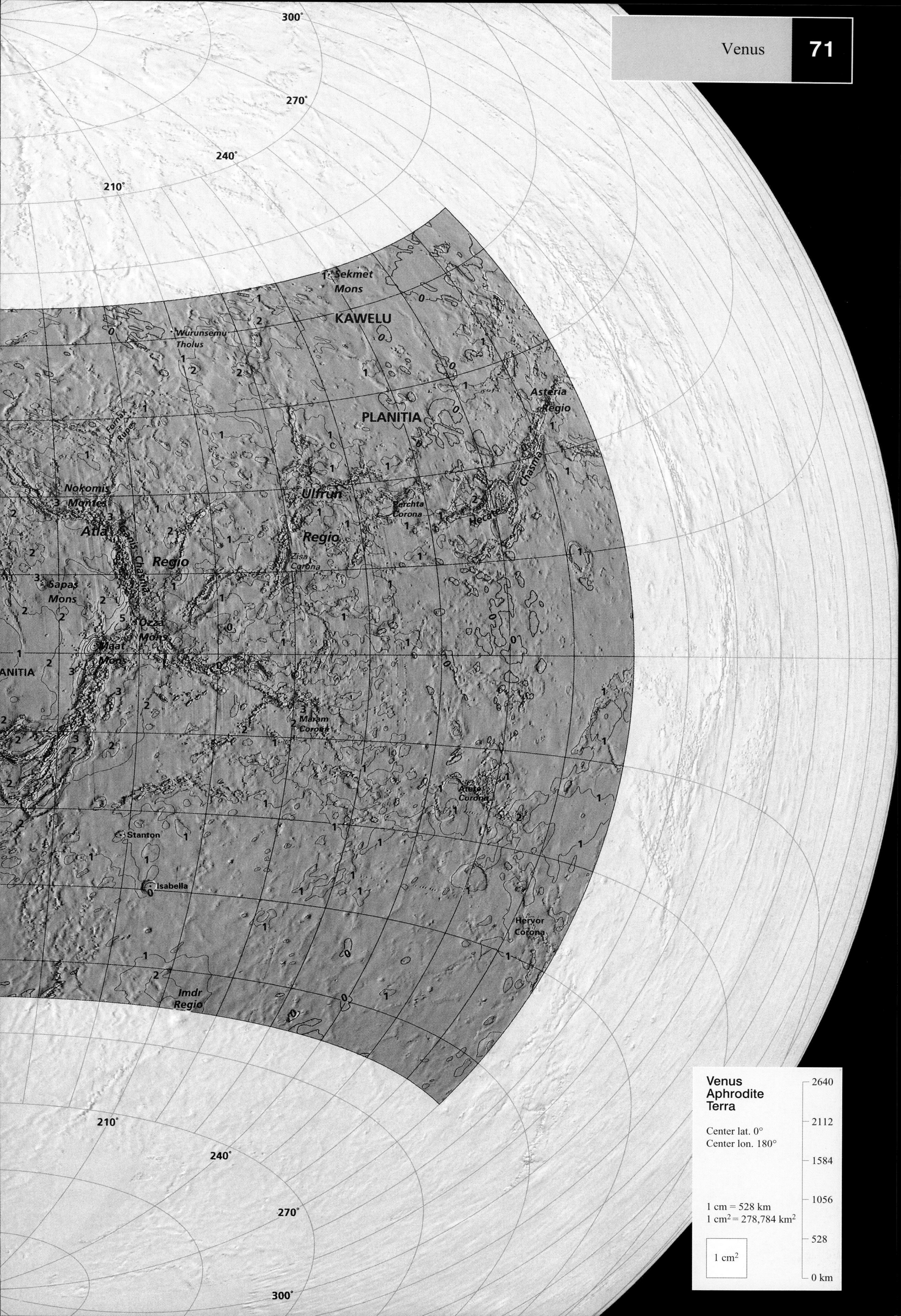
300°
270°
240°
210°
Sekmet Mons
KAWELU
PLANITIA
Wurunsemu Tholus
Fornax Rupes
Asteria Regio
Nokomis Montes
Atla
Ganis Chasma
Regio
Ulfrun
Regio
Parchta Corona
Hecate Chasma
Zisa Corona
Sapas Mons
Ozza Mons
Maat Mons
ANITIA
Maram Corona
Atete Corona
Stanton
Isabella
Hervor Corona
Imdr Regio
210°
240°
270°
300°
Venus
Aphrodite
Terra
Center lat. 0°
Center lon. 180°
1 cm = 528 km
1 cm² = 278,784 km²
1 cm²
2640
2112
1584
1056
528
0 km

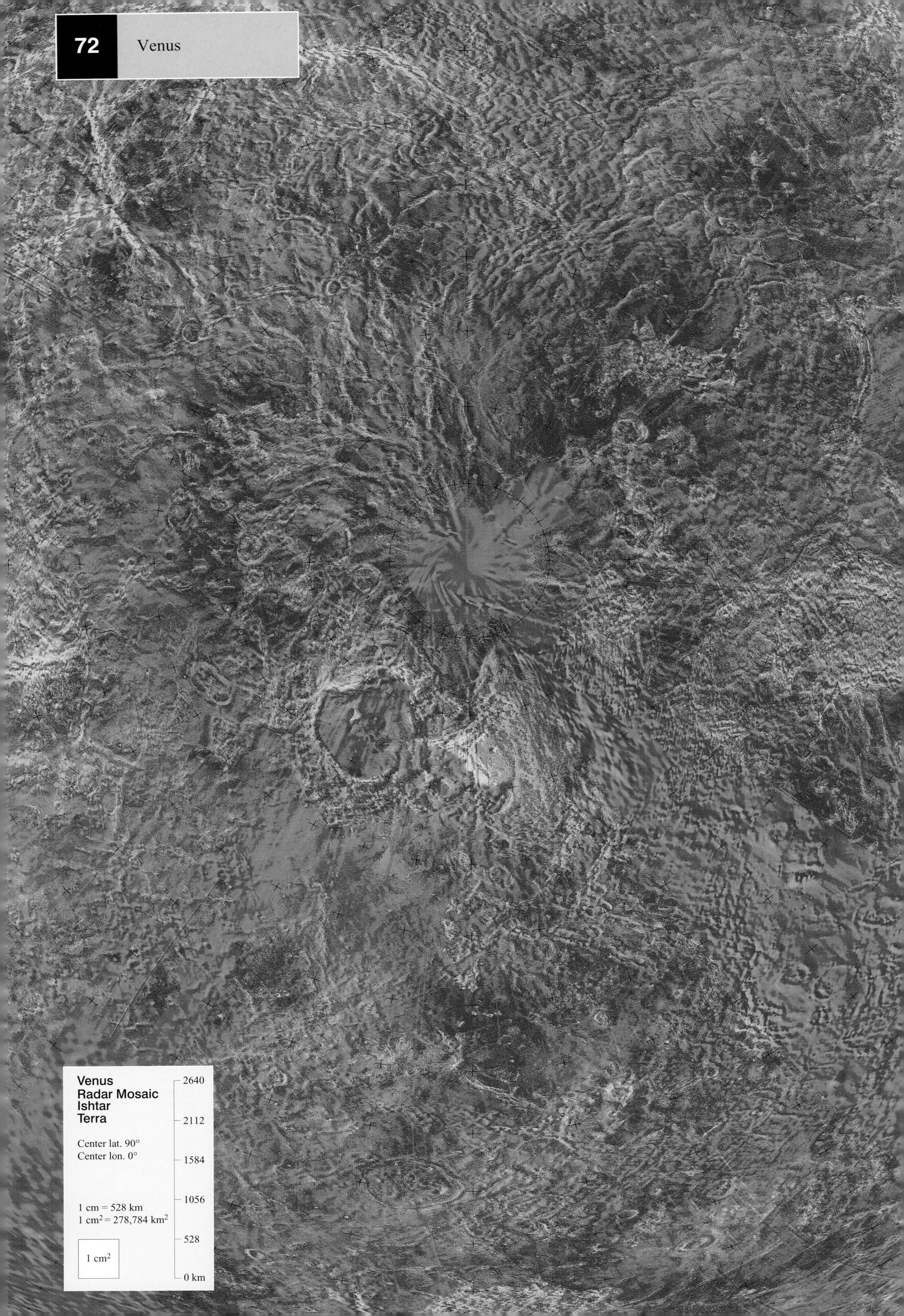
Venus
Radar Mosaic
Ishtar
Terra
Center lat. 90°
Center lon. 0°
1 cm = 528 km
1 cm² = 278,784 km²
1 cm²
2640
2112
1584
1056
528
0 km

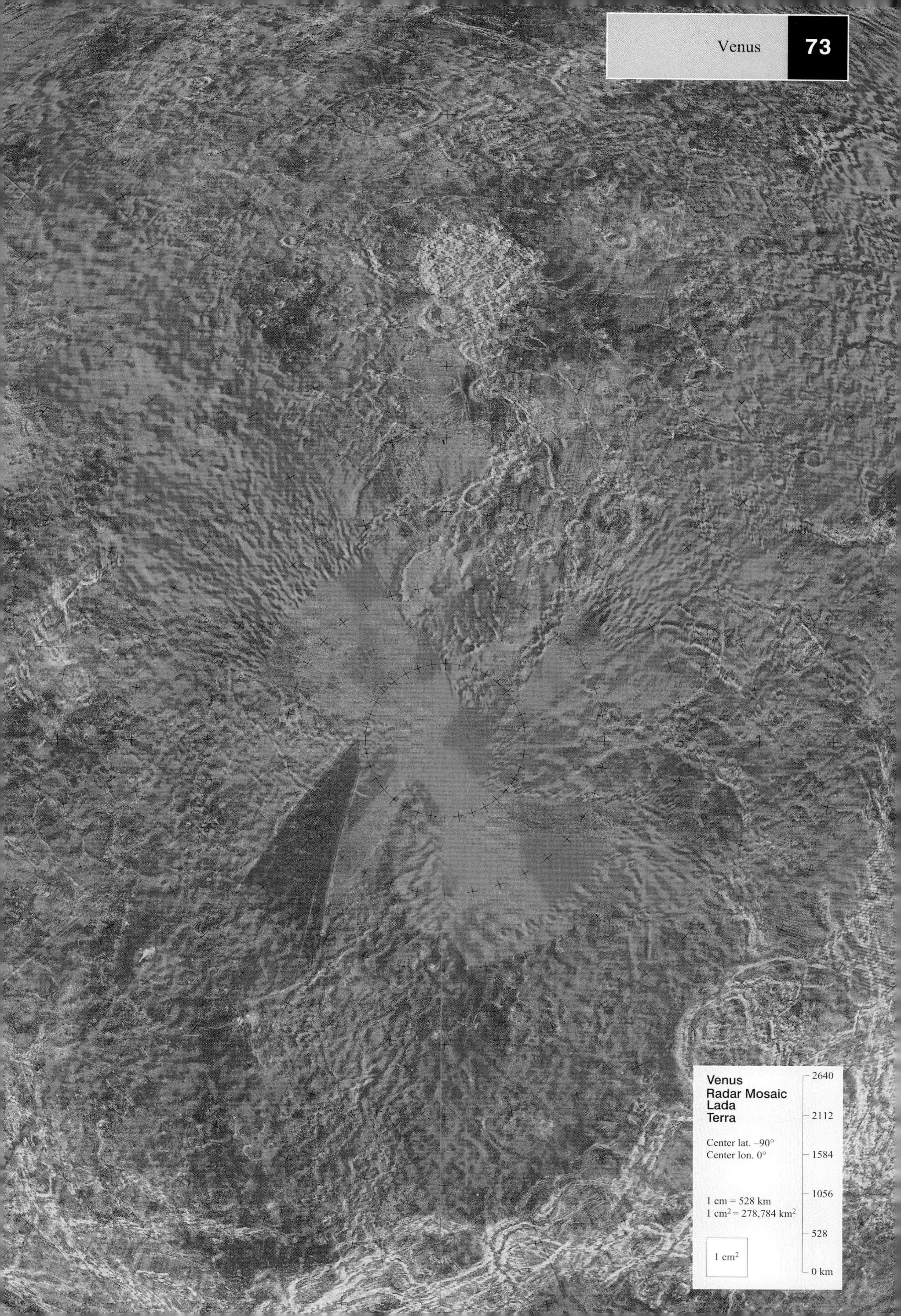
Venus
Radar Mosaic
Lada
Terra
Center lat. –90°
Center lon. 0°
1 cm = 528 km
1 cm² = 278,784 km²
1 cm²
2640
2112
1584
1056
528
0 km

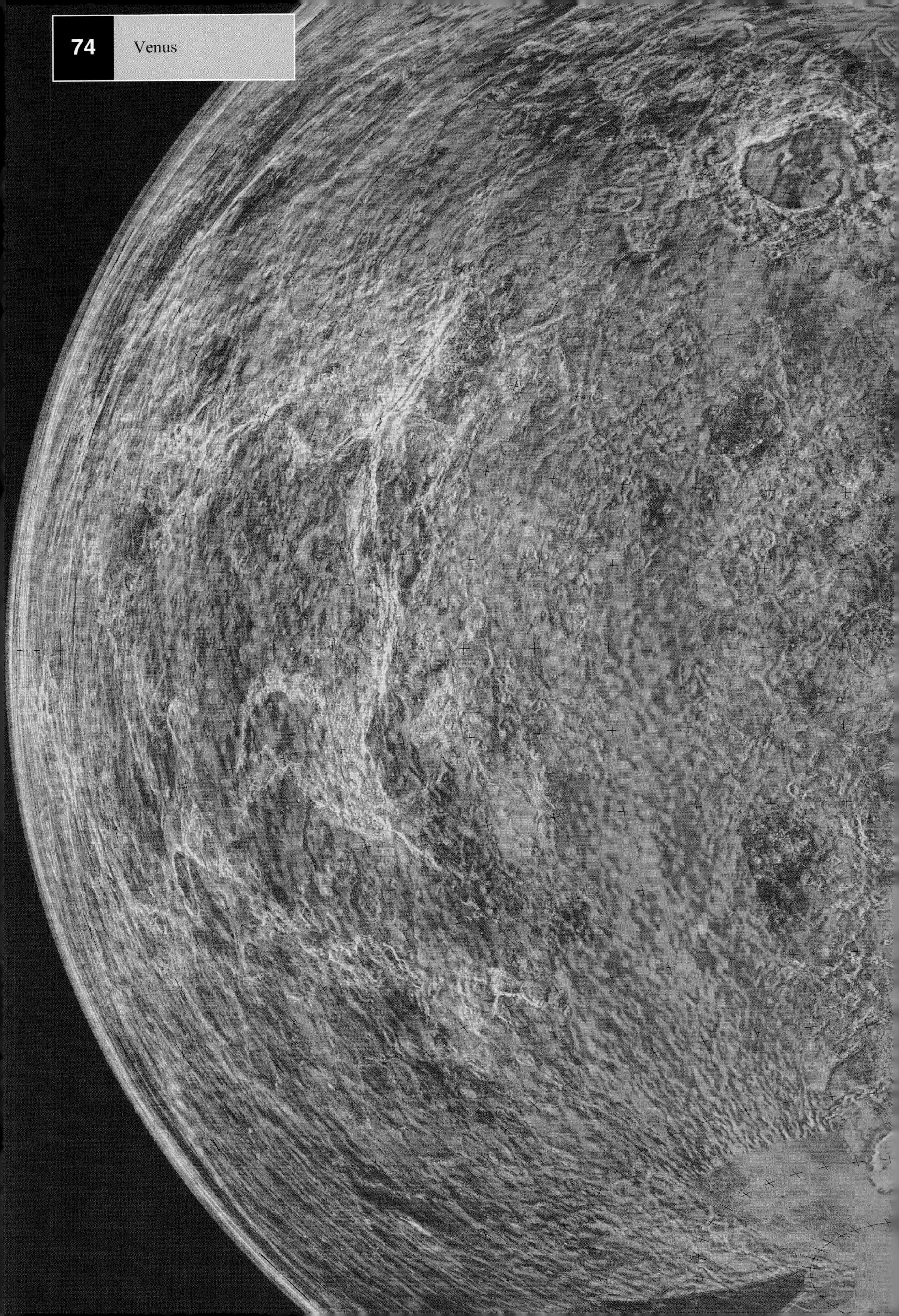

Venus
Radar Mosaic
Alpha
Regio
Center lat. 0°
Center lon. 0°
1 cm = 528 km
1 cm² = 278,784 km²
1 cm²
2640
2112
1584
1056
528
0 km

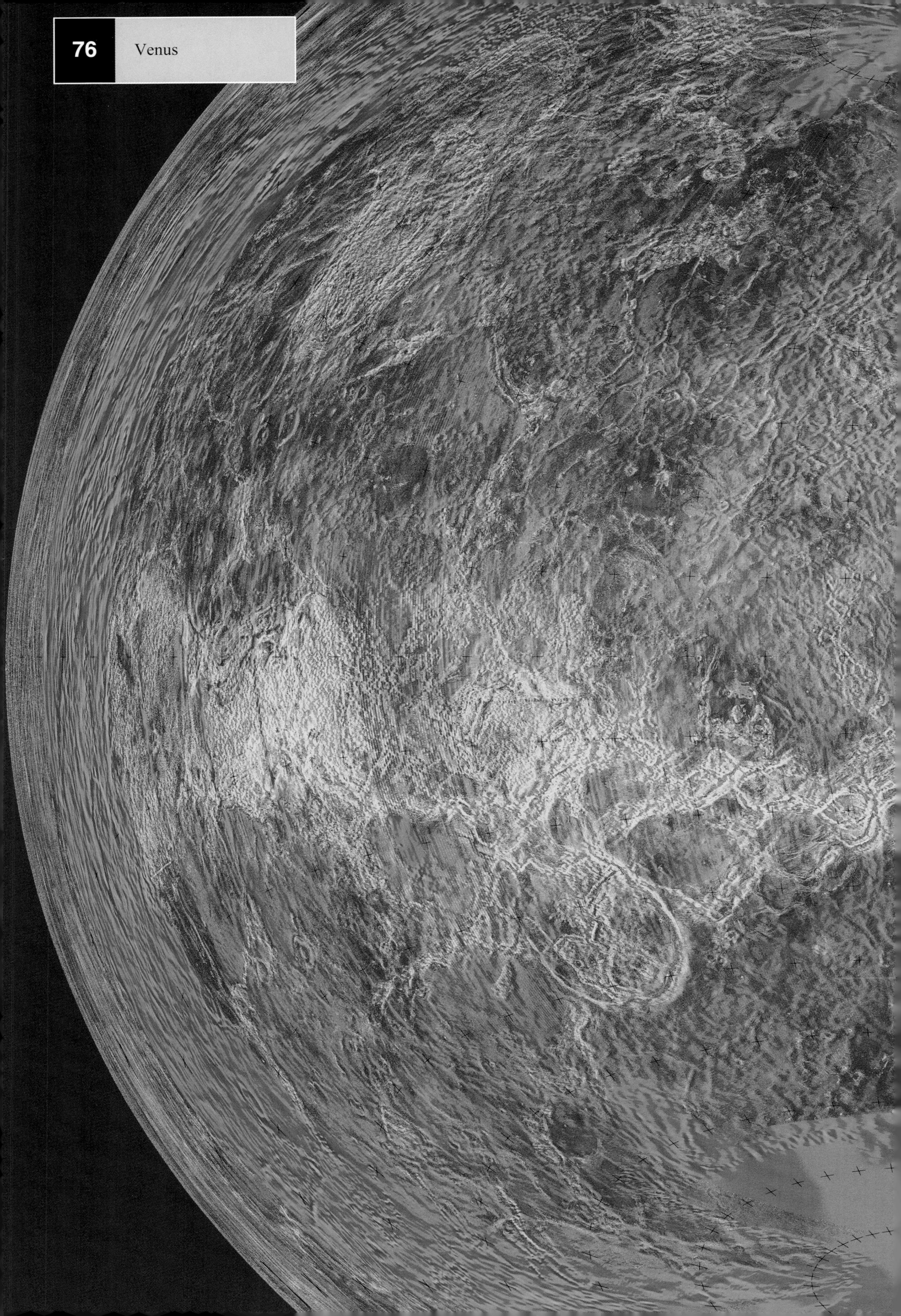

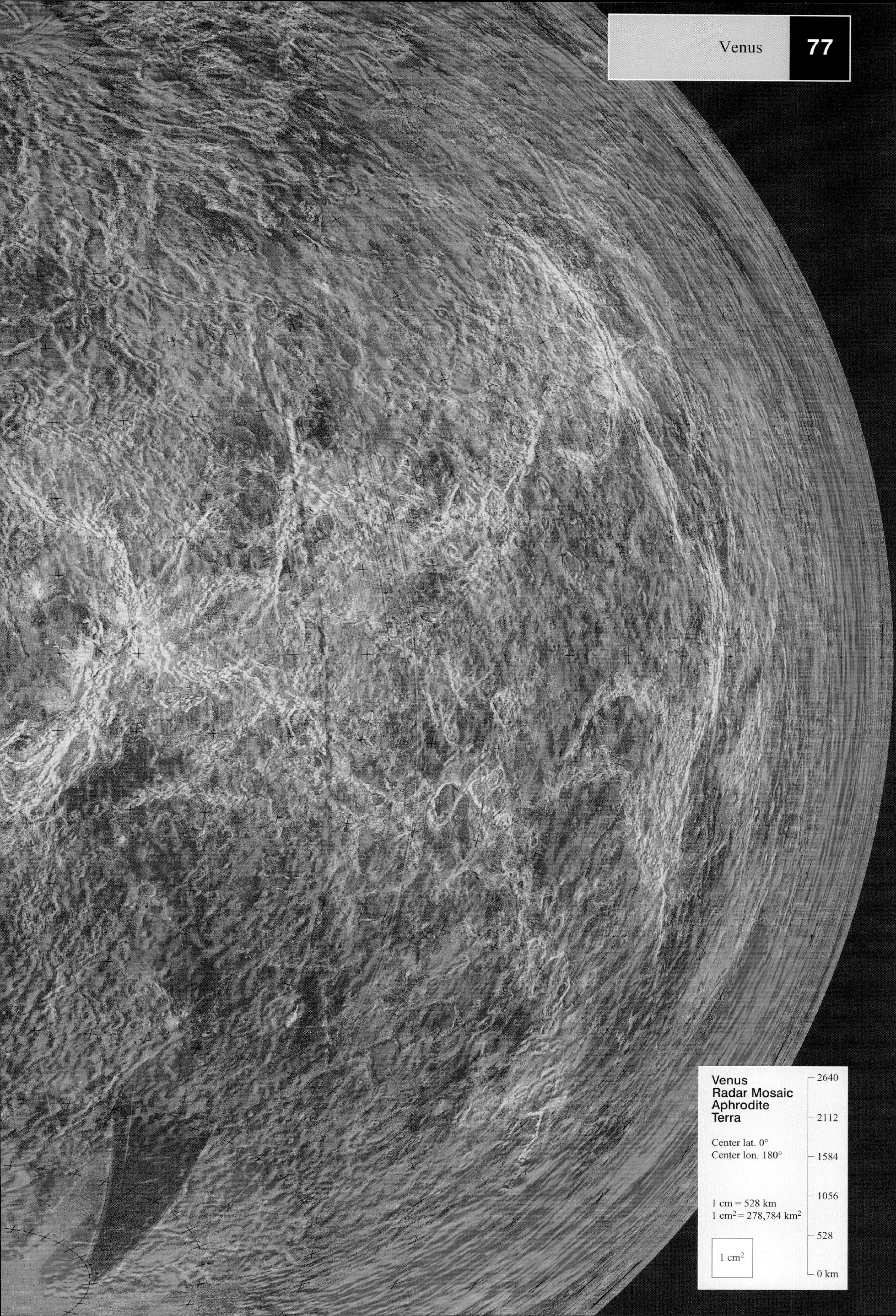
Venus
Radar Mosaic
Aphrodite
Terra
Center lat. 0°
Center lon. 180°
1 cm = 528 km
1 cm² = 278,784 km²
1 cm²
2640
2112
1584
1056
528
0 km

Earth-Moon System

In many ways, Earth is unique in our family of Solar System objects. So far as we know, Earth is the only planet that harbors life. Moreover, Earth's natural satellite, the Moon, is comparatively so large that some planetologists view the Earth and Moon as a binary planet. Unlike other planets, Earth's place in the Solar System allows abundant liquid water to exist on its surface, leading some people to suggest that our planet should have been named Hydra, Latin for "water." Certainly, many characteristics set Earth apart from all other planets.

On the other hand, the more we learn about the Solar System, the more similarities we see among the planets, including the Earth. For example, in the first decade and a half of Solar System exploration, Earth was the only planet known to experience active volcanism. With attention focused on the outer planets by the *Voyager* mission in the late 1970s and 1980s, volcanoes were first predicted and then discovered on Jupiter's moon, Io. Ten years later, when the *Voyager 2* spacecraft flew past the Neptune system, active geyserlike volcanoes were also found on Triton, Neptune's largest satellite.

Continuing exploration of the Solar System will reveal even more similarities and differences among the planets and satellites. Earth and its closest planetary neighbor, the Moon, are central to planetary studies because they are readily accessible and provide the scientific framework for comparison with other planets and satellites and for understanding planetary geologic processes.

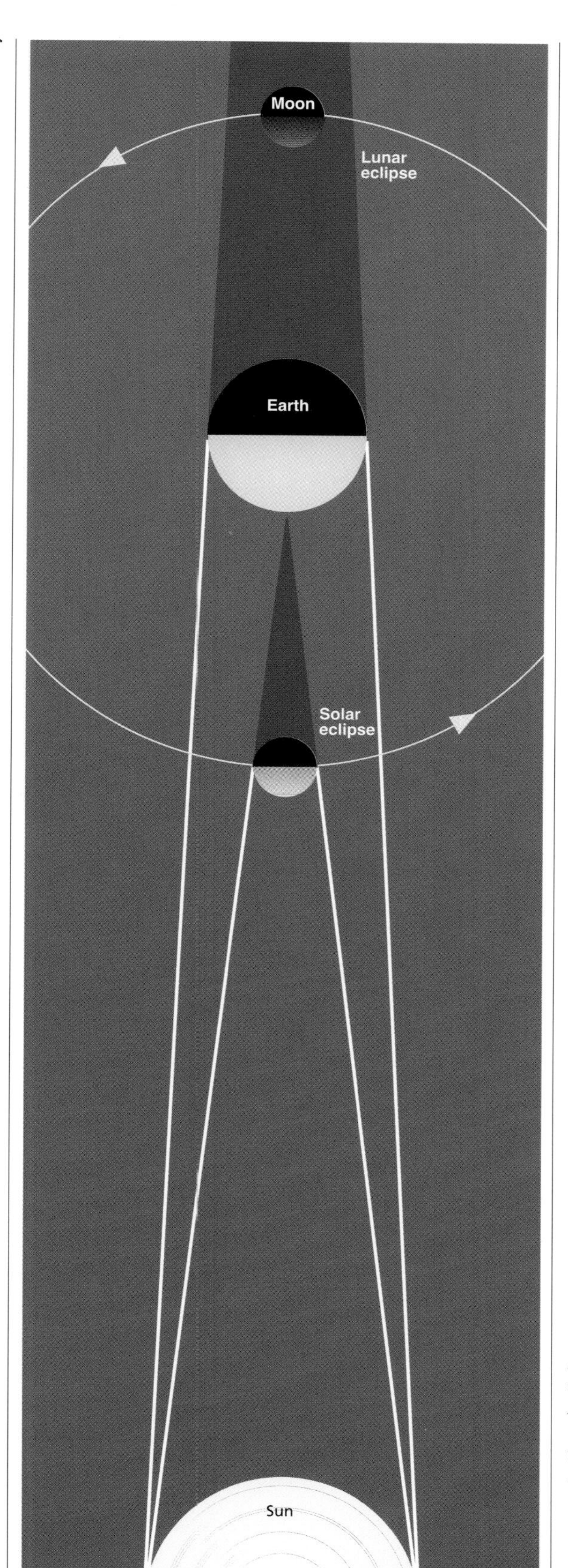

The Earth-Moon system (not to scale).

Facing page: view of Earth as it rises over the Moon as seen by the *Apollo 17* astronauts in December 1972 (*Apollo 17* photograph courtesy of Johnson Space Center, AS17-152-23274).

Earth

Global perspectives of planet Earth were greatly enhanced from photographs taken by the *Apollo* astronauts on their return home from the Moon. These and other photographs from Earth orbit show three dominant colors – reddish brown, blue, and white – representing the three principal components of our planet – rocks, water, and the atmosphere. In addition patches of green, representing the biosphere, can be seen from space. The atmosphere consists primarily of nitrogen (about 75 percent) and oxygen (nearly 25 percent). All other components of the atmosphere constitute less than

Meteosat photograph of Earth showing Africa; red and brown areas are land, blue is water, and the white patches are clouds; also visible is the dark green band across the continent that is predominantly vegetation (courtesy of G. Neukum).

1 percent and include carbon monoxide, carbon dioxide, water vapor, and solid particles such as dust.

Driven mainly by heat from the Sun – either directly or as reflected from land, ocean, and ice surfaces – the atmosphere is constantly in motion, forming complex circulation patterns. In recent years, the health and general vitality of the atmosphere has come to the forefront with concern for the ozone hole over the South Pole (and perhaps the North Pole) and for the effects of the depletion of tropical forests on oxygen production.

Water covers nearly three-fourths of the surface of our planet, making up the hydrosphere. As in the atmosphere, most water is also in motion and is capable of causing enormous changes on the surface of Earth. Running rivers and streams, slow but ever-moving glaciers, and ocean wave action are all familiar. Less well known are the huge oceanic currents that convey enormous volumes of water from one part of the globe to another.

View of Earth and Moon obtained by the *Voyager 1* spacecraft in 1977 as it began its journey to the outer Solar System (NASA photograph 78-HC-3).

Geology of the Earth

If all of the water on Earth were removed, the perspective would be similar to views of other planets and satellites from space. On a planetary scale, the rocky surface of Earth includes high-standing platforms – the continents – and lowland plains – the sea floors. Superimposed on both are mountains, mountain chains, and huge valleys.

Much of the Earth's surface is a manifestation of processes originating in the interior of our planet. What is known about the interior of the Earth? Although the deepest drilled holes and mine shafts barely puncture the skin of our planet, the deep interior can be explored using information from seismometers and other instruments. The core of iron and nickel contains a solid inner section about 2,460 kilometers across and an outer liquid section about 6,970 kilometers across. The mantle is composed mostly of hot, plastic, silicate materials and is subdivided into the mesosphere, the asthenosphere (a weak, plastic zone about 250 kilometers thick), and the lower part of the lithosphere. The lithosphere is a cool rigid layer about 100 kilometers thick and contains the

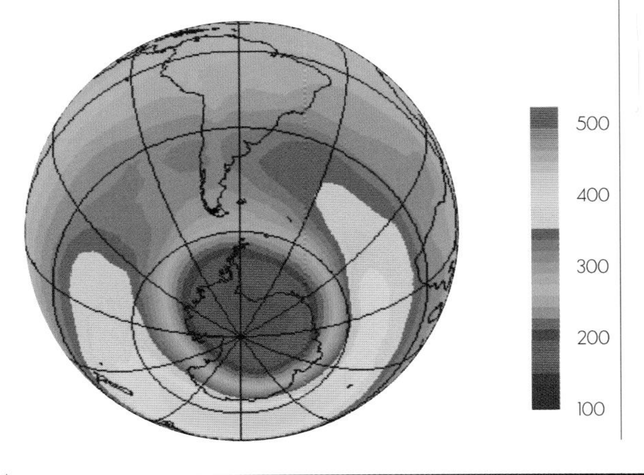

Image taken in October 1989 by the *Nimbus* satellite showing the ozone "hole" over the South Pole. Ozone is a molecule of oxygen composed of three atoms and is concentrated between 12 and 50 kilometers above the surface. This false-color image shows ozone in Dobson Units (1,000 units is equivalent to a 1 centimeter-thick layer of pure ozone) (image courtesy of M. Schoeberl, NASA – Goddard Space Flight Center).

Photograph of *Apollo 17* astronaut during the last manned mission to the Moon (Apollo 17 photograph AS17-152-23390).

outermost "skin" of our planet, the crust. Ranging in thickness up to 70 kilometers, continental crust is composed of silicate rocks rich in aluminum, silica, and calcium, with an average density of about 2.7 grams per cubic centimeter. Oceanic crust is composed of silicate rocks rich in iron and magnesium, with a density of about 3.0 grams per cubic centimeter and an average thickness of 8 kilometers.

Except for earthquakes and active volcanoes in certain areas, Earth's crust was once considered fixed and stable – until the concept of plate tectonics emerged to revolutionize the view of the world. Born out of the concept early in this century of slowly "drifting" continents, the instability of the crust and lithosphere on large scales was demonstrated by instruments that became available in the 1960s. Earth's lithosphere consists of seven major segments, or plates, and many smaller segments, all of which slide about on the underlying asthenosphere. In some places, plates move more than 20 centimeters per year. The energy that drives the plates comes from heat deep within the Earth derived from the decay of radioactive elements, plus heat remaining from planetary formation some 4.6 billion years ago.

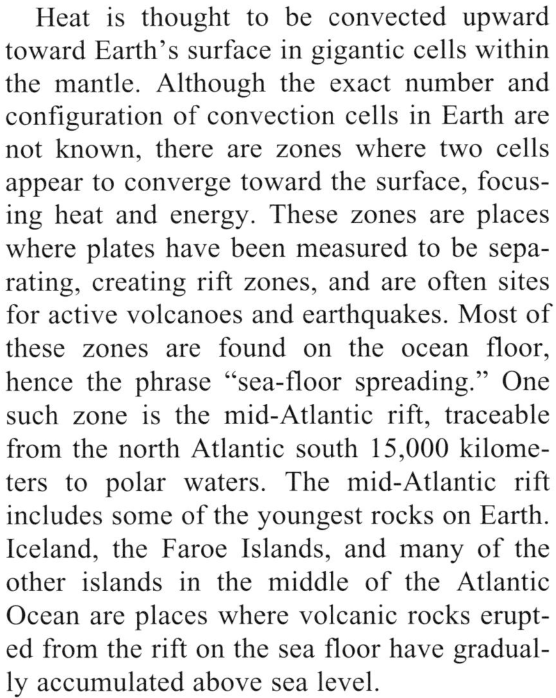

Heat is thought to be convected upward toward Earth's surface in gigantic cells within the mantle. Although the exact number and configuration of convection cells in Earth are not known, there are zones where two cells appear to converge toward the surface, focusing heat and energy. These zones are places where plates have been measured to be separating, creating rift zones, and are often sites for active volcanoes and earthquakes. Most of these zones are found on the ocean floor, hence the phrase "sea-floor spreading." One such zone is the mid-Atlantic rift, traceable from the north Atlantic south 15,000 kilometers to polar waters. The mid-Atlantic rift includes some of the youngest rocks on Earth. Iceland, the Faroe Islands, and many of the other islands in the middle of the Atlantic Ocean are places where volcanic rocks erupted from the rift on the sea floor have gradually accumulated above sea level.

In other places, plates meet head-on, forming convergent plate boundaries. In these highly complex regions, plates may be compressed, folding and faulting the crust. Some of the largest mountain ranges on Earth were formed by these tectonic processes, including the Himalayan Mountains where the Indian plate collides with the Asian plate. In other convergent boundary zones, segments of the crust are subducted, or pulled, into the mantle where they often melt and are recycled back to the surface by volcanism. The crust associated with collision zones may extend more than 100 kilometers into the mantle and provides for some of the most active volcanic and earthquake regions on Earth. In still other areas, plates slide laterally past one another, creating major tectonic systems such as the San Andreas fault zone of California.

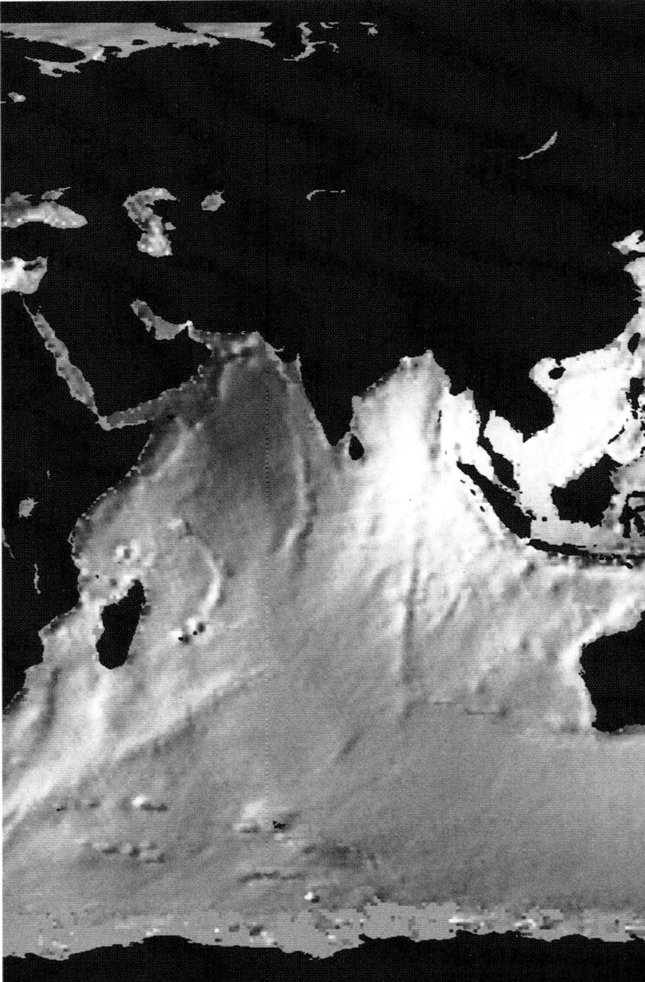

Some places on Earth are hot spots, or places where mantle plumes focus heat and generate magma. Hot spots appear to be relatively fixed

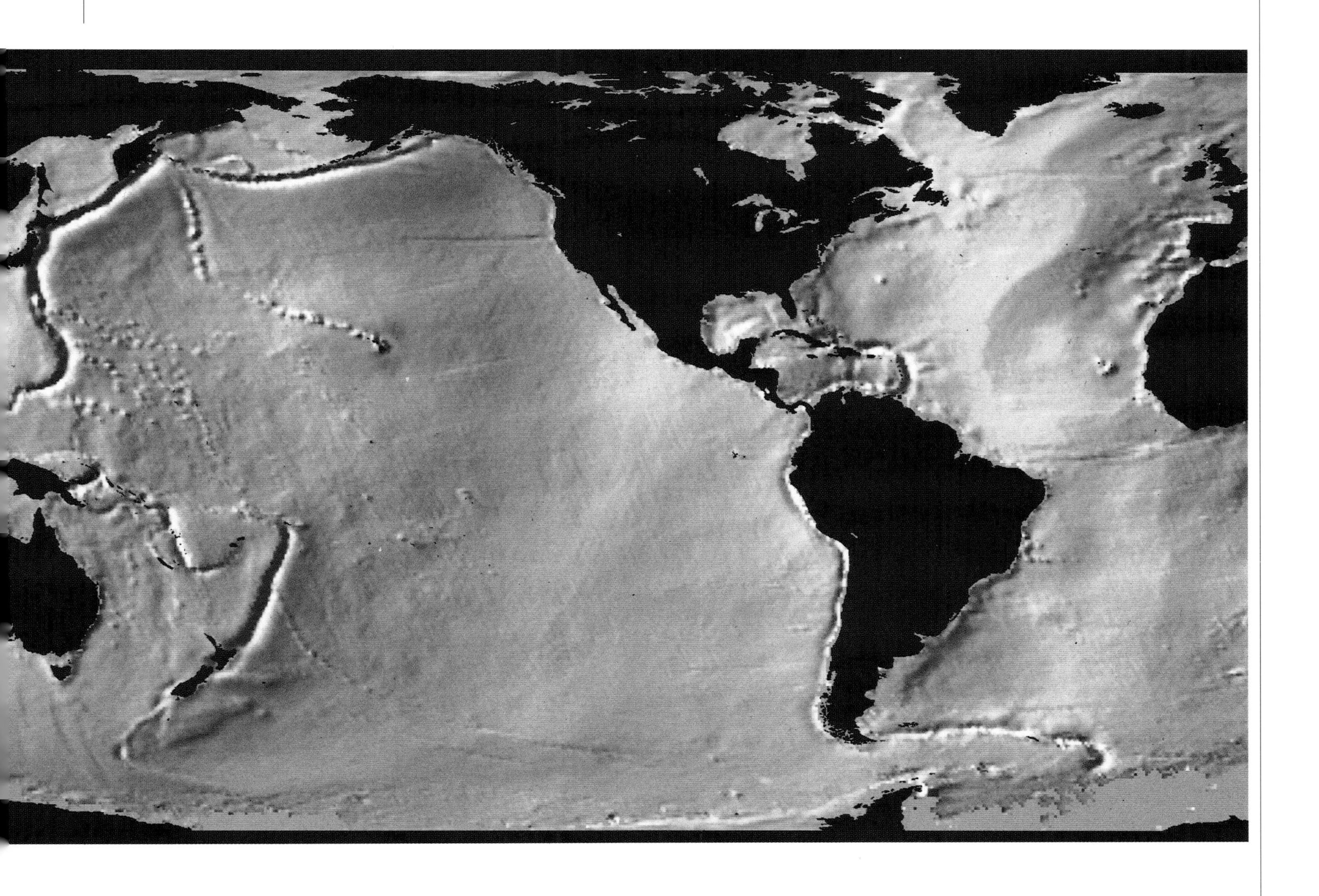

in place beneath lithospheric plates. The Hawaiian Islands are progressively older toward the northwest (the direction of plate motion) away from the currently active volcanoes on the Island of Hawaii. They are considered to have formed as the Pacific plate slid across a hot spot, producing volcanoes in a line of eruptions through time.

Magma generated within the mantle is rich in magnesium and iron, leading to the formation of the common igneous rock, basalt. Eruptions of basaltic magma on the sea floor or on land's surface tend to involve large quantities of very fluid lava spread as vast sheets called flood lavas, or as thin, multiple-flow accumulations called shield volcanoes, such as Mauna Loa in Hawaii. Sometimes the magma contains small amounts of gases that are contained under high pressure below the surface. As the magma approaches the surface, the lower pressures allow the release of these gases and can cause explosions, termed fire fountains, and the formation of cones from the accumulation of cinders and ash.

Magma derived from recycled crust commonly accompanies zones of subduction and often incorporates sediments and rocks rich in silicon and oxygen, which are combined as silica. Silica includes the mineral quartz. These silicic magmas tend to be less fluid than basaltic magmas and form viscous, pastelike lavas that can plug their vents. Consequently, eruptions of silicic magma often involve lavas that cannot flow very far from their vents, but rather form steep volcanic domes. In other cases, plugged vents may cause accumulation

Topographic relief radar image shown in false color, derived from *Seasat*, a radar system flown in orbit around Earth. This image shows that the ocean is higher over seamounts and ridges and lower over trenches (JPL photograph P-25621).

Folded rocks, such as these in the Andes of Argentina, result from compression of crustal rocks (photograph by Stephen J. Reynolds, Arizona State University).

Aerial view along the San Andreas fault zone in California. Faults fracture the crust, and rocks on one side of the fault are shifted or displaced with respect to rocks on the other side of the fault (U.S. Geological Survey photograph by Robert Wallace).

of enormous pressures that, on rupture, can produce some of the most explosive eruptions in the Solar System.

Tectonic deformation of the crust and volcanism commonly lead to the formation of mountains and valleys on Earth, producing topographic relief. At the same time, erosion from running water, blowing sand, and other geologic processes are continually reducing the high areas of our planet, transporting weathered sediments to the low places, and depositing sediments to form plains. Biologic activity, including the work of humans, can lead to significant weathering and erosion, as well as the formation of huge rock deposits, as evidenced by the existence of vast coral reefs and thick deposits of fossil-bearing limestone.

Impact cratering, volcanism, tectonic deformation, and erosion are geologic processes that have been operating for most of Earth's history. However, their relative importance has not been constant through time. For example, impact cratering was far more important early in Earth's history than it is now. One of the goals of Solar System exploration is to determine how these processes and their rates of activity differ among the planets and satellites in their respective geologic histories.

Geologic History of the Earth

Earth's geologic history has been pieced together from the record preserved in rocks exposed on or near the surface. Because of vigorous plate motion, most of the record earlier than about 600 million years ago has been lost through plate tectonics and crustal recycling. Consequently, some 80 percent of Earth's history is not well understood. Nevertheless, ancient rocks as old as about 4.0 billion years have been found in isolated exposures. These rocks, coupled with information gained from meteorites and comparisons with the early history of the Moon, enable the derivation of Earth's history and the development of a geologic time scale. The time scale includes four major segments, or eons, of time.

The Hadean Eon, "fiery home of the dead" – Hades – covers the time from the origin of the Solar System some 4.6 billion years ago to 4.0 billion years ago. The oldest rocks dated by radioactive methods are more than 4.0 billion

Vertical aerial photograph of part of Hualalai volcano, Hawaii, showing small cinder cones and a partly collapsed lava tube (U.S. Department of Agriculture).

years old and mark the close of the Hadean Eon. These rocks rest on top of still older rocks, including water-laid sediments, and show that oceans or lakes and surficial processes, such as weathering and erosion, were taking place very early in Earth's history.

In the early Hadean Eon, heat from planetary accretion melted most, if not all, of the mass. Most scientists believe that volatile elements such as oxygen, carbon, and nitrogen were "implanted" in the accreting Earth by planetesimals. Dense elements, such as iron, settled into the interior of the molten protoplanet to form the core, while lighter elements rose to the surface, leading to chemical differentiation of our planet.

Cooling of the surface of the protoplanet led to the formation of a thin crust. However, continued impact bombardment fractured the crust, allowing parts of the still-molten interior to flow to the surface. Gases trapped in the interior were released to the surface from these volcanic processes and formed Earth's early atmosphere. With time, rainfall accompanied the evolution of the early atmosphere and led to the accumulation of liquid water on the surface that eventually collected into sterile oceans.

The Archean Eon extends from 3.8 to 2.5 billion years ago. The name refers to "ancient" rocks and is applied to the development of widespread continental crust, remnants of which are preserved today. By the close of this eon, perhaps 50 to 90 percent of continental crust had developed. The Archean Eon also saw eruptions of extremely fluid flood lavas, remnants of which are in Canada, Australia, and other places on Earth.

Vast oceans covered much of the Earth during the Archean Eon and gave rise to a signal event in the history of the Solar System – the origin of life. Most planetary biologists believe that prior to this event millions of years of chemical evolution led to complex organic molecules, the precursors to life.

Early life forms probably were simple, single-celled plants, ultimately developing into blue-green algae. Plants expanded throughout the widespread oceans, releasing oxygen into the atmosphere. In turn, atmospheric oxygen reacted with sunlight to produce an enveloping ozone layer in our atmosphere.

Eruption of basaltic lava flows at Pu'u O'o on Kilauea's east rift zone in 1986, showing fire-fountaining and rivers of lava (U.S. Geological Survey photograph by J. D. Griggs).

Mauna Loa volcano, Hawaii (left side) and Mauna Kea volcano (right side), both typical shield volcanoes (U.S. Navy photograph).

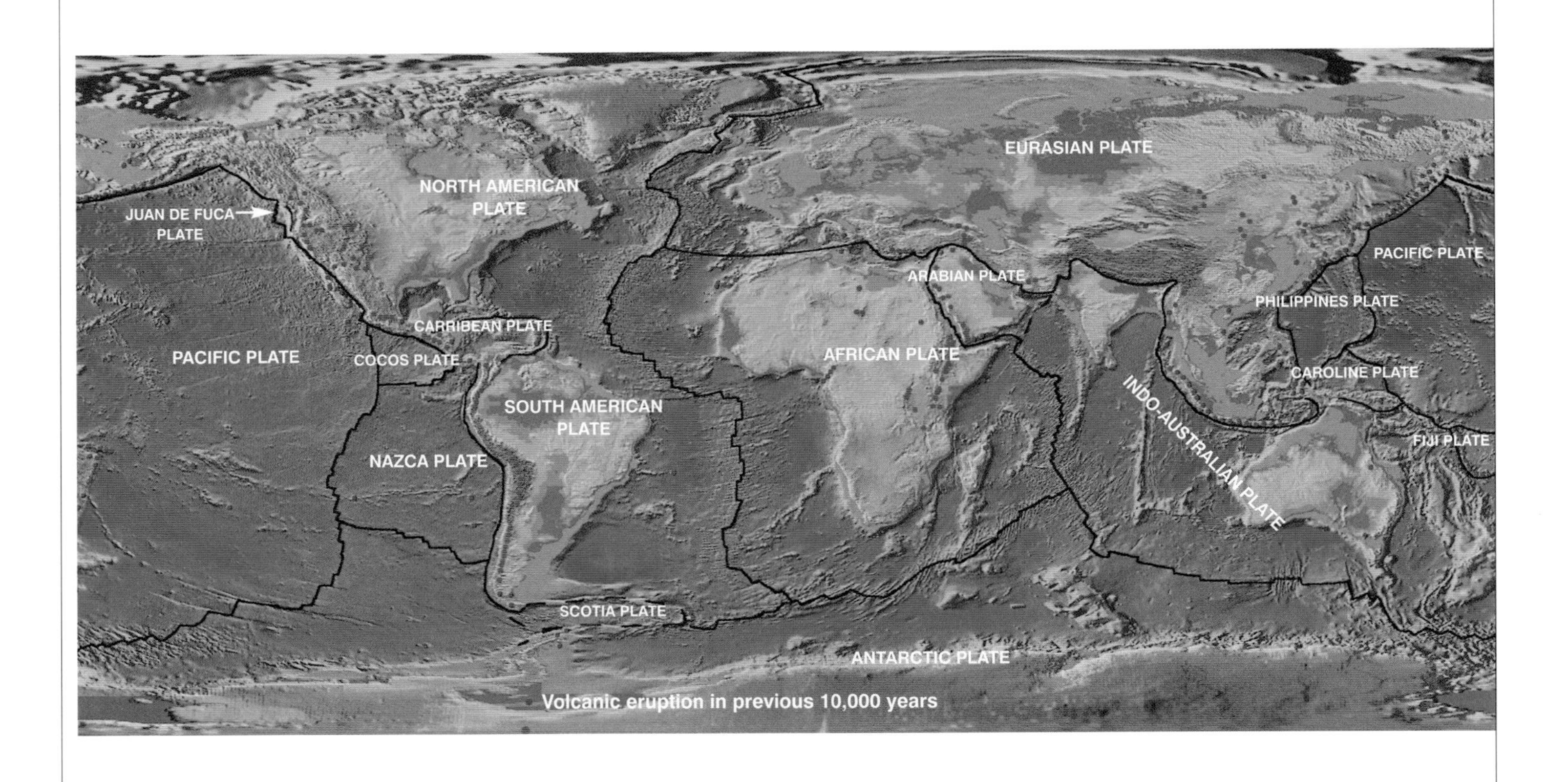

Map of Earth showing the major lithospheric plates, and smaller plates. Also shown are the locations of volcanic eruptions in the last 10,000 years and their relationship to major plate boundaries (U.S. Geological Survey).

Facing page: cross-section of Earth, showing core, mantle, lithosphere, and convection cells (Ralph Aeschliman, U.S. Geological Survey).

The Proterozoic Eon, named for "early life," extends from 2.5 to 0.6 billion years ago. Complex life forms evolved and, by the end of this eon, nearly all the major groups (phyla) of life were represented. Carbon dioxide was released through biologic activity, leading to the deposition of carbonate rocks, such as limestones. Oxygen in the atmosphere led to the oxidation of iron and the deposition of iron-rich sediments. Continued cooling of the interior of the Earth thickened the lithosphere, and the continents developed into relatively stable masses.

The Phanerozoic Eon (referring to abundant life) dates from 0.6 billion years ago to the present. This eon is marked by the development of "hardparts" in life forms, such as shells, which could be preserved as fossils. The continental land masses were repeatedly worn down to sea level, often experiencing flooding by shallow seas and mountain-building tectonic processes.

A vast "supercontinent," called Pangaea, existed about 300 million years ago and was subjected to a wide variety of erosional, volcanic, and regional tectonic processes. Thermal evolution and convection in the interior of the Earth developed the continents, and Pangaea began to break apart some 100 million years ago. Since that time, lithospheric plates and Earth's crust have continued to evolve through sea-floor spreading, subduction, and collision. Biologic evolution continued through geologic history until nearly every ecological niche on Earth was populated with life forms.

Earth is rich in geologic, atmospheric, hydrologic, and biologic activity. The delicate balance of these processes was not fully appreciated until the first pictures of Earth were returned from space in the 1960s, showing the thin, fragile shell of our planet. How does this image of Earth with its active volcanoes, plate tectonics, and biologic activity compare with our close companion, the Moon?

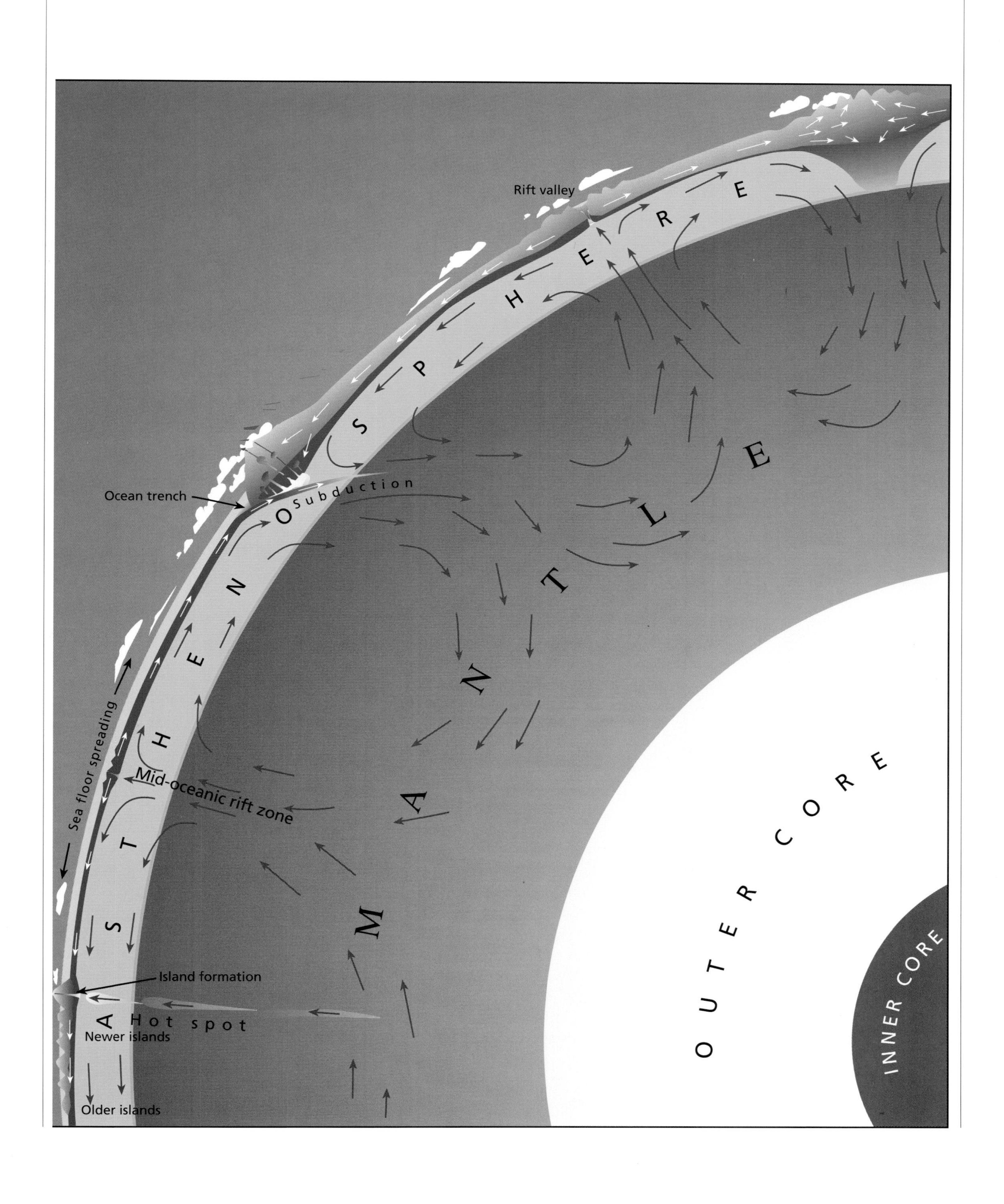
Rift valley
Ocean trench
Subduction
Sea floor spreading
Mid-oceanic rift zone
Island formation
Hot spot
Newer islands
Older islands
ASTHENOSPHERE
MANTLE
OUTER CORE
INNER CORE

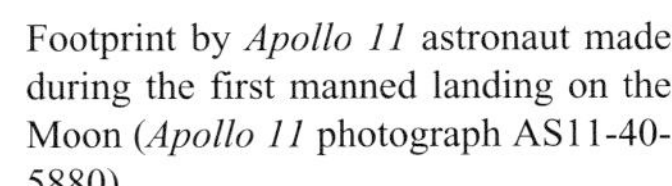

Footprint by *Apollo 11* astronaut made during the first manned landing on the Moon (*Apollo 11* photograph AS11-40-5880).

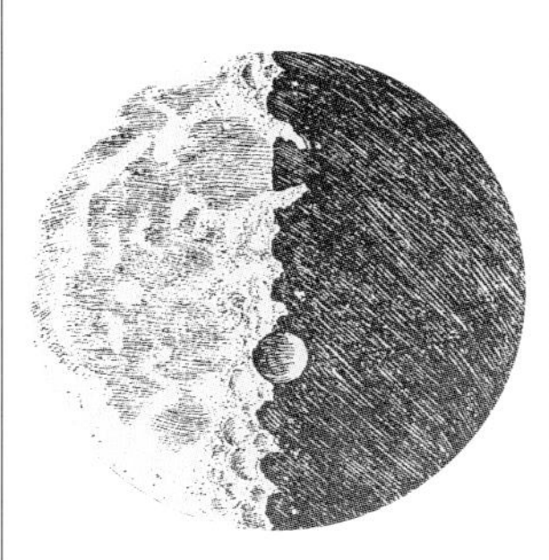

Galileo's sketch of the Moon, published in 1610.

Earth's Moon

Nearly every ancient culture references the Moon. Because the Moon is an imposing object in the nighttime sky, primitive notions of people becoming "moonstruck" still have vestiges in modern society. In recent decades, more has been learned about the Moon than any other object in the Solar System except the Earth. More than 40 spacecraft have been sent to the Moon by the United States and the Soviet Union, 12 men have walked on its surface, and nearly a half ton of lunar samples has been returned to Earth for study. In 1990, Japan initiated lunar exploration by launching a small spacecraft into orbit about the Moon. The Moon is a reference object in studying other planets and satellites and may well set the stage for expansion of humans into space.

The Moon was the first non-Earthly object studied scientifically. The use of the telescope to view the Moon in the early 1600s led to the first maps of the lunar surface. Early astronomers identified and named hundreds of craters and other landforms, including mountain chains, fractures, and huge sinuous channels called rilles. The Moon rotates on its axis at the same rate that it orbits about the Earth, so it always shows Earth the same hemisphere. Termed synchronous rotation, from Earth only the near side of the Moon can be seen, with the far side remaining hidden.

Before the Space Age, nearly all of the prominent features larger than about 10 kilometers on the near side of the Moon had been named or at least mapped. The far side remained unknown with only speculation about what would be found. The first photographs of the far side, taken by the Soviet *Luna 3* mission in 1959, showed some remarkable surprises. Most notable was the relative scarcity of the dark regions, or the maria, that are so prominent on the nearside.

Exploration of the Moon by spacecraft, the work of the *Apollo* astronauts, and the return of lunar samples to Earth have provided a great wealth of data on the origin and the evolution of the Moon. This information continues to be a valuable resource, not only for understanding

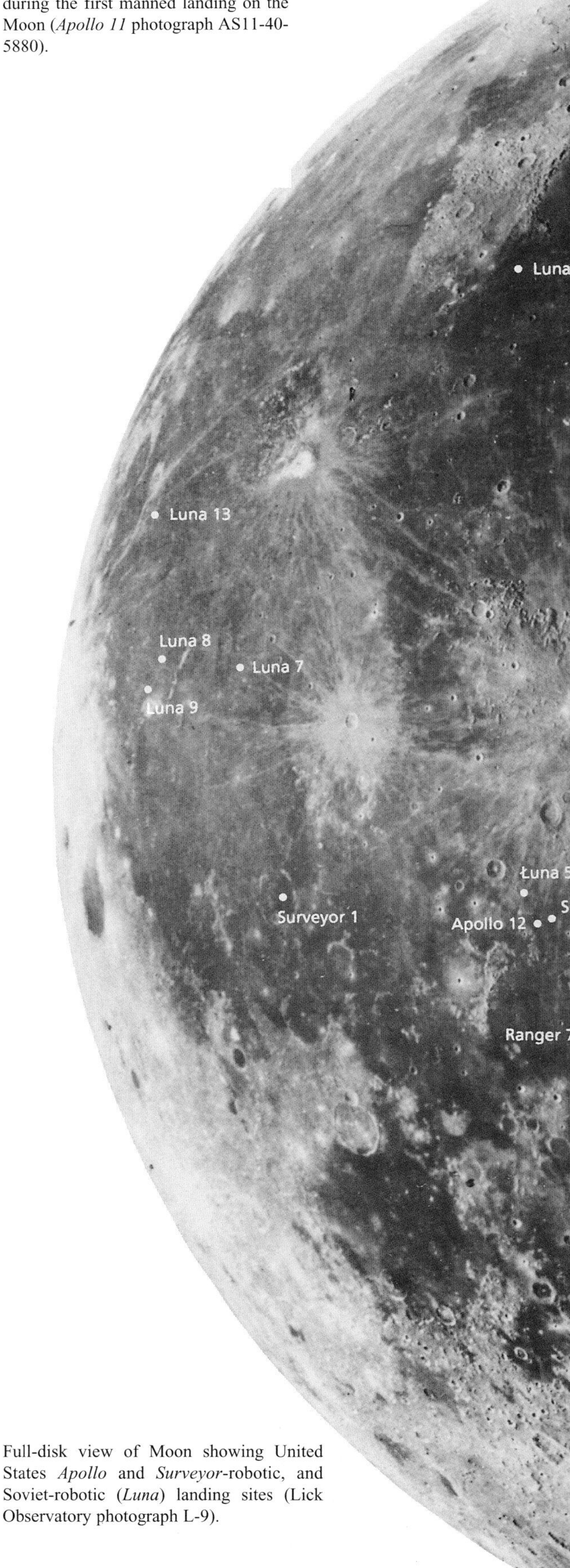

Full-disk view of Moon showing United States *Apollo* and *Surveyor*-robotic, and Soviet-robotic (*Luna*) landing sites (Lick Observatory photograph L-9).

Luna 2
Apollo 15
Luna 21
Apollo 17
Luna 24
Luna 23
Ranger 6
Luna 20
Surveyor 2
Ranger 8
Luna 16
Surveyor 6
Surveyor 5
Apollo 11
Surveyor 4
14
Apollo 16
Ranger
Surveyor 7

Top: views of the lunar far side (left) and near side (right), taken by the *Galileo* spacecraft in 1990 (JPL photograph P-37364).

Bottom: color composite views made from *Galileo* multispectral images. Blue colors correspond to basalts rich in titanium; greens, yellows, and light orange indicate basalts low in titanium; reds and deep oranges are highland rocks (JPL photograph P-37363).

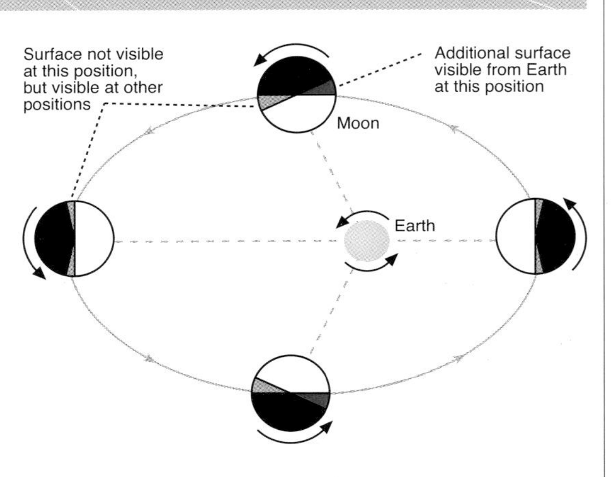

The Earth-Moon System, showing the orbit of the Moon in synchronous rotation, in which the same side of the Moon always faces Earth. Slight wobbles, or nutation, in the spin axis of the Moon allow slightly more than half of the Moon to be seen from Earth.

Apollo 17 astronaut on the Moon in the Taurus-Littrow Valley (*Apollo 17* photographs AS17-140-21497 and AS17-140-21493).

our nearest planetary neighbor, but also for the keys that are provided to understanding the rest of the Solar System.

Geology of the Moon

Even to the casual observer, the near side of the Moon consists of two dominant terrains – the dark maria (from the Latin word for seas) and light highlands, or terrae (from the Latin word for earth, or land). These terms were applied in the seventeenth century for their fanciful – but unfounded – resemblance to oceans and continents on Earth. Closer inspection reveals subdivisions of both maria and terrae. For example, light plains of several types were found in both maria and highland regions.

Craters are everywhere on the surface of the Moon. For centuries their origin was debated. Although some initial ideas were bizarre, eventually two principal hypotheses emerged: origin by volcanic processes or origin by impact processes. It was not until the Space Age and the return of data from lunar probes that the matter was finally put to rest with the recognition that the vast majority of the craters seen on the Moon – and by extension on most of the other planets and satellites – are of impact origin. This recognition has enabled a powerful technique to be developed for assessing the relative ages of planetary surfaces. Statistically, the longer a surface is exposed to space (that is, the older the surface), the more impact craters will appear. Thus, by measuring and counting the impact craters on surfaces and comparing the results with other surfaces, it is possible to determine the relative geologic ages of surfaces.

The lunar highlands are characterized by their bright surfaces, rugged terrain, and thousands upon thousands of overlapping craters of all sizes. The great number of craters in the highlands in comparison to the maria attest to their greater age. This fact has been amply verified by rock samples of the Moon that have been dated using radiometric "clocks" in laboratories on Earth. Lunar samples also show that the dominant rock in the highlands is composed of amalgamated rock fragments, collectively called breccia. Fragments include a wide variety of rocks, some of which are composed of still older breccias. These "breccias within breccias" reflect the continual breakup, mixing, and reassembly of the lunar crust by impact cratering. Detailed studies of highland rocks show the formation of the early crust, including rocks from magma, that cooled beneath the surface and were excavated by impact craters.

Apollo 15 photograph showing lunar sinuous rilles east of the Aristarchus Plateau. Aristarchus crater is shown to the right. Sinuous rilles were once rivers of flowing lava (*Apollo 15* photograph AS15-2606 M).

An important component found among the lunar samples is called KREEP. The term KREEP is derived from its elemental components: K (potassium), REE (the rare earth elements, such as samarium), and P (phosphorus). The formation age for the KREEP component of 4.35 billion years and its other characteristics suggest that this material came from an extensive molten rock or magma ocean that may have covered the entire Moon. KREEP probably represents the latest formation of the lunar crust as the magma ocean cooled.

The largest features on the Moon are the huge impact craters, termed basins, that exceed 300 kilometers in diameter. Most of the mountain chains identified by early lunar scientists are now recognized to be the rims of these

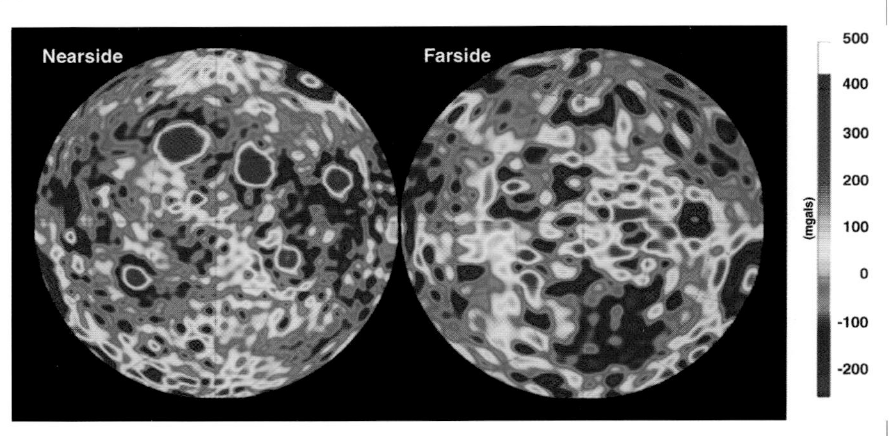

Gravity map from *Lunar Prospector* data, showing positive areas (red) over zones thought to contain high density materials (Jet Propulsion Laboratory).

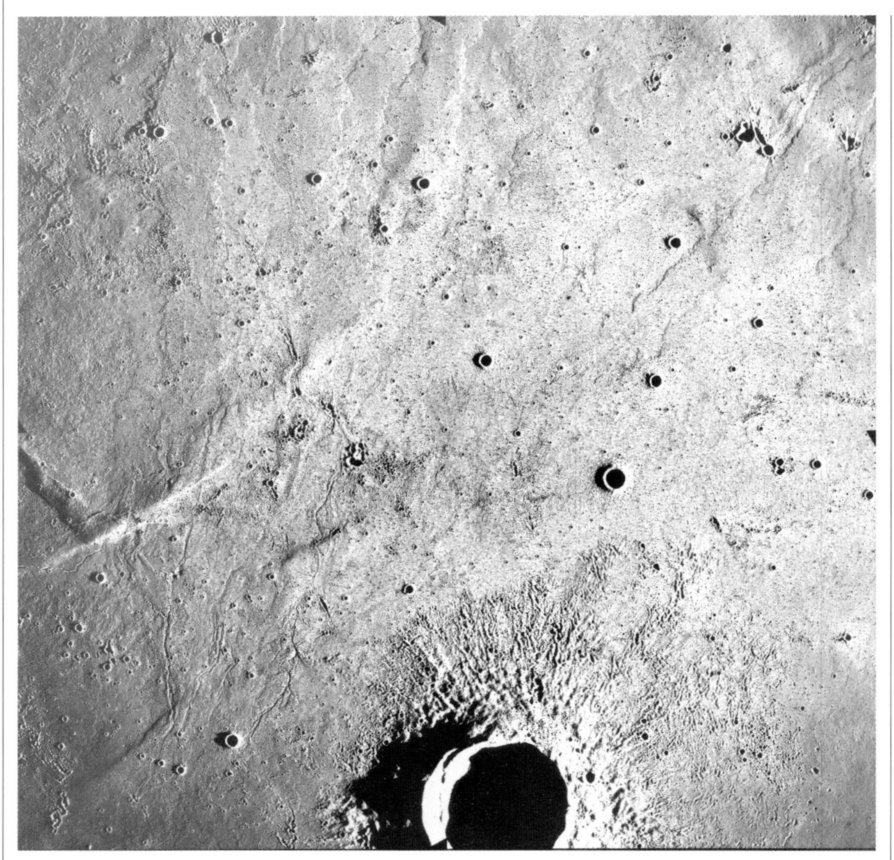

Lava flows in Mare Imbrium showing remnants of lava channels and individual lava flows, some of which can be traced 1,000 kilometers. Crater at the bottom of the picture is Euler (*Apollo 15* photograph M-1701).

impact basins, more than two dozen of which have been identified. The Orientale basin, found on the western limb of the Moon, is one of the youngest, best preserved of the larger structures. Basins range in size from 300 kilometers to more than 2,500 kilometers across. In general, the complexity of basins increases with size, with larger basins having two, three, or even more concentric rings of mountains. Basins are uniformly distributed on the Moon, but most basins on the near side have been filled or partly filled with lava flows.

In contrast to the highlands, lunar maria are dark and relatively smooth, with few impact craters, demonstrating their youth in comparison with the highlands. Exposed maria occupy about 17 percent of the lunar surface and are found mostly on the nearside. Most maria occur in the low parts of the lunar crust excavated by impact basins.

Even before the *Apollo* missions, photographs of flows in the Imbrium region led many scientists to speculate that maria were composed of lava flows. Some of the flows can be traced as thin sheets for more than 1,000 kilometers, indicating that the lavas must have been very fluid during eruption. Samples returned to Earth for analysis confirmed these speculations, showing that the flows consist of the iron-rich volcanic rock – basalt – and that during eruption they had the consistency of motor oil at room temperature. The samples also held some surprises, however. Because the maria are so sparsely cratered, it was thought that the flows were very young, yet radiogenic dating yielded ages in excess of 3 billion years for most of the lavas. While the dates indicate that these rocks are younger than the rocks in the highlands, most lavas on the Moon had erupted long before most rocks preserved on Earth were formed. Another surprise was the high titanium content in some lunar lavas. Some lunar basalts are richer in titanium than otherwise similar rocks on Earth by a factor of 10 or more.

Most of the volcanic eruptions on the Moon produced flows that formed thick "ponds" of lava on the floors of large craters, although some eruptions also emplaced thin lava flows. Flood lavas probably erupted as vast sheets from long fissures. Other flows were fed through open channels or underground tunnels of lava that transported magma to the flow front. Vestiges of lava channels can be traced hundreds of kilometers across the Moon.

Several areas on the Moon were centers of prolonged and complex volcanic eruptions. Parts of the Aristarchus Plateau, the eastern edge of Mare Serenitatas, isolated small craters, and other areas show dark patches covering the surface. These patches are considered to be pyroclastic (ash) deposits composed of glass beads formed during fire-fountaining eruptions.

Although we have no direct measurements of the thicknesses of mare lavas on the Moon, several indirect methods suggest that they range from 200 to 400 meters in thickness. Local deposits more than of 1,500 meters thick also probably occur at the centers of the impact structures, such as the Imbrium basin.

It is estimated that less than 1 percent of the lunar crust consists of mare basalts. In recent years, however, it has been suggested that more extensive mare lavas may remain hidden, buried by younger deposits such as ejec-

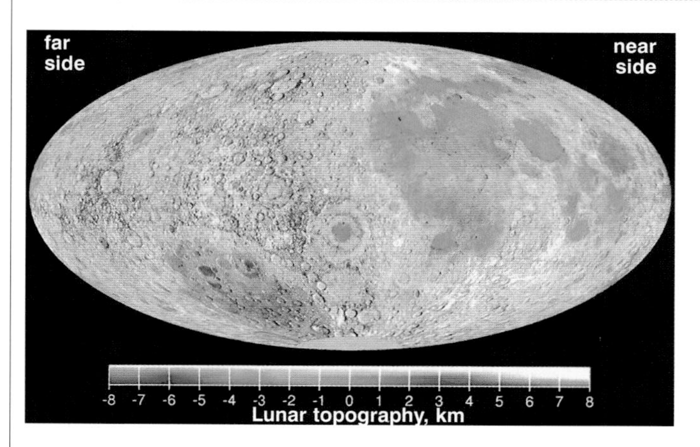

Lunar topography. The Clementine mission provided the first global altimetric data for the Moon. The enormous south pole-Aitken impact basin is the large low area in blue in the lower left. At 1,500 km across, the south pole-Aitken basin was suspected, but unproven, prior to Clementine (image courtesy David Smith and Maria Zuber).

ta from some impact craters. This proposal is supported by the discovery of fragments of basalts 4 billion years old found in highland rocks and detected by remote sensing data. The idea is complemented by the recognition of some mare flows with few superposed impact craters, suggesting a very young age. Thus, volcanism on the Moon may have been much more extensive and much longer-lived than earlier estimates.

Much more is known about the interior of the Moon than any other planet or satellite other than Earth. Data from the network of seismometers established by the *Apollo* astronauts and from other information provide a general view of the lunar interior. In addition, tracking of spacecraft in orbit around the Moon has revealed areas that are gravity anomalies, termed mascons, in reference to concentrations of mass that can be related to interior structure.

The Moon has a crust, mantle, and (possibly) a small core. The crust appears variable in thickness, ranging from a few tens of kilometers beneath some basins to more than 100 kilometers under some highland regions. The crust probably has been intensely fractured by impact cratering to depths of 20 kilometers or more. The mantle makes up some 90 percent of the total volume of the Moon and was the source for the lavas erupted on the surface. The Moon probably has an iron-rich core that may be partially molten and is perhaps 800 kilometers in diameter.

Alphonsus, a 125-kilometer impact crater, showing dark areas (arrows) thought to be volcanic pyroclastic deposits (*Apollo 16* photograph AS16-2478).

Geologic History of the Moon

The origin of the Moon has been a subject of much discussion. Was it "born" as a companion of Earth at the same time that the other planets were formed in the Solar System? Was it formed elsewhere in the Solar System, then captured into orbit by the Earth? Or was it derived from Earth itself? In recent years, a variation of this latter idea has gained popularity. Early in Solar System history when the planets were still accreting and evolving chemically, Earth may have been struck by a Mars-size planet, blasting a vaporous mixture of both the proto-Earth and the impacting object into orbit around the Earth. The vaporous mixture formed a disklike mass that was depleted in iron and volatile elements. In a few tens of millions of years, the disk condensed to form the "proto-Moon." Energy from condensation and the infall of smaller bodies, such as planetesimals, could have generated enough heat in the outer shell of the mass to form a "magma ocean." As the molten, chemical "soup" cooled and crystallized, the lower density materials such as calcium and aluminum-rich silicates floated to the surface to form the crust – seen as the highlands. Higher density minerals rich in iron tended to sink to the bottom of the magma ocean. The early Moon continued to sweep up Solar System debris, as recorded by the abundant impact craters in the highlands. With time, the number and size of impacting objects decreased markedly.

For convenience, the lunar geologic time scale has been subdivided into major intervals. The early history is designated the pre-Nectarian Period and is preserved in highland rocks and ancient cratered terrain found

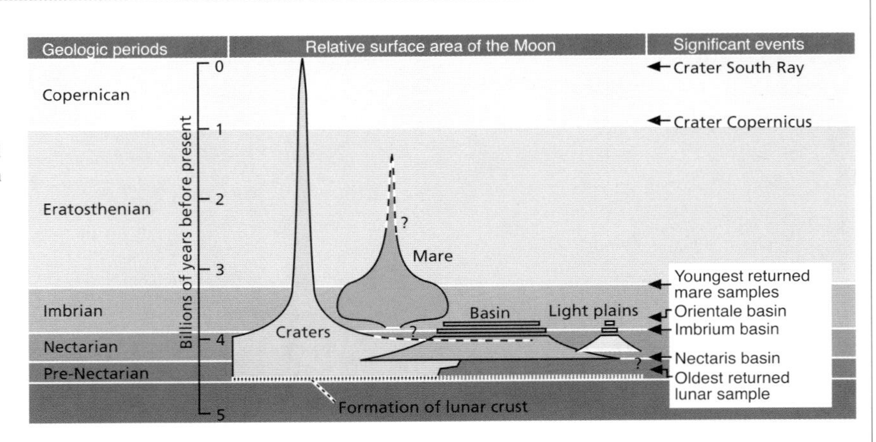

Lunar time scale, showing principal events in the history of the Moon (courtesy D. Wilhelms).

Color-ratio image of the Moon using data from the *Galileo* mission superposed on a shaded-relief base map. The Orientale basin, 900 kilometers in diameter, and parts of the farside highlands (left, red) are similar in composition to soils collected at the *Apollo 16* site. Several highland regions have enhanced iron content (yellow); mare regions (upper right) consist of basalts with relative high (blue) and low (orange) titanium dioxide content (U.S. Geological Survey Lunmap 12).

Facing page: *Apollo 17* astronaut using Lunar Rover in the Taurus-Littrow Valley (*Apollo 17* photograph AS17-147-22525).

mostly in the south polar region of the Moon and on the far side. The South Pole-Aitken basin, an ancient impact scar some 2000 kilometers across, formed in this early part of lunar history. This basin has been highly modified by subsequent impact events, but traces of its rim are visible in photographs and its presence is indicated by altimetric data obtained from orbit during the *Apollo* missions. Remote sensing data returned from the *Galileo* and *Clementine* missions suggest ancient mare deposits partly fill the basin. The Nectarian Period began with the formation of the Nectaris basin about 3.92 billion years ago. Rocks of this age consist mostly of impact-fractured highland crust and are exposed mostly on the eastern limb and the far side. Some volcanic eruptions of basalt and formation of KREEP also occurred during this period.

The last major basins to form on the Moon are the Imbrium basin (marking the start of the Imbrian Period 3.85 billion years ago) and the Orientale basin. The early Imbrian included the last stages of KREEP volcanism and the initiation of eruptions to form most of the mare lavas visible on the surface. The Eratosthenian Period, beginning about 3.2 billion years ago, saw the continued eruption of lavas, mostly in the western regions to form parts of Oceanus Procellarum and Mare Imbrium.

The Copernican Period, which includes deposits of rayed craters (craters with bright rays), including the crater Copernicus, began about 0.85 billion years ago. Volcanism of this relatively late stage (for the Moon) is suggested in some areas of the Moon, such as near the crater Lichtenberg. Except for these eruptions and occasional impacts to form small craters, most of the surface of the Moon has not changed for the past billion years.

In contrast to the history of the Moon, consider all of the events on Earth in the past billion years. The blossoming of life and the evolution of land masses through plate tectonics both contribute to the rich diversity that characterizes our planet, while the Moon has remained a relatively cold, inert body. Thus, the Moon is a small "fossil" planet, providing us with a window to the early geologic evolution of the Solar System.

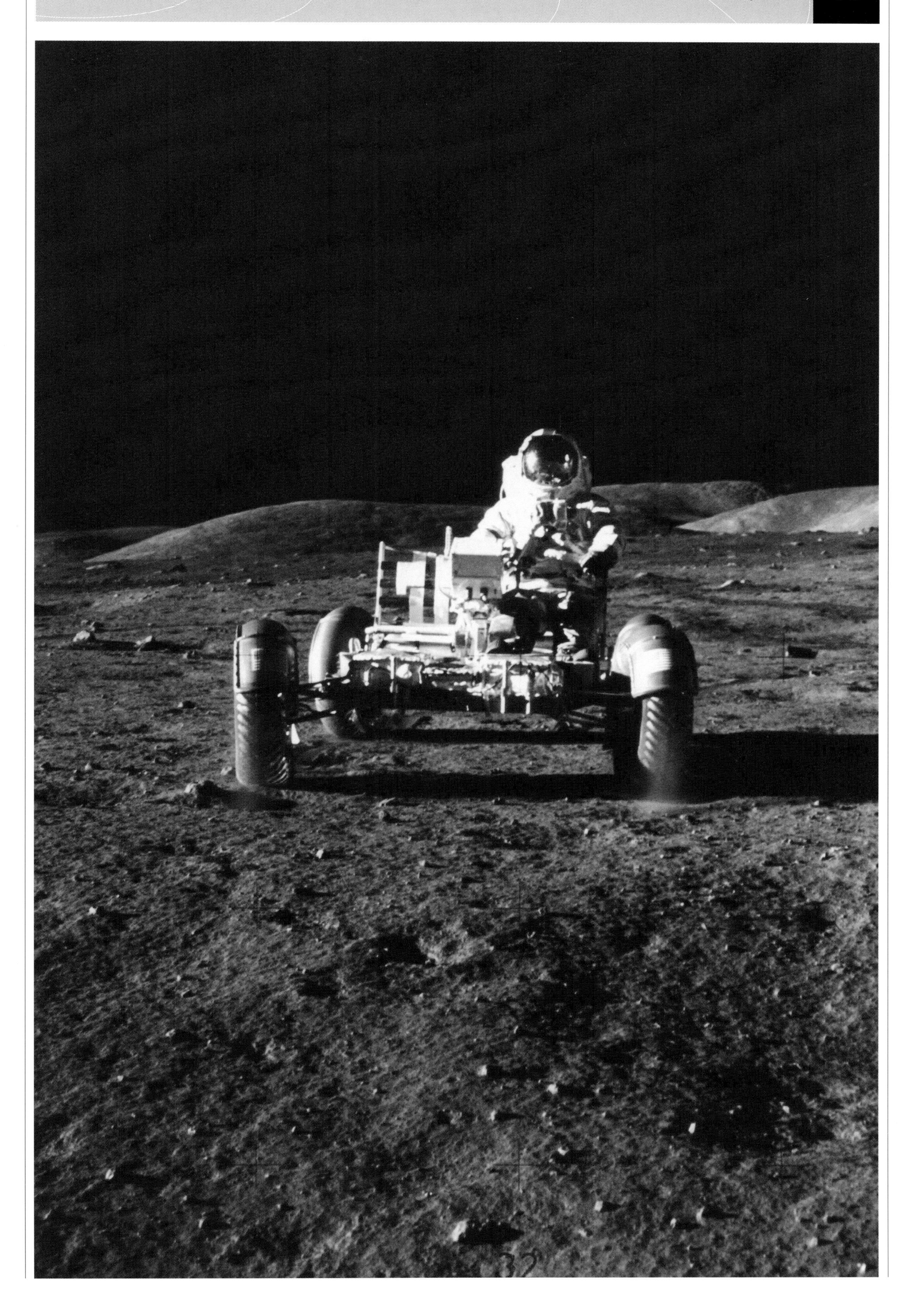

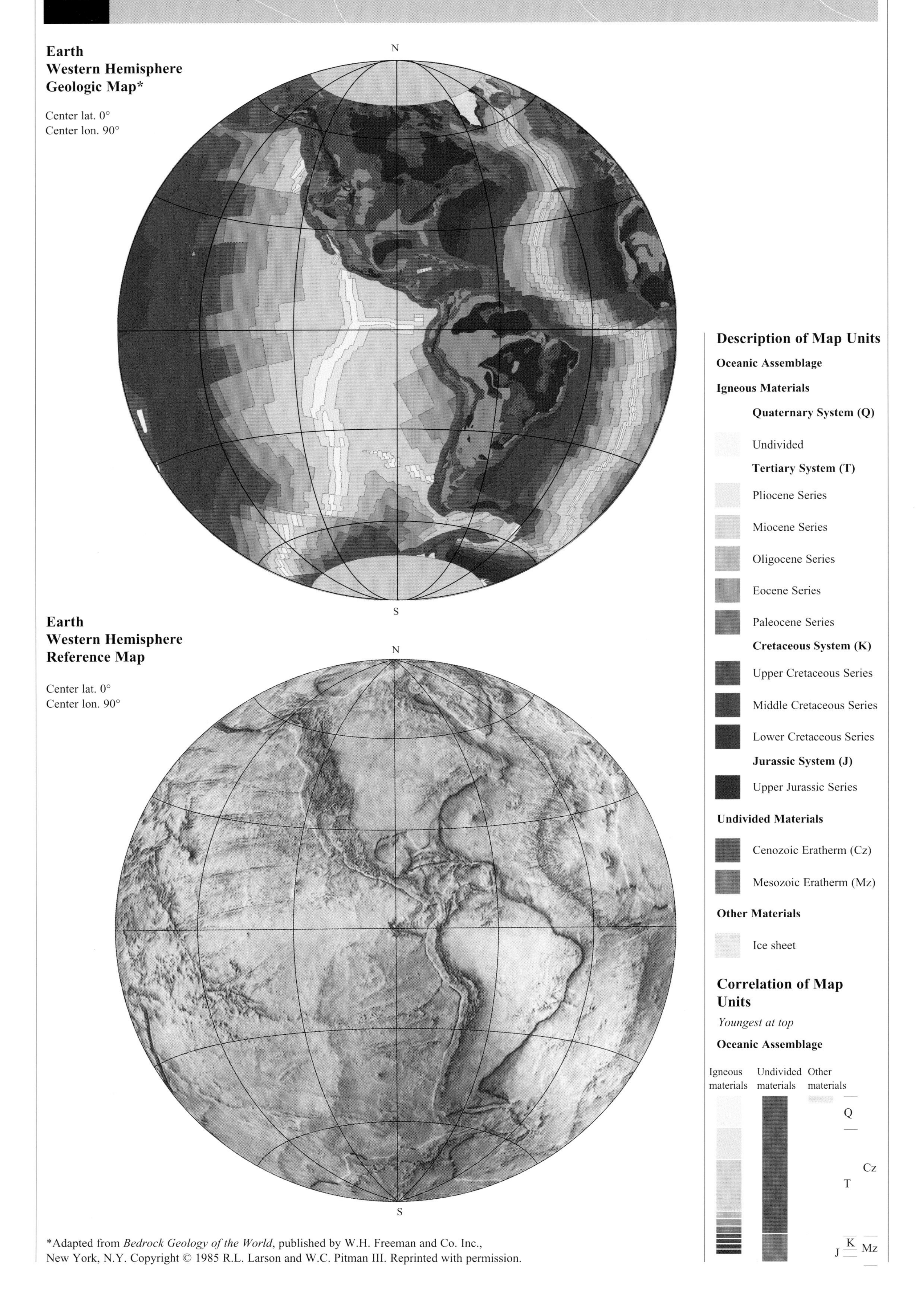

*Adapted from *Bedrock Geology of the World*, published by W.H. Freeman and Co. Inc., New York, N.Y. Copyright © 1985 R.L. Larson and W.C. Pitman III. Reprinted with permission.

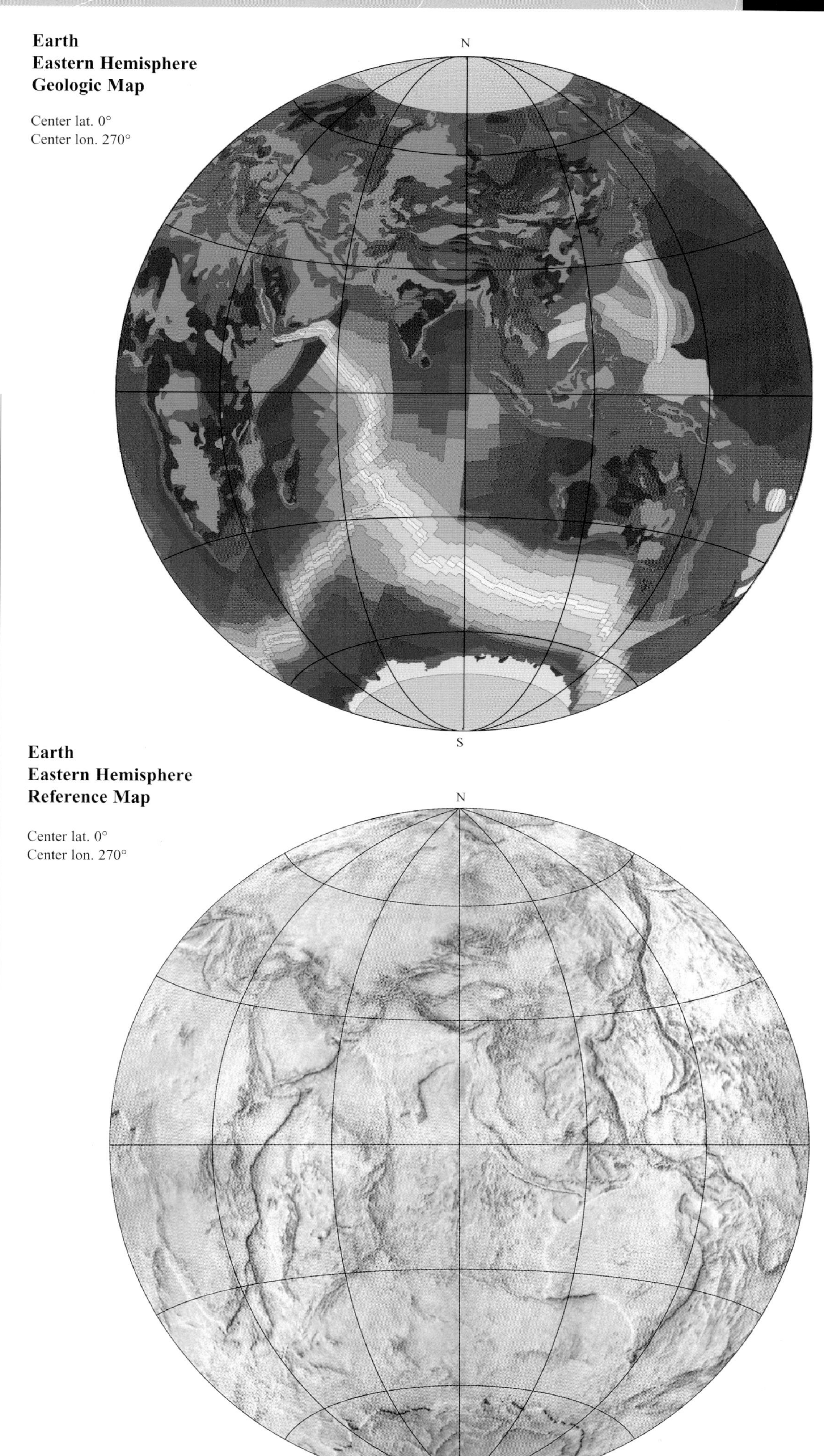

Description of Map Units

Continental Assemblage

Igneous Materials

- Cenozoic Eratherm (Cz)
- Mesozoic Eratherm (Mz)
- Paleozoic Eratherm (Pz)
- Phanerozoic undivided

Sedimentary and Metamorphic Materials

- Cenozoic Eratherm (Cz)
- Mesozoic (Mz)
- Paleozoic Eratherm (Pz)

Undivided Materials

- Precambrian (P€)

Other

- Unmapped

Correlation of Map Units

Youngest at top

Continental Assemblage

Igneous materials | Undivided igneous materials | Sedimentary and metamorphic materials

Cz

Mz

Pz

P€

Earth
Asian Hemisphere
Index Map

Center lat. –30°
Center lon. 100° E

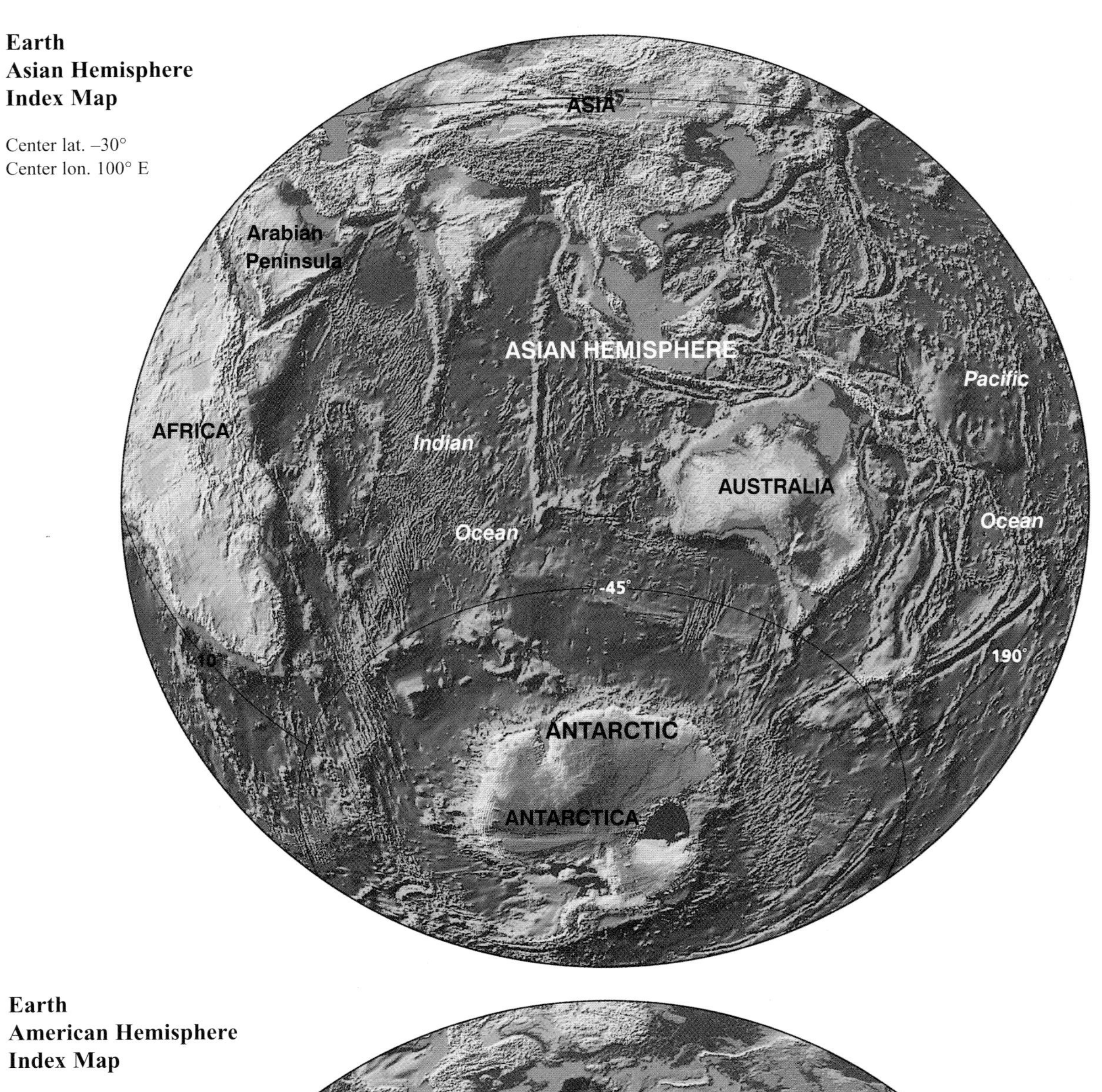

Earth
American Hemisphere
Index Map

Center lat. 30°
Center lon. 80° W

Arctic Ocean
EUROPE
ARCTIC
190°
10°
NORTH AMERICA
45°
AFRICA
Pacific
Atlantic
AMERICAN HEMISPHERE
Ocean
Ocean
SOUTH AMERICA
-45°

6,000 m
4,000 m
2,000 m
0
–2,000 m
–4,000 m
–6,000 m

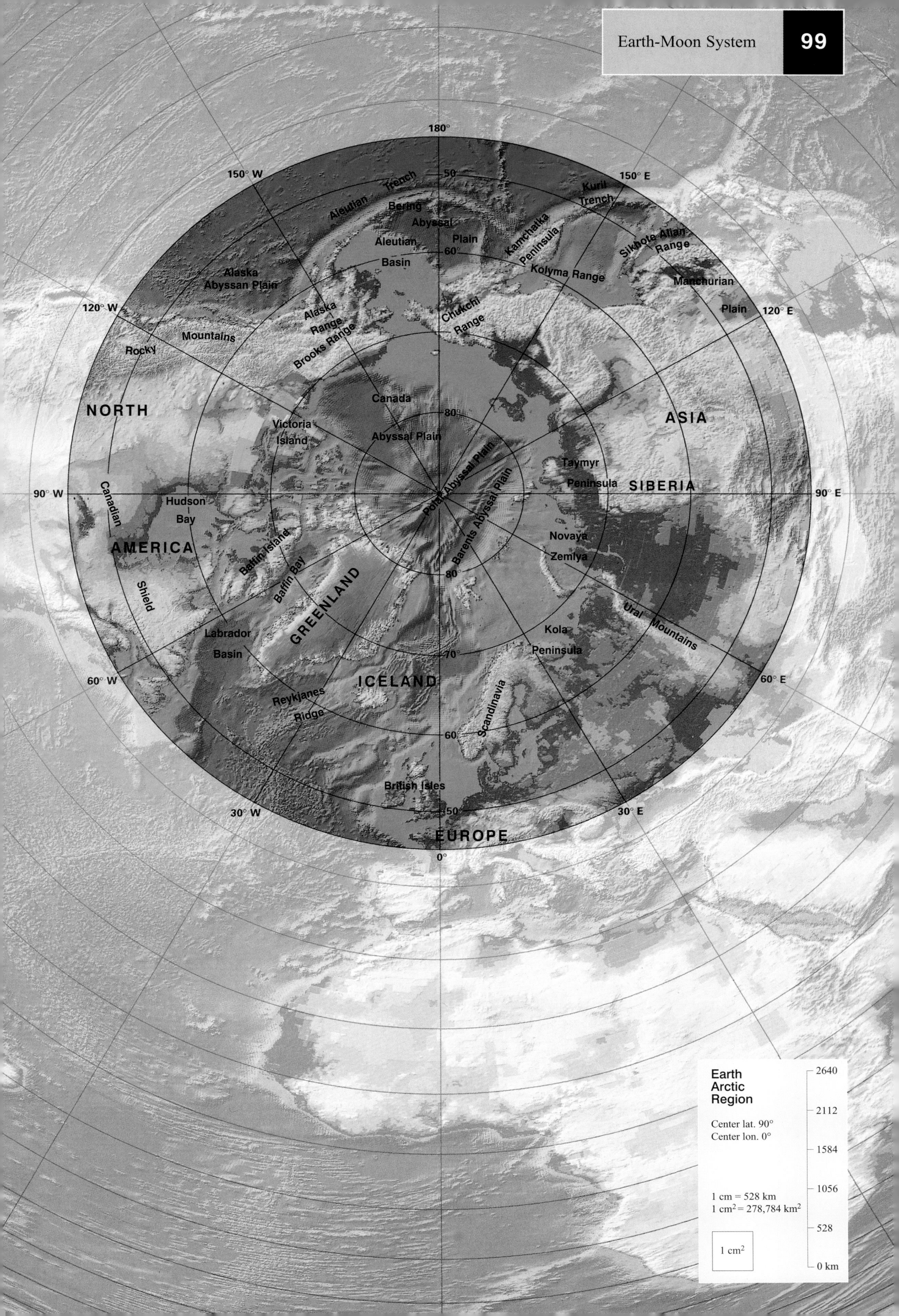
180°
150° W
150° E
120° W
120° E
90° W
90° E
60° W
60° E
30° W
30° E
0°
Aleutian Trench
Bering
Abyssal
Plain
Aleutian
Basin
Kuril
Trench
Kamchatka
Peninsula
Sikhote Alian
Range
Kolyma Range
Manchurian
Plain
Alaska
Abyssan Plain
Alaska
Range
Brooks Range
Chukchi
Range
Rocky
Mountains
NORTH
AMERICA
Canada
Abyssal Plain
Victoria
Island
ASIA
Taymyr
Peninsula
SIBERIA
Polar Abyssal Plain
Barents Abyssal Plain
Canadian
Shield
Hudson
Bay
Baffin Island
Baffin Bay
GREENLAND
Novaya
Zemlya
Ural Mountains
Kola
Peninsula
Labrador
Basin
ICELAND
Reykjanes
Ridge
Scandinavia
British Isles
EUROPE
50
60
70
80
Earth
Arctic
Region
Center lat. 90°
Center lon. 0°
1 cm = 528 km
1 cm² = 278,784 km²
1 cm²
2640
2112
1584
1056
528
0 km

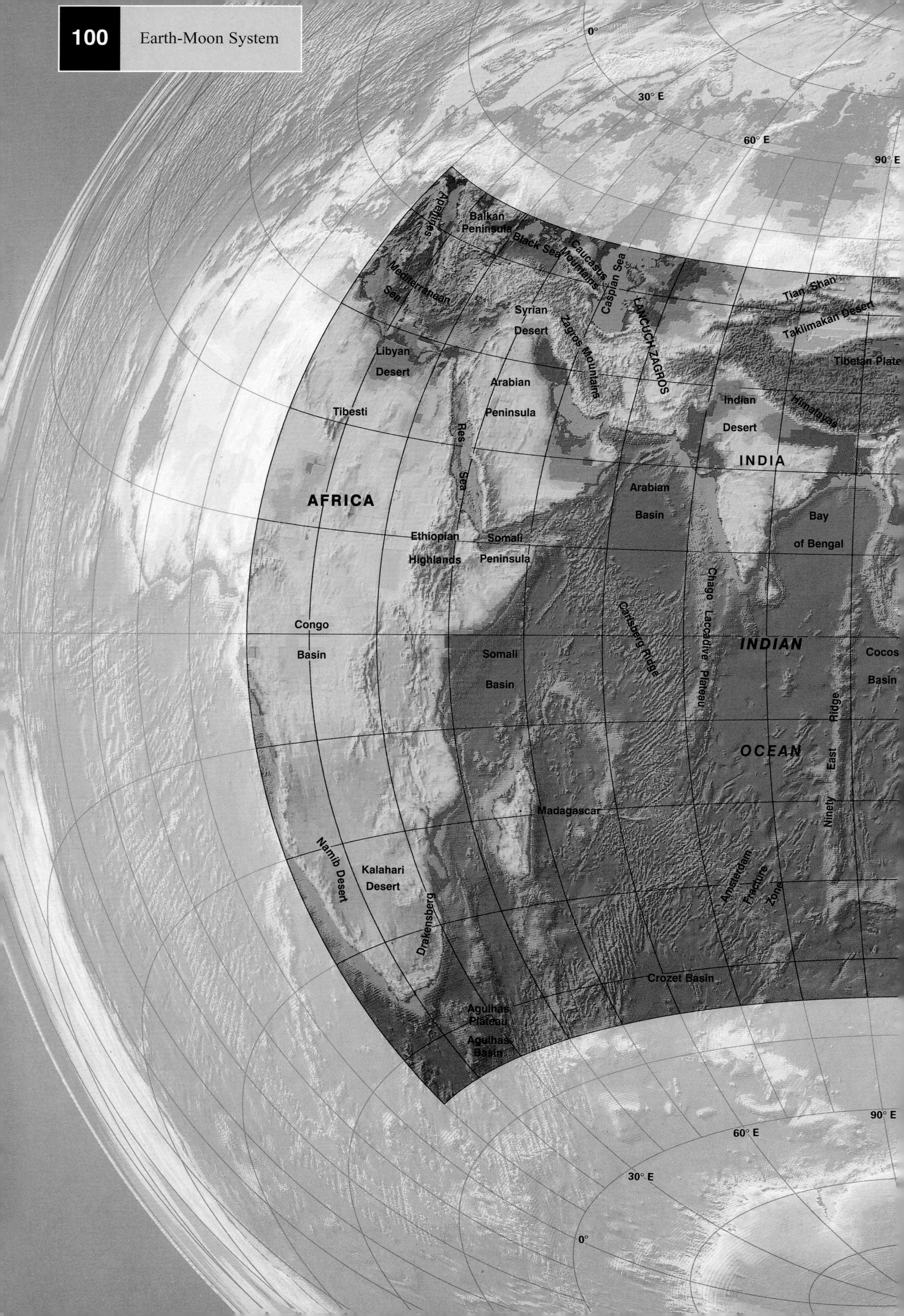

0°
30° E
60° E
90° E
Apennines
Balkan Peninsula
Black Sea
Caucasus Mountains
Caspian Sea
Mediterranean Sea
Syrian Desert
Zagros Mountains
LANCUCH ZAGROS
Tian Shan
Taklimakan Desert
Tibetan Plate
Libyan Desert
Arabian Peninsula
Tibesti
Res Sea
Indian Desert
Himalayas
INDIA
AFRICA
Arabian Basin
Bay of Bengal
Ethiopian Highlands
Somali Peninsula
Chago Laccadive Plateau
Carlsberg Ridge
Congo Basin
INDIAN
OCEAN
Cocos Basin
Somali Basin
Ninety East Ridge
Madagascar
Namib Desert
Kalahari Desert
Amsterdam Fracture Zone
Drakensberg
Crozet Basin
Agulhas Plateau
Agulhas Basin
90° E
60° E
30° E
0°

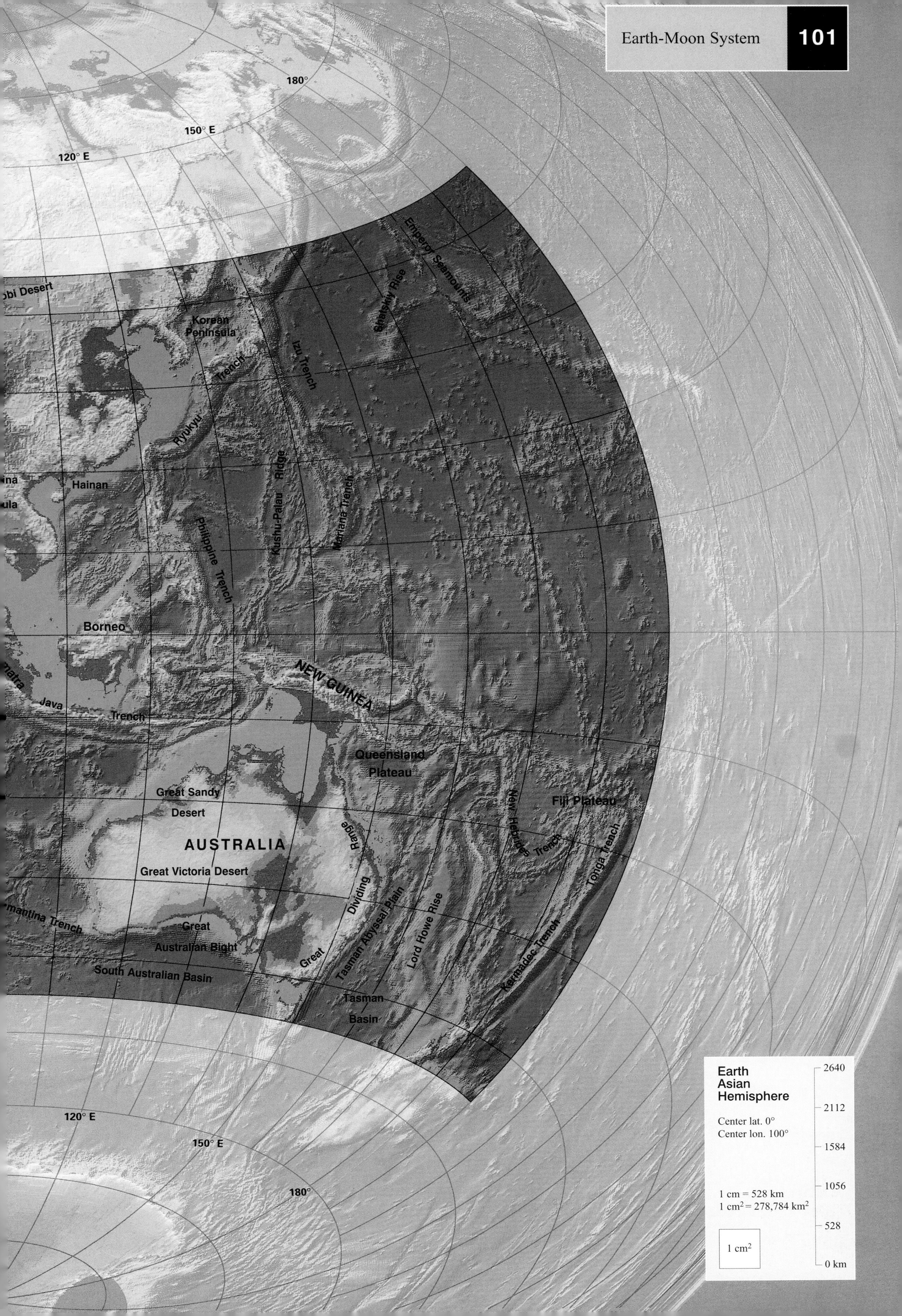

180°
150° E
120° E
bi Desert
Korean
Peninsula
Trench
Izu Trench
Emperor Seamounts
Shatsky Rise
Ryukyu
Kushu-Palau Ridge
Mariana Trench
Hainan
Philippine Trench
Borneo
NEW GUINEA
Java
Trench
Queensland
Plateau
Great Sandy
Desert
AUSTRALIA
Great Victoria Desert
Range
Dividing
Great
New Hebrides Trench
Fiji Plateau
Tonga Trench
Tasman Abyssal Plain
Lord Howe Rise
Kermadec Trench
mantina Trench
Great
Australian Bight
South Australian Basin
Tasman
Basin
120° E
150° E
180°
Earth
Asian
Hemisphere
Center lat. 0°
Center lon. 100°
1 cm = 528 km
1 cm² = 278,784 km²
1 cm²
2640
2112
1584
1056
528
0 km

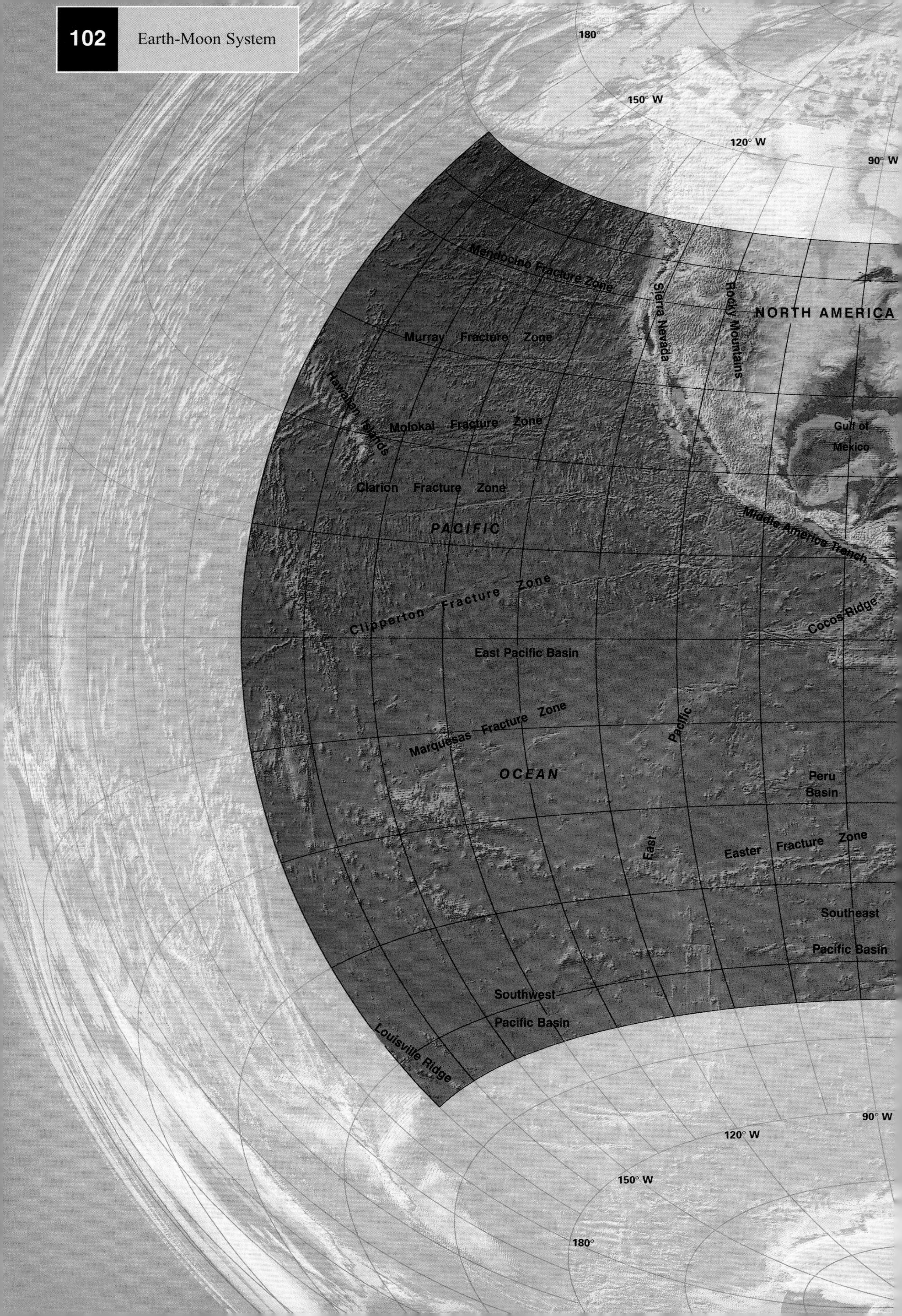

180°
150° W
120° W
90° W
Mendocino Fracture Zone
Sierra Nevada
Rocky Mountains
NORTH AMERICA
Murray Fracture Zone
Hawaiian Islands
Molokai Fracture Zone
Gulf of Mexico
Clarion Fracture Zone
PACIFIC
Middle America Trench
Clipperton Fracture Zone
Cocos Ridge
East Pacific Basin
Marquesas Fracture Zone
East Pacific
OCEAN
Peru Basin
Easter Fracture Zone
Southeast Pacific Basin
Southwest Pacific Basin
Louisville Ridge
90° W
120° W
150° W
180°

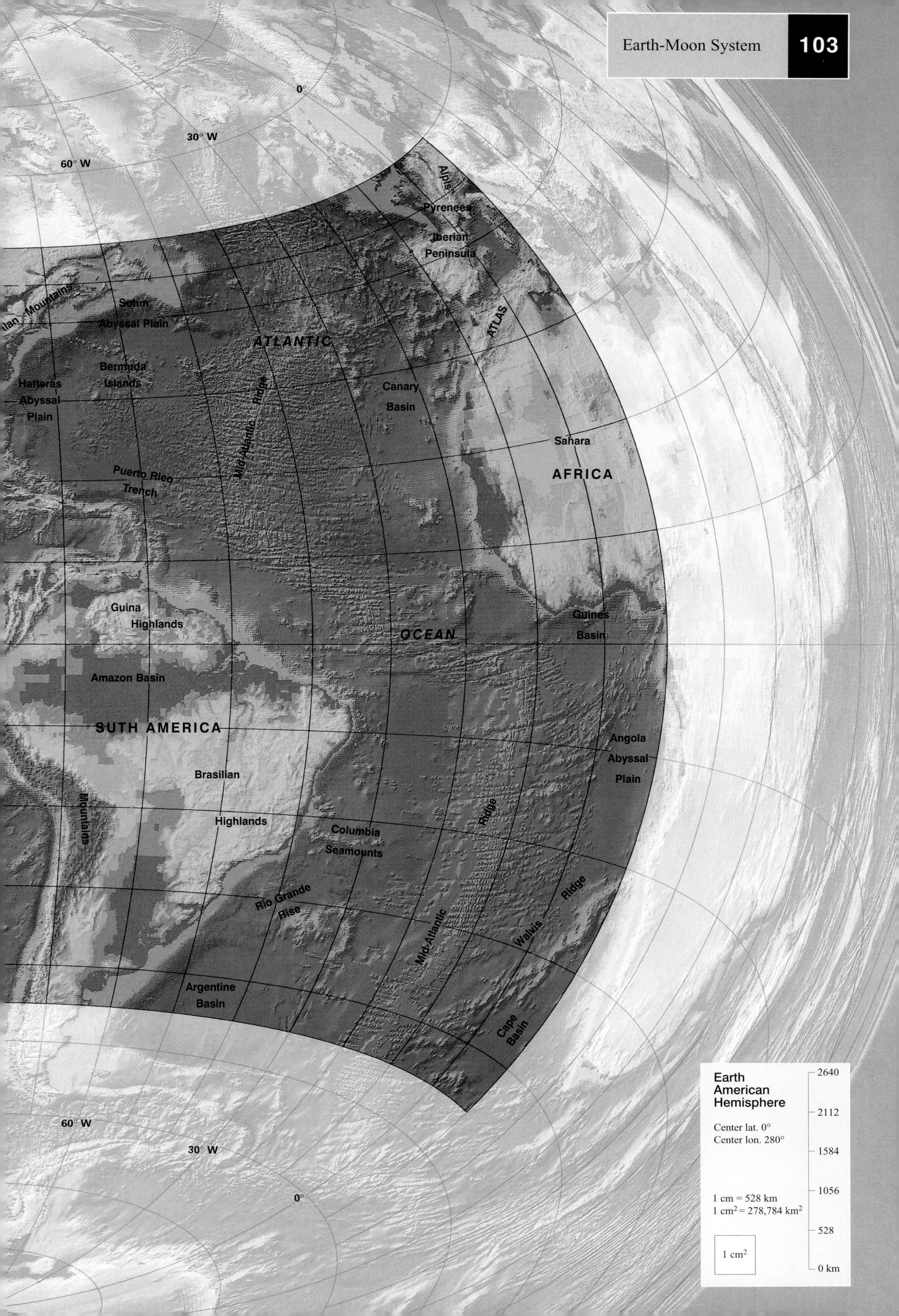
0°
30° W
60° W
Alpis
Pyrenees
Iberian
Peninsula
Mountains
Sohm
Abyssal Plain
ATLAS
ATLANTIC
Bermuda
Islands
Hatteras
Abyssal
Plain
Mid-Atlantic Ridge
Canary
Basin
Sahara
AFRICA
Puerto Rico
Trench
Guina
Highlands
OCEAN
Guines
Basin
Amazon Basin
SUTH AMERICA
Angola
Abyssal
Plain
Brasilian
Highlands
Mountains
Ridge
Columbia
Seamounts
Rio Grande
Rise
Mid-Atlantic
Walvis
Ridge
Argentine
Basin
Cape
Basin
60° W
30° W
0°
Earth
American
Hemisphere
Center lat. 0°
Center lon. 280°
1 cm = 528 km
1 cm² = 278,784 km²
1 cm²
2640
2112
1584
1056
528
0 km

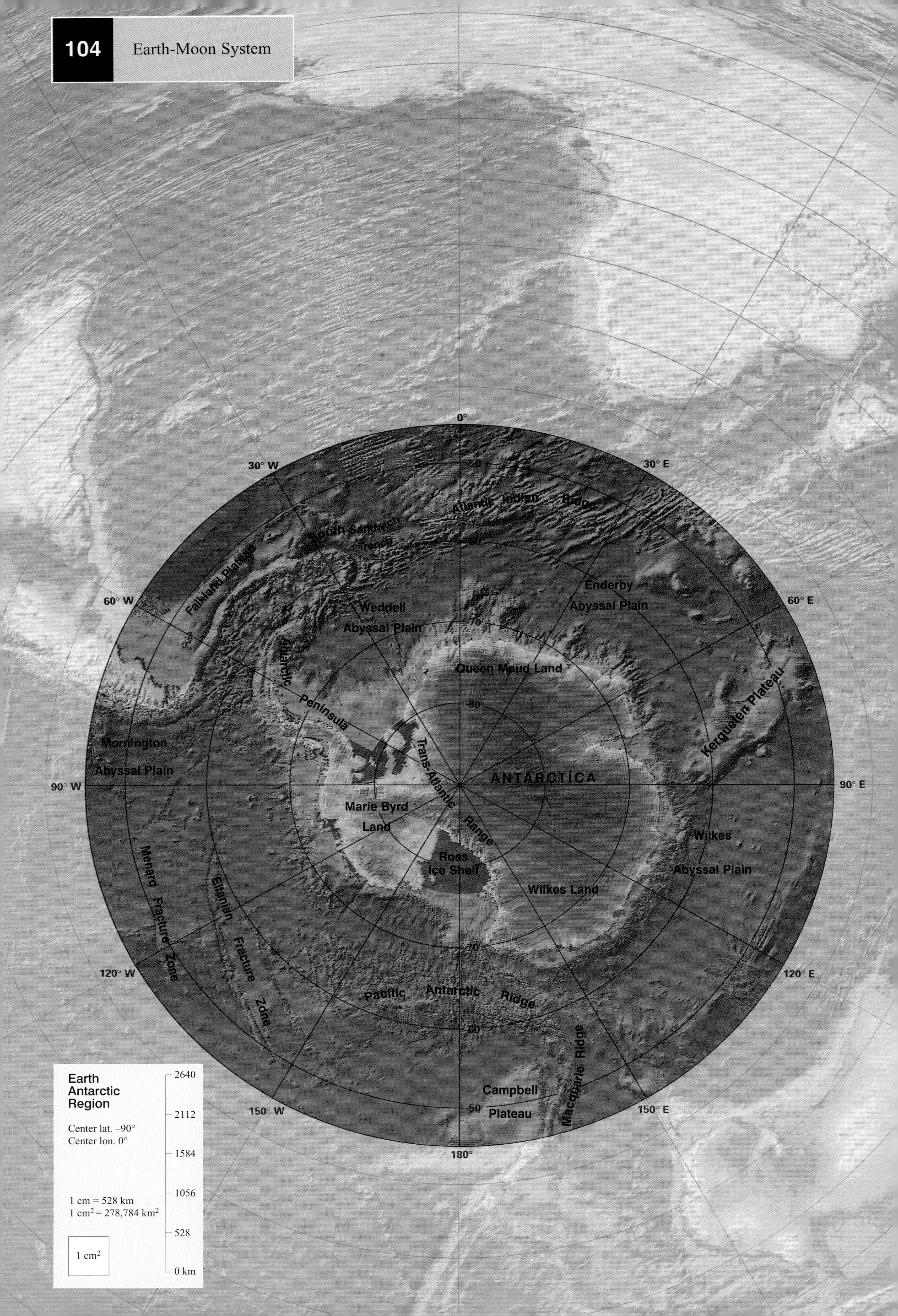

0°
30° W
30° E
-50°
Atlantic Indian Ridge
South Sandwich Trench
-60°
Falkland Plateau
Enderby Abyssal Plain
60° W
60° E
Weddell Abyssal Plain
-70°
Antarctic Peninsula
Queen Maud Land
Kerguelen Plateau
-80°
Mornington Abyssal Plain
Trans-Atlantic Range
ANTARCTICA
90° W
90° E
Marie Byrd Land
Ross Ice Shelf
Wilkes Abyssal Plain
Menard Fracture Zone
Eltanian Fracture Zone
Wilkes Land
-70
120° W
120° E
Pacific Antarctic Ridge
-60
Macquarie Ridge
Campbell Plateau
150° W
-50°
150° E
180°

Earth Antarctic Region

Center lat. –90°
Center lon. 0°

1 cm = 528 km
1 cm² = 278,784 km²

1 cm²

2640
2112
1584
1056
528
0 km

Moon
Near Side Hemisphere
Geologic Map

Center lat. 0°
Center lon. 0°

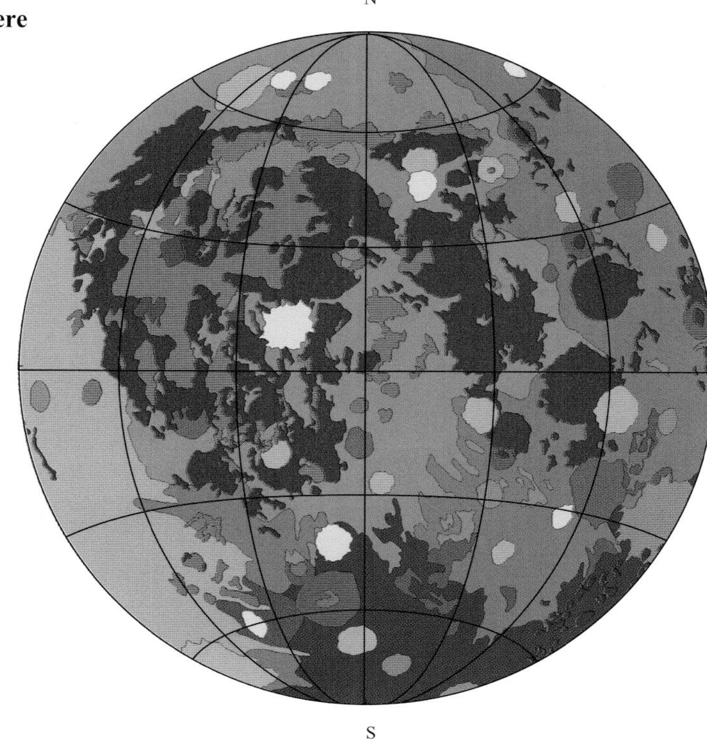

Moon
Near Side Hemisphere
Reference Map

Center lat. 0°
Center lon. 0°

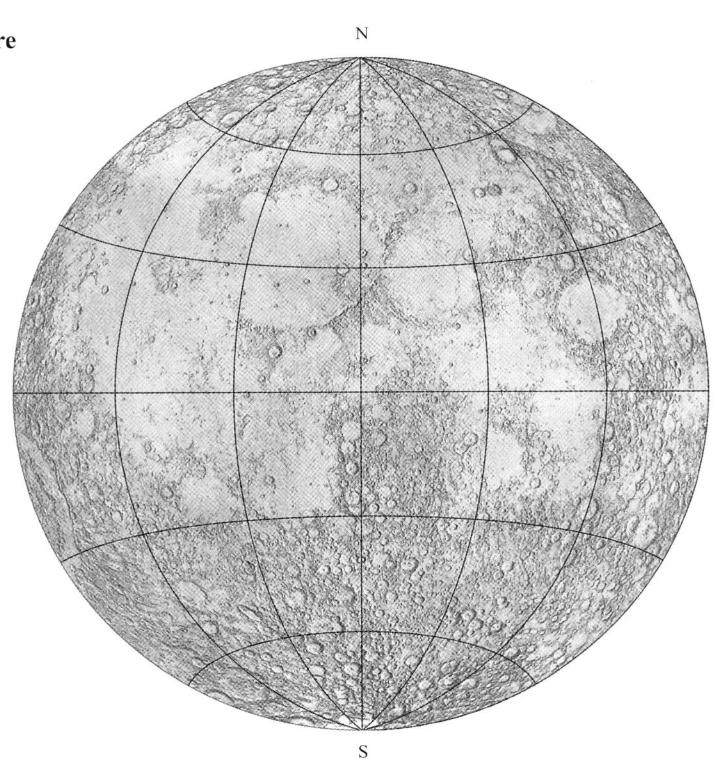

Description of Map Units

Igneous Plains Materials

- Copernican System (C), mare basalt
- Eratosthenian System (E), mare basalt
- Imbrian System (I), mare basalt
- Imbrian System (I), volcanic province
- Imbrian System (I), plains materials
- Nectarian System (N), plains materials

Impact Materials

Crater Materials

- Copernican System (C)
- Eratosthenian System (E)
- Upper Imbrian System (I)
- Lower Imbrian System (I)
- Nectarian System (N)

Basin Materials

- Orientale basin (I)
- Imbrian basin (I)
- Nectarian basin (N)
- Pre-Nectarian basin, undivided (pN)

Miscellaneous Materials

- Pre-Nectarian, undivided (pN)

Structure Symbols

- Graben
- Mare ridge

Correlation of Map Units

Youngest at top

Igneous plains materials

Impact materials

Crater

Basin

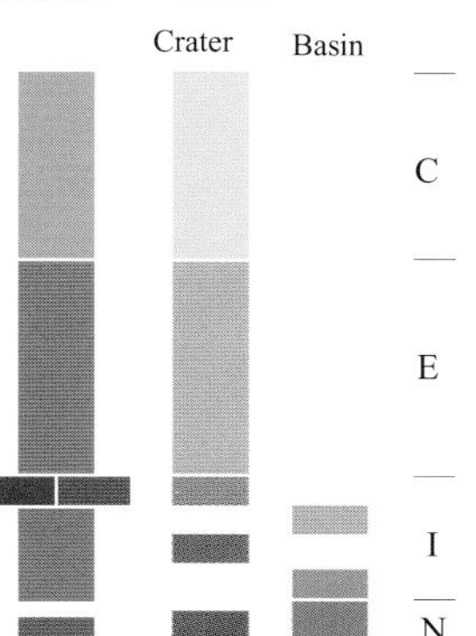

Moon
Far Side Hemisphere
Geologic Map

Center lat. 0°
Center lon. 180°

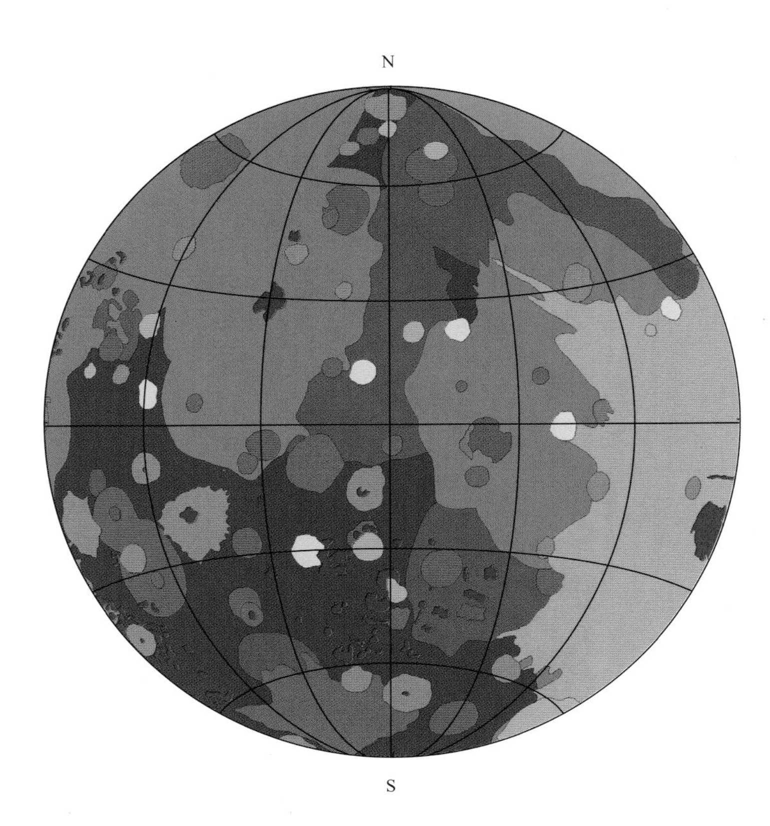

Description of Map Units

Igneous Plains Materials

- Copernican System (C), mare basalt
- Eratosthenian System (E), mare basalt
- Imbrian System (I), mare basalt
- Imbrian System (I), volcanic province
- Imbrian System (I), plains materials
- Nectarian System (N), plains materials

Impact Materials

Crater Materials

- Copernican System (C)
- Eratosthenian System (E)
- Upper Imbrian System (I)
- Lower Imbrian System (I)
- Nectarian System (N)

Basin Materials

- Orientale basin (I)
- Imbrian basin (I)
- Nectarian basin (N)
- Pre-Nectarian basin, undivided (pN)

Miscellaneous Materials

- Pre-Nectarian, undivided (pN)

Structure Symbols

- Graben
- Mare ridge

Monn
Far Side Hemisphere
Reference Map

Center lat. 0°
Center lon. 180°

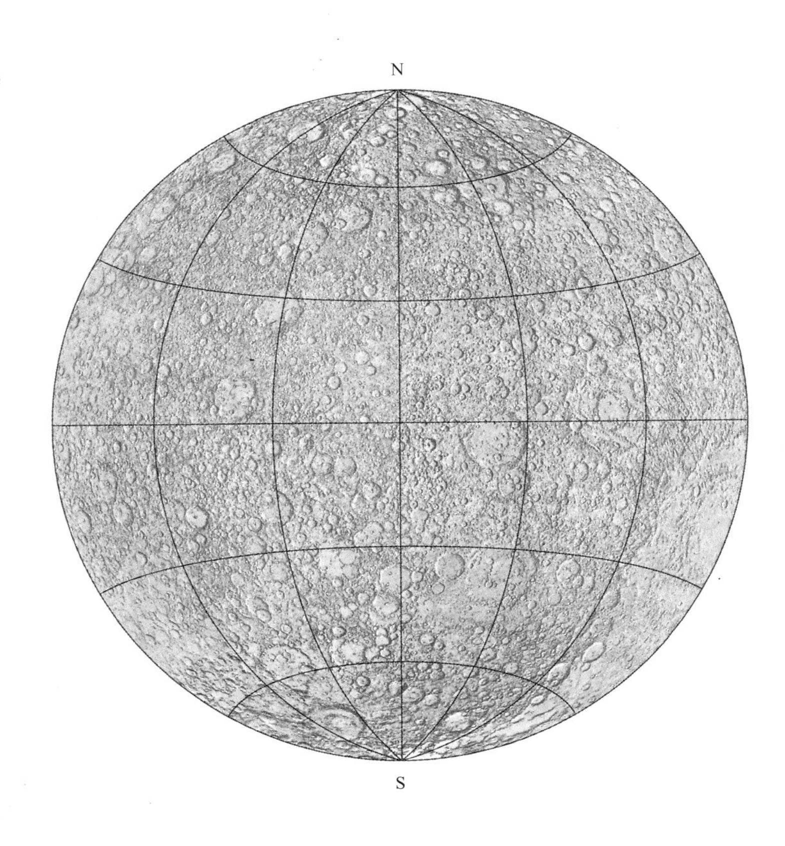

Correlation of Map Units

Youngest at top

Igneous plains materials
Impact materials
Crater
Basin
C
E
I
N
pN

Moon
Near Side
Index Map

Center lat. 30°
Center lon. 0°

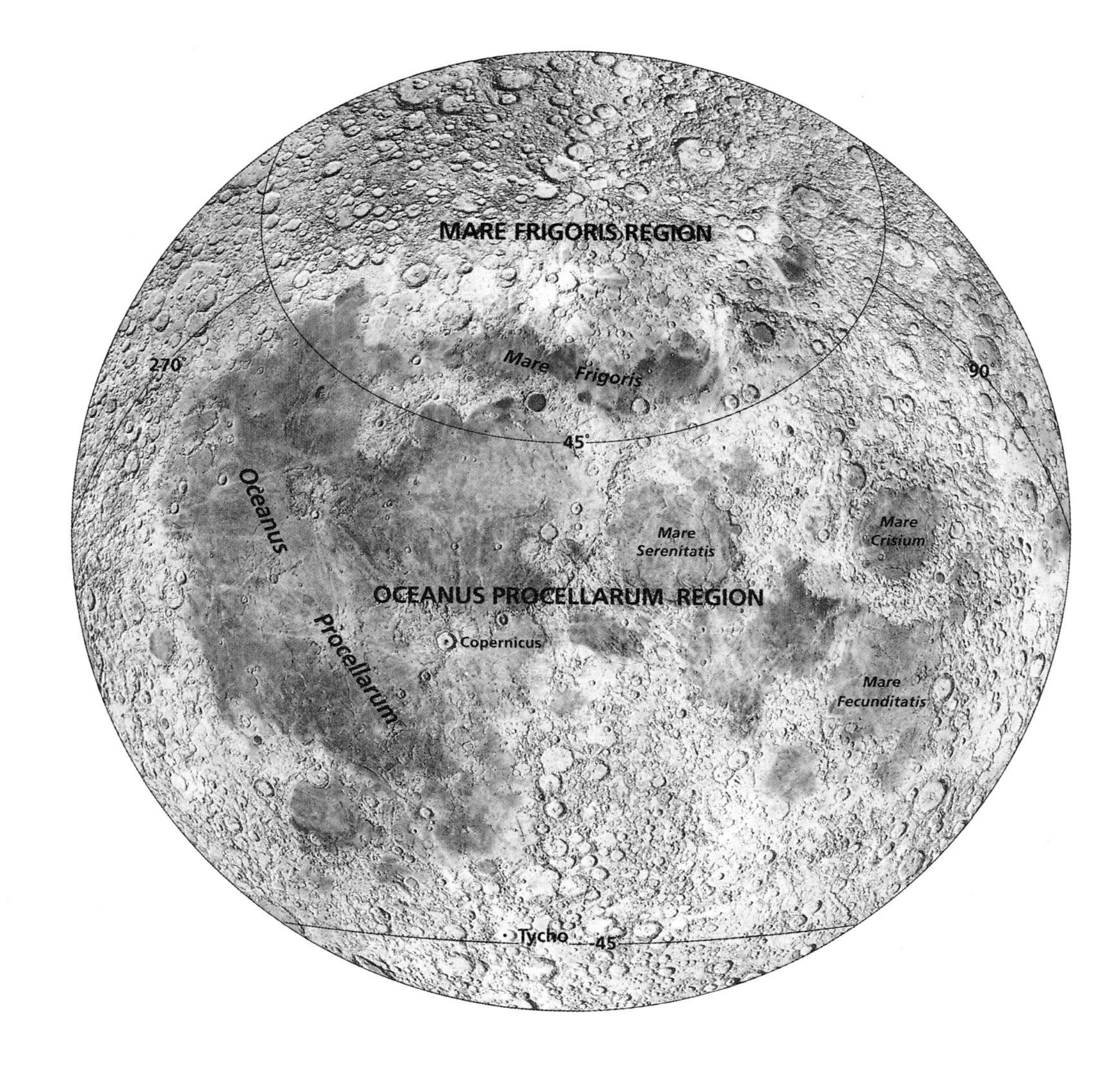

Moon
Far Side
Index Map

Center lat. –30°
Center lon. 180°

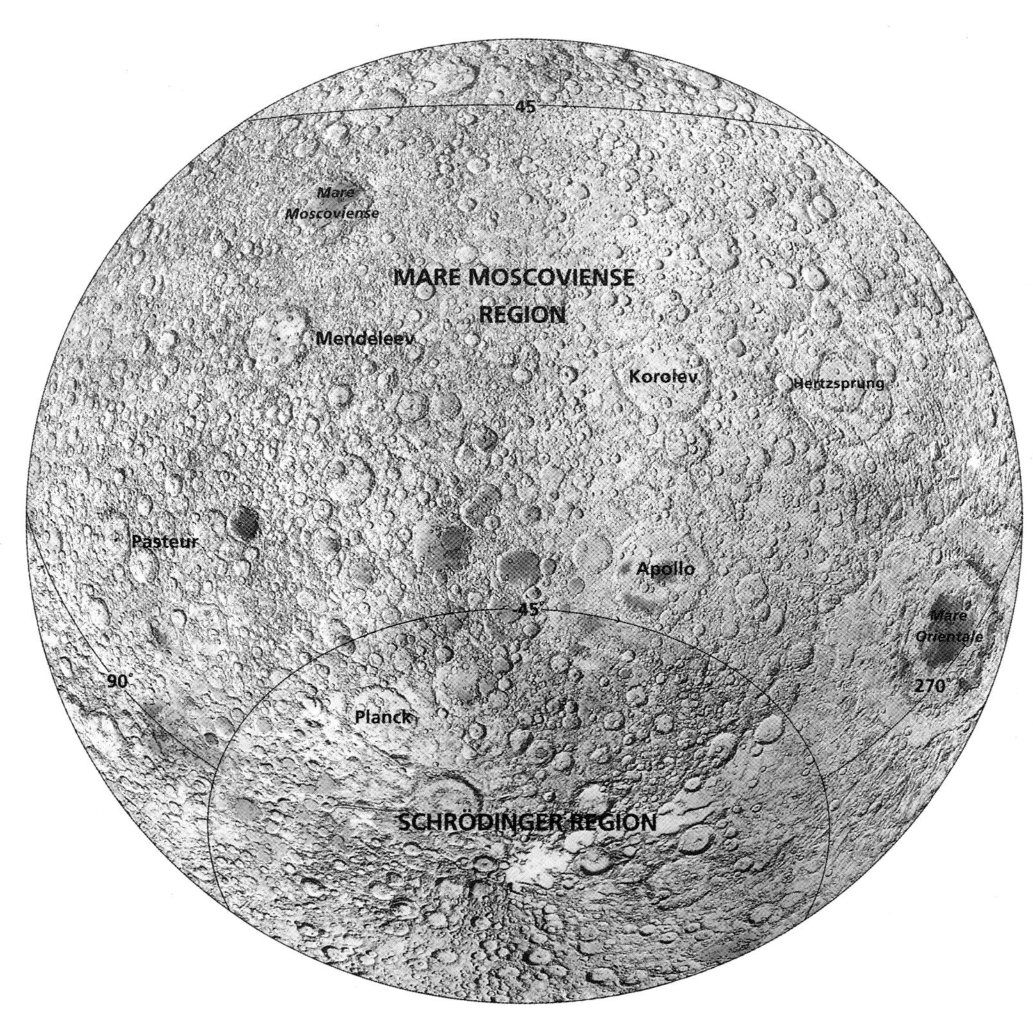

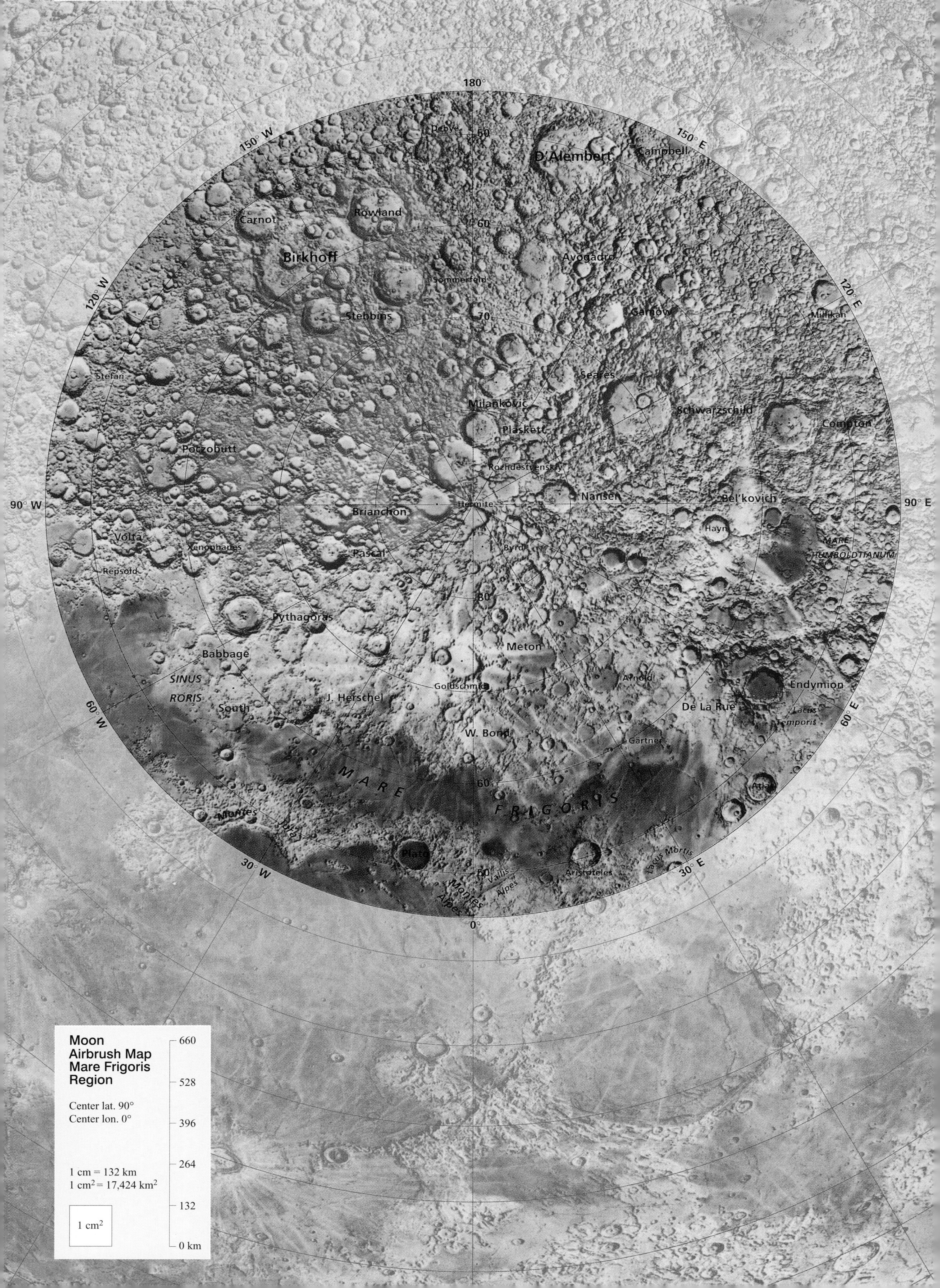

Moon
Airbrush Map
Mare Frigoris
Region

Center lat. 90°
Center lon. 0°

1 cm = 132 km
1 cm^2 = 17,424 km^2

1 cm^2

660
528
396
264
132
0 km

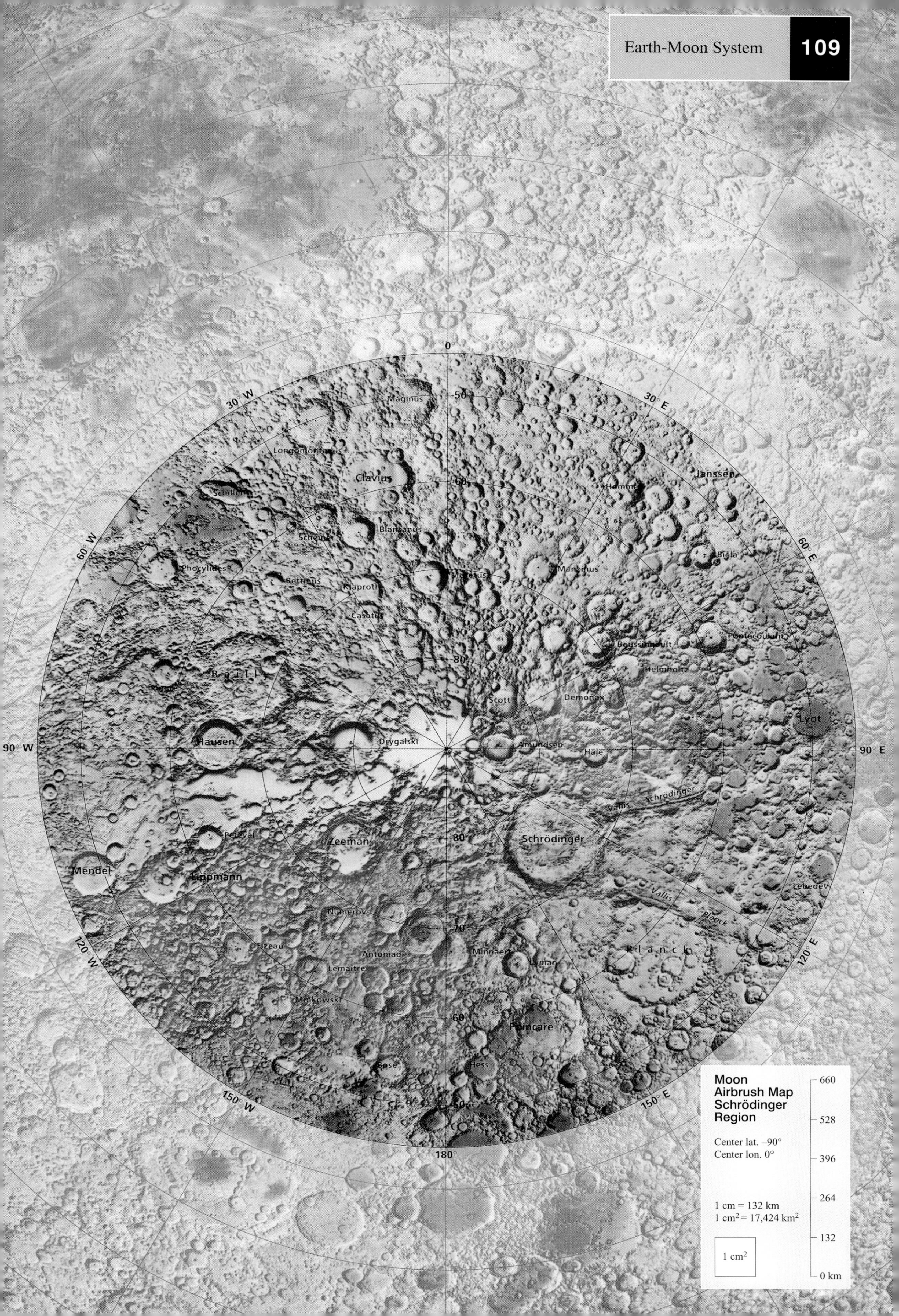
0°
30° W
30° E
60° W
60° E
90° W
90° E
120° W
120° E
150° W
150° E
180°
50
60
70
80
Maginus
Longomontanus
Clavius
Schiller
Scheiner
Blancanus
Phocylides
Bettinus
Klaproth
Casatus
Manzinus
Janssen
Hommel
Biela
Pontécoulant
Boussingault
Helmholtz
Demonax
Scott
Lyot
Amundsen
Hale
Drygalski
Hausen
Bailly
Vallis Schrödinger
Schrödinger
Zeeman
Mendel
Lippmann
Numerov
Fizeau
Antoniadi
Lemaître
Minkowski
Minnaert
Lyman
Planck
Vallis Planck
Lebedev
Poincaré
Hess
Bose
Moon
Airbrush Map
Schrödinger
Region
Center lat. –90°
Center lon. 0°
1 cm = 132 km
1 cm² = 17,424 km²
1 cm²
660
528
396
264
132
0 km

90° W
60° W
30° W
Sinus Iridum
MARE
IMBRIUM
Dorsum Zirkel
OCEANUS
PROCELLARUM
Russell
Struve
Eddington
Einstein
Vasco da Gama
Cardanus
Olbers
Hevelius
Riccioli
Grimaldi
Montes Carpatus
Eratosthenes
Montes Apenninus
SINUS AESTUUM
MARE INSULARUM
Kepler
Copernicus
Fra Mauro
Ptolemaeus
Montes Riphaeus
MARE COGNITUM
Letronne
Gassendi
MARE HUMORUM
MARE NUBIUM
Rupes Recta
Mersenius
Darwin
Byrgius
Vieta
Palus Epidemiarum
Regiomontanus
Pitatus
Deslandres
Purbach
Wilhelm
Tycho
Mee
Schickard
Vallis Inghirami
Vallis Bouvard
Lacus Autumni
Montes Cordillera
Rimae Sirsalis
Book
Montes Rook
30° W
60° W
90° W

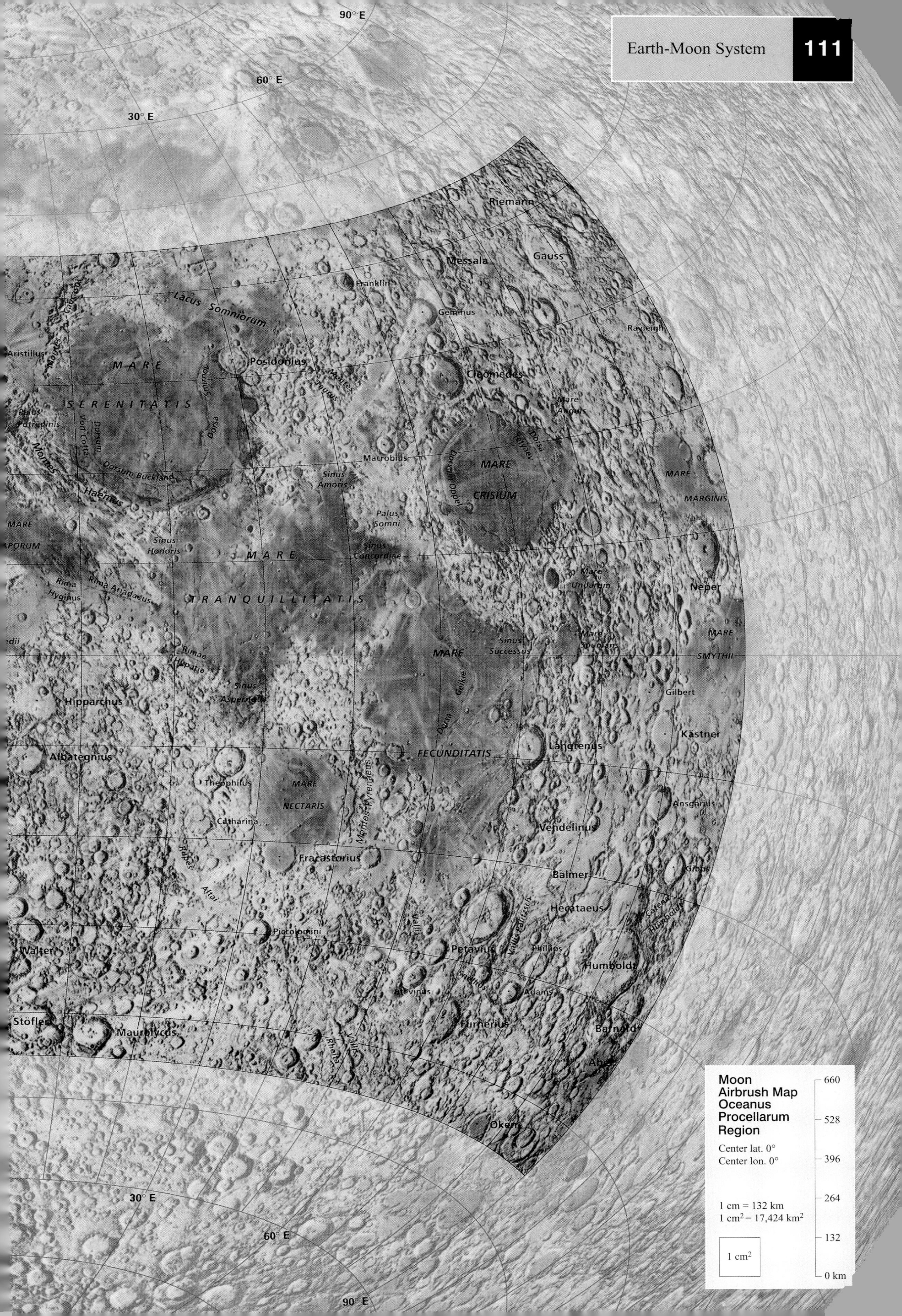
90° E
60° E
30° E
Riemann
Gauss
Messala
Franklin
Geminus
Rayleigh
Lacus Somniorum
Aristillus
Montes Caucasus
MARE SERENITATIS
Posidonius
Dorsa Smirnov
Montes Taurus
Cleomedes
Mare Anguis
Palus Putredinis
Dorsum Von Cotta
Montes Haemus
Dorsum Buckland
Macrobius
Sinus Amoris
Dorsa Tetyaev
Dorsum Oppel
MARE CRISIUM
MARE MARGINIS
Palus Somni
Sinus Honoris
Sinus Concordiae
MARE TRANQUILLITATIS
Rima Hyginus
Rima Ariadaeus
Mare Undarum
Neper
Mare Spumans
MARE SMYTHII
Sinus Successus
Rimae Hypatia
Sinus Asperitatis
Dorsa Geikie
Gilbert
Hipparchus
Kästner
Langrenus
MARE FECUNDITATIS
Albategnius
Theophilus
MARE NECTARIS
Montes Pyrenaeus
Catharina
Ansgarius
Vendelinus
Rupes Altai
Fracastorius
Balmer
Gibbs
Hecataeus
Catena Humboldt
Vallis Palitzsch
Piccolomini
Walter
Petavius
Phillips
Humboldt
Snellius
Stevinus
Adams
Stöfler
Maurolycus
Furnerius
Barnard
Vallis Rheita
Abel
Oken
Moon
Airbrush Map
Oceanus
Procellarum
Region
Center lat. 0°
Center lon. 0°
1 cm = 132 km
1 cm^2 = 17,424 km^2
1 cm^2
660
528
396
264
132
0 km

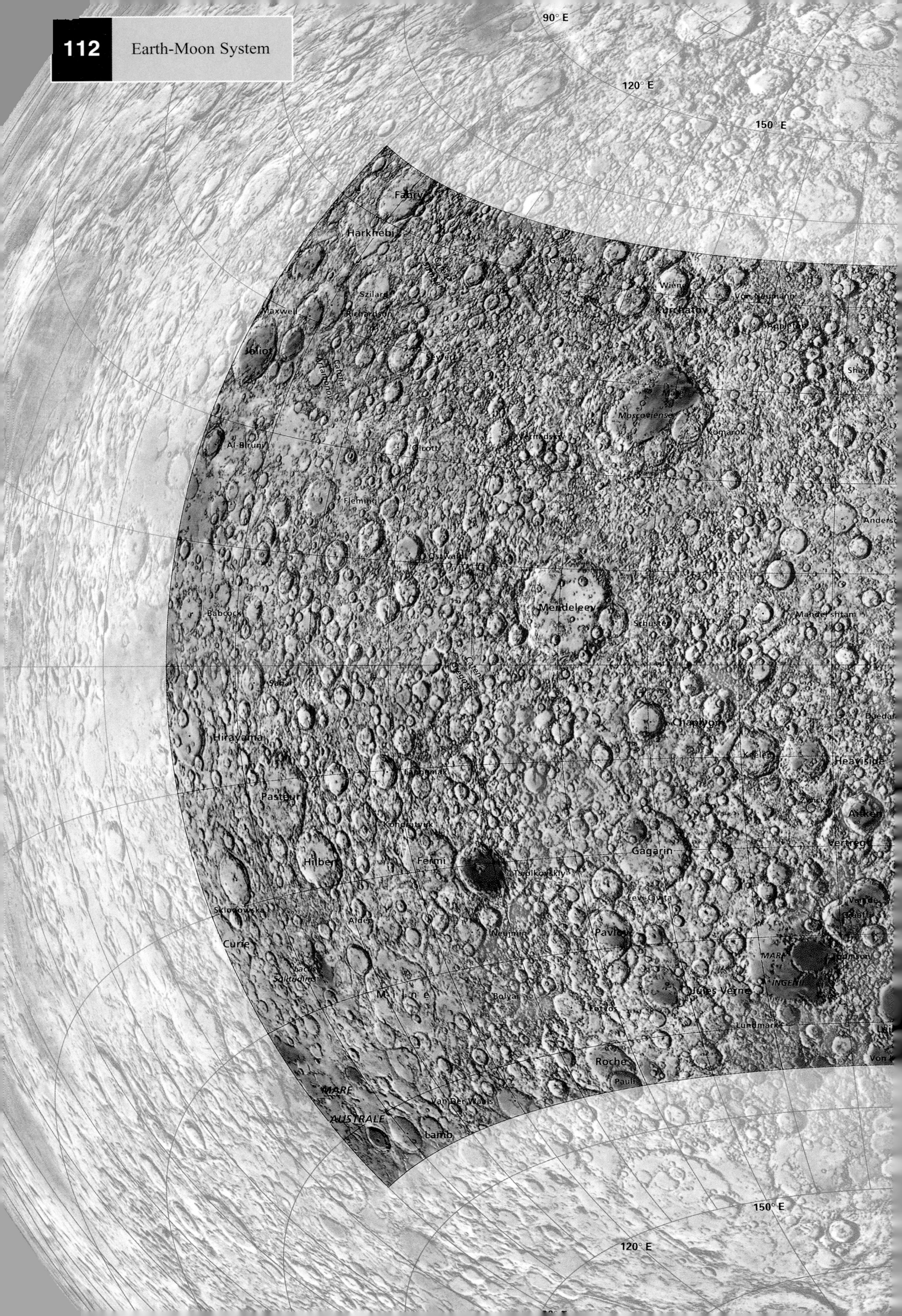
90° E
120° E
150° E
Fabry
Harkhebi
H.G. Wells
Wiener
Von Neumann
Kurchatov
Appleton
Szilard
Maxwell
Richardson
Joliot
Seyfert
Mare
Moscoviense
Komarov
Vernadskiy
Al-Biruni
Olcott
Fleming
Ostwald
Mendeleev
Schuster
Babcock
Mandel'shtam
Saha
Chaplygin
Hirayama
Keeler
Heaviside
Langemak
Pasteur
Kondratyuk
Aitken
Gagarin
Hilbert
Fermi
Tsiolkovskiy
Vertregt
Levi-Civita
Sklodowska
Alden
Neujmin
Pavlov
Curie
Mare
Ingenii
Milne
Bolyai
Jules Verne
Lundmark
Roche
Pauli
Mare
Australe
Van Der Waals
Lamb
150° E
120° E

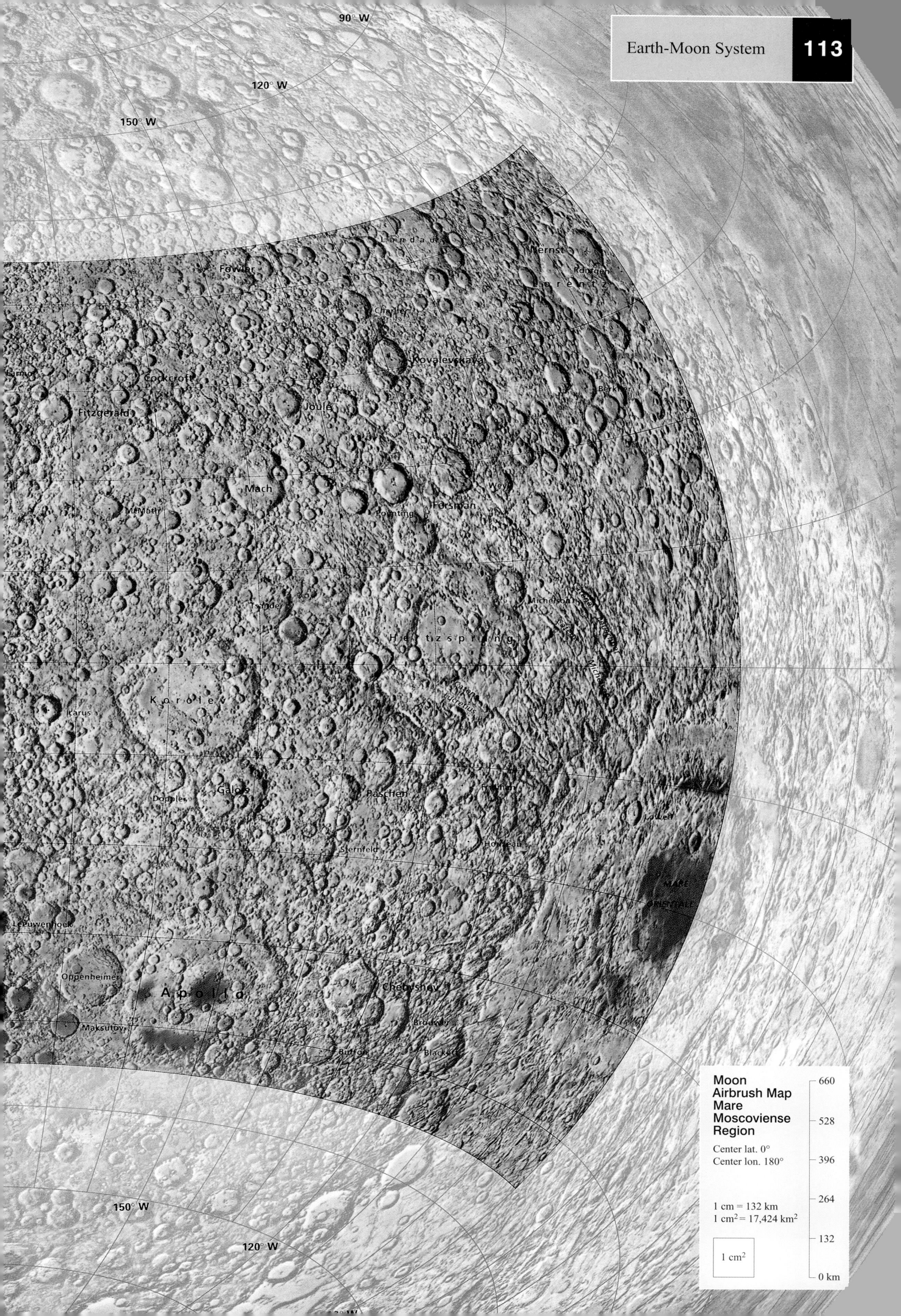
90° W
120° W
150° W
Landau
Nernst
Röntgen
Fowler
Kovalevskaya
Cockcroft
Fitzgerald
Joule
Mach
Fersman
Poynting
Weyl
Tsander
Hertzsprung
Michelson
Korolev
Icarus
Galois
Doppler
Paschen
Sternfeld
Lowell
MARE
ORIENTALE
Leeuwenhoek
Oppenheimer
Apollo
Chebyshev
Maksutov
Brouwer
Buffon
Blackett
150° W
120° W
Moon
Airbrush Map
Mare
Moscoviense
Region
Center lat. 0°
Center lon. 180°
1 cm = 132 km
1 cm² = 17,424 km²
1 cm²
660
528
396
264
132
0 km

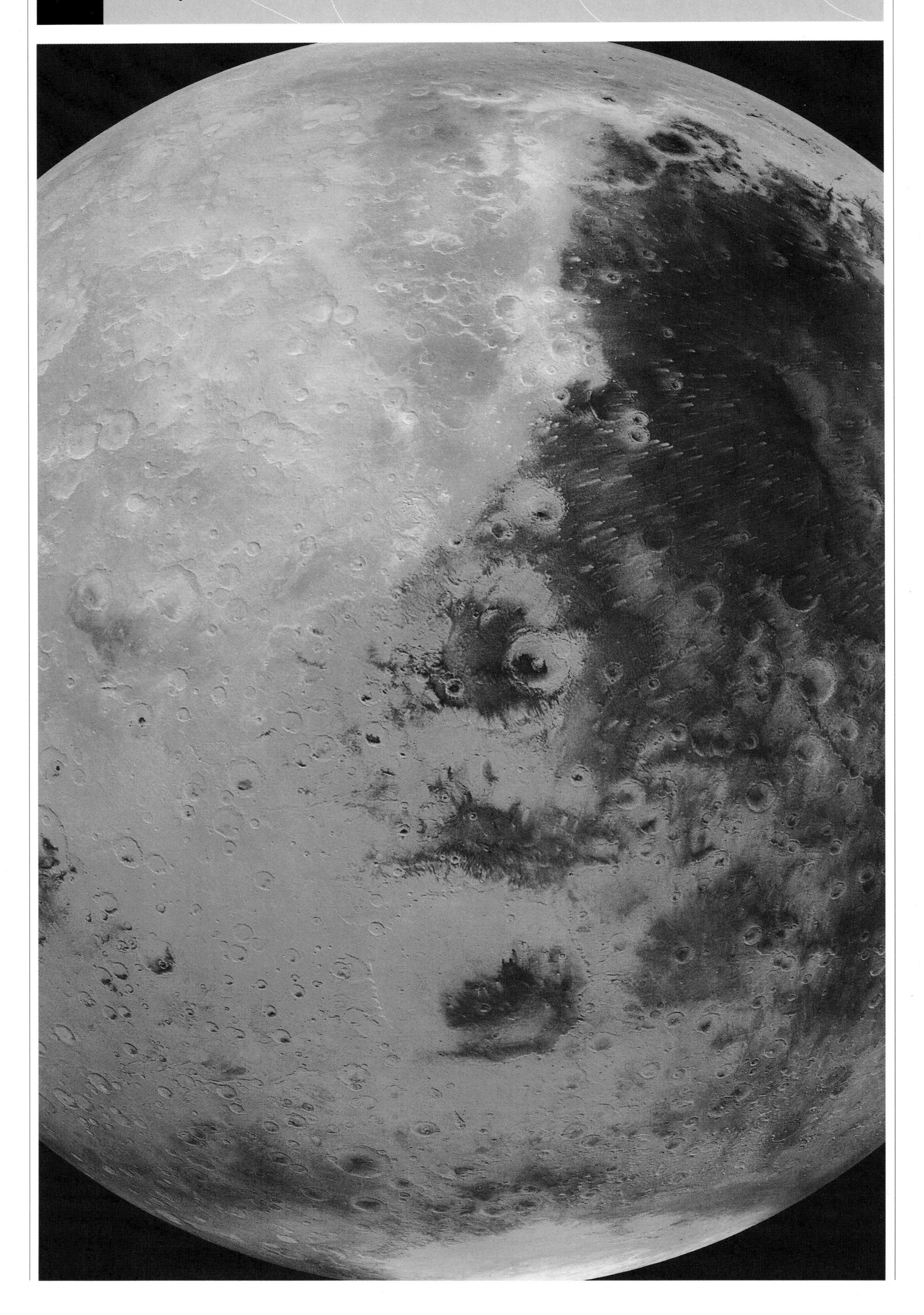

Mars System

Mars, often referred to as the Red Planet, has long been a subject of fascination. Speculation about the existence of "Martians" was widespread early in the twentieth century. For example, issues of *Scientific American* in 1920 described possible communication with Mars using searchlights and giant mirrors to reflect sunlight. Other proposals for communicating included digging huge trenches in the form of mathematical symbols in the Sahara Desert. These trenches were to be filled with kerosene and set aflame to signal Mars. Although these ideas were never carried out, they reflected intense public interest in Mars. Popular books, such as *The Gods of Mars* (1918) by Edgar Rice Burroughs, and the radio broadcast, *War of the Worlds* (1938), by H. G. Wells fired the imagination of a great many people.

Telescopic observations of Mars began in earnest in the late nineteenth century and provided fuel for speculation about the nature of Mars. For example, Giovanni Schiaparelli, the director of the Milan Observatory, described linear markings on Mars that were termed canali. This term was loosely translated to "canals," which conjured up images of intelligent construction. Percival Lowell, a wealthy Bostonian, became so intrigued with the possibility of intelligence on Mars that he established his own observatory in 1885 at Flagstaff, Arizona, to study Mars. These studies and Lowell's persistence dominated ideas about Mars for the succeeding 30 years. Telescopic observations showed that Mars has an atmosphere, clouds, polar caps, and shifting light and dark patterns that were correctly interpreted by some observers as dust storms.

Spacecraft exploration of Mars was initiated with the *Mariner 4* flyby in 1965. Although it took only 22 close-up pictures covering a tiny fraction of the surface, *Mariner 4* provided the first clues to the nature of the surface and showed the presence of lunarlike craters. The scientific community was disappointed that Mars was similar to the Earth's Moon. These thoughts were not greatly dispelled when *Mariners 6* and *7* flew past Mars in 1969 and again returned images limited to the cratered terrain similar to the lunar highlands.

The geologic diversity of Mars was finally revealed by the *Mariner 9* spacecraft. This mis-

The Mars system compared to the Earth/Moon system.

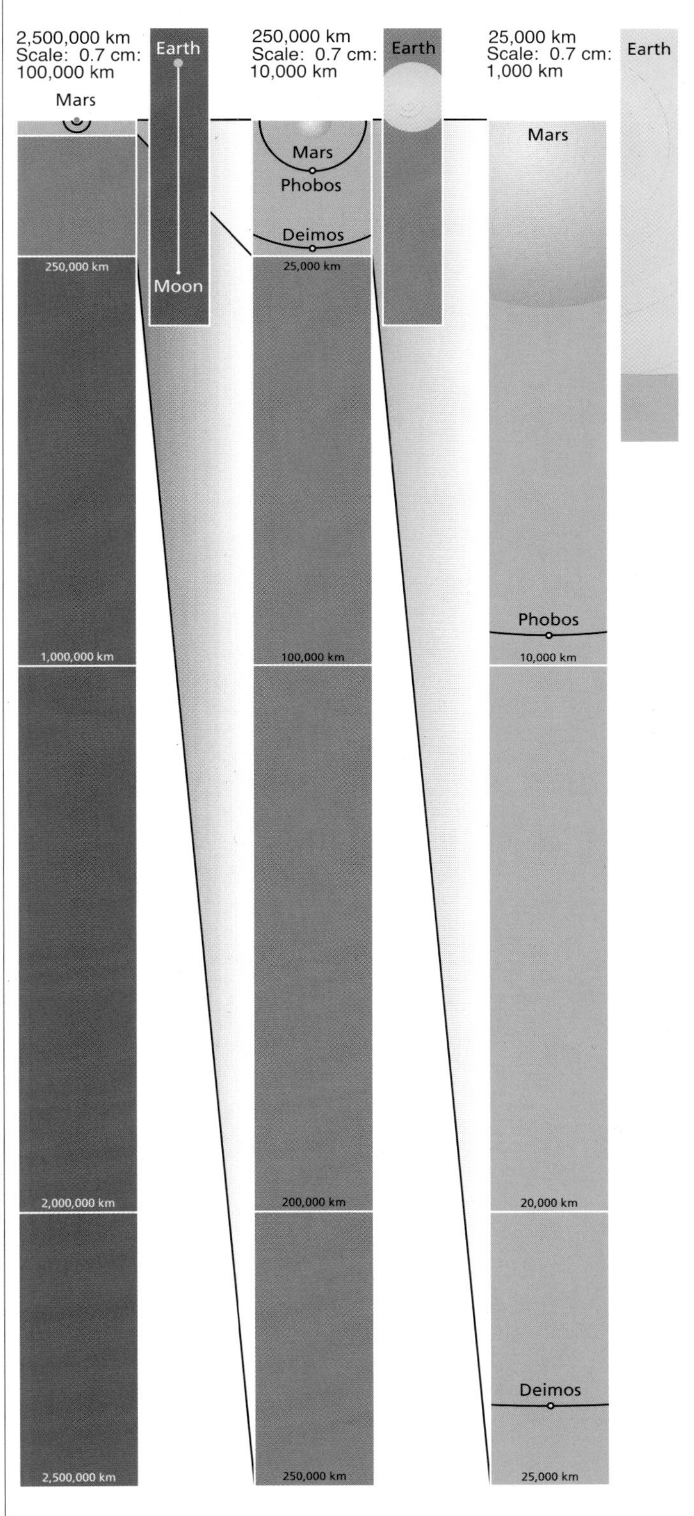

Facing page: view of Mars synthesized from *Viking Orbiter* images (courtesy of Alfred McEwen, U.S. Geological Survey).

Four views of Mars synthesized from *Viking Orbiter* images. Each view was generated from more than 100 separate photographs, mosaicked to produce a seamless view in color (courtesy of Alfred McEwen, U.S. Geological Survey).

Left top: mosaic centered at about the equator and 305° W, showing Syrtis Major and Isidis Planitia (dark area on the right), Sinus Sabaeus (dark area on the left), the Hellas Basin (bright area at bottom of mosaic), and cratered uplands.

Left middle: mosaic centered at about the equator and 190° W, showing the low-albedo Cerberus region, prominent bright and dark wind streaks, and the bright plains of Elysium (upper left part of the mosaic).

Left bottom: mosaic centered at about the equator and 335° W, showing Sinus Meridiani and Sinus Sabaeus (two prominent dark areas visible from Earth-based telescopes) and Schiaparelli, a crater 450 kilometers in diameter near the center of the mosaic.

Right: mosaic centered at about the equator and 80° W, showing the Valles Marineris canyonlands, two of the prominent Tharsis volcanoes, Ascreaus Mons (upper left) and Pavonis Mons (middle left), and channel systems that trend north toward the top of the mosaic.

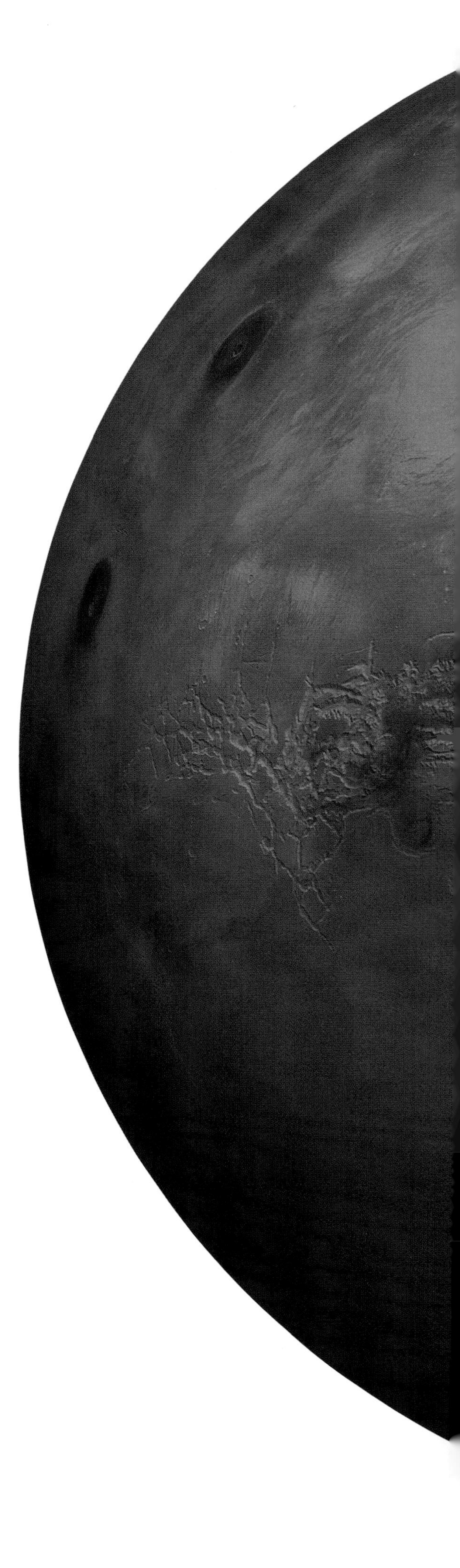

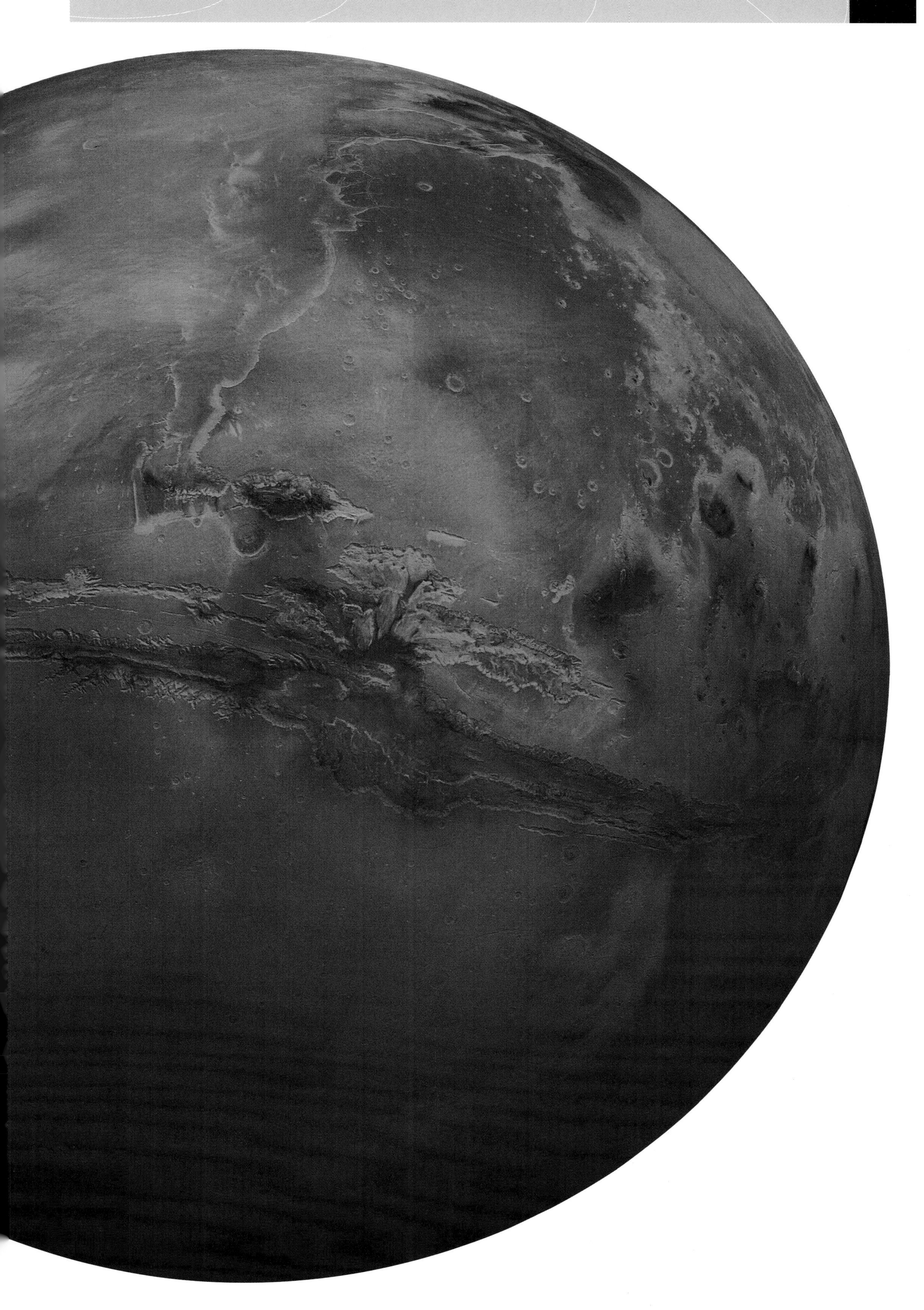

Artist's rendition of the sequence leading to the landing by the *Viking* spacecraft on Mars (left side) and of the *Viking Orbiter* as it photographed the surface.

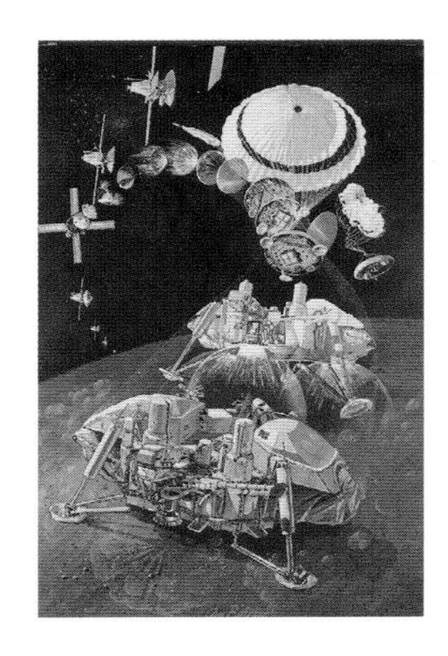

Viking Orbiter image taken across the Argyre impact basin (foreground) and cratered terrain on Mars toward the horizon. Haze, thought to be crystals of carbon dioxide, is visible in the atmosphere above the horizon (JPL photograph P-23692).

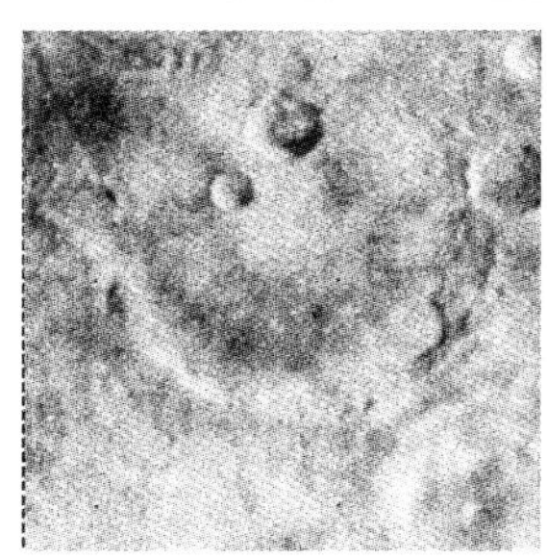

Mariner 4 photograph showing southern cratered terrain on Mars. Images like this and similar ones obtained by *Mariners 6* and *7* gave the erroneous impression that Mars was similar to Earth's Moon. Area shown is 105 kilometers across (*Mariner 4*, photograph 11).

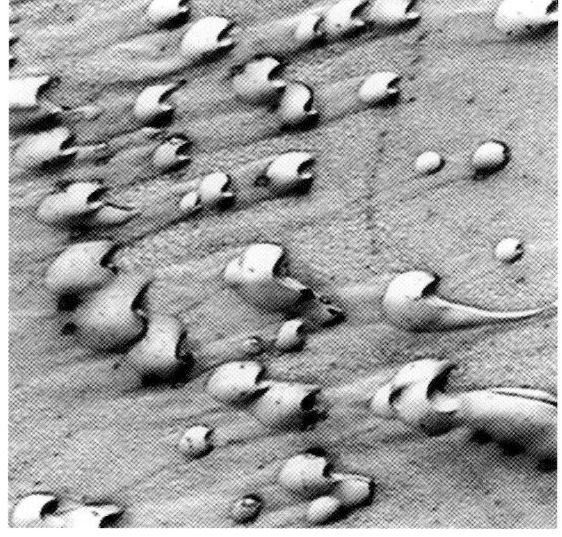

Large dunes of the north polar region have a form similar to barchan dunes on Earth. From their shape, the dunes appear to have formed from winds blowing generally from the west (left). Area shown is 1.5 kilometers across. (*Mars Global Surveyor* photograph, Malin Space Science Systems, NASA, PIA 02069).

sion, however, did not begin very auspiciously. Two spacecraft, *Mariners 8* and *9*, were scheduled to orbit Mars, but *Mariner 8* was lost after launch. Consequently, *Mariner 9* had to be reprogrammed to incorporate *Mariner 8*'s mission. Although *Mariner 9* was launched successfully, when it arrived and was placed in orbit one of the largest storms ever observed enveloped the planet in thick clouds of dust, hiding the surface from view. *Mariner 9* was placed in a semidormant mode in orbit until the dust storm concluded. Pictures were taken periodically and, as the dust slowly settled, a remarkable array of surface features began to emerge. Before the end of the mission in 1972, *Mariner 9* had taken more than 7,300 images covering most of the planet.

The *Viking* mission was the most complicated unmanned mission flown in Solar System exploration. Consisting of two orbiters and two landers (all operating simultaneously), the spacecraft arrived at Mars in 1976. All four spacecraft carried an array of instruments. The search for evidence of life was the primary scientific objective, but results failed to show that Mars harbors extant life. Nonetheless, a wealth of data was returned on the characteristics of Mars that fueled a research effort for more than two decades.

It would be 20 years after the *Viking* project before the next successful spacecraft visited Mars. In 1997 *Mars Pathfinder* and its rover, *Sojourner*, landed and returned information for a third site on the Red Planet. Nearly concurrently, the orbiter *Mars Global Surveyor* began operations, returning a wealth of new data on the topography, composition, remanent magnetism, and surface features. These missions have set the stage for more complex future missions, including the return of samples to Earth.

Mars

More than 100 years of observations of Mars from Earth and by spacecraft have shown the Red Planet to be a fascinating object. After Earth, it is the most hospitable planet in the Solar System from the human perspective. Mars has an atmosphere composed mostly of carbon dioxide, although its atmosphere is considerably less dense than the Earth's (the

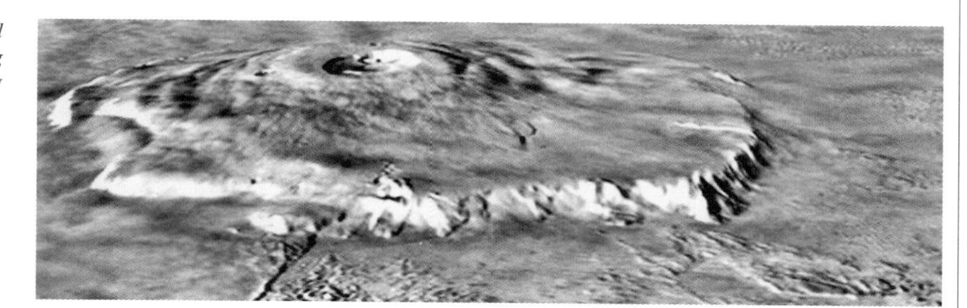

Topographic data from *Mars Global Surveyor* were combined with *Viking Orbiter* images for this perspective view of Olympus Mons (NASA PIA 02806).

atmospheric pressure on the surface of Mars is about the same as being at an altitude of 35 kilometers above the Earth). The polar caps, visible even in telescopic views of Mars, consist of frozen carbon dioxide (the material commonly called "dry ice"), condensed from the atmosphere together with a small amount of water ice. Despite its low density, the atmosphere is sufficiently active to generate winds capable of moving sand and dust to form features such as sand dunes. The reddish-orange appearance of Mars results primarily from iron oxides in the dust and soils that mantle most of the planet.

The rotational axis of Mars is tilted about 25° from its orbital plane, nearly the same as Earth. This means that, like Earth, Mars experiences seasons. In the winter hemisphere, the polar cap grows in size as the carbon dioxide freezes from the atmosphere. In the summer hemisphere, the cap shrinks, leading to intense temperature contrasts between the polar caps and the heated surface around the caps. Local dust storms are common and can grow to a global scale. At present, summers in the southern hemisphere are hotter than in the northern hemisphere. However, because Mars "wobbles" on its spin axis, this relationship reverses every 50,000 years, with the northern summer becoming hotter than the southern.

Mars has probably experienced major changes in climate in the past, as evidenced by dry river channels once carved by flowing water. Because liquid water cannot exist on the surface of Mars under present conditions (the atmosphere is too thin and cold), Mars probably had a warmer, denser atmosphere at the point in its history when the river channels were eroded. Was the change in climate on Mars to its present cold state unique to the Red Planet, or did all the terrestrial planets, including Earth, experience similar changes? Answering this question is fundamental to understanding the history of the Solar System.

Very little is known about the interior of Mars. Most information about planetary interiors is derived from the study of seismic events such as earthquakes (or in this case, "marsquakes"). Although seismometers were carried to Mars by both Viking landers, one

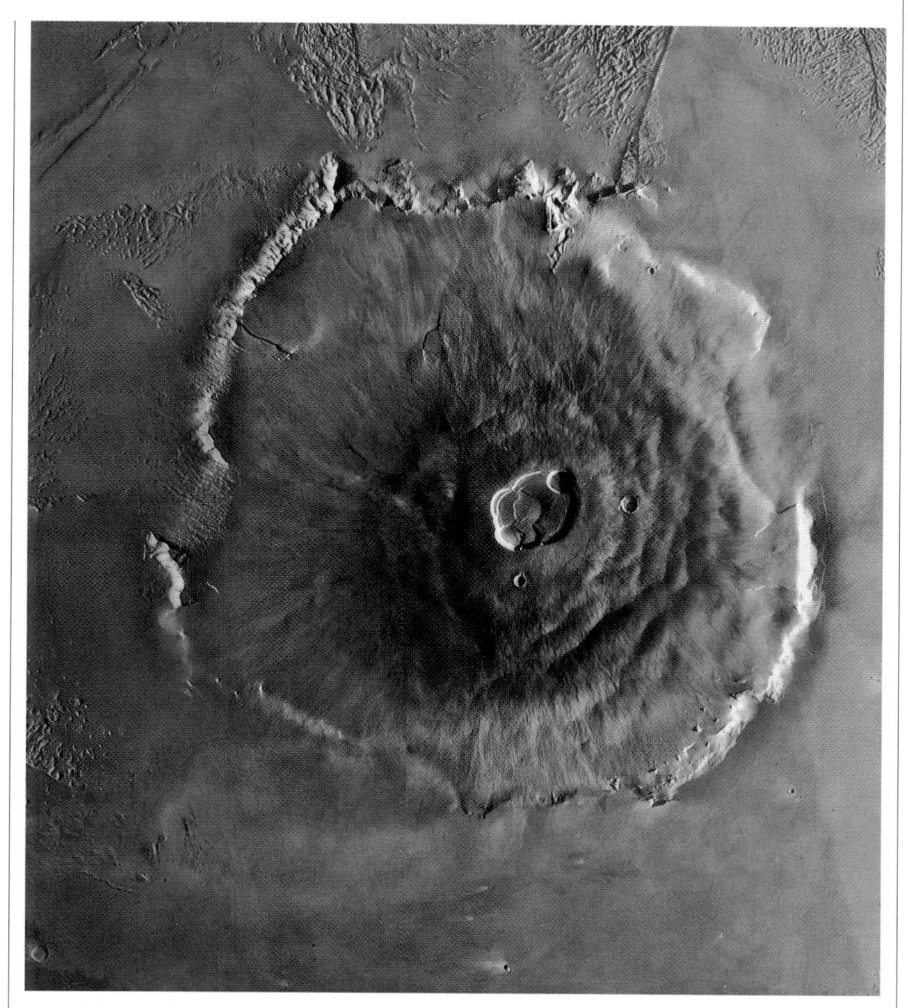

Photograph of Olympus Mons, one of the largest shield volcanoes known in the Solar System, taken during clear atmospheric conditions. Area shown is 700 kilometers across. (*Viking Orbiter* photograph, U.S. Geological Survey, Flagstaff).

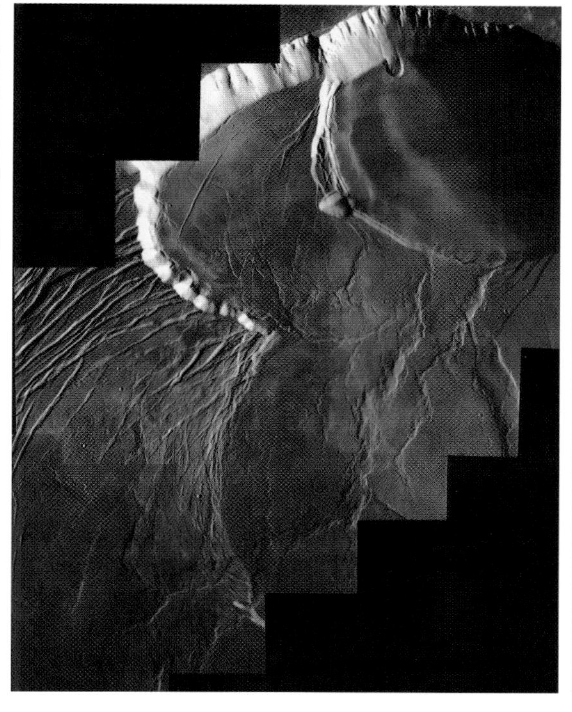

Mosaic of high-resolution frames of part of the Olympus Mons summit caldera, showing extensive ridges in one of the floor units. The scalloped caldera and multiple, nested craters indicate that the caldera resulted from several episodes of eruption and collapse. Area shown is about 30 by 60 kilometers (JPL photograph P-23694).

View of Mars in Utopia Planitia taken by *Viking Lander 2* (*Viking Lander* image 76/10/07/125702).

Image taken of the martian surface in morning illumination by *Viking Lander 1*, showing drifts of windblown sediments that appear to have been sculpted or eroded by the wind. The large boulder at the left is about 8 meters from the spacecraft and measures about 1 by 3 meters (JPL photograph P-17430).

View of the martian surface from *Viking Lander 1* in Chryse Planitia, showing pitted rocks, soil, and drifts of wind-sculpted dust (courtesy of R. Arvidson, Washington University).

failed to operate and the other provided limited information. Simplified models of the interior, based mostly on theory, suggest that Mars has a small core, but its properties and questions regarding marsquakes must await the future establishment of a network of geophysical instruments.

Geology

Mars displays a wide variety of terrains, all of which signal diverse geologic processes. From the windswept sand plains of the north polar region to the summit caldera of the huge volcano Olympus Mons, Mars is a cornucopia of geologic delights and puzzles.

On a global scale, Mars can be divided roughly into two terrains – the northern lowlands and the southern highlands. In places, the two terrains are separated by a scarp as high as 2 kilometers. Many ideas have been proposed to explain this basic dichotomy. Some planetary scientists have suggested that the lowlands represent an enormous impact basin or a series of basins. Other investigators have proposed that the difference reflects internal processes. In any case, after the establishment of the highland-lowland boundary, the scarp separating the two terrains was eroded by a variety of processes, including slope failure and the formation of debris flows.

The northern lowlands are covered by diverse plains and include lava flows, sedimentary deposits, and materials reflecting periglacial (cold-region) processes. Both *Vikings* and *Mars Pathfinder* landed in the northern plains, and images from the surface show rocks up to a meter or more across, soils, and wind-sculpted dust deposits. Chemical analysis of the soils obtained by the landers indicate that they probably formed from the weathering of volcanic basalts. Basalts are rich in iron that, when chemically weathered, produces the red soils characteristic of Mars.

Although samples of martian rocks have not yet been returned to Earth for analysis, there is a small, unique class of meteorites on Earth that may, in fact, have come from Mars. Good arguments suggest that these meteorites are rocks that were ejected from Mars by impact and were carried in trajectories to Earth. Analyzing traces of gas within these rocks shows similarities of the gas to the martian atmosphere measured by the *Viking* landers.

The southern highlands of Mars are heavily cratered and superficially resemble the lunar highlands. Closer inspection, however, reveals that the craters have been eroded by wind and water and that the floors of many of the larger craters have been filled with sediments. Areas between some of the craters have also been blanketed with windblown dust and stream deposits, or covered with lava flows.

Large impact basins are also present on Mars, including the Hellas basin, which is nearly 2,000 kilometers across, and the Argyre

Panoramic view from the *Mars Pathfinder* lander showing "Twin Peaks" on the horizon (NASA PIA 01466).

basin. Although many other impact basins can be recognized on Mars, most have been altered severely by erosion and other processes and have been nearly obliterated.

Ejecta from many of the martian impact craters shows flowlike patterns unlike those seen around lunar craters. One theory holds that the impact occurred in ground that contained water or ice and that a mixture of rock, soil, and water was ejected as a slurrylike mixture. Alternatively, the flowlike patterns may result from interactions of the ejecta with the atmosphere during the ejection phase of the cratering process. Resolving this problem may be important for future exploration of Mars; if the first idea is correct, these craters could provide clues to the search for subsurface water that might still be present.

Former river channels and stream valleys of several types have been identified on Mars. One type may have resulted from the catastrophic release of large volumes of water from the subsurface. Another forms networks of valleys and tributaries that might have been eroded by surface water. Other channels were probably formed by subsurface water that reached the surface through hundreds of small springs scattered over thousands of square kilometers. In some areas, the channels empty into large craters and are thought to have formed ancient lakes where sediments were deposited.

Some of the larger martian channels flow across the northern lowland plains. For example, channel remnants can be traced through the general area of the *Viking 1* landing site northward more than 1,000 kilometers across Chryse Planitia. High-resolution images of this general area show that vast quantities of water must have flowed on the surface of Mars. Detailed mapping indicates that such flow occurred repeatedly and was not an isolated event.

The youngest areas on Mars are in the north and south polar regions. The south polar area contains a series of layered deposits presumed to be composed of ice and dust settled from the atmosphere. The layers may reflect cyclic patterns of deposition and erosion related to the 50,000-year wobble in the spin axis of Mars described earlier in this chapter. Similar deposits are found in the north polar region, but they are not as extensive. Instead, vast fields of sand dunes surround the north polar cap like a discontinuous collar. High-resolution photographs show that the dunes are similar in size and form to sand dunes on Earth. The area covered by the martian north polar dune field is greater than a half million square kilometers, equal in size to the dunes of the North African Sahara. In addition to the north polar area, small dune fields and individal dunes are found in many other parts of Mars. All of the dunes probably formed late in Mars's history.

View from *Viking Lander 2* in Utopia Planitia, showing rocky surface; tops of many of the rocks have a thin mantle of windblown dust (courtesy of R. Arvidson, Washington University).

Evidence for active aeolian, or wind, processes is shown by various bright and dark wind streaks. These are patterns on the surface of Mars that change their size, shape, and ori-

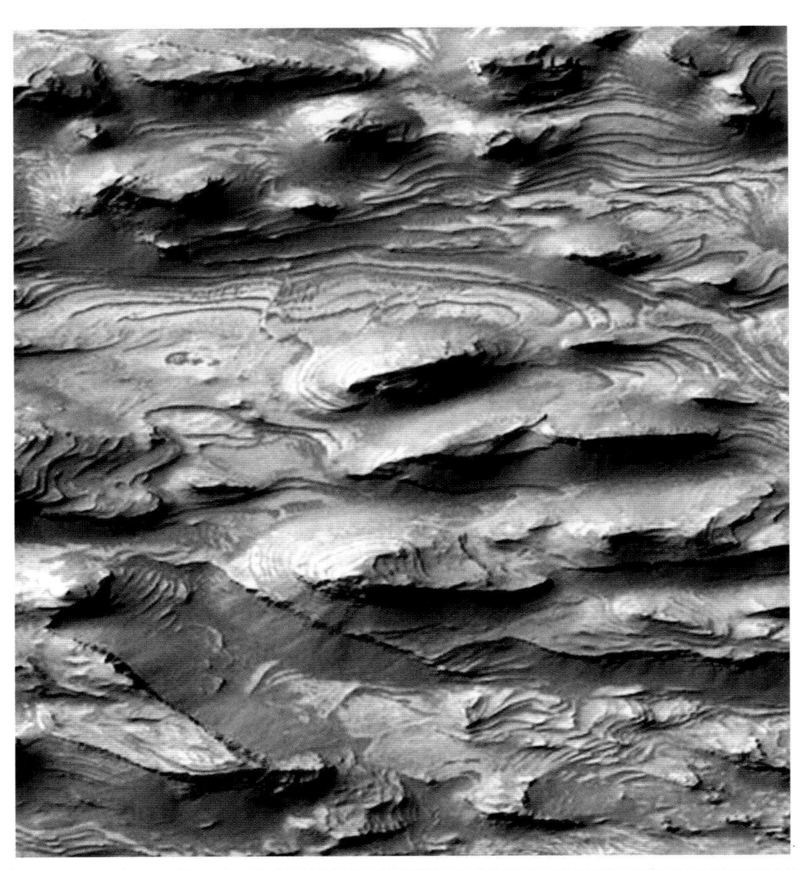

Layered deposits in Valles Marineris, photographed in high resolution by *Mars Global Surveyor*. Each layer is estimated to be about 30 meters thick. Area shown is 1.5 kilometers across (NASA PIA 02839, Malin Space Science Systems).

Martian sunset over Chryse Planitia from *Viking Lander 1*. Blue to red colors in the sky result from scattering and absorption of sunlight by particles in the atmosphere, estimated to be from one to a few microns in diameter. "Layered" sky colors are image artifacts (JPL photograph P-17704).

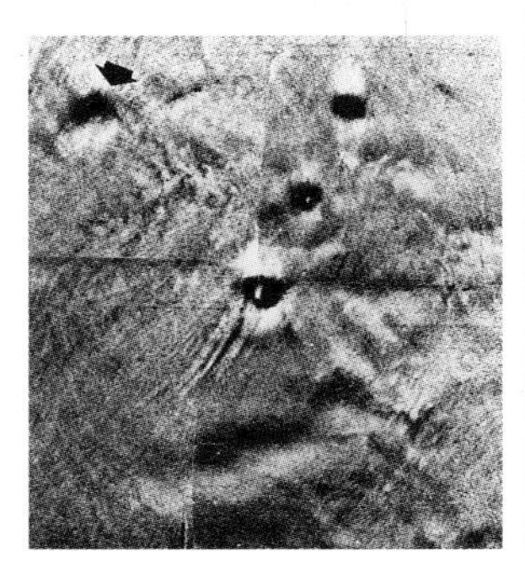

Mosaic of *Mariner 9* images of the Tharsis region taken during the Great Dust Storm of 1971; nearly all of the surface was obscured by dust. Arrow points to Olympus Mons; three dark spots to the right side of Olympus Mons correspond to the volcanoes (from top to bottom), Ascraeus Mons, Pavonis Mons, and Arsia Mons (JPL photograph P-12676-6).

entation with time and serve as local wind vanes. Maps of these features have allowed global wind patterns to be defined and used for comparison with theoretical models of atmospheric circulation.

One of the geologic surprises presented by the *Mariner 9* photographs was the existence of enormous volcanoes on Mars. The more impressive structures are found in the Tharsis region, a broad, elevated area in the western hemisphere of Mars. Olympus Mons and other shield volcanoes rise as high as 27 kilometers above the average elevation of the planet and are among the largest mountains found in the Solar System. Except for their great size, they are very similar to the shield volcanoes of the Hawaiian islands. Both the martian and Hawaiian volcanoes have complex summit calderas, are built from thousands of individual lava flows, and appear to be composed of iron-rich silicate rocks, such as basalt.

Other volcanoes on Mars are in the Elysium region, in Syrtis Major, and on the margins of the Hellas basin. Although these volcanoes are impressive, by far the greatest volumes of volcanic materials on Mars occur as vast sheets of flood lavas, similar to those of the maria on Earth's Moon. In all, nearly half the surface of Mars has been covered with various volcanic deposits.

Stretching more than 4,000 kilometers eastward from the Tharsis volcanic province is the "Grand Canyon" of Mars, named Valles Marineris. This canyon system is one of the major tectonic features on the planet and may have resulted from crustal extension, or rifting. However, it has also been enlarged by landslides, wind and water erosion, and by collapse, which has formed chains of pit craters. Parts of the canyon have been filled with sediments, which some planetary scientists have suggested were deposited in ancient lakes contained in the canyonlands.

In addition to Valles Marineris, other tectonic features on Mars include sets of faults that have disrupted thousands of square kilometers. The most extensive are sets of fractures that are roughly radial to the Tharsis region. The fractures probably represent deformation of the martian crust in response to uplift and crustal

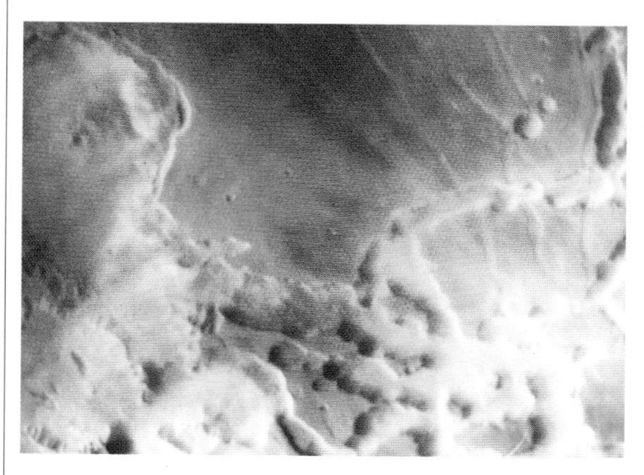

False-color composite image showing white patches of early morning fog in the Noctis Labyrinthus region of western Valles Marineris of Mars. Area shown is about 100 kilometers across (*Viking Orbiter 1* photograph, NASA 76-HC-791).

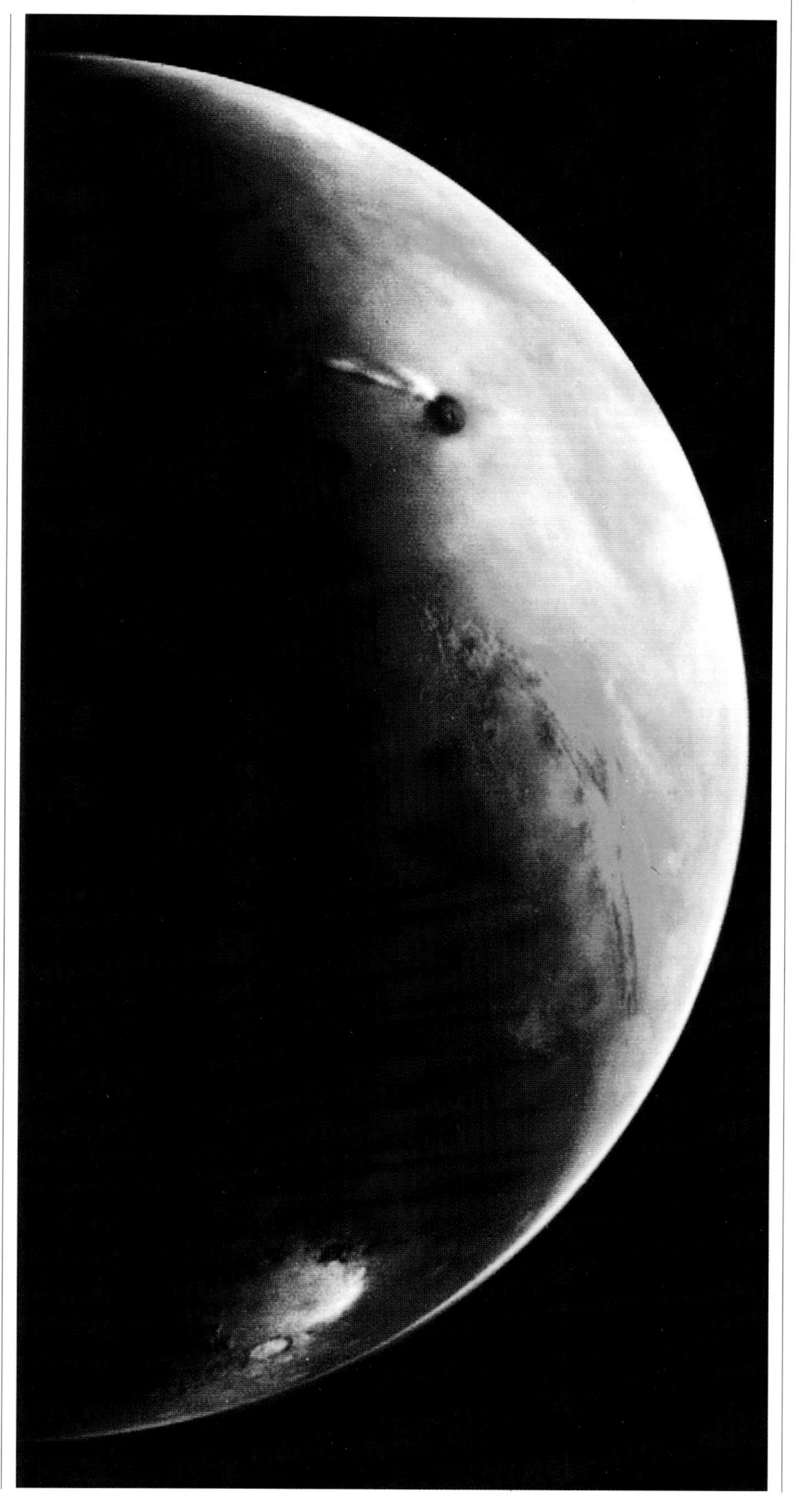

Mars viewed by the *Viking Orbiter 2* spacecraft during its approach, showing bright plumes of water ice clouds extending northwest from Ascraeus Mons, the northernmost of large, young Tharsis volcanoes. The bright area toward the bottom of the image marks frost deposits in the Argyre impact basin (JPL photograph P-19009).

extension. However, some disruption is also related to the growth of the volcanic region; the weight of the accumulated lava flows would have pressed the supporting crust downward, causing it to fracture.

Geologic History

By mapping the age and distribution of the features seen on Mars, it is possible to derive a general history. The earliest history on Mars, termed the Noachian Period, is recorded in the heavily cratered terrain and the ancient impact basins, such as Hellas. These reflect the final stages of crustal solidification and the end of the period of heavy bombardment seen throughout the inner Solar System. The development of the crustal dichotomy to form the northern lowlands occurred in this period and was accompanied by extensive erosion by wind and water, and the eruption of vast sheets of lava in many areas of Mars. Sediments and volcanic materials (lavas and possibly ash deposits) blanketed parts of Mars but were then eroded in many places, leaving the networks of valleys seen today.

The Hesperian Period marks Mars's "middle age." Extensive flood lavas were erupted from long fissures, creating many of the plateau regions, such as Hesperia Planum. Some of the eruptions evolved to a different type, in which magma reached the surface through local vents, such as craters, in contrast to the earlier fissure eruptions. Earliest representations of this style of volcanism are the highland patera, seen around the Hellas basin, and the Elysium volcanoes, such as Albor Tholus. The older volcanoes of the Tharsis area also began to grow during the Hesperian Period. This growth was accompanied by extensive crustal deformation along the equator, which opened up Valles Marineris. The enormous outflow channels were carved in this period of martian history and released tremendous floods of water that washed across the surface of Mars and might have collected as local seas.

The Amazonian Period is the youngest subdivision of time on Mars. It is represented by the late-stage lava flows in the Tharsis region, including those that make up most of the volcano Olympus Mons, and the younger lavas of

False-color photograph of layered deposits in the north polar cap, showing layers that are thought to be made of ice and dust. The regularity of the layering suggests that it results from periodic changes in erosion and deposition; area shown is 60 by 30 kilometers. Dunelike features (dark areas with a rippled texture) possibly formed from material eroded from the layered terrain (center and right) (JPL photograph P-18459).

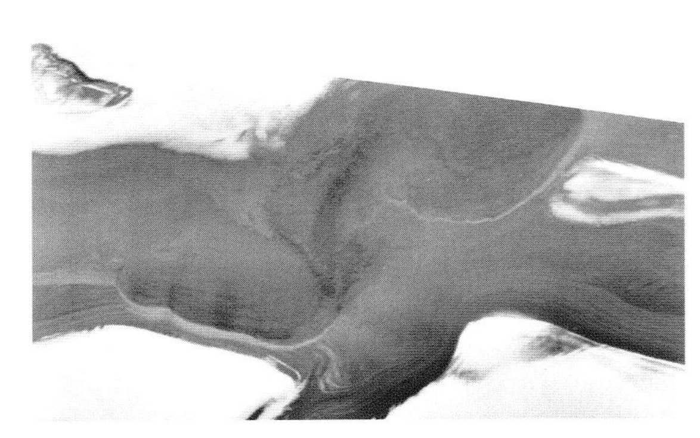

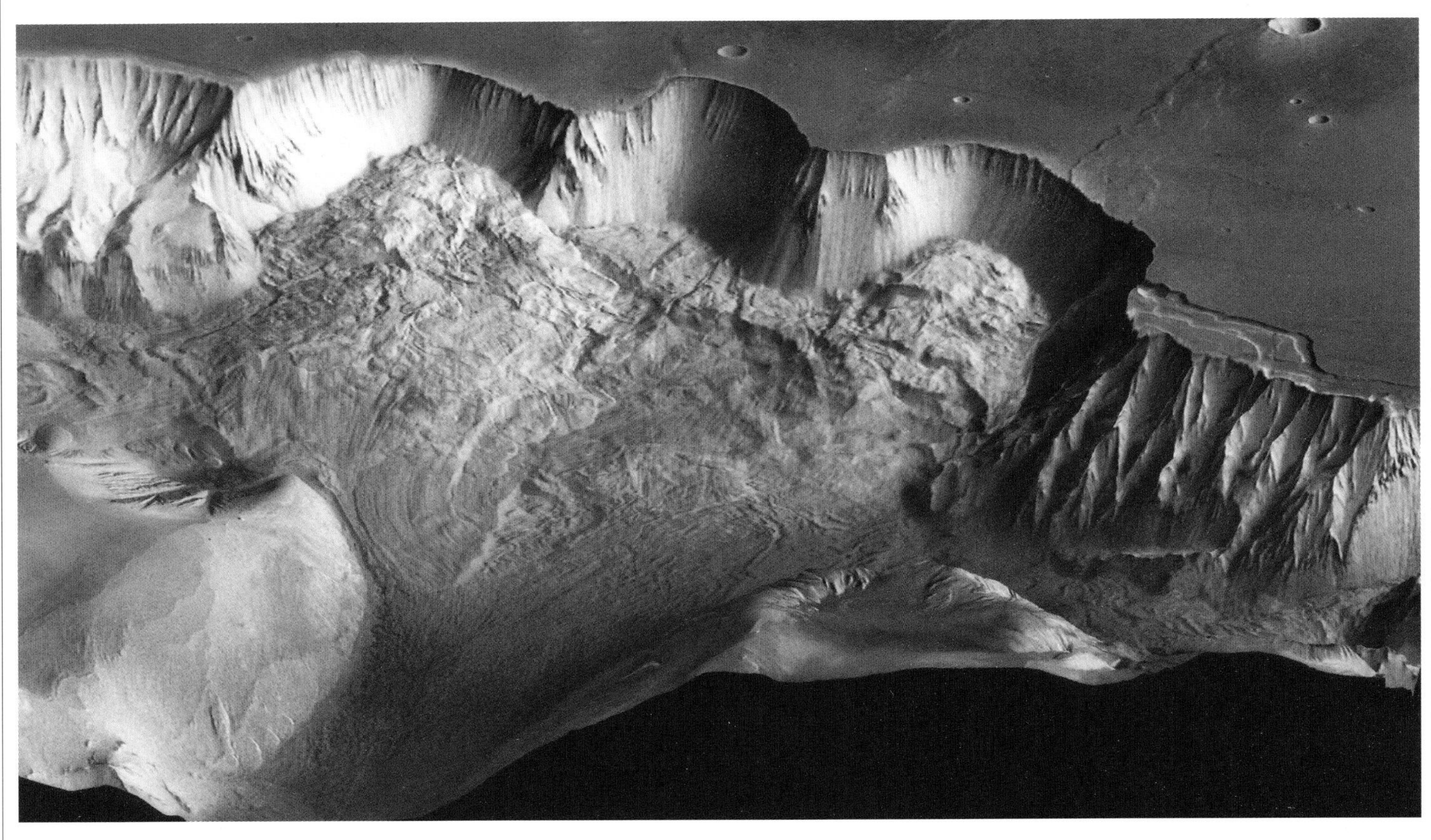

Oblique view shows huge landslides in the Valles Marineris canyonlands; the exposed cliff is several kilometers high (courtesy of A. McEwen, U.S. Geological Survey).

Mars Global Surveyor image showing gullies in the southern hemisphere thought to represent erosion by running water in geologically recent times (NASA–Malin Space Science Systems frame M07–01873).

the Elysium region. Wind and water sediments of this age fill parts of the Hellas, Argyre, and Isidis impact basins, as well as parts of the northern lowlands, and show that extensive erosion and deposition were taking place.

Currently, the surface of Mars is shaped primarily by wind and landslides, punctuated by the occasional formation of impact craters. Whether Mars experiences active volcanism or marsquakes is unknown at present but will be an important question to be addressed by future missions to the Red Planet.

Satellites

The moons of Mars were discovered by Asaph Hall in 1877 and consist of two tiny chunks of rock and debris. Named Phobos and Deimos for the Roman gods of fear and dread, respectively, these satellites are very irregular-shaped objects. Both are best described as "triaxial ellipsoids" in shape; Phobos measures 27 by 22 by 18 kilometers, whereas Deimos is 15 by 12 by 11 kilometers in size. Remote sensing of their surfaces suggests compositions that are

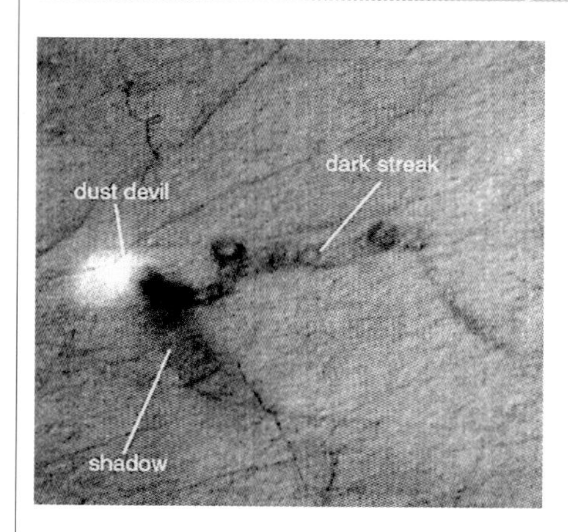

Mars Global Surveyor captured this view of an active dust devil and the dark streak left in its path (NASA PIA 02378, Malin Space Science Systems).

similar to the class of meteorites known as carbonaceous chondrites, which are rich in organic materials and volatiles.

Mariner 9 provided the first images of the moons and showed that they are nonspherical and have cratered surfaces. The *Viking* orbiters were programmed to obtain close-up views of both moons. In fact, *Viking Orbiter 2* flew within 28 kilometers of Deimos and there was some concern about crashing into the moon. Fortunately, this did not happen and some of the highest resolution (1 meter) photographs of the mission were obtained.

Outwardly, Phobos and Deimos are similar; both are small, "lumpy" objects, have low densities, and are heavily cratered. Close inspection shows important differences between the two satellites, however. Phobos has three large impact craters named Stickney, Hall, and Roche. Radiating away from Stickney is a series of grooves that wrap nearly all the way around Phobos. These grooves average 200 meters wide, 5 to 30 meters deep, and 15 to 20 kilometers long and tend to become wider toward Stickney crater. In some sections, grooves appear to consist of coalescing depressions. Although most investigators agree that the grooves are somehow related to the impact that caused Stickney, the mechanism of formation is not known. Some proposals include drainage of surface debris into subsurface cavities, while others have suggested that the grooves were formed by outward flow of gases released by heat generated from the impact.

In addition to the grooves, Phobos shows sets of ridges. The largest is about 5 kilometers wide by 15 kilometers long. Some ridges may be the remnants of degraded crater rims. Others may be scars left when large impacts caused chunks of the satellite to be blasted into space.

In contrast, Deimos lacks grooves, ridges, and craters larger than about 2.5 kilometers in diameter. It does, however, show ample evidence for downslope movement of loose surface material. Numerous bright streaks oriented downslope are seen, despite the extremely low gravitational acceleration – the driving force in downslope movements. The low gravity probably explains the lack of clearly identifiable ejecta deposits around the craters. Except for building-size (up to 200 meters wide) blocks on

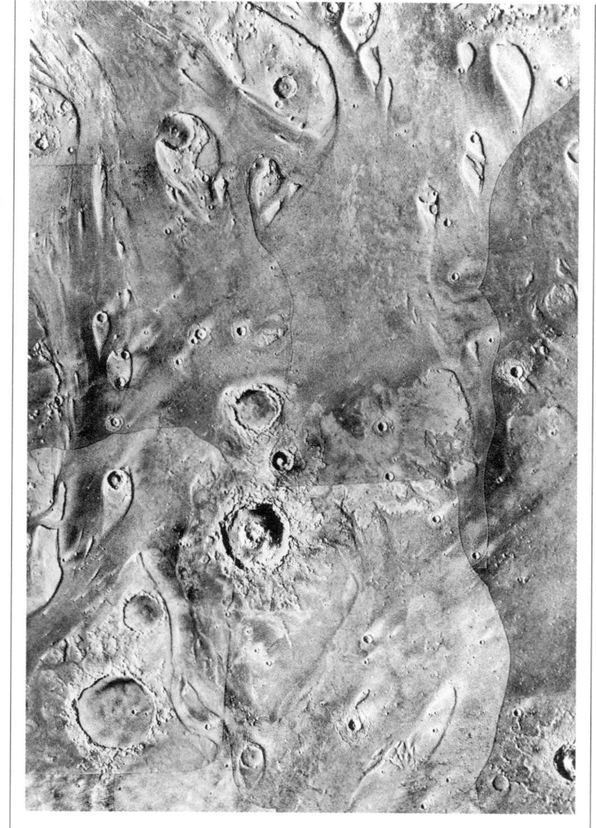

Mosaic of *Viking Orbiter* photographs showing ancient river system in Chryse Planitia and eroded, teardrop-shaped islands. Flow was toward the top of the image. Area shown is 550 by 550 kilometers (from MC 11-NW, U.S. Geological Survey).

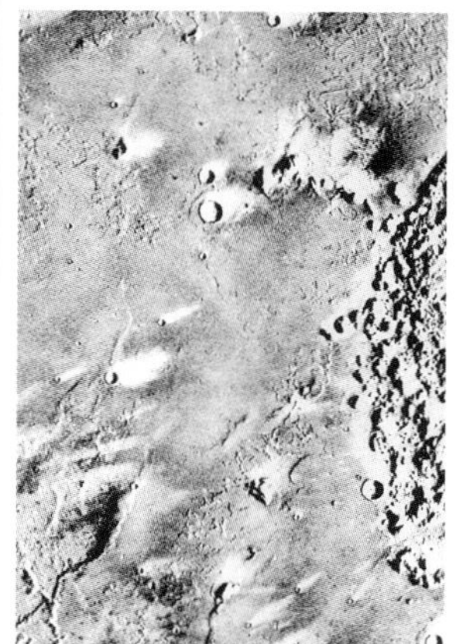

Bright "wind streaks" associated with craters. Bright streaks, thought to be deposits of dust, indicate the direction of winds at the time of their formation. Largest streaks shown here are about 18 kilometers long (*Viking Orbiter* photograph 545A53).

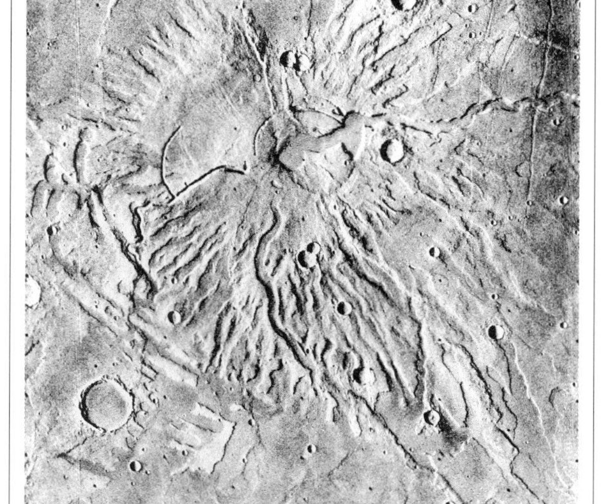

View of Tyrrhena Patera, one of the ancient volcanoes near the Hellas basin. Area shown is 200 kilometers across (*Viking Orbiter* photograph 87A14).

"Ejecta flow" crater (30 kilometers in diameter), showing thin flow lobes typical of many impact craters on Mars (*Viking Orbiter* photograph 608A45).

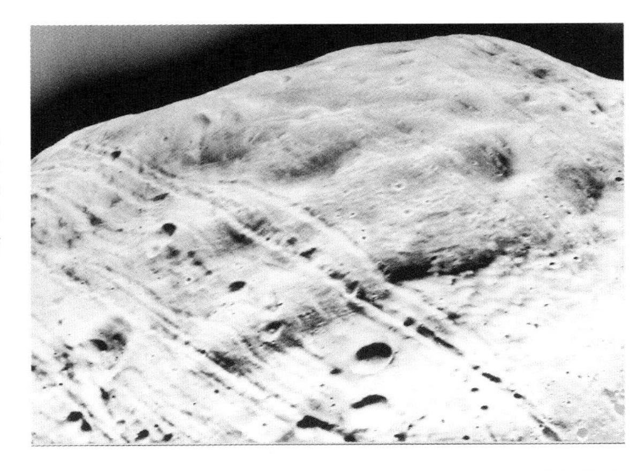

These 100- to 200-meter wide grooves on Phobos appear to be related to Stickney crater. Picture covers area about 12 kilometers across in the foreground (*Viking Orbiter* photograph 343A13).

False-color image of the martian moon, Phobos, with Mars in the background (Soviet Phobos mission image, processed by U.S. Geological Survey).

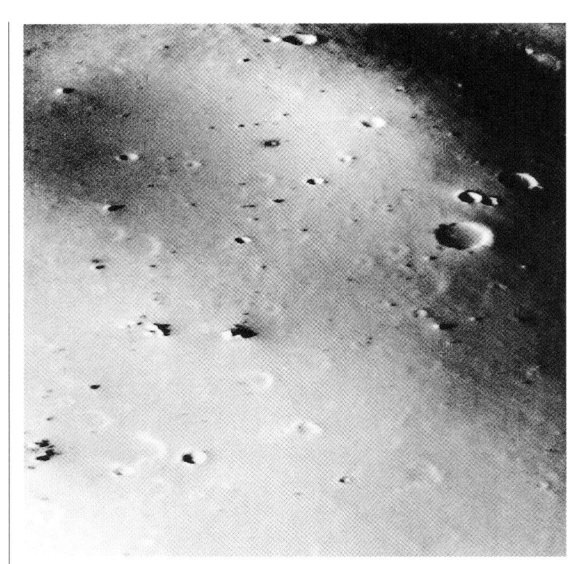

High-resolution *Viking Orbiter* photograph showing area about 2 kilometers by 2 kilometers on Deimos and the presence of blocks on the surface (*Viking Orbiter* photograph 423B63).

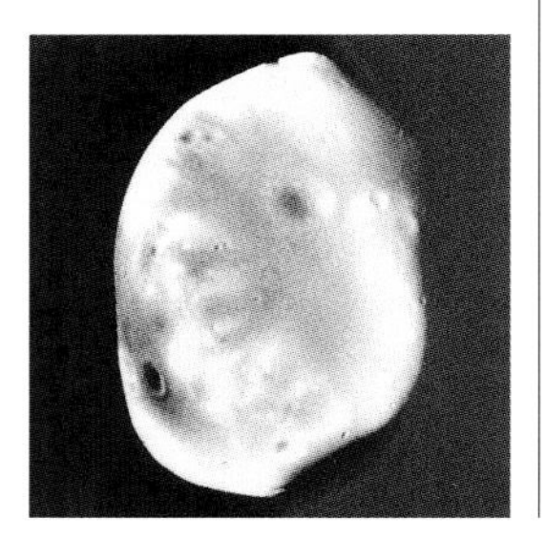

View of the martian moon, Deimos, showing surface markings that suggest the movement of surface material downslope (*Viking Orbiter* photograph 428B22).

the rims of some craters and elsewhere, much of the ejecta from cratering probably exceeded the gravitational escape velocity of both satellites and was lost to space.

Despite the apparent lack of crater ejecta, both Phobos and Deimos have extensive regolith deposits. The estimated thickness of regolith on Deimos is about 10 meters, whereas it may be as great as 100 meters on Phobos.

What is the origin of the martian satellites? Because of their size, shape, and apparent composition, both have been suggested to be asteroidlike objects. Consequently, some investigators have suggested that perhaps they were captured from the asteroid belt. But because of the equatorial orbital geometry of the satellites around Mars, other investigators have argued that capture is unlikely and that Phobos and Deimos evolved contemporaneously with Mars. Still other planetologists have drawn on the heavily cratered surface to suggest that both objects were part of a much larger body that was fragmented by collision early in the history of Mars.

Mars
Key to Geologic Maps

Description of Map Units

Volcanic Materials

- Amazonian lava flows of Tharsis Montes, Alba Patera, Ceraunius Fossae, and Olympus Mons Formations (A)
- Aureole deposits of the Olympus Mons Formation (A)
- Lava and volcaniclastic flows of the Elysium Formation (A)
- Hesperian lava flows of the Tharsis Montes, Alba Patera, and Syria Planum Formations (H)
- Ridged plains and lava flows of the Syrtis Major Formation (H)
- Highland paterae and associated dissected material (H)
- Volcanic shields and domes, undivided (A,H,N)

Channel-system Materials

- Amazonian channel and flood-plain deposits (A)
- Hesperian channel and flood-plain deposits (H)
- Hesperian chaotic terrain material (H)

Surficial Materials

- Aeolian, dune, and mantle materials (A)
- Landslide materials (A)
- Polar ice (A)
- Polar layered material (A)

Lowland Terrain Materials

- Arcadia Formation – lowland lava flows and other plains units (A)
- Medusae Fossae Formation – friable, layered, light-colored deposits (A)
- Undivided smooth, etched, knobby, and intracrater plains materials (A,H)
- Vastitas Borealis Formation – older knobby, grooved, ridged, and mottled plains material (H)

Highland Terrain Materials

Hellas Assemblage

- Amazonian floor and rim units of Hellas basin (A)
- Hesperian floor units of Hellas basin (H)

Plateau and High Plains Assemblage

- Layered and floor deposits of Valles Marineris (A,H)
- Lava flows and smooth plains of the Tempe Terra and Dorsa Argentea Formations (H)
- Smooth and mottled units and intercrater plains material of the plateau sequence (H,N)
- Rugged, dissected, etched, and ridged heavily cratered material of the plateau sequence (N)
- Ancient hilly, basin-rim, and mountainous materials, undivided (N)
- Highly deformed terrain materials (H,N)
- Ridged plains materials (H,N)

Other Materials

- Undivided cratered terrain materials (H,N)
- Superposed impact-crater material (A,H)
- Partly buried impact-crater material (N)

Structure Symbols

- Scarp
- Graben
- Ridge

Correlation of Map Units

Youngest at top

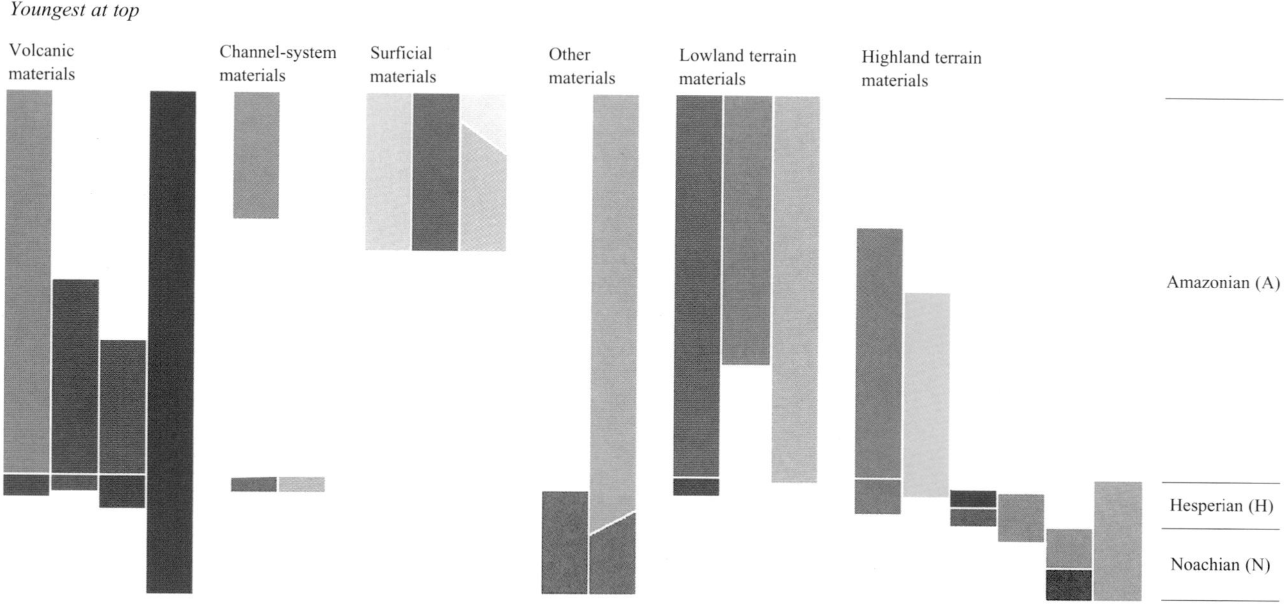

Mars
Eastern Hemisphere
Geologic Map

Center lat. 0°
Center lon. 270°

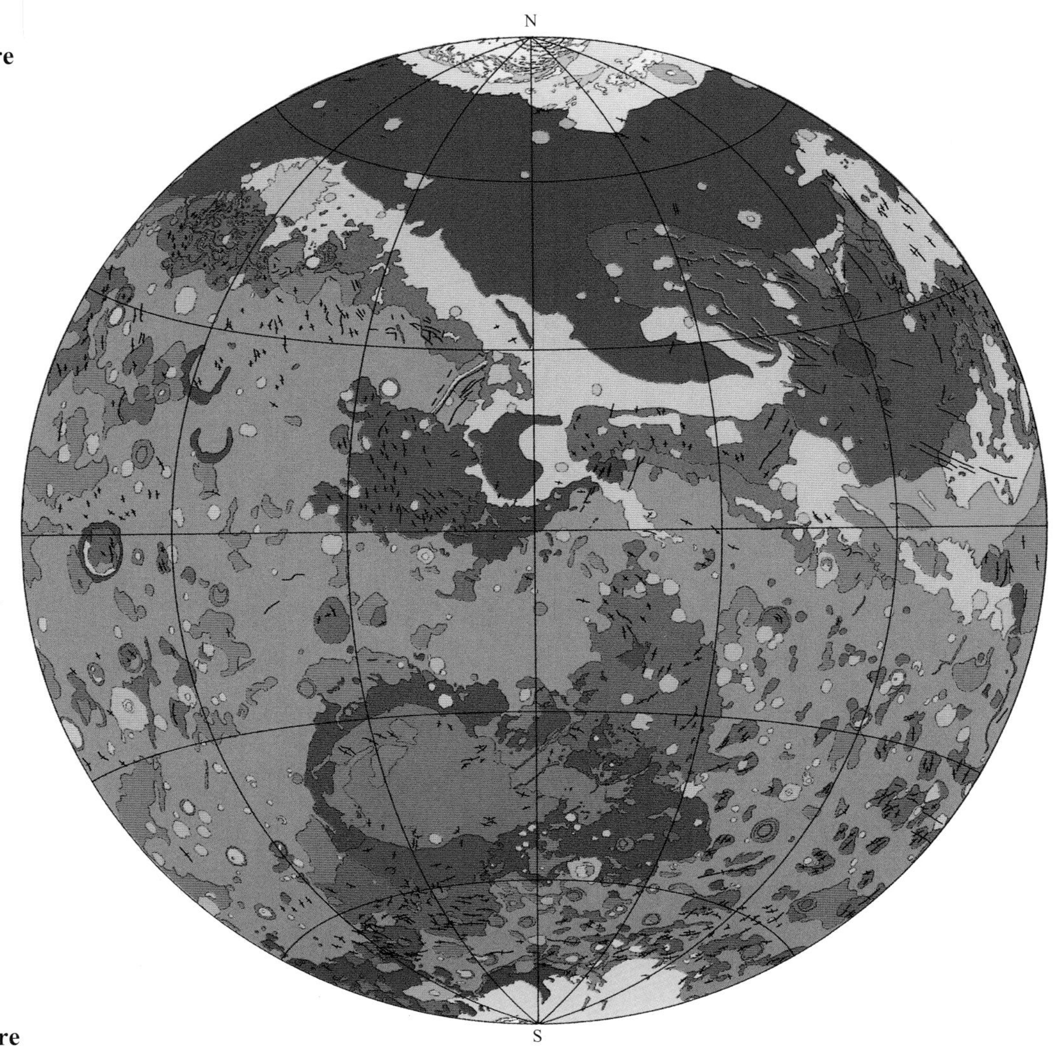

Mars
Eastern Hemisphere
Reference Map

Center lat. 0°
Center lon. 270°

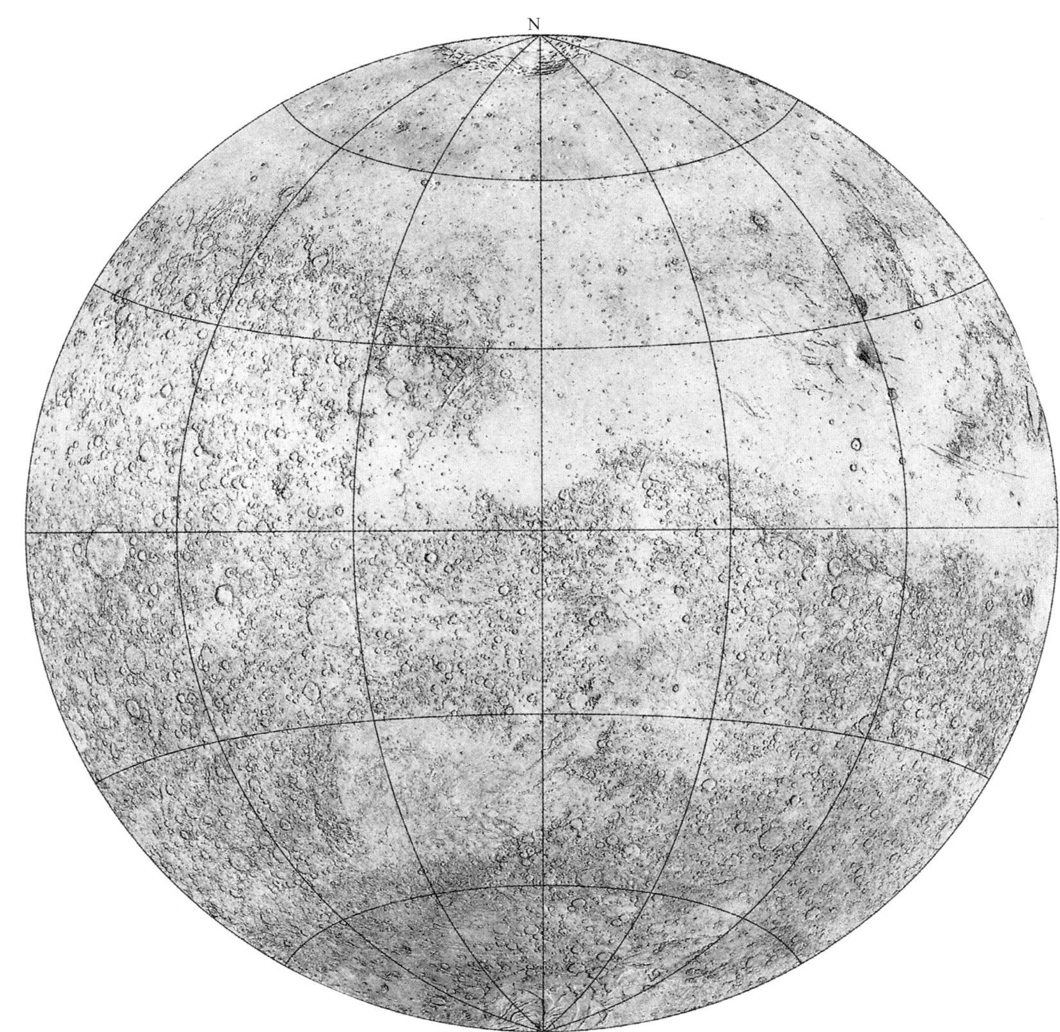

Mars
Western Hemisphere
Geologic Map

Center lat. 0°
Center lon. 90°

Mars
Western Hemisphere
Reference Map

Center lat. 0°
Center lon. 90°

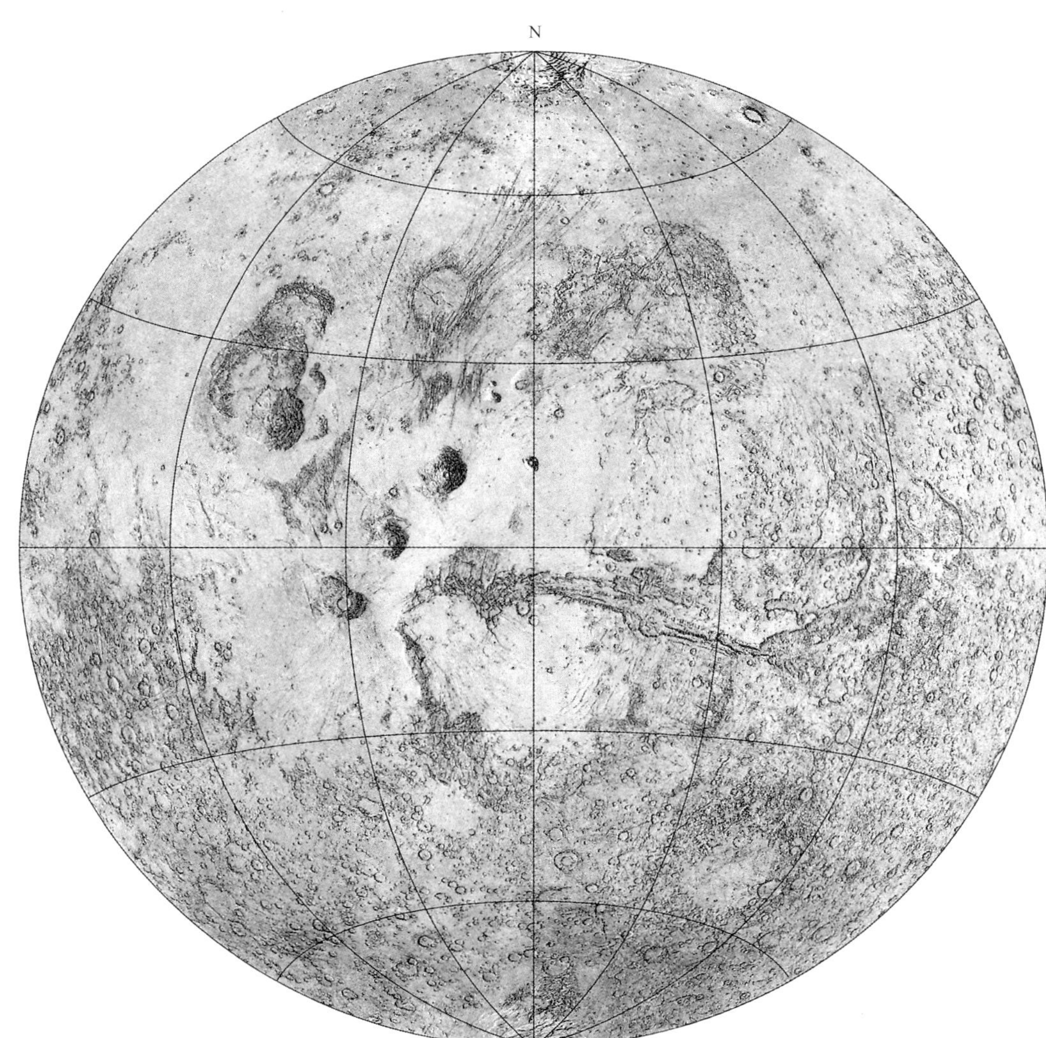

Mars
Index Map

Center lat. 45°
Center lon. 180°

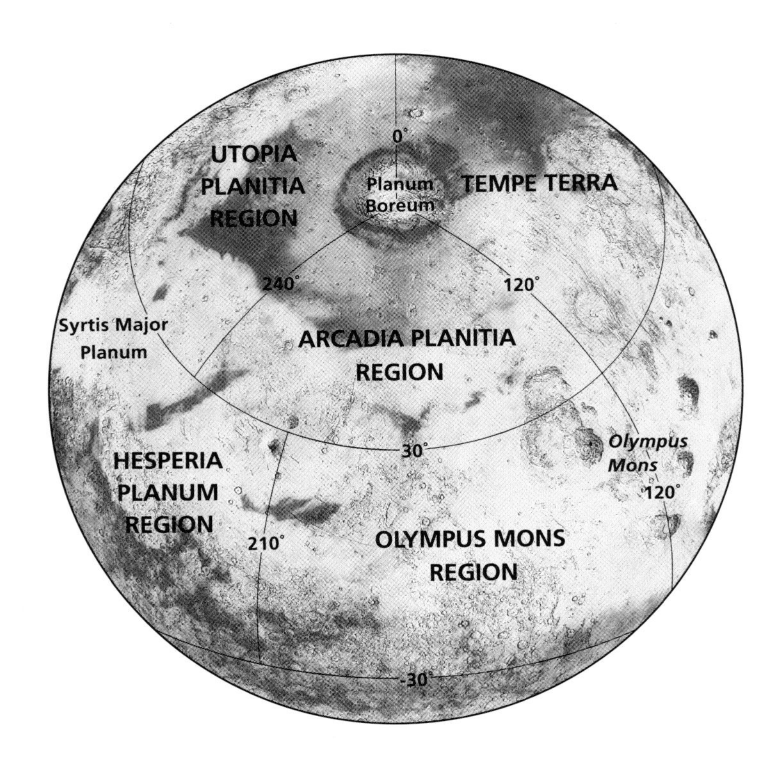

Mars
Index Map

Center lat. –45°
Center lon. 180°

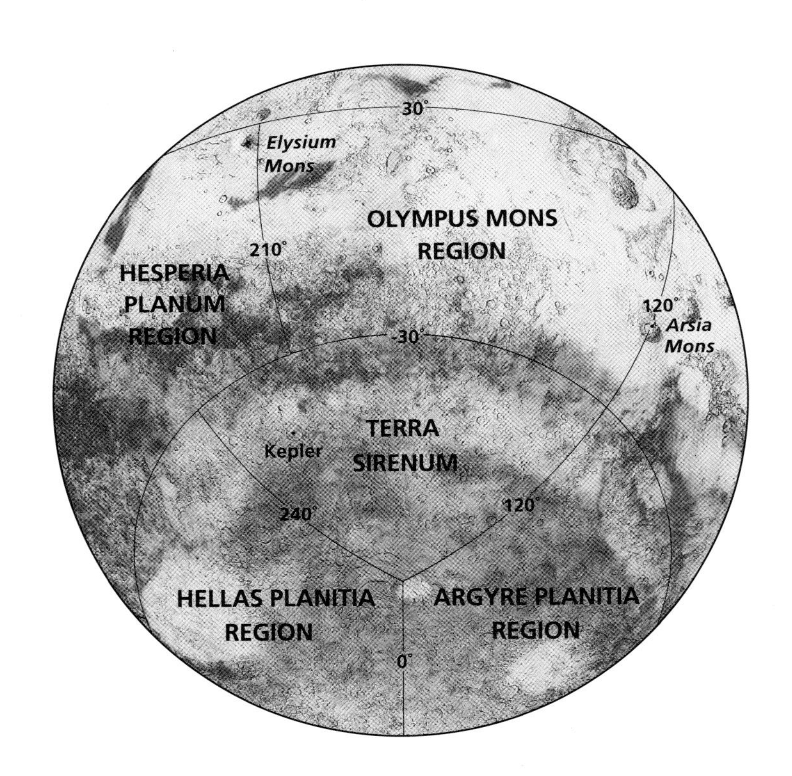

Mars
Index Map

Center lat. 0°
Center lon. 0°

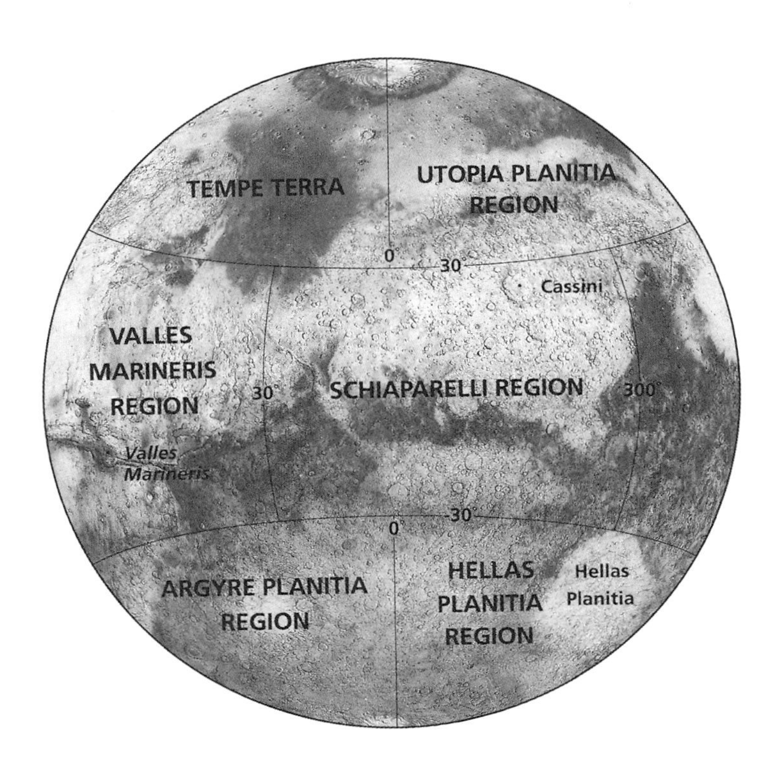

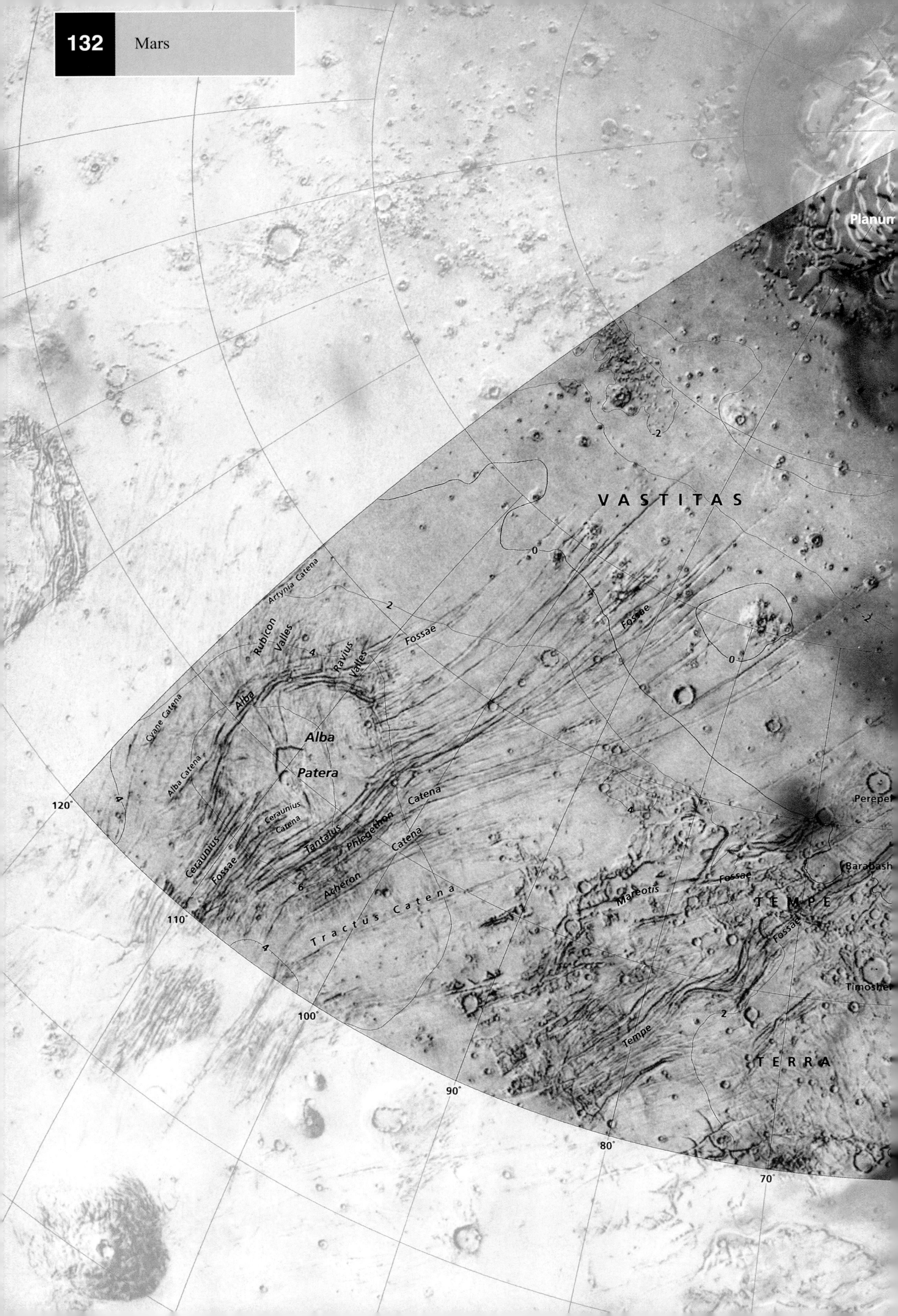
Planum
VASTITAS
-2
0
Artynia Catena
2
Rubicon
Valles
Fossae
Fossae
4
Ravius
Valles
0
Cyane Catena
Alba
Alba
Patera
Alba Catena
Ceraunius
Catena
Catena
120°
4
Tantalus
Phlegethon
Catena
4
Perepel
Ceraunius
Fossae
6
Acheron
Barabash
Fossae
Mareotis
TEMPE
Tractus Catena
Fossae
110°
4
Timoshe
100°
2
Tempe
TERRA
90°
80°
70°

Boreum
Chasma Boreale
Hyperboreae
Undae
Lomonosov
BOREALIS
Kunowsky
ACIDALIA
PLANITIA
Bamberg
ARABIA
Sklodowska
Cydonia
TERRA
Mensae
Sytinskaya
Nilokeras
Mensae
Scopulus
keras
0°
10°
20°
30°
40°
50°
Mars
Airbrush Map
Tempe
Terra
Center lat. 60°
Center lon. 60°
Contour interval
2 km
1 cm = 132 km
1 cm² = 17,424 km²
1 cm²
660
528
396
264
132
0 km

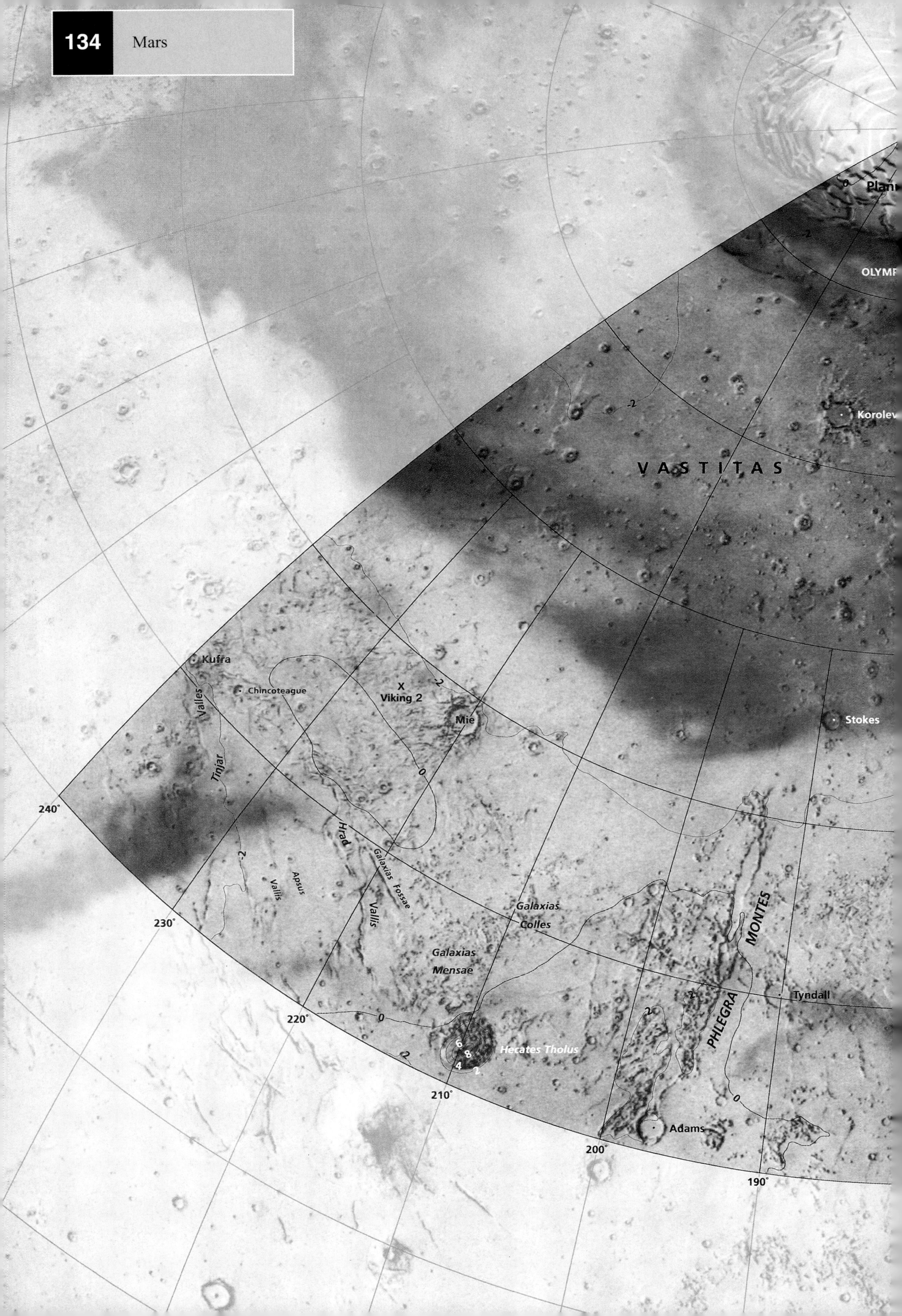

Plan
OLYMP
Korolev
VASTITAS
Kufra
Chincoteague
X
Viking 2
Mie
Stokes
Valles
Tinjar
240°
Hrad
Galaxias Fossae
Apsus
Vallis
Vallis
230°
Galaxias
Colles
MONTES
Galaxias
Mensae
Tyndall
PHLEGRA
220°
Hecates Tholus
210°
Adams
200°
190°

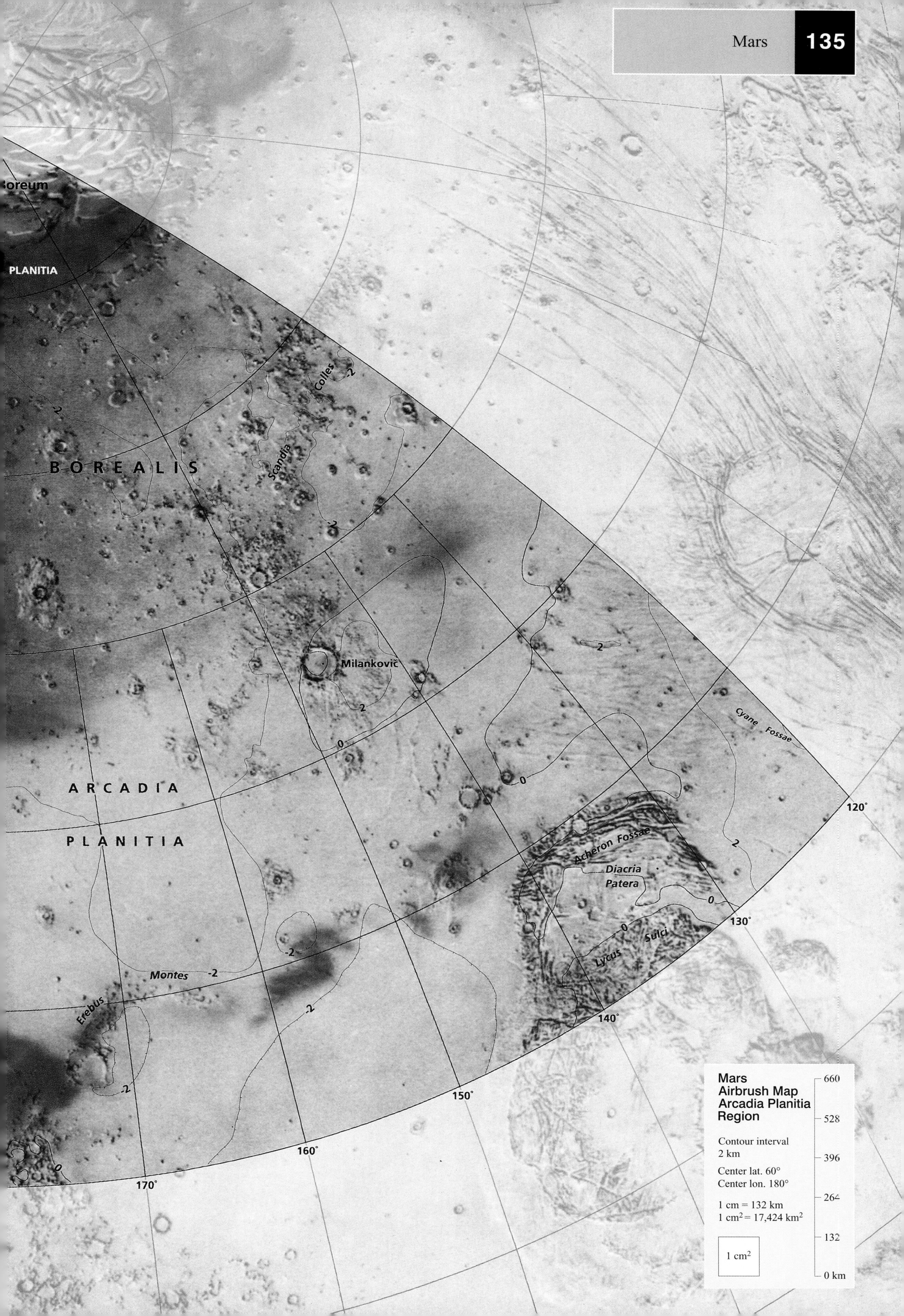

oreum
PLANITIA
BOREALIS
Colles
Scandia
-2
ARCADIA
PLANITIA
Milankovič
2
0
Cyane Fossae
Acheron Fossae
Diacria
Patera
Lycus Sulci
Erebus
Montes
120°
130°
140°
150°
160°
170°
Mars
Airbrush Map
Arcadia Planitia
Region
Contour interval
2 km
Center lat. 60°
Center lon. 180°
1 cm = 132 km
1 cm² = 17,424 km²
1 cm²
660
528
396
264
132
0 km

Planum
VASTITAS
Semeykin
Vallis
Deuteronilus
Mensae
Mamers
Deuteronilus
Colles
Lyot
Focas
Protonilus
Mensae
Moreux
Ismeniae
Fossae
Cerulli
Budaux
Colloe
Quenisset
Clasia
Vallis
0°
350°
340°
330°
320°
310°

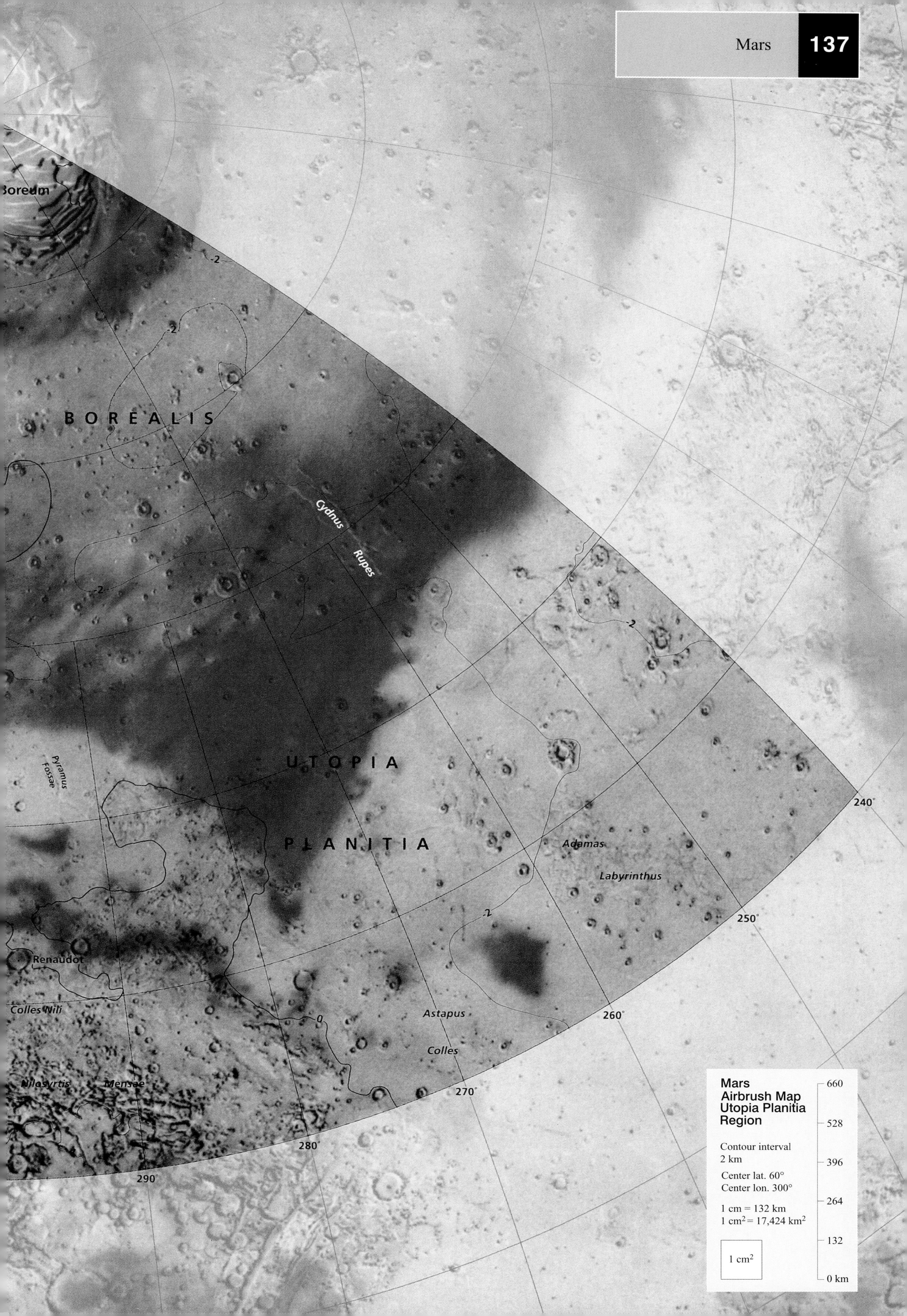
Boreum
BOREALIS
Cydnus
Rupes
Pyramus
Fossae
UTOPIA
PLANITIA
Adamas
Labyrinthus
Renaudot
Colles Nili
Astapus
Colles
Nilosyrtis
Mensae
-2
0
240°
250°
260°
270°
280°
290°
Mars
Airbrush Map
Utopia Planitia
Region
Contour interval
2 km
Center lat. 60°
Center lon. 300°
1 cm = 132 km
1 cm2 = 17,424 km2
1 cm2
660
528
396
264
132
0 km

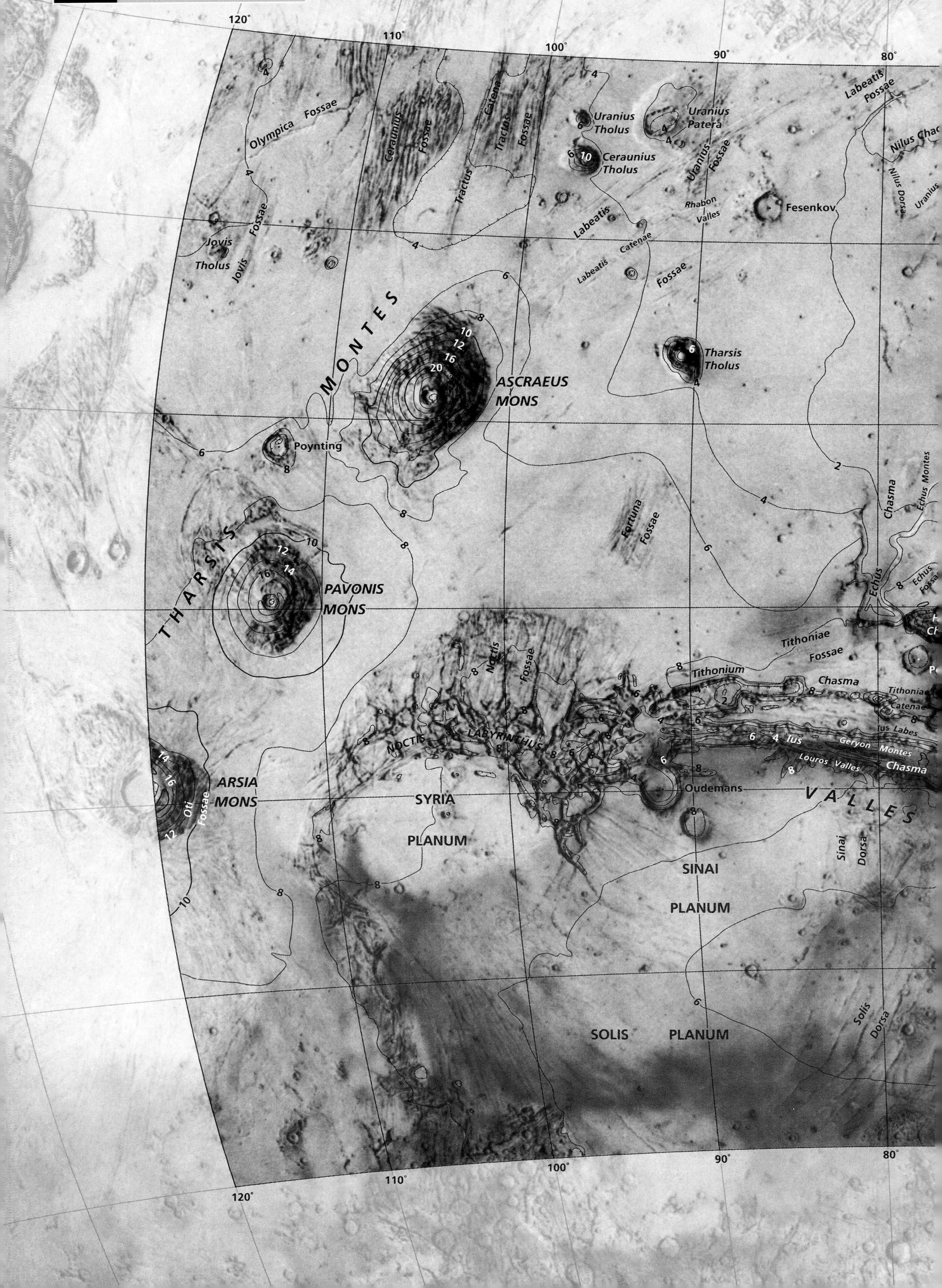

120° 110° 100° 90° 80°

Olympica Fossae
Ceraunius Fossae
Catena
Tractus
Tractus Fossae
Uranius Tholus
Uranius Patera
Labeatis Fossae
Nilus Chaos
Nilus Dorsa
Uranius
Ceraunius Tholus
Uranius Fossae
Rhabon Valles
Fesenkov
Jovis Tholus
Jovis Fossae
Labeatis
Labeatis Catenae
Fossae
THARSIS MONTES
ASCRAEUS MONS
Tharsis Tholus
Poynting
Fortuna Fossae
Echus Chasma
Echus Montes
Echus Fossae
PAVONIS MONS
Noctis Fossae
Tithoniae Fossae
Tithonium Chasma
Tithoniae Catenae
Ius Labes
NOCTIS LABYRINTHUS
Ius Chasma
Geryon Montes
Louros Valles
ARSIA MONS
Oti Fossae
SYRIA PLANUM
Oudemans
VALLES
Sinai Dorsa
SINAI PLANUM
SOLIS PLANUM
Solis Dorsa

120° 110° 100° 90° 80°

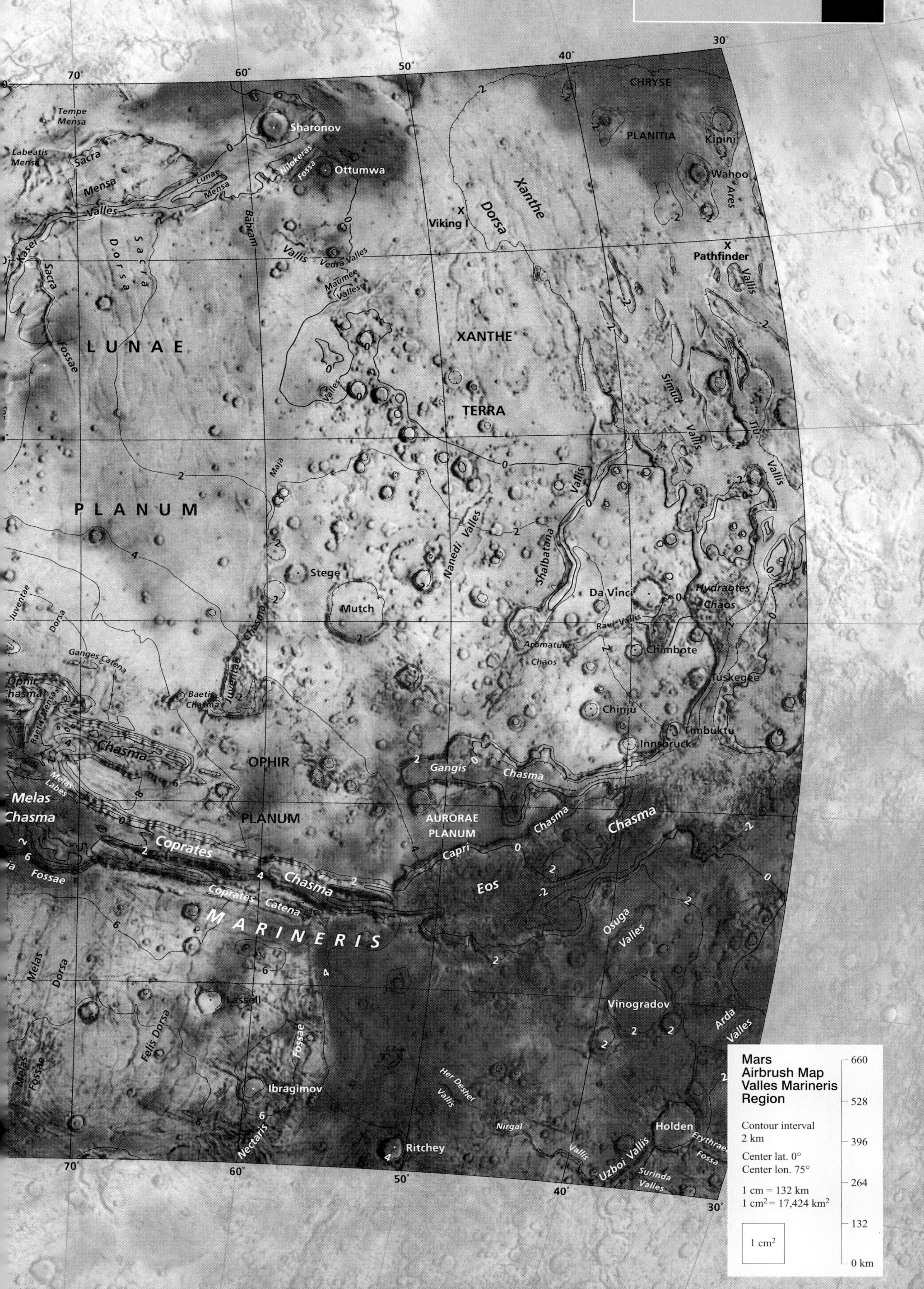

Mars
Airbrush Map
Valles Marineris
Region
Contour interval
2 km
Center lat. 0°
Center lon. 75°
1 cm = 132 km
1 cm² = 17,424 km²
1 cm²
660
528
396
264
132
0 km
CHRYSE
PLANITIA
Kipini
Wahoo
Ares
Pathfinder
Vallis
Sharonov
Ottumwa
Nilokeras
Fossa
Lunae
Mensa
Tempe
Mensa
Labeatis
Mensa
Sacra
Mensa
Valles
Kasei
Sacra
Dorsa
Sacra
Fossae
Bahram
Vallis
Vedra Valles
Maumee
Valles
Viking I
Dorsa
Xanthe
XANTHE
TERRA
LUNAE
PLANUM
Simud
Vallis
Tiu
Vallis
Maja
Valles
Nanedi Valles
Shalbatana
Vallis
Stege
Mutch
Da Vinci
Hydraotes
Chaos
Ravi Vallis
Aromatum
Chaos
Chimbote
Tuskegee
Chinju
Timbuktu
Innsbruck
Juventae
Dorsa
Juventae
Chasma
Ganges Catena
Ophir
Chasma
Baetis
Chasma
Baetis Mensa
Chasma
Melas
Labes
Melas
Chasma
OPHIR
PLANUM
Gangis
Chasma
AURORAE
PLANUM
Chasma
Chasma
Capri
Eos
Coprates
Chasma
Coprates
Catena
Fossae
MARINERIS
Osuga
Valles
Vinogradov
Arda
Valles
Melas
Dorsa
Lassell
Felis Dorsa
Fossae
Melas
Fossae
Ibragimov
Nectaris
Her Desher
Vallis
Nirgal
Vallis
Ritchey
Holden
Uzboi Vallis
Erythraea
Fossa
Surinda
Valles
70°
60°
50°
40°
30°

210° 200° 190° 180° 170°

Lockyer

Ituxi Vallis

Patapsco Vallis

Stygis Catena

Iberus Vallis

Albor Tholus

Albor Fossae

Tartarus Montes

Tartarus Colles

Orcus Patera

Pettit

Cerberus Rupes

ELYSIUM PLANITIA

Hibes Montes

Avernus Dorsa

Tartarus Rupes

Apollinaris Patera

Memnonia Sulci

Reuyl

Zephyria Mensae

TERRA CIMMERIA

Gusev

Boeddicker

Al-Qahira Vallis

Hadley

Graff

Ma'adim Vallis

Eirikson

210° 200° 190° 180° 170°

AMAZONIS
PLANITIA
Lycus Sulci
Halex Fossae
Cyane Sulci
Olympus Rupes
OLYMPUS MONS
Sulci Gordii
Ulysses Fossae
Gigas Sulci
Gigas Fossae
Ulysses Patera
Biblis Patera
Eumenides Dorsum
Gordii Dorsum
Nicholson
Medusae Fossae
Medusae Sulci
Amazonis Sulci
Minio Vallis
Abus Vallis
Labou Vallis
Mangala Valles
Arimanes Rupes
Marca
Cobres
Burton
DAEDALIA
PLANUM
Aganippe Fossa
ARSIA MONS
Williams
Comas Sola
Memnonia Fossae
Bernard
Dejnev
Columbus
Koval'skiy
Mars
Airbrush Map
Olympus Mons
Region
Contour interval
2 km
Center lat. 0°
Center lon. 165°
1 cm = 132 km
1 cm^2 = 17,424 km^2
1 cm^2
660
528
396
264
132
0 km

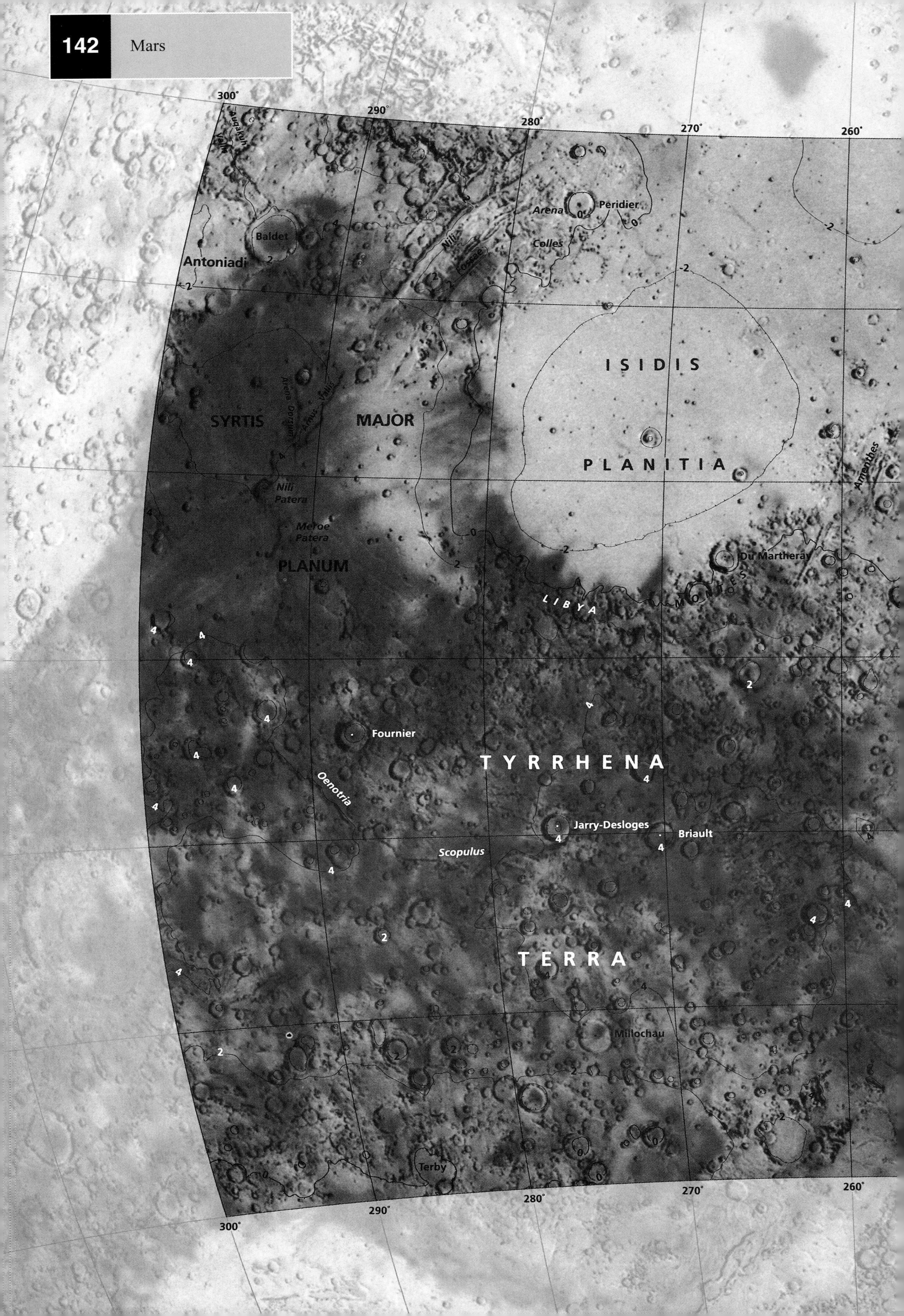

300°
290°
280°
270°
260°
Baldet
Antoniadi
Peridier
Arena
Colles
Nili Fossae
ISIDIS
PLANITIA
SYRTIS
MAJOR
Arena Dorsum
Amus Vallis
Nili Patera
Meroe Patera
PLANUM
Du Martheray
LIBYA
MONTES
Amenthes
Fournier
TYRRHENA
Oenotria
Jarry-Desloges
Briault
Scopulus
TERRA
Millochau
Terby

ELYSIUM
PLANITIA
ESPERIA
PLANUM
Granicus Valles
Elysium Fossae
Stygis Fossae
Elysium Rupes
Zephyrus Fossae
ELYSIUM MONS
Hyblaeus Chasma
Elysium Chasma
Hyblaeus Fossae
Albor Tholus
Elysium Catena
Hephaestus Fossae
Hebrus Valles
Eddie
Nepenthes Mensae
Amenthes Rupes
Escalante
Aeolis Mensae
Gale
Knobel
Lasswitz
Wien
Herschel
Müller
Molesworth
250°
240°
230°
220°
210°
Mars
Airbrush Map
Hesperia
Planum
Region
Contour interval
2 km
Center lat. 0°
Center lon. 255°
1 cm = 132 km
1 cm^2 = 17,424 km^2
1 cm^2
660
528
396
264
132
0 km

20°
10°
0°
350°
Curie
Maggini
Mawrth
Vallis
McLaughlin
Colles
Oxia
Becquerel
Rutherford
Trouvelot
Radau
Gill
Marth
Ares
ARABIA
Vallis
Galilaei
Crommelin
Hydaspis
Chaos
Aram
Chaos
Iani
Chaos
Aureum
Chaos
Airy
MARGARITIFER
TERRA
MERIDIANI
Arsinoes
Chaos
Margaritifer
Chaos
Pyrrhae
Chaos
Mädler
TERRA
Beer
Loire
Jones
Wislicenus
Lorica
Valles
Paraná
Valles
NOACHIS
Ladon Valles
Himera Valles
Samara
Oltis
Valles
Bakhuys
Kantang
Chekalin
Kasimov
Clota
Vallis
Valles
Ostrov
Kashira
30°
20°
10°
0°
350°

340° 330° 320° 310°

Luzin
Cassini
Phison Rupes
Flammarion
Schöner
Pasteur
Indus Vallis
Scamander Vallis
Tikhonravov
Henry
Arago
TERRA
Valles Locras
Naktong
Janssen
Vallis
Teisserenc de Bort
Schiaparelli
Schroeter
Valles
TERRA SABAEA
Dawes
Liris Valles
Tisia Valles
Huygens
Mosa Vallis
Flaugergues
Denning
Bouguer
Lambert
TERRA
Scopulus
Scopulus
Scylla
Charybdis
Schaeberle
Niesten

340° 330° 320° 310° 300°

Mars Airbrush Map Schiaparelli Region

Contour interval 2 km

Center lat. 0°
Center lon. 345°

1 cm = 132 km
1 cm^2 = 17,424 km^2

1 cm^2

660
528
396
264
132
0 km

70°
80°
90°
100°
110°
120°
Aniak
BOSPOROS
PLANUM
Voeykov
Babakin
Lampland
Pulawy
Coracis
Fossae
Claritas
Fossae
Warrego
Valles
AONIA
Slipher
ICARIA
PLANUM
Douglass
Lowell
Thaumasia
Fossae
Coblentz
TERRA
Argyre Rupes
Fontana
Porter
Ross
Brashear
Lamont
Bianchini
Smith
Agassiz
Schmidt
Heaviside
Steno
PLANUM
Cavi
Angusti
Lau
ANGUSTUM

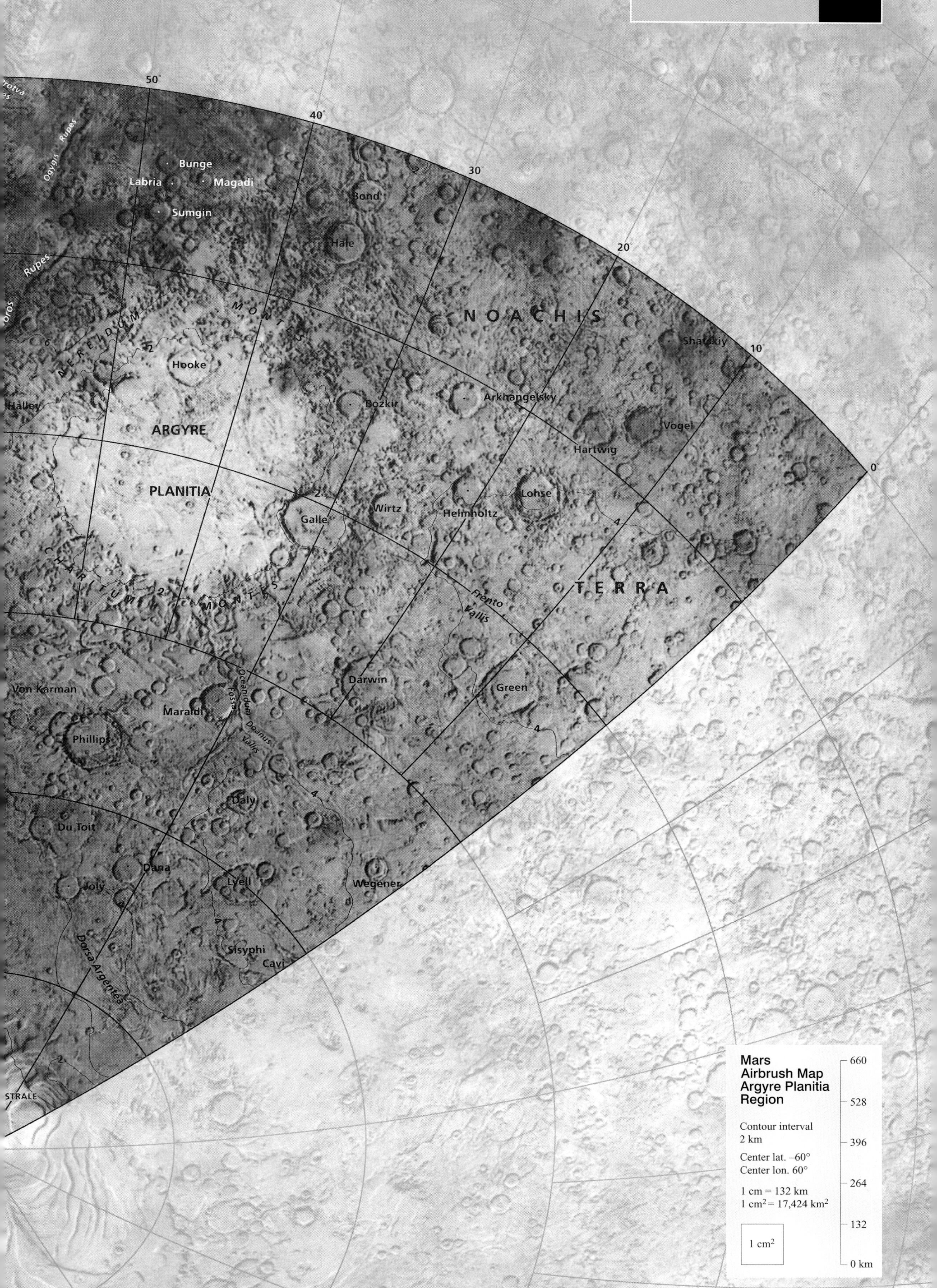

50°
40°
30°
20°
10°
0°
Ogygis Rupes
Bunge
Labria
Magadi
Sumgin
Bond
Hale
NEREIDUM MONTES
NOACHIS
Shatskiy
Hooke
Halley
Arkhangelsky
Bozkir
Vogel
Hartwig
ARGYRE
PLANITIA
Lohse
Wirtz
Helmholtz
Galle
CHARITUM MONTES
TERRA
Frento Vallis
Von Karman
Darwin
Green
Maraldi
Oceanidum Fossa
Doanus Vallis
Phillips
Daly
Du Toit
Dana
Joly
Lyell
Wegener
Dorsa Argentea
Sisyphi Cavi
Mars
Airbrush Map
Argyre Planitia
Region
Contour interval
2 km
Center lat. –60°
Center lon. 60°
1 cm = 132 km
1 cm² = 17,424 km²
1 cm²
660
528
396
264
132
0 km

190°
200°
210°
220°
230°
240°
Ariadnes Colles
Martz
Bjerknes
Cruls
TERRA
Rossby
Morpheos Rupes
Huggins
CIMMERIA
Kepler
Tycho Brahe
Campbell
Arrhenius
Scopulus
Mendel
Eridania
Vinogradsky
Alexey Tolstoy
Haldane
Priestley
Planum
Chronium
Ulyxis Rupes
Wells
Jeans
Byrd
Thyles Rupes
4

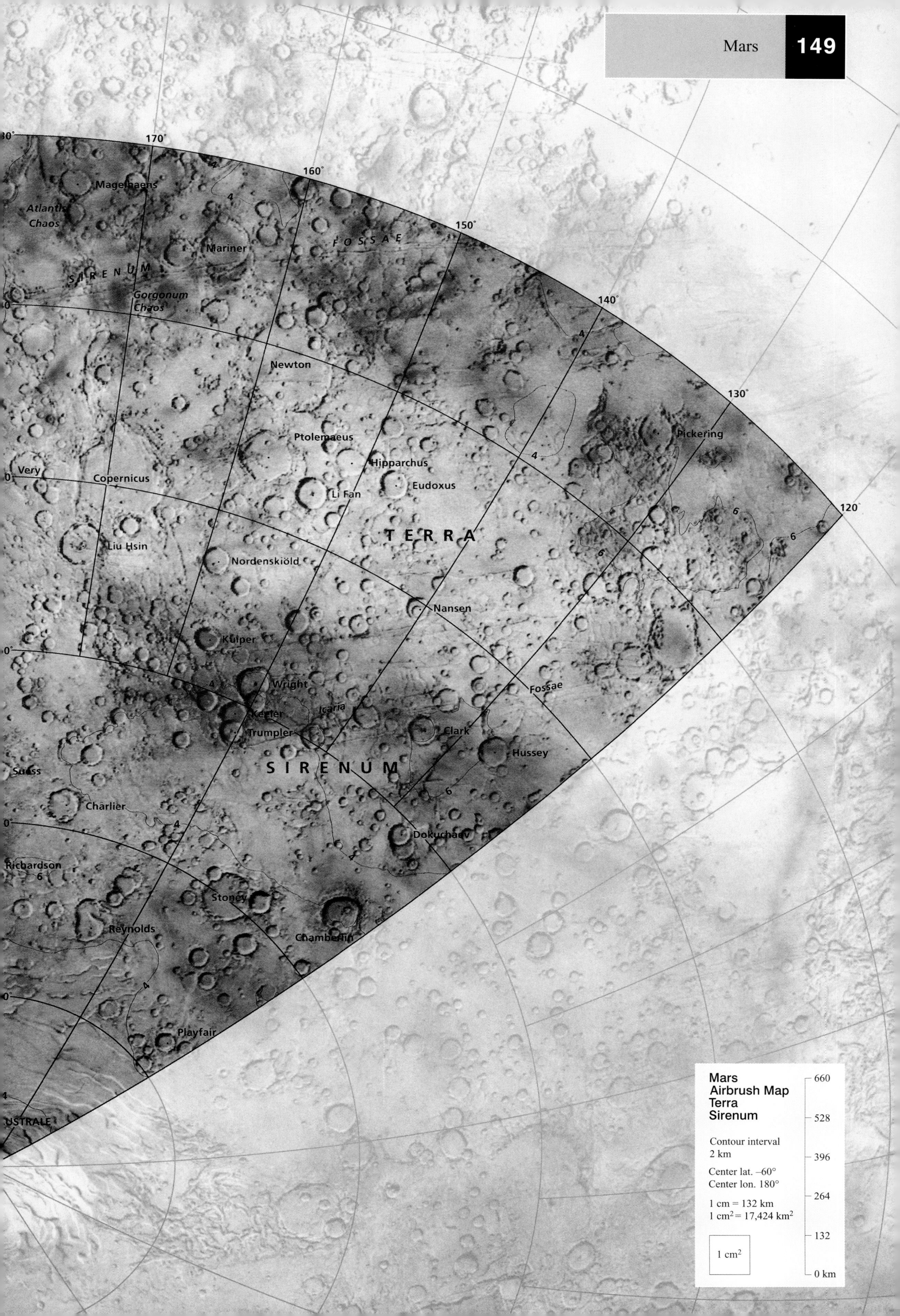
170°
160°
150°
140°
130°
120°
Magelhaens
Atlantis Chaos
Mariner
FOSSAE
SIRENUM
Gorgonum Chaos
Newton
Ptolemaeus
Hipparchus
Very
Copernicus
Eudoxus
Li Fan
Pickering
TERRA
Liu Hsin
Nordenskiöld
Nansen
Kuiper
Wright
Fossae
Kepler
Icaria
Trumpler
Clark
Hussey
SIRENUM
Suess
Charlier
Dokuchaev
Richardson
Stoney
Reynolds
Chamberlin
Playfair
USTRALE
Mars
Airbrush Map
Terra
Sirenum
Contour interval
2 km
Center lat. –60°
Center lon. 180°
1 cm = 132 km
1 cm² = 17,424 km²
1 cm²
660
528
396
264
132
0 km

310°
320°
330°
340°
350°
0°
NOACHIS
TERRA
Le Verrier
Rabe
Proctor
Kaiser
Russell
Maunder
Hellespontus Montes
Chalcoporos Rupes
Pityusa Rupes
Sisyphi Montes
South
Main
PLAN

290°
280°
270°
260°
250°
240°
Coronae
Scopulus
Hadriaca Patera
HELLAS
PLANITIA
Dao Vallis
Harmakhis Vallis
Reull Vallis
Zea Dorsa
Axius Valles
Amphitrites Patera
Barnard
Gledhill
Krishtofovich
PROMETHEI
TERRA
Spallanzani
Tikhov
Redi
Secchi
Wallace
MALEA
LANUM
Mitchel
Huxley
Gilbert
Brevia
Holmes
Weinbaum
Hutton
Vishniac
Burroughs
Liais
Rayleigh
Australe
Chasma
ISTRALE
Mars
Airbrush Map
Hellas Planitia
Region
Contour interval
2 km
Center lat. –60°
Center lon. 300°
1 cm = 132 km
1 cm² = 17,424 km²
1 cm²
660
528
396
264
132
0 km

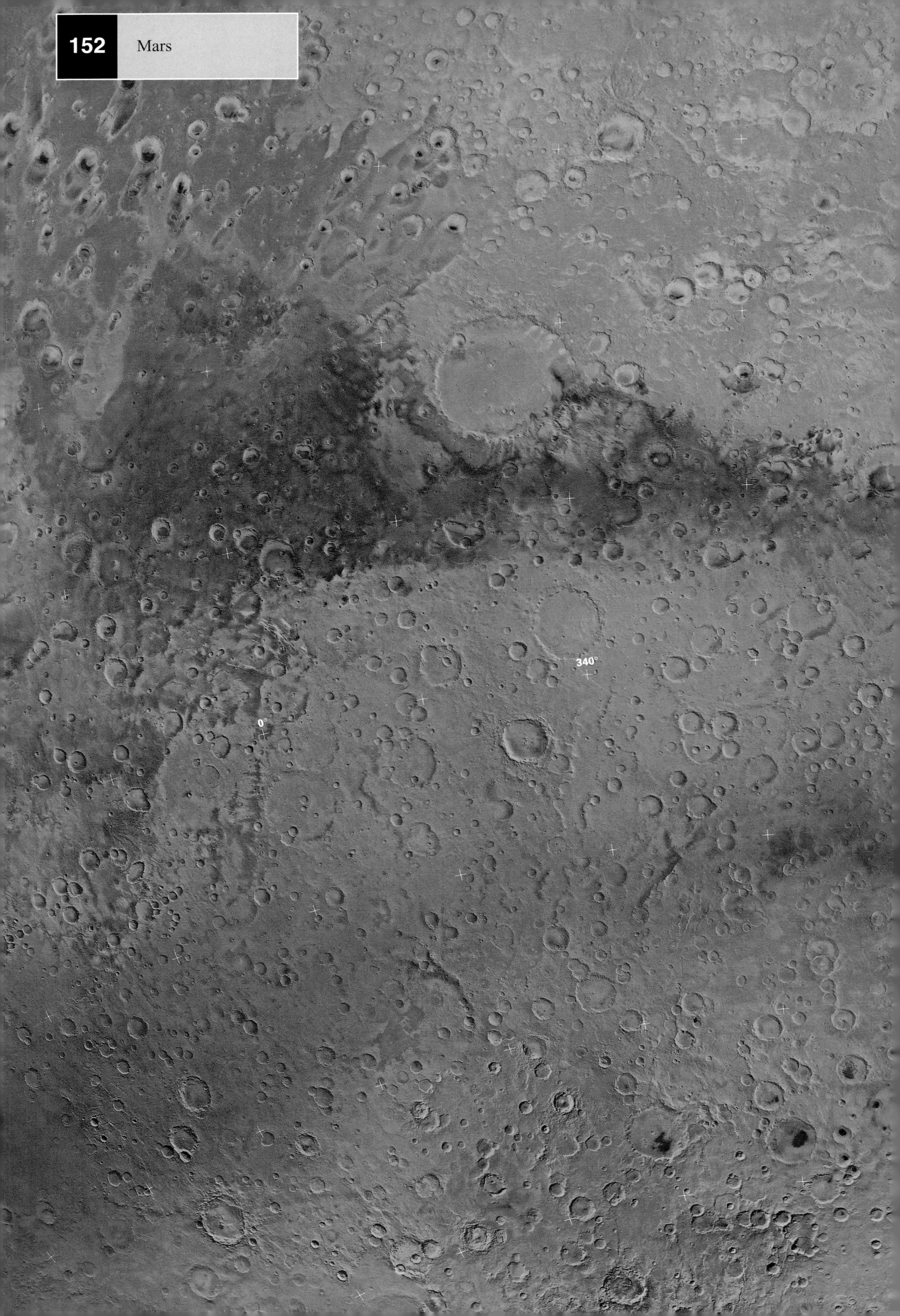
340°
0°

300°
280°

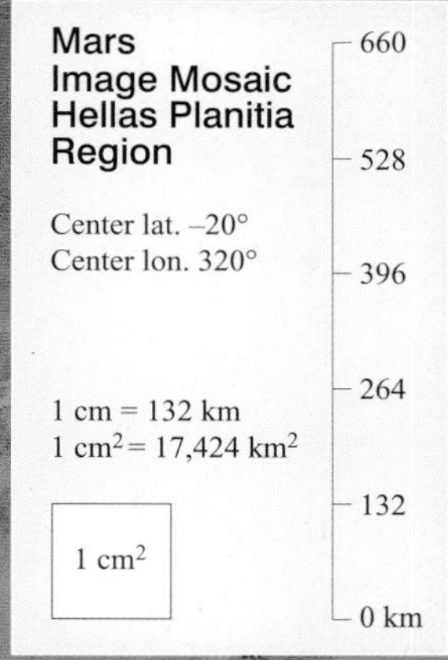
Mars
Image Mosaic
Hellas Planitia
Region
Center lat. –20°
Center lon. 320°
1 cm = 132 km
1 cm² = 17,424 km²
1 cm²
660
528
396
264
132
0 km

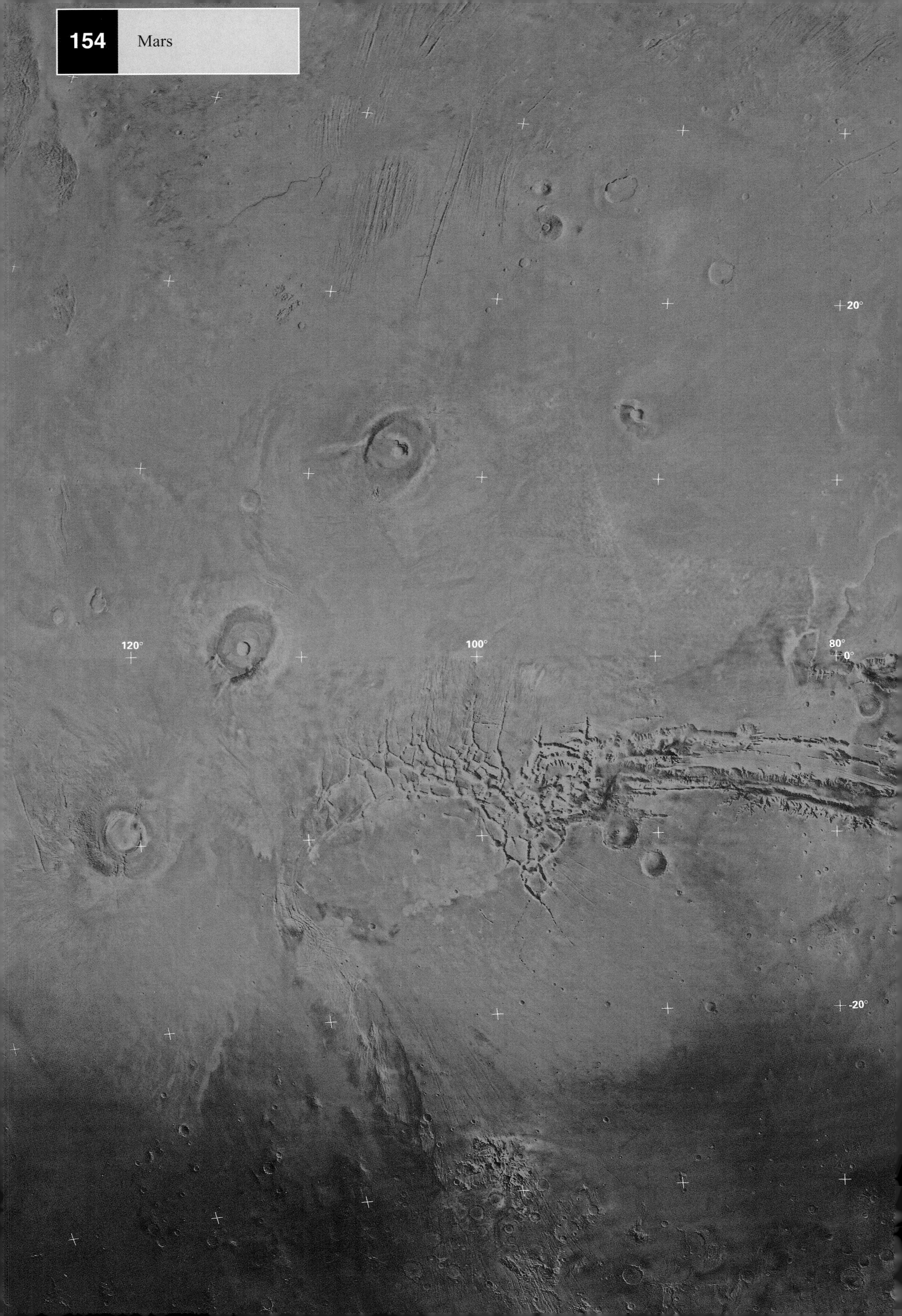

20°
120°
100°
80°
0°
-20°

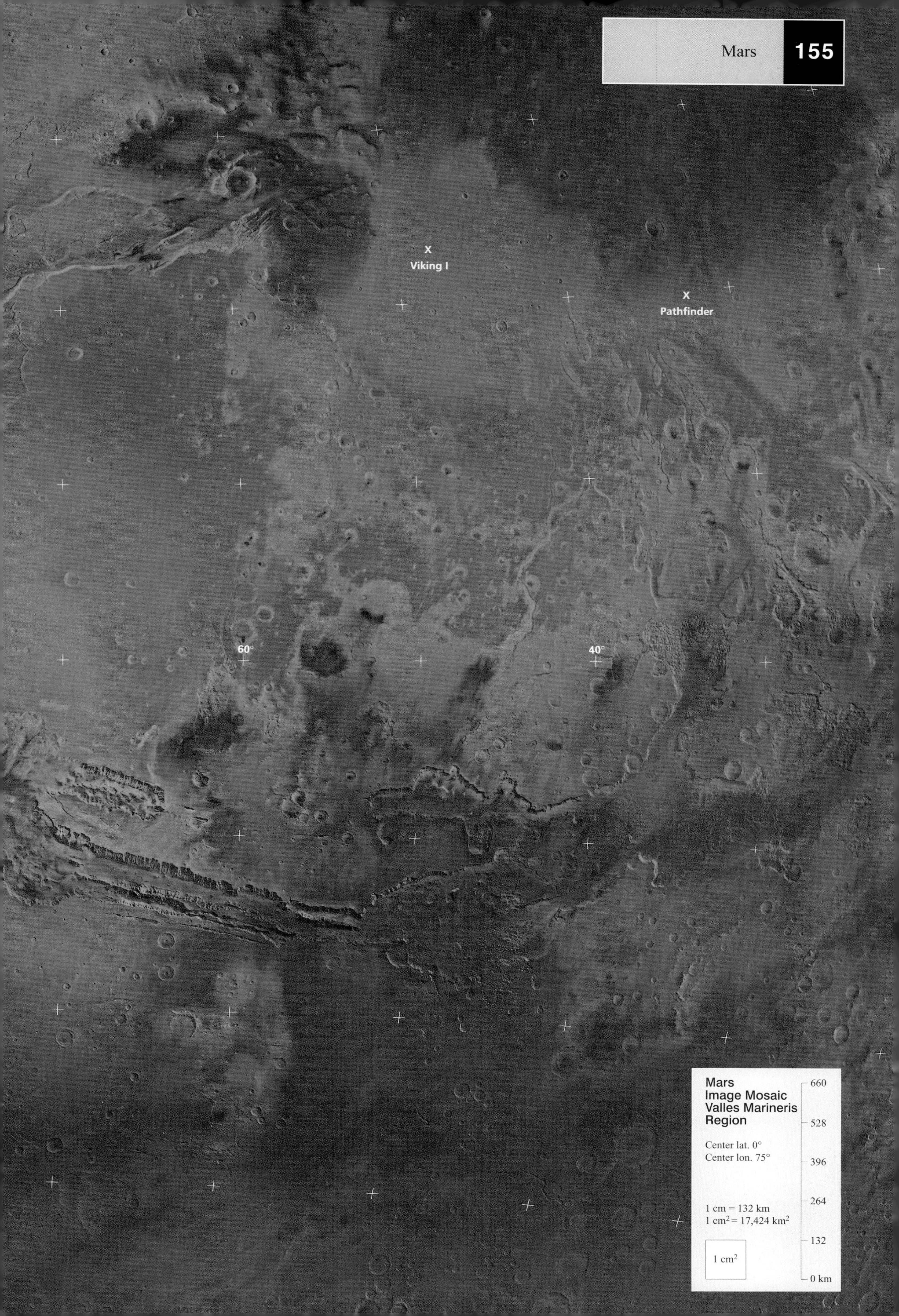
X
Viking I
X
Pathfinder
60°
40°
Mars
Image Mosaic
Valles Marineris
Region
Center lat. 0°
Center lon. 75°
1 cm = 132 km
1 cm² = 17,424 km²
1 cm²
660
528
396
264
132
0 km

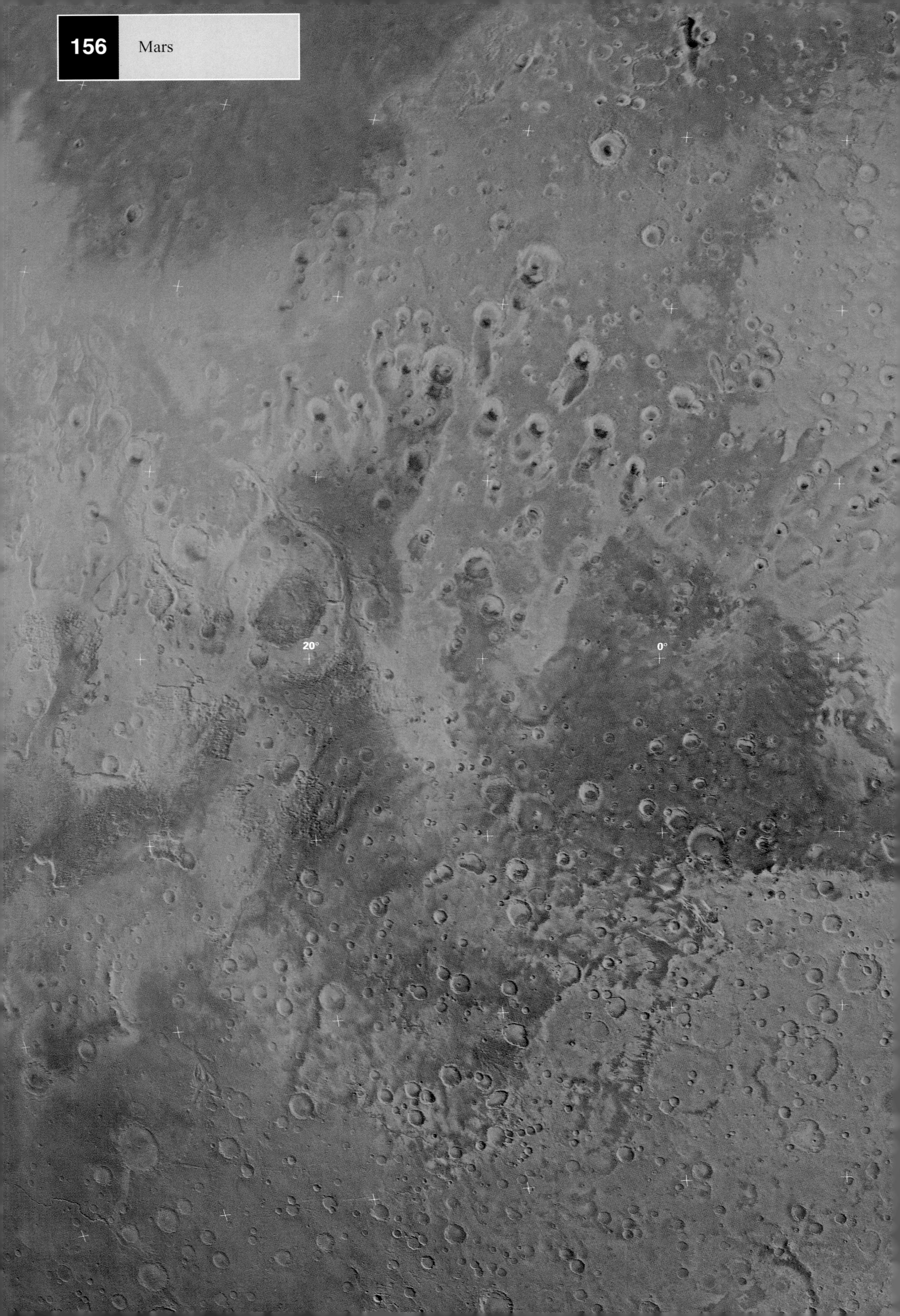
20°
0°

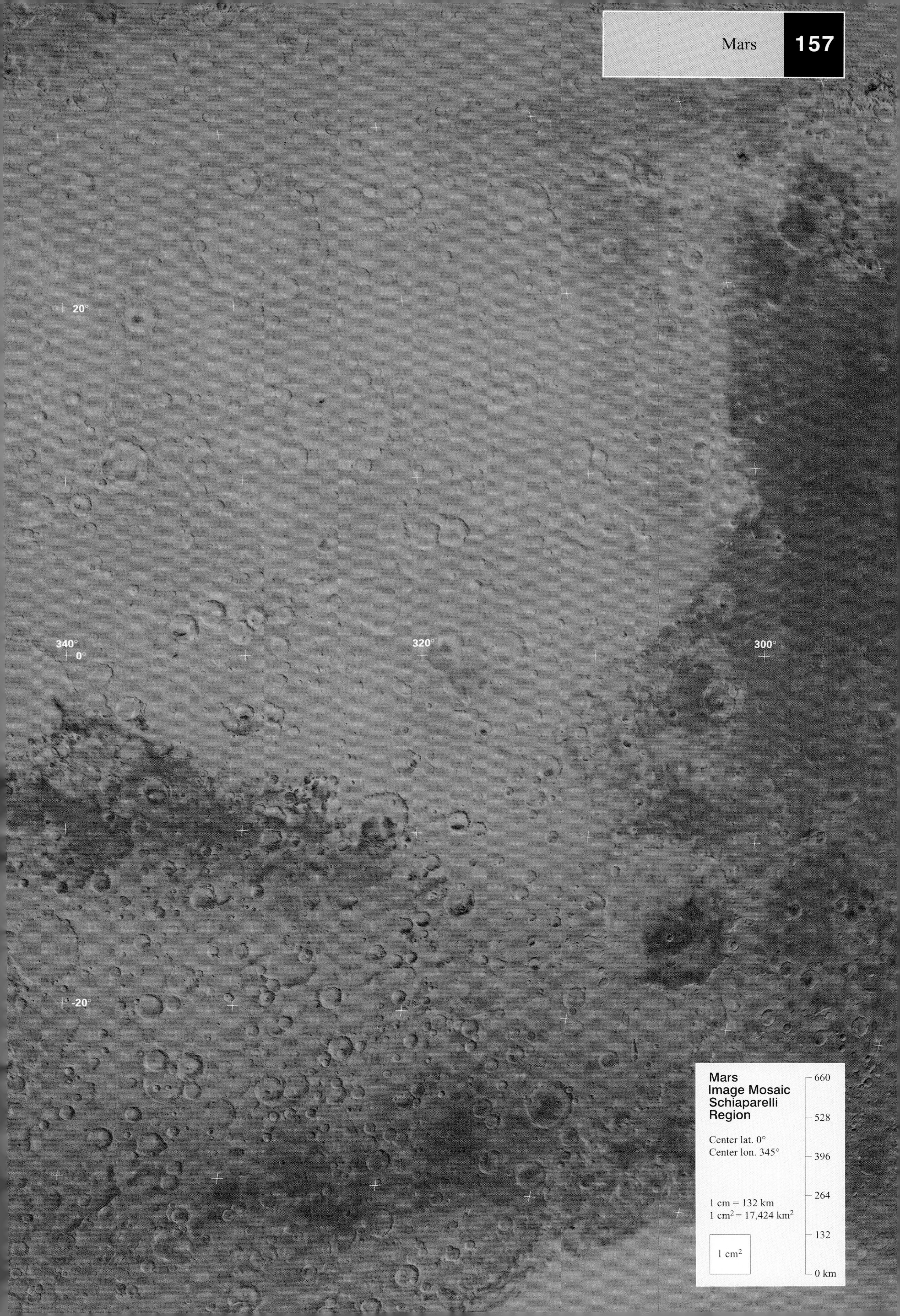

20°
340°
0°
320°
300°
-20°

Mars
Image Mosaic
Schiaparelli
Region

Center lat. 0°
Center lon. 345°

1 cm = 132 km
1 cm^2 = 17,424 km^2

1 cm^2

660
528
396
264
132
0 km

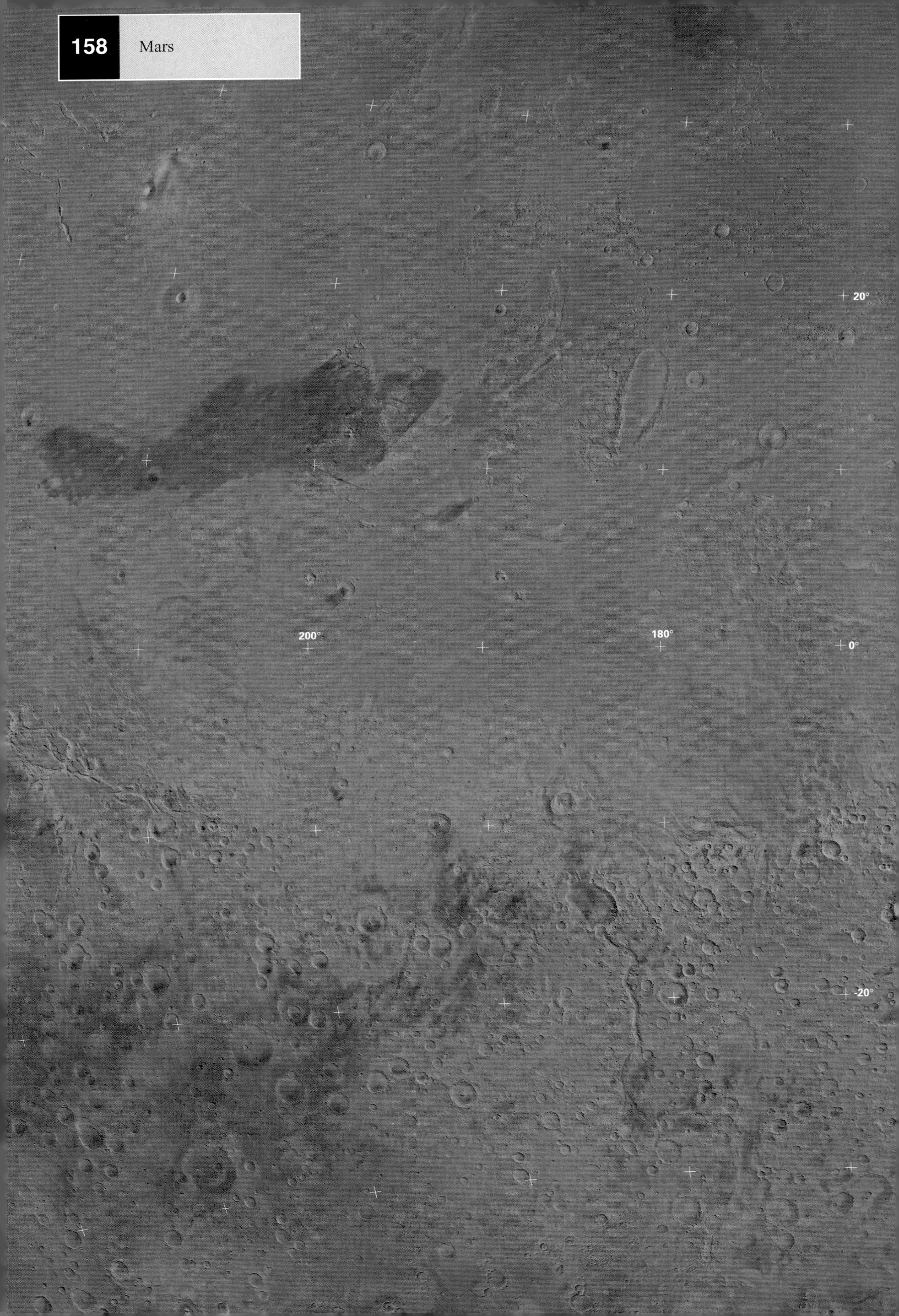

20°
200°
180°
0°
-20°

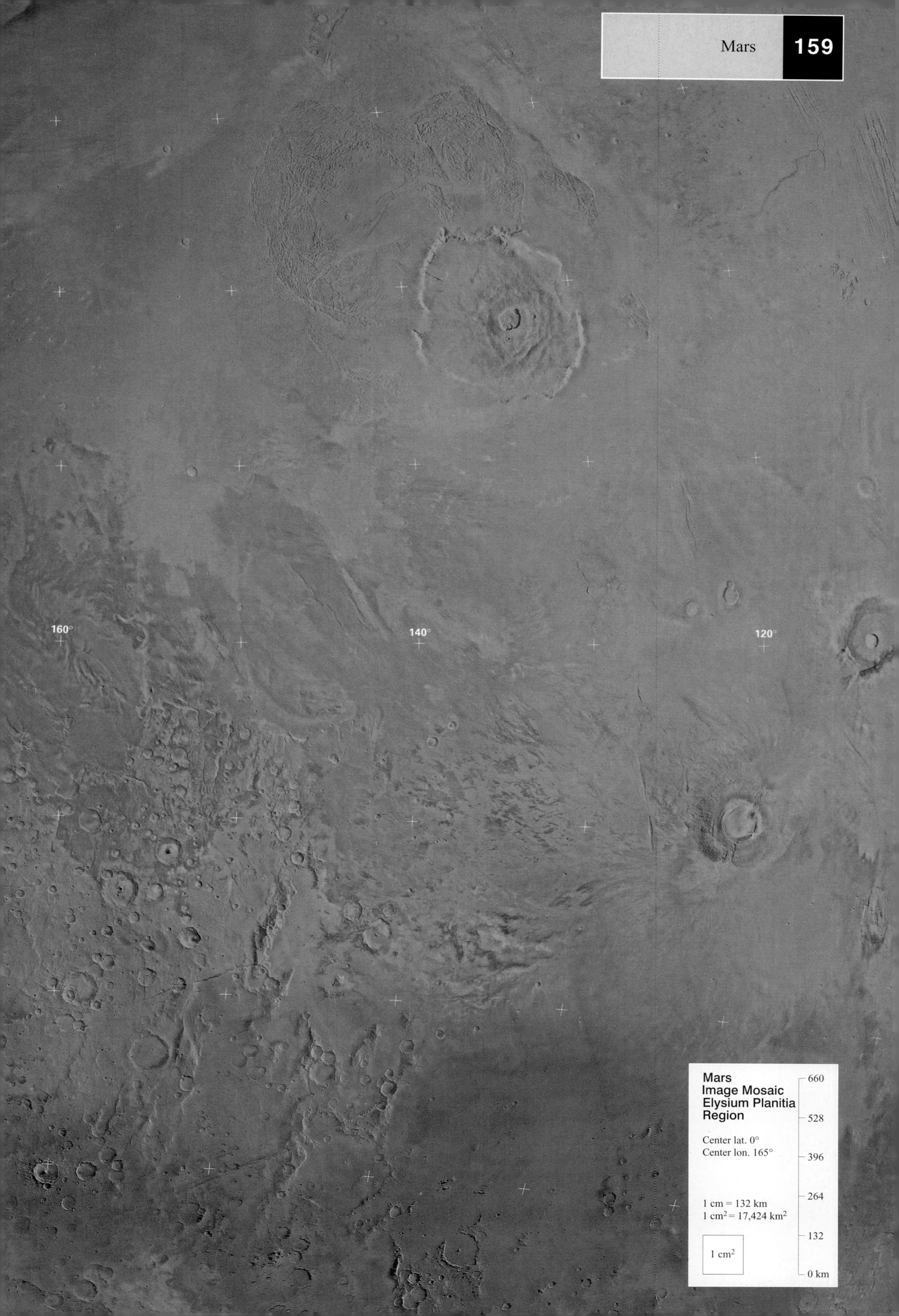

160°
140°
120°

Mars
Image Mosaic
Elysium Planitia
Region

Center lat. 0°
Center lon. 165°

1 cm = 132 km
1 cm^2 = 17,424 km^2

1 cm^2

660
528
396
264
132
0 km

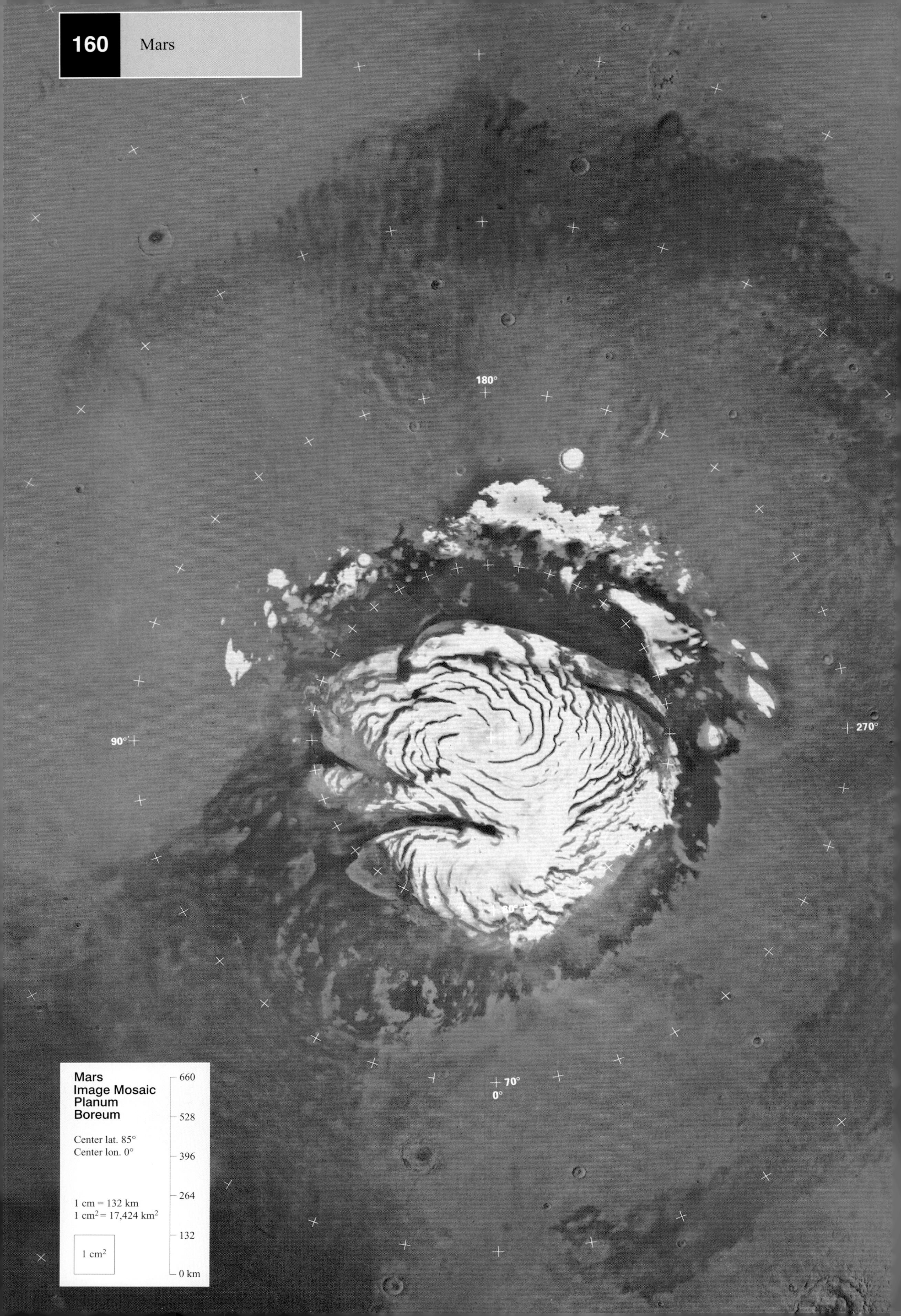

180°

90°

270°

80°

70°
0°

Mars
Image Mosaic
Planum
Boreum

Center lat. 85°
Center lon. 0°

1 cm = 132 km
1 cm² = 17,424 km²

1 cm²

660
528
396
264
132
0 km

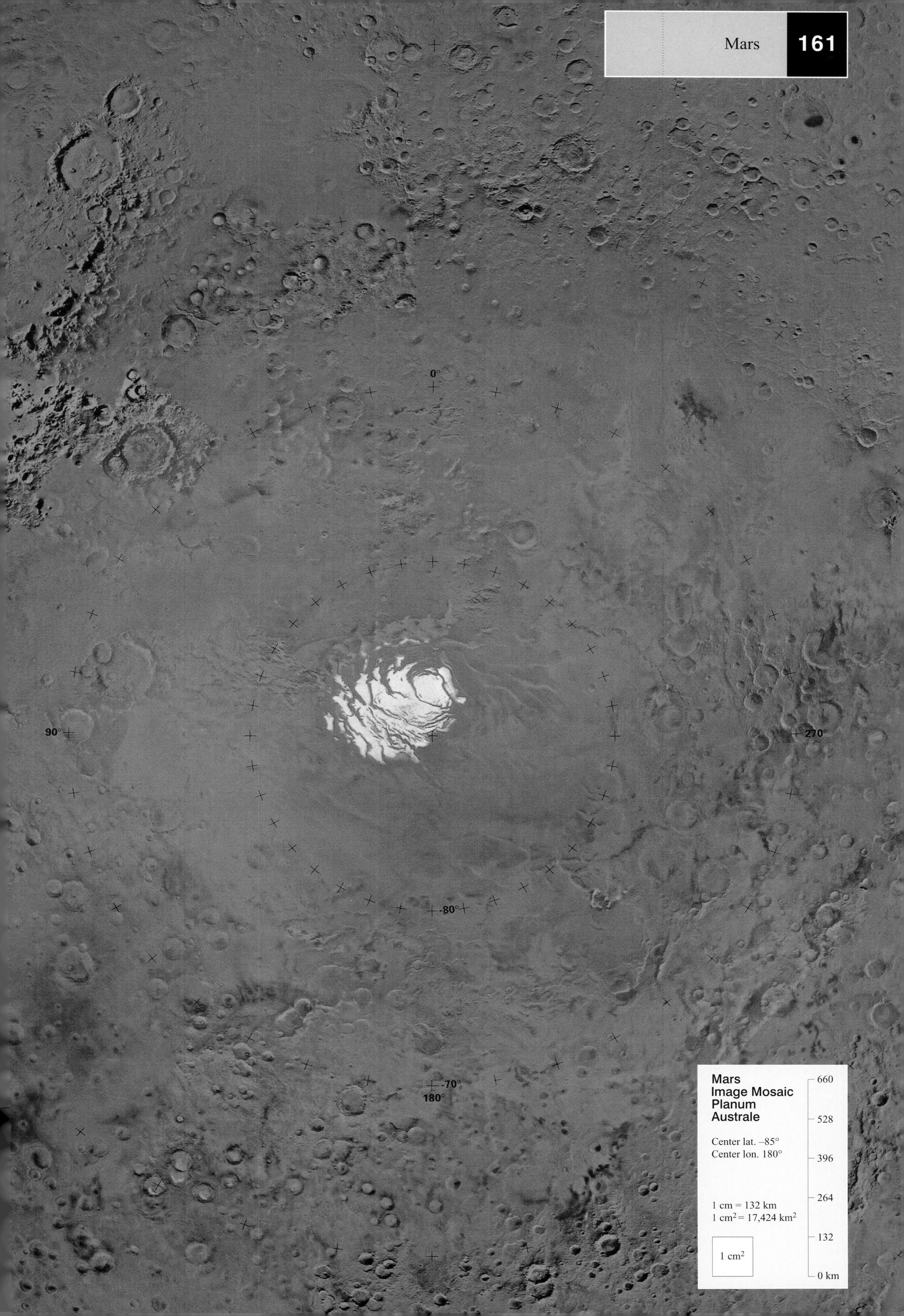
0°
90°
270°
-80°
-70°
180°
Mars
Image Mosaic
Planum
Australe
Center lat. –85°
Center lon. 180°
1 cm = 132 km
1 cm² = 17,424 km²
1 cm²
660
528
396
264
132
0 km

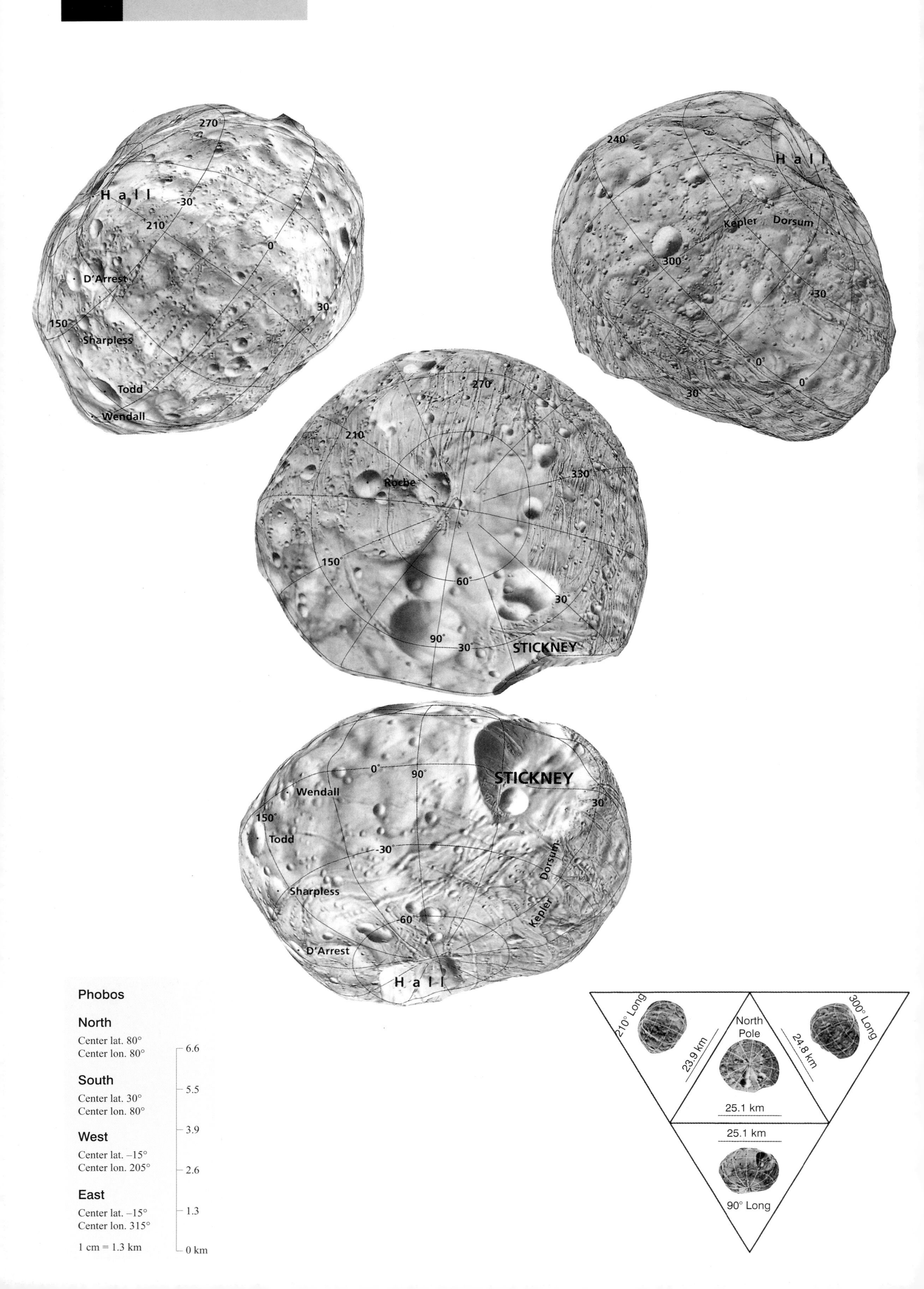
270°
Hall
-30°
210°
0°
D'Arrest
30°
150°
Sharpless
Todd
Wendall
240°
Hall
Kepler
Dorsum
300°
30°
0°
30
0°
270°
210
Roche
330°
150°
60°
30°
90°
30°
STICKNEY
0°
90°
STICKNEY
Wendall
30°
150°
Todd
-30°
Dorsum
Sharpless
Kepler
-60°
D'Arrest
Hall
Phobos
North
Center lat. 80°
Center lon. 80°
South
Center lat. 30°
Center lon. 80°
West
Center lat. –15°
Center lon. 205°
East
Center lat. –15°
Center lon. 315°
1 cm = 1.3 km
6.6
5.5
3.9
2.6
1.3
0 km
210° Long
North Pole
300° Long
23.9 km
24.8 km
25.1 km
25.1 km
90° Long

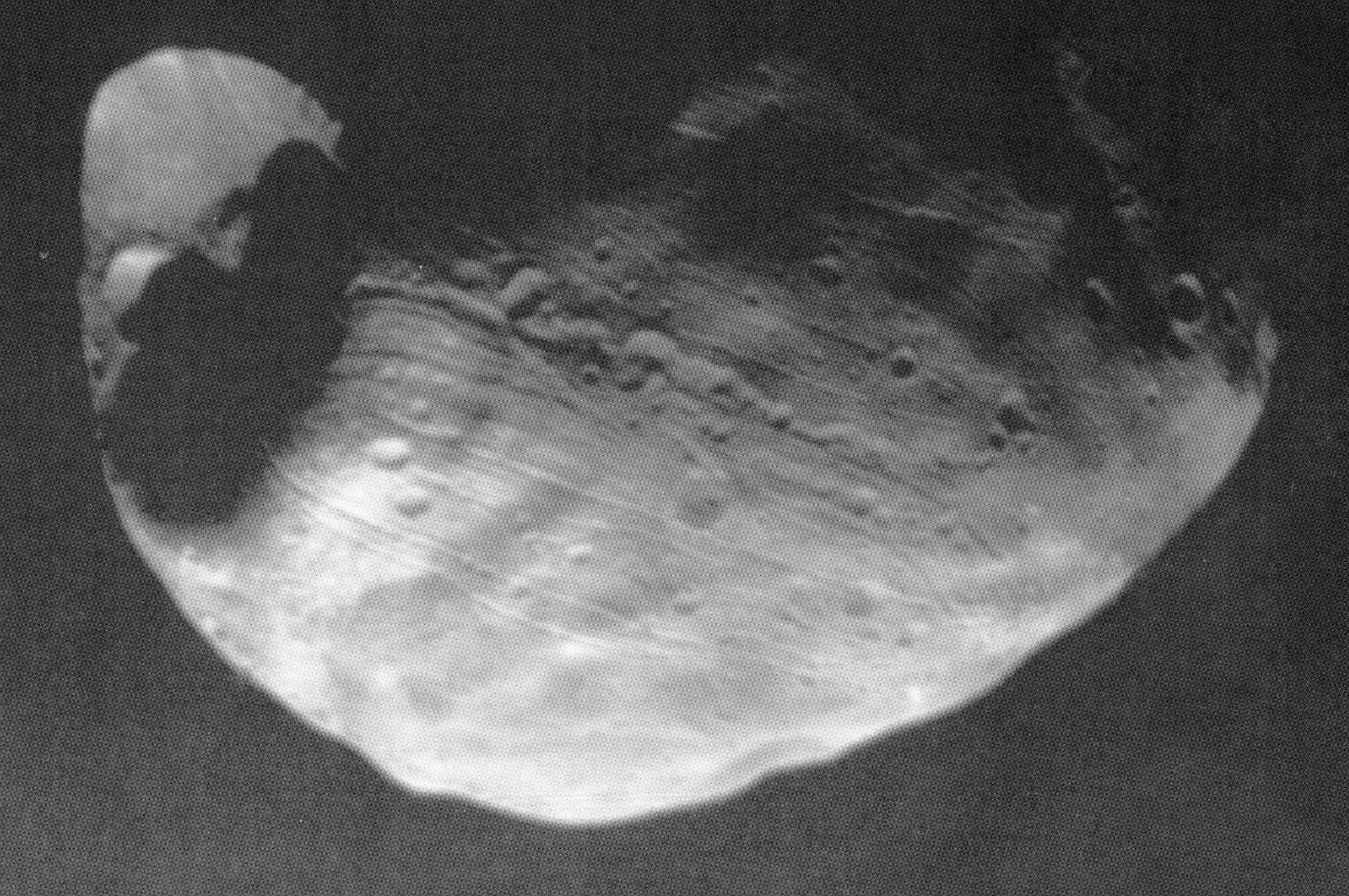

Phobos

Viking Orbiter photograph.

Jupiter System

Certainly ancient star watchers would have noticed one of the brightest objects in the sky. Because of its dominance, this object was named Jupiter after the mightiest of the Roman gods. More massive than the other planets combined, Jupiter and its rings and satellites form a "miniature Solar System." More than 300 years of telescopic observations and, more importantly, data returned from the *Pioneer*, *Voyager, Galileo,* and *Cassini* missions have yielded views of the Jovian system that are perhaps the most spectacular images of Solar System exploration to this time.

Jupiter

The Jupiter System was formed in the colder, outer part of the protosolar cloud 4.6 billion years ago. Jupiter is composed mostly of hydrogen and helium, plus small amounts of methane, ammonia, and ethane. The planet emits more than twice the amount of energy that it receives from the Sun, indicating an internal heat source. During the initial phase of formation, Jupiter was probably 10 times its present diameter, but after 10 million years the gaseous mass shrank to its present size. As the mass collapsed gravitationally, frictional heat in the gas raised temperatures to perhaps 50,000 K. Over the last 4.6 billion years, Jupiter has been slowly discharging heat into space from beneath its clouds.

The characteristics of Jupiter's interior are highly speculative. Present technology is inadequate to allow space probes to penetrate far into the clouds, nor is it easy to simulate in the laboratory the interior conditions. However, calculations based on theory can be used to surmise what the interior may be like. Traveling from the upper clouds into the interior, the density of the gas steadily increases, generating high pressures and temperatures. Toward the center of the planet, the pressure is estimated to be 100 million times greater than the atmospheric pressure on the surface of the Earth, and the temperature may reach 30,000 K. Under these conditions, hydrogen would have a density of about 4 grams per cubic centimeter and would form a "metallic" phase. Some models suggest a central solid core as large as 20,000 kilometers across, which may contain

The Jupiter system compared to the Earth/Moon system.

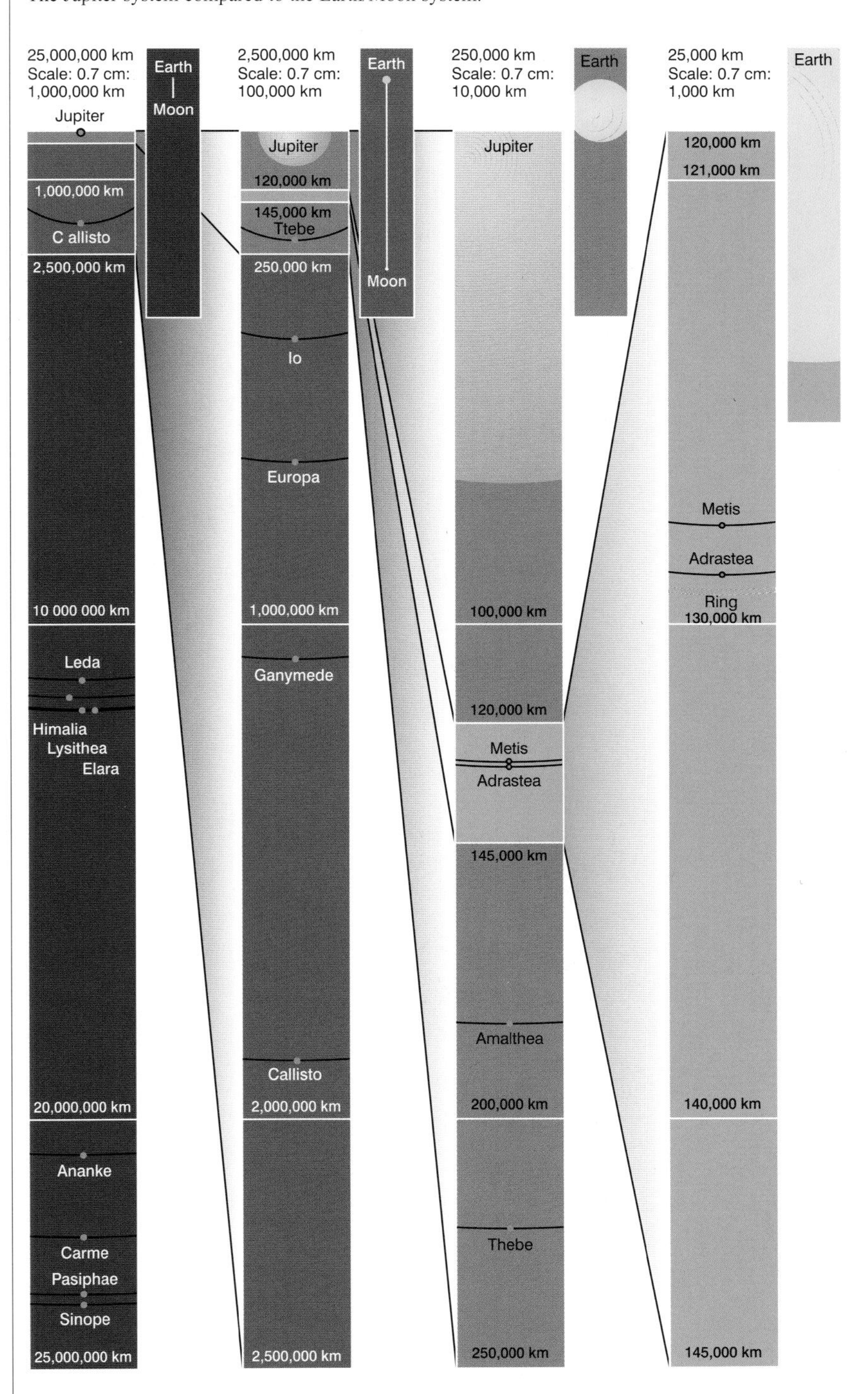

Facing page: *Voyager 2* image of the southern hemisphere of Jupiter, showing area extending from the Great Red Spot to the south pole. The presence of the white oval beneath the Great Red Spot causes the flow streamlines to distort (JPL photograph P-21737).

Voyager 1 image of the Great Red Spot showing an area about 25,000 kilometers wide. The north (top) side is partly obscured by a thin layer of ammonia clouds. South of the Great Red Spot is one of the three white ovals which first appeared in 1938, as discovered by Earth-based telescopic observations. Sequential photographs show a 6-day counterclockwise rotation for clouds on the outer edges of the Great Red Spot, whereas the center contains little apparent motion (JPL photograph P-21430C).

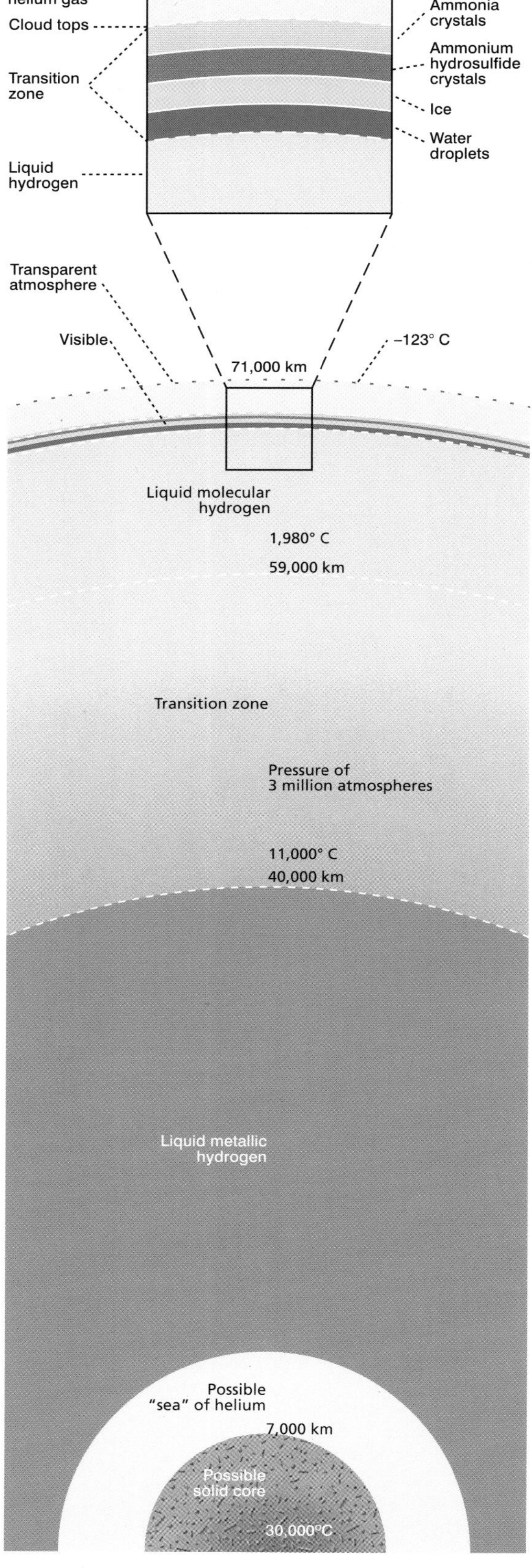

A possible "plug" taken of Jupiter from the upper clouds to its center. Although Jupiter is composed primarily of hydrogen and helium, the tremendous pressures in the interior of the planet, generated by its great mass, cause the hydrogen to be compressed to first a liquid phase, then a possible "metallic" phase, and finally a solid inner core. Heat energy is constantly transferred outward from the interior and sets up convection cells within the clouds.

small amounts of rocky material. There is probably no clearly defined surface but a gradual phase change from liquid to gas with decreasing pressure from the center of Jupiter outward into its clouds.

Despite its large size, Jupiter spins rapidly on its axis, making a complete rotation in less than 10 hours. This rapid spin causes the equator to bulge and, in combination with the presence of a "metallic" inner zone, may explain the huge magnetic field that surrounds Jupiter. It may also explain Jupiter's fascinating cloud patterns.

The cloud bands are far brighter and more intricate than suspected prior to spacecraft imaging. Sequential photographs from *Voyager* and *Cassini* taken over many weeks permitted cloud motions and speeds to be assessed. Although the major zonal "jets" (bands of westward winds and counter-flowing eastward winds) have not changed their general configuration in more than 100 years of observations, spacecraft images showed that they travel at speeds in excess of 50 meters per second and that there is incredible turbulence in the shear regions between the zones. Eddies appear, disappear, and blend together over 1 to 2 days. In addition, there are various filamentary curls, spirals, and feathery clouds that change on similar time scales. These motions apparently result from convection cells that are driven by the escape of heat from Jupiter and from solar heating.

Cloud colors span nearly the entire spectrum and include reds, yellows, browns, blues, and various shades of white. Because hydrogen and helium, the principal components of the atmosphere, are colorless, the cloud colors are attributed to the presence of small amounts of other elements and compounds such as sulfur and organic materials. The colors also seem to depend on altitude, as determined by temperature measurements. Blues represent the deepest levels and are seen only through holes in the upper clouds. Brown colors dominate the next higher level, followed by upper white clouds. At the highest and coldest levels are the red clouds, including the Great Red Spot.

The Great Red Spot was discovered by the English scientist, Robert Hooke, in the mid-seventeenth century. It varies somewhat in size and can be as long as 40,000 kilometers

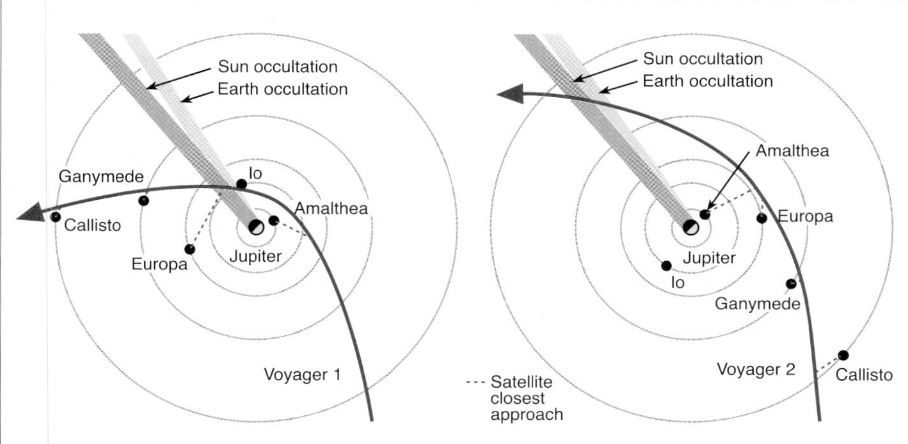

The passage of *Voyager 1* and *Voyager 2* through the Jupiter system and their "encounters" with the Galilean satellites. Also shown is one of the smaller inner satellites, Amalthea.

Voyager 1 color composite image showing detail in atmospheric structure near the Great Red Spot, part of which is seen in the upper right of the image. This view was obtained when the spacecraft was 5 million kilometers from the planet (JPL photograph P-21182).

Cassini and *Voyager* images of Jupiter (full disk, northern pole, southern pole). Distinctive banded structures have letter codes. The Great Red Spot is the hurricanelike pattern located in the southern hemisphere. Velocities in the atmosphere and the directions of motion are indicated for the latitudes in degrees where the motion was assessed.

Voyager images of Jupiter showing its appearance on March 1, 1979 (top) and early July (bottom). The two pictures show relative motions of the various bands and vortices (JPL photograph P-21771C).

NTB = North temperate belt,
NTrZ = North tropic zone,
NEB = North equatorial belt,
EZ = equatorial zone,
SEB = South equatorial belt,
STrZ = South tropic zone,
STB = South temperate belt.

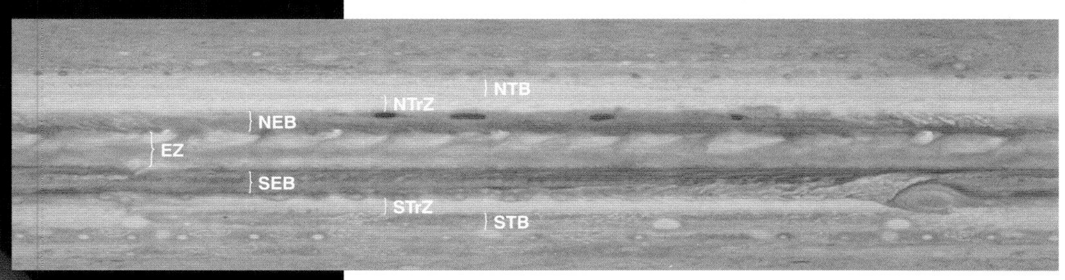

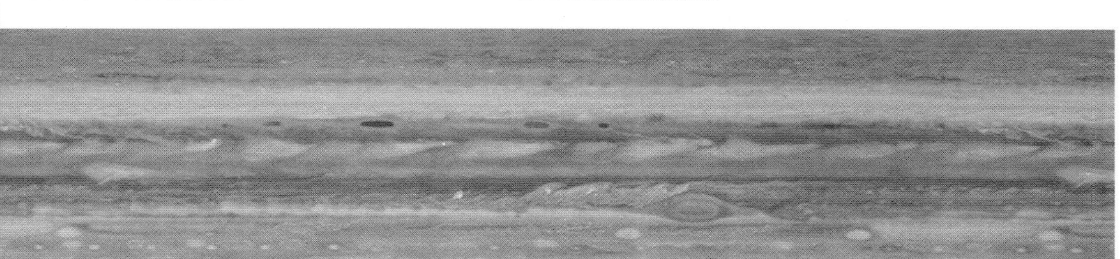

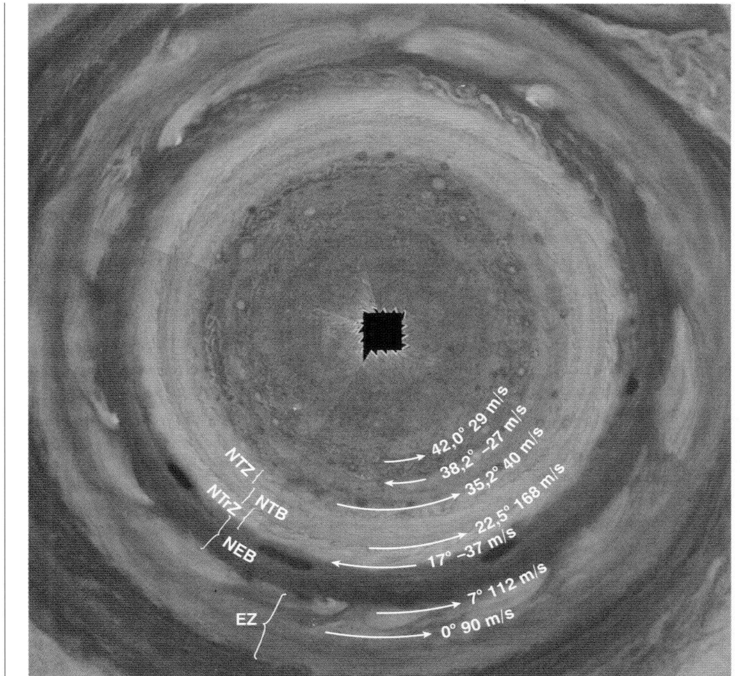

Jupiter, northern pole.

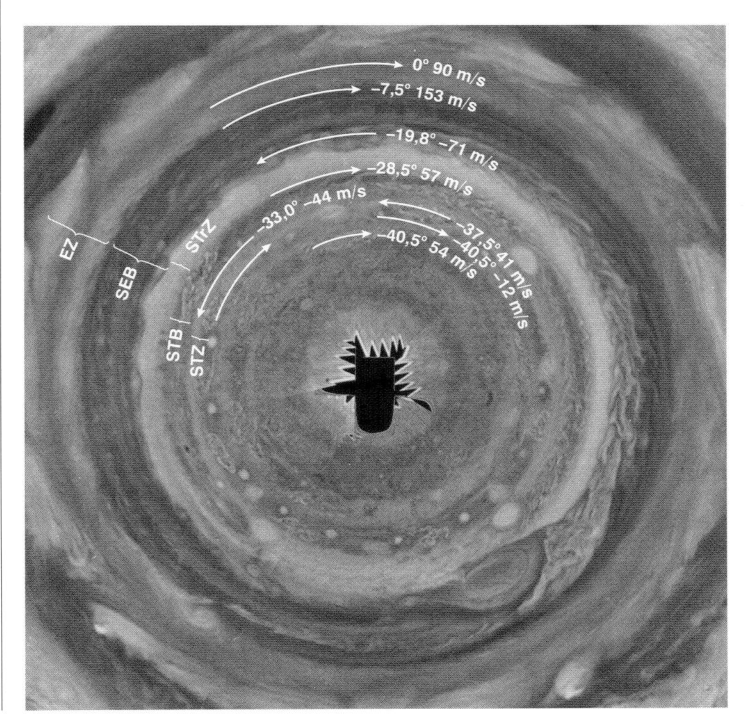

Jupiter, southern pole.

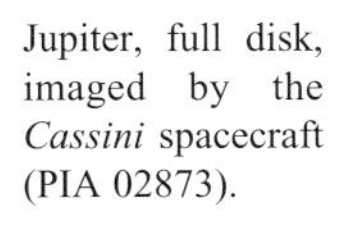

Jupiter, full disk, imaged by the *Cassini* spacecraft (PIA 02873).

Composite of *Voyager 2* images taken from inside the shadow of Jupiter and showing the planet and the position of the ring system (JPL photograph P-21774B/W).

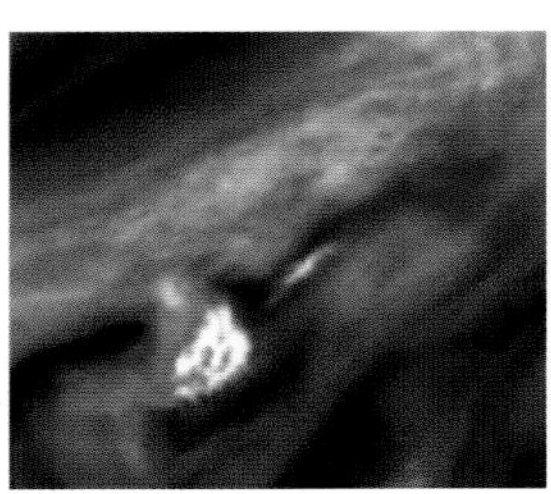

Water cloud thunderstorm near the Great Red Spot, seen from *Galileo*. The white cloud is 1000 kilometers across (NASA PIA 01639).

and as wide as 14,000 kilometers – several times larger than the Earth. It rises some 24 kilometers above the surrounding clouds. The Great Red Spot has been described as an enormous maelstrom that consumes smaller storms and cloud vortices. A vast convective upwelling of the atmosphere, it may involve a "weather cycle" of about 6 years, during which it changes its size and colors. For example, during the *Pioneer* spacecraft encounters in 1973 and 1974, the Great Red Spot was enveloped in a white diffuse cloud that was mostly absent during the *Voyager* encounters in 1979.

There is considerable debate about the trace chemical coloration of the Great Red Spot, but phosphorous probably composes its upper clouds. One model suggests that an oxide of phosphorous (phosphine) is drawn upward into the Great Red Spot from the interior of Jupiter by convection. Ultraviolet light from the Sun then causes a chemical reaction that splits the phosphine apart and generates pure phosphorous, which is red in the cold environment of the upper clouds of Jupiter. Minor amounts of organic compounds may account for some of the other visible colors.

In addition to the Great Red Spot, numerous smaller "spots" occur within the atmosphere of Jupiter, many of which may have short lifetimes. For example, in 1938, three white spots, each about the diameter of Earth's Moon, were discovered telescopically near the Great Red Spot where no previous structures had been observed. These and similar features evidently are zones of local atmospheric turbulence.

Galileo false-color image showing Jupiter's oval clouds at 30° south latitude and 100° west longitude (NASA PIA 00700).

Rings

Prior to the Space Age, only Saturn was known to have rings. Data from the reconnaissance mission of *Pioneer 11* suggested that a ring might be present around Jupiter. *Voyagers 1* and *2* were programmed to photograph the area, and through careful processing, the pictures did show the presence of a faint ring disk. The ring system was again imaged by *Galileo* during its multi-year mission. The system consists of three main parts: a bright ring, a diffuse disk, and a halo. The bright ring has a width of about 6,000 kilometers and occurs at a distance of 58,000 kilometers above Jupiter. The diffuse disk extends downward from the bright ring toward Jupiter. Both the bright ring and the diffuse disk are very thin; viewed on edge, they appear to be less than 30 kilometers thick. The halo component extends vertically above and below the two rings at a thickness of possibly 20,000 kilometers.

Particles in the rings travel very rapidly around Jupiter; one complete orbit takes 5 to 7 hours. The optical properties of the ring system suggest that the rings are composed of tiny dust-size grains about the size of smoke particles. The particles are very dark and are probably composed of silicate materials. Because of their small size and rapid motion, the particles should erode quickly and be pulled out of orbit into the atmosphere. The rings are thought to be constantly replenished with dust, generated by impacts on the small Jovian satellites. For example, spacecraft images revealed that two small satellites, 40-kilometer-diameter Metis and 25-kilometer-diameter Adrastea, bound the bright ring on either side. Erosion of these satellites by meteoroidal impacts adds dust particles to the rings.

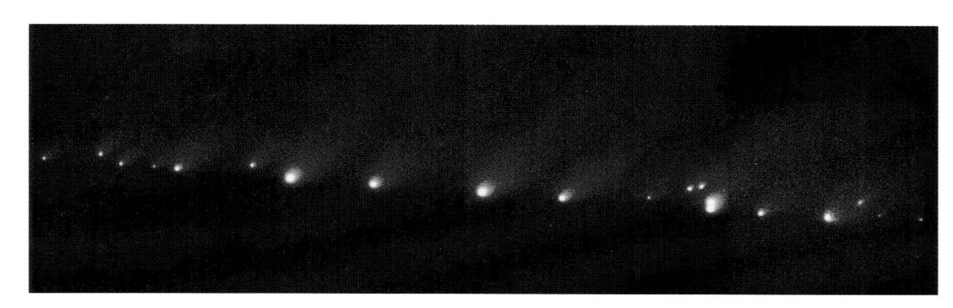

Hubble Space Telescope view (red filter) of Comet Shoemaker-Levy 9 taken May 1994, 22 months after being captured by Jupiter, and 2 months before hitting Jupiter. Stresses on the comet during the capture caused it to fragment into many pieces (Space Telescope Science Institute image).

Comet Shoemaker-Levy 9 Impact

Sometime in the first third of the 20th Century, one of the many comets that reside beyond the orbit of Pluto was jogged out of its normal path and drawn toward the sun. The comet began a series of chaotic orbits that carried it past Jupiter about every two years. Named for co-discoverers Eugene and Carolyn Shoemaker and David Levy, in July 1992 it came so close to Jupiter that it was ripped into more than 23 fragments, the largest of which was about 700 meters across. The chunks of rock and ice were strung into a chain some 7 million kilometers long. Their path carried them on a collision course with Jupiter and during six days in July, 1994 the chunks began slamming into the atmosphere of Jupiter at speeds of 60 kilometers per second.

Although the area of impact on Jupiter was not visible from Earth, Jupiter and its atmosphere rotated into view so that the effects of the collision could be observed. Fortuitously, at the time of the impacts, the *Galileo* spacecraft was on its way to Jupiter and in a position to observe the impact zone. *Galileo* data, combined with images from the *Hubble Space Telescope* and Earth-based observatories, provided new information on what happens during a major impact.

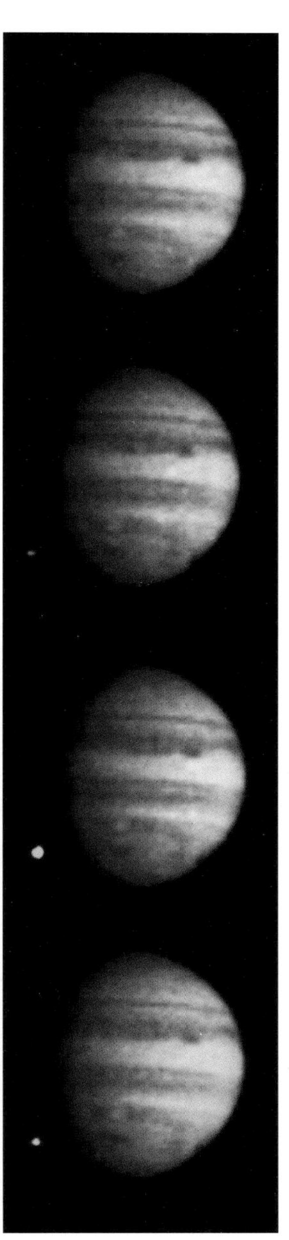

A direct view of the impact of fragment W was afforded by the *Galileo* spacecraft, en route to Jupiter, while still some 238 million kilkometers away. This sequence of images was obtained July 22, 1994. Images were taken at intervals of 2 1/3 seconds (JPL photograph P-44542).

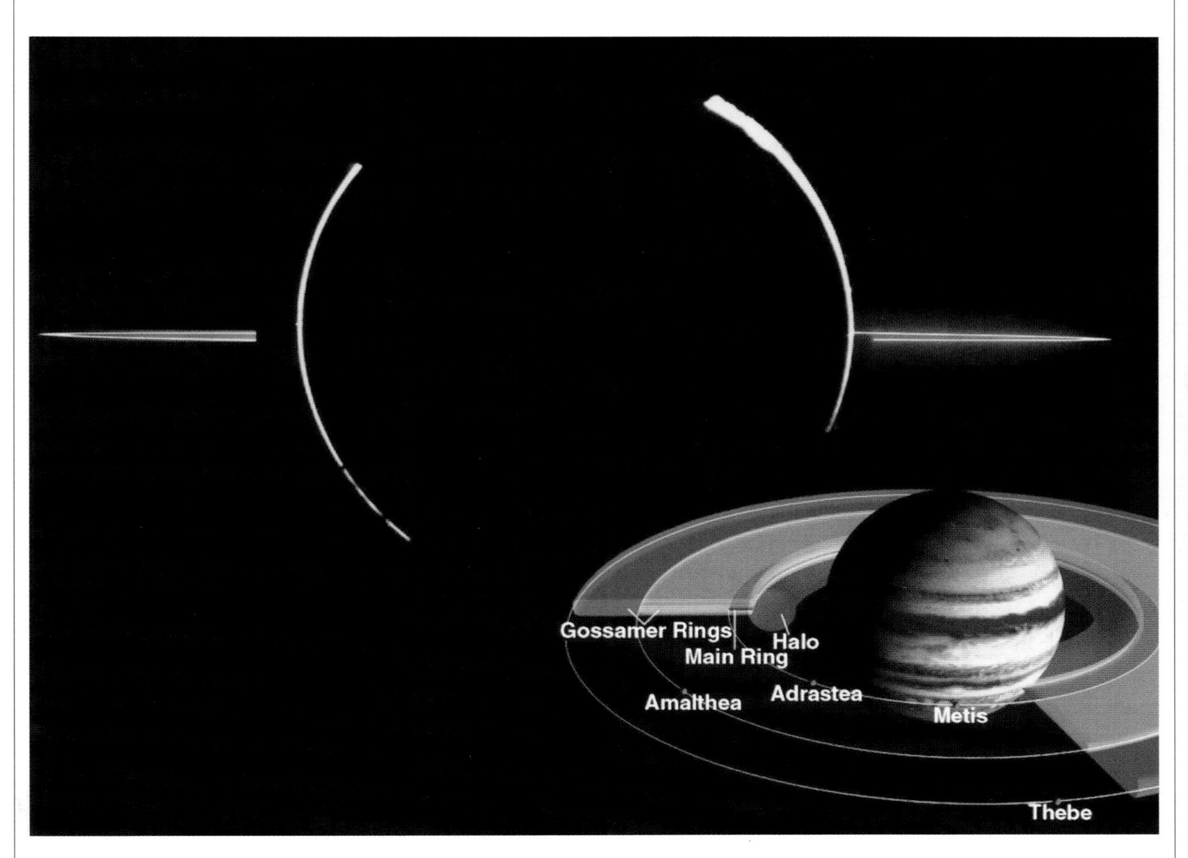

Galileo images showing Jupiter's faint ring system and diagram of relationships with small satellites (NASA PIA 03001).

Diagrams of the Galilean satellites drawn to the same scale as the terrestrial planets for comparison. The curved zones represent Jupiter and the Sun, respectively.

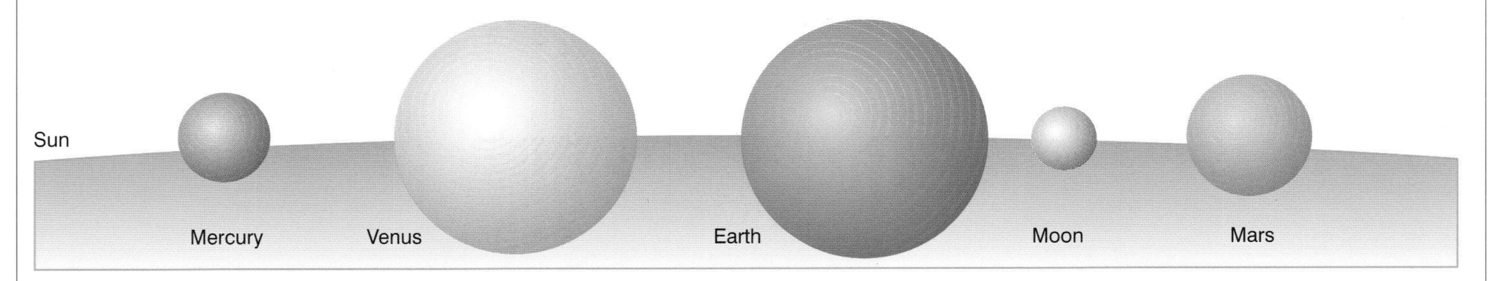

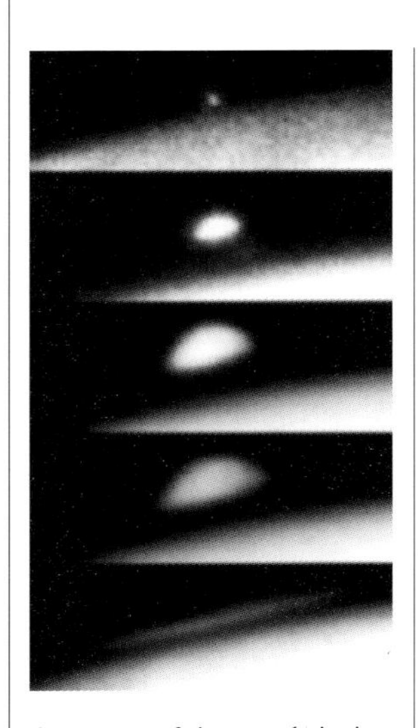

Sequence of images beinning five minutes after the impact of fragment G of Comet Shoemaker-Levy 9 into Jupiter. A plume of material rises over Jupiter's limb in this sequence taken over a period of 18 minutes. Because the impact was just beyond the limb of the planet, the lower part of the plume is in Jupiter's shadow. Images taken by the Hubble Space Telescope. (Space Telescope Science Intsitute).

As each cometary fragment plunged through the atmosphere it exploded into an enormous fireball, leaving a rising column of hot gases from Jupiter and cometary debris in its wake. This column formed a plume that rose some 3500 kilometers above the cloud tops. As the energy within the plume was spent, the material splashed back into the atmosphere, covering an area the size of the Earth, heating more gas, and creating another, smaller fireball. The zones where each impact took place were marked by dark blotches in the clouds of Jupiter, some of which remained for many months.

Based on the Shoemaker-Levy 9 impact, there is speculation that some of the spots seen in pictures of Jupiter's atmosphere may represent former impacts by comets and asteroids, and that some of these features could remain as atmospheric disturbances for centuries. What would happen if such an impact occurred on Earth? The effects would be devastating. Among the lessons learned from the Shoemaker-Levy 9 impact is the realization that huge fireballs are created by impacts, especially on planets with atmospheres.

Small Satellites

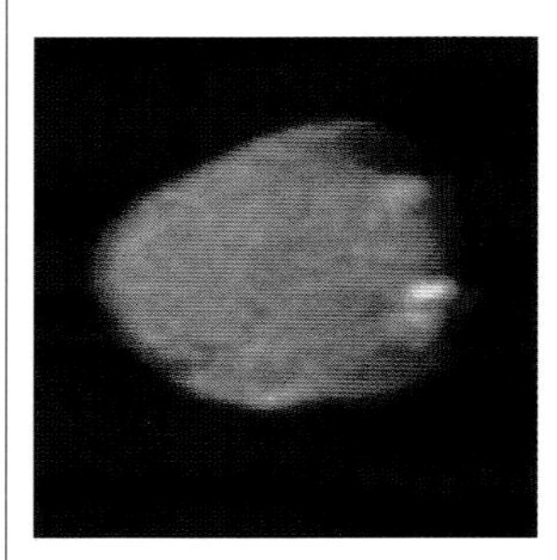

Voyager 1 image of Amalthea showing its dark red appearance and the presence of craters, presumably formed by impact collision (NASA P-21223C).

Adrastea and Metis are but two of a host of small satellites orbiting Jupiter. Amalthea, perhaps best known, is a dark reddish, cratered object measuring 270 by 165 by 150 kilometers and was first imaged by *Voyager 1*. Its largest crater, Pan, is more than 90 kilometers wide. In addition to craters, numerous grooves and ridges tens of kilometers long were found on the surface. Its red color may result from sulfur volcanically erupted from Io, the innermost of the Galilean satellites. Amalthea orbits between Jupiter and Io, and material exploded from Io is gravitationally drawn toward Jupiter and into the path of Amalthea. Thebe (~50 kilometers diameter), the other small inner satellite, appears between the orbits of Amalthea and the larger moons, called the Galilean satellites.

The Galilean satellites separate the inner and outer small moons of Jupiter. The outer small satellites can be divided into two groups. Himalia, Elara, Lysithea, and Leda are all in prograde orbits (traveling in the same direction as the rotation of Jupiter), about 11.5 million kilometers from Jupiter. Their orbits are inclined to the equator of Jupiter between 26° and 29°. The other group is composed of Pasiphae, Carme, Sinope, and Ananke. These satellites are about 22 million kilometers from Jupiter and are all in retrograde orbits (traveling in the direction opposite to Jupiter's rotation) that are inclined between 16° and 33°. All eight satellites are dark and have properties suggestive of carbonaceous materials similar to the Trojan asteroids, discussed in the last chapter. Consequently, they might have been gravitationally captured by Jupiter. Alternatively, the two groups may be fragments derived from two larger parent bodies, one at 11.5 million kilometers in prograde orbit around Jupiter, the other at 22 million kilometers in retrograde orbit.

In late 2000, ten additional small moons were discovered by telescopic observations. All of these satellites are small, with diameters ranging from 3 to 8 kilometers. These discoveries bring the total number of Jovian satellites to 28.

The Galilean Satellites

On January 7, 1610, Galileo turned his primitive telescope toward Jupiter and noted "three little stars, small but bright." Later observations revealed a fourth object in proximity. From their motion in relation to Jupiter, Galileo deduced that they were satellites, a revelation that contributed to the demise of the "Earth-centered" notion of the Solar System. The satellites were later named by Simon Marius, a German astronomer, for the various lovers of Jupiter in Greco-Roman mythology. Io is closest to Jupiter, followed by Europa, Ganymede, and Callisto. Marius claimed to have sighted the satellites one month earlier than Galileo, but these observations were not documented at the time.

The moons are known collectively as the Galilean satellites, except in Italy where they are often referred to as the Medician satellites. Galileo suggested the name in recognition of the financial support that he received from the powerful Medici family.

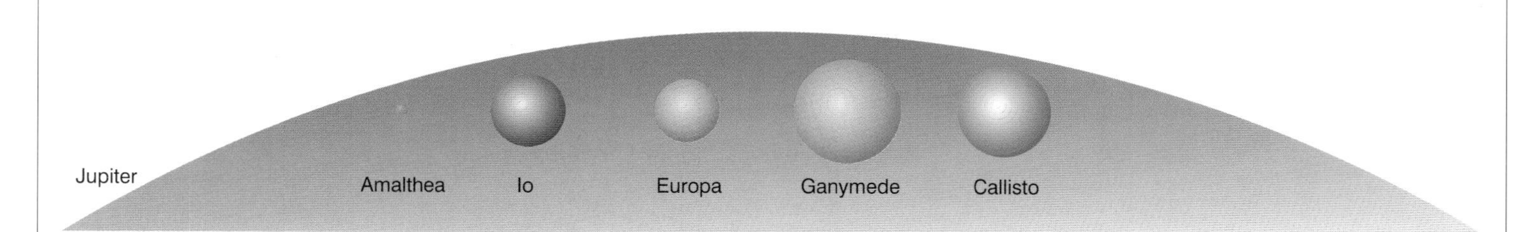

Like Earth's Moon, all four Galilean satellites are in synchronous rotation, always oriented with the same side (the Jovian hemisphere) facing Jupiter. The opposite side is referred to as the anti-Jovian hemisphere. In addition, the satellites have a leading hemisphere, the forward-facing side in orbit, and a trailing hemisphere, the aft-facing side in orbit.

The Galilean satellites can be divided into two groups. Io and Europa are the innermost objects. Both are about the size and density of Earth's Moon and are composed predominantly of rocky material, although Europa is encased in a water-rich shell 150 kilometers thick. Ganymede and Callisto, the outer two moons, are about the size of Mercury. Their low density suggests water or water-ice composition.

One of the most spectacular discoveries in Solar System exploration is the existence of active volcanoes on Io. The volcanoes were predicted shortly before their discovery from *Voyager 1* images. Planetary scientists had calculated that frictional heat should be generated within Io as a consequence of its being constantly "push-pulled" by the gravity fields of Jupiter and Io's neighboring satellite, Europa. They suggested that enough heat would be generated in Io to melt the interior and that active volcanoes could be present on the surface. Two weeks after their prediction was published, *Voyager 1* returned images that showed volcanic eruptions sending explosive debris more than 200 kilometers above the surface of Io. Subsequent images from *Voyager* and *Galileo* revealed that the surface of Io is dominated by volcanic features, including vast lava flows, shield volcanoes, and calderas more than 65 kilometers wide. Io is the only solid-surface planetary object thus far to lack impact craters. The vigorous volcanic activity probably "resurfaced" and buried the impact craters that may have once existed.

Galileo images for Europa show a bizarre surface with a complex pattern of interweaving ridges and grooves and areas where the icy crust has been disrupted. With few impact craters identified, the surface is relatively young, like that of Io. A layer of water ice or liquid water probably lies beneath the surface. Fracturing and disruption of the surface crust is probably induced by tidal flexing.

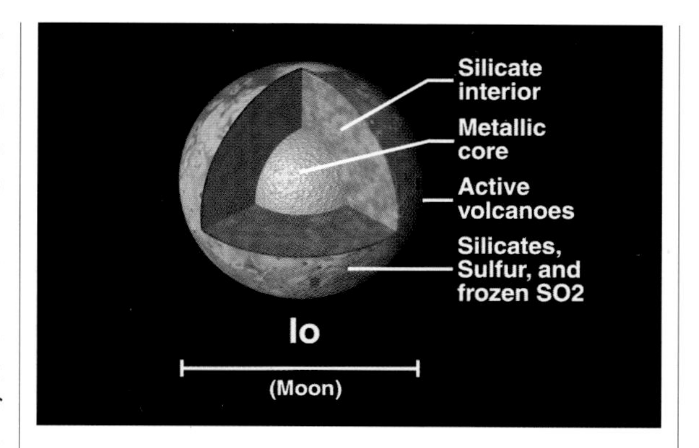

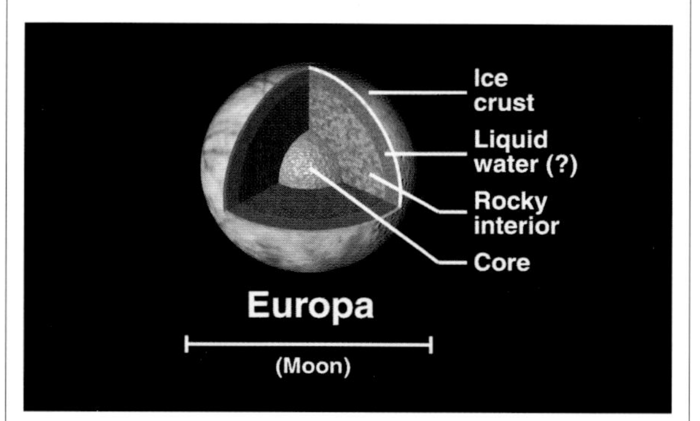

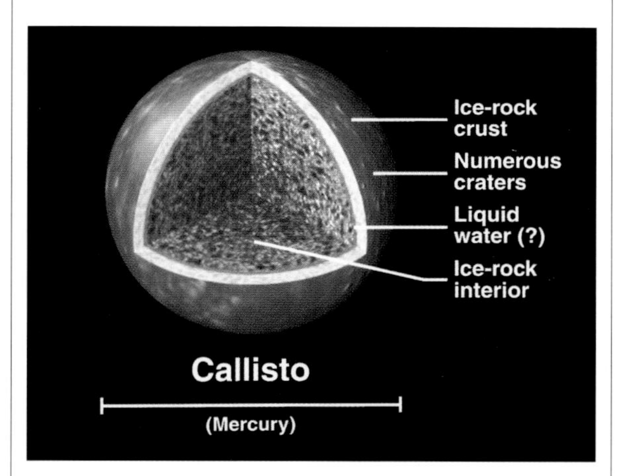

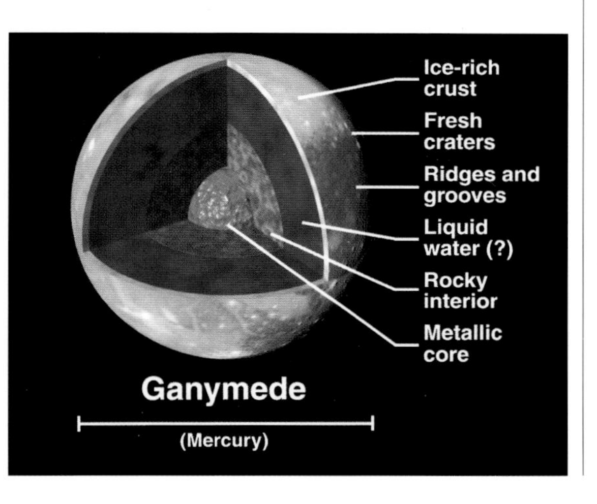

The interiors of the Galilean satellites modeled on the basis of their sizes and densities (courtesy Torrence Johnson).

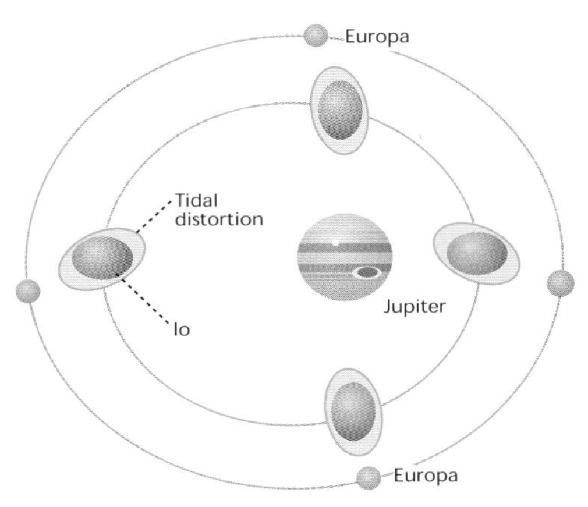

The orbit of Io and Europa in relation to Jupiter (not to scale). The varying gravitational "pull" of Jupiter and Europa due to the noncircular orbit of Io causes tidal flexing of the satellite. This, in turn, generates internal friction and heat, leading to the active volcanoes seen on Io via Voyager images (Io not drawn to scale).

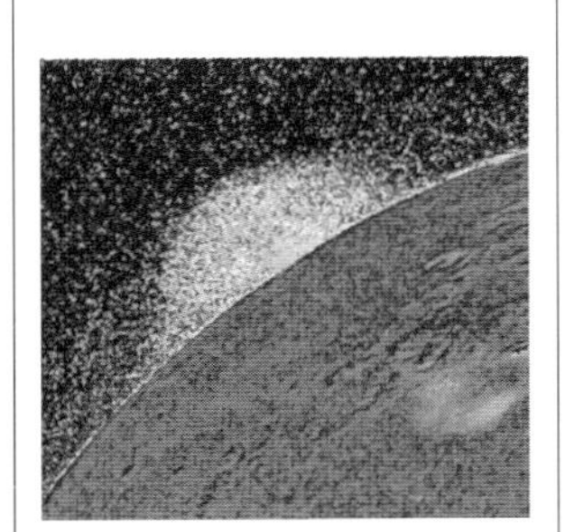

Computer-enhanced enlargement of *Voyager 1* image showing the volcano Loki on Io (JPL photograph P-21334).

Despite their similar size and density, Ganymede and Callisto display strikingly disparate surfaces that reflect different evolutionary histories. Ganymede includes both ancient, heavily cratered terrain and younger, highly fractured regions indicative of internal activity and tectonic processes. In contrast, the surface of Callisto consists entirely of geologically old, cratered crust essentially inactive since its formation some 4.6 billion years ago. Although the explanations for the disparity between the objects are speculative, Ganymede probably received a greater proportion of radioactive elements during its early formation than Callisto; hence, it generated more internal heat. It may also have been subjected to tidal stresses similar to those exerted within Io. Such tidal stresses would have been absent in Callisto and may have contributed sufficient heat to Ganymede to permit internal melting and disruption of its surface.

The Jupiter system was the first of the outer planets to be explored by spacecraft. The revelations provided by the *Pioneer*, *Voyager,* and *Galileo* missions only previewed discoveries to be made in the journeys of exploration outward from the Sun. Each of the four Galilean satellites is presented in detail.

Io

Since its discovery nearly four centuries ago, Io has been studied intensively by telescope to determine its general properties. Io has an overall density of 3.53 grams per cubic centimeter and a diameter of 3,640 kilometers, suggesting that Io is composed mostly of rock, probably of silicate materials. Io is the reddest object seen in the Solar System. Its poles are generally dark, while the equatorial region is bright. In addition, the leading hemisphere is brighter than the trailing hemisphere. The surface of Io displays a wide variety of colors – pale shades of red, yellow, orange, and brown. Many of the colors are attributed to sulfur or mixtures of sulfur, sulphur dioxide frost, and sulfurous salts of sodium and potassium on the surface.

Much of the surface of Io was photographed with good resolution by the *Voyager* and *Galileo* spacecraft. The surface can be subdi-

Facing page: *Cassini* image of Jupiter and one of its Galilean satellites, Io and the dark shadow cast by Io on Jupiter's cloud deck. Io's orange color is attributed to the presence of sulfur and sulfur compounds on its surface.

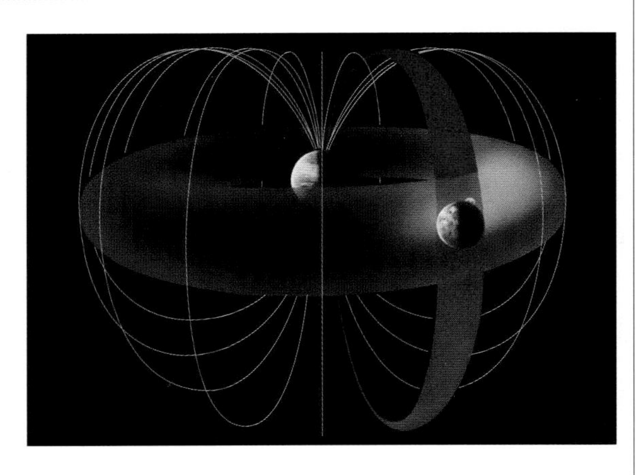

Io's sulfurous volcanoes spew material into Jupiter's immense magnetosphere, resulting in a glowing sodium cloud and the globe-encircling sulfur/oxygen torus. Strong radio bursts, energetic currents in a magnetic flux tube, and Jovian auroras are among the fascinating physical and chemical phenomena influenced by Io (University of Arizona Press and NASA).

Galileo image of Io showing the erupting plume of the volcano, Masubi, on the horizon (JPL photograph PIA 02502).

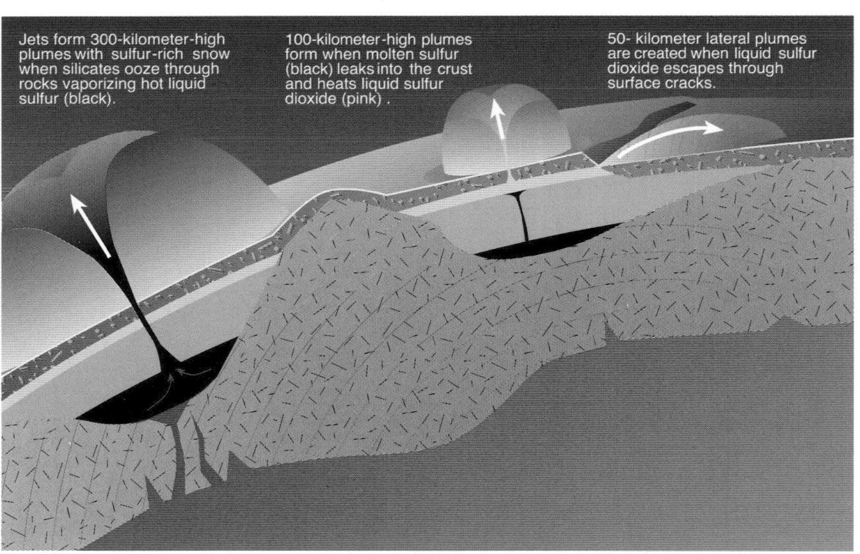

Jets form 300-kilometer-high plumes with sulfur-rich snow when silicates ooze through rocks vaporizing hot liquid sulfur (black).
100-kilometer-high plumes form when molten sulfur (black) leaks into the crust and heats liquid sulfur dioxide (pink) .
50- kilometer lateral plumes are created when liquid sulfur dioxide escapes through surface cracks.

Typical surface features on Io (including an active volcanic plume) and an idealized cross--section of the interior of Io.

Facing page: false-color *Voyager* image of the Ionian volcano, Pele, as it sprays particles 300 kilometers above the surface (NASA P-26276, courtesy of U.S. Geological Survey, Flagstaff).

vided into four principal terrains and landforms: mountains, plains, vent areas, and flows. The plains, vents, and flows are all related to volcanic processes, but the mountains may not be. The mountains form high-standing, rugged blocks, some of which are more than 100 kilometers wide and rise 16 kilometers above the surrounding plains.

Plains of several types are widespread on Io. Although most plains are pale yellows and brown, those in the polar areas are dark brown to black. Smooth plains with low scarps are thought to be composed of volcanic ash interbedded with lava flows and sulfurous salts deposited from volcanic gases. Mesas, cliffs, and pits in some plains probably represent periods of erosion and deposition. Because Io lacks an atmosphere and is very cold, the erosion cannot result from running water or wind, but may be a result of "sapping" processes in which liquid sulfur dioxide is the main erosional agent. Subsurface sulphur dioxide under pressure can be released by fractures in the crust. This release may have produced droplets of sulphur dioxide snow sprayed tens of kilometers from the fracture, possibly explaining the numerous bright patches in many areas. Withdrawal of support from beneath the solid crust as the sulphur dioxide is released would cause the surface to collapse, forming irregular pits and "etched" surfaces.

More than 300 active and inactive volcanic vents have been identified on Io. Vents include huge calderas as wide as 65 kilometers in the south polar region. The calderas show histories involving multiple eruptions and enlargement by collapse. Many of the floors are very dark, suggesting the presence of molten material, perhaps liquid silicates or sulfur. Pictures of some calderas show changes that occurred in the 4-month interval between the encounters of *Voyager 1* and *Voyager 2* and between *Galileo* flybys.

The distribution of volcanoes on Io is much more random than on Earth, Earth's Moon, or Mars, where tectonic patterns influence the location of volcanism. Volcanism on Io is not controlled by simple convection cells in the interior but is driven by a more complex process involving frictional heat generated within Io.

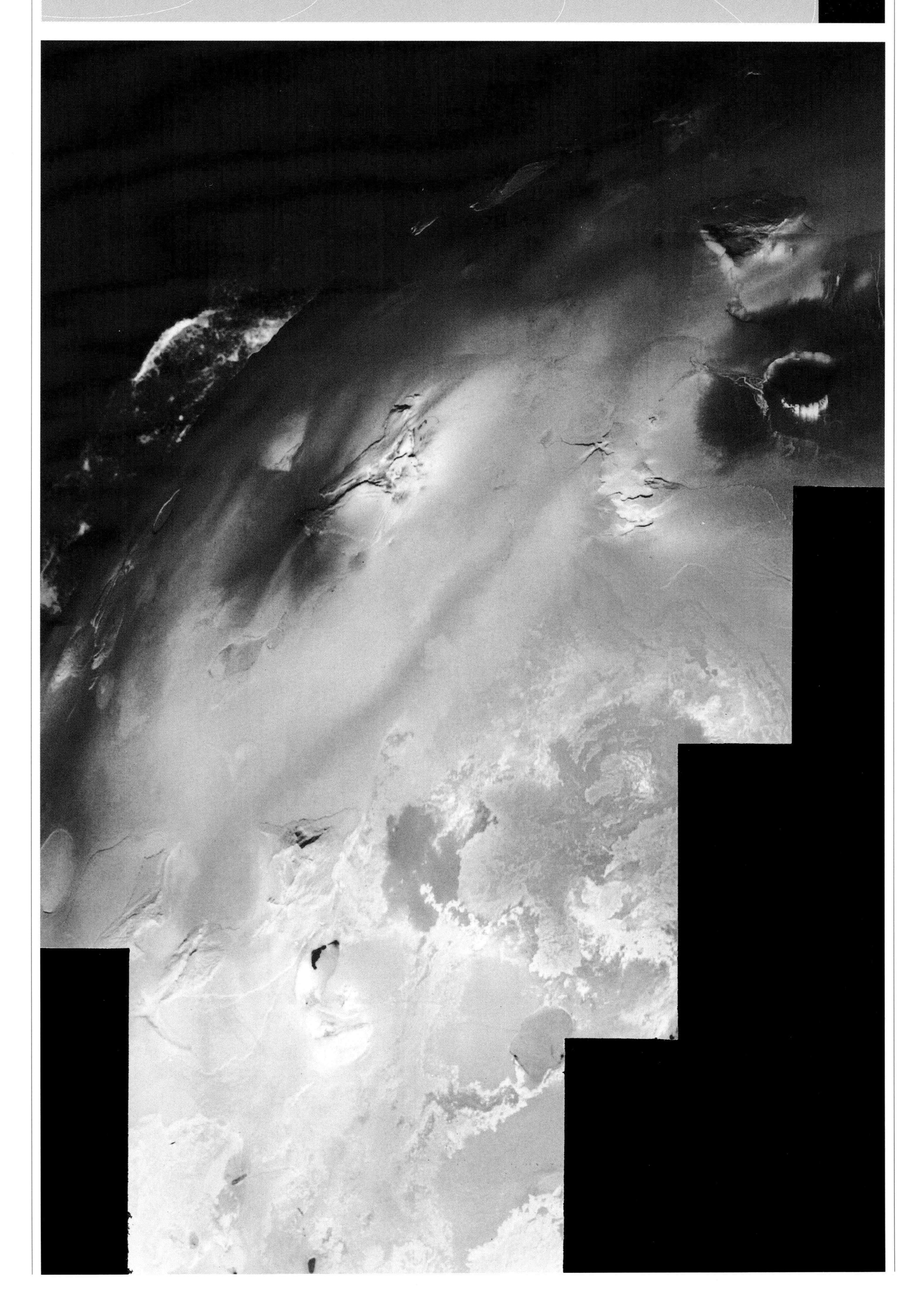

Voyager image (top) and *Galileo* image (bottom) of Io showing its mottled, reddish surface. Most of the features seen on the surface are attributed to volcanic processes; north is to the top (top: U.S. Geological Survey; bottom: NASA PIA 02309).

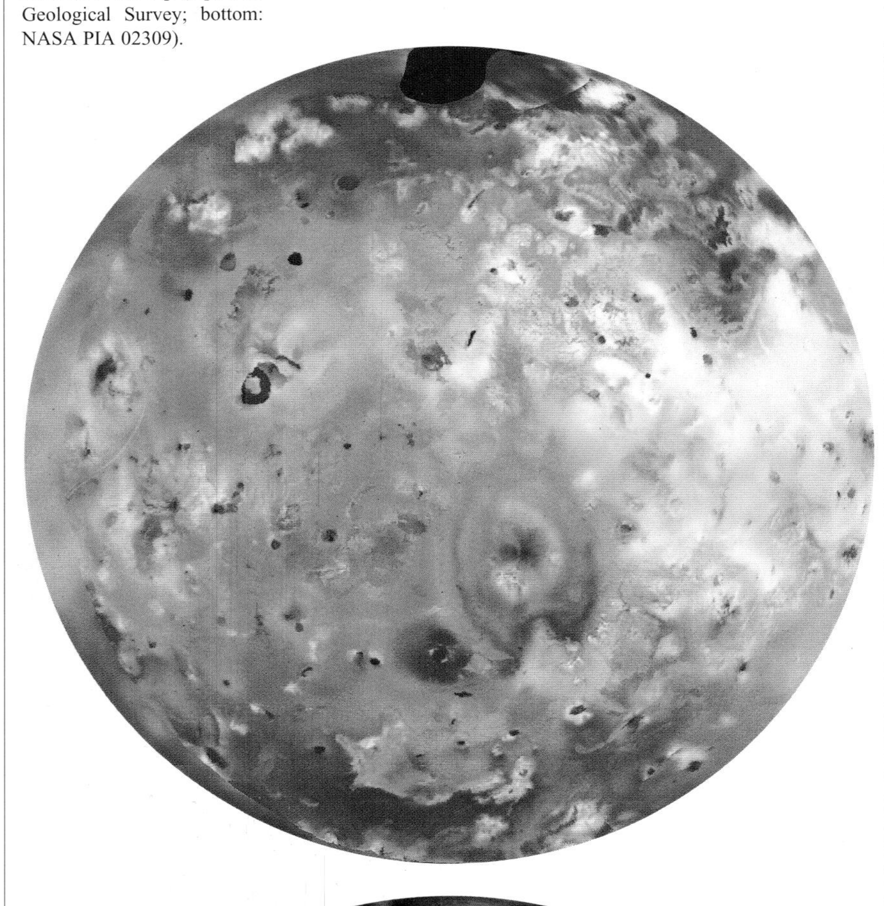

Voyager spacecraft completed the first reconnaissance of the outer Solar System, visiting all of the outer planets except Pluto. Images taken by *Voyager 1* and 2 enabled the first maps to be made for the outer planet satellites (NASA P-19727).

Nine active volcanoes were seen during the *Voyager 1* flyby, and many more were detected by *Galileo*. The most vigorous eruptions sprayed sulfurous gases and solids as high as 300 kilometers above the surface at velocities up to 1 kilometer per second. Ash and other particles rained onto the surface as far as 600 kilometers from the vents.

In addition to explosive volcanoes, several different types of volcanic flows exist, including some as hot us 1600ºC. Such high temperature lavas are thought to have occured on Earth 2–3 billion years ago, but not since then. Some flows are narrow sinuous features as long as 300 kilometers, which radiate from small vents. Others, emanating from calderas, form sheet-like flows 700 kilometers long. Considerable controversy has arisen regarding the composition of the lava flows on Io. Remote-sensing data indicate that sulfur and sulfur-rich compounds are present, but these may be only thin surface deposits overlying flows of silicate rocks. The forms of the vents and flow materials are typical of silicate volcanoes seen on Earth; however, some flows could be composed entirely of sulfur. It is unlikely that the controversy regarding the flows will be resolved until measurements of their compositions can be made directly on Io, unlikely for many decades. Regardless of composition, the volcanic flows have accumulated to form several types of volcanoes including shields, domes, fissure volcanoes, and small cones.

For a planetary object that is so volcanically active, Io displays few features that can be attributed to tectonic processes. Some cliffs, such as in Nemea Planum, may be faults. However, these cliffs and similar scarps are not associated with any obvious volcanic vents, nor do they form patterns diagnostic of large-scale deformation such as terrestrial plate tectonics.

Because most of the surface is relatively young (indicated by the lack of impact craters), it is difficult to derive more than a general geologic history of Io based on information available in 2001. Io is assumed to have formed at the same time as the other satellites in the Jupiter system. Because Io is a highly active planetary object, it undoubtedly has been chemically differentiated through geologic time.

Based on current volcanic activity, the entire mass of Io may have been recycled, or turned "inside-out." Volatiles, such as water and carbon dioxide, have long been lost, while most heavier materials have sunk to form a core. Sulfur and various sulfur compounds, aided perhaps by silicate magmas, are constantly being recycled, forming the complex surface.

Europa

Europa has an overall density of 3.0 grams per cubic centimeter. Its interior is separated into a silicate core surrounded by a 150 kilometer-thick shell composed predominantly of soft ice or water. The surface of Europa is the brightest of the Galilean satellites and is considered to be mostly water ice. The brightness and colors of the surface are not uniform, which suggests that the ice includes various impurities. These impurities could include various salts and silicates erupted onto the surface from the interior or derived from exterior sources, such as meteoritic material.

Europa has little topography, such as mountains, but has a variety of colorful terrains, along with networks of grooves and ridges. The polar areas are rather bright with a relatively dark trailing hemisphere. The dark appearance might result from the bombardment of the surface by ions originating from Jupiter's magnetosphere. Plasma within the orbital path of Europa travels faster than the satellite, ultimately attaching to Europa, implanting ions in the ice and darkening the surface of the trailing hemisphere.

Although not numerous, impact craters are seen on Europa; some, such as Pwyll crater, are as large as a few tens of kilometers across. They have raised rims, central peaks, and traces of ejecta deposits. Other impact structures include large, flat, circular areas 100 kilometers or wider, such as Tyre.

As noted on Io, the scarcity of impact craters on Europa denotes a young surface. Like Io, Europa experiences tidal stresses in its orbit between Ganymede and Io that generate heat and may lead to volcanism. In fact, the ridges could represent ice-filled fractures in the crust, formed in response to tidal stresses. However, some calculations show that the tidal heating

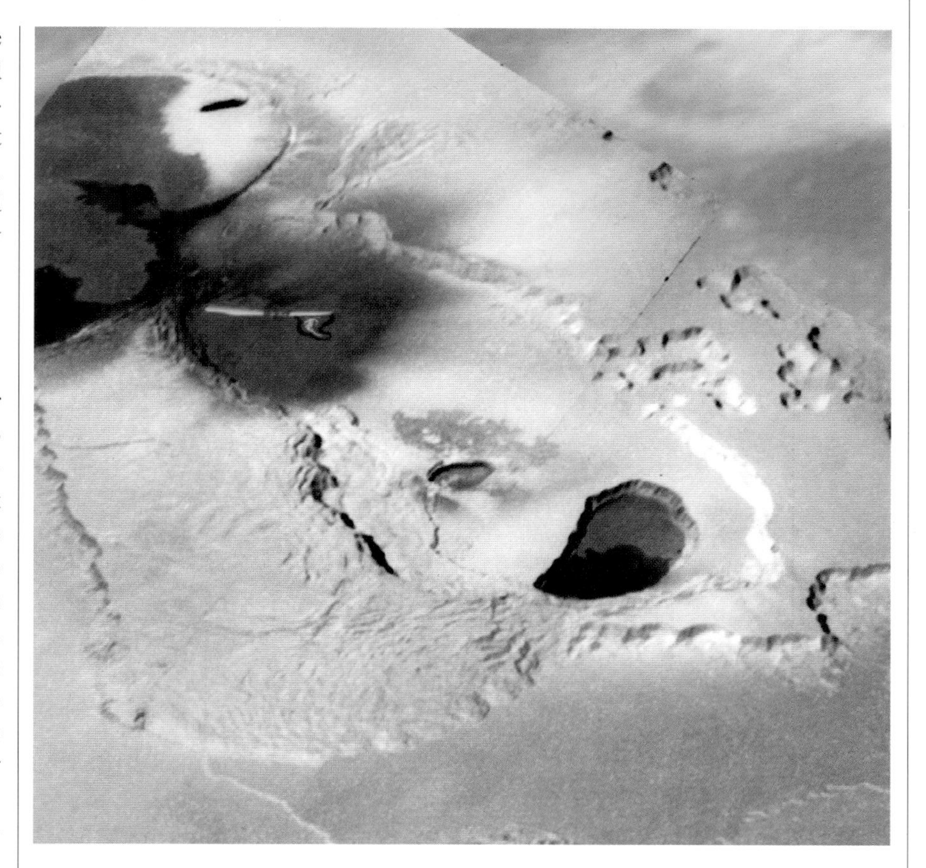

Galileo image taken in November 1999 showing an active eruption on the caldera floor of Tvashtar volcano. The active zone is about 45 kilometers long and is colorized in red to correspond to the very high temperatures derived from the image. The very dark flows are probably the most recent, but have cooled and solidified (JPL PIA 02545).

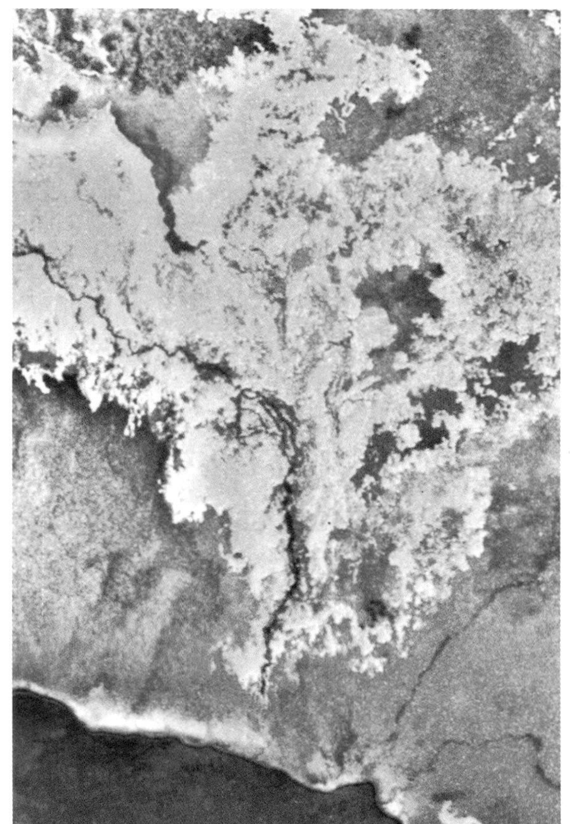

Galileo images of Emakong Patera on Io, showing its caldera (dark area at bottom of picture) and series of lava flows, some of which contain channels; area shown is 80 km wide (NASA PIA 02539).

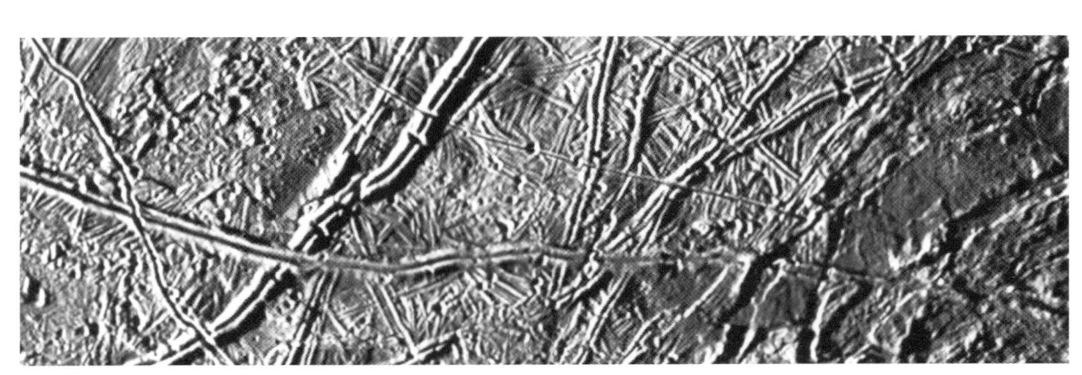

High resolution *Galileo* image of ridges on Europa (NASA PIA 00542).

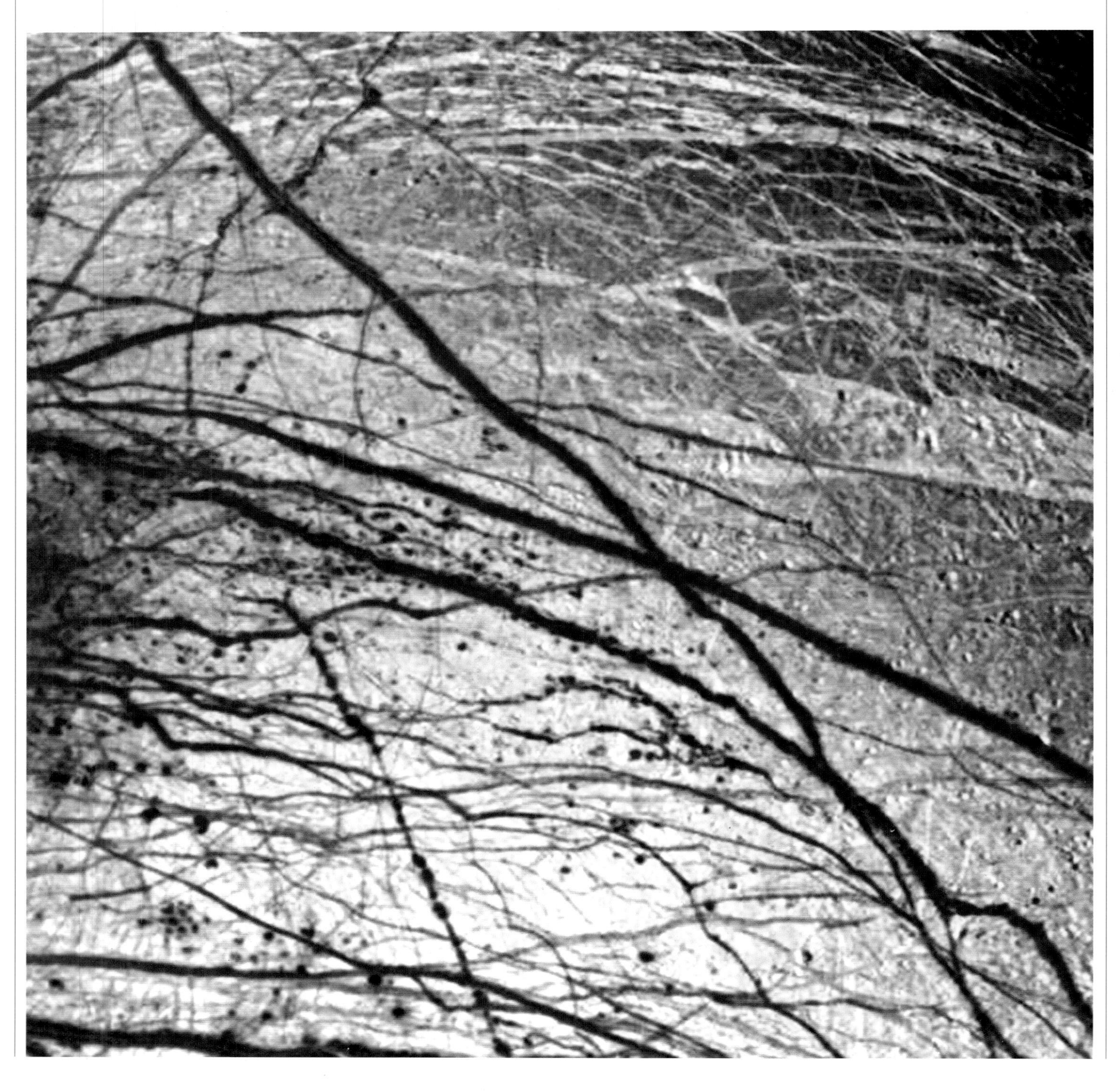

Topographic image of Pwyll crater on Europa. The crater is 26 kilometers in diameter. Red shows high areas, blue represents low areas (NASA PIA 01175).

may be too low to melt ice in the interior of Europa.

Global views of Europa's surface show numerous linear features, most of which are complex ridges. Some parts of Europa's crust have been broken apart into icy slabs which have been rafted into new positions. Much like the pieces of a jig-saw puzzle, the pieces look as though they could fit back together. In some places, chunks of the crust resemble partly submerged icebergs.

The various styles of disruption seen on Europa could mean that the ice crust was relatively thin and underlain by liquid water when the features seen in images formed. Although this does not mean that liquid sub-ice water exists today, the geological youth of the surface indicated by the paucity of impact craters supports this notion. Moreover, the magnetometer data from the *Galileo* mission are consistent with the presence of salt water beneath the surface.

The possibility of liquid water, along with tidal heating in the rocky interior and the probable presence of organic materials implanted by cometary impacts make Europa a high priority in the search for life in the Solar System beyond Earth. Future missions are being planned to carry out additional exploration of Europa. First would be an orbiter designed to determine if liquid water is present and, if so, where it could be found and at what depth below the surface. Imaging in high resolution would show the best places for the next stage of exploration, including surface landers, some of which might have the capability to burrow through the ice.

Ganymede

Ganymede is the largest of the satellites in the Solar System. The first spacecraft image of Ganymede was taken by *Pioneer 10*. Although low resolution, the image shows dark and light zones and demonstrates that the surface is not homogeneous. Ganymede was seen in detail during the *Voyagers 1* and *2* flybys in 1979 and by *Galileo* beginning in 1996. In all, about 80 percent of the surface of Ganymede was photographed. Two prominent terrains are seen: a dark, heavily cratered terrain considered to be ancient and a bright terrain characterized by sets of ridges and grooves.

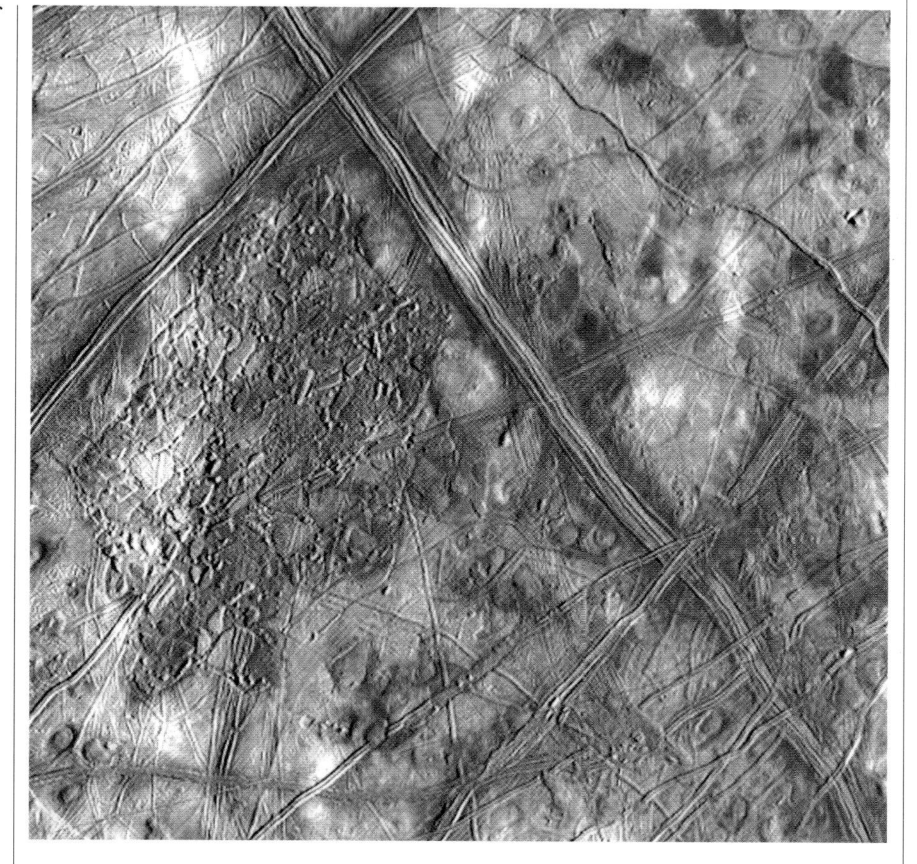

The Conamara Chaos region of Europa is shown in this false-color *Galileo* image. Fracture systems identified in lower resolution images are seen to consist of parallel ridges and grooves. Area shown is 200 kilometers wide (NASA PIA 01296).

Ganymede is about the same size as Mercury, but its low density (1.9 grams per cubic centimeter) suggests that it is composed of about 60 percent water and 40 percent rocky materials. Overall, the satellite is a very bright object; even the so-called dark terrains are actually brighter than the lunar highlands. The surface also appears to be spectrally "red," with the dark terrain being somewhat redder than the bright, grooved terrain. Moreover, as in the case of Europa, the trailing hemisphere of Ganymede is darker than the leading hemisphere.

Dark terrain constitutes about 40 percent of the surface of Ganymede and includes large polygonal regions separated by grooved terrain. The high number of impact craters superposed on the dark terrain reflects the period of intense bombardment in the final stages of Solar System formation. However, when craters on Ganymede are compared with those on the inner planets, few craters are larger than 60 kilometers in diameter. Perhaps the larger craters have been obliterated by some

Facing page: false-color *Galileo* image of Europa. Icy plains, shown in tans and blues, are cut by fractures which appear to have brought non-ice material to the surface, leaving stains shown here in reds and browns. Area shown is about 1,260 kilometers across (NASA PIA 00275).

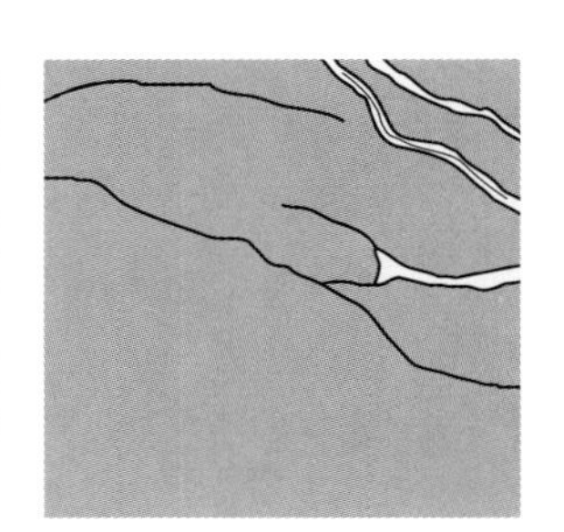
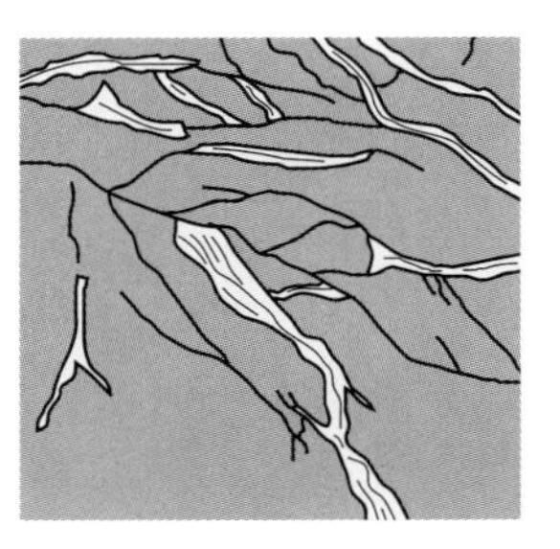

Possible sequence of grooved-terrain formation: (left) formation of primary grooves; (center) formation of more primary grooves, development of secondary grooves, and flooding by "clean" ice; (right) addition of third-level grooves and extensive emplacement of clear ice (courtesy of M. Golombek).

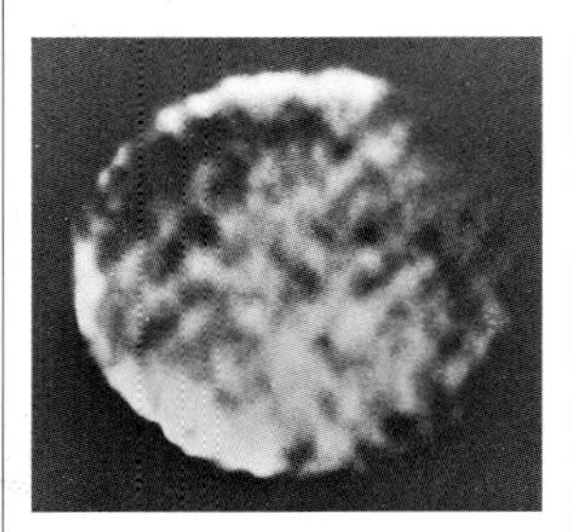

Pioneer 10 image of Ganymede (taken in 1973) is the first spacecraft photograph taken of the satellite and shows its mottled surface (courtesy of B. R. Frieden).

Above and facing page: *Voyager 2* and *Galileo* color composite photographs of Ganymede showing ancient dark terrain and younger bright (grooved) terrain. The large circular dark area (opposite page) is Galileo Regio. Light circular patches are impact scars called palimpsests. Very bright areas are ice-rich ejecta deposits from relatively young impact craters (JPL photograph P-21749 above and *Galileo* photograph PIA 00716 on facing page).

unknown process, or perhaps the cratering record is different for the outer and inner parts of the Solar System.

The most extensive region of ancient, dark terrain includes Galileo Regio, a semicircular area more than 3,200 kilometers across on the leading hemisphere. Furrowlike depressions occur in many dark areas and are among the oldest features found on the satellite. They may represent surface deformation from internal processes, but more likely they are the relics of impact cratering processes.

Bright terrain forms by the conversion of dark terrain primarily through various tectonic processes. Bright terrain appears to be about 1 kilometer thick and occupies most of the polar areas. It also occurs as a zone hundreds of kilometers wide around Galileo Regio and forms large patches in the southern hemisphere. Linear bands of grooved terrain, called sulci, dissect many parts of the older, dark terrain. Bright, grooved terrain is characterized by sets, or bundles, of ridges and grooves 100 kilometers or more long by tens of kilometers in width. Individual ridges are several kilometers wide and may stand as high as 700 meters.

In general, bright, grooved terrain cuts across craters and other topographic features of the older, dark terrain, suggesting tectonic processes superficially analogous to plate tectonics on Earth, in which crustal segments are in lateral motion. However, bright terrain on Ganymede apparently results primarily from extensive fracturing of the icy crust. The fracturing may have occurred through normal faulting of the lithosphere, as parts of the crust sank or subsided. Once thought to have also involved a type of "ice-slush" volcanism, very few areas show evidence of this process in high resolution *Galileo* images.

Impact craters range in diameter from the limit in resolution (less than about 20 meters) to hundreds of kilometers, with the larger craters being more shallow. The relationship between crater size and depth is considered to be a consequence of impacts into ice. With time, the high, ice-rich rims of craters "relax" by viscous flow. The larger craters are more affected by this process than small craters. Alternatively, the craters may be close to their original form, reflecting the properties of the target surface at the time of impact.

Craters about 5 to 35 kilometers in diameter usually have central peaks on their floors, similar to craters found on the terrestrial planets. Larger craters often have central pits, or depressions, on their floors, perhaps as a consequence of the crustal ice melting from heat generated by the impact, or of the impacting object punching through to a liquid zone beneath the icy crust.

Among the many remarkable discoveries made by the *Voyager* mission were the large craterlike features on Ganymede termed palimpsests. These are circular patches of bright terrain as wide as several hundred kilometers, found principally in the dark terrain. They are considered to be scars left by impact into the evolving icy surface of Ganymede early in its history.

Ganymede grew from the collapse of the evolving gas and dust nebulae that formed Jupiter and its satellites. Calculations show that, at the distance of Ganymede from the proto-Jupiter core, the temperatures should have been low enough for water ice and silicate materials to condense to form planetesimals. In less than a half million years, these planetesimals accreted and formed the satellite Ganymede. Heat generated from accretion would have then melted parts of the satellite to form an outer water layer, an inner silicate zone, and an ice-silicate core. These zones, however, would not be stable. As cooling occurred, the denser silicate materials would have sunk to form a core, leaving an intermediate ice-silicate zone between the core and the outer ice layer. *Galileo* data suggest that a liquid water zone lies beneath the ice crust.

Callisto

Callisto is the darkest of the Galilean satellites, yet it is still twice as bright as Earth's Moon. Its density is only about 1.8 grams per cubic centimeter, making it the least dense of the Galilean satellites and suggesting that it has the greatest proportion of water. Images from the *Voyager* and *Galileo* spacecraft show that the surface of Callisto is

Voyager 2 view of an impact crater (80 kilometer-diameter) containing a central pit. Crater is found in the western equatorial region of Ganymede (*Voyager 2* photograph 20638.39).

Below: Moderate-resolution *Voyager 2* image of Ganymede showing older dark terrain cut by younger, bright grooved terrain (JPL photograph P-21769).

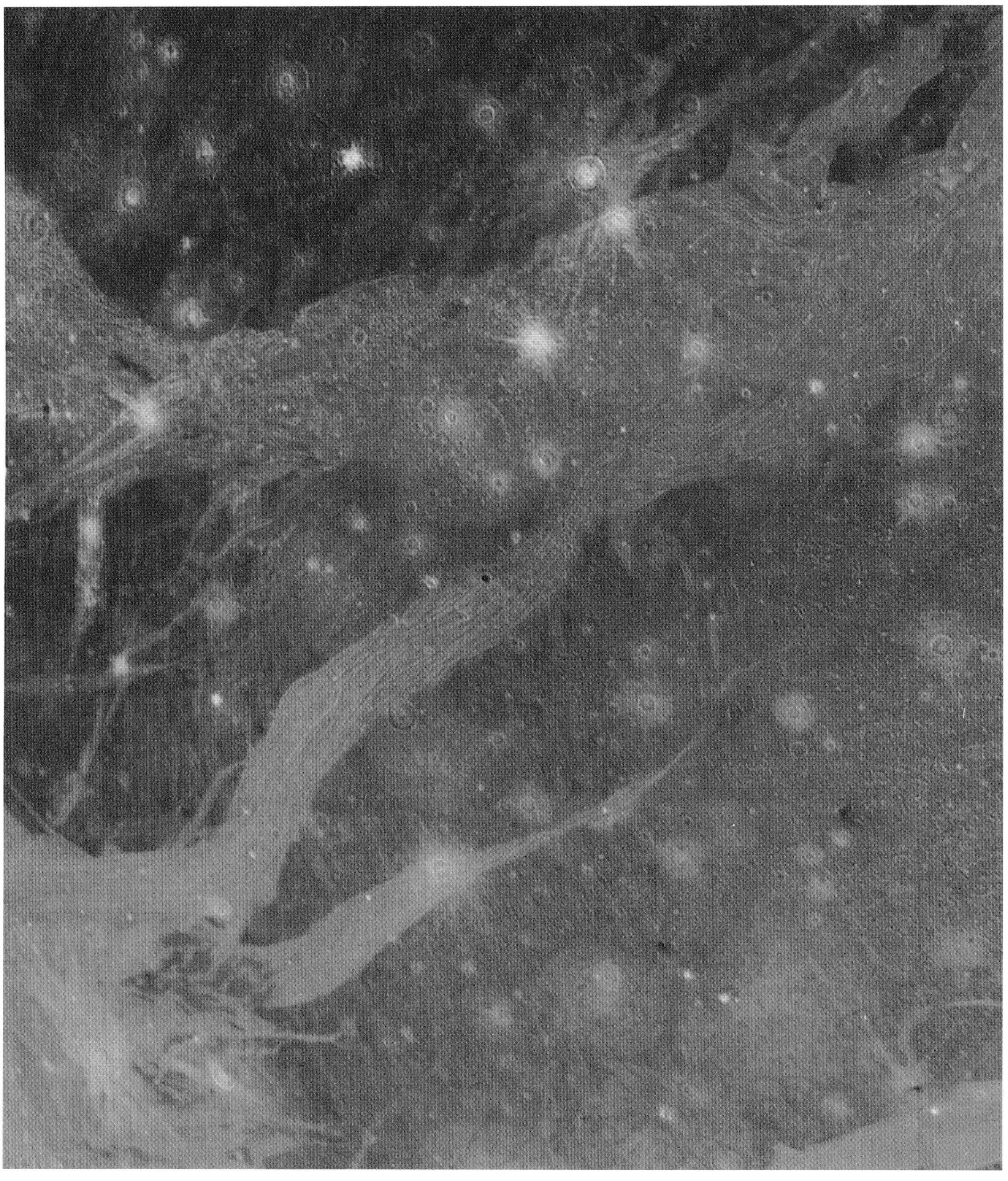

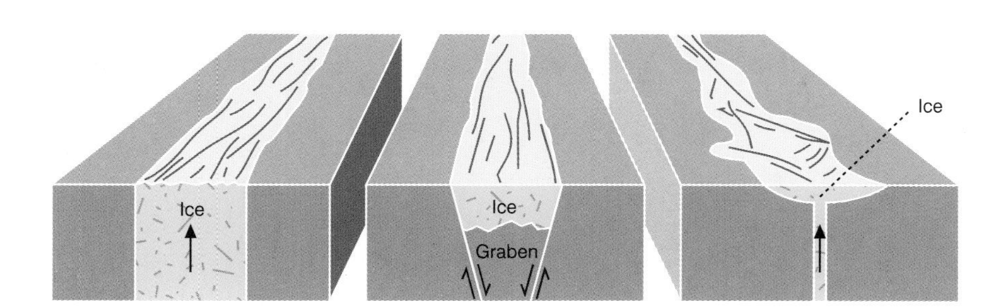

Three possible ways in which bright terrain could form: (left) dark terrain fractures are filled with fresh ice from below; (center) dark terrain is cut by normal faults to form a graben that is then filled with ice; (right) low-lying areas of the dark terrain are flooded with thin deposits of ice erupted as liquid water or slush from below the surface (courtesy of M. Parmentier).

heavily cratered but has relatively little relief. Except for terrains related to large impact events, the surface viewed by the spacecraft is essentially uniform, consisting of relatively dark, heavily cratered terrain.

Callisto is more heavily cratered than even the oldest terrain on Ganymede, and the surface of Callisto probably records a longer geologic history than that of Ganymede. Many planetary scientists see Callisto as essentially devoid of most internal activity since its initial formation, in contrast to Ganymede.

The most prominent feature on Callisto is a multiring structure of presumed impact origin named Valhalla. The Valhalla structure consists of a bright central zone about 600 kilometers wide, surrounded by concentric rings extending outward for nearly 2,000 kilometers from the center of the structure. The bright central zone presumably marks the position of the initial crater immediately following impact. Impact may have occurred in a rather thin, rigid crust overlying a fluidlike zone. The impact penetrated into the fluid mass, which rapidly filled the initial crater bowl. At least seven other multiring features have been recognized on Callisto, including Asgard with its central bright zone 230 kilometers wide.

One of the biggest suprises from the *Galileo* mission was the discovery in some parts of Callisto of terrains which lack craters smaller than a few kilometers across. The surface appears very smooth, and it is thought that some process unique to Callisto effectively erases the small craters in these areas.

Ganymede and Callisto are about the same size and density. Prior to the *Voyager* flybys, it was assumed that the two objects would show surfaces reflecting similar geologic histories. However, spacecraft images revealed strikingly different worlds and generated considerable debate regarding the comparative evolution of Ganymede and Callisto. Nonetheless, some agreement has been reached on the general events in their history, if not on the detailed evolution. In the early history of the Jupiter system, both objects formed from materials leading to water-silicate bodies. Heating from various sources, including impact cratering, radioactive decay, and tidal stresses, led to dif-

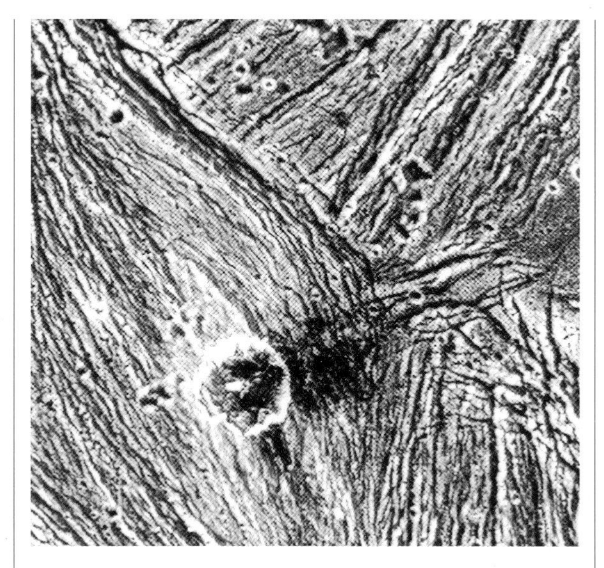

Galileo image of ridged and grooved terrain in the *Uruk Sulcus* region of Ganymede, and a 6.5 kilometer in diameter impact crater (*Galileo* image PIA 00280).

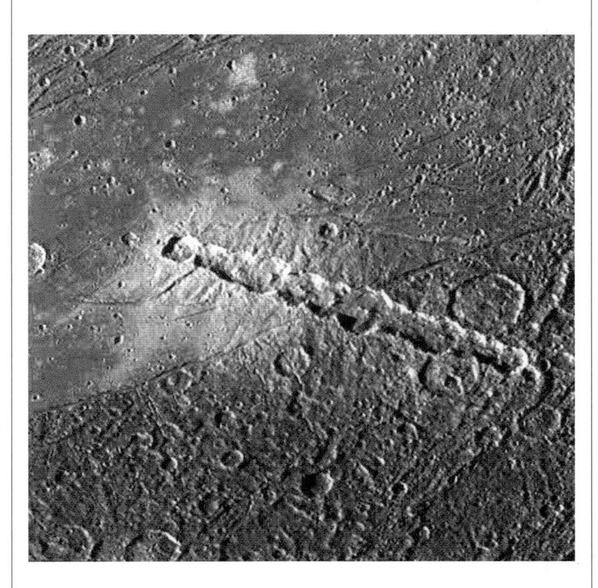

This chain of craters is thought to have formed from the collisions of cometary fragments striking Ganymede, similar to the impact of comet Shoemaker-Levy 9 on Jupiter. Area shown is 214 by 217 kilometers (*Galileo* image PIA 01610).

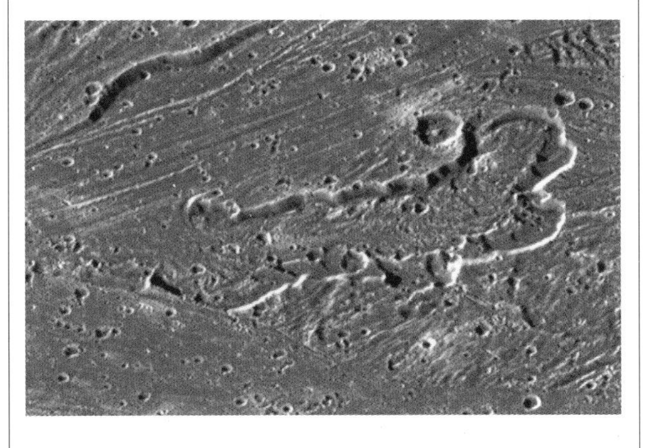

Possible evidence of water-ice eruptions on Ganymede is seen in the form of this caldera-like feature 55 kilometers long by 17–20 kilometers wide (*Galileo* image PIA 01614).

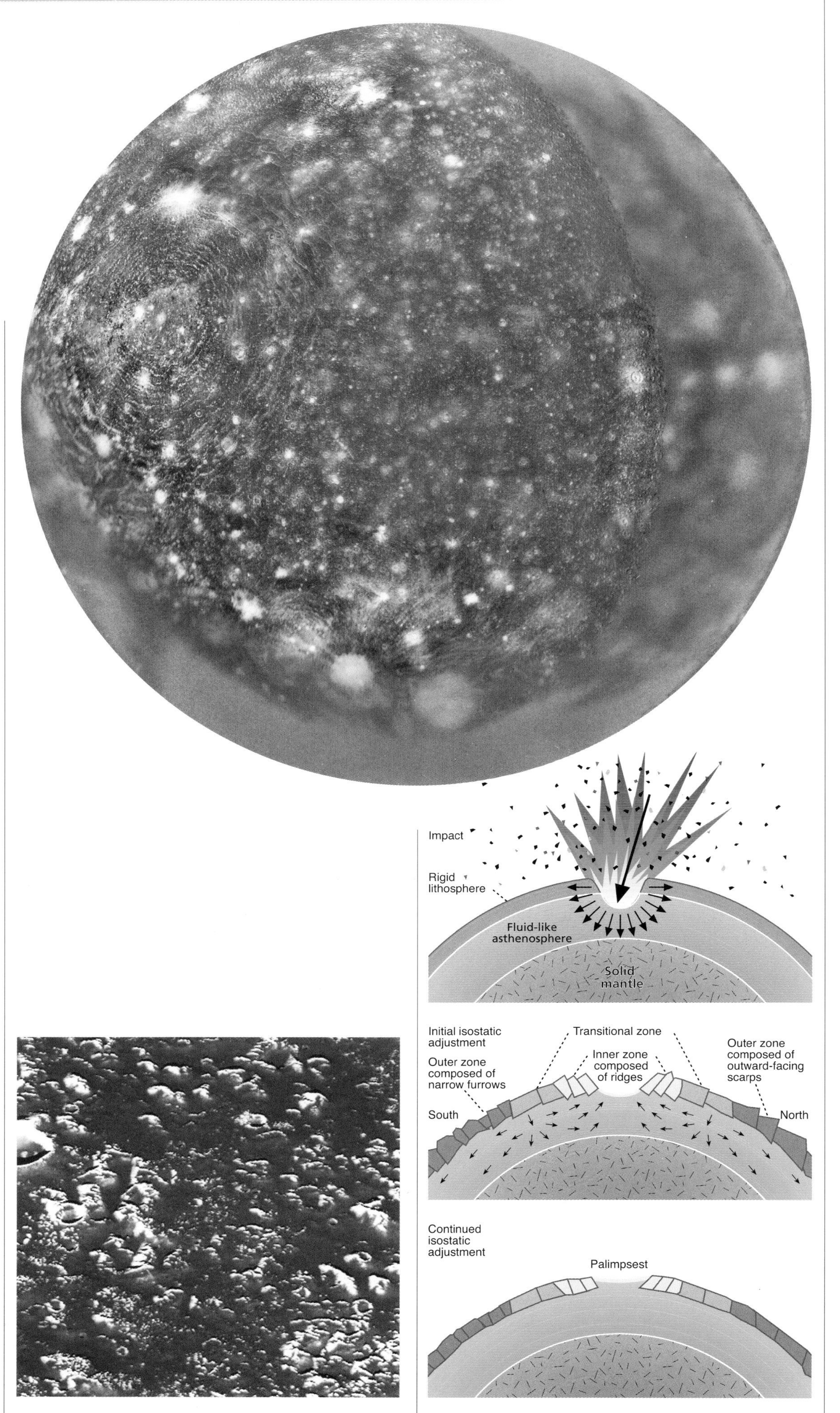

Voyager mosaic showing the Valhalla basin, defined by a central bright spot about 600 kilometers in diameter surrounded by numerous concentric rings (*Voyager* mosaic U.S. Geological Survey).

High resolution *Galileo* view of the central zone of the Valhalla structure on Callisto, showing the paucity of small impact craters. Area shown is about 11 kilometers across (NASA PIA 00516).

Cross-section through the Valhalla basin of Callisto, showing inferred deformation of the rigid crust following the impact event.

ferentiation and the formation of a "mud" core. However, there are several differences that would result in less heat on Callisto, including its smaller size (meaning fewer radioactive elements and, hence, less heat from radioactive decay), and significantly less tidal stresses because there is no large body outside its orbit.

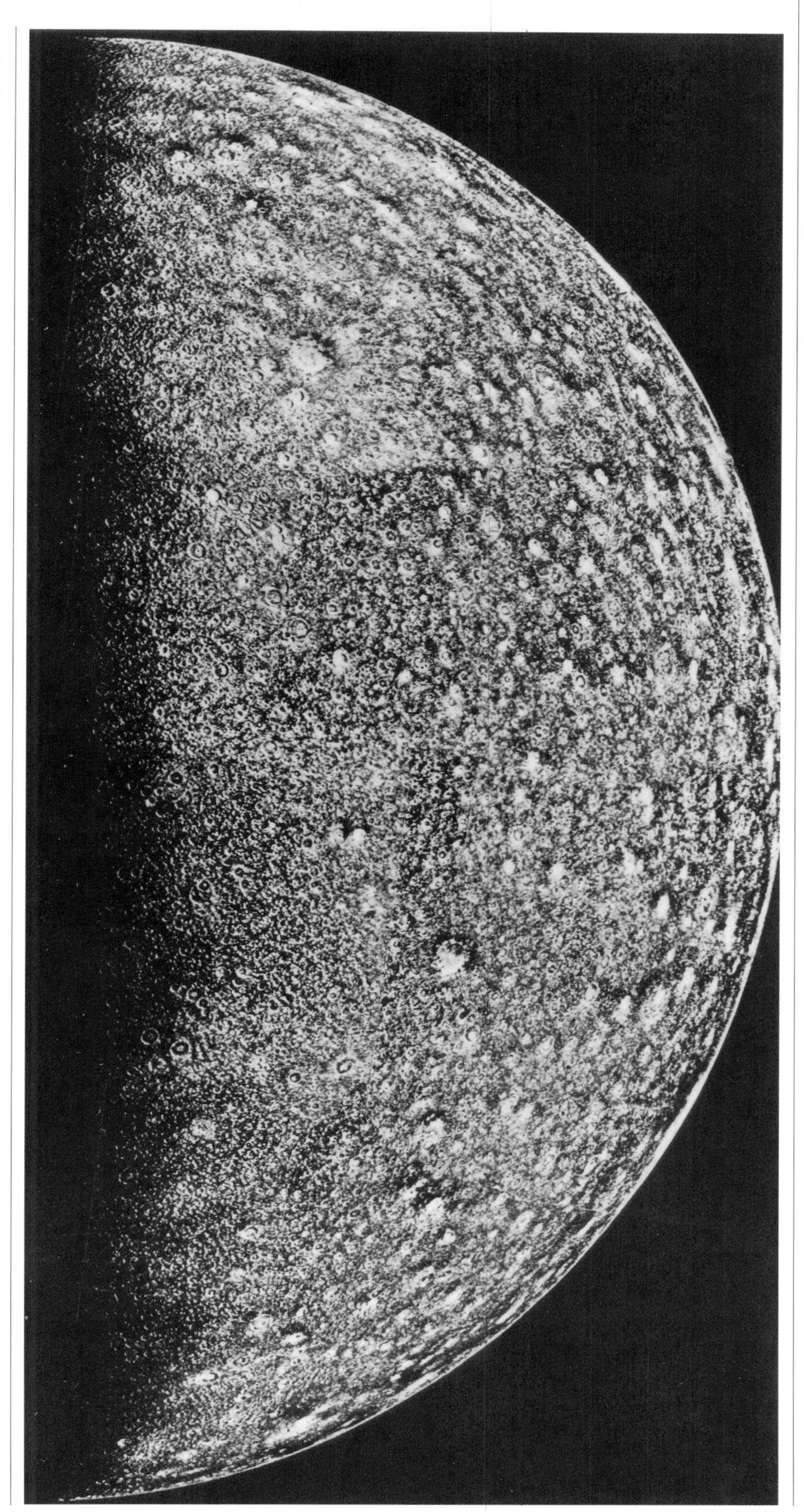

Mosaic showing the heavily cratered terrain of Callisto. The general lack of smooth, uncratered terrain indicates that most of the surface dates from the early history of the Jupiter system (*Voyager* mosaic 260-586).

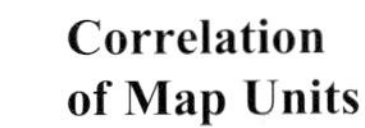

Correlation of Map Units

Youngest at top

Vent floor materials | Vent cone materials | Flow materials | Plains materials | Mountain materials

Io Sub-Jovian Hemisphere Geologic Map

Center lat. 0°
Center lon. 0°

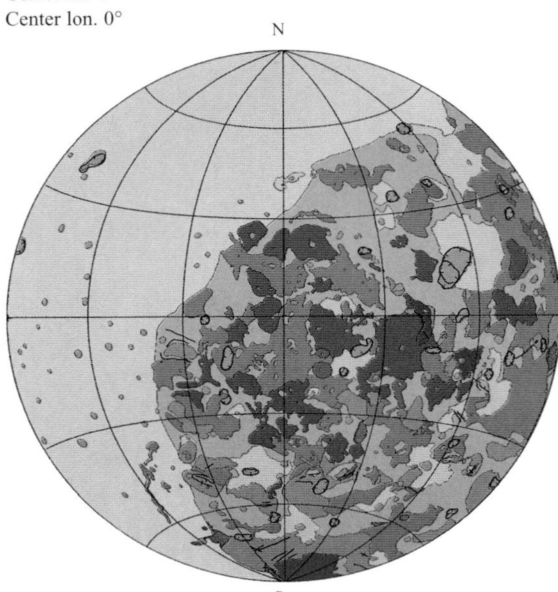

Io Anti-Jovian Hemisphere Geologic Map

Center lat. 0°
Center lon. 180°

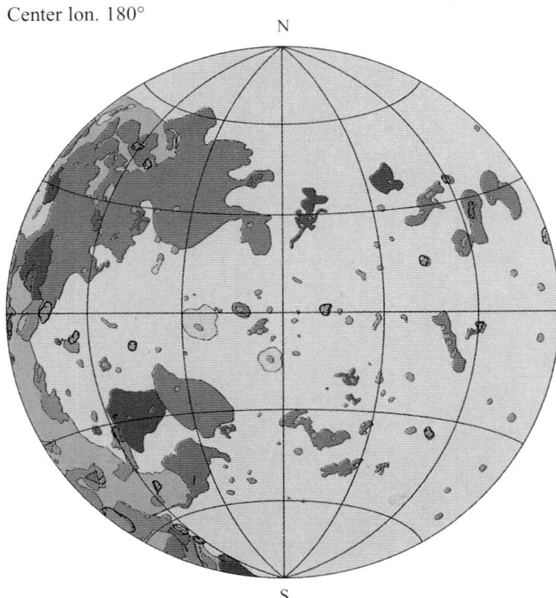

Io Sub-Jovian Hemisphere Reference Map

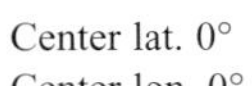

Center lat. 0°
Center lon. 0°

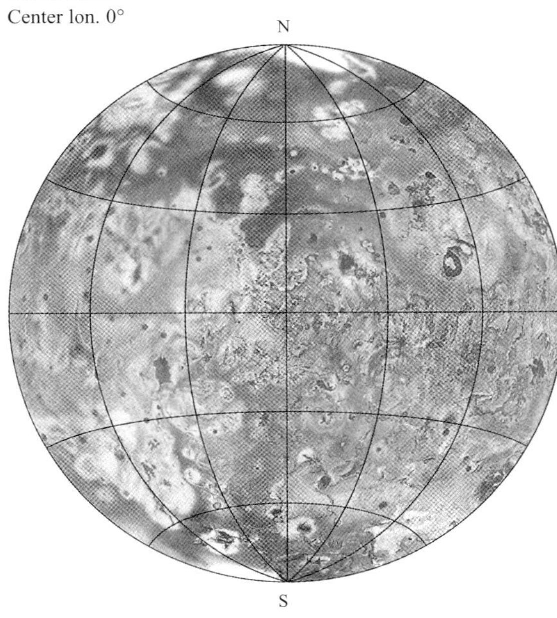

Io Anti-Jovian Hemisphere Reference Map

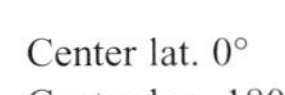

Center lat. 0°
Center lon. 180°

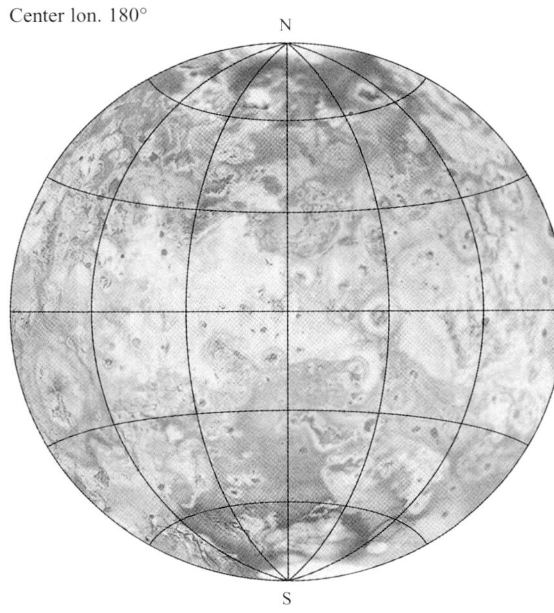

Description of Map Units

Igneous Materials

Vent Floor Materials

- Undivided

Vent Cone Materials

- Cone materials

Flow Materials

- Fissure
- Breakout
- Hummocky
- Patera
- Lobate
- Plains-forming
- Tholus
- Shield
- Undivided

Plains Materials

- Intervent
- Layered

Mountain Materials

- Grooved
- Smooth
- Undivided

Unmapped

- Data resolution insufficient

Structure Symbols

- Fault
- Graben
- Ridge

Europa
Sub-Jovian Hemisphere
Geologic Map

Center lat. 0°
Center lon. 0°

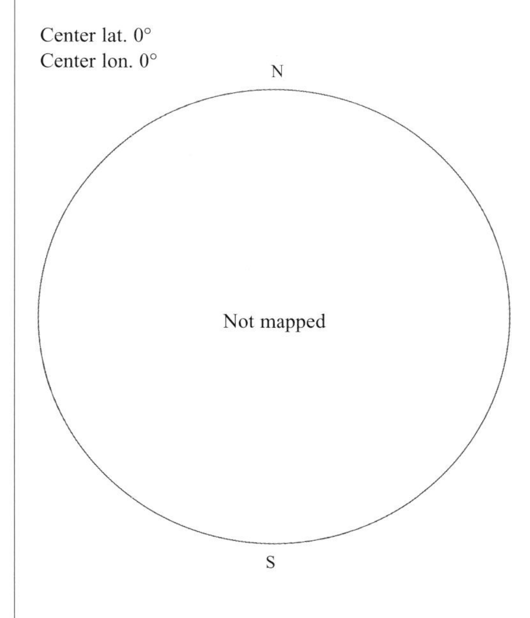

Europa
Anti-Jovian Hemisphere
Geologic Map

Center lat. 0°
Center lon. 180°

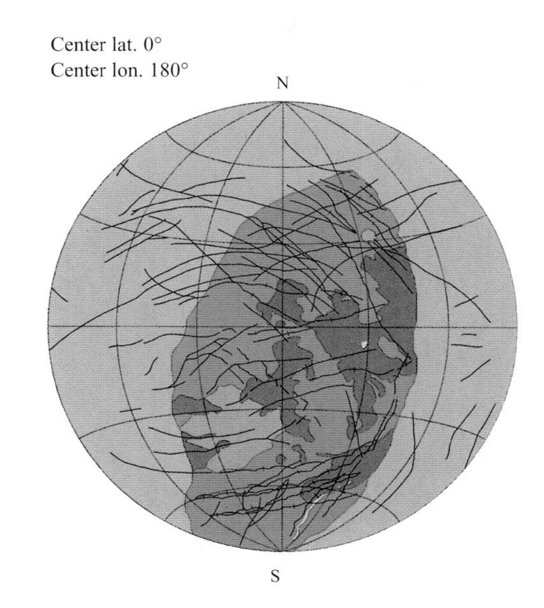

Description of Map Units

Ice Plains Materials

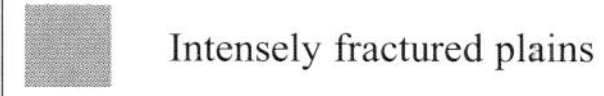
Intensely fractured plains

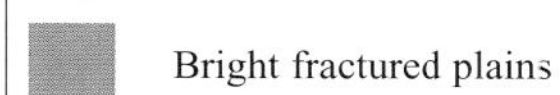
Bright fractured plains

Impact Materials

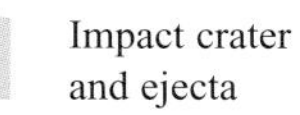
Impact crater and ejecta

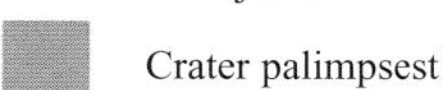
Crater palimpsest

Undetermined Materials

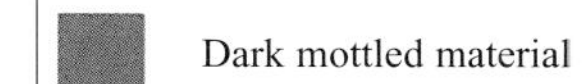
Dark mottled material

Lineated terrain

Unmapped

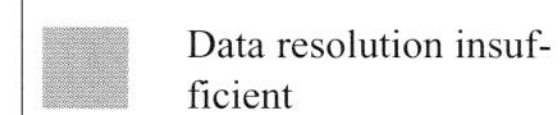
Data resolution insufficient

Structure Symbols

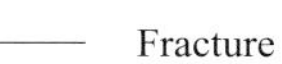
Fracture

Ridge

Correlation of Map Units

Youngest at top

Ice plains materials Impact materials Undetermined materials

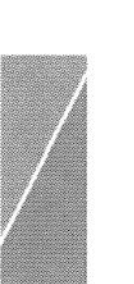

Europa
Sub-Jovian Hemisphere
Reference Map

Center lat. 0°
Center lon. 0°

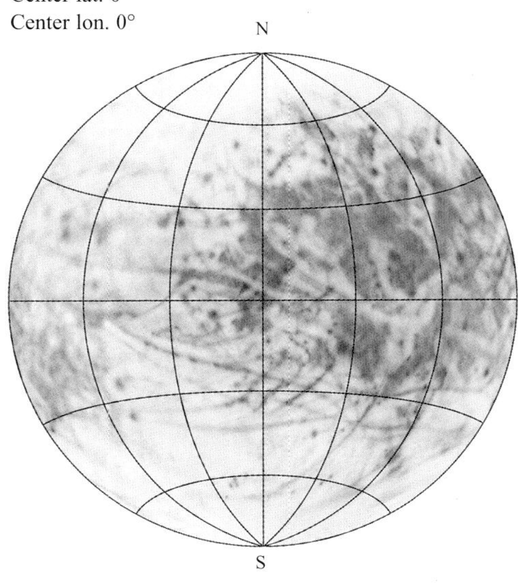

Europa
Anti-Jovian Hemisphere
Reference Map

Center lat. 0°
Center lon. 180°

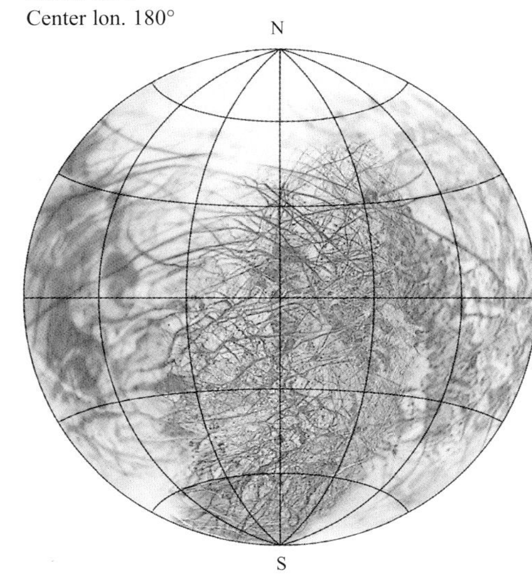

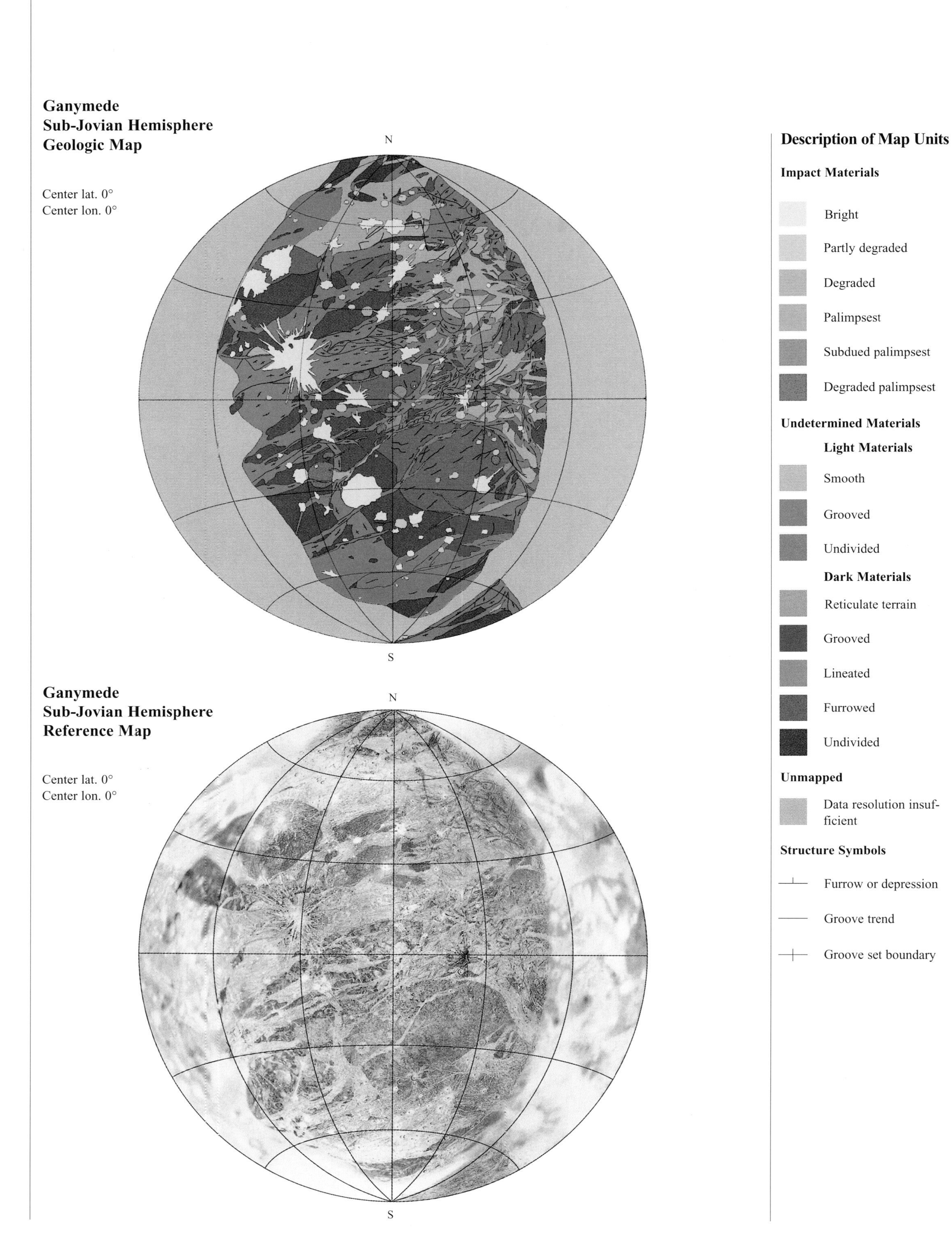
Ganymede
Sub-Jovian Hemisphere
Geologic Map
Center lat. 0°
Center lon. 0°
N
S
Ganymede
Sub-Jovian Hemisphere
Reference Map
Center lat. 0°
Center lon. 0°
N
S
Description of Map Units
Impact Materials
Bright
Partly degraded
Degraded
Palimpsest
Subdued palimpsest
Degraded palimpsest
Undetermined Materials
Light Materials
Smooth
Grooved
Undivided
Dark Materials
Reticulate terrain
Grooved
Lineated
Furrowed
Undivided
Unmapped
Data resolution insufficient
Structure Symbols
Furrow or depression
Groove trend
Groove set boundary

Correlation of Map Units

Youngest at top

Impact materials

Undetermined materials

Light

Dark

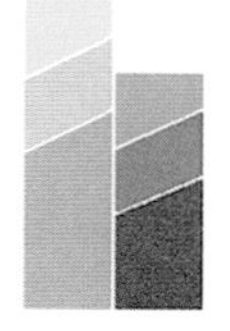

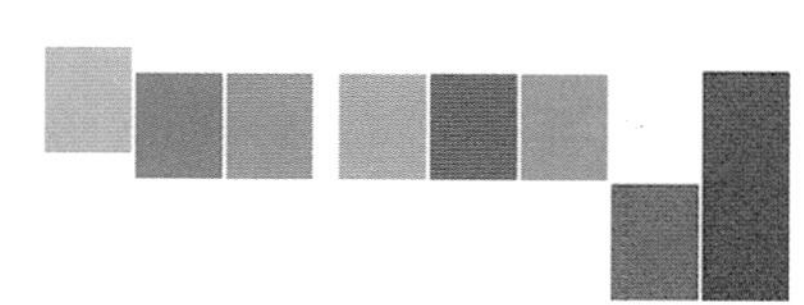

Ganymede
Anti-Jovian Hemisphere
Geologic Map

Center lat. 0°
Center lon. 180°

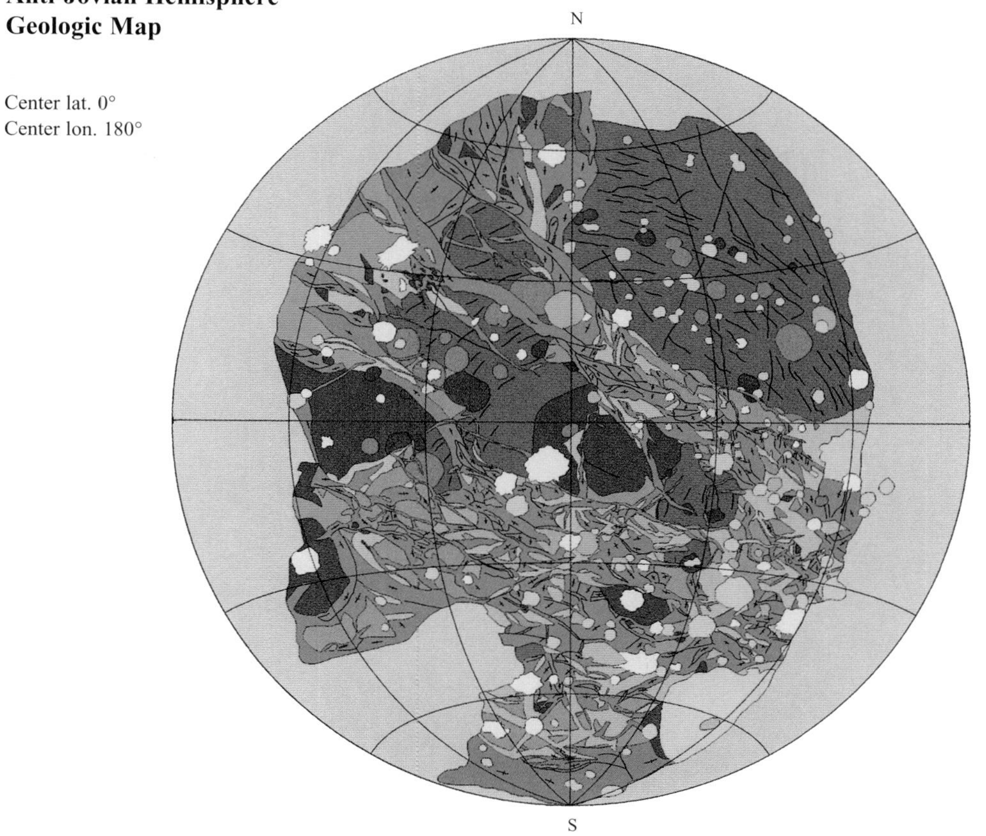

Ganymede
Anti-Jovian Hemisphere
Reference Map

Center lat. 0°
Center lon. 180°

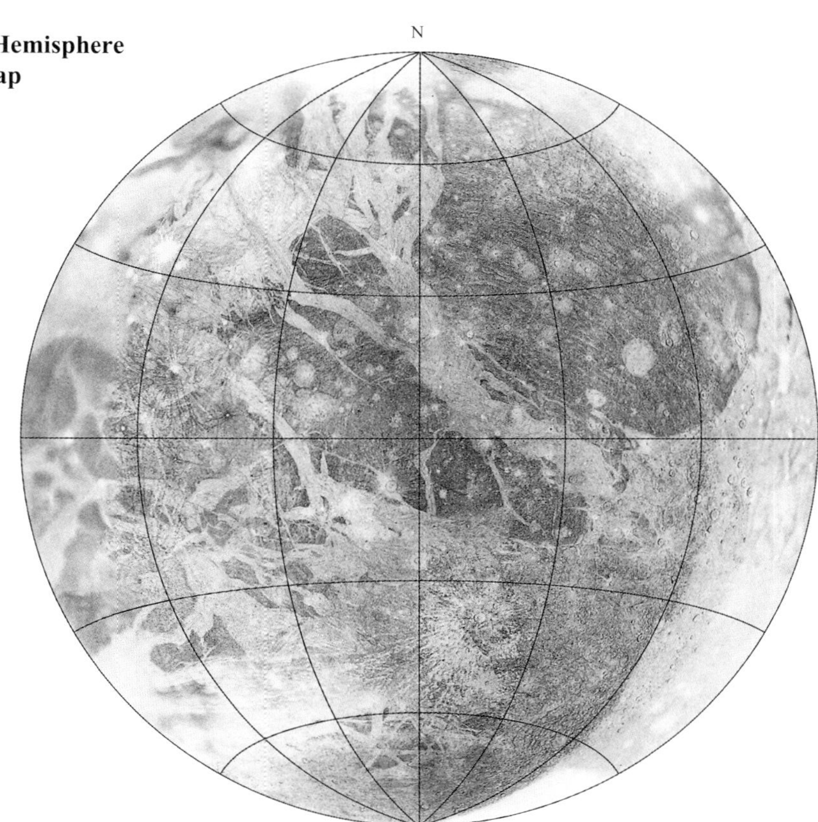

Callisto Sub-Jovian Hemisphere Geologic Map

Center lat. 0°
Center lon. 0°

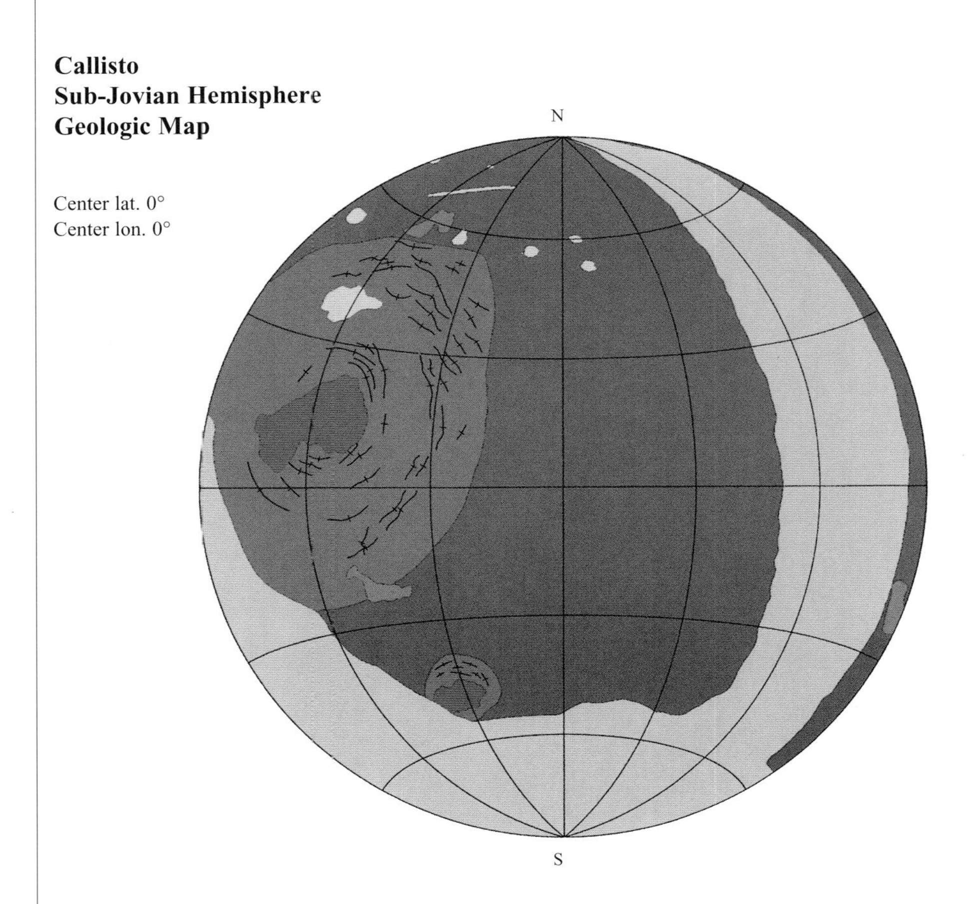

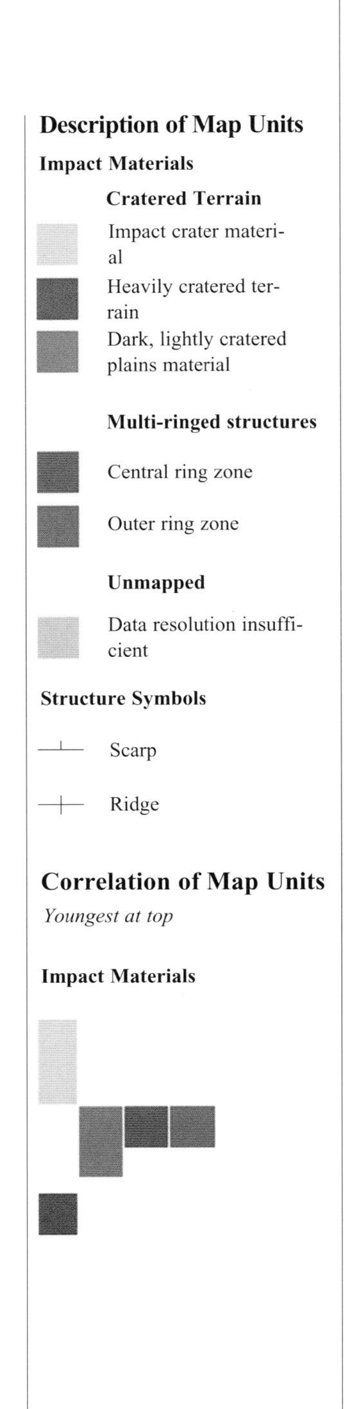

Callisto Sub-Jovian Hemisphere Reference Map

Center lat. 0°
Center lon. 0°

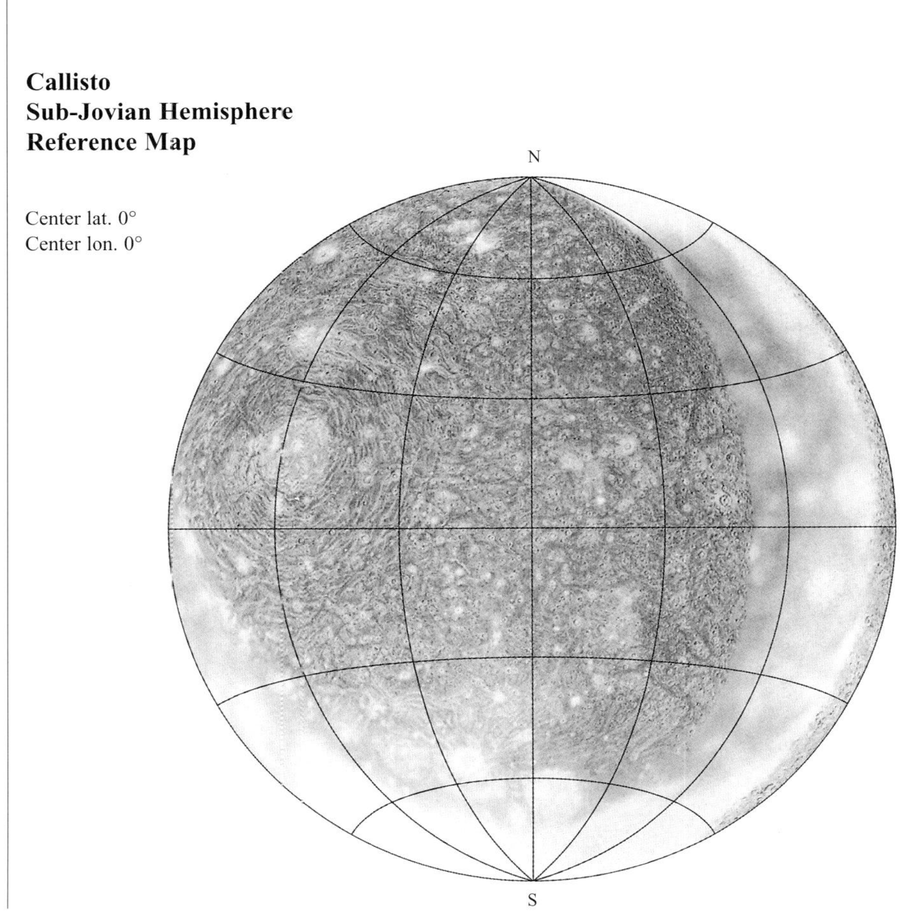

Callisto
Anti-Jovian Hemisphere
Geologic Map

Center lat. 0°
Center lon. 180°

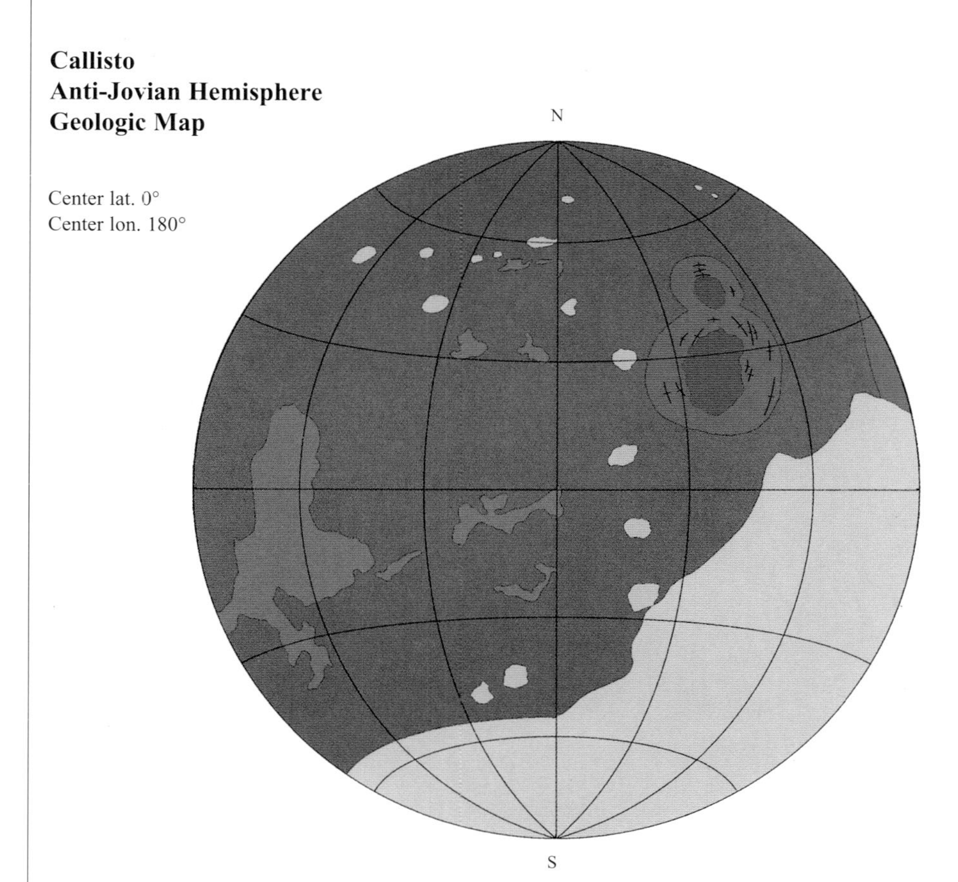

Callisto
Anti-Jovian Hemisphere
Reference Map

Center lat. 0°
Center lon. 180°

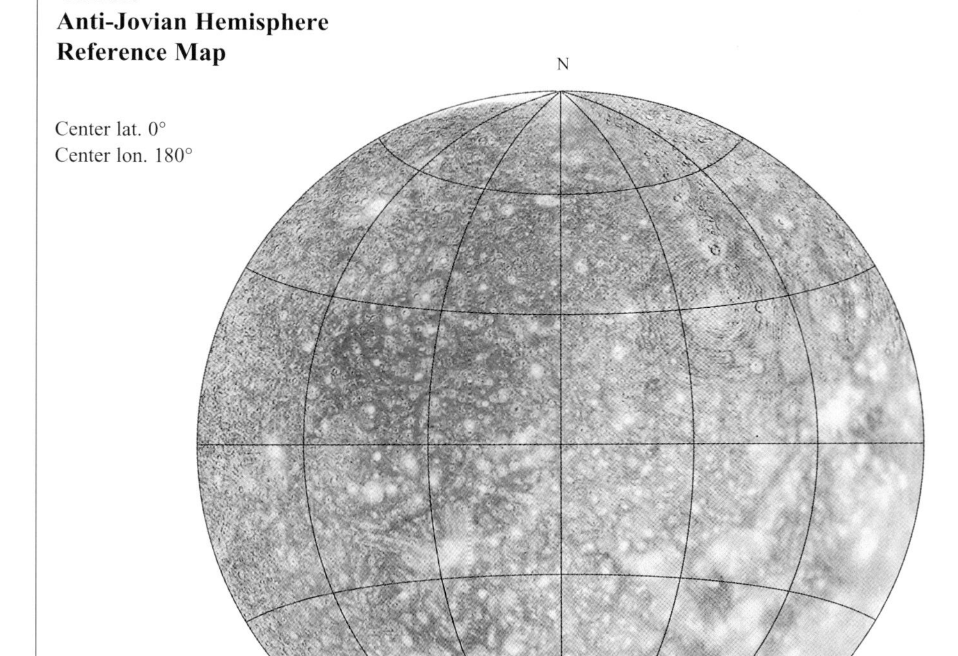

Io
Index Map

Center lat. 30°
Center lon. 0°

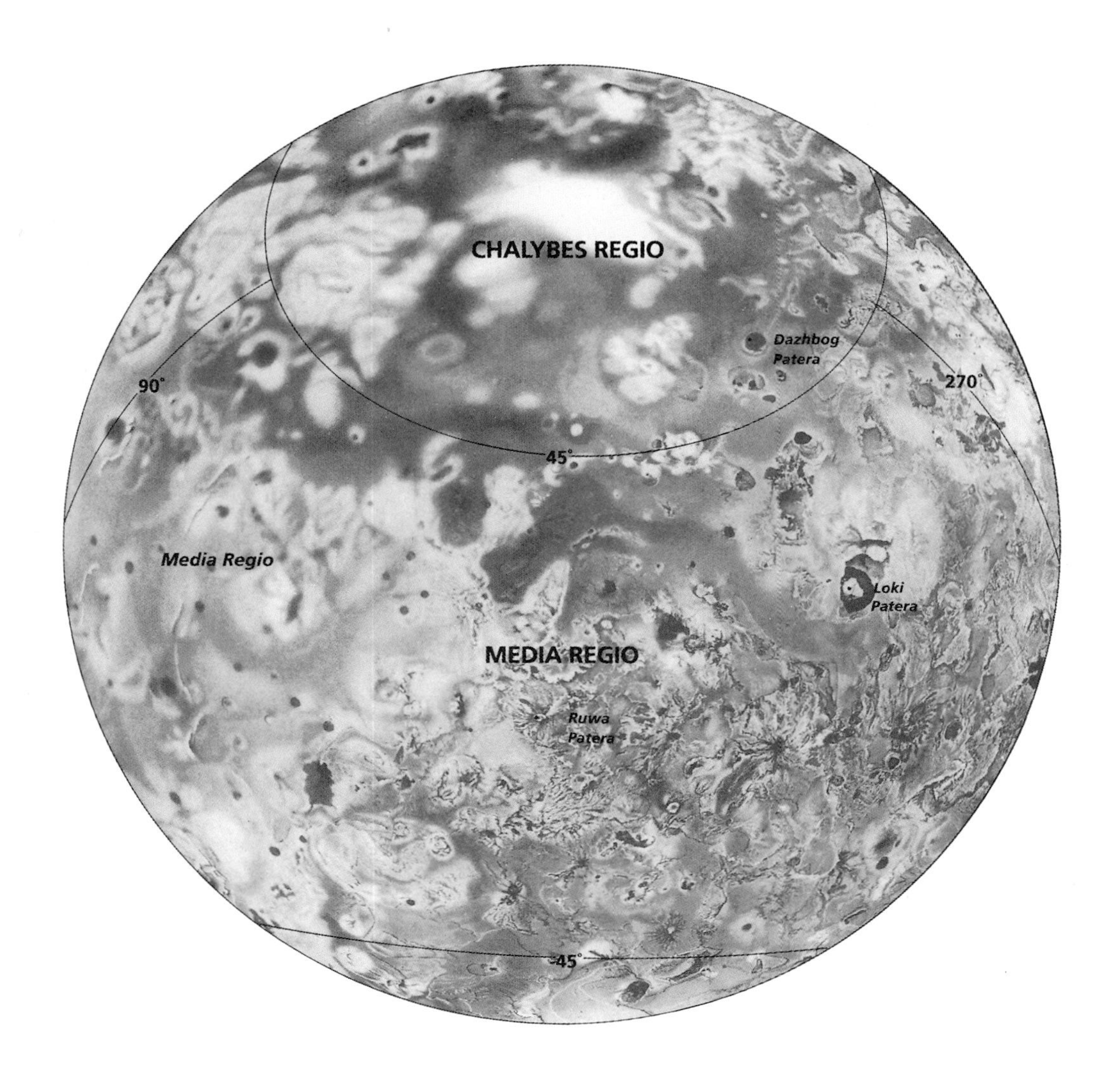

Io
Index Map

Center lat. –30°
Center lon. 180°

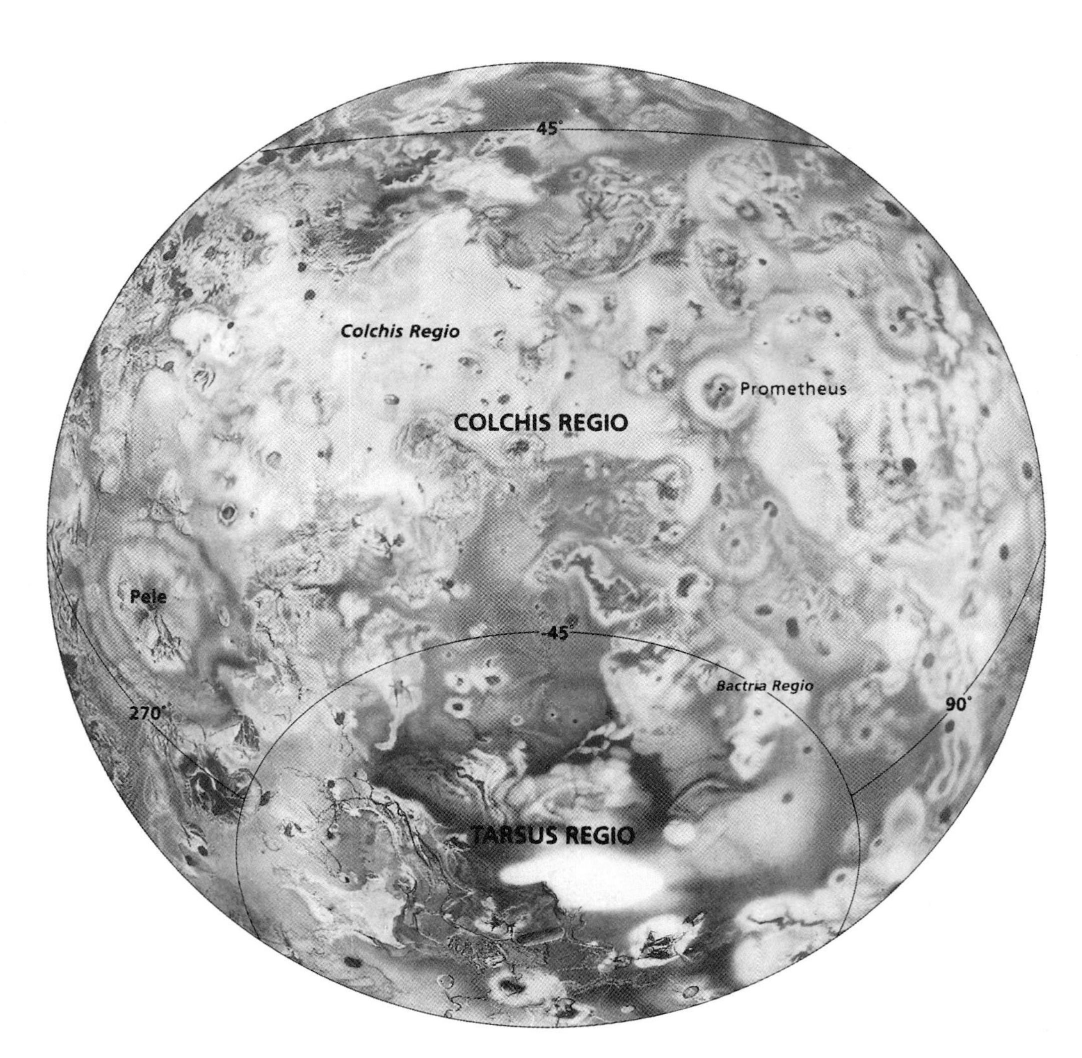

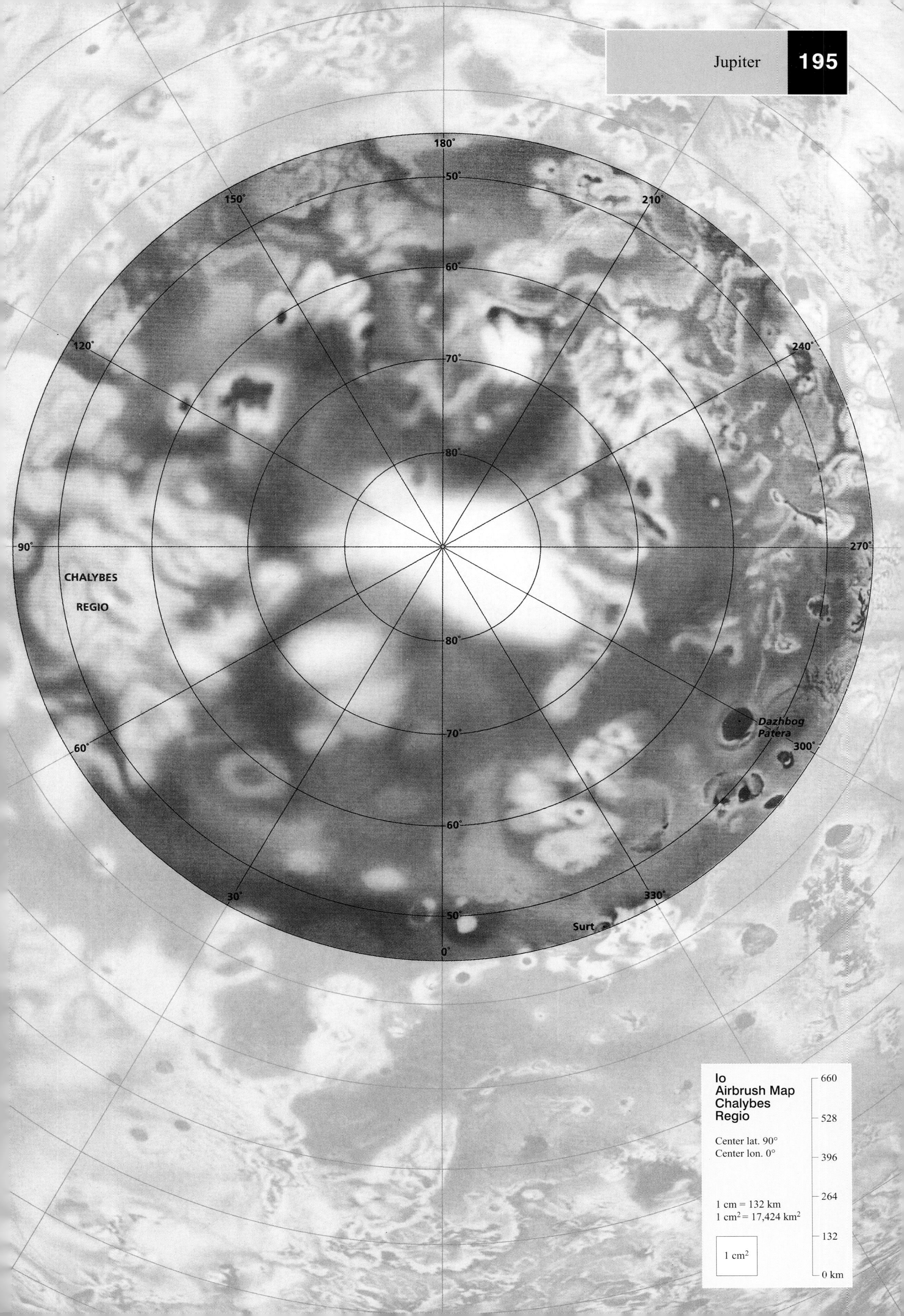
180°
50°
150°
210°
60°
120°
240°
70°
80°
90°
270°
CHALYBES
REGIO
80°
70°
Dazhbog
Patera
60°
300°
60°
30°
330°
50°
Surt
0°
Io
Airbrush Map
Chalybes
Regio
Center lat. 90°
Center lon. 0°
1 cm = 132 km
1 cm² = 17,424 km²
1 cm²
660
528
396
264
132
0 km

60°
30°
MEDIA REGIO
Ruwa
Ilmarinen
Tung Yo
Sui Jen
Angpetu Patera
Cataquil
Patera
Uta
Fluctus
Mbali Pa
Uta
Patera
TARSUS REGIO
Ethiopia Planum
30°
60°

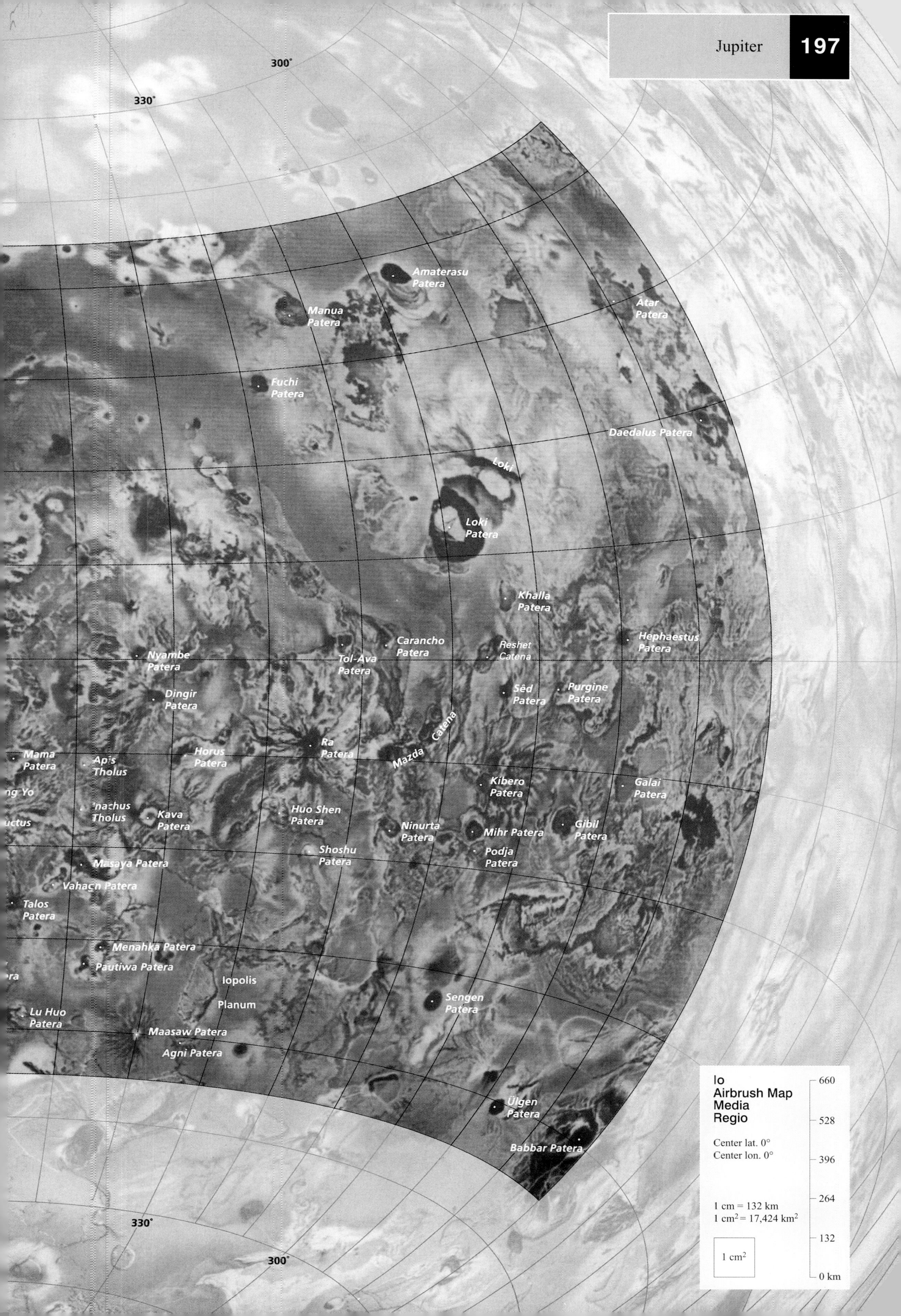
300°
330°
Amaterasu Patera
Manua Patera
Ātar Patera
Fuchi Patera
Daedalus Patera
Loki
Loki Patera
Khalla Patera
Carancho Patera
Reshet Catena
Hephaestus Patera
Nyambe Patera
Tol-Ava Patera
Dingir Patera
Sêd Patera
Purgine Patera
Ra Patera
Mazda Catena
Mama Patera
Apis Tholus
Horus Patera
Kibero Patera
Galai Patera
Inachus Tholus
Kava Patera
Huo Shen Patera
Ninurta Patera
Mihr Patera
Gibil Patera
Shoshu Patera
Podja Patera
Masaya Patera
Vahagn Patera
Talos Patera
Menahka Patera
Pautiwa Patera
Iopolis
Planum
Sengen Patera
Lu Huo Patera
Maasaw Patera
Agni Patera
Ülgen Patera
Babbar Patera
330°
300°
Io
Airbrush Map
Media
Regio
Center lat. 0°
Center lon. 0°
1 cm = 132 km
1 cm² = 17,424 km²
1 cm²
660
528
396
264
132
0 km

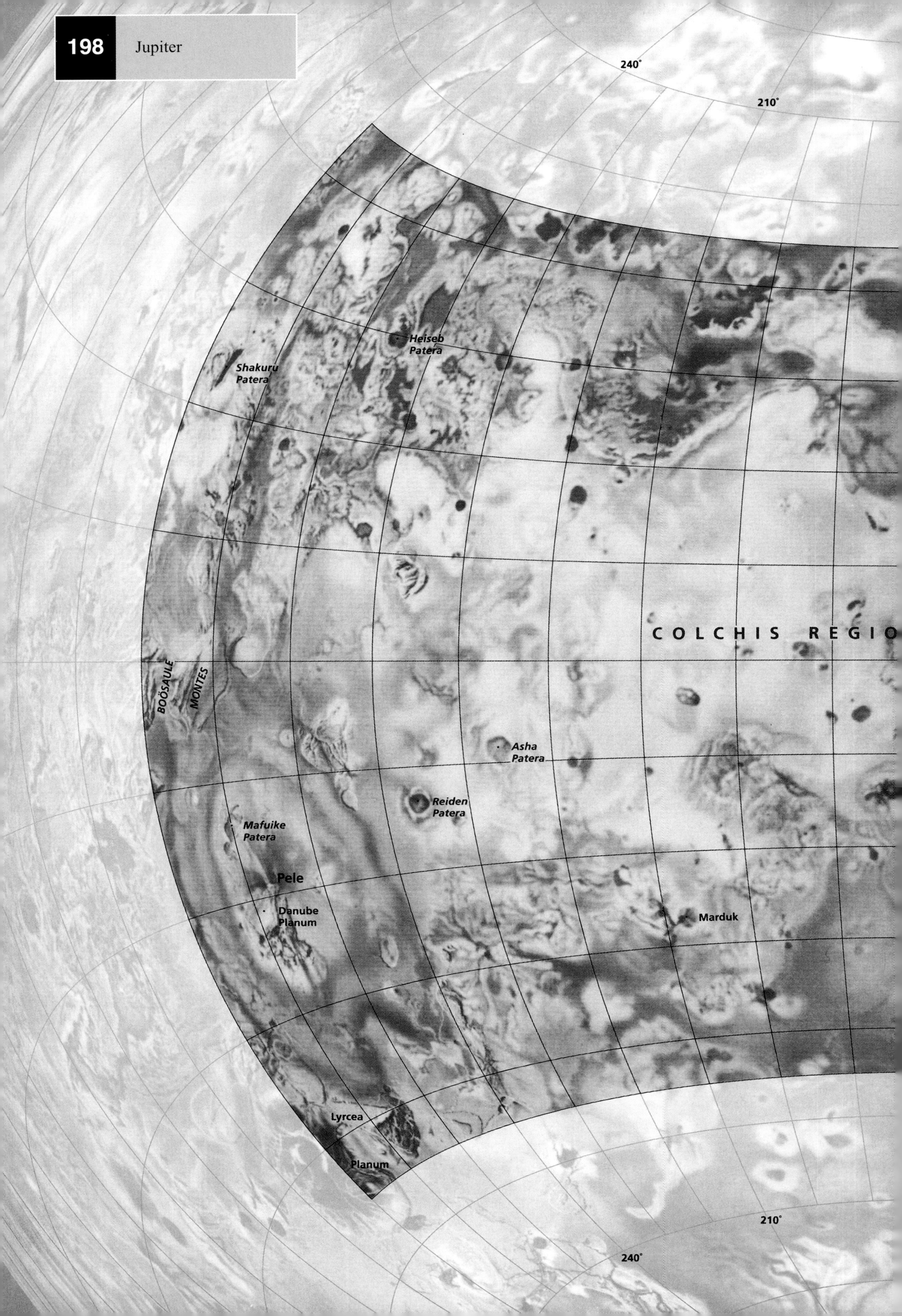

240°
210°
Heiseb Patera
Shakuru Patera
COLCHIS REGIO
BOÖSAULE MONTES
Asha Patera
Reiden Patera
Mafuike Patera
Pele
Danube Planum
Marduk
Lyrcea
Planum
210°
240°

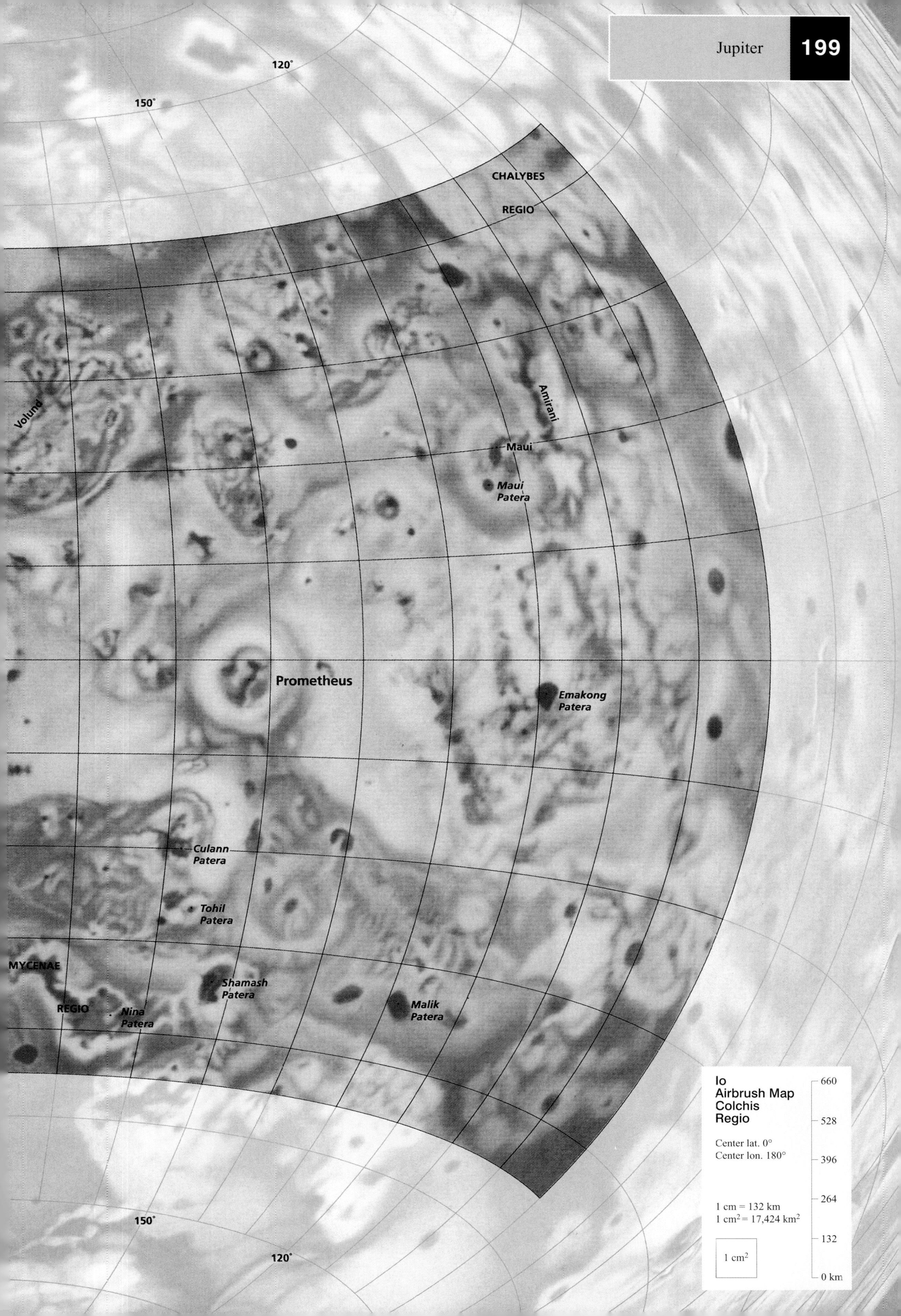

Io
Airbrush Map
Colchis
Regio

Center lat. 0°
Center lon. 180°

1 cm = 132 km
1 cm² = 17,424 km²

Paive Patera
Kane Patera
Siun Patera
Euboea
Montes
Creidne Patera
Hybristes Planum
Pan Mensa
DODONA PLANUM
Argos Planum
Bochica Patera
Hatchawa Patera
Nusku Patera
Aten Patera
Inti Patera
Hiruko Patera
Heno Patera
Taranis Patera
Aramazd Patera
Lynx Mensa
Echo Mensa
LERNA REGIO
NEMEA PLANUM
Viracocha Patera
Silpium Mons
Mithra Patera
Svarog Patera
Crimea Mons
Pyerun Patera
Epaphus Mensa
BACTRIA REGIO
0°
30°
60°
90°
120°
150°
180°
210°
240°
300°
330°
-50°
-60°
-70°
-80°
Io
Airbrush Map
Tarsus
Regio
Center lat. –90°
Center lon. 0°
1 cm = 132 km
1 cm² = 17,424 km²
1 cm²
660
528
396
264
132
0 km

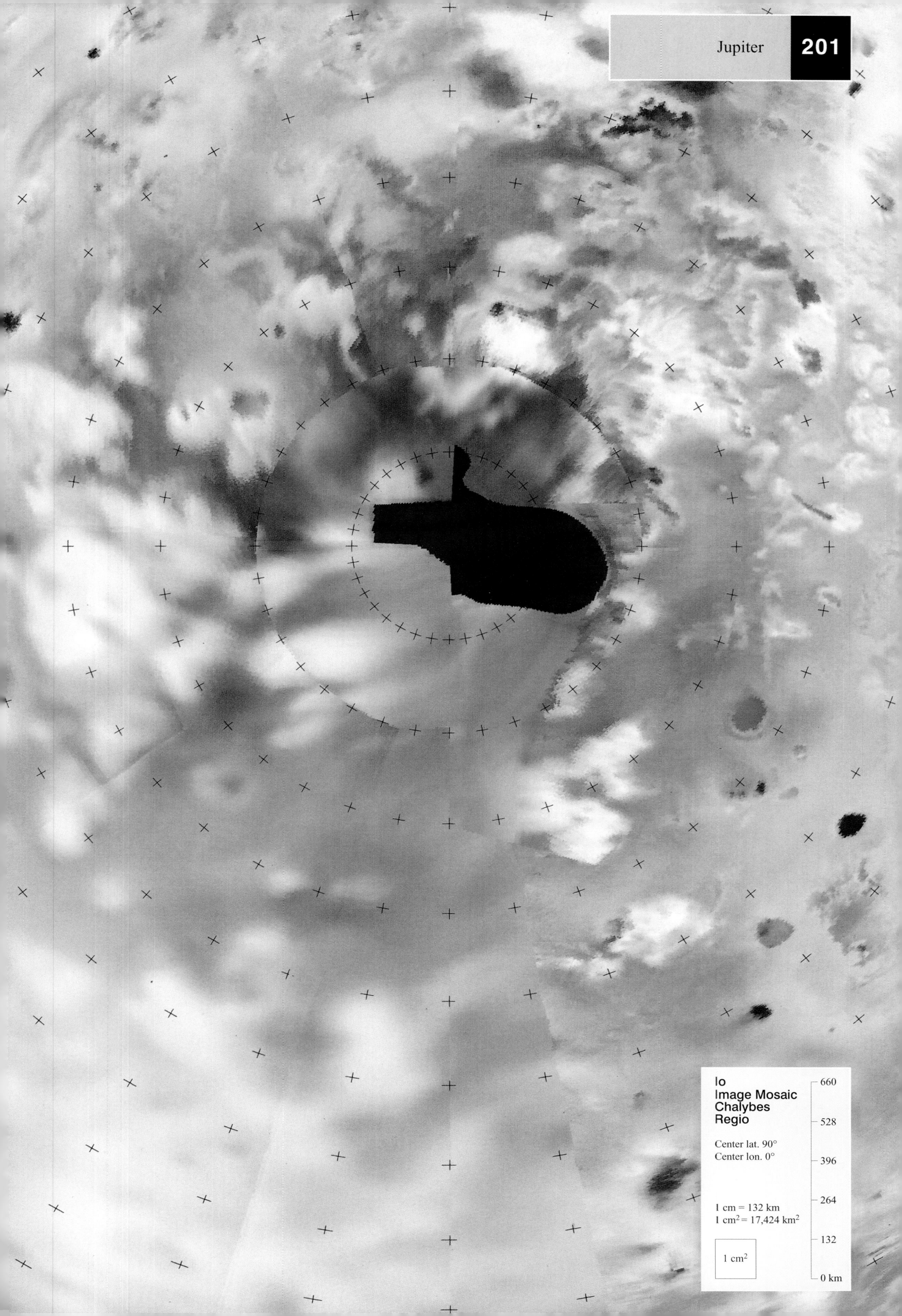
Io
Image Mosaic
Chalybes
Regio
Center lat. 90°
Center lon. 0°
1 cm = 132 km
1 cm² = 17,424 km²
1 cm²
660
528
396
264
132
0 km

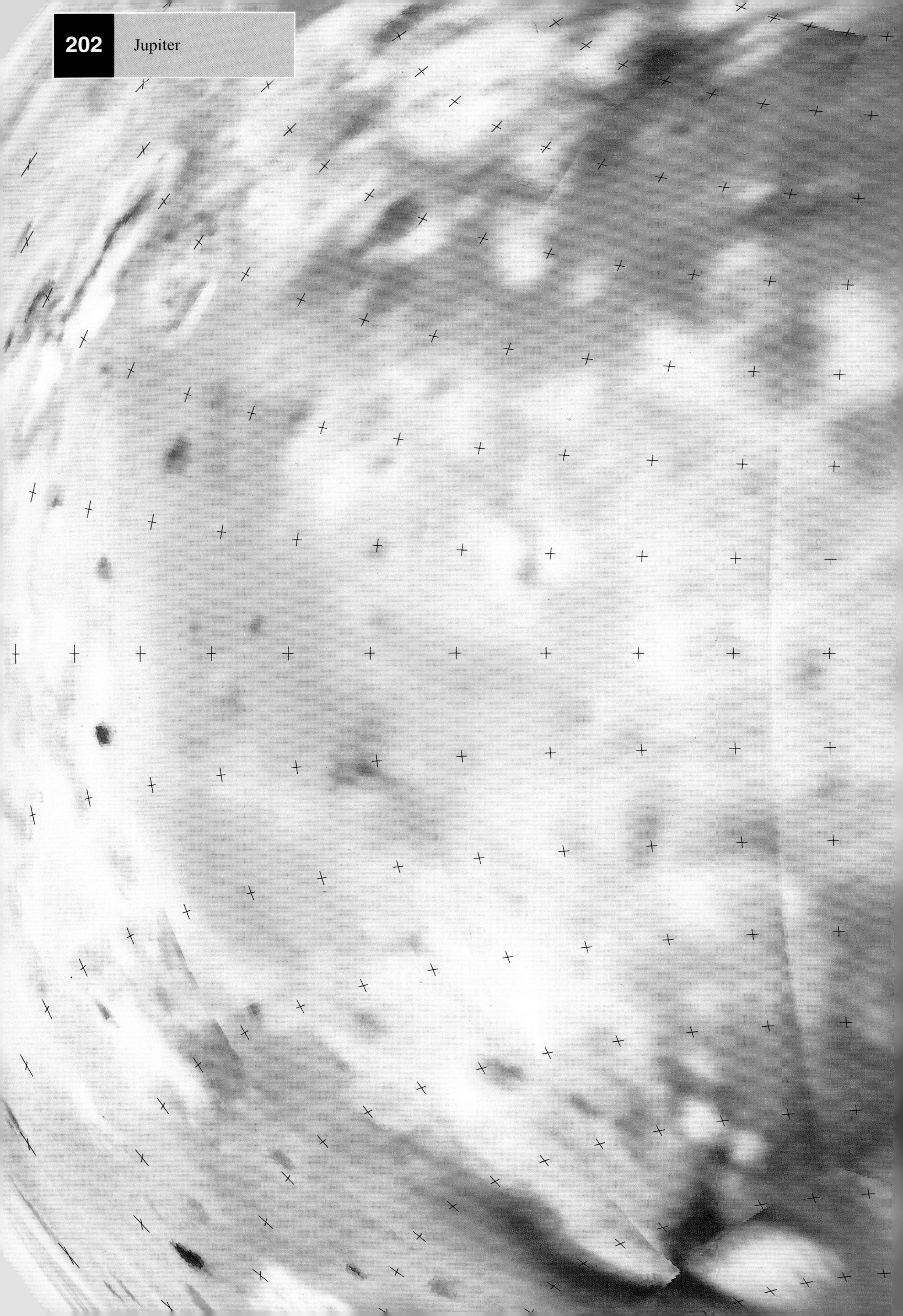

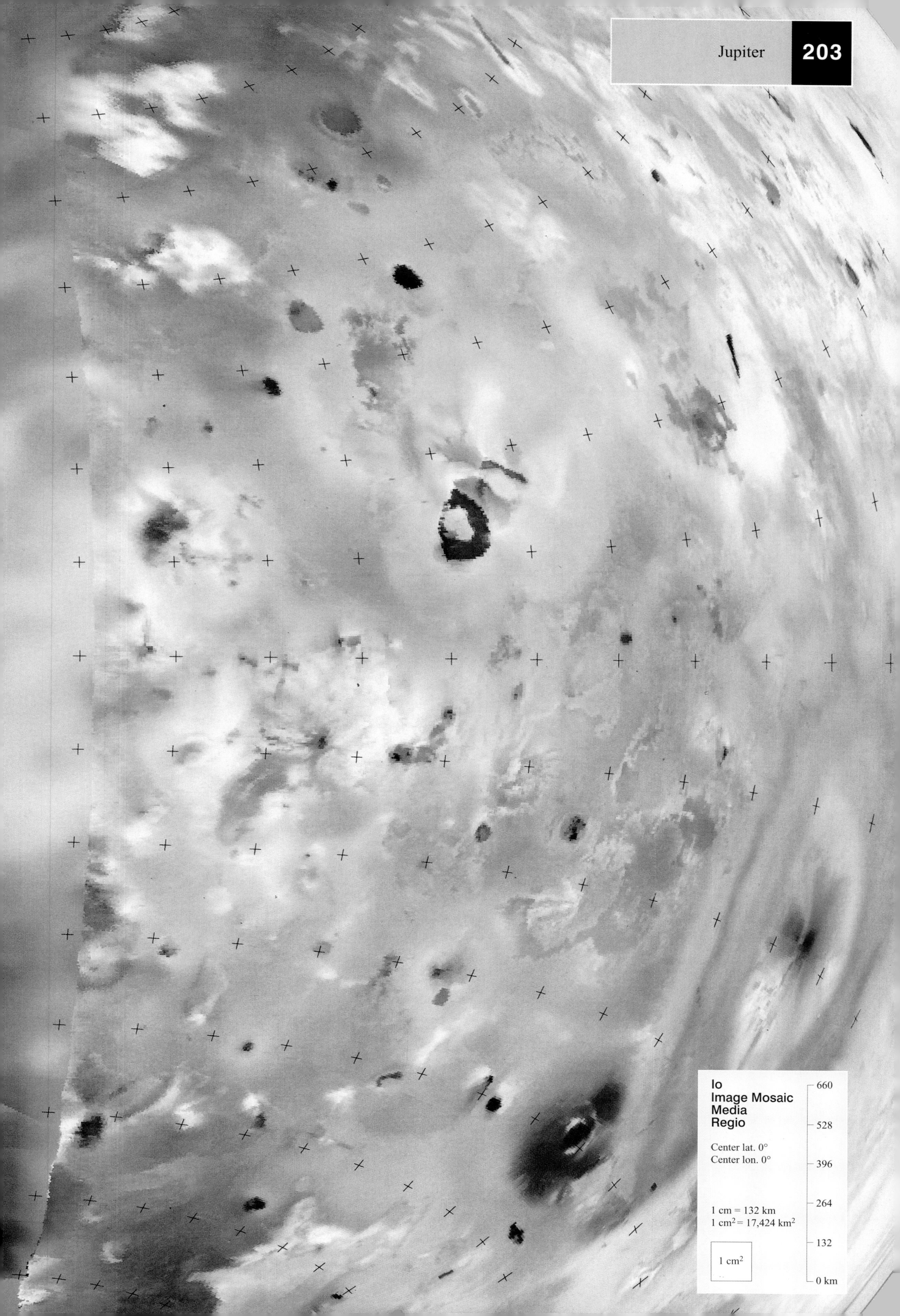
Io
Image Mosaic
Media
Regio
Center lat. 0°
Center lon. 0°
1 cm = 132 km
1 cm² = 17,424 km²
1 cm²
660
528
396
264
132
0 km

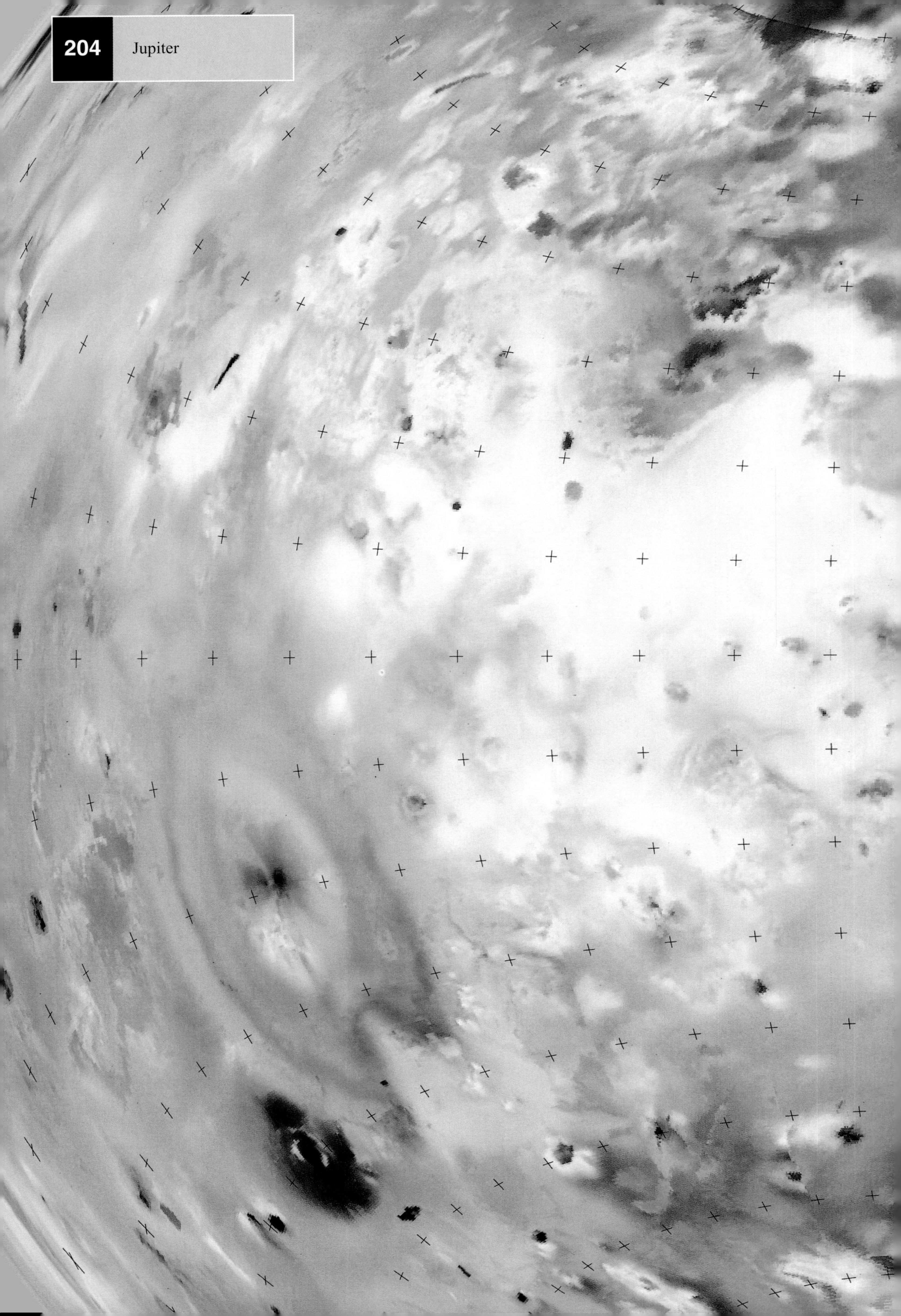

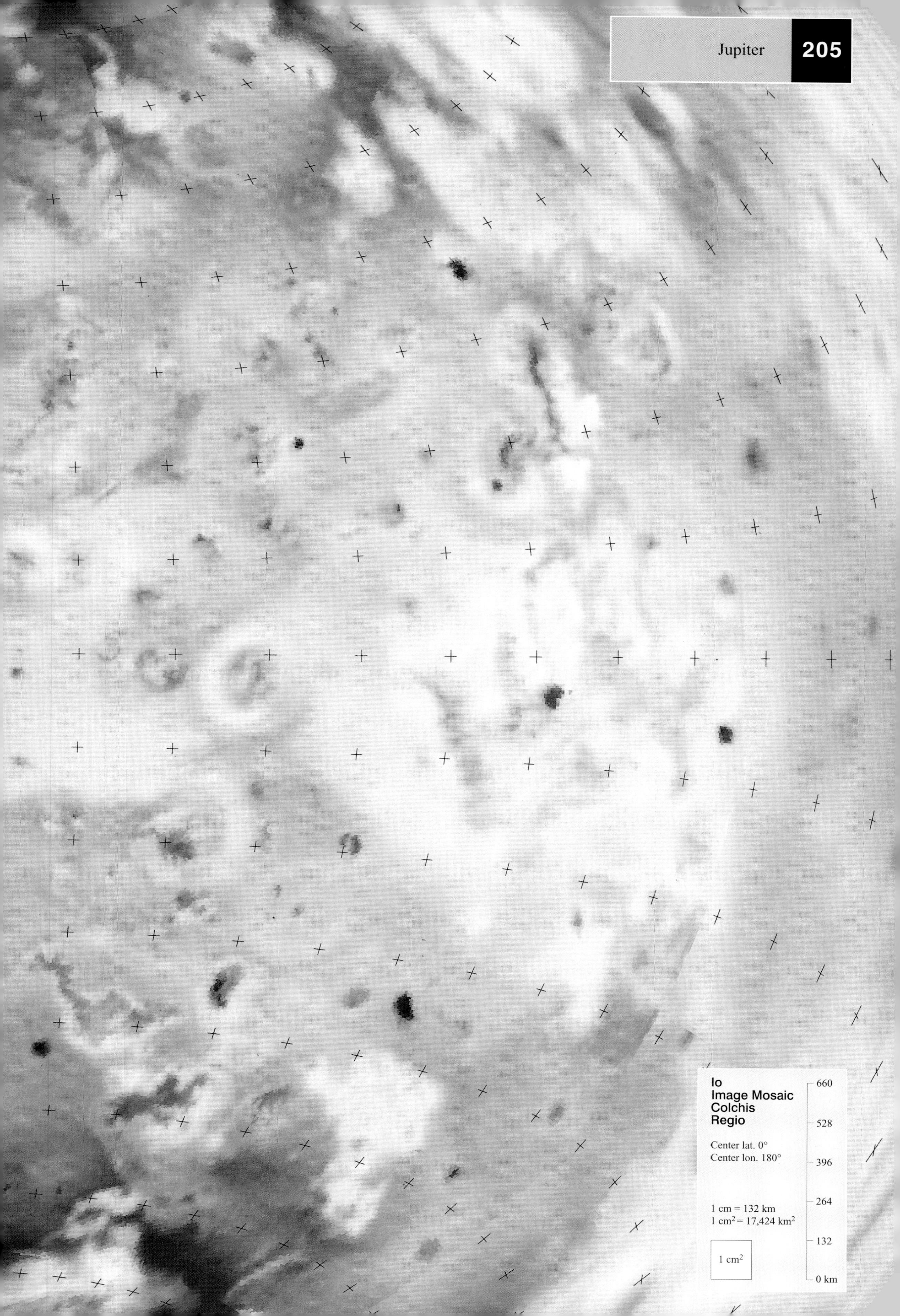
Io
Image Mosaic
Colchis
Regio
Center lat. 0°
Center lon. 180°
1 cm = 132 km
1 cm² = 17,424 km²
1 cm²
660
528
396
264
132
0 km

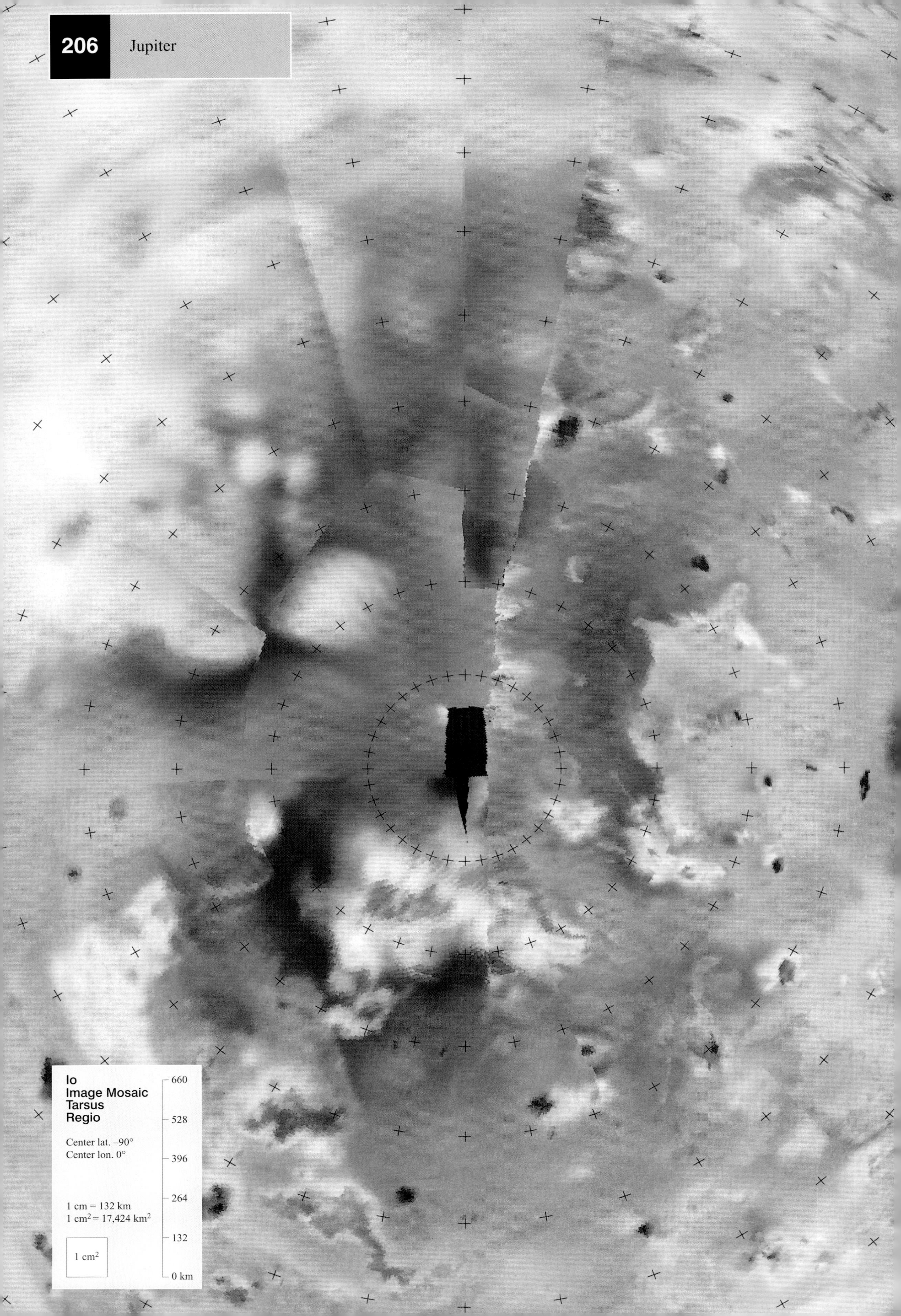
Io
Image Mosaic
Tarsus
Regio
Center lat. –90°
Center lon. 0°
1 cm = 132 km
1 cm² = 17,424 km²
1 cm²
660
528
396
264
132
0 km

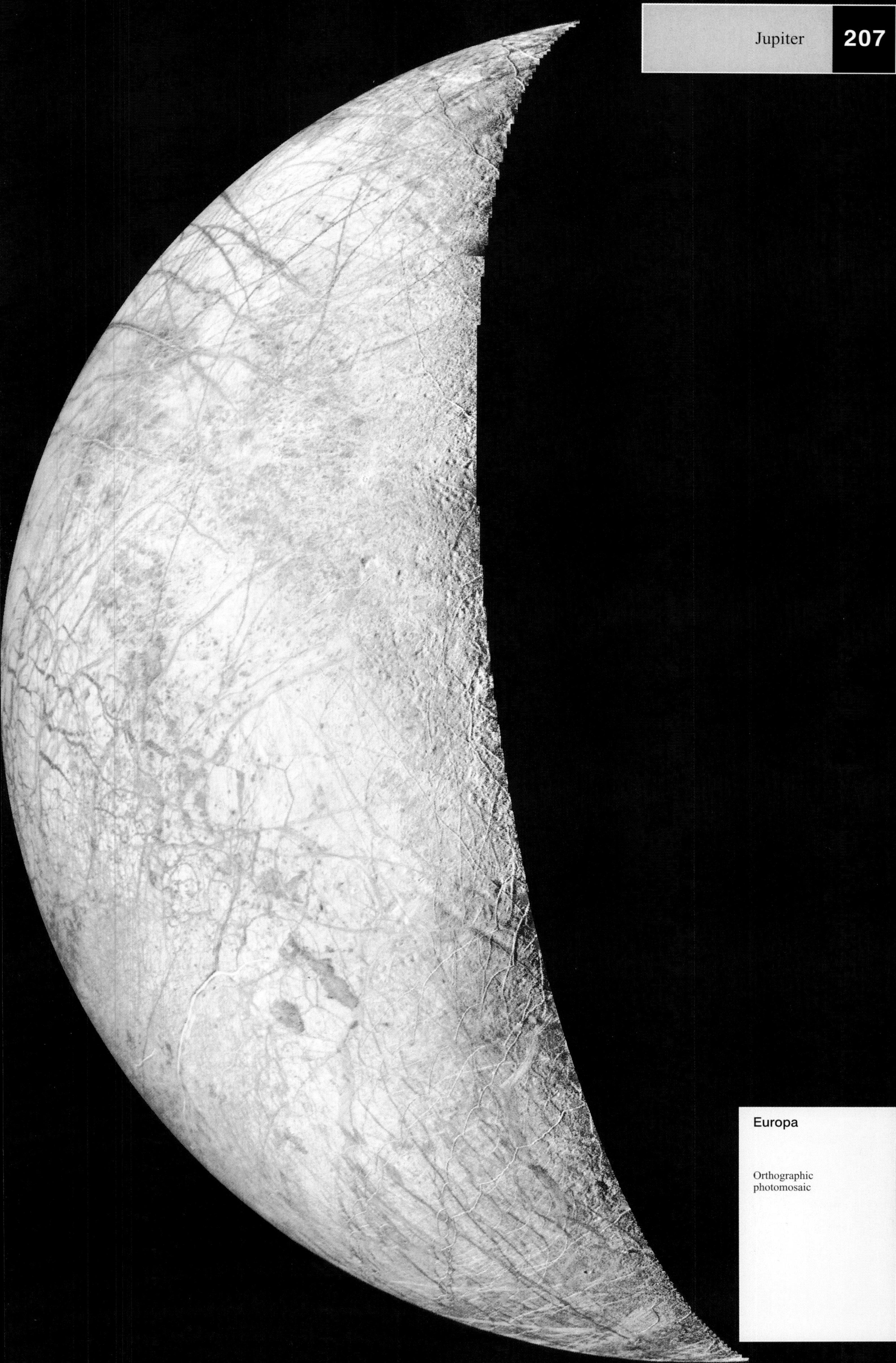

Europa

Orthographic photomosaic

270°
240°
210°
MINOS
ASTERIUS LINEA
BELUS LINEA
TECTAMU
PHOENIX
RHADAMANTHY
TELEPH
ARGIOPE LINEA
ECHION
PELORUS
ALPHESIBOE
AGENOR LINEA
PHOCIS FLEXUS
PHINEUS LINEA
LIBYA LINE
210°
240°
THASUS LINE
270°
Rhiann
300°

90°
120°
150°
TYRE
MACULA
LINEA
LINEA
Tegid
LINEA
Morvran
Taliesin
FLEXUS
THRACE
MACULA
GORTYNA
FLEXUS
FLEXUS
LINEA
FLEXUS
ADONIS LINEA
SARPEDON LINEA
120°
90°
60°
Europa
Center lat. –20°
Center lon. 180°
1 cm = 132 km
1 cm² = 17,424 km²
1 cm²
660
528
396
264
132
0 km

Ganymede Index Map

Center lat. 45°
Center lon. 180°

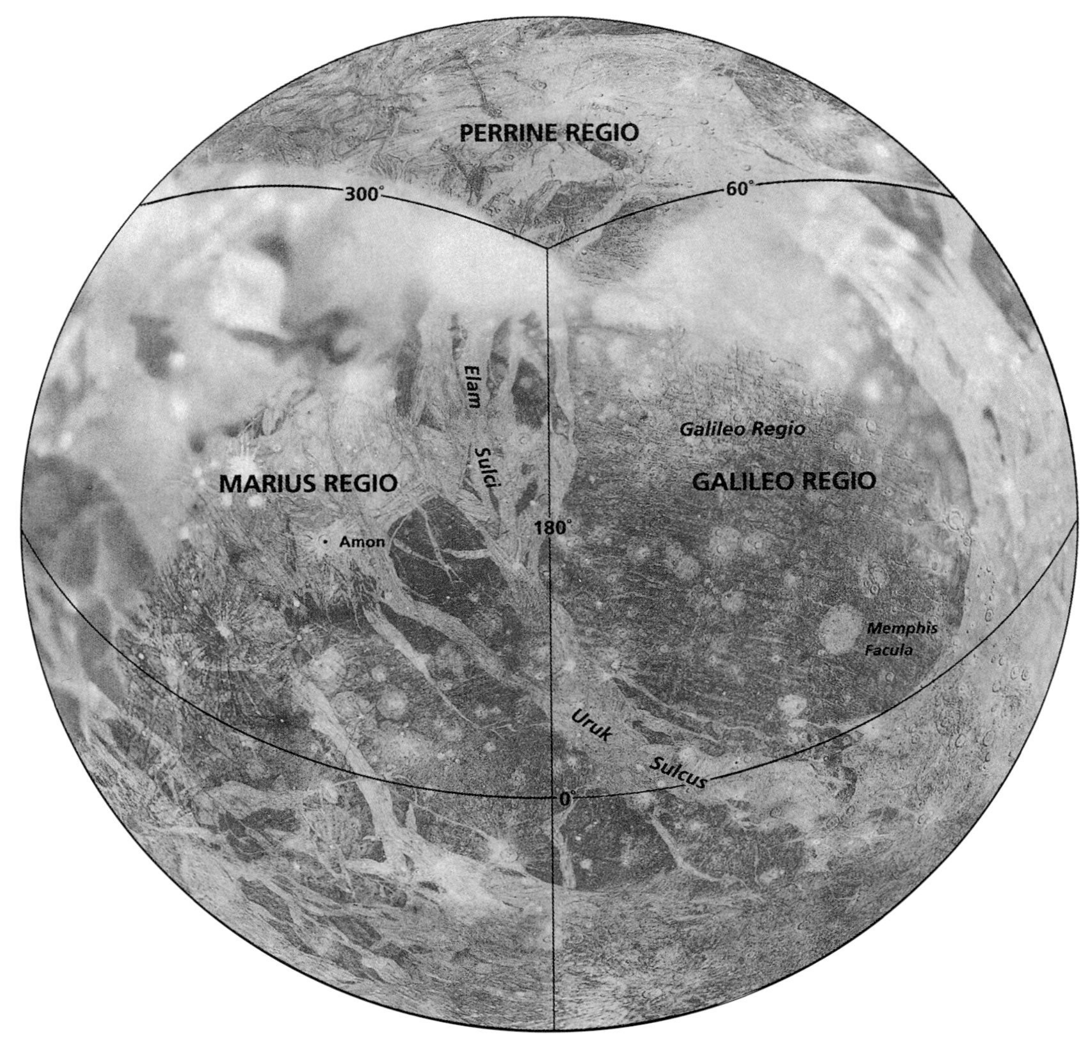

Ganymede Index Map

Center lat. −45°
Center lon. 180°

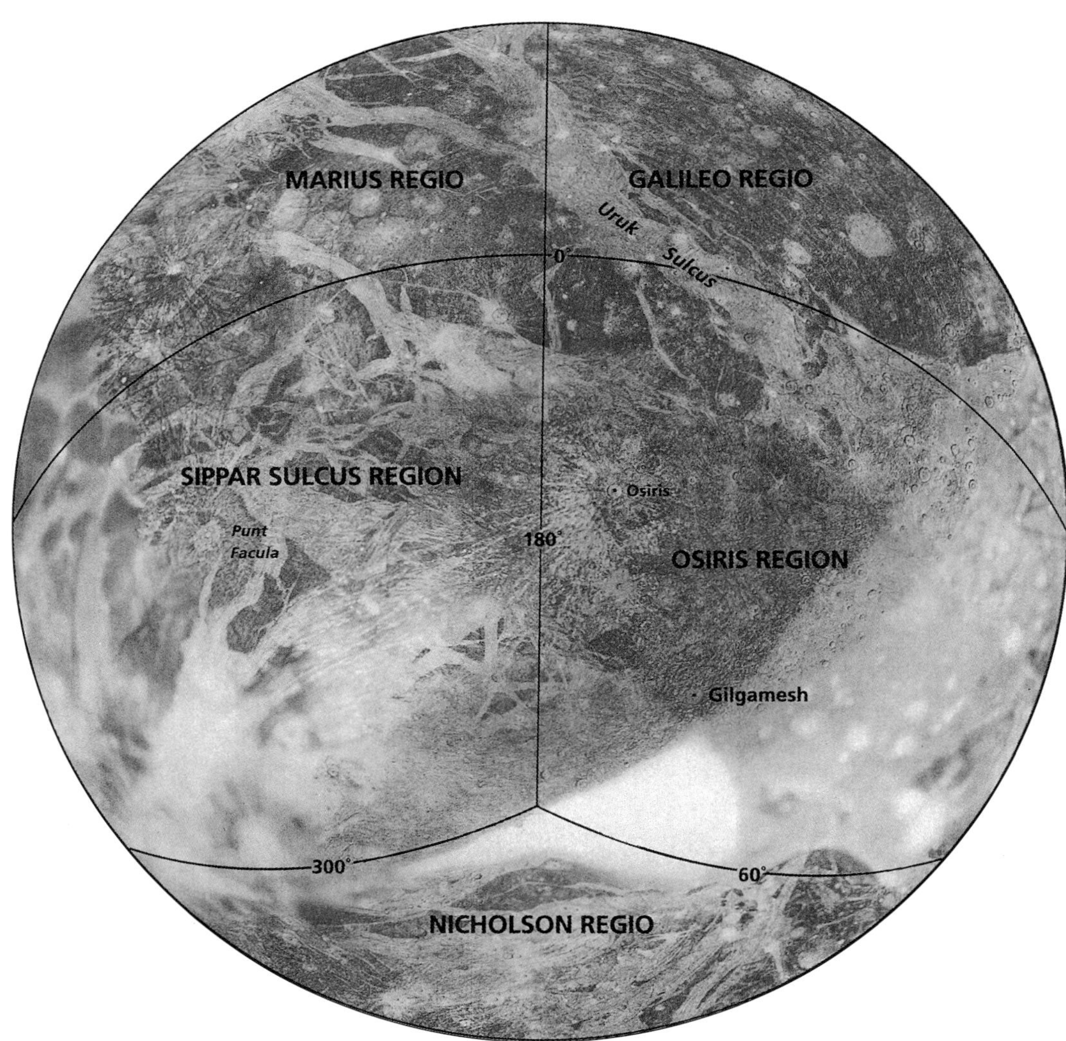

Ganymede Index Map

Center lat. 0°
Center lon. 0°

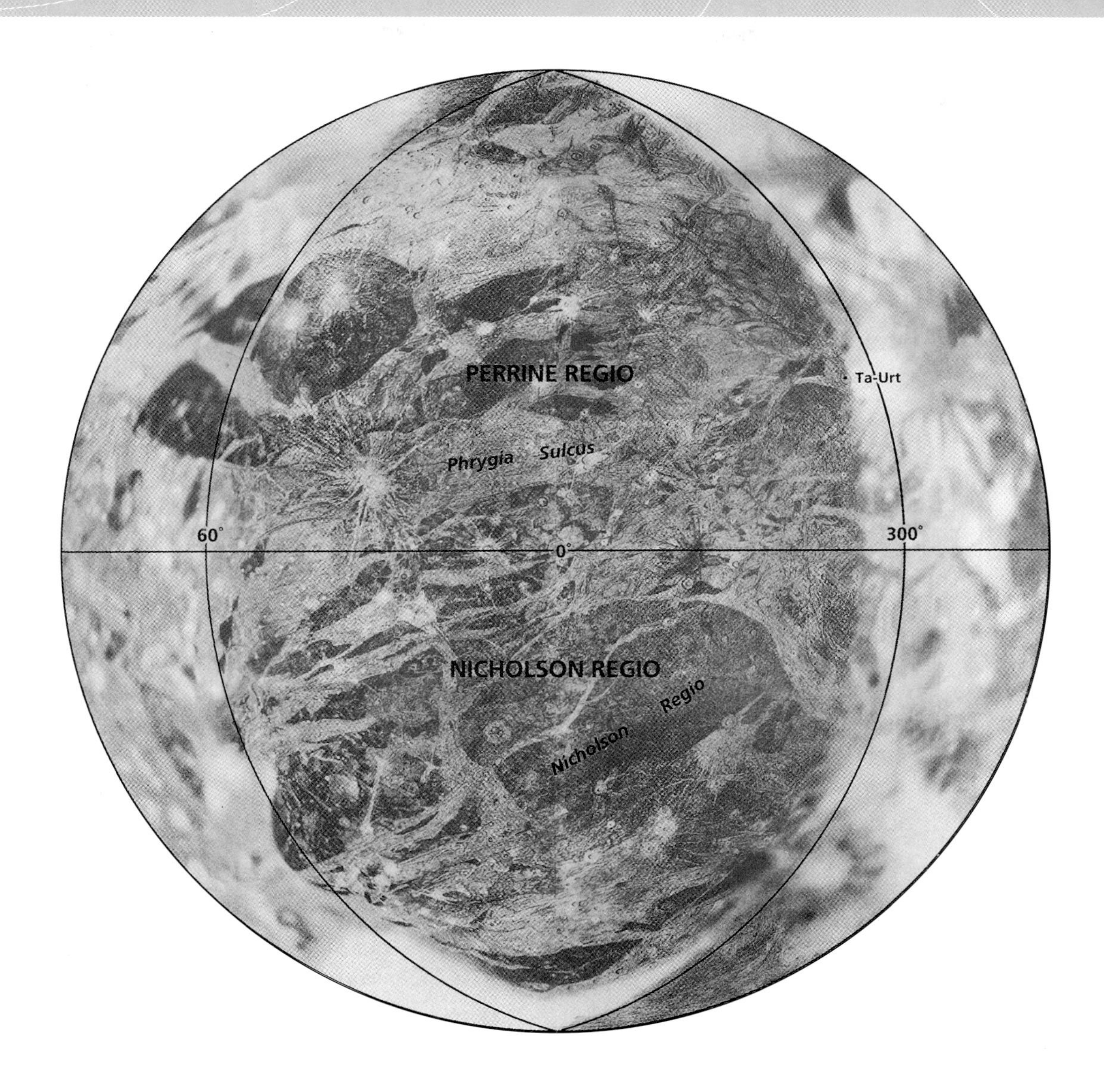

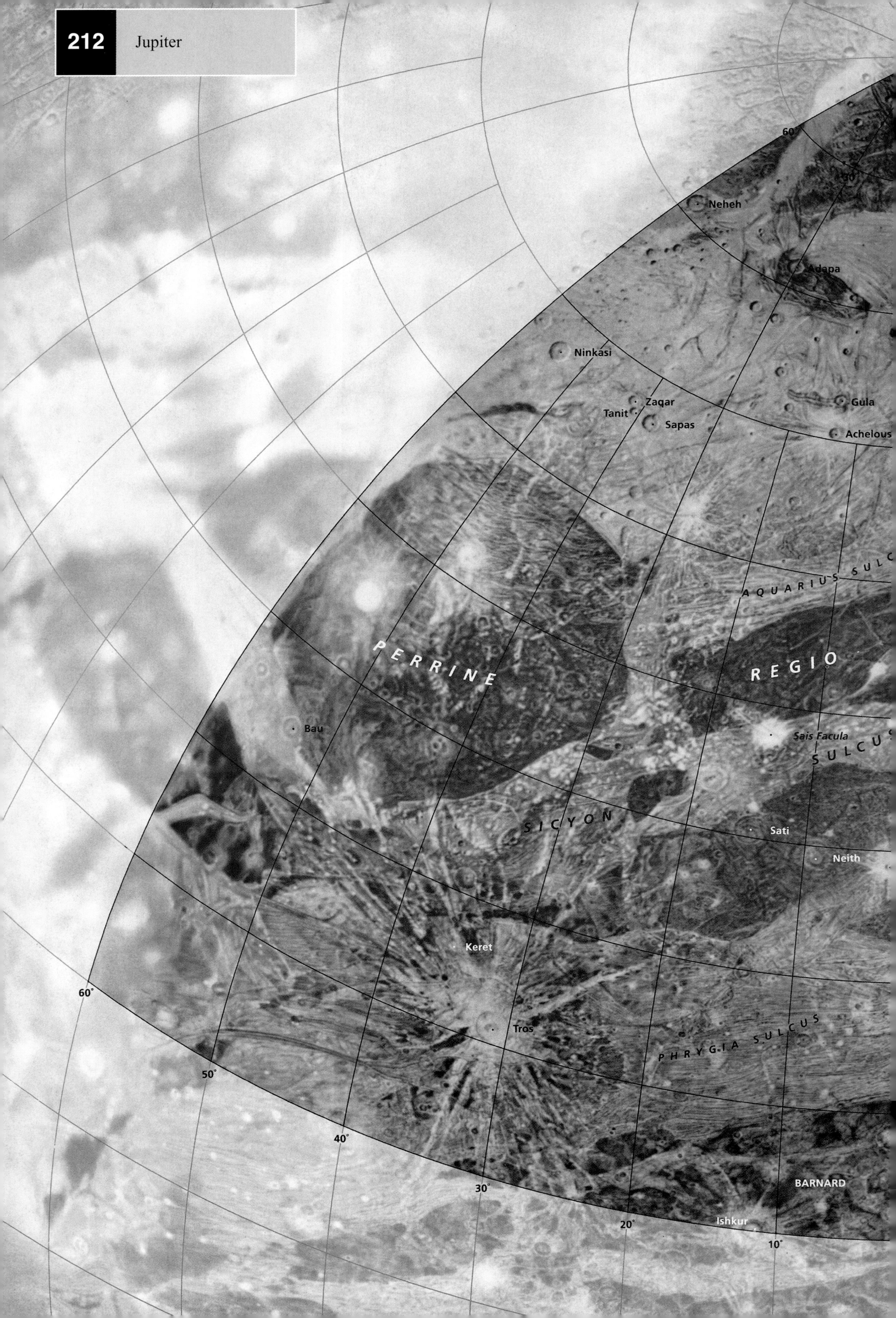

Neheh
Adapa
Ninkasi
Zaqar
Tanit
Sapas
Gula
Achelous
AQUARIUS SULC
PERRINE
REGIO
Bau
Sais Facula
SULCUS
SICYON
Sati
Neith
Keret
Tros
PHRYGIA SULCUS
BARNARD
Ishkur
60°
50°
40°
30°
20°
10°

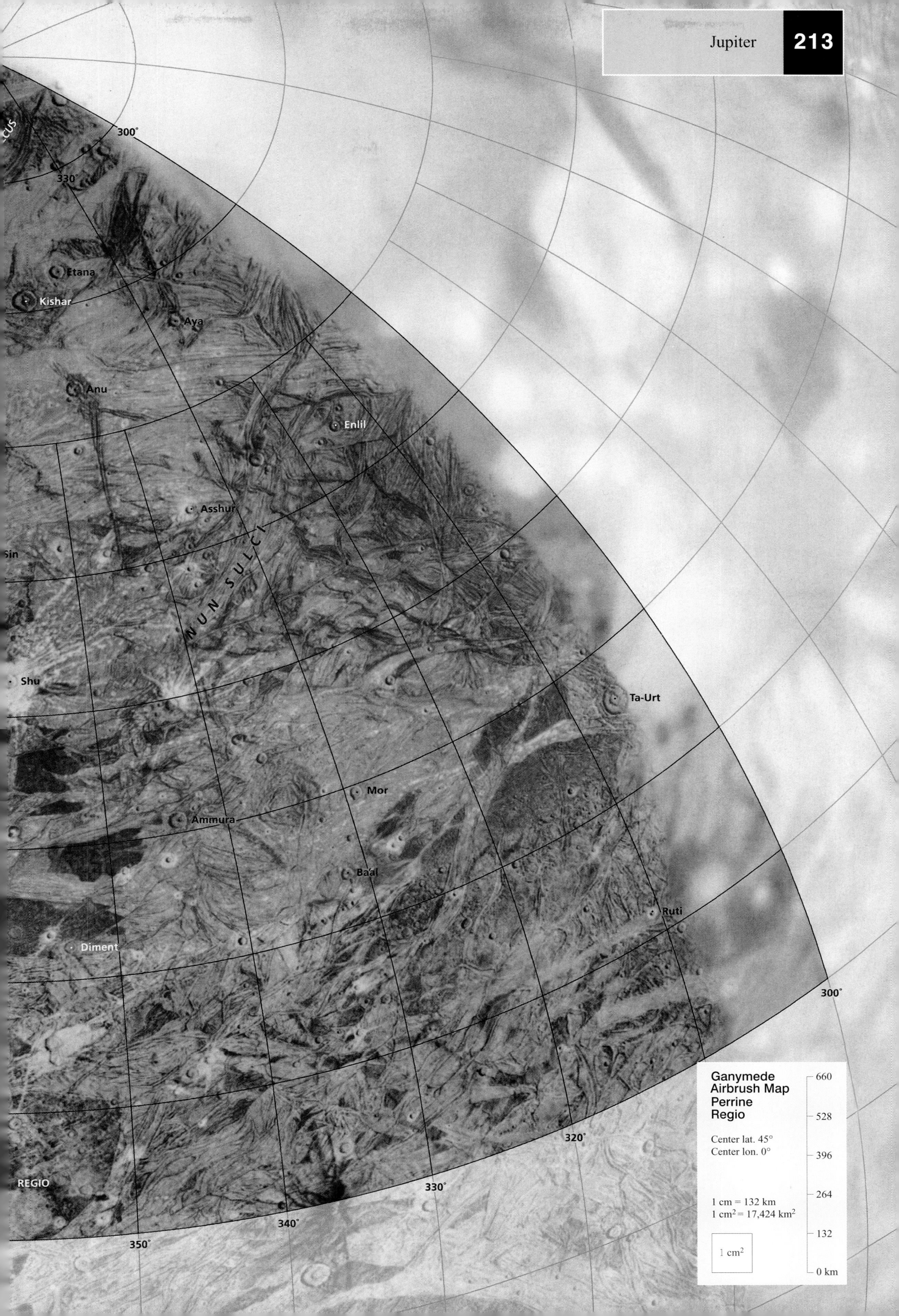
300°
330°
Etana
Kishar
Aya
Anu
Enlil
Asshur
Sin
NUN SULCI
Shu
Ta-Urt
Mor
Ammura
Baal
Ruti
Diment
300°
REGIO
320°
330°
340°
350°
Ganymede
Airbrush Map
Perrine
Regio
Center lat. 45°
Center lon. 0°
1 cm = 132 km
1 cm² = 17,424 km²
1 cm²
660
528
396
264
132
0 km

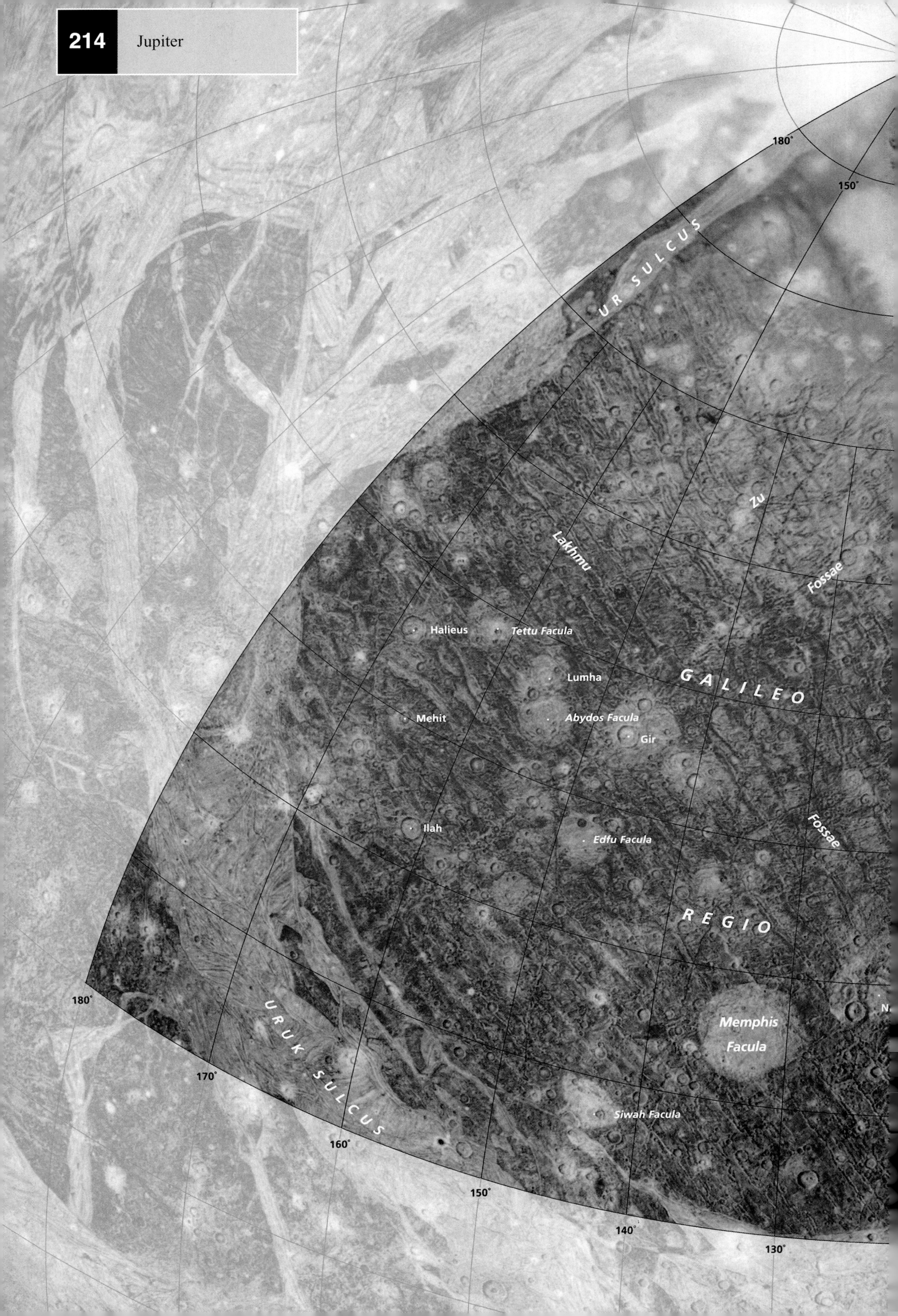

180°
150°
UR SULCUS
Zu
Lakhmu
Fossae
Halieus
Tettu Facula
Lumha
GALILEO
Mehit
Abydos Facula
Gir
Ilah
Edfu Facula
Fossae
REGIO
180°
Memphis Facula
URUK SULCUS
170°
Siwah Facula
160°
150°
140°
130°

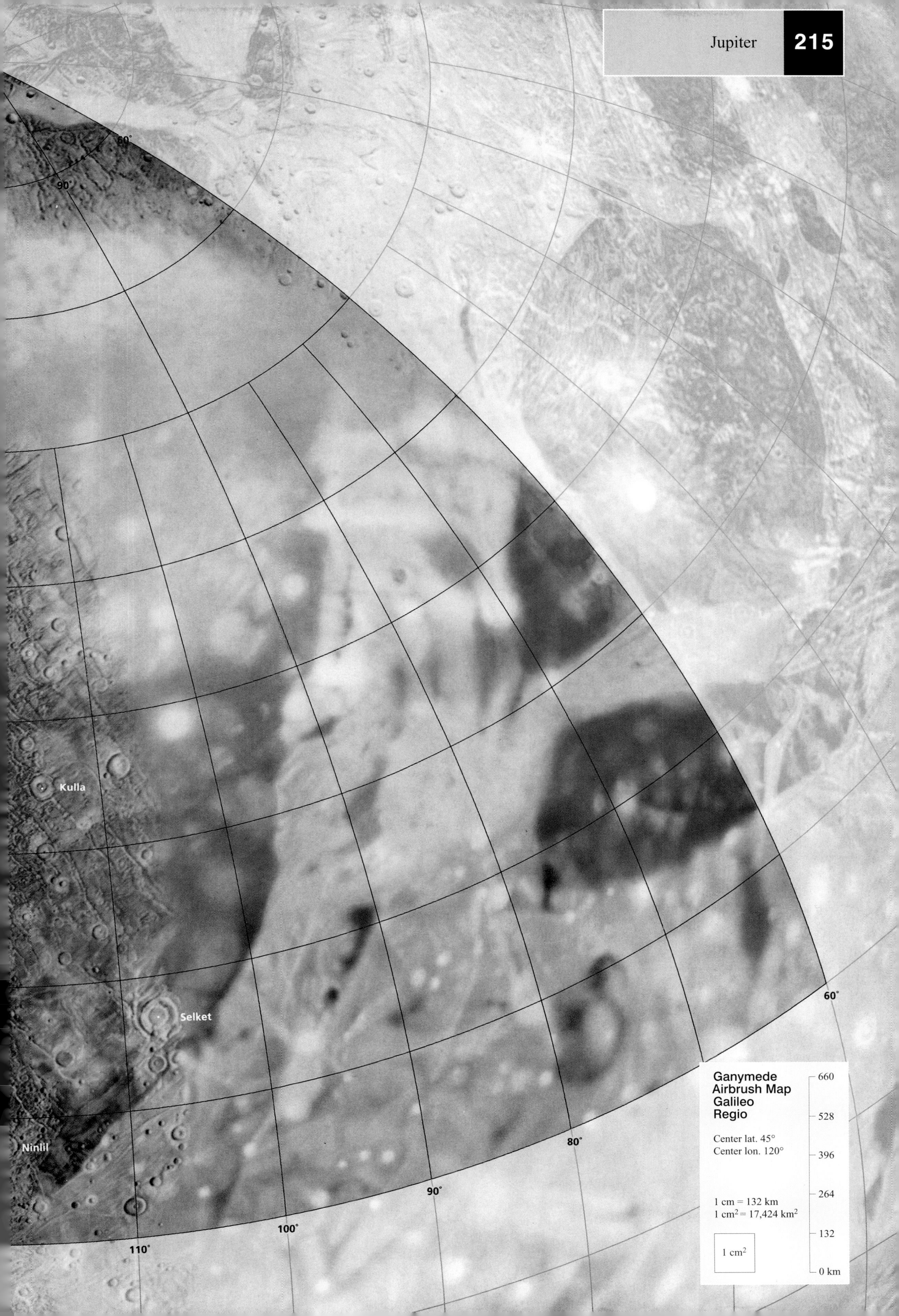
60°
90°
Kulla
Selket
Ninlil
60°
80°
90°
100°
110°
Ganymede
Airbrush Map
Galileo
Regio
Center lat. 45°
Center lon. 120°
1 cm = 132 km
1 cm² = 17,424 km²
1 cm²
660
528
396
264
132
0 km

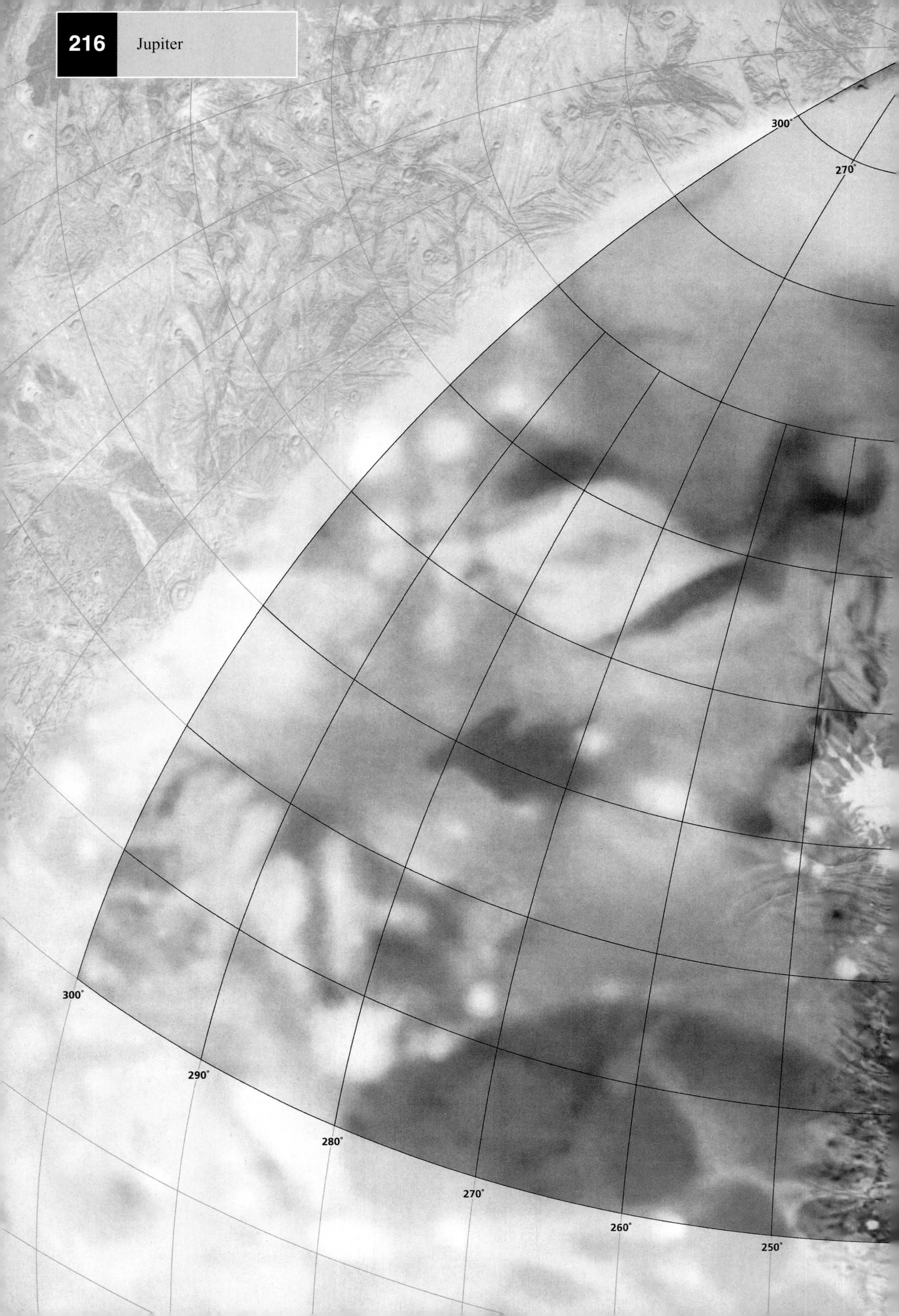
300°
270°
300°
290°
280°
270°
260°
250°

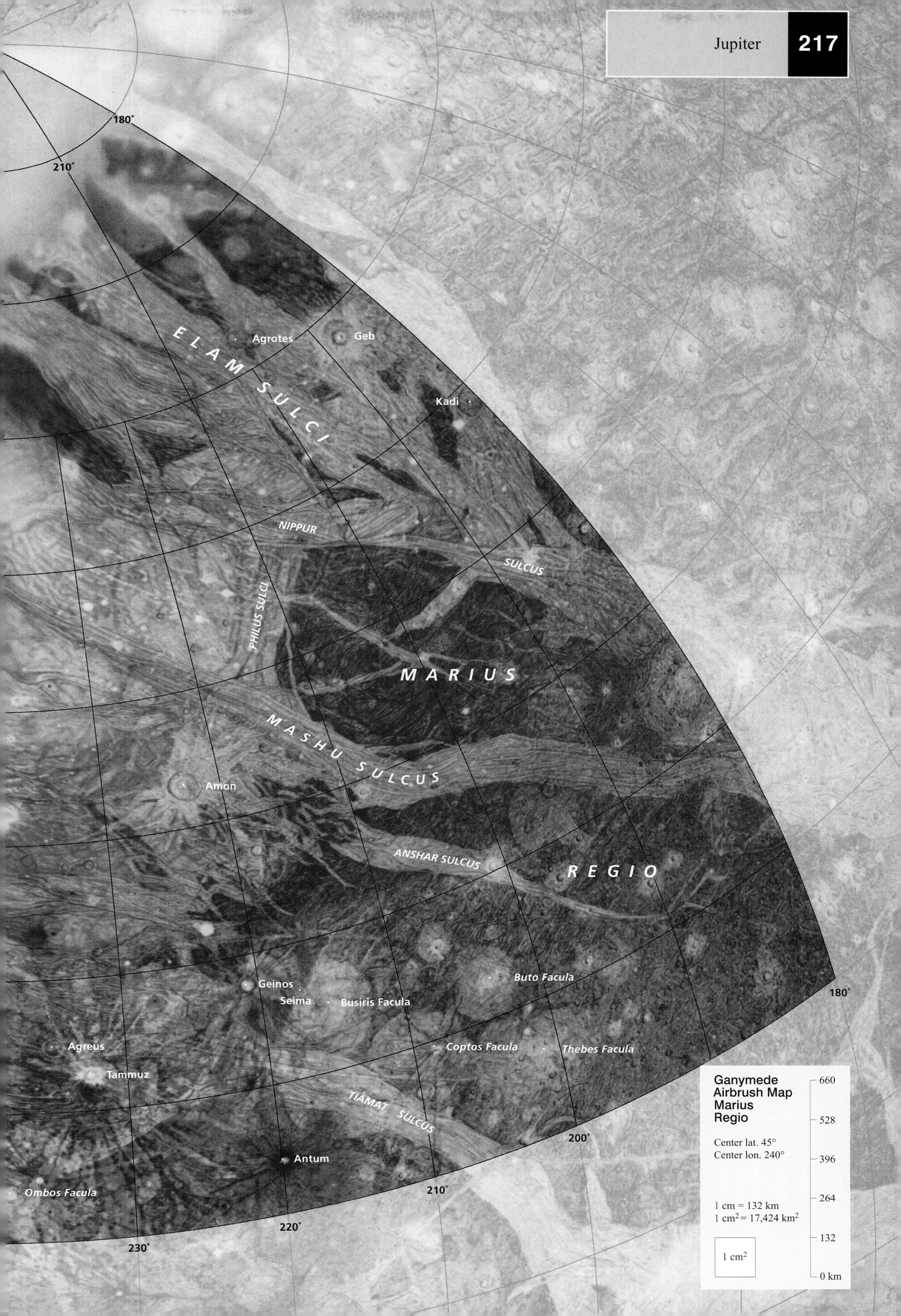

180°
210°
ELAM SULCI
Agrotes
Geb
Kadi
NIPPUR
SULCUS
PHILUS SULCI
MARIUS
MASHU SULCUS
Amon
ANSHAR SULCUS
REGIO
Buto Facula
Geinos
Seima
Busiris Facula
180°
Agreus
Tammuz
Coptos Facula
Thebes Facula
TIAMAT SULCUS
200°
Antum
Ombos Facula
210°
220°
230°
Ganymede
Airbrush Map
Marius
Regio
Center lat. 45°
Center lon. 240°
1 cm = 132 km
1 cm² = 17,424 km²
1 cm²
660
528
396
264
132
0 km

10°
20°
30°
40°
50°
60°
BARNARD REGIO
Danel
SULCI
MYSIA
SULCUS
Nabu
DARDANUS
30°
60°
90°

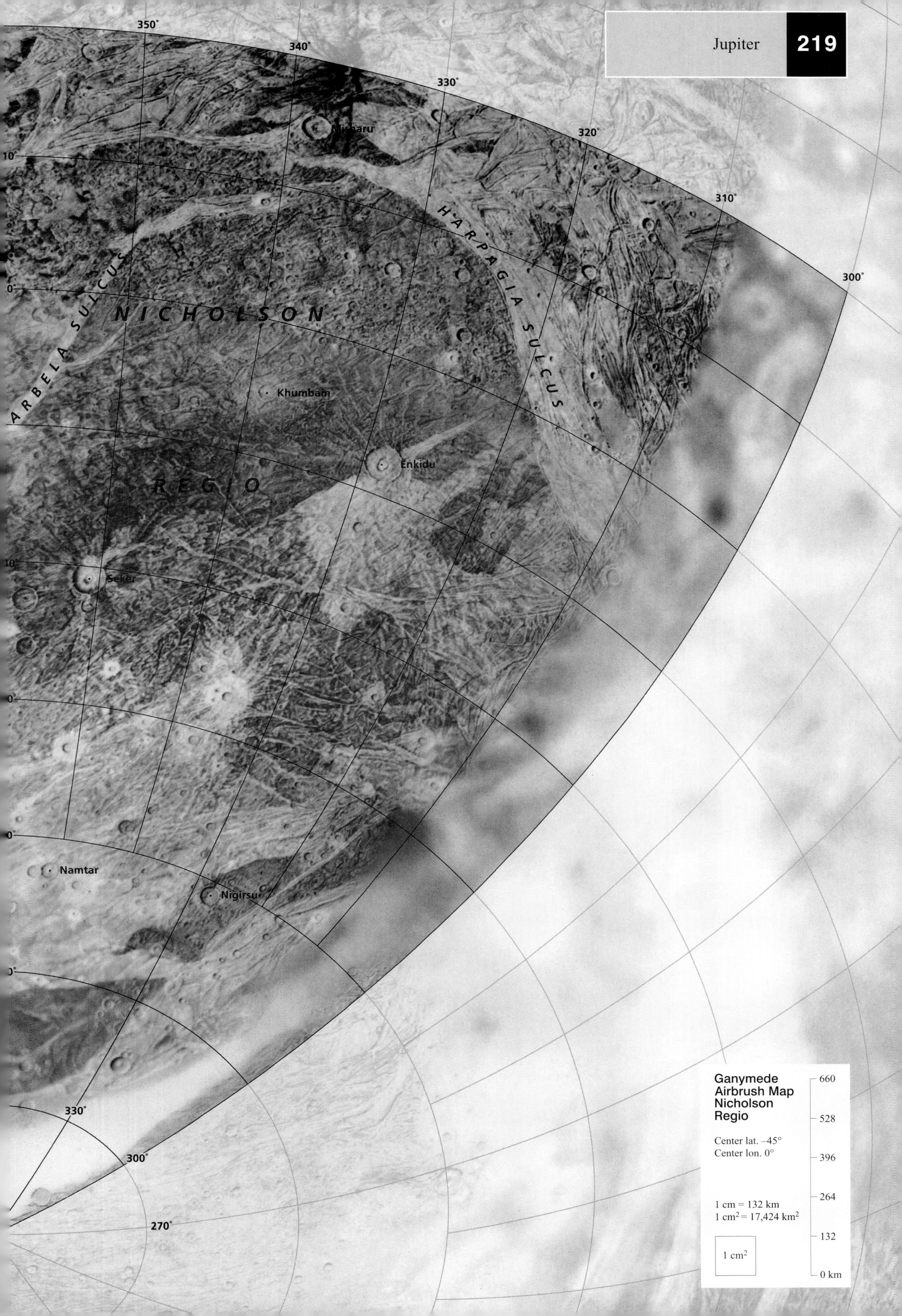
350°
340°
330°
320°
310°
300°
Misharu
HARPAGIA SULCUS
ARBELA SULCUS
NICHOLSON
REGIO
Khumbam
Enkidu
Seker
Namtar
Nigirsu
330°
300°
270°
Ganymede
Airbrush Map
Nicholson
Regio
Center lat. –45°
Center lon. 0°
1 cm = 132 km
1 cm² = 17,424 km²
1 cm²
660
528
396
264
132
0 km

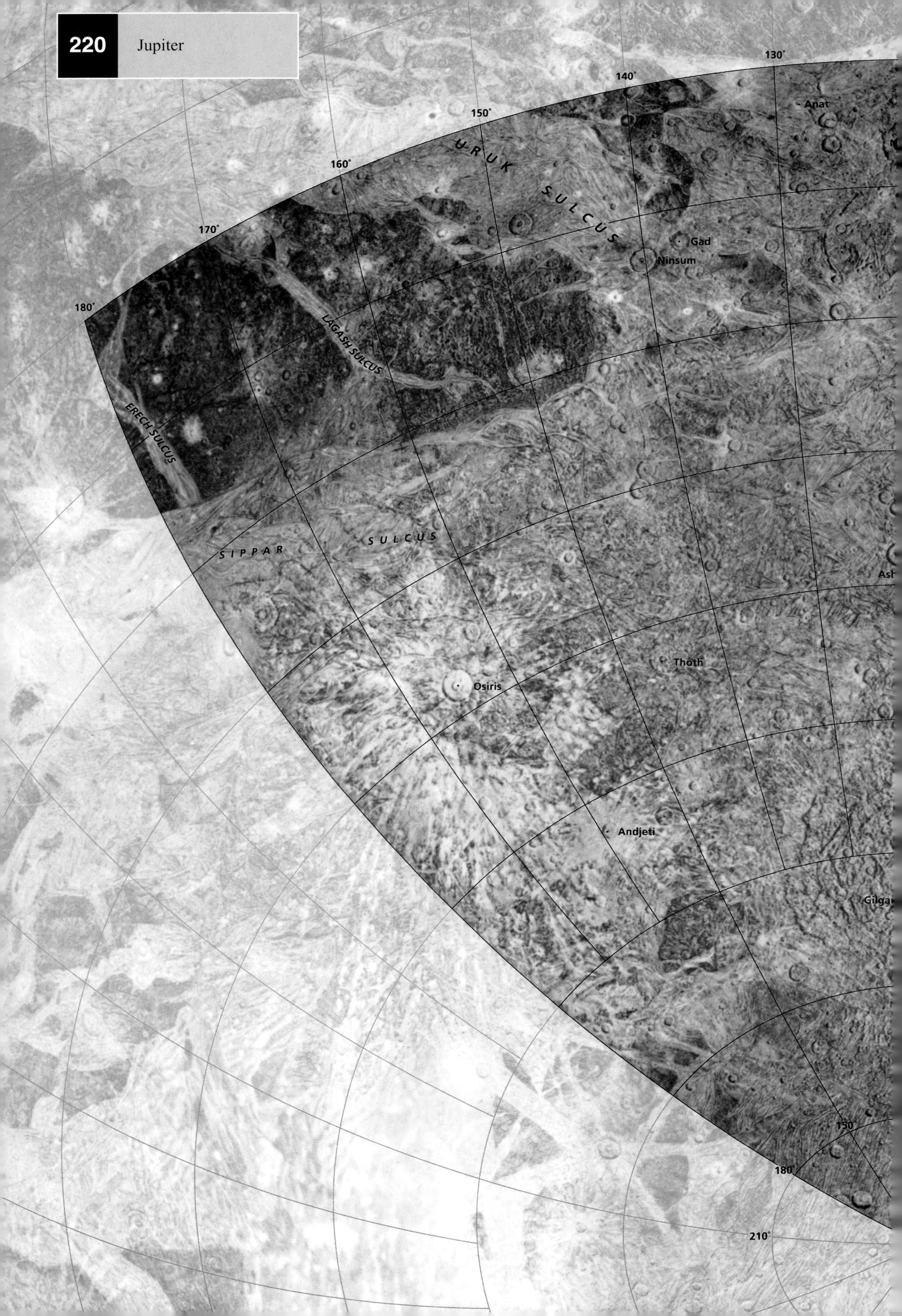
130°
140°
150°
160°
170°
180°
Anat
URUK SULCUS
Gad
Ninsum
LAGASH SULCUS
ERECH SULCUS
SIPPAR SULCUS
Thoth
Osiris
Andjeti
150°
180°
210°

110°
100°
90°
80°
70°
60°
Ilus
Mush
Irkalla
90°
ubis
60°
30°
Ganymede
Airbrush Map
Osiris
Region
Center lat. –45°
Center lon. 120°
1 cm = 132 km
1 cm² = 17,424 km²
1 cm²
660
528
396
264
132
0 km

250°
260°
Dendera Facula
270°
280°
290°
300°
Nut
BUBASTIS
Hathor
270°
300°
330°

230°
220°
210°
200°
190°
180°
Mir
SULCUS
HURSAG SULCUS
Lakhamu
Fossa
KISHAR
SIPPAR
SULCUS
Melkart
Eshmun
SU SULCI
Kingu
Hapi
Bes
Khonsu
Ptah
Isis
210°
180°
150°
Ganymede
Airbrush Map
Sippar Sulcus
Region
Center lat. –45°
Center lon. 240°
1 cm = 132 km
1 cm² = 17,424 km²
1 cm²
660
528
396
264
132
0 km

Callisto Index Map

Center lat. –45°
Center lon. 180°

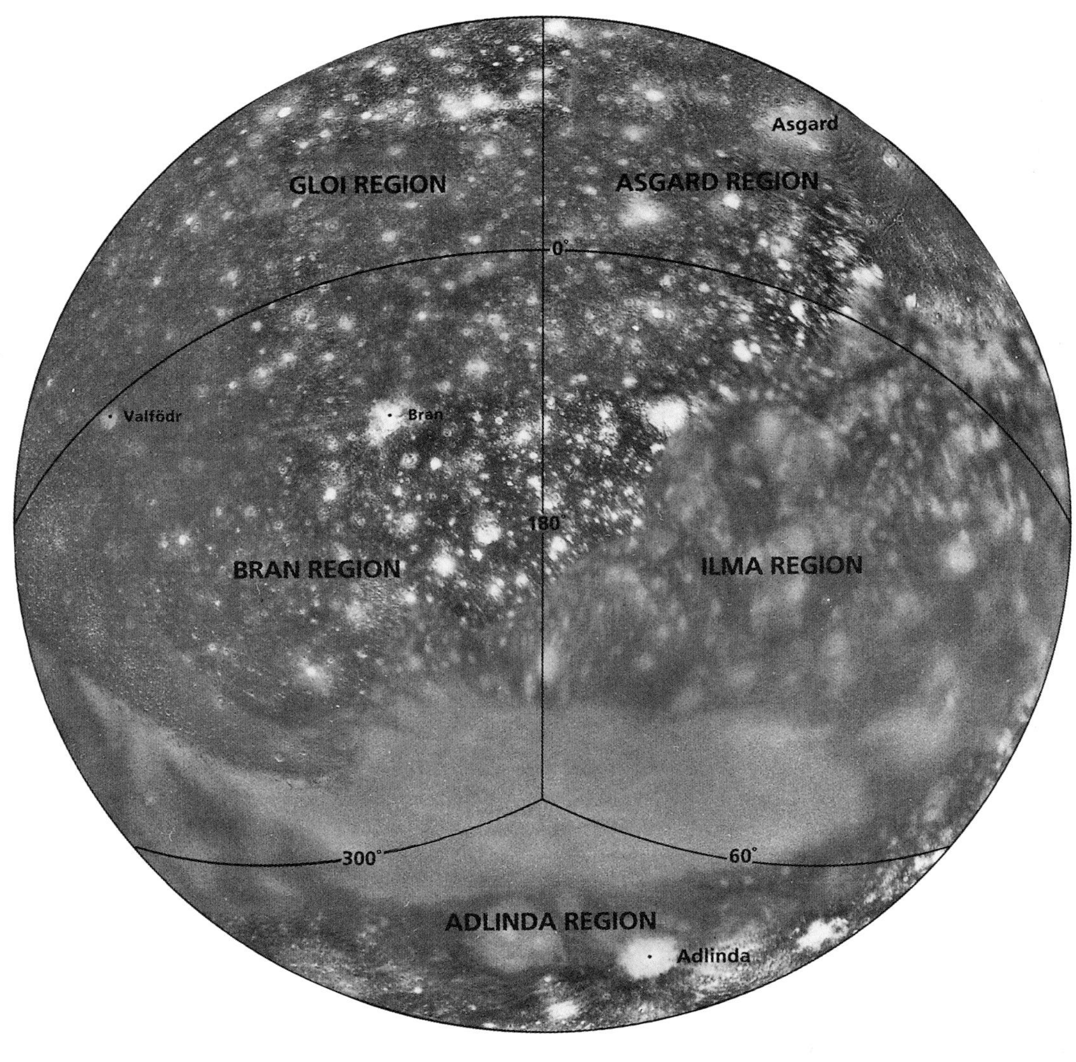

Callisto Index Map

Center lat. 45°
Center lon. 180°

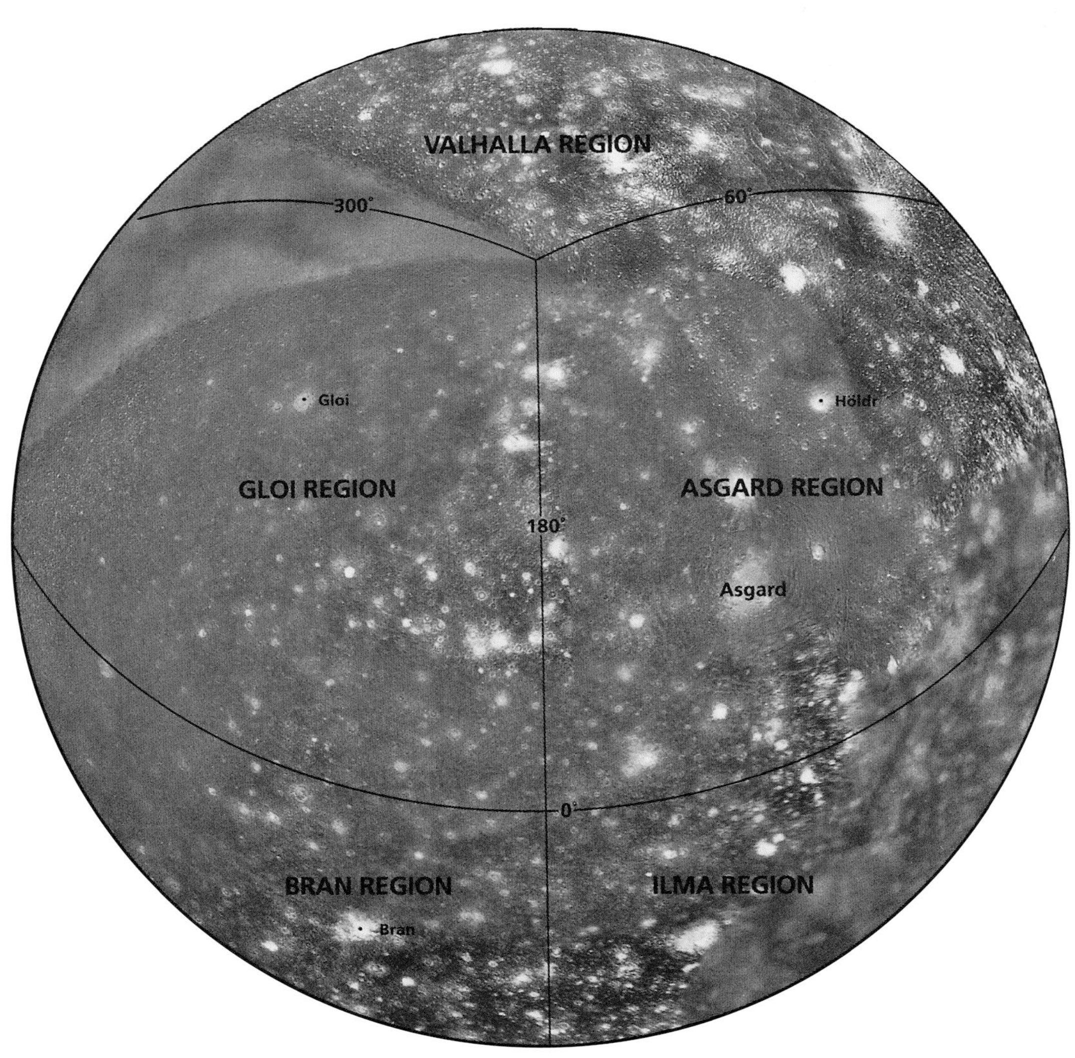

Callisto
Index Map

Center lat. 0°
Center lon. 0°

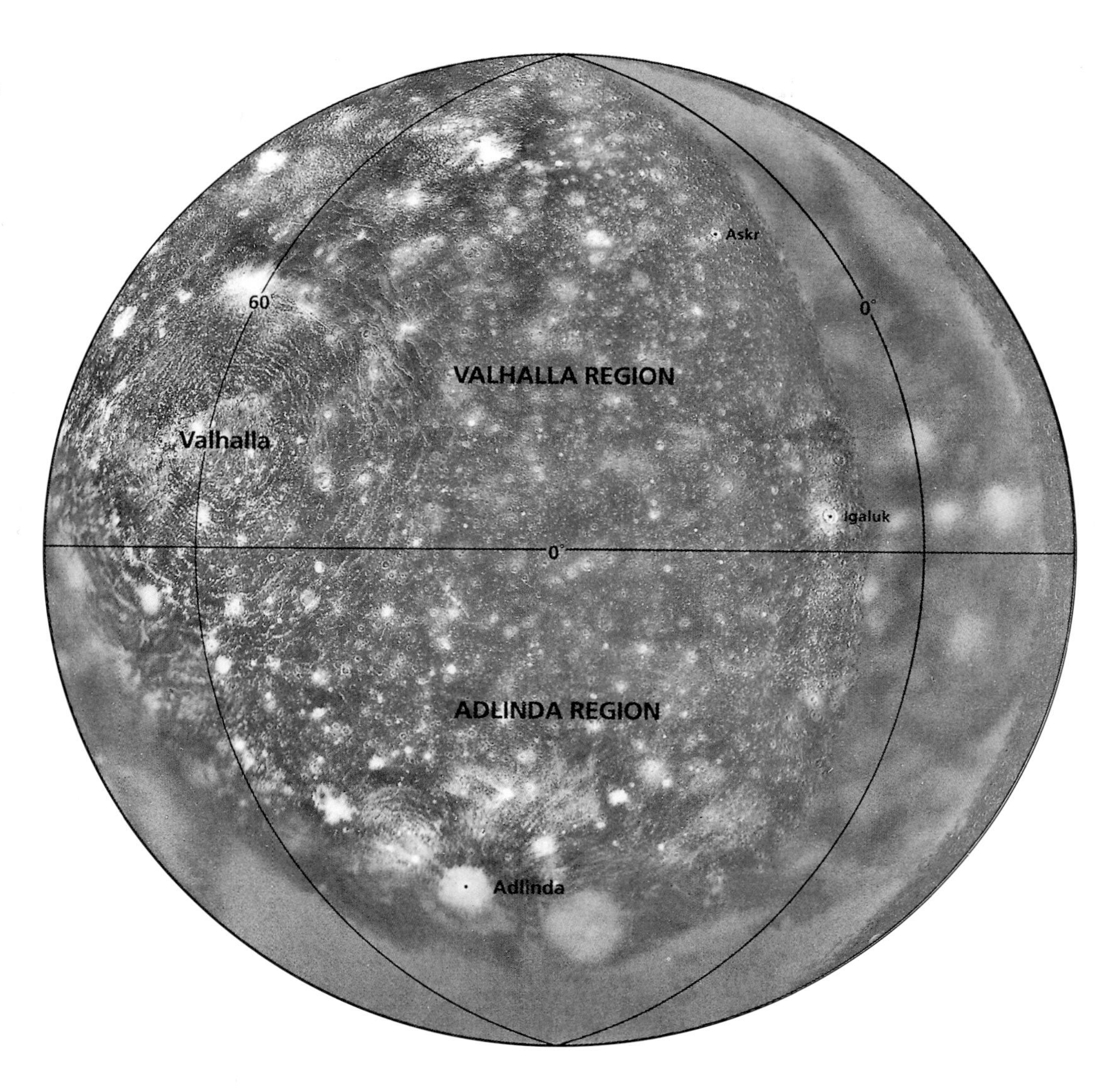

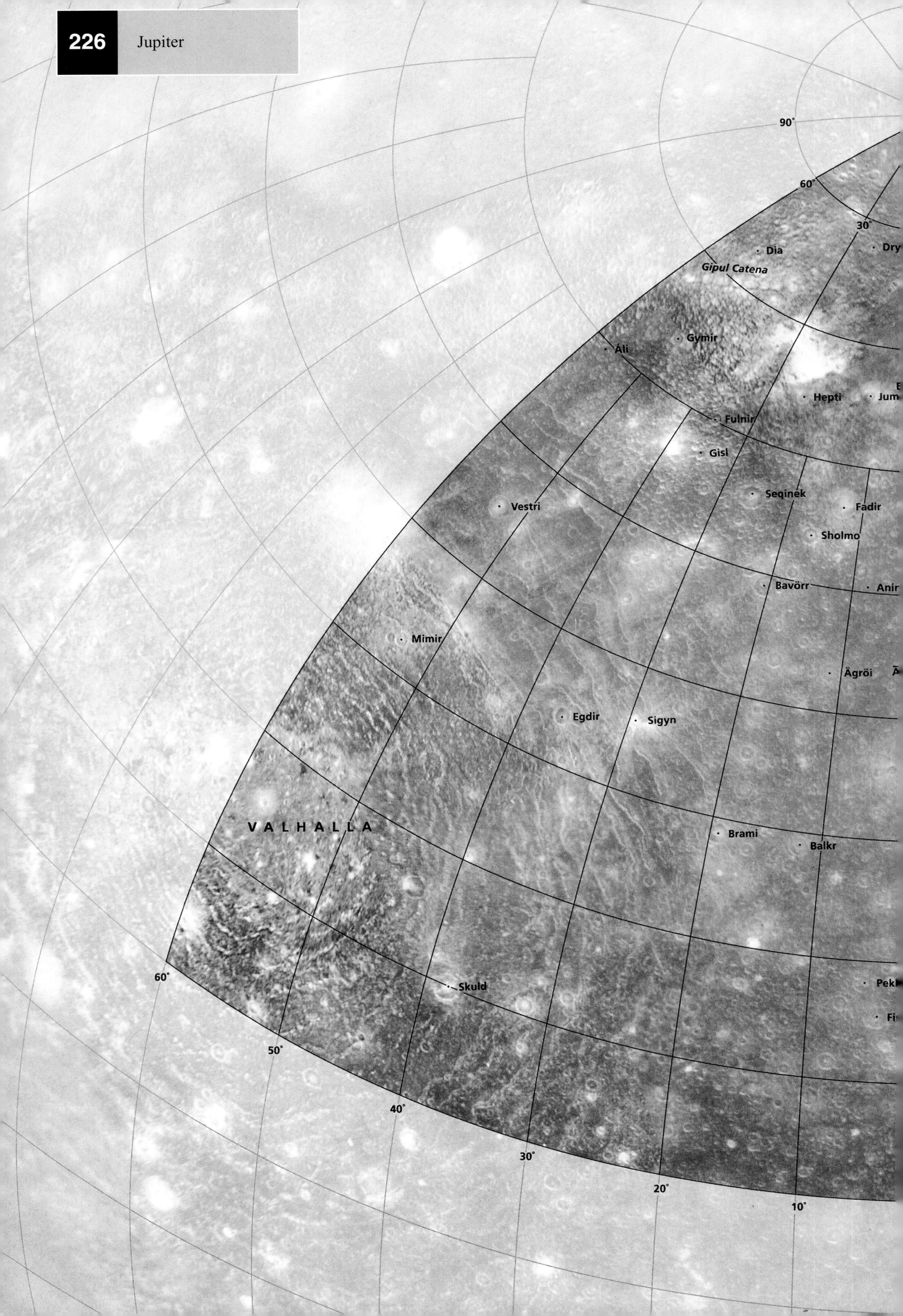
90°
60°
30°
Dia
Dry
Gipul Catena
Gymir
Áli
Hepti
Jum
Fulnir
Gisl
Seqinek
Vestri
Fadir
Sholmo
Bavörr
Anir
Mimir
Ägröi
Egdir
Sigyn
VALHALLA
Brami
Balkr
Skuld
Pek
Fi
60°
50°
40°
30°
20°
10°

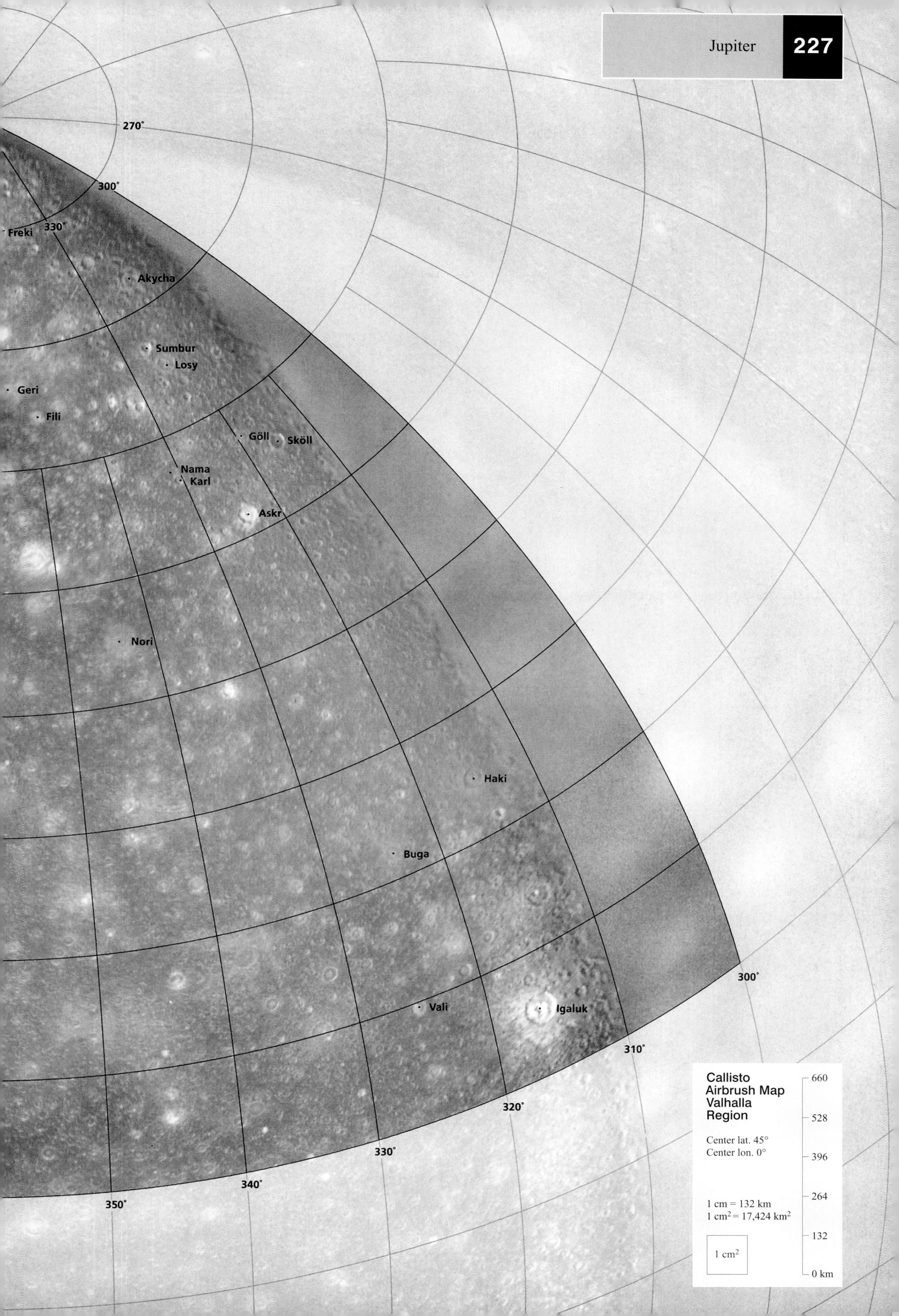

270°
300°
330°
Freki
Akycha
Sumbur
Losy
Geri
Fili
Göll
Sköll
Nama
Karl
Askr
Nori
Haki
Buga
Vali
Igaluk
300°
310°
320°
330°
340°
350°
Callisto
Airbrush Map
Valhalla
Region
Center lat. 45°
Center lon. 0°
1 cm = 132 km
1 cm² = 17,424 km²
1 cm²
660
528
396
264
132
0 km

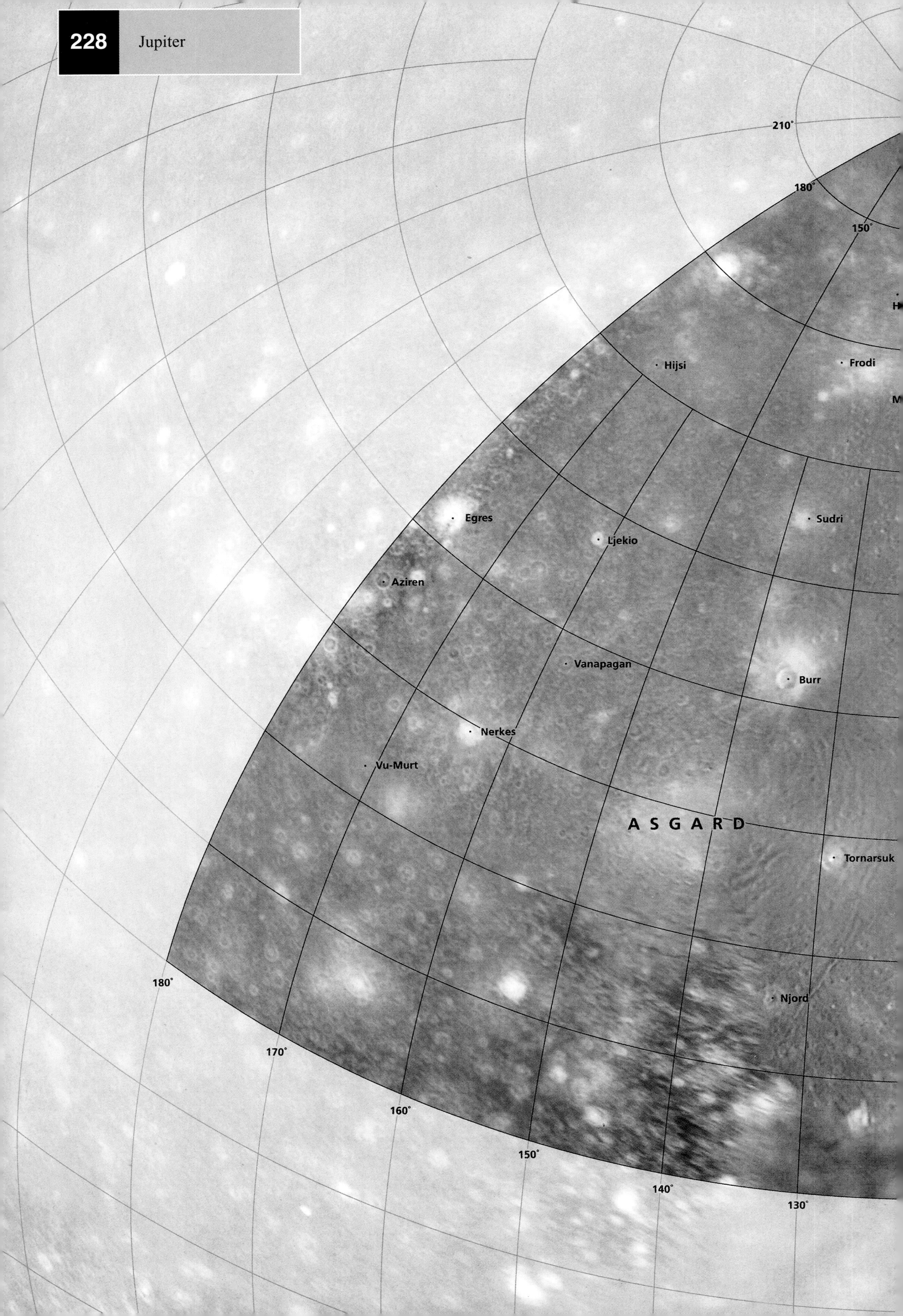

210°
180°
150°
Hijsi
Frodi
Egres
Ljekio
Sudri
Aziren
Vanapagan
Burr
Nerkes
Vu-Murt
ASGARD
Tornarsuk
Njord
180°
170°
160°
150°
140°
130°

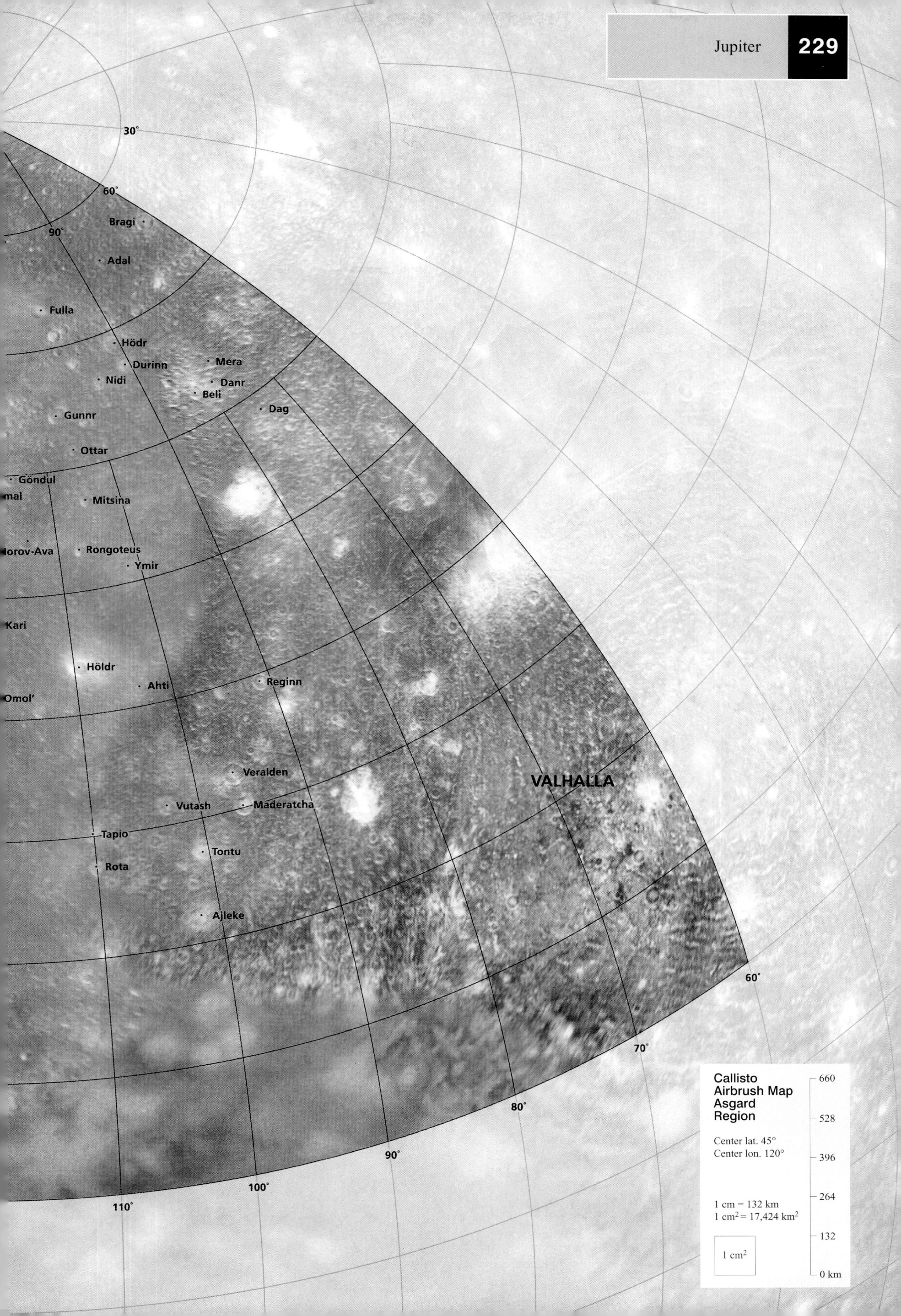
30°
60°
90°
Bragi
Adal
Fulla
Hödr
Durinn
Mera
Nidi
Danr
Beli
Dag
Gunnr
Ottar
Göndul
mal
Mitsina
orov-Ava
Rongoteus
Ymir
Kari
Höldr
Ahti
Reginn
Omol'
Veralden
VALHALLA
Vutash
Maderatcha
Tapio
Tontu
Rota
Ajleke
60°
70°
80°
90°
100°
110°
Callisto
Airbrush Map
Asgard
Region
Center lat. 45°
Center lon. 120°
1 cm = 132 km
1 cm² = 17,424 km²
1 cm²
660
528
396
264
132
0 km

330°
300°
270°
Nuada
Oski
300°
290°
280°
270°
260°
250°

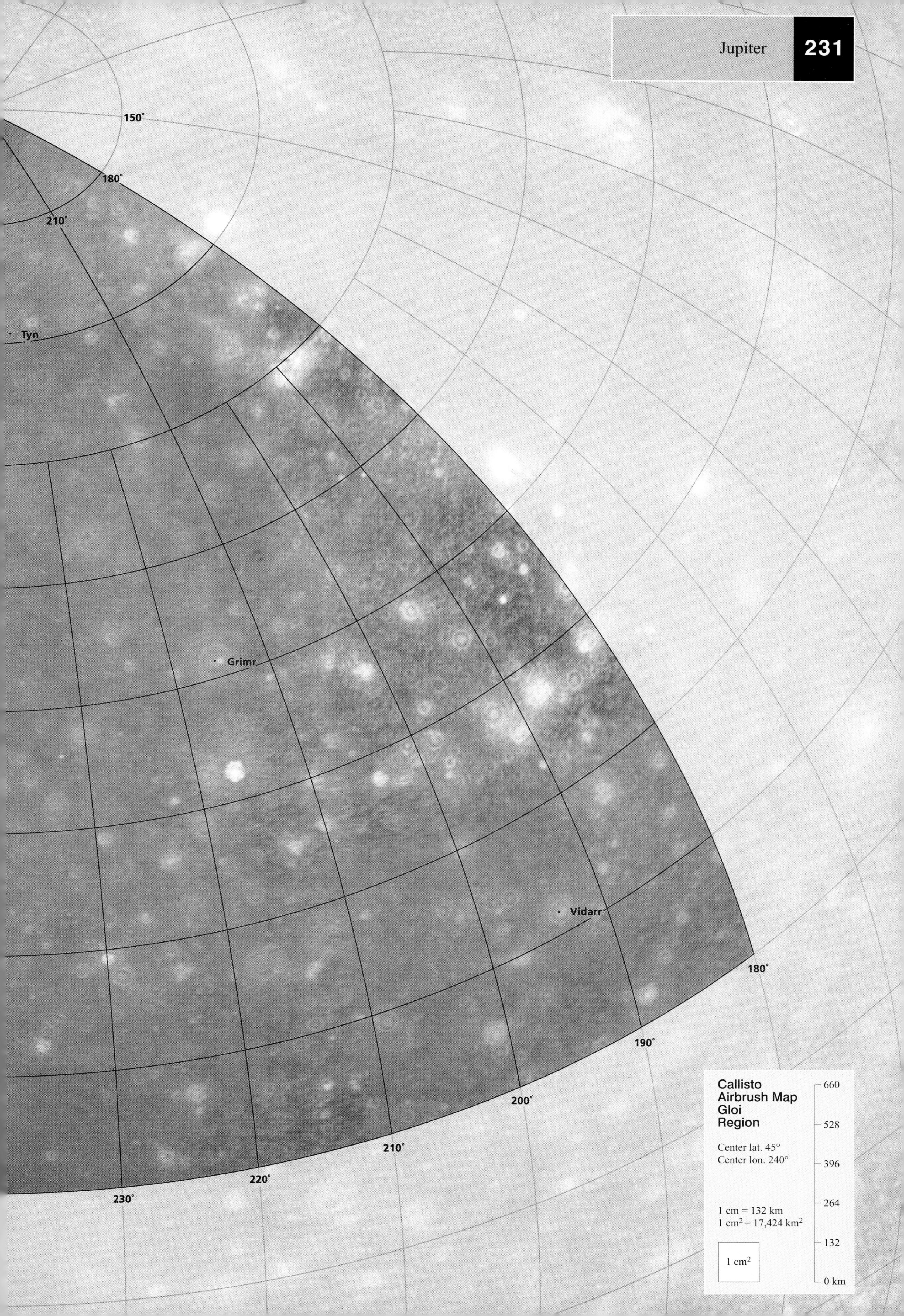
150°
180°
210°
Tyn
Grimr
Vidarr
180°
190°
200°
210°
220°
230°
Callisto
Airbrush Map
Gloi
Region
Center lat. 45°
Center lon. 240°
1 cm = 132 km
1 cm² = 17,424 km²
1 cm²
660
528
396
264
132
0 km

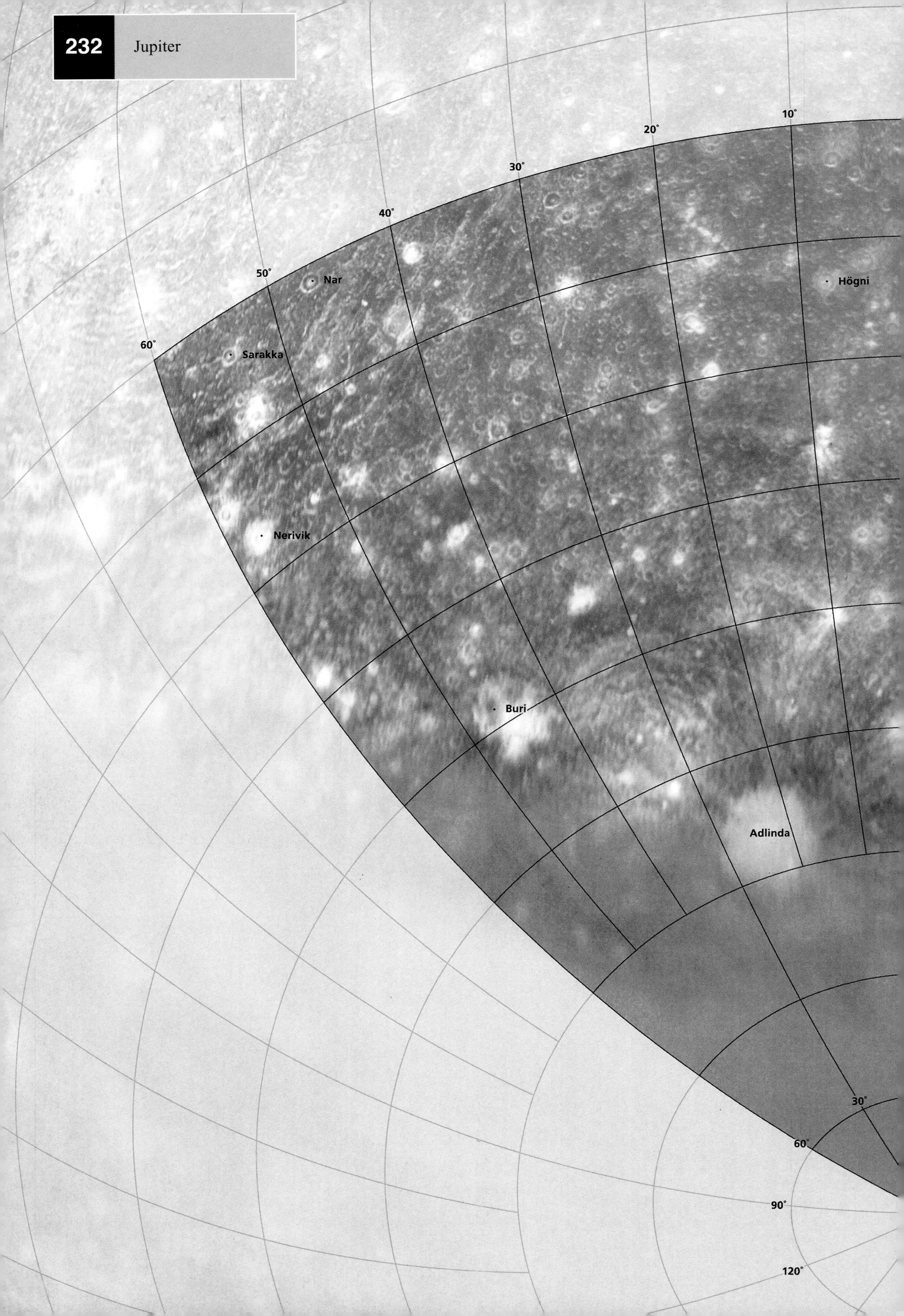

10°
20°
30°
40°
50°
60°
Nar
Högni
Sarakka
Nerivik
Buri
Adlinda
30°
60°
90°
120°

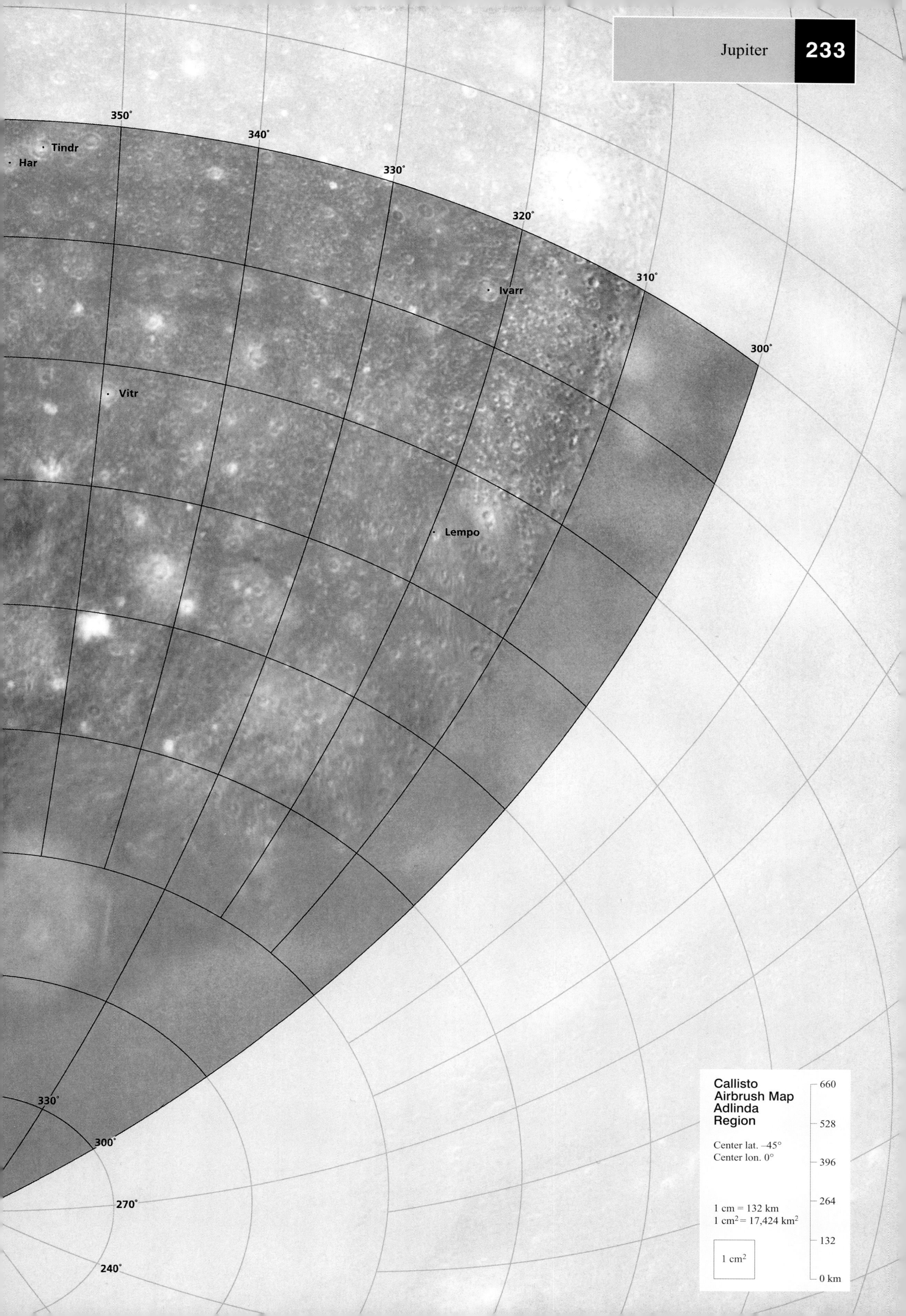
350°
340°
330°
320°
310°
300°
Tindr
Har
Ivarr
Vitr
Lempo
330°
300°
270°
240°
Callisto
Airbrush Map
Adlinda
Region
Center lat. –45°
Center lon. 0°
1 cm = 132 km
1 cm² = 17,424 km²
1 cm²
660
528
396
264
132
0 km

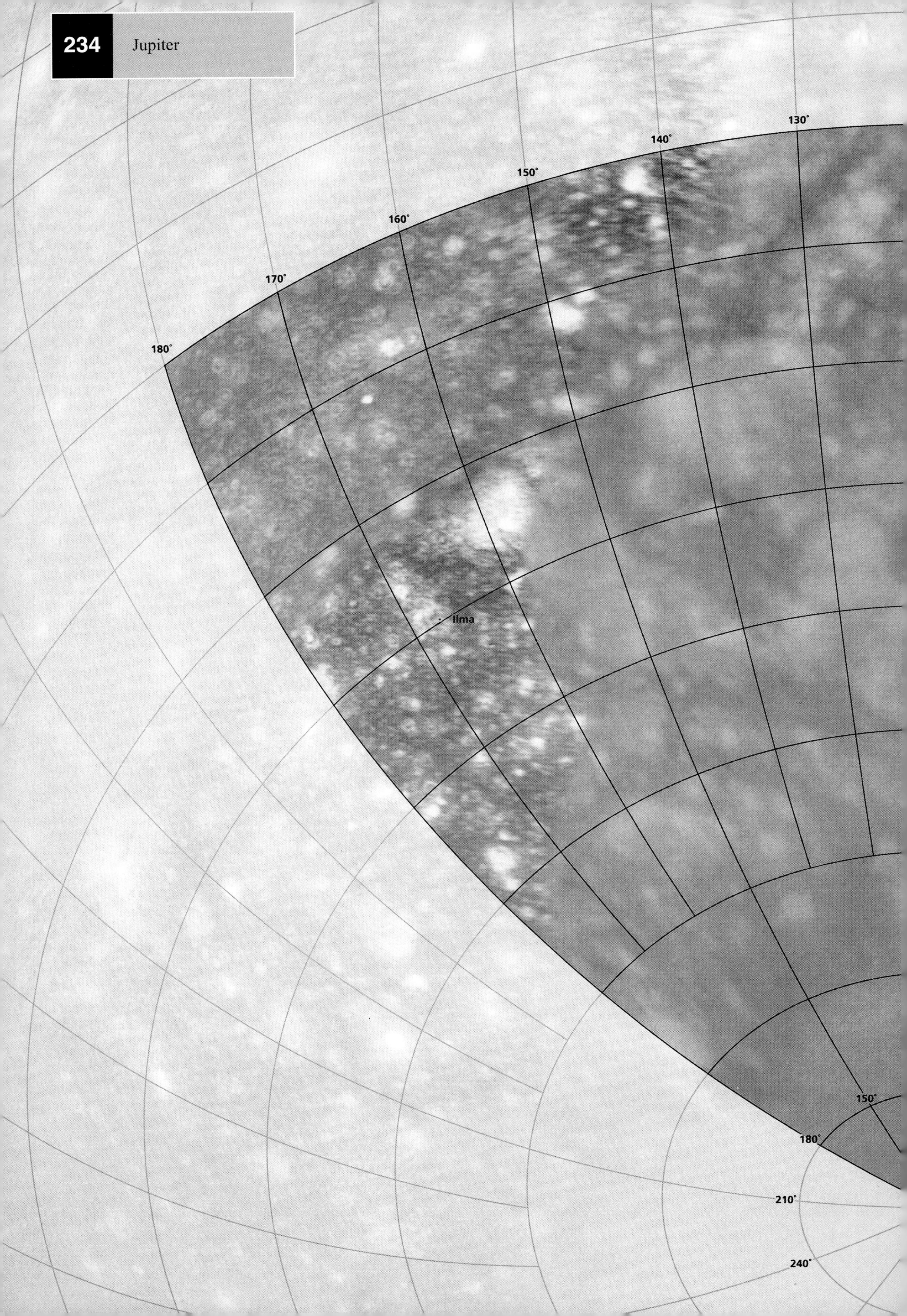

130°
140°
150°
160°
170°
180°
Ilma
150°
180°
210°
240°

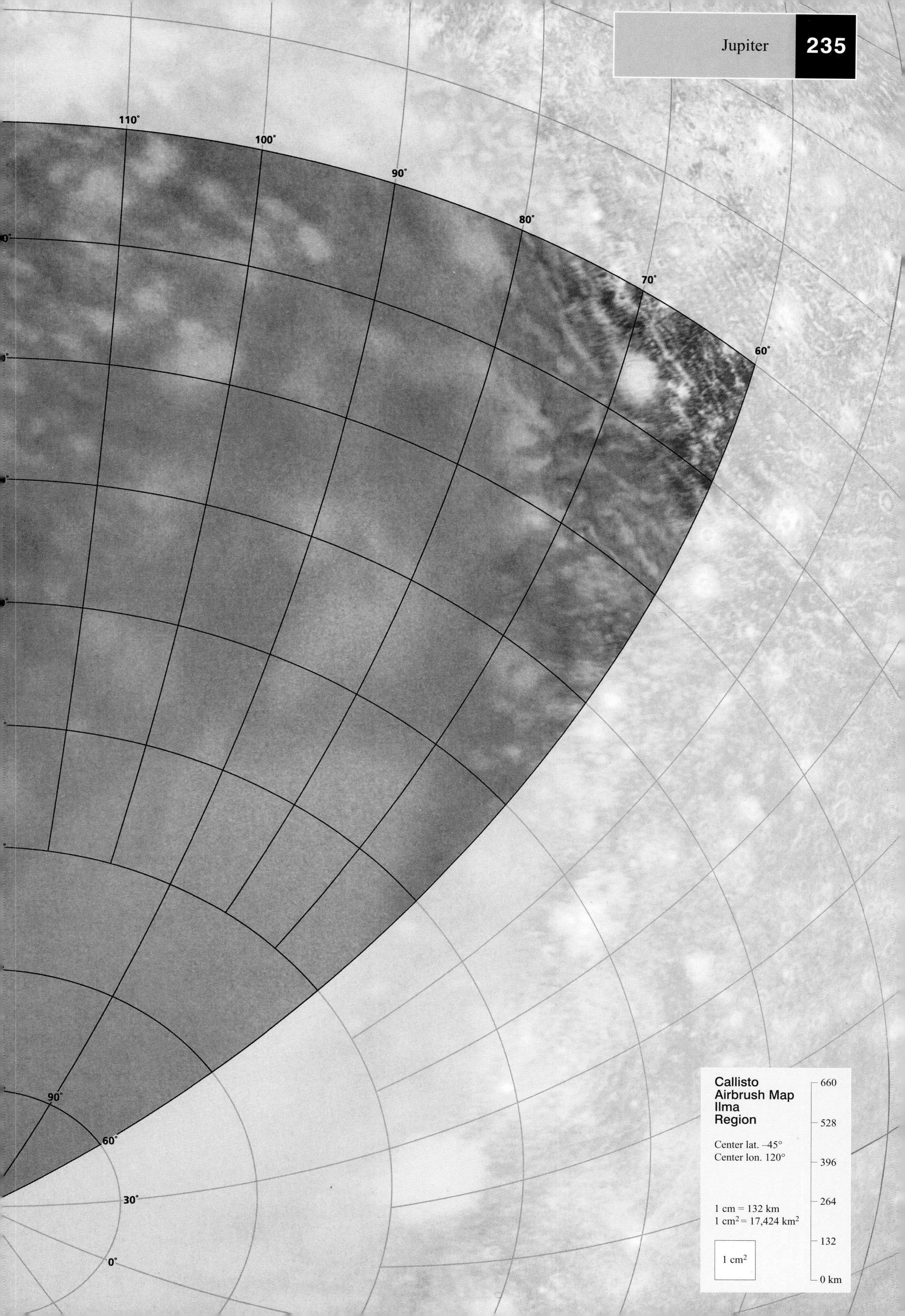
110°
100°
90°
80°
70°
60°
90°
60°
30°
0°
Callisto
Airbrush Map
Ilma
Region
Center lat. –45°
Center lon. 120°
1 cm = 132 km
1 cm² = 17,424 km²
1 cm²
660
528
396
264
132
0 km

250°
260°
270°
280°
290°
300°
Valfödr
Hoenir
Lodurr
270°
300°
330°
0°

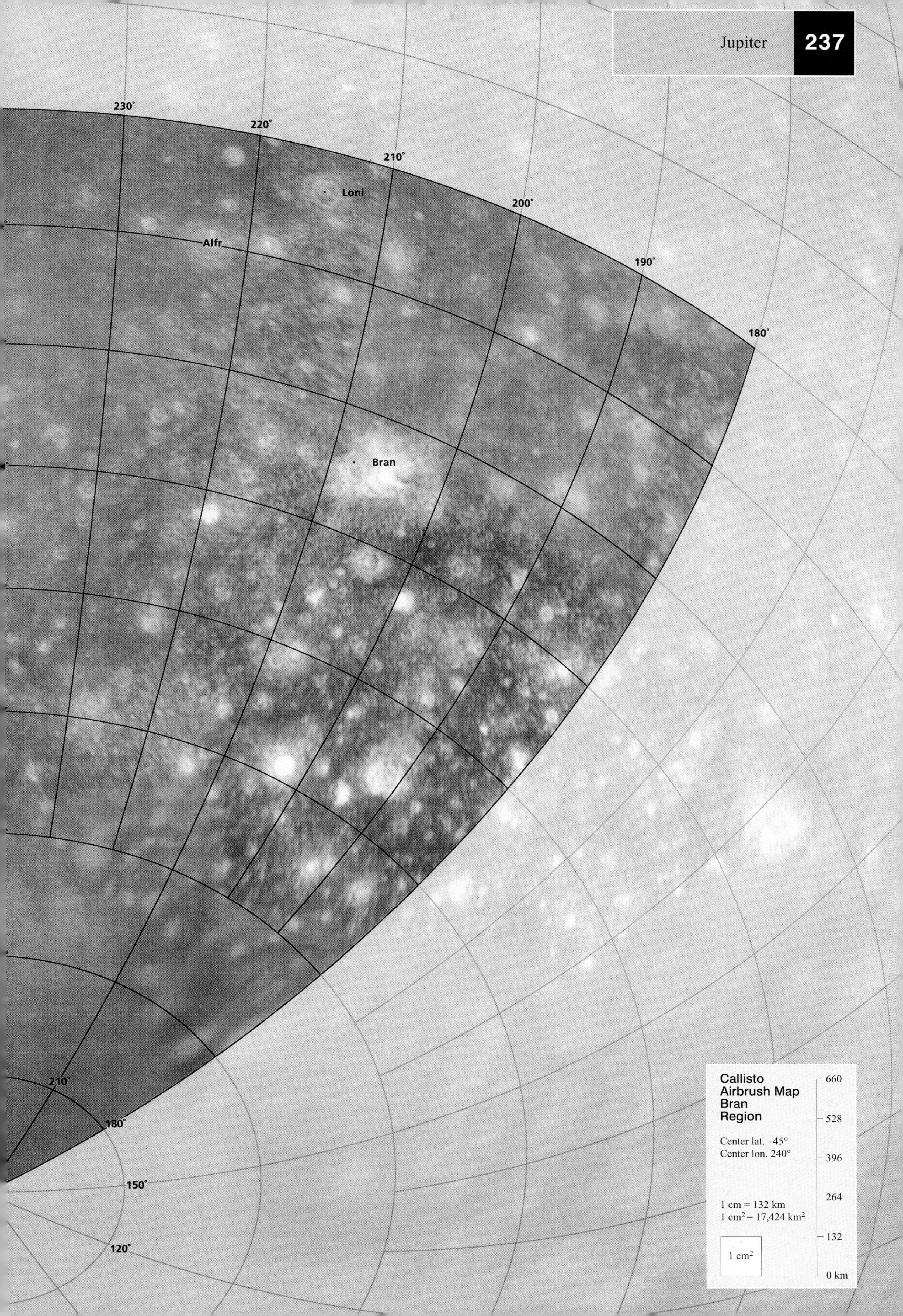
230°
220°
210°
200°
190°
180°
Loni
Alfr
Bran
210°
180°
150°
120°
Callisto
Airbrush Map
Bran
Region
Center lat. –45°
Center lon. 240°
1 cm = 132 km
1 cm² = 17,424 km²
1 cm²
660
528
396
264
132
0 km

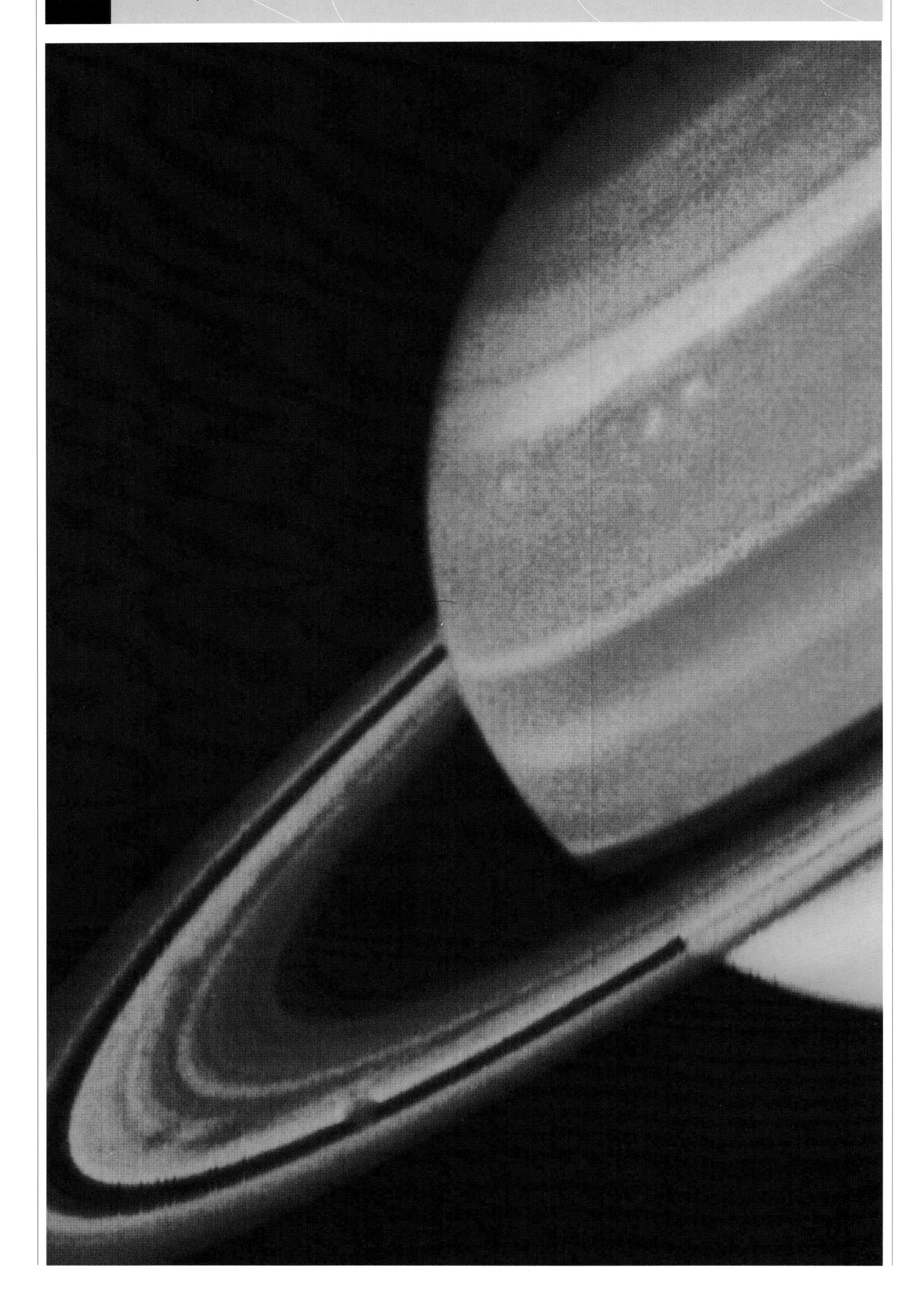

Saturn System

Mention "planet" and most people immediately picture Saturn and its beautiful array of rings. Second in diameter only to Jupiter, Saturn is an enormous planet. From the discovery of rings around Saturn nearly 400 years ago until the last three decades when rings were found around the other giant gas planets, Saturn was thought to be unique in the Solar System. Despite these discoveries, Saturn remains the grand champion for the size, complexity, and sheer beauty of its ring system. Moreover, Saturn has more satellites and a greater variety of satellites than any other planet. Consequently, Saturn holds a special place in our view of the Solar System.

Scientific study of Saturn was initiated in the early 1600s by Galileo. When he viewed Saturn through his 20-power telescope for the first time in July 1610, Galileo apparently thought he was seeing three separate objects. Later observations led to his publishing a sketch in 1616 that clearly showed Saturn and its ring system. Rapid improvements in telescopes and their application to planetary observations resulted in more detailed descriptions of Saturn and prompted wide speculation on the origin and characteristics of its system of rings.

Exploration of the Saturn system by spacecraft began with the *Pioneer 11* flyby in 1979, followed by *Voyager 1* and *Voyager 2* in 1980 and 1981, respectively. Data from the *Pioneer 11* mission yielded new insight into the complex magnetic field surrounding Saturn and enabled discovery of the F Ring. The *Voyager* spacecraft, carrying sophisticated camera systems, returned clear photographs that revealed the great geologic diversity of Saturn's satellites – as well as many newly found satellites – and showed the complexity of both the ring system and cloud patterns on Saturn. More recently, cameras on the *Hubble Space Telescope* returned new information on features in Saturn's atmosphere, such as the Great White Spot that seems to appear and disappear on a 30-year cycle.

The Saturn system compared to the Earth–Moon system

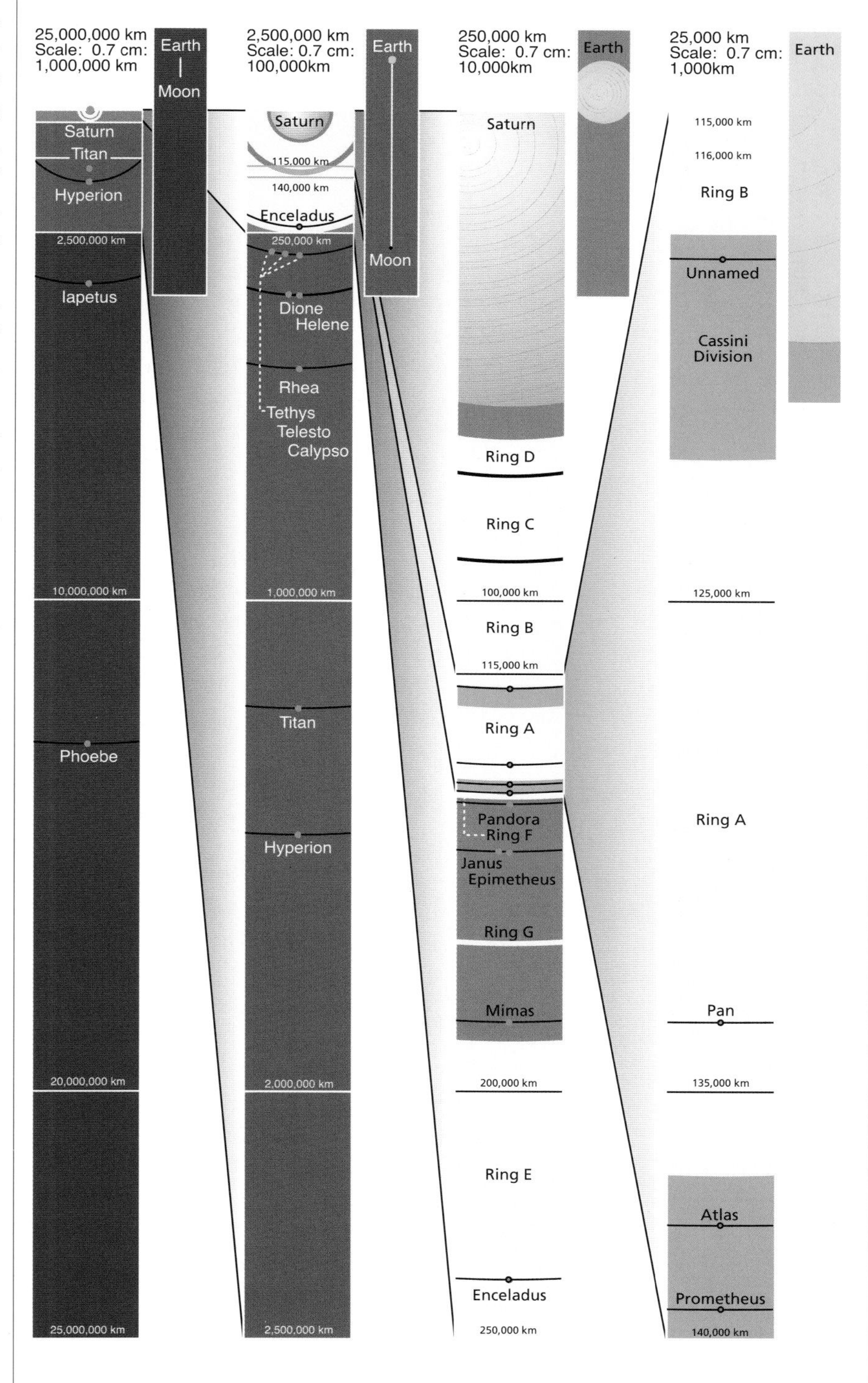

Facing page: False–color photograph of Saturn taken by *Voyager* generated from images taken through the ultraviolet, violet, and green filters (JPL photograph P-23876).

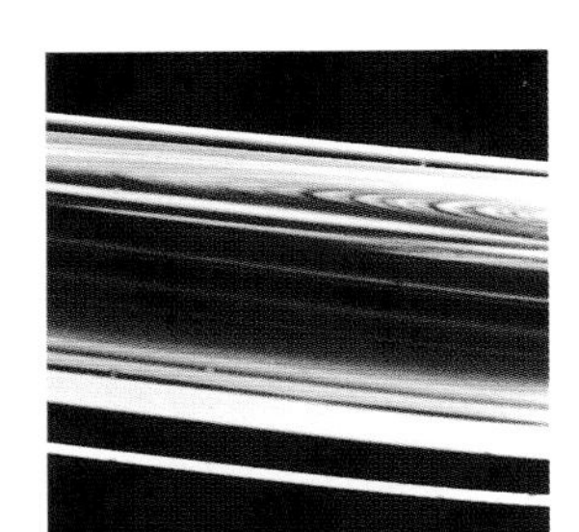

Wide-angle camera view of Saturn's rings; starting from the bottom, visible are the F Ring, the A Ring with the Encke Gap, the Cassini Division (narrow dark band at center), the B Ring, and the C Ring (JPL photograph P-23964).

Saturn

Saturn is estimated to have a density of about 0.7 grams per cubic centimeter – the lowest of all the planets and significantly less than the density of water. Like Jupiter, Saturn consists of about 80 percent hydrogen, 18 percent helium, and oxygen, iron, neon, nitrogen, and sulfur, all totaling less than 2 percent. The interiors of Saturn and Jupiter are similar, with both probably having rocky cores. However, Saturn's core is proportionately larger than Jupiter's core. The cores of both planets are overlain by a mantle of metallic hydrogen and a layer of liquid molecular hydrogen that grades into the gaseous atmosphere. Like Jupiter, Saturn also releases more energy (nearly three times as much) than it receives from the Sun. A major source of this internal energy is probably related to the gravitational separation of helium from hydrogen, somewhat similar to the process that occurs in the Sun but on a much smaller scale.

Stylized diagram of Galileo's first observations of Saturn, made in 1610, showing three objects. Evidently, his telescope was inadequate to distinguish the rings and the view was interpreted to include two satellites.

Like the other giant planets, Saturn spins on its axis extremely rapidly, taking only 10 hours to complete one rotation. This causes the equator to bulge outward and the polar regions to flatten. In fact, the equator is some 13,000 kilometers greater in diameter than the pole-to-pole distance. The high-spin rate, coupled with the interior configuration of core and mantle probably account for the large magnetic field generated by Saturn, based on the dynamo theory.

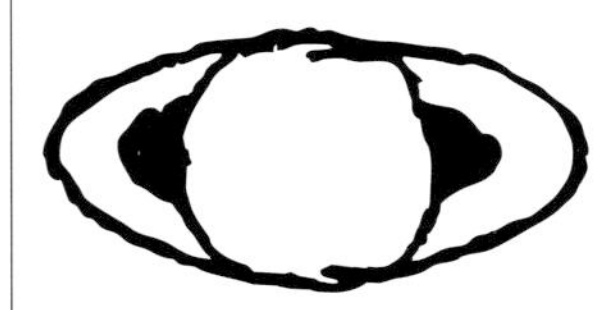

Galileo's sketch of Saturn made in 1616, clearly showing the ring system.

Although the atmosphere and cloud patterns on Saturn are not as spectacular as those on Jupiter, they are nonetheless interesting. A thin zone of reddish-brown and pale-orange streaky clouds tops a clear atmosphere of hydrogen. Both the colors and patterns in Saturn's clouds are more muted and less complex than those of Jupiter. However, wind velocities are much greater – sequential pictures taken by the *Voyager* spacecraft show that the clouds are moving at a speed of 1,500 kilometers per hour – four times faster than those of Jupiter.

Oval clouds of various colors are features common to the atmospheres of Saturn and Jupiter. These clouds last from a few days to several years and apparently indicate the presence of storm systems. Unlike storms on Earth, driven primarily by heating from the Sun, sat-

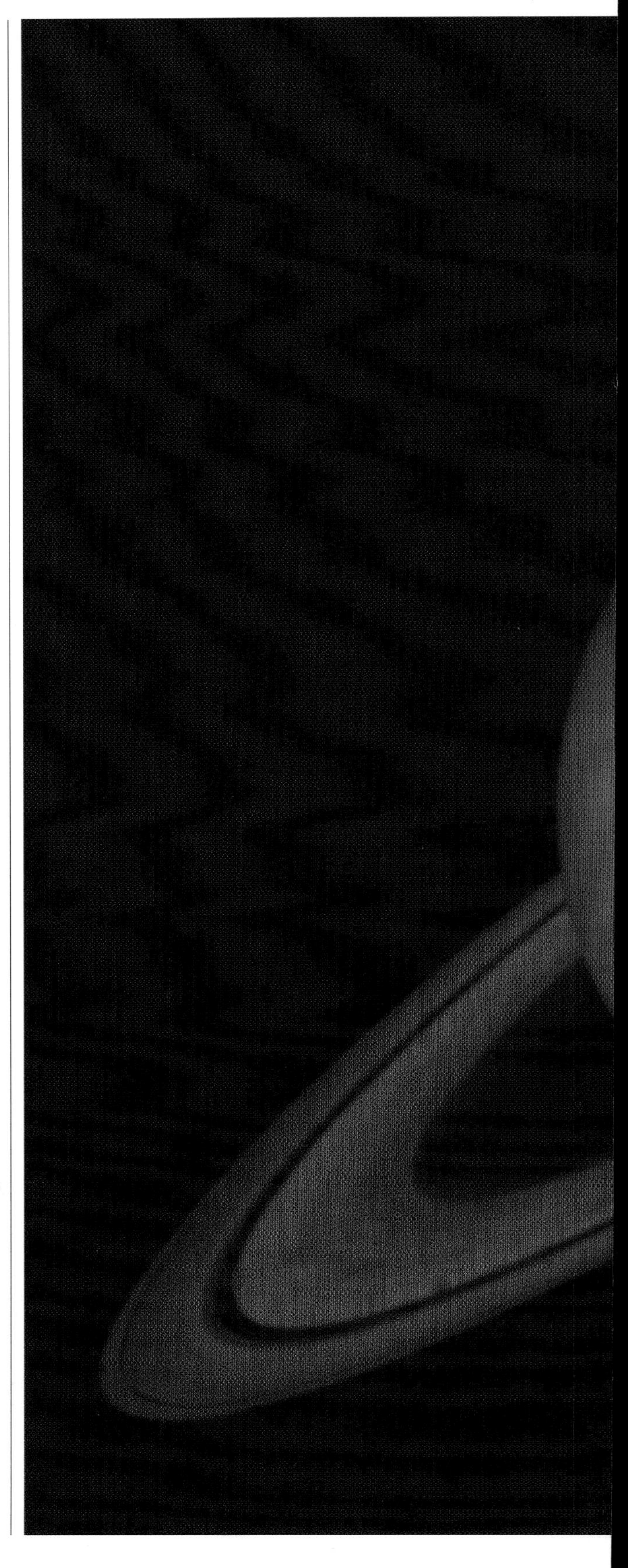

Facing page: false-color image of Saturn taken by the *Voyager 2* spacecraft. Notice two bright, presumably convective, cloud patterns in the northern (top) hemisphere and spokelike features in the B Ring (left of planet). Rhea and Dione, two satellites of Saturn, are visible as dots to the lower right of Saturn (JPL photograph P-23883, courtesy of R. Beebe).

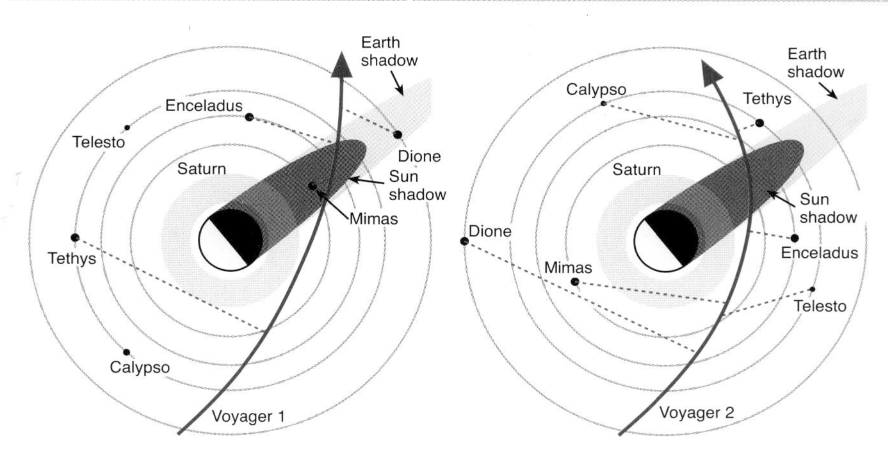

Diagrams showing the paths of *Voyager 1* and *2* through the Saturn system.

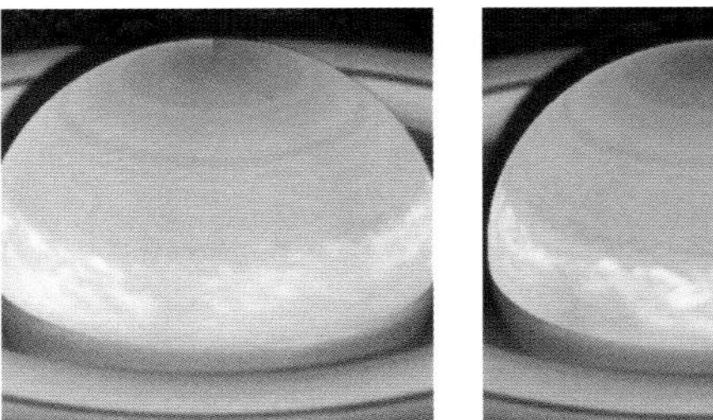
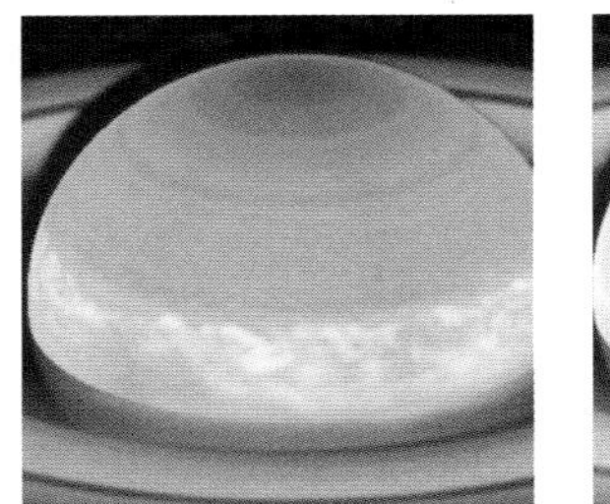

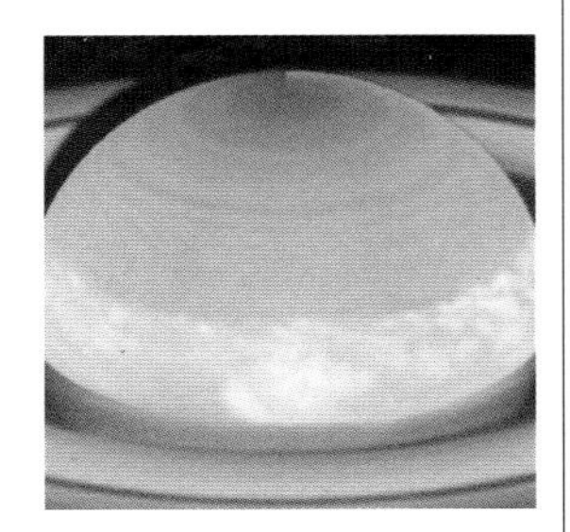

Four images taken as part of a photographic sequence on November 17, 1990, with the *Hubble Space Telescope*. The images show a rare mid-latitude storm that completely encircled Saturn (see detail in image shown below; NASA Goddard Space Flight Center photograph).

Cross-sections of Jupiter and Saturn showing a rock-ice core, a mantle composed of metallic hydrogen, and an outer liquid molecular hydrogen zone that grades into the gaseous atmosphere.

False-color image produced by the *Hubble Space Telescope* in 1990 showing the development of the Great White Spot (arrow). This image was made from two images (one taken in blue light, the other in infrared) from the Planetary Camera that were combined in a computer so that the lower parts of the cloud are shown in blue and the high clouds are shown in red (*Hubble Space Telescope*, released November 20, 1990).

urnian storm systems are probably generated from heat sources derived from the interior of the planet.

One recurring storm rivals the Great Red Spot of Jupiter. Termed the Great White Spot, it was first discovered on Saturn in 1876 by Asaph Hall of the U.S. Naval Observatory. It was later seen in 1903, 1933, and 1960, leading to predictions of its appearance in 1990. The Great White Spot was found independently by two amateur astronomers in the fall of 1990. It was photographed, and the evolution of the Great White Spot was followed by both Earth-based and *Hubble Space Telescope* observations. Within a couple of weeks after its initial appearance as an intense, bright white spot 10,000 kilometers across, the spot grew in complexity and size to more than 50,000 kilometers. As it grew, it also diminished in brilliance.

The Ring System

Even simple telescopes show that the rings around Saturn occur in sets. Earth-based observations using visible, infrared, and microwave (radar) instruments, coupled with spacecraft observations and analytical models, provide a great deal of information about the rings of Saturn. The rings are composed predominantly of water-ice particles that range in diameter from tiny dust specks to boulders 10 meters or more. Each particle is a tiny satellite, traveling at the speed appropriate to its radial distance from the planet. From a distance, they give the appearance of solid disks encircling Saturn.

The subtle colors in the ice particles include reddish-tan and brown-tones, due to impurities, such as iron-oxide rust, or to structural damage in the ice crystals caused by ultraviolet radiation from the Sun.

The main ring system extends from about 700 kilometers above Saturn's clouds to greater than 74,000 kilometers. The ring system is very thin – perhaps averaging only 100 meters thick – so that when viewed from the side, the ring system looks like a knife edge. Despite the enormous extent of the rings, if all of the material in the ring system were accumulated, it would form a satellite-size object only about 500 kilometers in diameter.

View of Saturn taken by *Voyager 1* 4 days after the spacecraft flew past the planet (JPL photograph P-23254).

These "spokes" in the B Ring probably consist of dust clouds driven by electrostatic forces and are as long as 10,000 kilometers, oriented away from Saturn (JPL photograph P-23881BW).

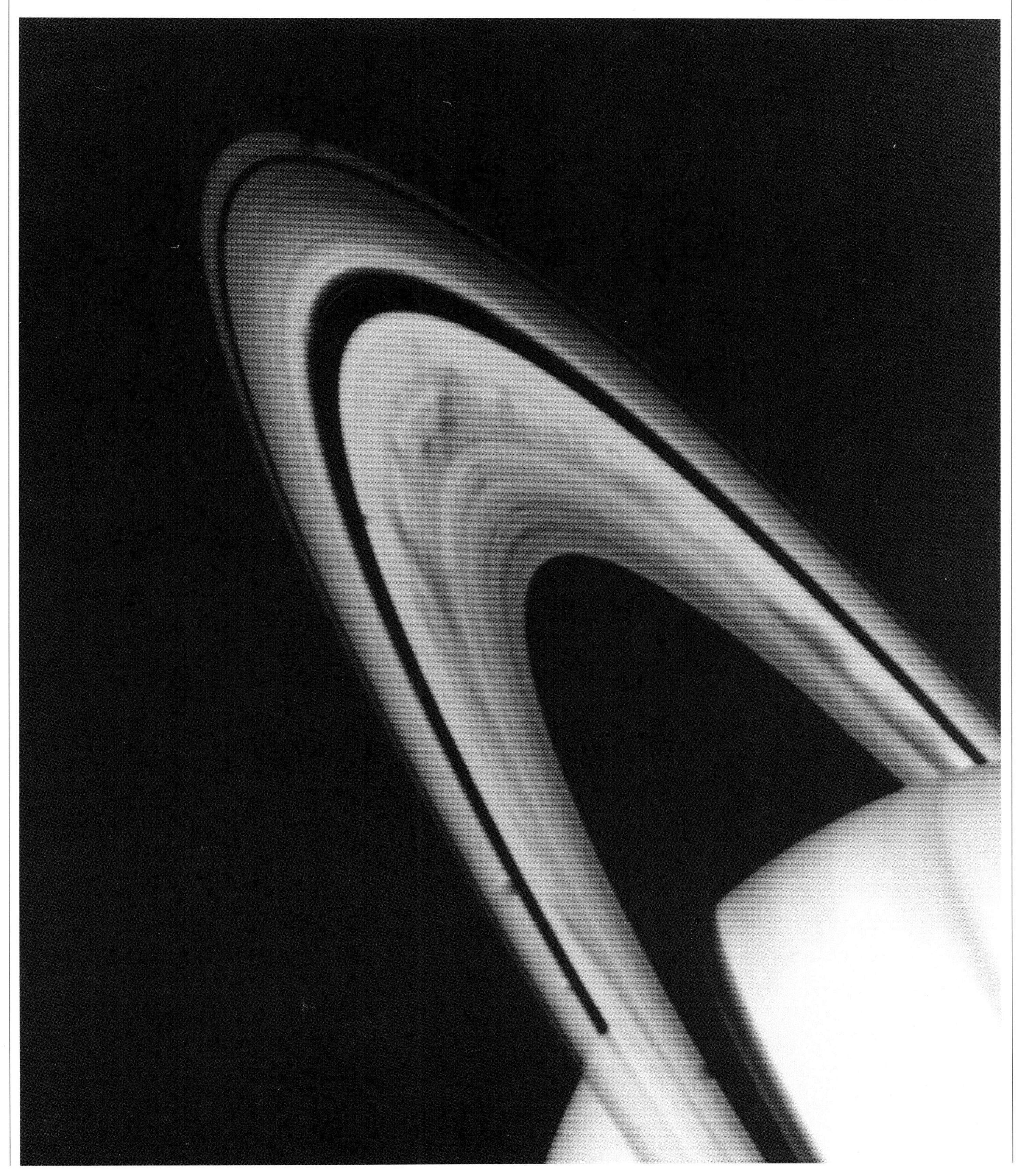

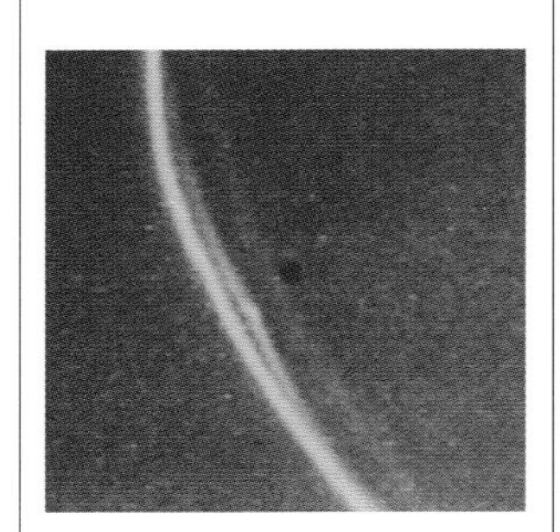

Saturn's F Ring photographed by *Voyager 1* showing two narrow, braided bright ring components and a broader diffuse component about 35 kilometers wide (JPL photograph P-23099).

The various parts of the ring system are designated by letters, but the order of the letters is based on the sequence of their discovery and not their position in relation to Saturn. *Voyager* observations show that each major ring has considerable structure, with each being composed of numerous "ringlets." The innermost ring, designated the D Ring, is the most difficult to see. It consists of very narrow ringlets composed of fine-grained particles that appear to be "leaking" inward from the C Ring toward Saturn. The C Ring consists of light and dark bands and is thought to be dominated by meter-sized particles spaced about 20 meters apart. It may also include some objects that are 10 meters to perhaps 1 kilometer in diameter. The C Ring contains little or no dust, most of it evidently having been swept away to form the D Ring.

The B Ring is the brightest of the saturnian rings. Radar observations show that it is about 2 kilometers thick, much thicker than any other ring. The B Ring apparently is composed mostly of particles about 10 centimeters in diameter, although some may be as large as 10 meters. Viewed by *Voyager* cameras, transient spokelike patterns appear superposed on the B Ring. The spokes are 10,000 kilometers long and range from 100 to 1,000 kilometers in width. They radiate outward through the B Ring and appear and disappear. Although there has been much speculation about the characteristics of the spokes, most planetary scientists conclude that they are clouds of dust particles, perhaps levitated by electrostatic forces generated as the ring particles collide with each other.

The A Ring is composed of ice particles ranging in size from dust specks to 10-meter boulders. The ring consists of several narrow ringlets, with the outermost ringlet having a sharp, well-defined edge. A small satellite, Atlas, discovered during the *Voyager* mission, is in orbit just outside the A Ring. It is called a "shepherding" satellite, because it appears to bound material in the A Ring by gravitational forces. Separating the A Ring from the B Ring is the Cassini Division, noted in 1675 by the Italian astronomer Gian Domenico Cassini. This gap in the rings is some 4,000 kilometers wide and was thought to be an area relatively

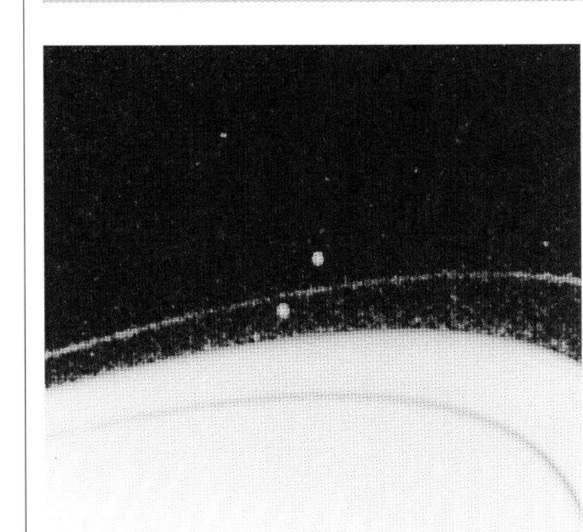

Saturn's thin F Ring bracketed by two "shepherding" satellites, Pandora and Prometheus (JPL photograph P-2391 BW).

Voyager false-color image showing the major ring systems of Saturn. Blue colors correspond to the C Ring (lower left and the Cassini Division (JPL photograph P-23953C).

free of particles. *Voyager* photographs, however, show that some narrow rings are present.

The F Ring was discovered using data collected from the Pioneer mission. This ring is about 700 kilometers wide and orbits some 4,000 kilometers beyond the A Ring. Two "shepherding" satellites on either side of the F Ring, Pandora and Prometheus, appear to keep the ring particles constrained to a very narrow orbit. Close inspection of the F Ring reveals an irregular form, including clumps, strands, and braids, that changes with time and that may indicate electrostatic or magnetic forces at work. The E Ring is a very faint, barely visible strand that orbits far beyond the main rings.

Many ideas have been proposed to explain the origin and evolution of the rings around Saturn and the other Jovian planets. Some planetary scientists have proposed that Saturn's rings represent one or more satellites that were broken apart by tidal forces generated from Saturn. Others have suggested that the rings represent the breakup of one or more small satellites through collisions. Alternatively, the rings may represent material that never accreted to form a satellite in the early stages of Solar System formation. All three proposals are viable and will continue to be the subject of discussion.

Satellites

The Saturn system displays a fascinating collection of moons. The 30 known satellites range in size from less than 10 kilometers across to a Mercury-size object – Titan – that has an atmosphere 1.7 times as dense as that of Earth. All of the satellites have densities that suggest they are ice-rich objects with different amounts of rocky material composing their interiors.

The nine largest satellites were discovered telescopically, some as early as 1655. Most of their names were proposed by Sir John Herschel in the 1700s and are taken from Greek mythology. Dione, Rhea, Tethys, Mimas, Enceladus, Titan, and Phoebe were giants, while Hyperion and Iapetus were brothers of Saturn.

All of the satellites but two, Hyperion and Phoebe, are in synchronous rotation, keeping the same side facing Saturn. All but two travel in circular, prograde orbits in the equatorial plane, the same as Saturn's rings. The exceptions, Iapetus and Phoebe, travel in paths inclined to the equatorial plane, and Phoebe travels in a retrograde direction. The innermost satellites of Saturn are in orbit between some of the prominent rings.

Nearly all of the information about the surfaces of Saturn's satellites has come from the *Voyager* mission. The paths of *Voyagers 1* and *2* were planned to use a gravitational "boost" from Jupiter to speed them on their way to Saturn. Voyager 1 flew closer to Jupiter and received a greater boost in speed that allowed it to arrive at Saturn in November 1980, 9 months earlier than *Voyager 2*. Aimed to make a close pass by Saturn's giant moon, Titan, *Voyager 1* also took high-resolution pictures of Mimas, Dione, and Rhea. Programming the flight path for a close flyby of Titan meant that the trajectory would carry *Voyager 1* out of the ecliptic plane of the Solar System and prohibit it from journeying onward to the other outer planets.

The path of the *Voyager 2* spacecraft was designed so that it would be able to fly past Uranus in 1986 and Neptune in 1989. Although this design placed some restrictions on its passage through the Saturn system, the trajectory nicely complemented that of *Voyager 1*. *Voyager 2* provided close views of the satellites Iapetus, Hyperion, Enceladus, and Tethys.

There are no obvious patterns relating the sizes or densities of the satellites to their distance from Saturn or to the degree of geologic activity among them. For simplicity, the satellites are discussed here in groups according to size: intermediate-size satellites (400 to 1,500 kilometers in diameter), small satellites (less than 400 kilometers in diameter), and Titan, the largest satellite.

Intermediate-Size Satellites

These objects are intermediate in size between the large Galilean moons and the tiny satellites of Mars and Jupiter. As such, they afford the opportunity to study processes and geologic histories that may be unique in the Solar System. Within this group the satellites are discussed in pairs for comparison.

Mimas and Enceladus

Mimas and Enceladus are 394 and 502 kilometers in diameter, respectively. Their orbits are the closest to Saturn of the intermediate-size satellites. Aside from these characteristics, the similarity to other intermediate-size satellites ends. *Voyager 1* images show the surface of Mimas to be heavily cratered with deep, bowl-shaped depressions. Many of the craters larger than about 30 kilometers have central peaks. Perhaps the most striking aspect of Mimas is the presence of a huge impact crater found on the leading hemisphere. Named Herschel, after the English astronomer who discovered many of Saturn's satellites, the crater is 130 kilometers across, or one-third the size of Mimas. This crater is nearly 10 kilometers deep and has a central peak that rises some 6 kilometers from the crater floor.

In addition to craters, the surface of Mimas is cut by grooves up to 90 kilometers long, 10 kilometers wide, and 1 to 2 kilometers deep. These grooves are fractures in the lithosphere that could have developed in response to the formation of the crater Herschel, some other impact event, or to internal radioactive heating.

Like Mimas, Enceladus was discovered by Herschel in 1789. Because of their similar sizes and positions relative to Saturn, it was assumed before the *Voyager* mission that the surfaces of Mimas and Enceladus would be similar, yet they represent two extremes. While Mimas preserves an ancient, cratered surface dating back to the time of its formation and shows little signs of internal activity, the satellite Enceladus displays a much younger surface with a complex geologic history.

Geologic mapping of Enceladus shows a variety of terrains, including smooth plains, ridges, grabens, and cratered regions. Some of the larger craters are rather shallow and may represent the deformation of icy crust similar to the shallow craters seen on the Galilean satellite, Ganymede. The grooves, ridges, and plains are evidence for tectonic processes and repeated resurfacing events, unexpected for such a small object. The resurfacing events probably involved eruptions of water, perhaps mixed with ammonia, or slurries of water ice

High-resolution image of Mimas showing the south polar region and prominent grooves (*Voyager 1*, JPL photograph P-23112).

Voyager image of Mimas showing impact crater (132 kilometers in diameter) Herschel (*Voyager 1*, JPL photograph P-23210).

The floor of Tethys' impact crater, Odysseus (400 kilometers in diameter), matches the curvature of the satellite (*Voyager 2* photograph 43980.27, ASU IPF-78).

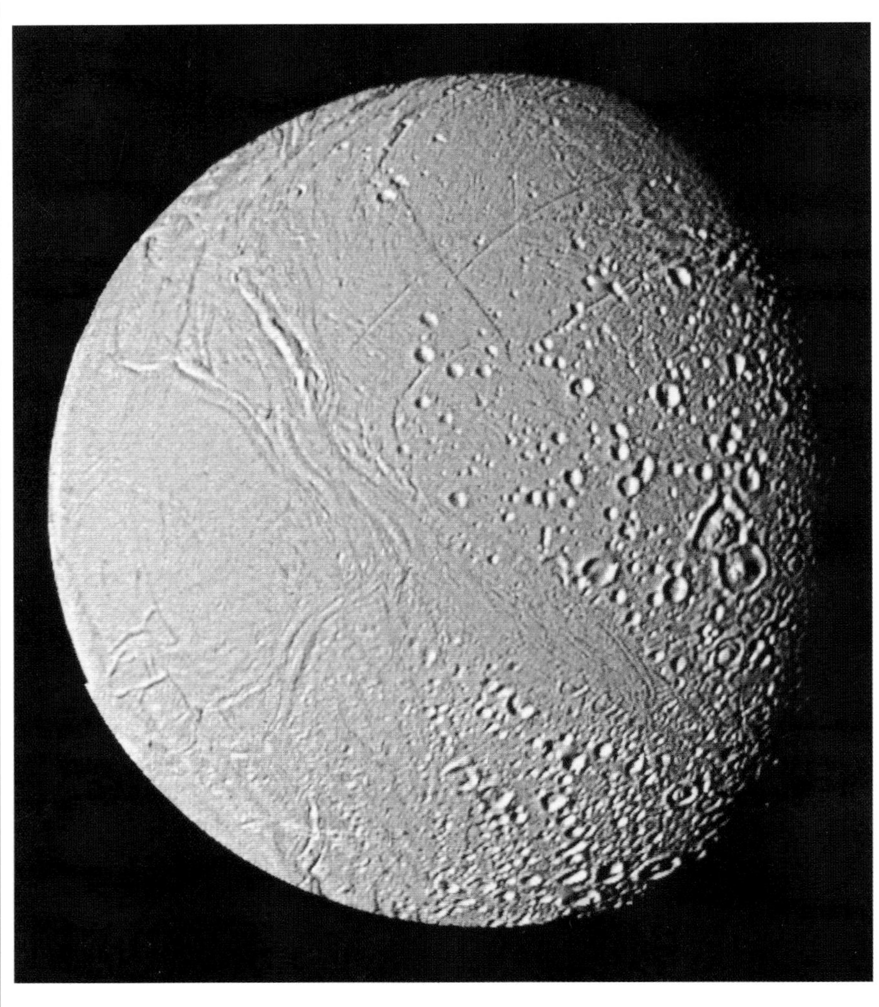

False-color image of Enceladus showing cratered, grooved, and smooth terrain. The smooth terrain was probably formed by eruption of slush from the interior (JPL photograph P-24308).

onto the surface from melt zones in the interior. The grabenlike grooves probably resulted from brittle failure of the icy lithosphere. The ridges, similar to those seen on Ganymede, could have formed either from compression of the crust, from upwelling of plastic ice from the interior, or from expansion of freezing water that pushed upward and intruded through fractures in the crust.

What could have caused the tectonic deformation of Enceladus's crust and the resulting heat to release water onto its surface? Because of its small size, neither primordial heat left over from impacts early in the history of Enceladus nor heating from radioactive decay could account for the activity evidenced on the surface – calculations of heat loss show that Enceladus would have frozen solid very early in its history. The most likely source of heat is internal friction from tidal stresses, similar to that which drives active volcanoes on Io. In this model, the orbit of Enceladus in relation to Dione "push-pulls" the satellite, and that, in turn, generates tidal heating. Although greater eccentricities of orbits than present values are required to melt parts of Enceladus, especially if pure water ice were involved, impurities, such as ammonia, in the ice could lower the melting temperature by 100°C or more. It is less likely that the interior of Enceladus could include concentrations of radioactive elements, leading to local "hot spots"; this would require unusual compositions in comparison to the rest of the Solar System and most planetary scientists discount this possibility.

Tethys and Dione

Tethys, at 1,048 kilometers in diameter, and Dione, at 1,120 kilometers in diameter, constitute the next pair of intermediate-size satellites. They are found in orbits outside the E Ring. Both are unusual in that they have tiny companion satellites. Tethys' two companions, Telesto and Calypso, are irregular-shaped objects about 20 kilometers across. One companion leads Tethys and the other trails behind, with all three in the same orbital path. Dione has a single companion, Helene, found 60° in front of Dione. All three companion satellites are apparently in gravitational balance with their larger neighbors, as discussed further in this chapter.

Tethys is more than twice the diameter of Enceladus and also displays evidence of crustal deformation and resurfacing. However, its ridges, grooves, and smooth plains are not as extensive as those on Enceladus. The two most striking features on the surface of Tethys are Odysseus, an impact crater 400 kilometers in diameter (about 40 percent of the diameter of the satellite), and an enormous canyonland named Ithaca Chasma, which wraps at least three-quarters of the way around the globe. Odysseus is a shallow crater with a ringlike central peak complex. Its low relief may have resulted from viscous flow of the icy lithosphere. The impact that formed Odysseus could have occurred when Tethys was partly liquid and the ice in its interior was soft and plastic.

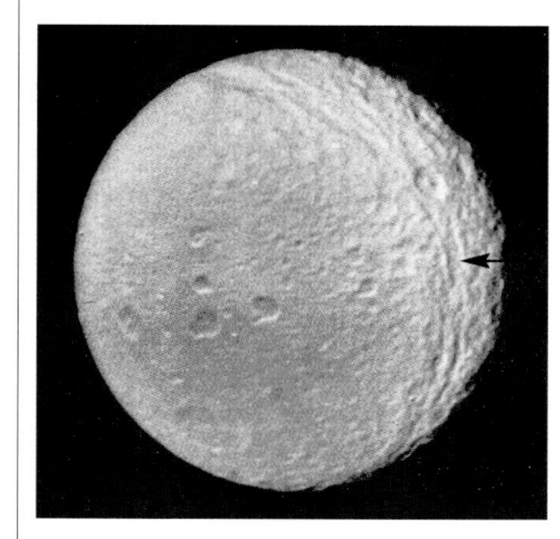

Voyager view of Tethys showing Ithaca Chasma (arrow), a plains unit (lower left), and heavily cratered terrain (*Voyager 2* photograph 43995.58, ASU IPF-84).

Ithaca Chasma is a branching canyon system more than 1,000 kilometers long. Terraces on the canyon walls could have formed by enormous landslides of the icy lithosphere. The canyon system may have been created as some sort of crustal response to the impact that formed Odysseus, perhaps by seismic disruption.

Dione, discovered in 1684 by Cassini, is about the same size as Tethys but is considerably more dense. At 1.43 grams per cubic centimeter, it is second only to Titan in density, indicating that it may have more rocky material in its interior than the other satellites of Saturn. The largest craters on Dione are found on the trailing hemisphere and are as large as 200 kilometers across. The trailing hemisphere also shows a network of bright streaks on a dark background. The bright wispy streaks, however, do not resemble the typical bright rays found with impact craters but follow irregular paths. These markings could be frost deposited by explosive releases of gas from the interior through fractures in the crust.

Rhea and Iapetus

Rhea and Iapetus are the largest of the intermediate-size satellites of Saturn. Like the pairs of satellites discussed above, except for their similarity in size (1,530 kilometers and 1,435 kilometers in diameter, respectively) they seem to have little else in common. Except for Titan, Rhea is the largest of the saturnian satellites. Like Dione, its trailing hemisphere is dark and has bright wispy markings, whereas the leading hemisphere is uniformly bright. Parts of the surface of Rhea are dominated by large, degraded craters and resemble the lunar highlands. None of the craters show the "flattening" to the degree of those on Ganymede. Some craters show very bright patches on the walls; these may be relatively fresh ice deposits exposed by landslides.

Rhea displays an impact basin some 325 kilometers in diameter. The outer ring consists of two concentric, inward-facing scarps that are discontinuous in places, similar to Odysseus on Tethys. These outer scarps may be similar to giant terraces found around some

Color composite mosaic of Dione showing impact-cratered surface; the largest crater in this view (toward the top of the photograph) is just less than 100 kilometers in diameter and has a central peak (JPL photograph P-23113).

Voyager image of the trailing hemisphere of Dione showing heavily cratered terrain, the apparent association of bright wisps with troughs, and the parallel groups of some of the bright wisps' related lineaments (*Voyager 1* image 62S1+000).

Mosaic of *Voyager* images for north polar region of Rhea showing both ancient cratered terrain (on which occur coalescing pits and linear troughs) and resurfaced terrain (*Voyager 1*, ASU photomosaic 4216-H, courtesy J. Moore).

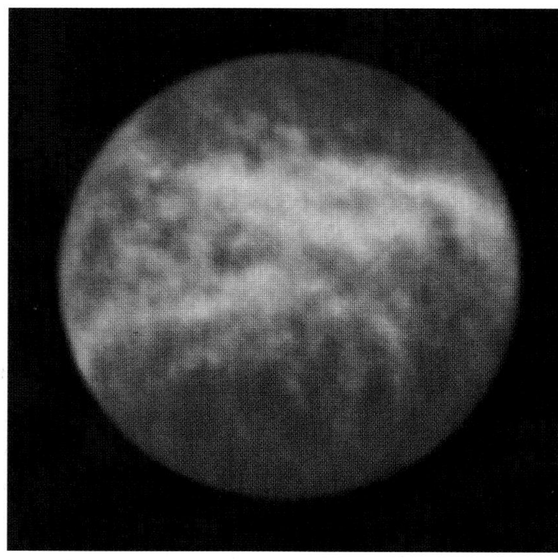

Color composite photograph of Rhea constructed from *Voyager 1* images taken through violet, blue, and orange filters. The hemisphere shown here trails Rhea in orbit, and shows light, wispy markings on the surface (JPL photograph P-23093C).

impact basins on the Moon. These types of terraces probably formed by enormous collapses of the basin rim.

Iapetus is nearly the twin of Rhea in its size but has a lower density, about 1.2 grams per cubic centimeter, suggesting that it is composed of more ice than that proposed for Rhea. Since the satellite's discovery in the seventeenth century, the contrasts in surface brightness on Iapetus have been recognized as distinctive in the Solar System. While the leading hemisphere is extremely dark, the trailing hemisphere is very bright. Surface features are difficult to see in the dark area, Cassini Regio, but craters as wide as 120 kilometers are clearly visible in the north polar area and in the bright terrain of the trailing hemisphere. The large craters have central peaks and well-defined rims. Some of the craters in the bright areas near the boundary with the dark terrain have floors covered with dark material that has the same spectral properties as the dark terrain, suggesting a common origin.

The origin of the dark material on Iapetas is controversial. The dark material is about as black as coal tar or asphalt, and the only known materials this dark that are common in the Solar System are those that make up the carbonaceous chondrites. Two general ideas, each with variations, have been proposed to explain the origin of Iapetus' dark material. One model suggests that the dark material came from external sources that accumulated on the surface, and the other suggests that it was erupted from the interior. In either case, most planetary scientists agree that the dark material forms a mantlelike deposit that is draped over the bright terrain.

Because the dark material is found on the leading hemisphere, it was proposed, even before the Voyager flyby, that impact craters eroded the ice and left a lag deposit of dark, rocky materials. However, this idea seems to require a large proportion of rocky material in relation to ice that does not fit the overall low density of Iapetus. Alternatively, the dark material may originate on Phoebe, which is in the next orbit outward from Saturn. Phoebe is very dark, and bombardment of its surface could eject dark debris, which then would be pulled by gravity inward toward Saturn. Some of this

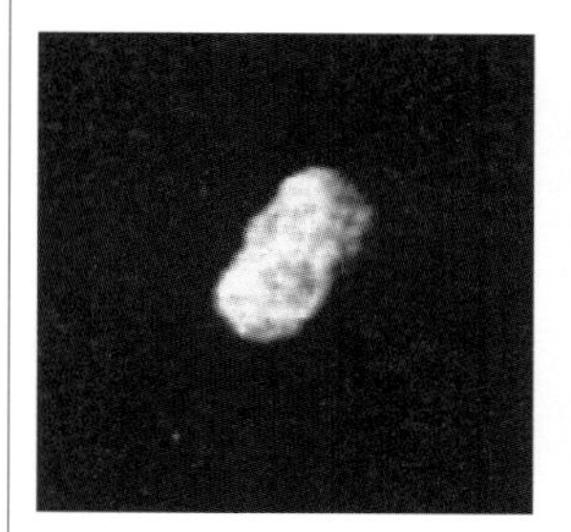
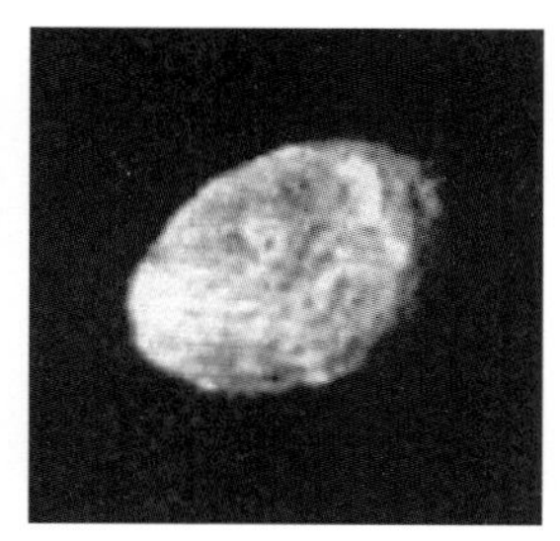
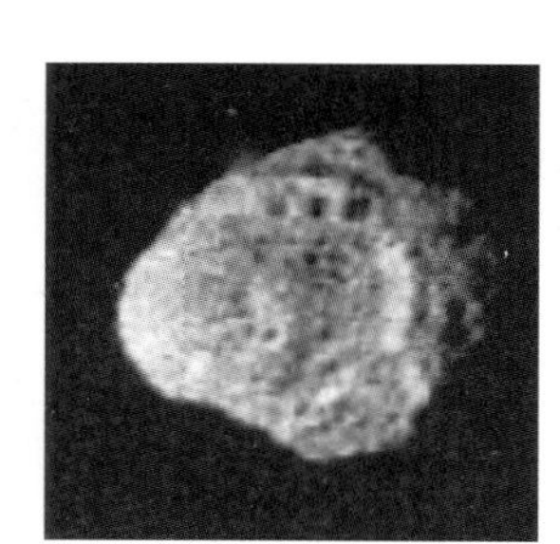

Voyager views of Hyperion showing its irregular shape and the presence of large degraded craters and scarps (*Voyager 2*, JPL photograph P-23932).

material would be swept up by Iapetus and deposited on its leading hemisphere. The main problem with this proposal is that the spectral properties of the surface of Phoebe and the dark terrain on Iapetus do not match. Iapetus is redder, suggesting that the two are not made of the same material.

The second general idea involves an internal origin for the dark material. Many of the low areas in the bright terrain on Iapetus appear to be flooded with smooth, dark material, while the high-standing areas are free of dark deposits. This suggests that, at least locally, the dark material erupted on the surface from the interior of Iapetus. However, this model requires some mechanism to concentrate dark material in the interior and erupt it as a mass onto the surface.

Despite the additional information provided by the *Voyager* mission, the origin of the dark material on Iapetus remains an enigma.

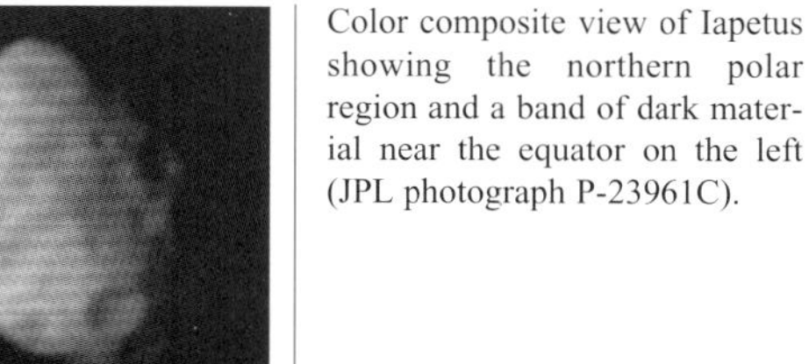

Color composite view of Iapetus showing the northern polar region and a band of dark material near the equator on the left (JPL photograph P-23961C).

Small Satellites

"Small" satellites range in size from 25 to 220 kilometers in diameter. Phoebe, the outermost saturnian satellite, is in a retrograde orbit inclined to the equatorial plane of Saturn. This unusual orbit suggests that Phoebe may have been captured by Saturn. Estimates of the mass of this 220 kilometer-size moon could not be made by *Voyager* because the spacecraft did not pass close enough to the satellite to determine its density. The low albedo and spectral properties of Phoebe suggest that its surface is composed of carbonaceous material. Although *Voyager* images are very poor, Phoebe appears spherical and the darkest object in the saturnian system. Circular markings and indentations suggest that craters are on the surface.

Five of the small satellites are found inside the orbit of Mimas. Two, Janus and Epimetheus, are called co-orbital satellites because their orbits are within 50 kilometers of each other. They are thought to be remnants of a larger moon broken apart by an impact collision.

Two satellites, Pandora and Prometheus, are called ring shepherds because they "bracket" the F Ring and are considered to confine the orbits of the particles that make up the ring.

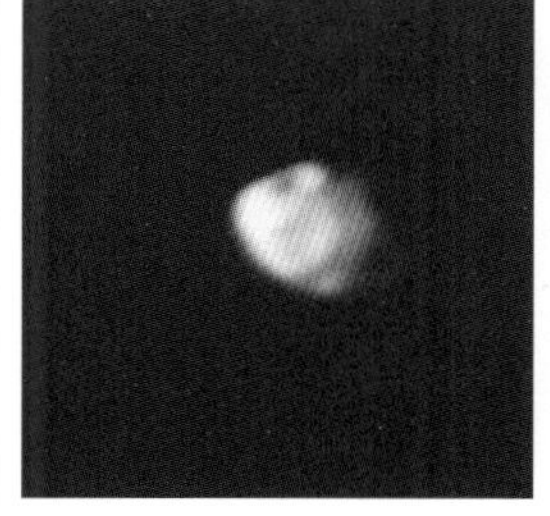
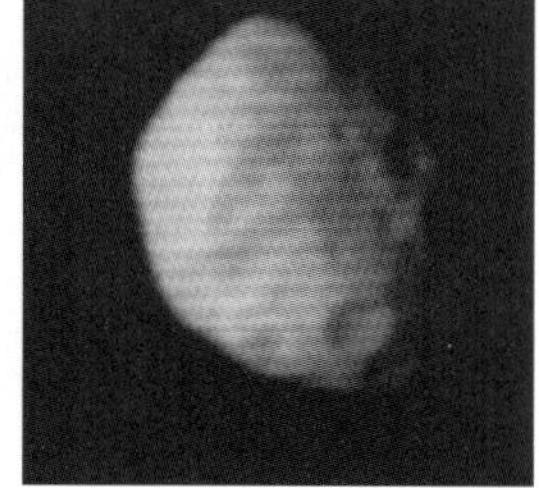
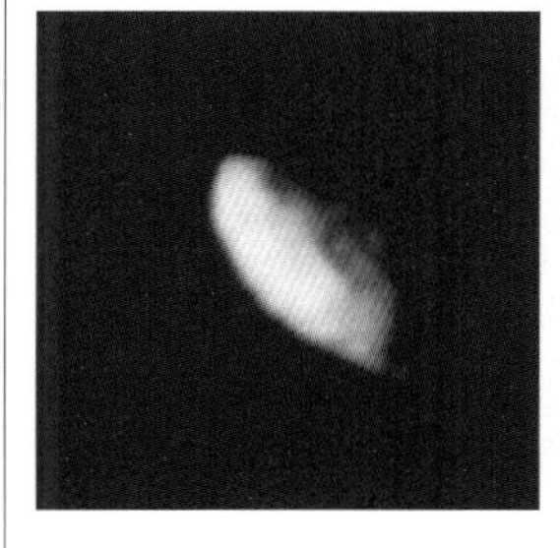

Mosaic of *Voyager* images showing some of the small saturnian satellites. Upper left: Pandora (outer F Ring shepherd), 120 by 100 kilometers; lower left: Prometheus (inner F Ring shepherd), 145 by 70 kilometers; upper right: Janus, 220 by 160 kilometers; lower right: Epimetheus, 140 by 100 kilometers (*Voyager 2*, JPL photograph P-24061).

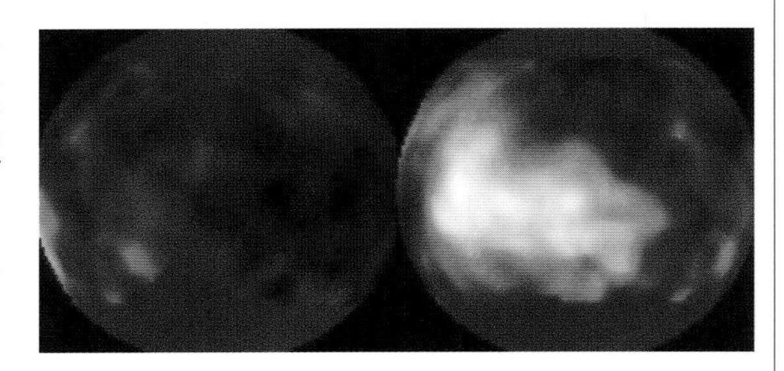

Titan, as viewed by the *Hubble Space Telescope* Planetary Camera at near-infrared wavelengths. Dark areas could represent oceans of hydrocarbons while bright Australia-size regions could be chunks of frozen water and ammonia ice. Alternatively, dark areas could be hydrocarbon tars covering solid surfaces (courtesy of Peter Smith and Mark Lemmon, University of Arizona, and the Space Telescope Institute).

False-color *Voyager* photograph showing the limb of Titan. Bands correspond to layers in the atmosphere (*Voyager 1*, JPL photograph P-23108C).

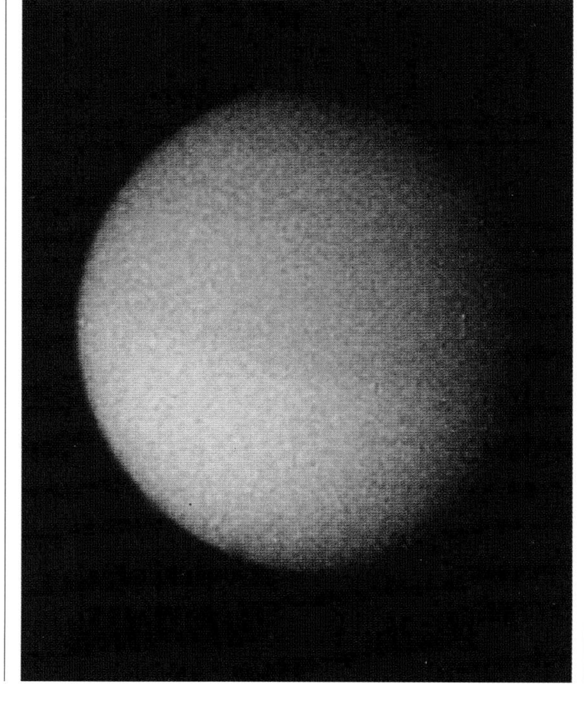

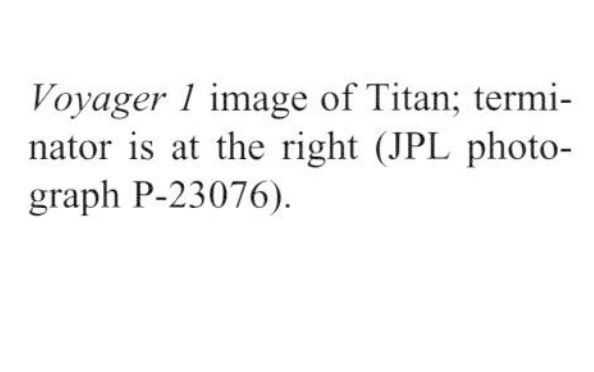

Voyager 1 image of Titan; terminator is at the right (JPL photograph P-23076).

A third ring shepherd, Atlas, is about 30 kilometers across and one of the smallest known saturnian satellites.

Three of the satellites, Calypso, Telesto, and Helene, are Lagrangian satellites. It has long been known that a small object can have the same orbit and speed as a larger object, so long as it maintains a 60° arc ahead of or behind the larger object in positions called Lagrangian points. Tethys has two such small objects in its orbit; Dione has one.

Finally, Hyperion is a small, irregular-shaped object whose elliptical orbit lies just outside that of Titan. Measuring 350 by 235 by 200 kilometers, its shape includes both angular and rounded parts. This satellite is very dark and, although its mass has not been determined, is presumed to be composed of ice and rock. The surface of Hyperion is heavily cratered. Large craters – some as large as 120 kilometers in diameter and 10 kilometers deep – are scattered over the surface. In addition, the surface displays scarps that may be part of a connected, arcuate feature nearly 300 kilometers long. This feature could be the remnant of a crater that would have a diameter greater than 200 kilometers.

Because of its irregular shape, heavily cratered surface, and a possible impact scar more than half its size, Hyperion is considered by most investigators to be the remnant of a much larger object shattered by collisional or impact processes.

Titan

Titan is a remarkable satellite. Larger than Mercury at 5,120 kilometers in diameter, it is in the same size class as Ganymede and Callisto. Like those satellites of Jupiter, its mass suggests that it is composed of about 45 percent water ice and 55 percent rocky material. What makes it remarkable is its dense atmosphere. Although the satellite was the first to be discovered in the saturnian system (found by Christian Huygens in 1655), the presence of an atmosphere was not suspected until the twentieth century. In 1944, the Dutch-American astronomer, Gerard Kuiper, identified methane on Titan, which initiated speculation on the various chemical reactions that might be taking place in its atmosphere.

The presence of extensive clouds in the atmosphere of Titan was well established by the time the *Voyager* spacecraft arrived. *Voyager* scientists hoped that the clouds would be broken and sufficiently scattered to allow glimpses of the surface. Such was not the case, and the *Voyager* pictures are disappointingly bland, showing a relatively uniform orange-brown sphere with only hints of variations in the clouds. Limb views, however, show layers in the atmosphere that are thought to be smoglike photochemical hazes. It is likely that very little sunlight penetrates the clouds, and the surface must be a frigid, gloomy place.

On the other hand, the nonimaging experiments on board *Voyager* returned results that show Titan to be a fascinating object. Nitrogen is the main component in the atmosphere, along with ethane, acetylene, ethylene, hydrogen cyanide, and other carbon-nitrogen components. It may be that when Titan first formed methane and ammonia ices were caught up in the accreting mass. With heating and subsequent chemical differentiation these compounds would have been released from the interior and formed an early-stage atmosphere. The action of sunlight on the ammonia would have released the nitrogen and allowed the escape of most of the hydrogen to deep space, while methane would have remained as a vast ocean that covered the entire satellite. Calculations show that some of the methane would have evaporated from the surface of this ocean and would be dissociated by sunlight. These chemicals would then recombine to form complex organic chemicals, resulting in the orangish smog seen in the atmosphere. These and other molecules would then condense to form aerosols that would rain down on Titan's surface, perhaps forming puddles of tar-like ooze.

Surface temperatures are about -180°C. This temperature is near the point where ethane and methane could coexist, and it has been suggested that Titan may be covered with a liquid ethane-methane ocean to a depth of 1 kilometer. Earth-based radar observations suggest that both liquids and solids exist on the surface. Thus, there could be islands of solid water ice rising from the ocean floor. In this model, liquid ethane and methane could play the same role on Titan as liquid water does on Earth, so far as surface processes are concerned.

From this description, we can only speculate about geologic events on Titan, either at present or in the past. Fluid processes may modify solid parts of the surface and could involve rivers of liquid ethane-methane, tidal activity by an ocean, or perhaps even wind erosion by ice particles. Whether these or other geologic processes, such as volcanism, tectonics, or impact cratering, have occurred and left landforms is open to question. Just as for cloud-covered Venus, a radar-imaging system is required to view the surface of Titan. Such a system was developed for a joint NASA-European Space Agency mission named *Cassini*. The *Cassini* spacecraft, launched in 1997, will include an orbiter around Saturn and a probe to descend to the surface of Titan in 2004. In many respects Titan resembles a primordial Earth, and the presence of organic compounds in a chemically rich atmosphere makes Titan a target of high priority for exploration.

Enceladus
Sub-Saturnian Hemisphere
Geologic Map

Center lat. 0°
Center lon. 0°

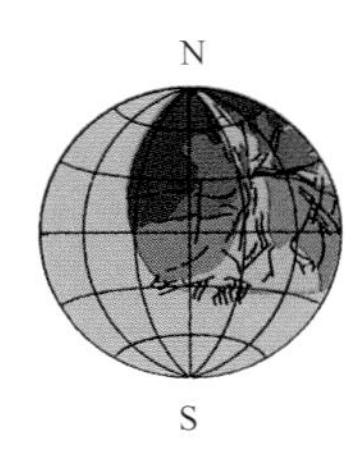

Enceladus
Anti-Saturnian Hemisphere
Geologic Map

Center lat. 0°
Center lon. 180°

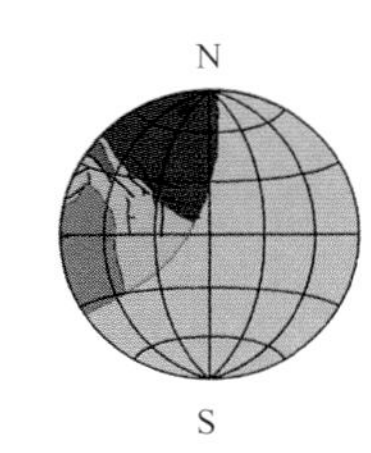

Enceladus
Sub-Saturnian Hemisphere
Reference Map

Center lat. 0°
Center lon. 0°

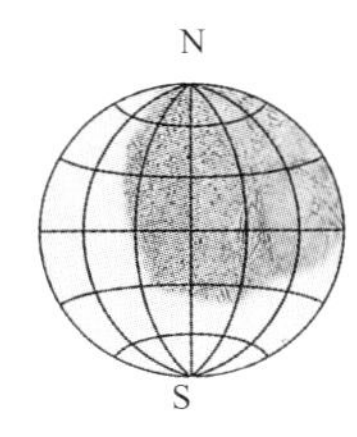

Enceladus
Anti-Saturnian Hemisphere
Reference Map

Center lat. 0°
Center lon. 180°

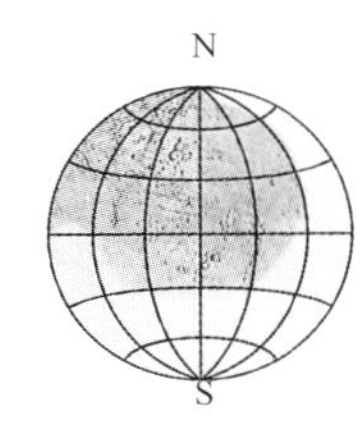

Tethys
Sub-Saturnian Hemisphere
Geologic Map

Center lat. 0°
Center lon. 0°

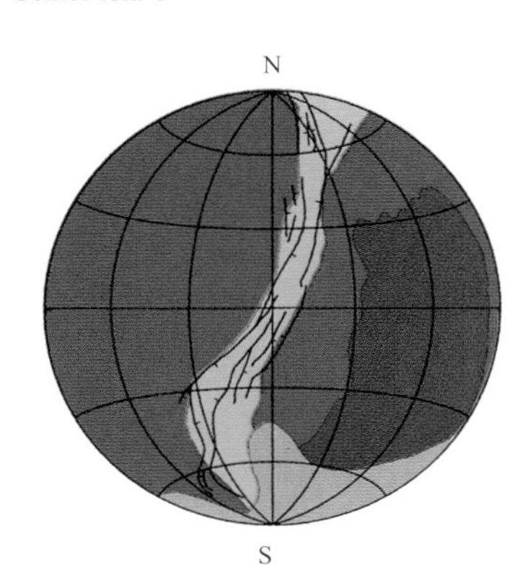

Tethys
Anti-Saturnian Hemisphere
Geologic Map

Center lat. 0°
Center lon. 180°

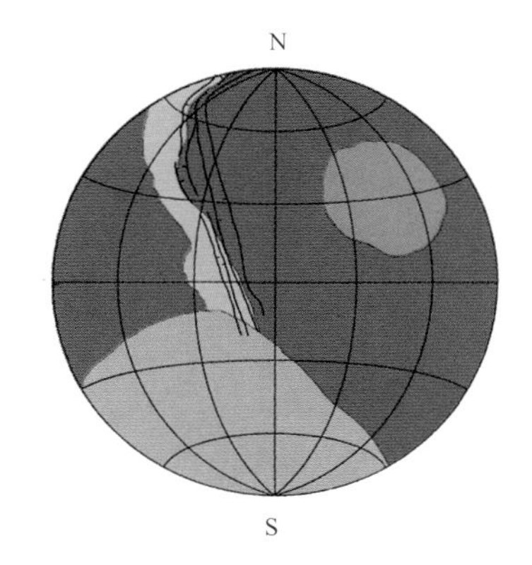

Tethys
Sub-Saturnian Hemisphere
Reference Map

Center lat. 0°
Center lon. 0°

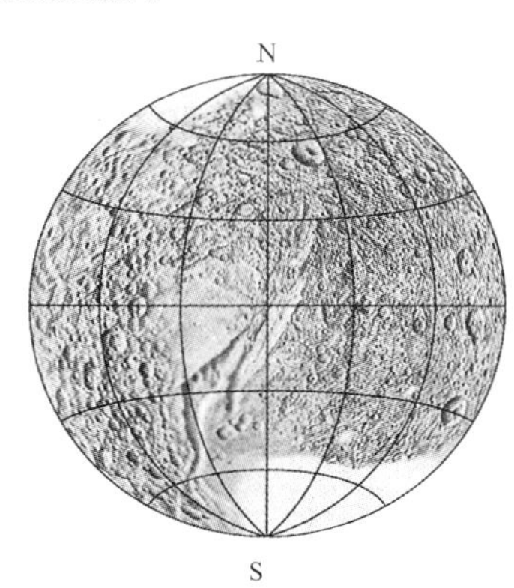

Tethys
Anti-Saturnian Hemisphere
Reference Map

Center lat. 0°
Center lon. 180°

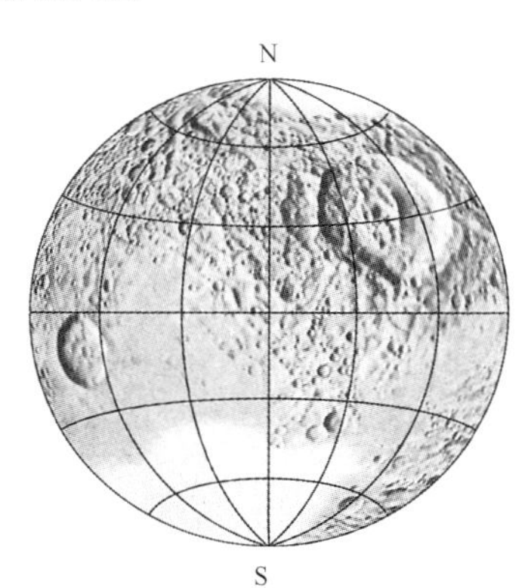

Enceladus

Description of Map Units

Cratered Terrain

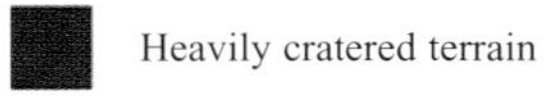
Heavily cratered terrain

Plains Materials

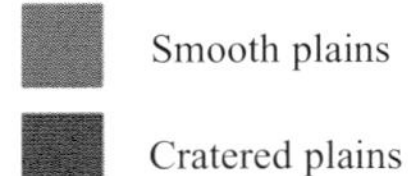
Smooth plains

Cratered plains

Grooved plains

Unmapped

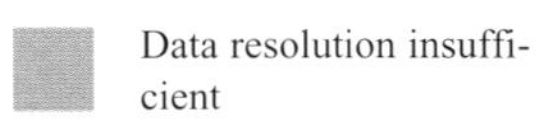
Data resolution insufficient

Structure Symbols

Fault scarp

Graben

Enceladus

Correlation of Map Units

Youngest at top

Cratered terrain | Plains materials

Tethys

Description of Map Units

Impact Materials

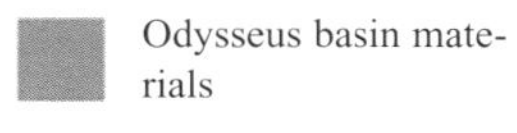
Odysseus basin materials

Undetermined Materials

Plains

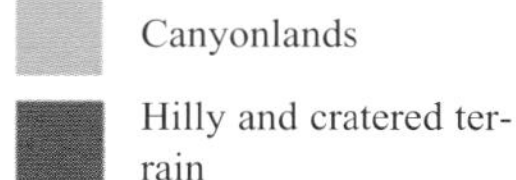
Canyonlands

Hilly and cratered terrain

Unmapped

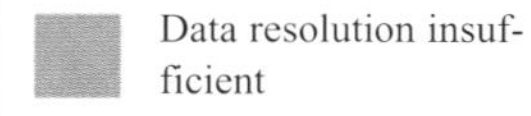
Data resolution insufficient

Structure Symbols

Fault scarp

Tethys

Correlation of Map Units

Youngest at top

Impact materials | Undetermined materials

Dione
Sub-Saturnian Hemisphere
Geologic Map

Center lat. 0°
Center lon. 0°

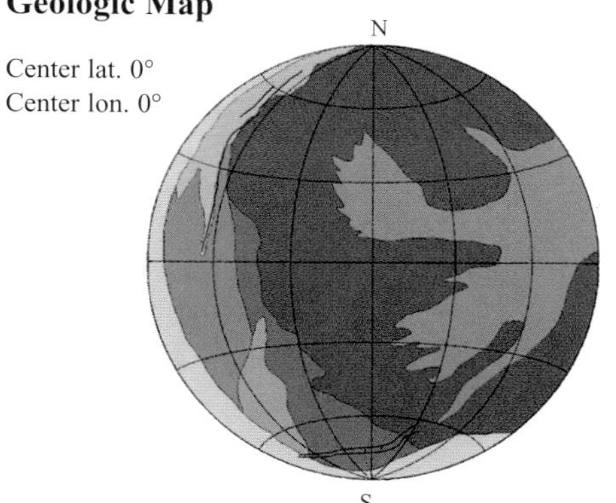

Dione
Anti-Saturnian Hemisphere
Geologic Map

Center lat. 0°
Center lon. 180°

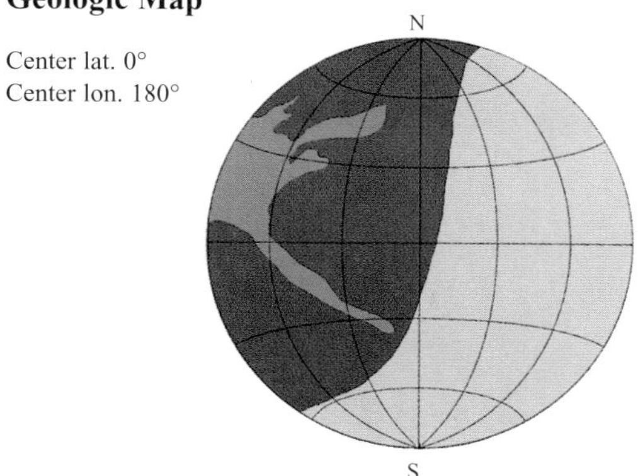

Dione
Sub-Saturnian Hemisphere
Reference Map

Center lat. 0°
Center lon. 0°

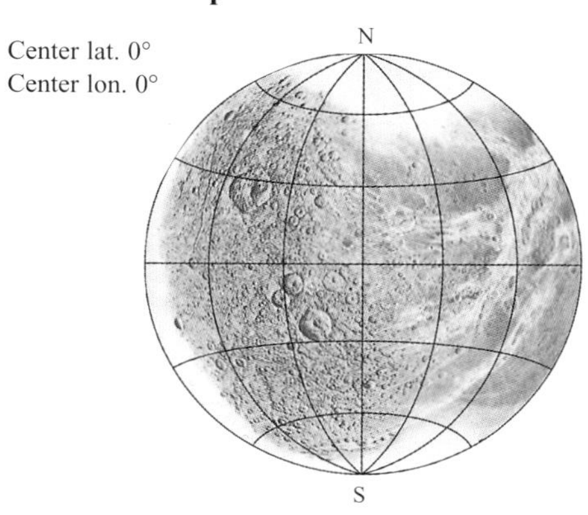

Dione
Anti-Saturnian Hemisphere
Reference Map

Center lat. 0°
Center lon. 180°

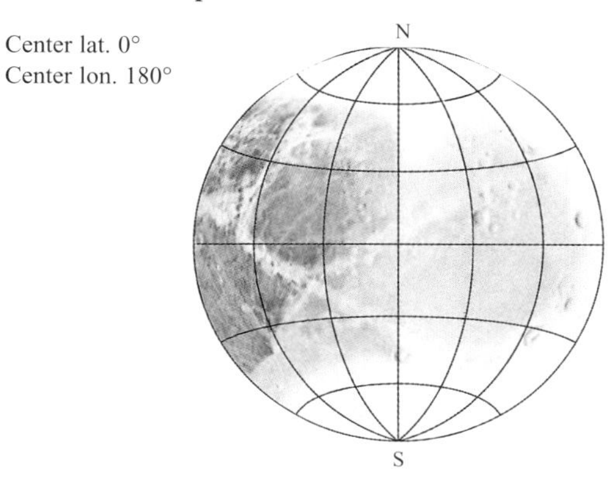

Rhea
Sub-Saturnian Hemisphere
Geologic Map

Center lat. 0°
Center lon. 0°

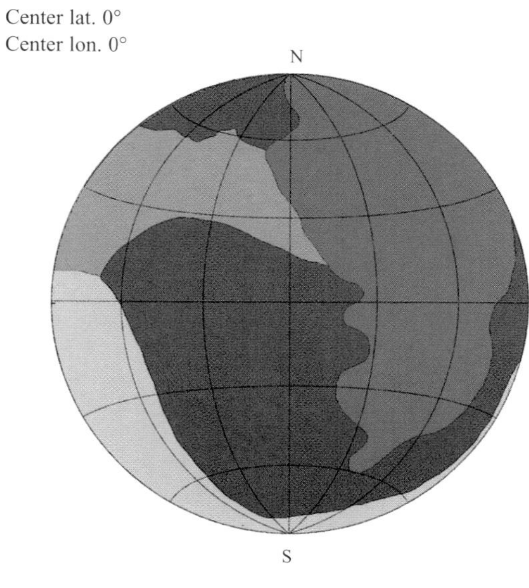

Rhea
Anti-Saturnian Hemisphere
Geologic Map

Center lat. 0°
Center lon. 180°

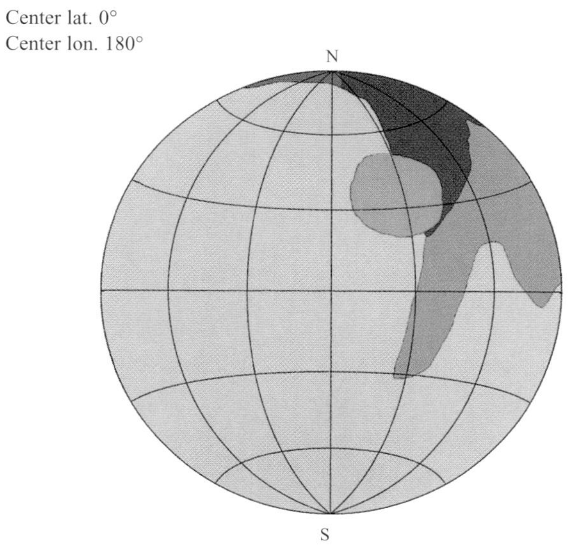

Rhea
Sub-Saturnian Hemisphere
Reference Map

Center lat. 0°
Center lon. 0°

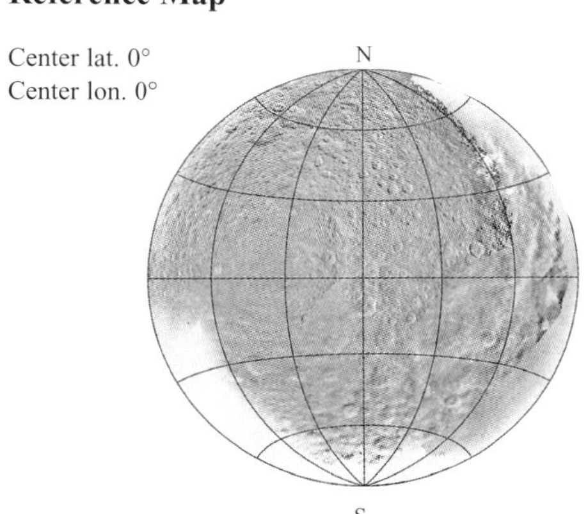

Rhea
Anti-Saturnian Hemisphere
Reference Map

Center lat. 0°
Center lon. 180°

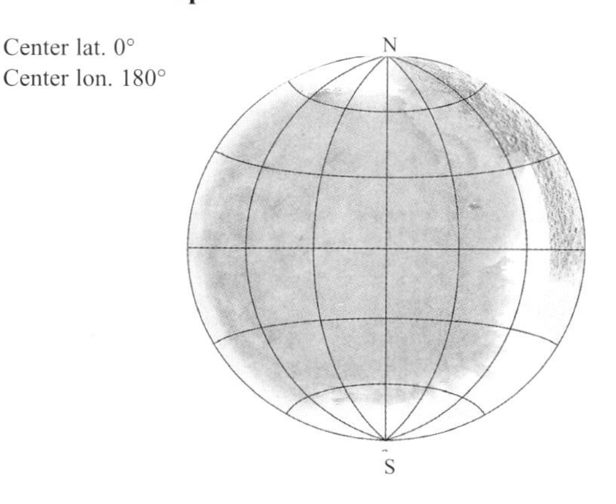

Dione

Description of Map Units

Plains Materials

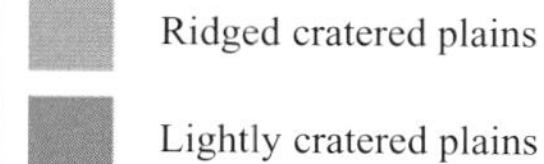
Ridged cratered plains

Lightly cratered plains

Cratered plains

Miscellaneous Units

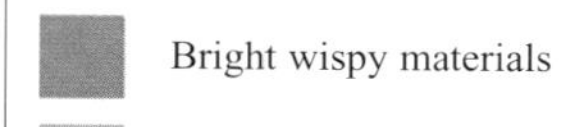
Bright wispy materials

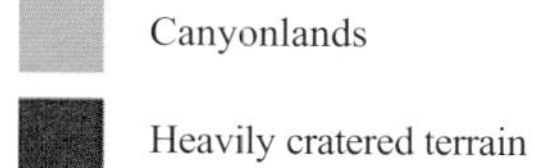
Canyonlands

Heavily cratered terrain

Unmapped

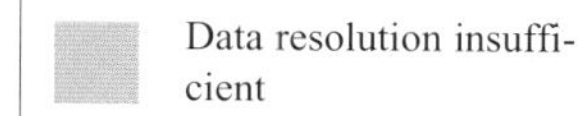
Data resolution insufficient

Structure Symbols

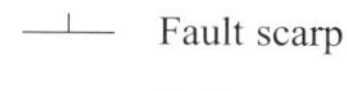
Fault scarp

Graben

Dione

Correlation of Map Units

Youngest at top

Plains materials | Miscellaneous units

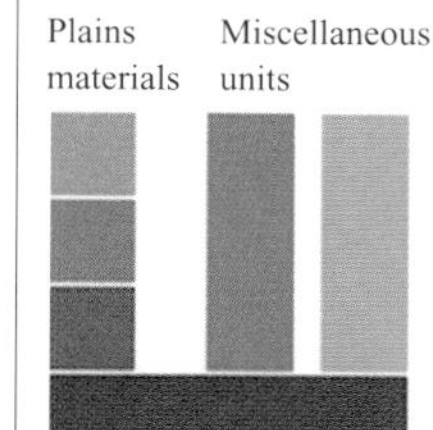

Rhea

Description of Map Units

Impact Materials

Tirawa basin materials

Cratered Terrains

Younger

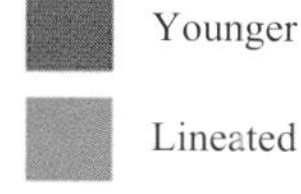
Lineated

Undivided

Unmapped

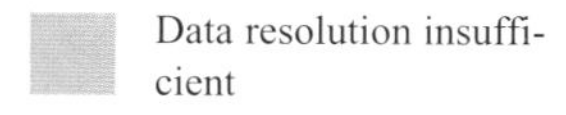
Data resolution insufficient

Rhea

Correlation of Map Units

Youngest at top

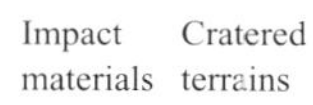
Impact materials | Cratered terrains

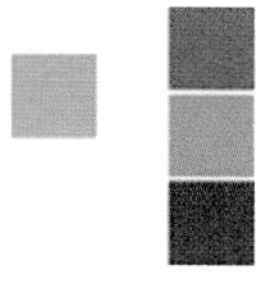

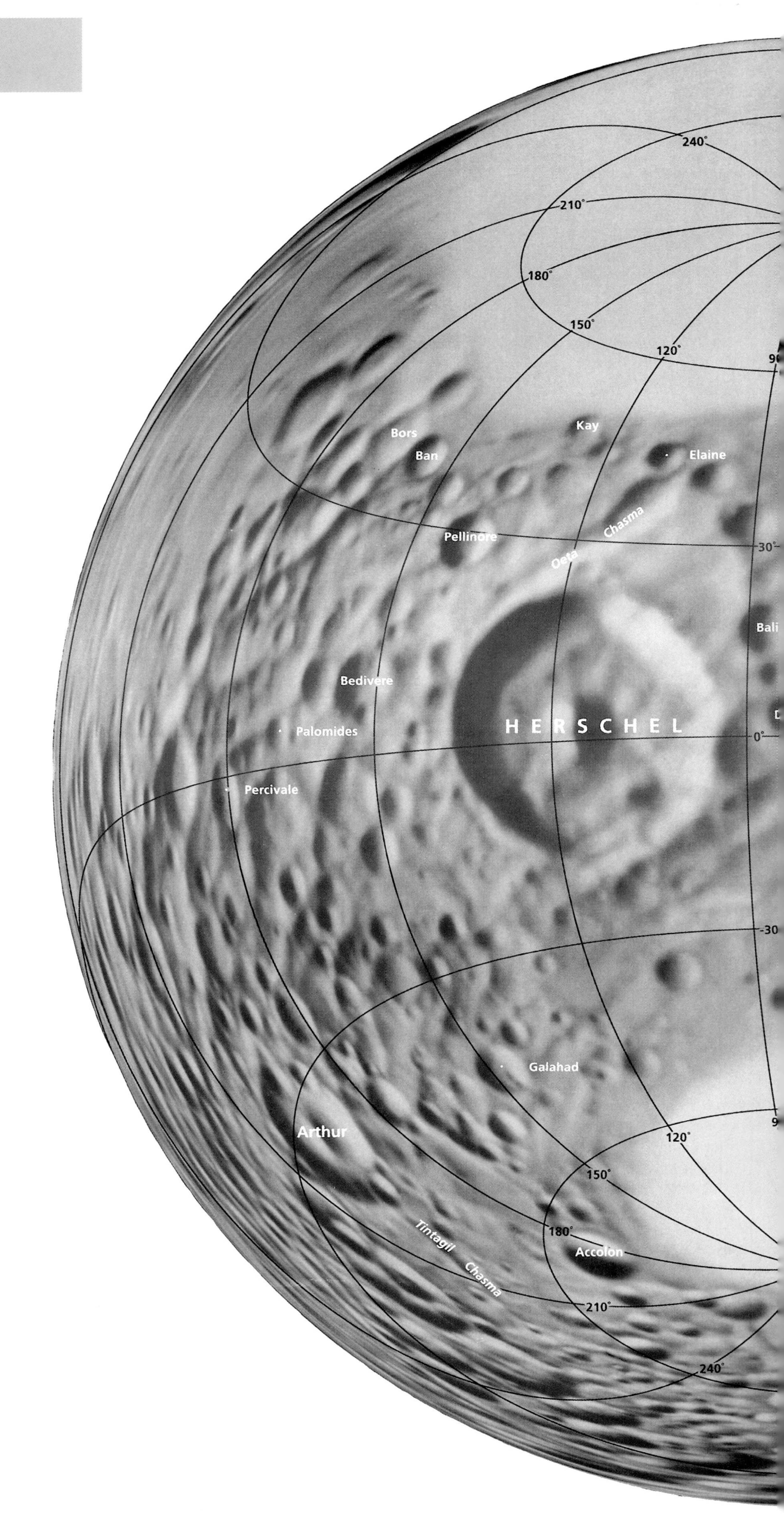

240°
210°
180°
150°
120°
Bors
Ban
Kay
Elaine
Chasma
Oeta
Pellinore
30°
Bali
Bedivere
HERSCHEL
Palomides
0°
Percivale
-30
Galahad
Arthur
120°
150°
Tintagil
Chasma
180°
Accolon
210°
240°

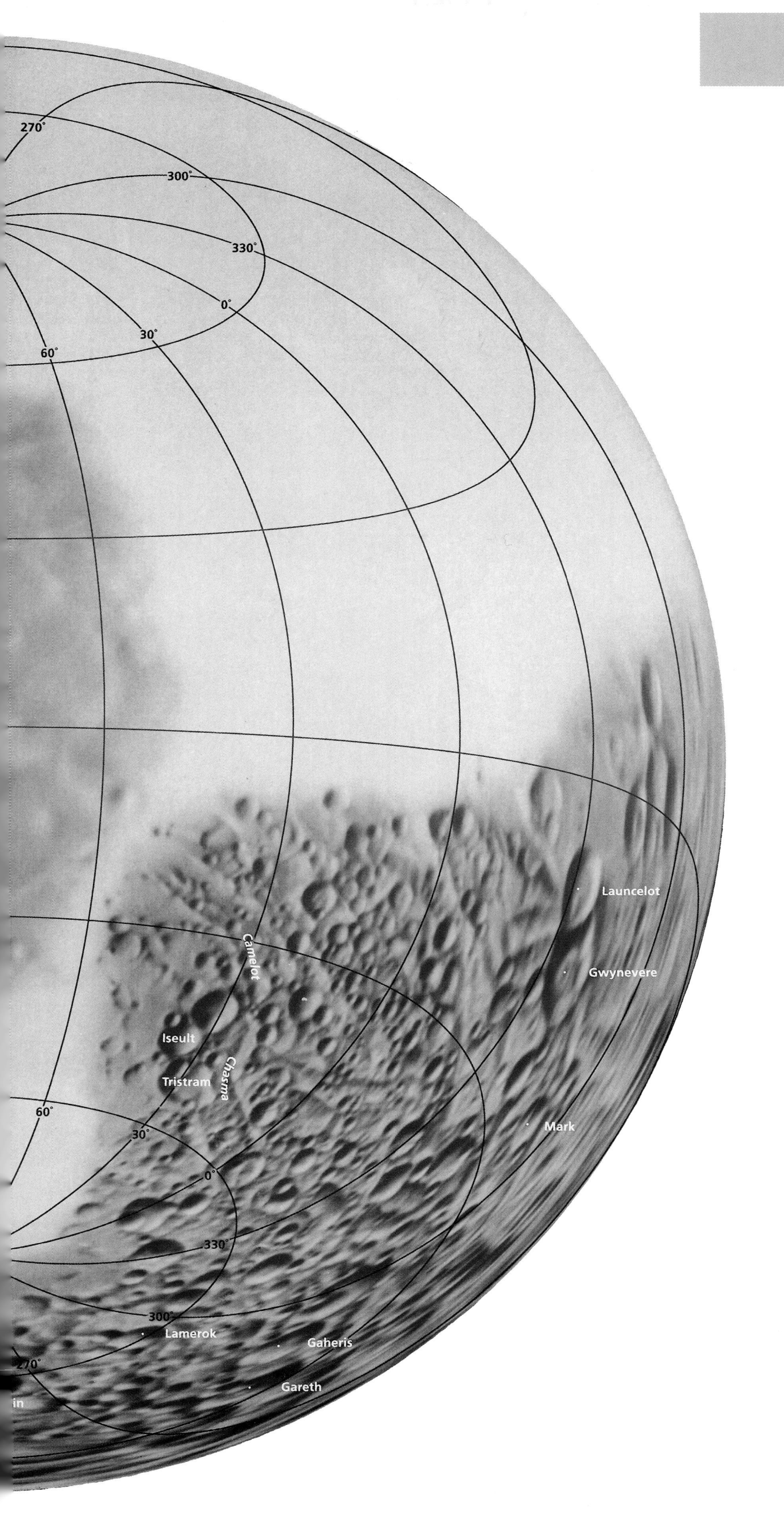

Mimas Airbrush Map Herschel Region

Center lat. –5°
Center lon. 80°

1 cm = 26 km
1 cm² = 676 km²

1 cm²

130
104
78
52
26
0 km

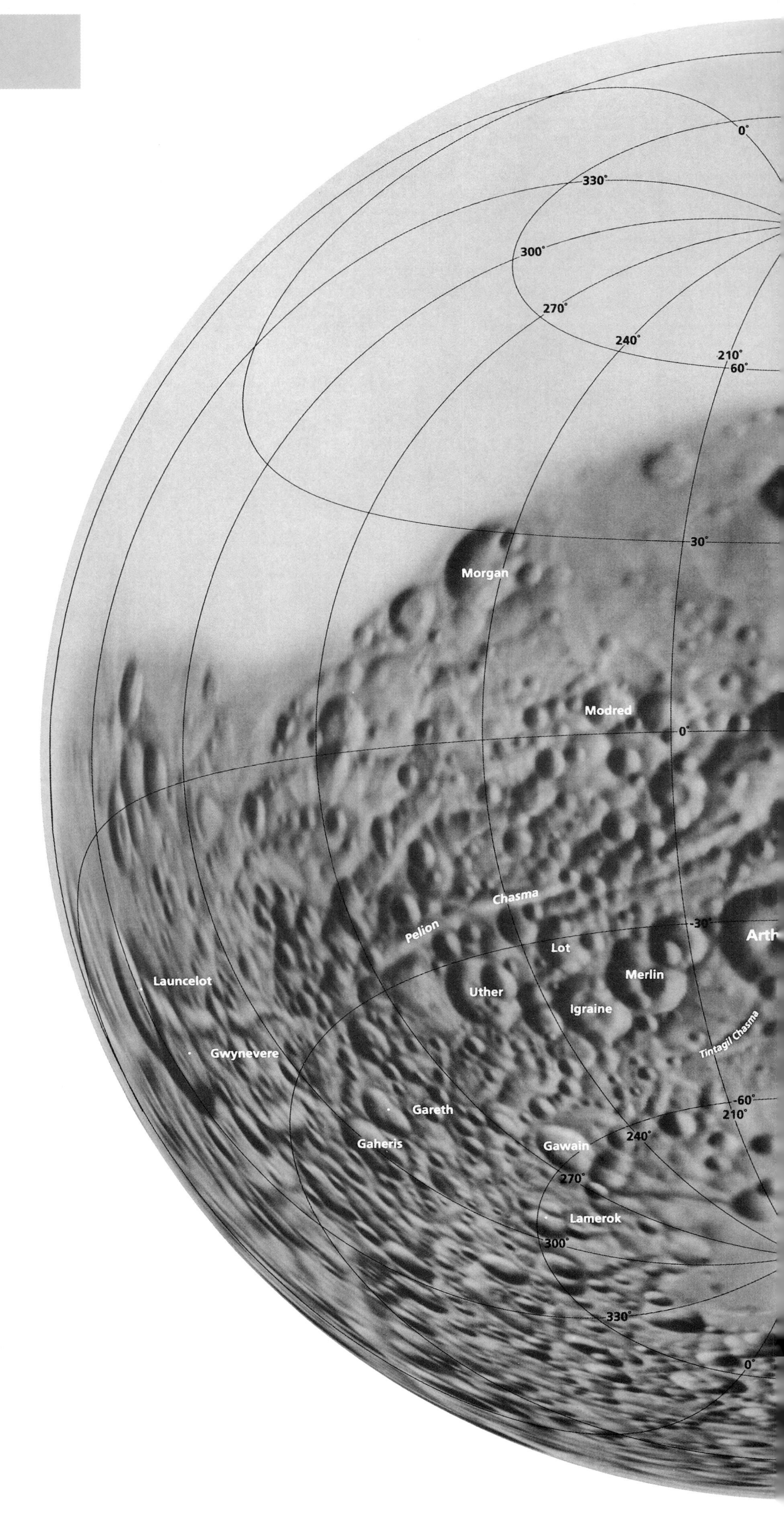

0°
330°
300°
270°
240°
210°
60°
30°
Morgan
Modred
0°
Chasma
Pelion
-30°
Arth
Lot
Merlin
Launcelot
Uther
Igraine
Tintagil Chasma
Gwynevere
-60°
210°
Gareth
240°
Gaheris
Gawain
270°
Lamerok
300°
330°
0°

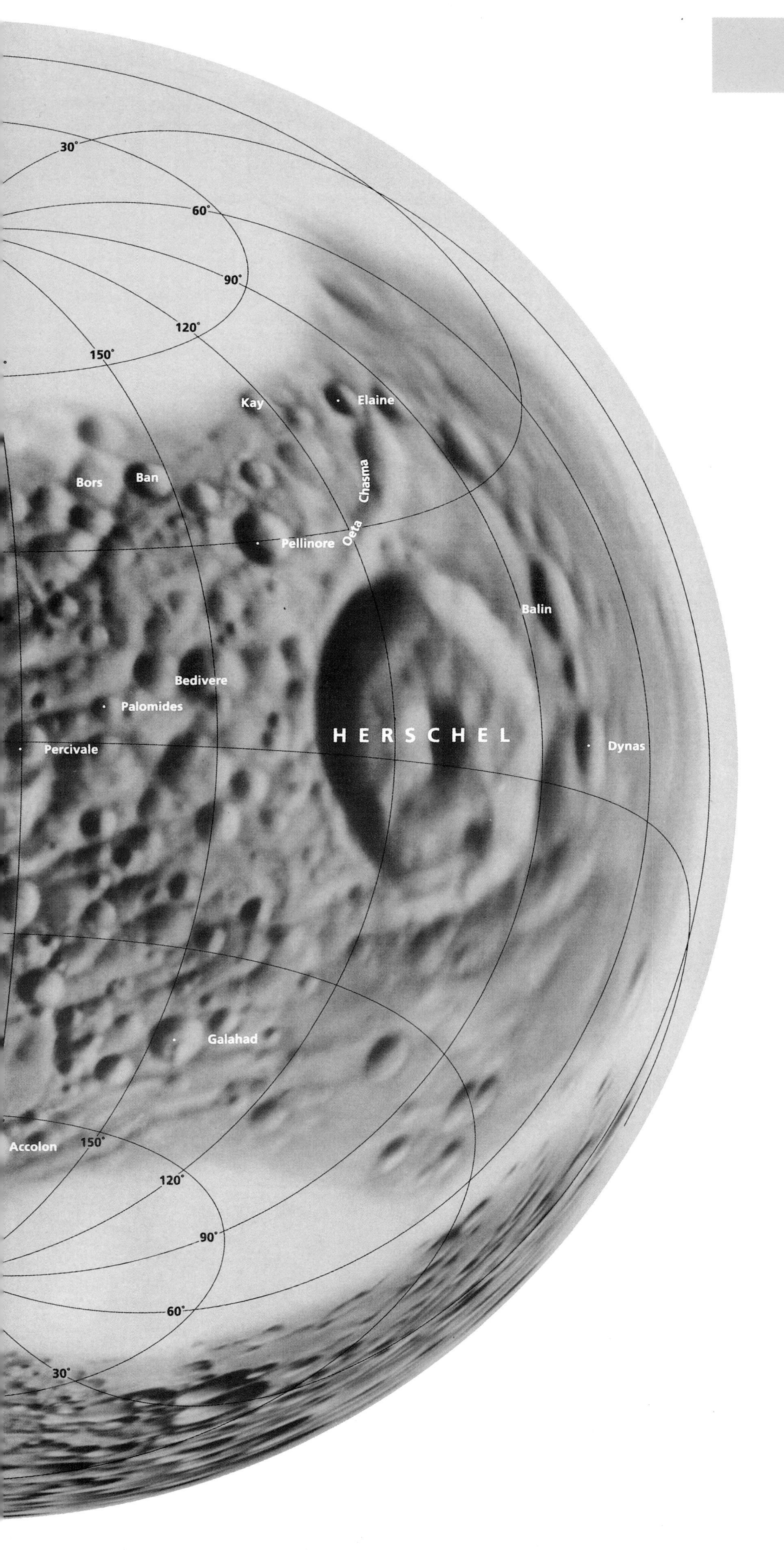

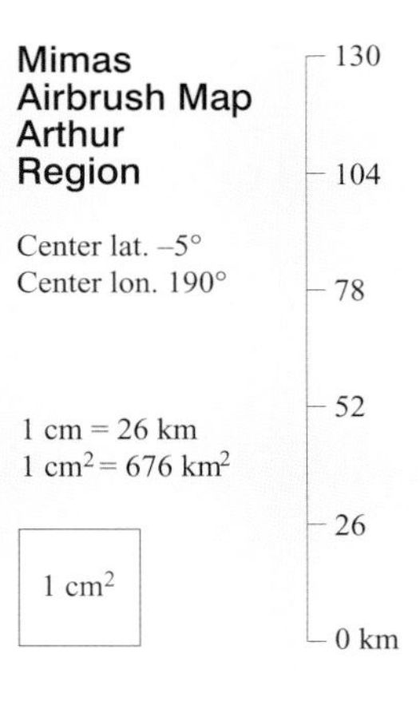

Mimas
Airbrush Map
Arthur
Region

Center lat. –5°
Center lon. 190°

1 cm = 26 km
1 cm² = 676 km²

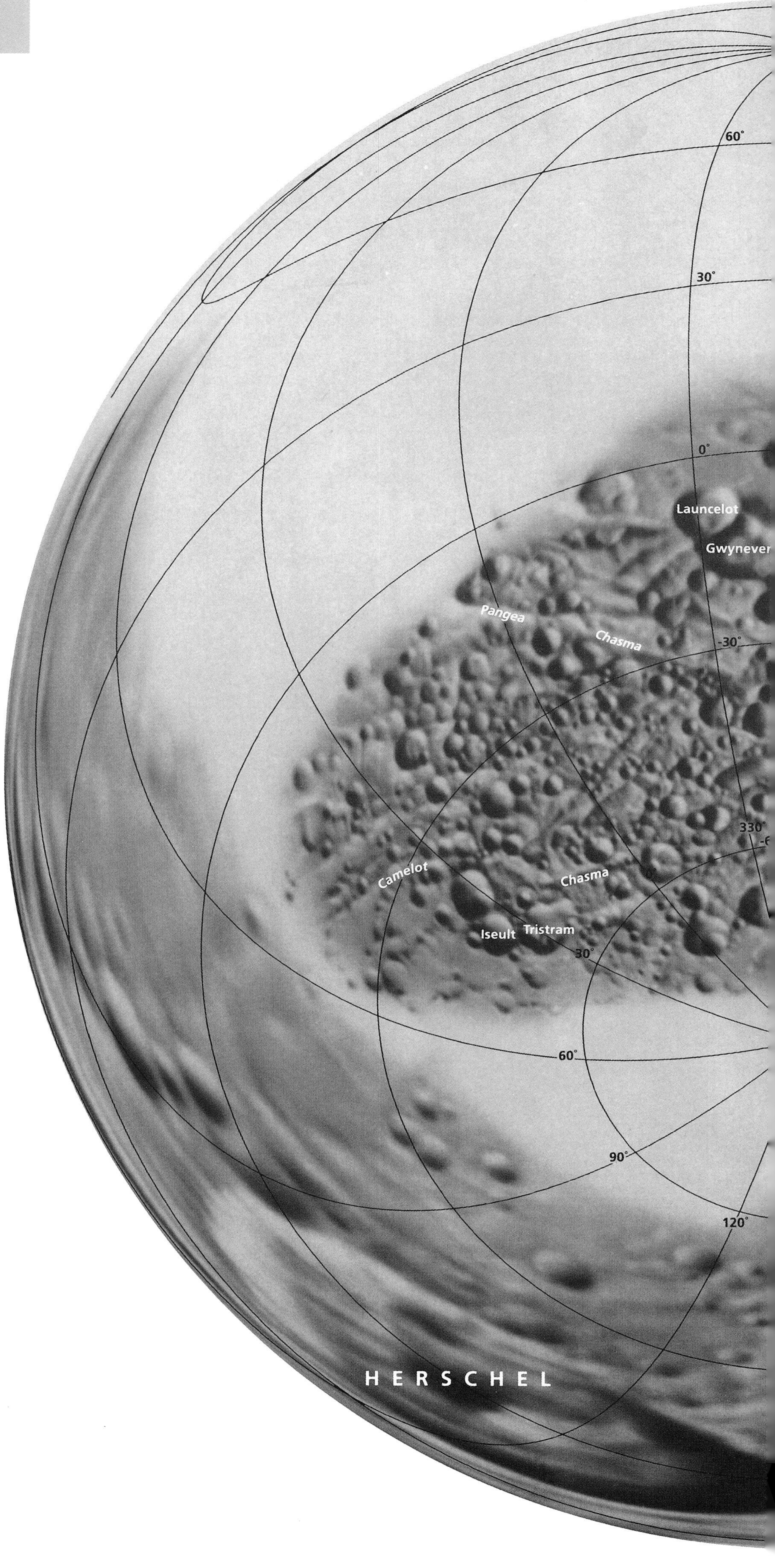

Launcelot
Gwynever
Pangea
Chasma
Camelot
Chasma
Iseult
Tristram
HERSCHEL

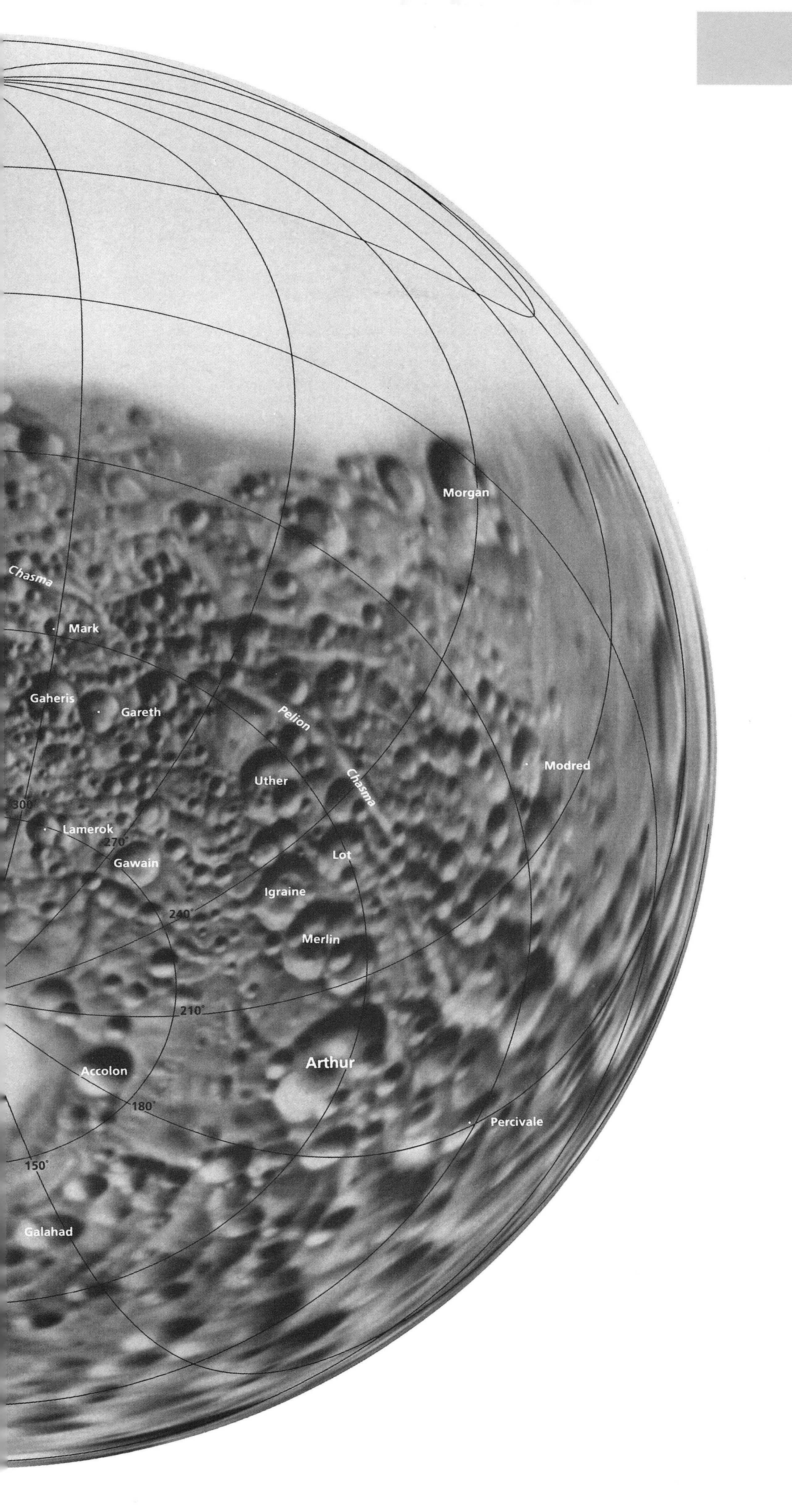

Mimas
Airbrush Map
Gwynevere
Region

Center lat. –50°
Center lon. 315°

1 cm = 26 km
1 cm^2 = 676 km^2

1 cm^2

130
104
78
52
26
0 km

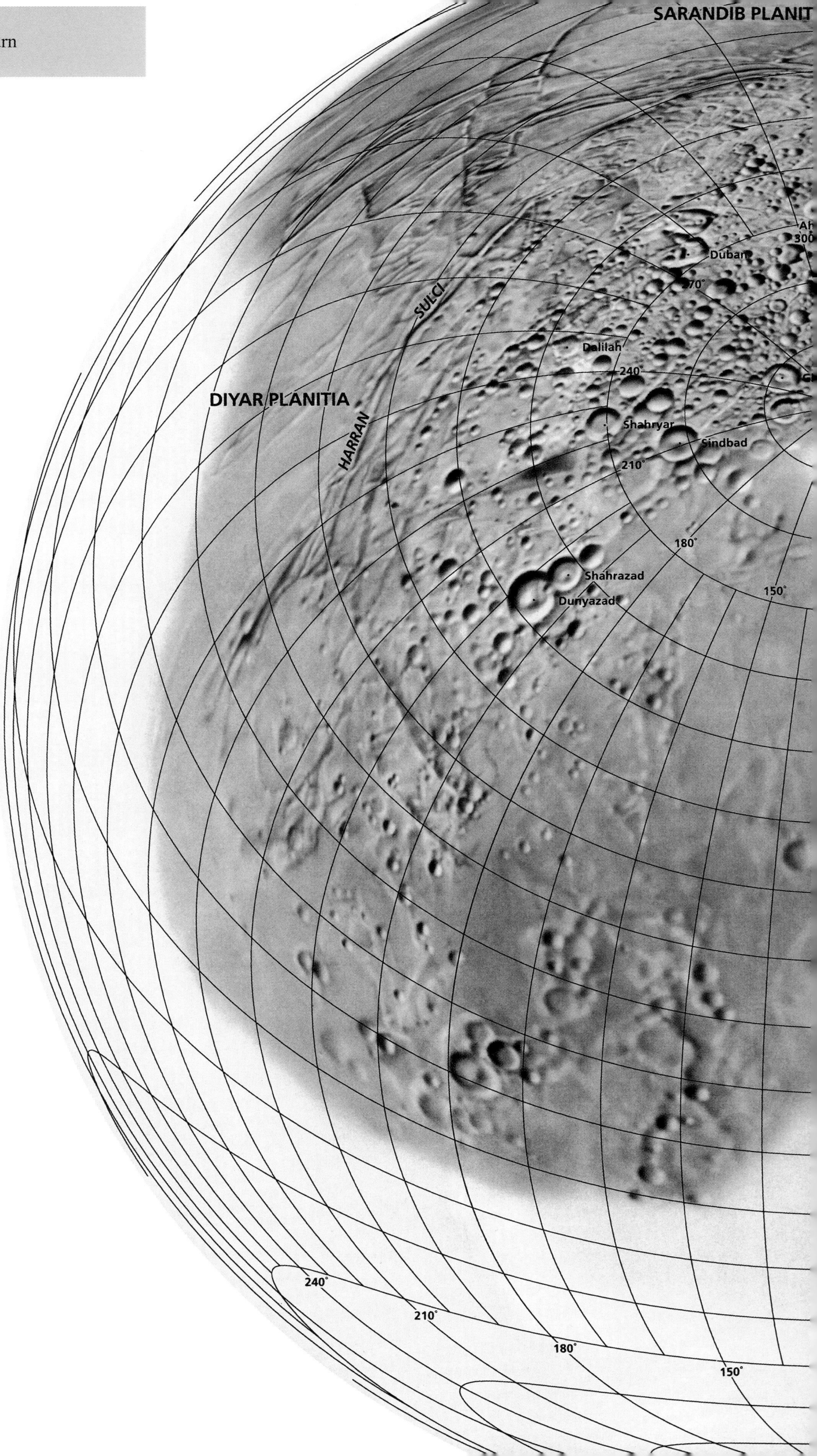

SARANDIB PLANIT

Duban

270°

300

Dalilah

240°

DIYAR PLANITIA

HARRAN SULCI

Shahryar

Sindbad

210°

180°

Shahrazad

Dunyazad

150°

240°

210°

180°

150°

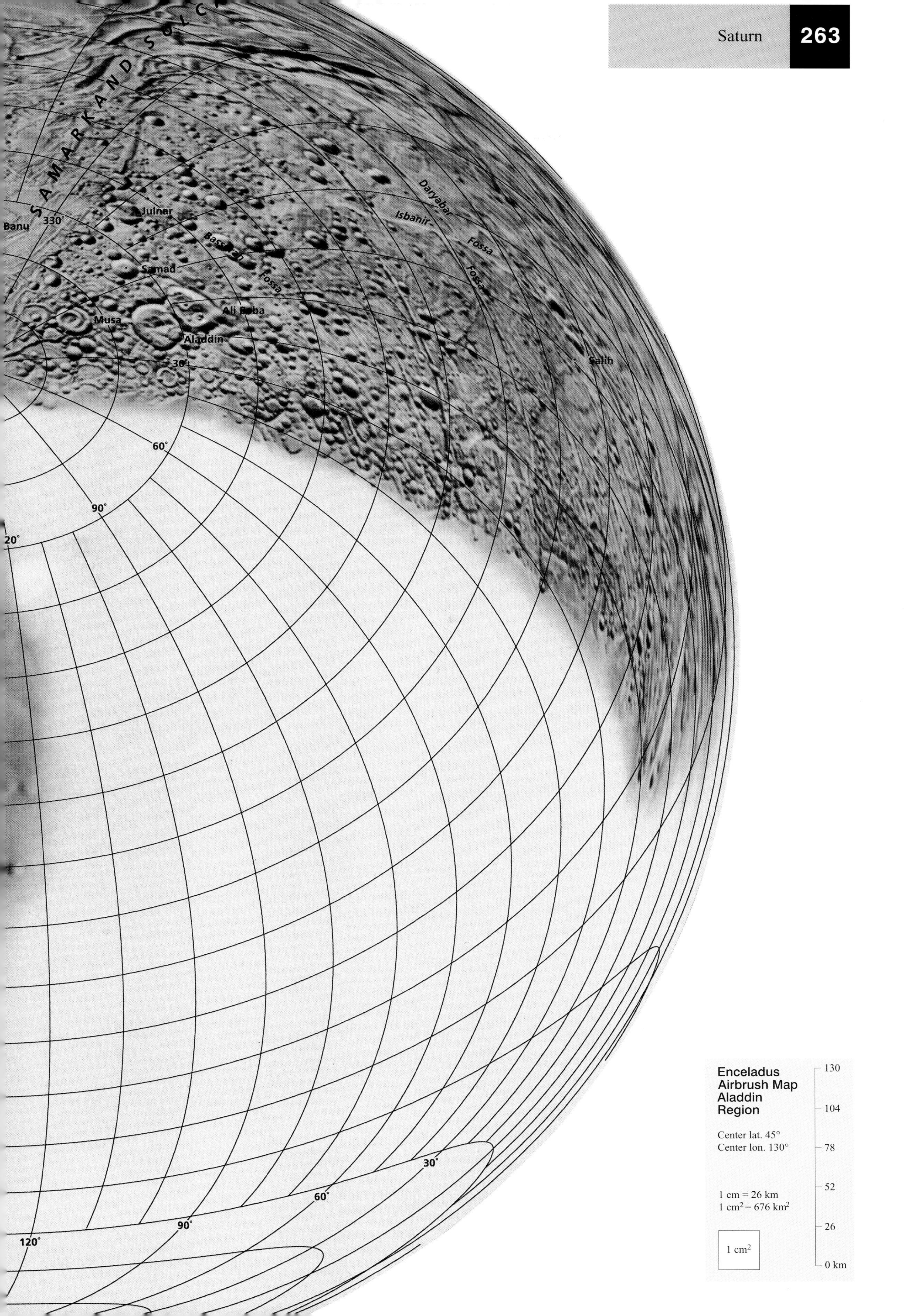

Enceladus Airbrush Map Aladdin Region

Center lat. 45°
Center lon. 130°

1 cm = 26 km
$1\ cm^2 = 676\ km^2$

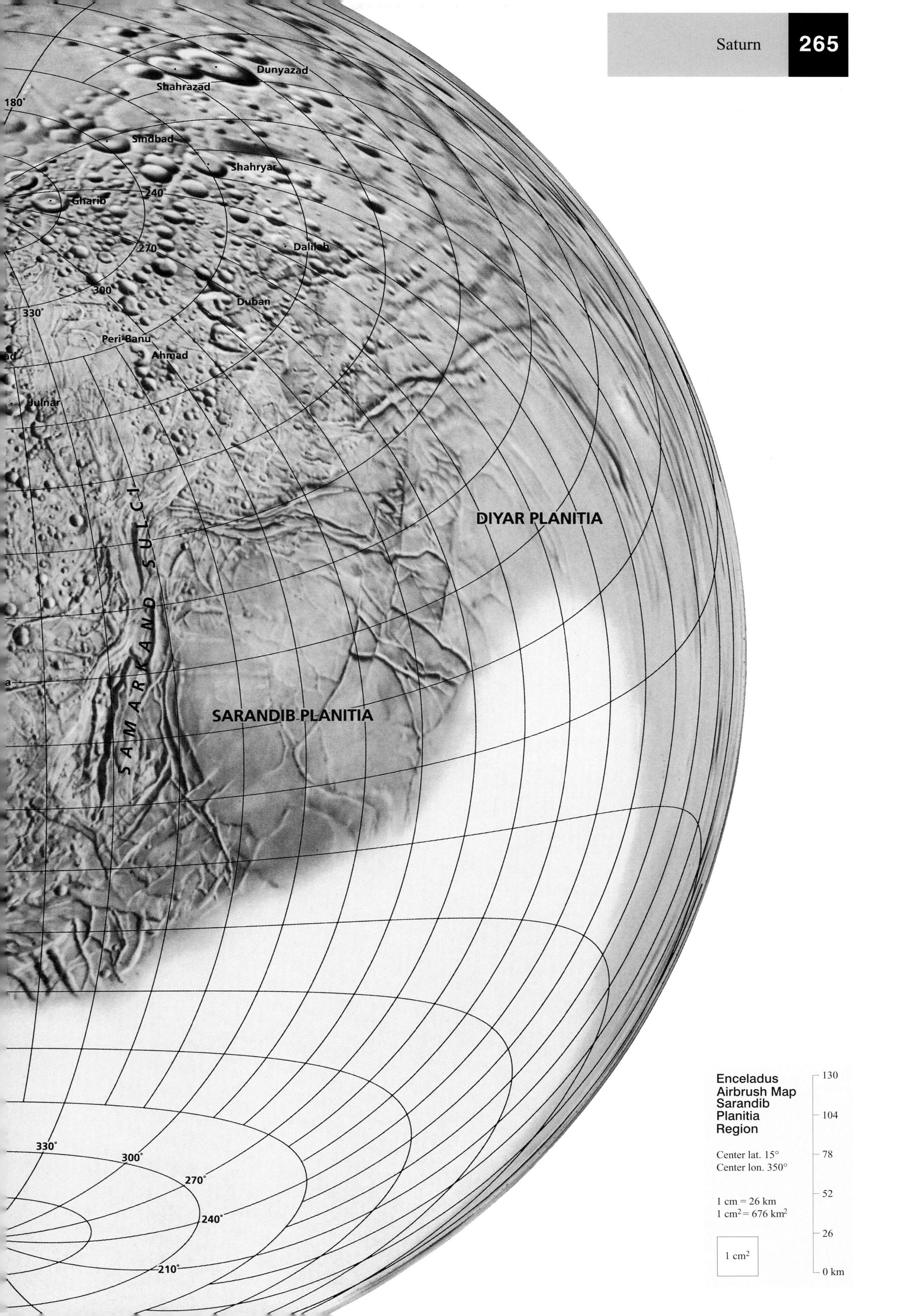

Enceladus Airbrush Map Sarandib Planitia Region

Center lat. 15°
Center lon. 350°

1 cm = 26 km
1 cm² = 676 km²

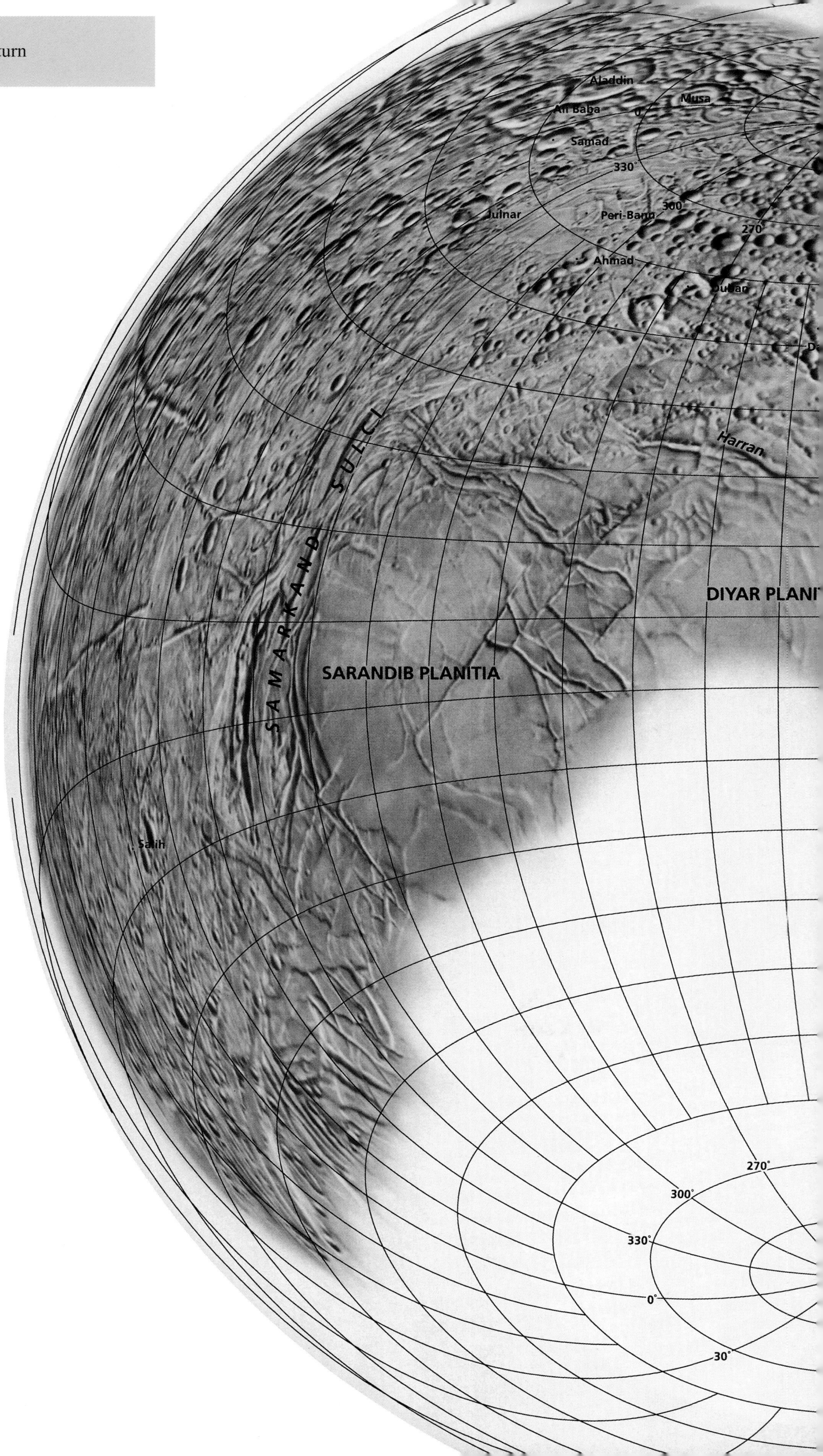

Aladdin
Musa
Ali Baba
0°
Samad
330°
Julnar
Peri-Banu
300°
270°
Ahmad
Duban
Harran
SAMARKAND SULCI
DIYAR PLANI
SARANDIB PLANITIA
Salih
270°
300°
330°
0°
30°

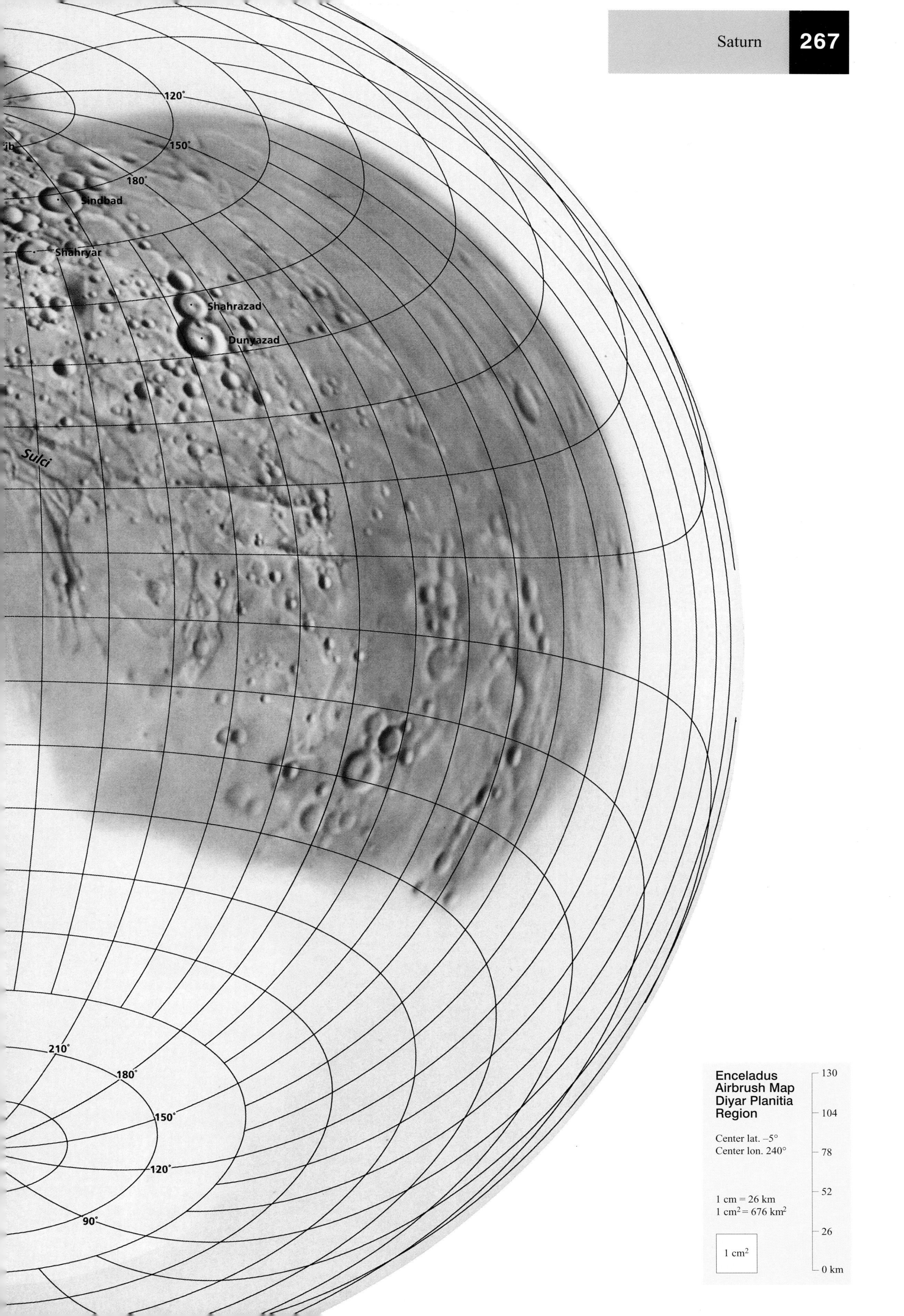

Enceladus Airbrush Map Diyar Planitia Region

Center lat. –5°
Center lon. 240°

1 cm = 26 km
1 cm^2 = 676 km^2

1 cm^2

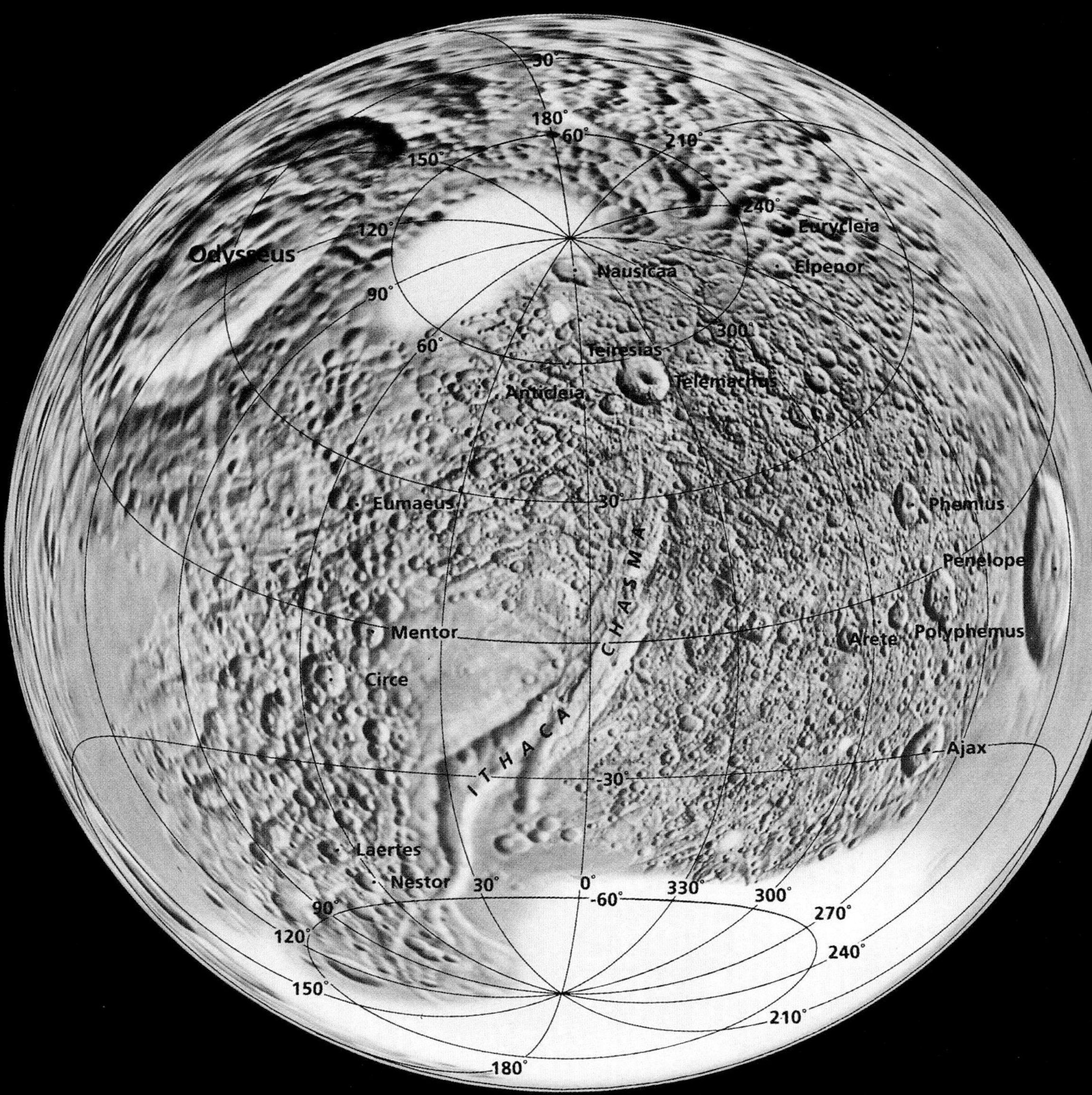
Odysseus
Eurycleia
Elpenor
Nausicaa
Teiresias
Telemachus
Anticleia
Eumaeus
Phemius
Penelope
ITHACA CHASMA
Mentor
Arete
Polyphemus
Circe
Ajax
Laertes
Nestor

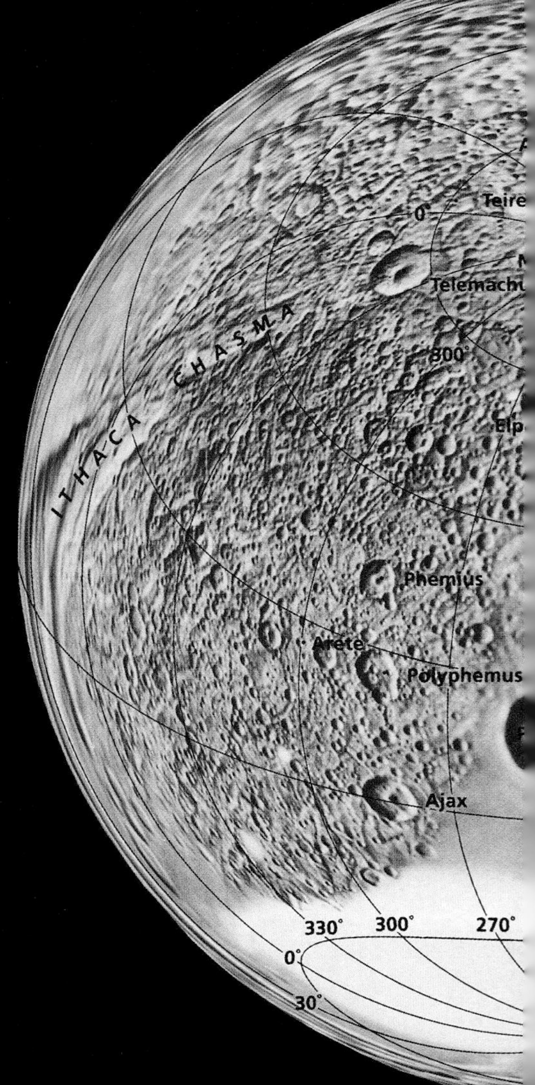
ITHACA CHASMA
Telemachu
Phemius
Arete
Polyphemus
Ajax

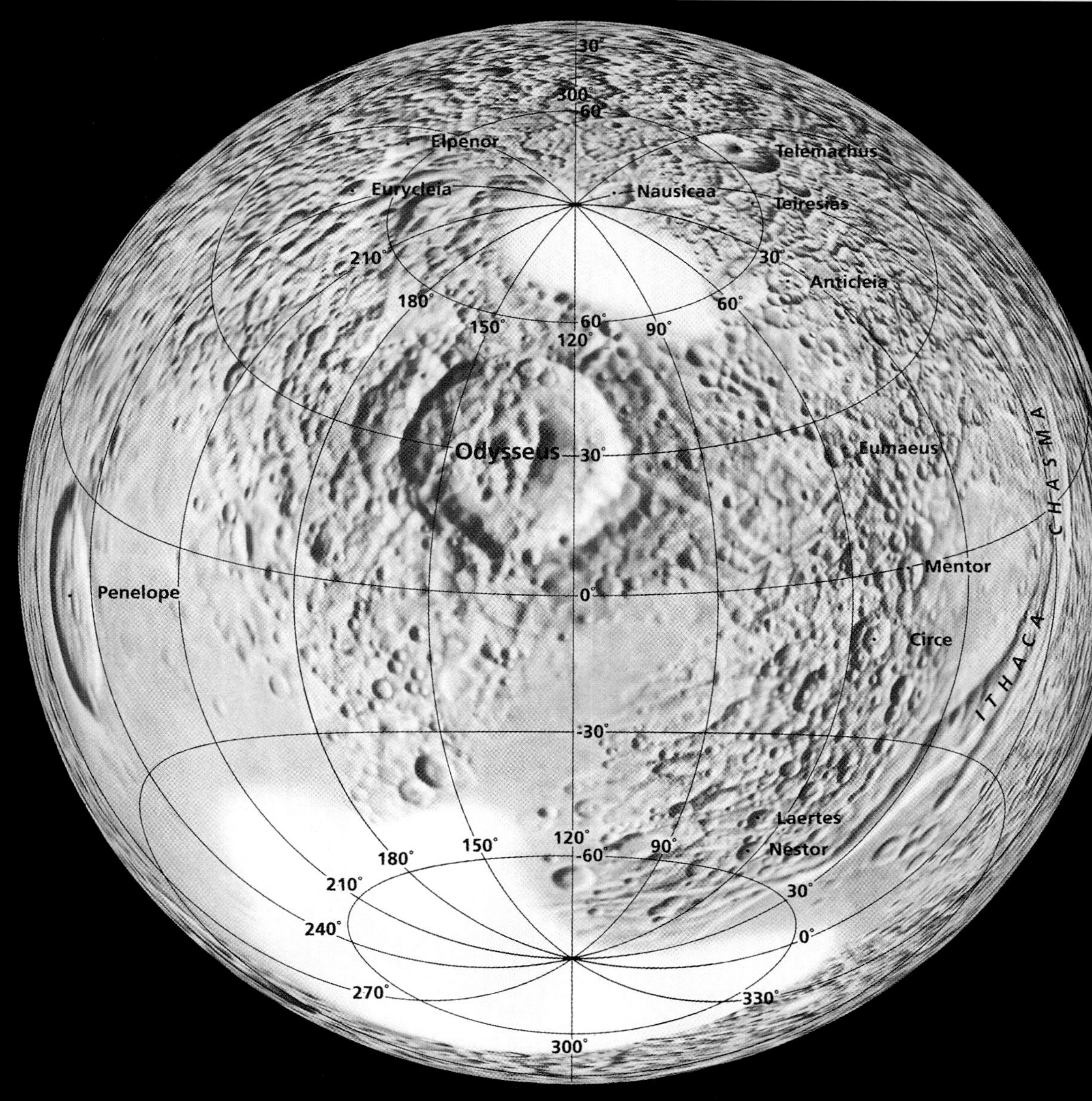

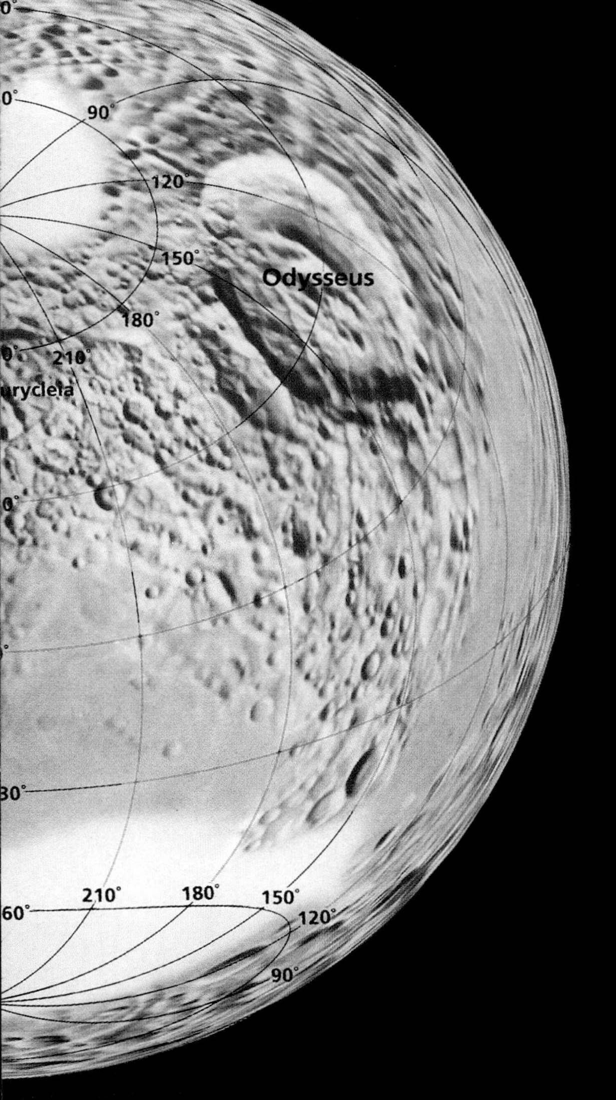

Tethys Airbrush Map

Ithaca Chasma Region

Center lat. 20°
Center lon. 5°

Odysseus Region

Center lat. 10°
Center lon. 120°

Penelope Region

Center lat. 30°
Center lon. 240°

1 cm = 132 km
1 cm^2 = 17,424 km^2

1 cm^2

660
528
396
264
132
0 km

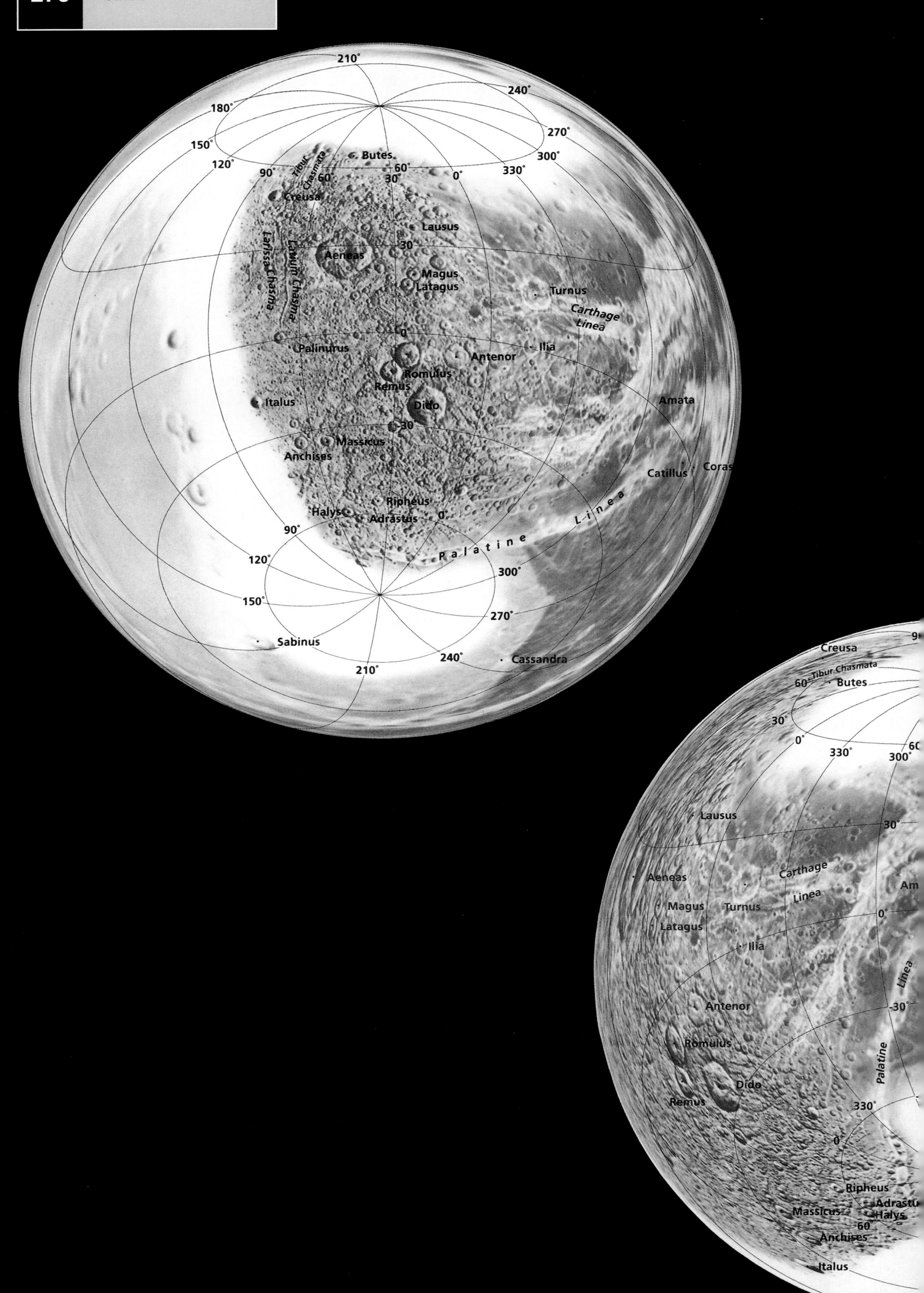
210°
240°
180°
270°
150°
300°
120°
330°
Butes
Tibur Chasmata
Creusa
Lausus
Larissa Chasma
Latium Chasma
Aeneas
Magus
Latagus
Turnus
Carthage Linea
Palinurus
Ilia
Antenor
Romulus
Remus
Italus
Dido
Amata
Massicus
Anchises
Catillus
Coras
Ripheus
Halys
Adrastus
Palatine Linea
Sabinus
Cassandra
Creusa
Tibur Chasmata
Butes
Lausus
Aeneas
Carthage Linea
Magus
Turnus
Latagus
Ilia
Antenor
Romulus
Dido
Remus
Palatine
Ripheus
Massicus
Anchises
Italus

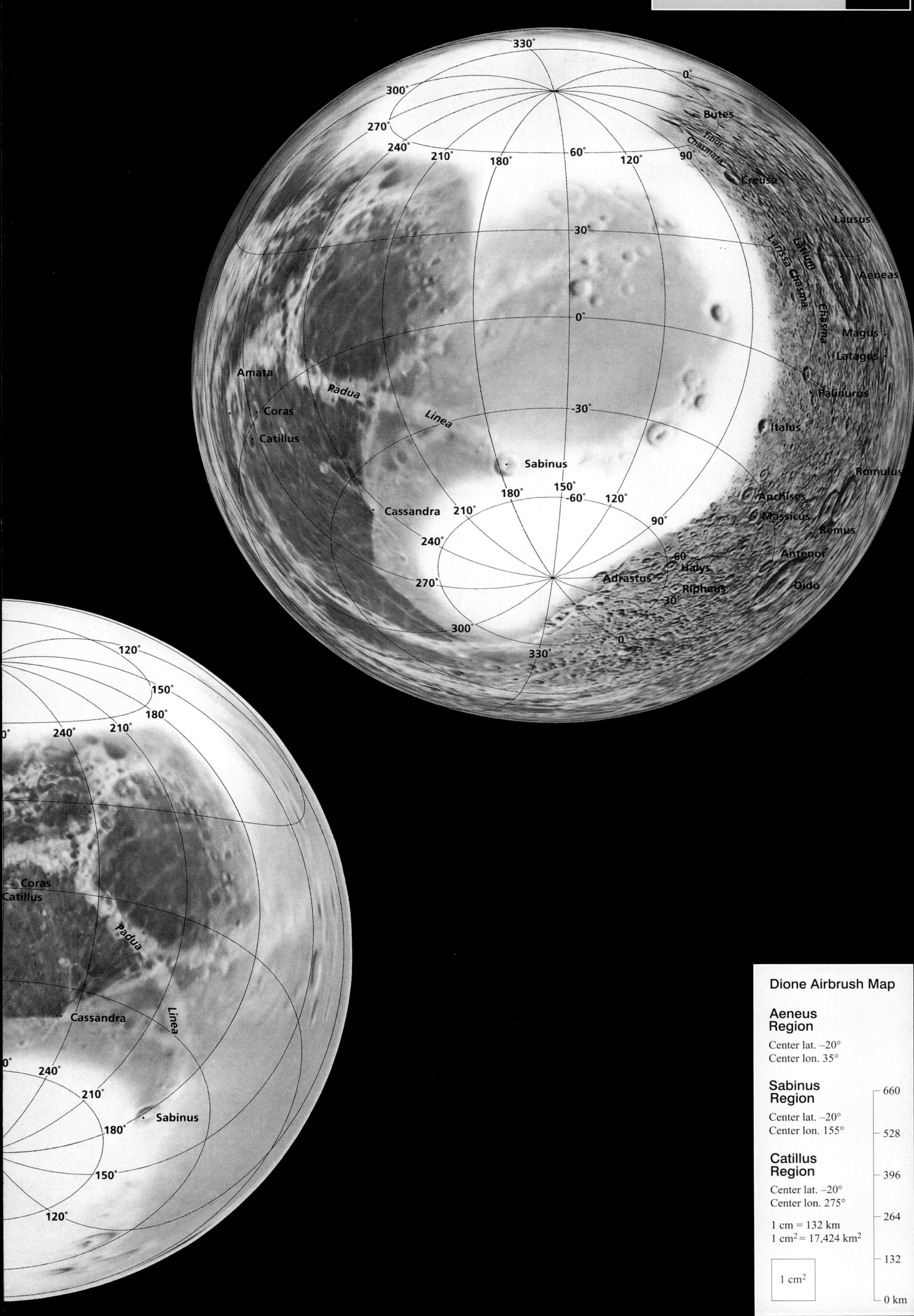
330°
300°
270°
240°
210°
180°
60°
120°
90°
0°
Butes
Tibur Chasmata
Creusa
Lausus
Latium
Larissa Chasma
Chasma
Aeneas
Magus
Latagus
Palinurus
Italus
Romulus
30°
0°
-30°
Amata
Coras
Catillus
Padua
Linea
Sabinus
150°
-60°
Cassandra
Anchises
Massicus
Remus
Antenor
Dido
Halys
Ripheus
Adrastus
Dione Airbrush Map
Aeneus Region
Center lat. –20°
Center lon. 35°
Sabinus Region
Center lat. –20°
Center lon. 155°
Catillus Region
Center lat. –20°
Center lon. 275°
1 cm = 132 km
1 cm² = 17,424 km²
1 cm²
660
528
396
264
132
0 km

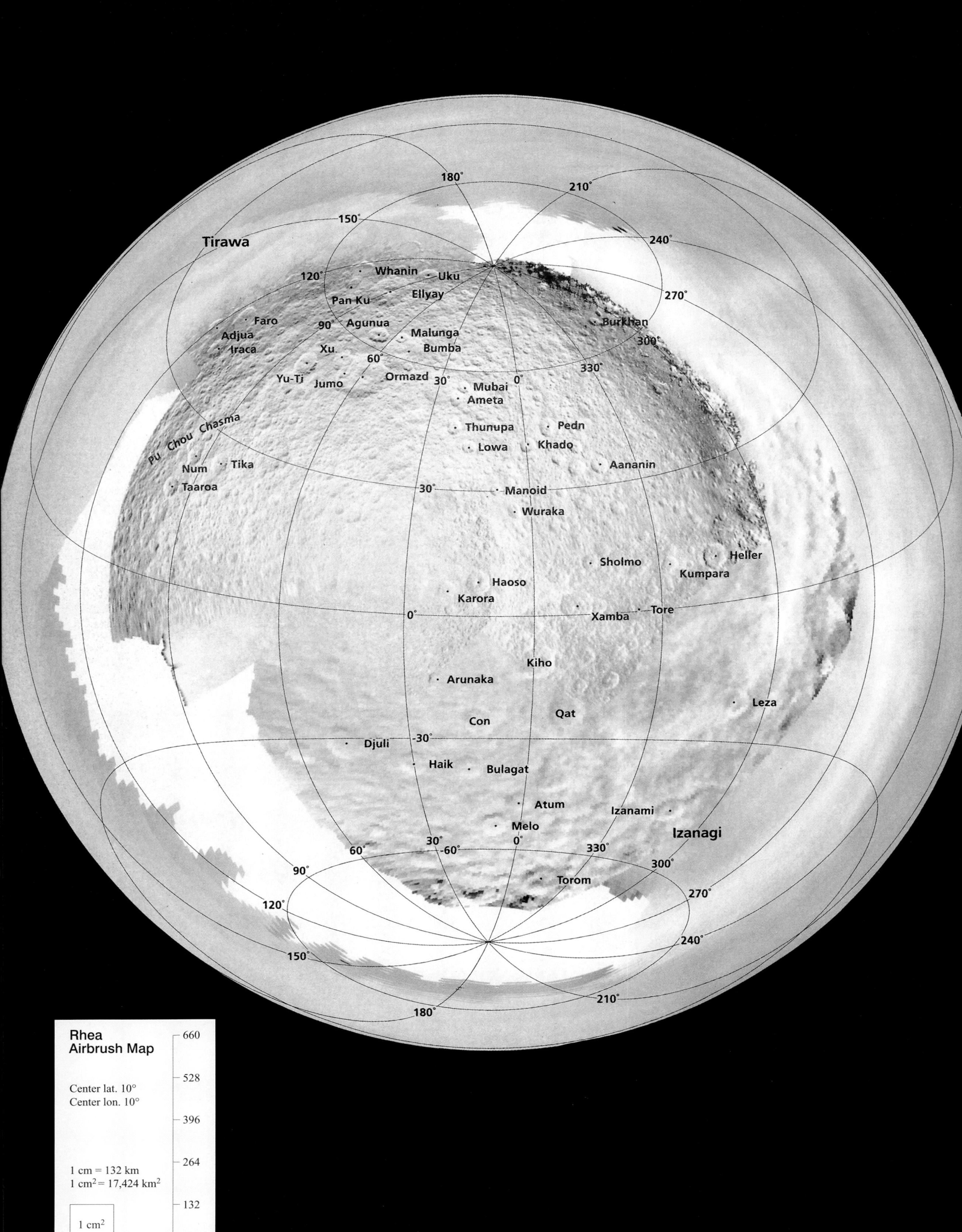

Rhea
Airbrush Map

Center lat. 10°
Center lon. 10°

1 cm = 132 km
1 cm^2 = 17,424 km^2

1 cm^2

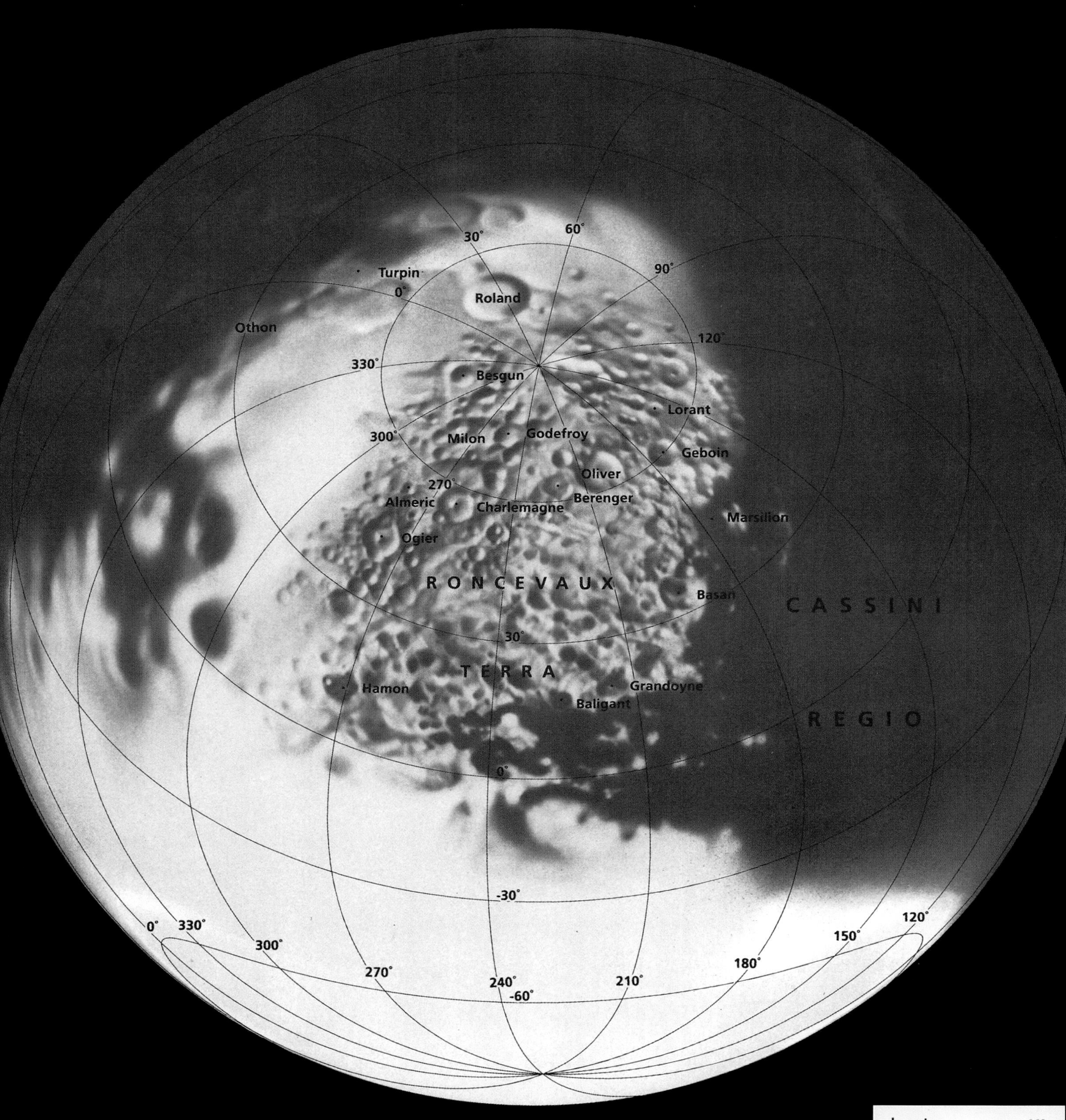
30°
60°
Turpin
90°
0°
Roland
Othon
120°
330°
Besgun
Lorant
300°
Milon
Godefroy
Geboin
Oliver
270°
Almeric
Berenger
Charlemagne
Marsilion
Ogier
RONCEVAUX
Basan
CASSINI
30°
TERRA
Hamon
Grandoyne
Baligant
REGIO
0°
-30°
0°
330°
120°
300°
150°
270°
180°
240°
210°
-60°

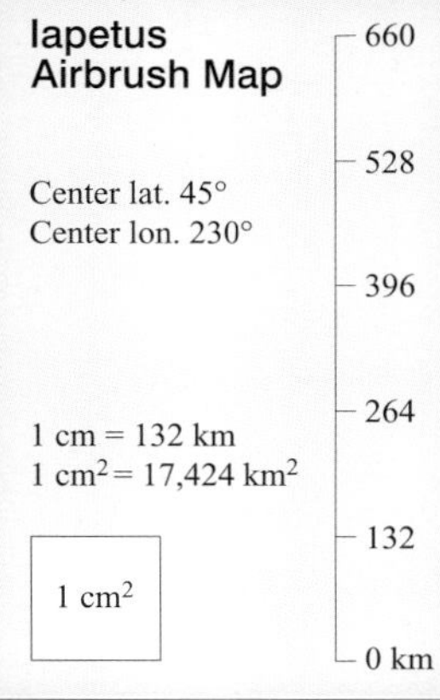
Iapetus
Airbrush Map
Center lat. 45°
Center lon. 230°
1 cm = 132 km
1 cm² = 17,424 km²
1 cm²
660
528
396
264
132
0 km

Uranus System

Traveling more than 20 kilometers per second, in early 1986 the Voyager 2 spacecraft streaked past Uranus on its continuing journey through the outer Solar System. In its brief encounter, *Voyager 2* returned more information on Uranus, its rings, and its satellites than had been accumulated in the 205 years since the discovery of this seventh planet from the Sun. With the return of these data and the discovery of 10 new satellites, the Uranus system was then incorporated into the family of objects available for geologic study.

Uranus was the first planet to be found in historical times. Its discoverer, William Herschel, earned his living as a musician in Bath, England, but was also an amateur astronomer who enjoyed building telescopes for observing stars. In March 1781, while using a small telescope to search for double stars, Herschel found a curious object that did not behave like a normal star. Although first described as some sort of nebular star or a comet, additional observations revealed the object to be a planet. Herschel wanted to name the planet "Georgium Sidus" after the reigning King George III, but instead it was suggested that the newly found object be named Uranus after the Roman god of the heavens.

Herschel continued viewing Uranus, and in 1787 he discovered its two largest satellites, later named Titania and Oberon. Another British amateur astronomer, William Lassell, found two additional satellites, Umbriel and Ariel, in 1851. The fifth large satellite, Miranda, was not detected until nearly 100 years later by the late planetary astronomer, Gerard Kuiper.

Rings around Uranus were found by accident in 1977. A team of astronomers was observing star occultations (blocking of light) by Uranus, using NASA's special airborne observatory. They noted several brief occultations immediately before and after the passage of Uranus in front of the star and, with subsequent observations, confirmed the existence of nine thin, dark rings encircling Uranus. Beginning with the outermost ring, they are designated epsilon, delta, gamma, eta, beta, alpha, 4, 5, and 6 (the innermost ring).

Uranus is the third largest planet in the Solar System, after Jupiter and Saturn. Like those two giants, Uranus has a thick atmosphere of hydro-

The Uranus system compared to the Earth/Moon system.

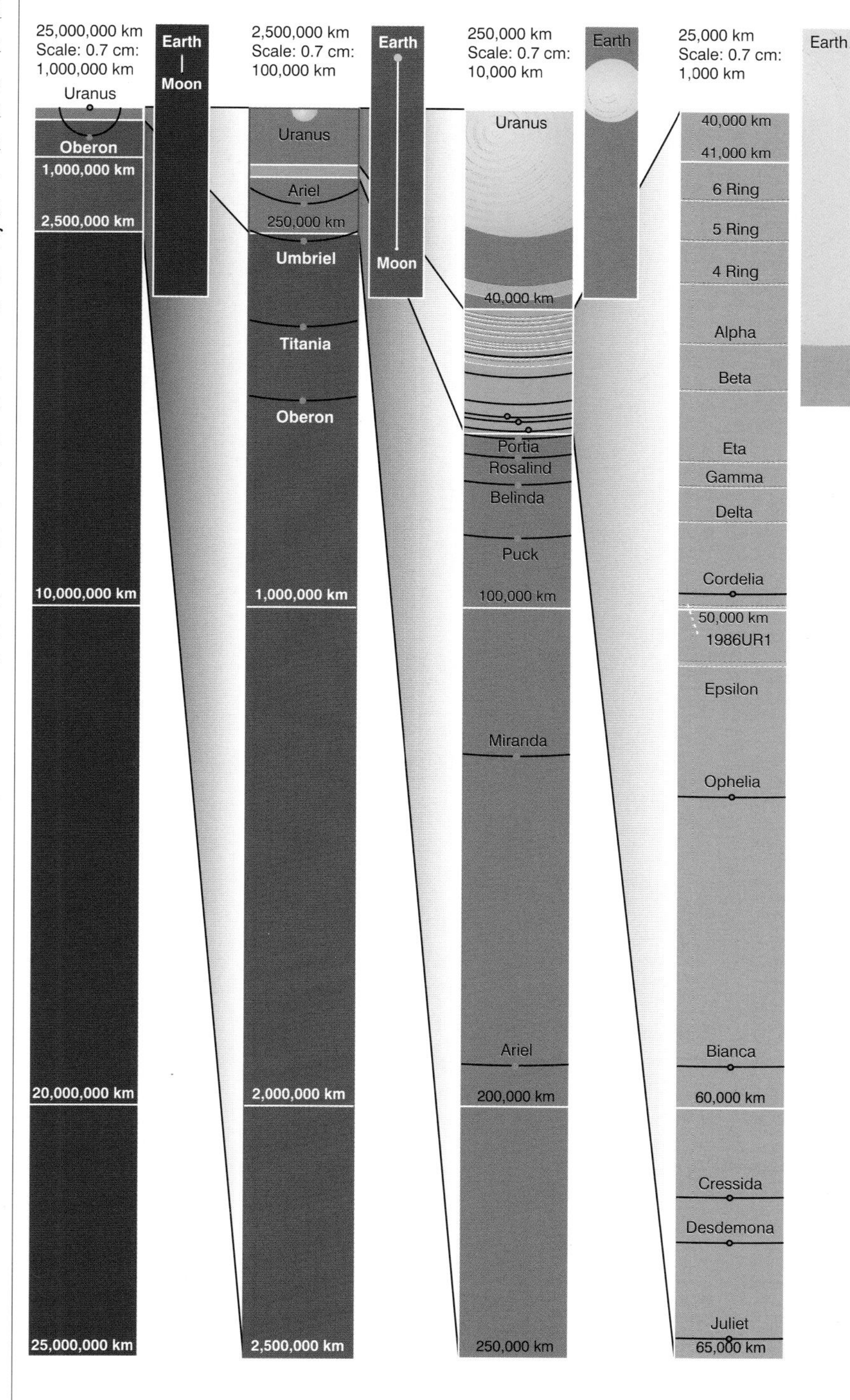

Facing page: *Voyager 2* false-color image of Uranus taken by the wide-angle camera after the spacecraft had made its closest approach to the planet (JPL photograph P-29539C).

Two views of the southern hemisphere of Uranus taken by the *Voyager 2* spacecraft. The image on the left shows Uranus as it appears to the human eye. The image on the right is a false-color view to enhance detail and to show the dark polar area, indicated by the dark orange colors. The donut-shaped rings are artifacts of the camera system. The pink band in the lower right is a processing artifact (JPL photograph P-29478C).

Diagram of the magnetic field associated with Uranus and its offset from the axis of rotation of the planet.

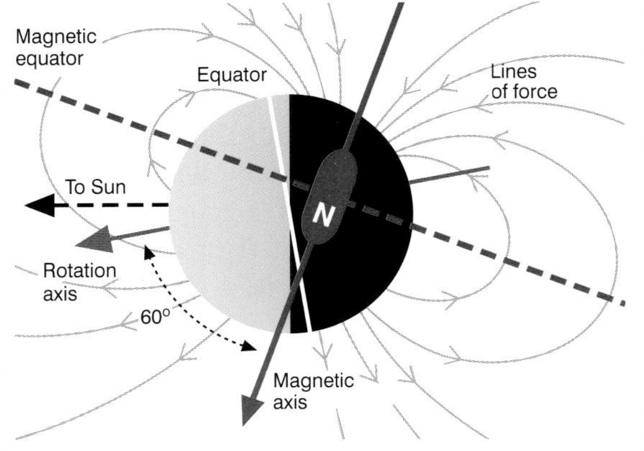

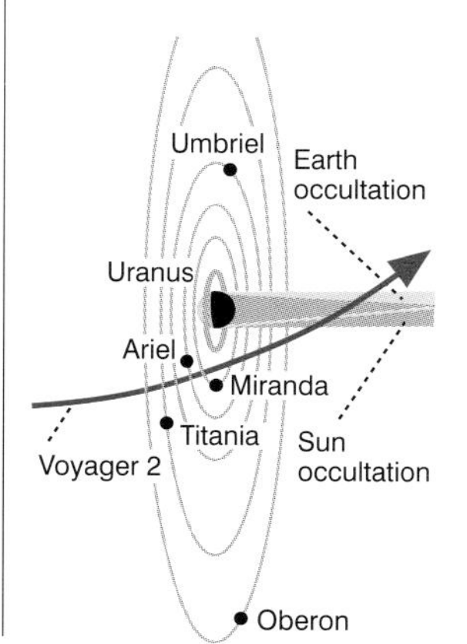

Uranus system showing the "bull's eye" pattern formed by the rings and satellites, with respect to the rest of the Solar System and the *Voyager 2* trajectory.

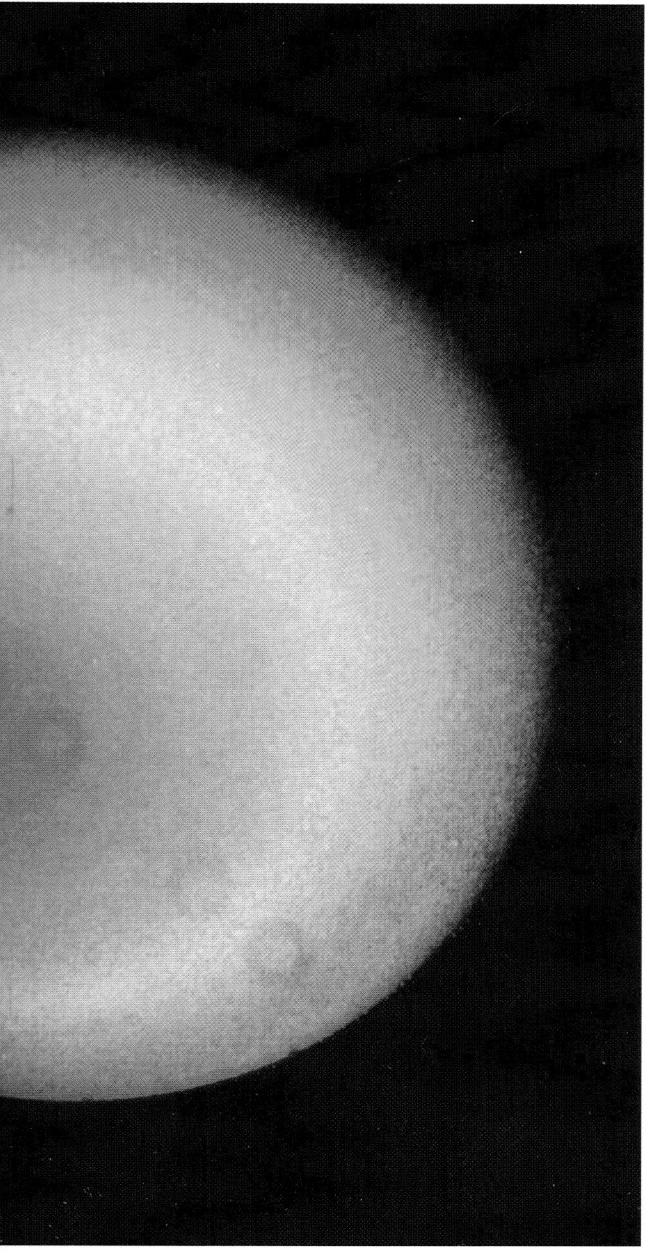

26,000 km
Liquid molecular hydrogen
18,000 km
Ice
8,000 km
Rock

The interior of Uranus. Although shown with a rocky interior core, alternative models suggest a liquid, low-density core for the gas-rich planet.

gen and helium. However, its density indicates that most of Uranus's mass consists of water, ammonia, and methane. A small, silicate-rich core probably exists at Uranus's center. The atmosphere of Uranus has a distinctive blue-green appearance because of the presence of methane, a gas which absorbs red light.

Uranus

Unlike any other planet except Pluto, the axis of rotation for Uranus is tipped 98° from the vertical, so that it lies nearly horizontal on the ecliptic plane. Although there is no wholly satisfactory explanation for this orientation, it is plausible that Uranus was struck early in its history by an Earth-size object that knocked it askew. At present, the south pole points toward the Sun. Thus, as the rings and satellites rotate about Uranus's equator, their paths inscribe patterns that appear similar to a bull's eye.

Even before the *Voyager 2* encounter, models of Uranus divided the planet into three regions – a rocky core, an "ocean," and a thick atmospheric envelope. Spacecraft data allowed further refinement and development of these models and revealed some surprises. Prior to the *Voyager* flyby, it was predicted that Uranus would have a magnetic field. The magnetometer on board the spacecraft verified the presence of a strong field – comparable to that of Saturn and Earth – but also showed, unexpectedly, that the field is not only offset from the center of Uranus but is tilted from the axis of rotation by some 60°. For comparison, the magnetic field of Earth is centered on the planet and tilted by only 11°. Perhaps, like Earth, Uranus experiences periodic reversals of the north and south magnetic poles and such a reversal may be occurring now. Alternatively, if Uranus were struck early in its history and its spin axis reoriented, the magnetic field orientation might be a remnant from that time.

In any event, the magnetic field may indicate some type of "dynamo," perhaps related to a uranian hydrogen ocean. The ocean is thought to be thousands of kilometers thick, and at depth the high pressures would raise the temperature in excess of 1,000°C. At these

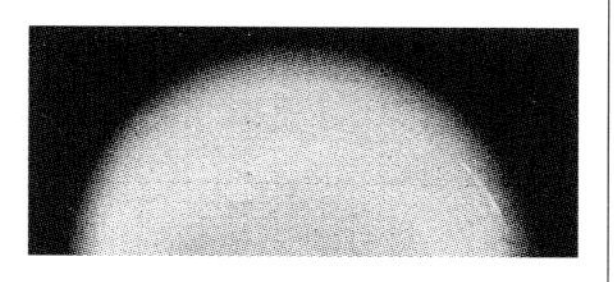
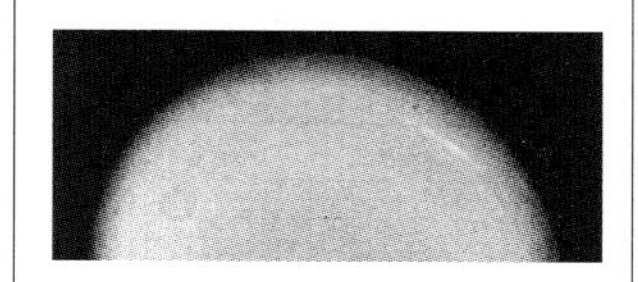
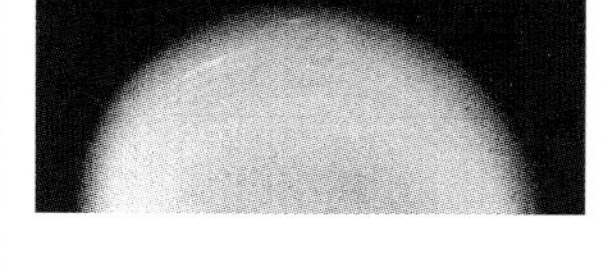
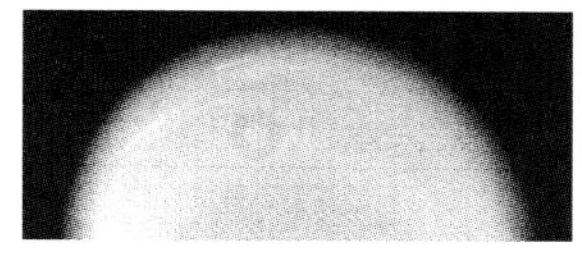

Time-lapse sequence of *Voyager 2* images showing movement of two bright, streaky clouds on Uranus. Tracking the clouds shows that they have periods of rotation of 16.2 and 16.9 hours. The dark circles visible in some images are camera artifacts (JPL photograph P-29467).

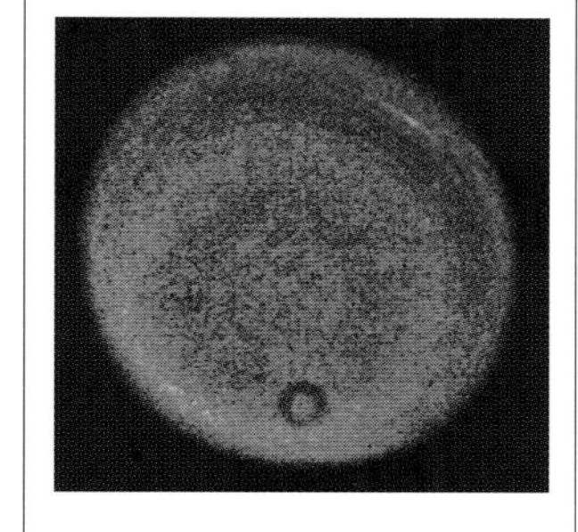

Composite *Voyager 2* images that have been computer enhanced to show clouds (white streaks in the blue zone) in the upper atmosphere of Uranus. Black rings are artifacts (JPL photograph P-29468).

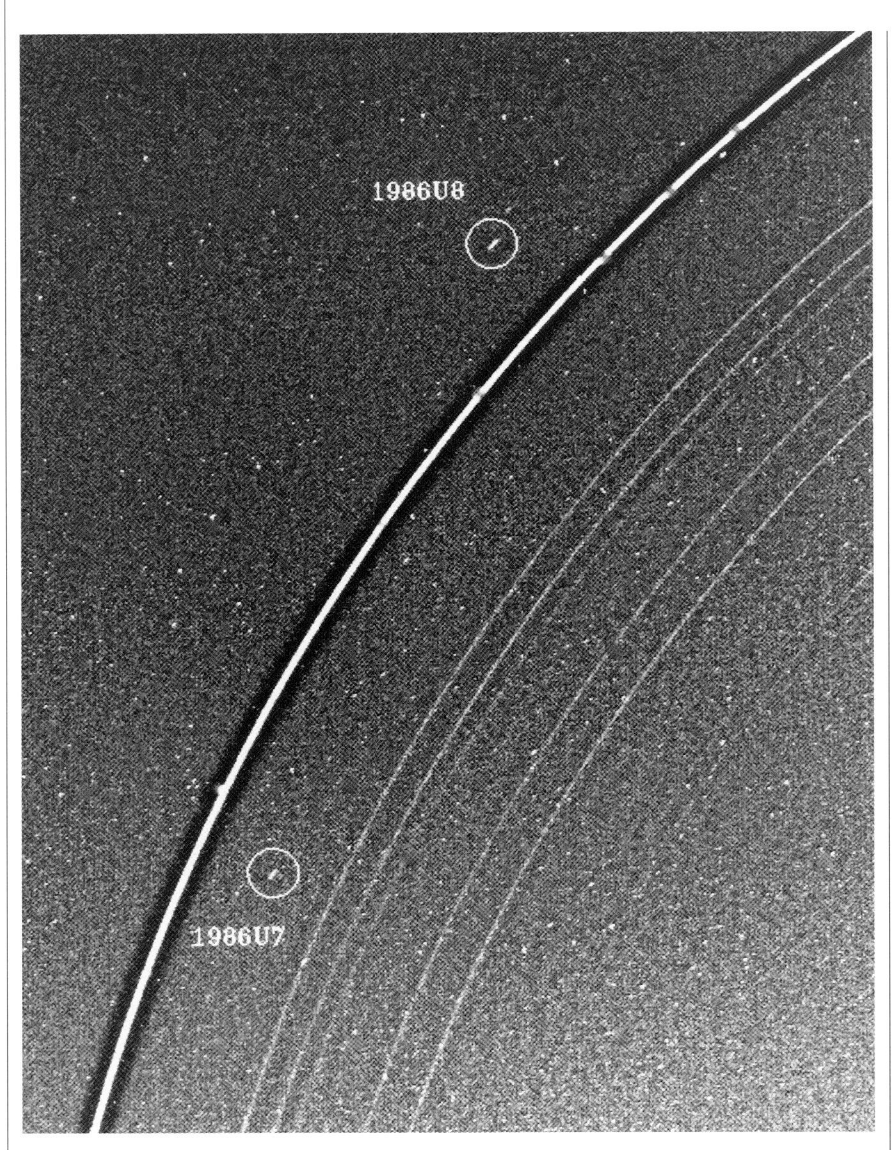

Voyager 2 image showing two "shepherd" satellites, Ophelia (1986U8) and Cordelia (1986U7), for the epsilon ring and the other rings of the Uranus system (JPL photograph P-29466).

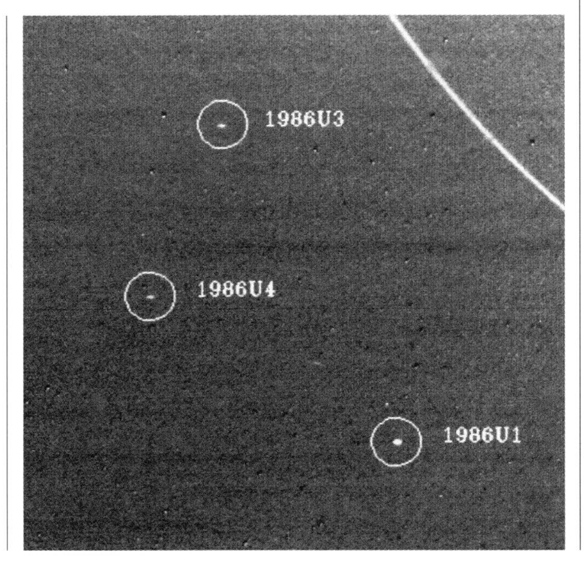

Voyager 2 image showing three satellites (1986U1 = Portia, 1986U3 = Cressida, and 1986U4 = Rosalinda) discovered during the flyby in 1986. Also visible is the outermost, or epsilon, ring (JPL photograph P-29465).

temperatures, hydrogen would break down into ions that would conduct electricity, thus permitting the generation of a magnetic field as the planet rotates.

The atmosphere of Uranus has long been of interest to planetary scientists. It is composed mostly of hydrogen. A key question prior to the *Voyager* flyby was the amount of helium present in the atmosphere. Although predictions of helium ran as high as 40 percent, *Voyager* showed that only about 12 percent is present. In addition to hydrogen and helium, small amounts of methane, ammonia, and other gases were found.

Computer processing of the pictures taken of Uranus reveals a brownish haze in the south polar region (the north pole, pointing away from the Sun, has not been seen). The haze could be some complex organic compound, perhaps produced by the alteration of methane and other materials in the atmosphere by the Sun.

Voyager pictures show a few white, cloud-like features in the upper atmosphere and faint but distinctive bands suggesting wind patterns. These features were recorded during several days of observations, which enabled detection of high-speed winds in the upper atmosphere. The winds move in the same direction as the rotation of the planet, the opposite of predictions made prior to the *Voyager* flyby. Thus, despite its bland appearance, the atmosphere of Uranus has posed a number of intriguing questions not yet answered.

Rings

Several Voyager observations were designed to gather information on the rings of Uranus. Scientific objectives for the *Voyager* flyby included characterizing the known rings, searching for new rings, and looking for "shepherd" satellites like those known to accompany some of the rings of Saturn. As was known from Earth-based observations, Voyager showed the uranian rings to be very dark – similar to soot. This is in contrast to the lighter and redder rings of Jupiter and the very bright rings of Saturn. It has been suggested that the uranian rings may be composed of carbonaceous materials. In addition, the amount of micron-size dust in the uranian main rings is much less than in the rings of Saturn and Jupiter, suggesting different processes of ring-particle interactions.

Uranus has at least 11 rings or partial rings. The largest, the epsilon ring, ranges from about 22 to 93 kilometers in width. Composed of two "strands" of particles separated by a dark region, it shows complex patterns. Although an intense search using *Voyager* data was conducted throughout the ring system, only the epsilon ring appears to have shepherding satellites – the outside satellite named Ophelia and the inside satellite Cordelia. *Voyager* also discovered a broad band of dust inside rings 4, 5, and 6. The band is very faint and extends to within 11,000 kilometers of Uranus. This dust band, the distribution of particle sizes in the main rings, and the presence of possible incomplete rings all suggest that the rings are evolving. Perhaps repeated impacts cause fragmentation of small moons, with the fragments being swept up later or incorporated into newly formed and "incomplete" rings.

Satellites

The five largest moons, Miranda, Ariel, Umbriel, Titania, and Oberon are named after characters in Shakespeare's *A Midsummer Night's Dream* and *The Tempest* and *Pope's The Rape of the Lock*. All the satellites are in synchronous rotation, always showing the same hemisphere toward Uranus. In addition, they are dark and have densities of about 1.5 to 1.7 grams per cubic centimeter. These densities are less than the density of the Galilean satellite, Ganymede, but are slightly higher than the saturnian satellites, suggestive of a higher proportion of rock to ice.

Voyager pictures show that all five satellites have experienced impact cratering, tectonic deformation, and resurfacing, although the intensity of these processes varies among the satellites. The outermost moons appear to have been modified after the period of heavy impact bombardment, whereas the innermost satellites – Miranda and Ariel – were extensively modified by internal processes.

Sixteen new moons were discovered through the analysis of *Voyager* and other data. All are

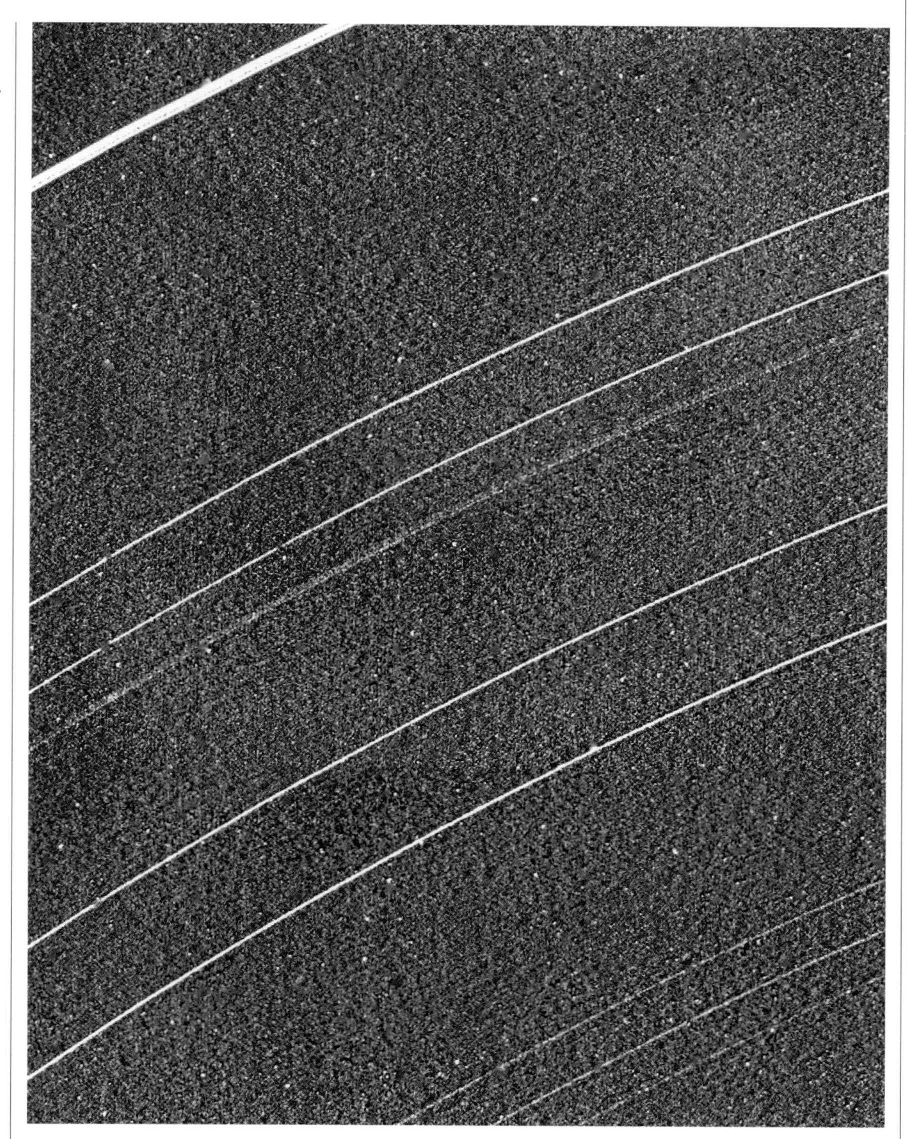

Voyager 2 mosaic showing rings of Uranus, including one discovered during the spacecraft flyby; the newly discovered ring is about midway between the bright, outermost epsilon ring and the next ring, or delta ring (JPL photograph P-29507).

Voyager 2 photograph taken while the spacecraft was in the shadow of Uranus; the combination of viewing geometry and long camera exposure enabled the continuous distribution of small particles to be detected throughout the ring system. Short bright streaks are stars "smeared" by the long exposure (JPL photograph P-29525).

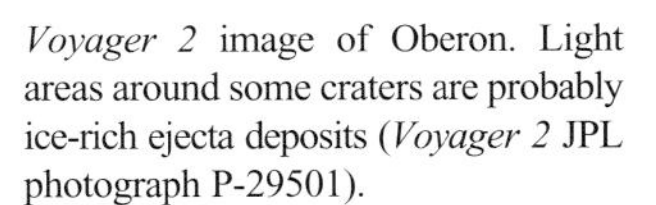
Voyager 2 image of Oberon. Light areas around some craters are probably ice-rich ejecta deposits (*Voyager 2* JPL photograph P-29501).

Mosaic of *Voyager* images showing complex terrains on Ariel. The largest fault-valleys, or grabens, have been partly filled with ice-rich materials erupted from the interior (*Voyager 2*, JPL photograph P-29520).

Umbriel, darkest satellite of Uranus, is seen in this *Voyager 2* image. The prominent crater on the terminator (named Vuver, upper right) is about 110 kilometers in diameter and has a large central peak. Also visible is a bright ring (named Wunda) near the equator (top of image) that may be a frost deposit (JPL photograph P-29521).

small (30 to 150 kilometers in diameter) and somewhat dark. The masses of the satellites are difficult to determine, and the data from the *Voyager* mission are inadequate to allow estimates of their densities to be obtained. It is presumed, however, that the densities of the newly discovered satellites are similar to those of their larger neighbors.

Remote sensing observations suggest that water ice is present on all the large satellite surfaces, but, like the rings, the moons are rather dark and gray in color. This implies either that they are composed of dark material or that they are mantled by dark material that has been swept up from orbit. Alternatively, a model derived from laboratory experiments has been proposed to explain the dark surfaces seen on many of the outer planet satellites, including those of Uranus. These experiments show that when photons (as would be generated from the Sun) impact methane ice for a long period of time, they drive out hydrogen molecules and leave a high proportion of carbon at the surface causing a general darkening. Although it has not yet been shown that methane ice is present in the ring or satellite materials or that such irradiation would occur in the uranian system, it is a reasonable explanation of the dark surfaces.

Oberon

Oberon, the outermost satellite, is the second largest moon in the uranian system. Its surface is dominated by large (50 to 100 kilometers) impact craters, some of which exhibit radial patterns of bright ejecta. Some linear and curved scarps appear to be faults. Several very dark patches are found on Oberon, particularly on the floors of large craters. These areas may represent the extrusion of slushy material onto the surface from the interior following the period of heavy impact cratering, perhaps similar to the dark terrain on Saturn's moon, Iapetus.

Titania

With a diameter of 1,610 kilometers, Titania is the largest satellite of Uranus. Like Oberon, much of its surface is heavily cratered and

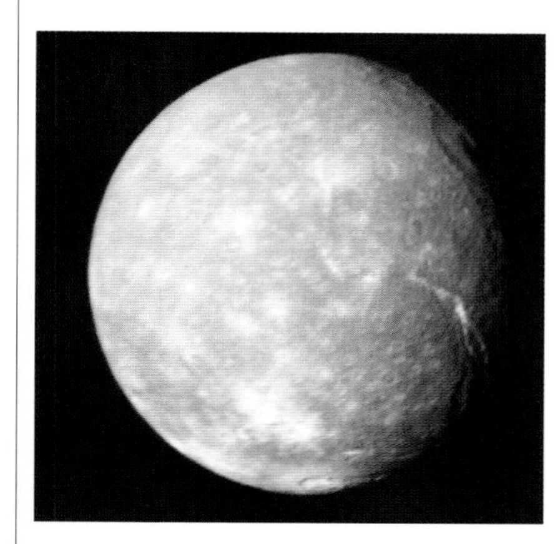

Titania, largest of the uranian satellites, seen in this *Voyager 2* image, has numerous impact craters and grabenlike troughs (*Voyager 2* JPL photograph P-29509).

includes multiring impact basins. Titania has more faults than Oberon, including features 20 to 50 kilometers wide that appear to be grabens. High-albedo materials – possibly ice or frost deposits – are exposed on some of the cliff faces. The faults may represent global extension of Titania's crust, perhaps in response to expansion that occurred as ice froze in the interior of the satellite. Smooth, relatively uncratered zones look like areas that have been resurfaced by the extrusion of fluid-like materials, such as ice slush.

Umbriel

Umbriel is the darkest of the large satellites and appears uniform in color, albedo, and general surface features. Although *Voyager* was unable to take photographs of the entire surface of Umbriel, the parts that were seen are uniformly cratered with a high proportion of large impact scars, indicative of an ancient surface. Although it is heavily cratered, few ejecta rays are visible and small-scale surface features are not apparent, suggesting that Umbriel has been blanketed with dark material from external sources. Alternatively, the material excavated during impact cratering may be of a composition whose optical properties are little affected by impact, inhibiting the formation of bright ejecta rays and deposits.

Ariel

The surface features revealed by *Voyager* for Ariel are much more complex than those seen on the outer, larger satellites. The older, cratered terrain on Ariel has been extensively fractured by faults and grabens. However, very old, large craters appear to be missing, perhaps having been obliterated by an early period of resurfacing. Smaller (less than 60 kilometers in diameter) craters are found in some areas; other areas consist of smooth plains, some of which may have been emplaced as flows of ice or water which buried the older terrain. Some of the grabens on Ariel show sinuous valleys on their floors that may represent fluid, or glacier-like flows. Alternatively, these features could be sinuous faults.

Miranda

Although Miranda is the smallest of the major uranian satellites, it displays a complex geologic history. A wide variety of terrains can be identified on Miranda, including old, heavily cratered surfaces and geologically younger, complex terrains. The younger terrains are found in at least three areas, each of which has been named: Inverness Corona, Arden Corona, and Elsinore Corona.

These younger units show complex patterns of grooves and ridges – similar in many respects to the tectonically deformed bright terrain on the Galilean satellite, Ganymede, and the saturnian satellite, Enceladus. Patches of smooth, uniform-albedo material within the complex terrain may be areas that have been resurfaced by eruptions of icy slush or glacial-like flows.

Miranda also displays enormous fault scarps, some of which cut across the entire globe. The largest of these include grabens that are 10 to 15 kilometers deep and hundreds of kilometers long. Bright material is exposed in the walls of some fault scarps and craters. This bright material is apparently part of a layer perhaps 1 kilometer thick.

One of the 10 moons of Uranus discovered during the flyby of *Voyager 2*. This 150-kilometer-diameter satellite, named Puck, shows a heavily cratered surface (*Voyager 2* JPL photograph P-29519).

Geologic Processes

The record of impact cratering that is exposed on planetary surfaces is a fundamental key to deriving geologic histories. It has been suggested that different "populations" of craters can be recognized on the outer planet satellites, and this idea was applied to the satellites of Uranus. The concept is based on crater statistics in which the size-frequency distribution per unit area for one part of an object is compared with statistics for other parts of the object and with values derived for other planets and satellites. From these data, sets of craters, or "populations," appear to be recognizable.

Population I craters may represent a period of heavy impact cratering following the time of planetary formation. This population of craters dominates the surfaces of Oberon and Umbriel. Population II craters have proportionately fewer large craters and are prevalent on Titania and

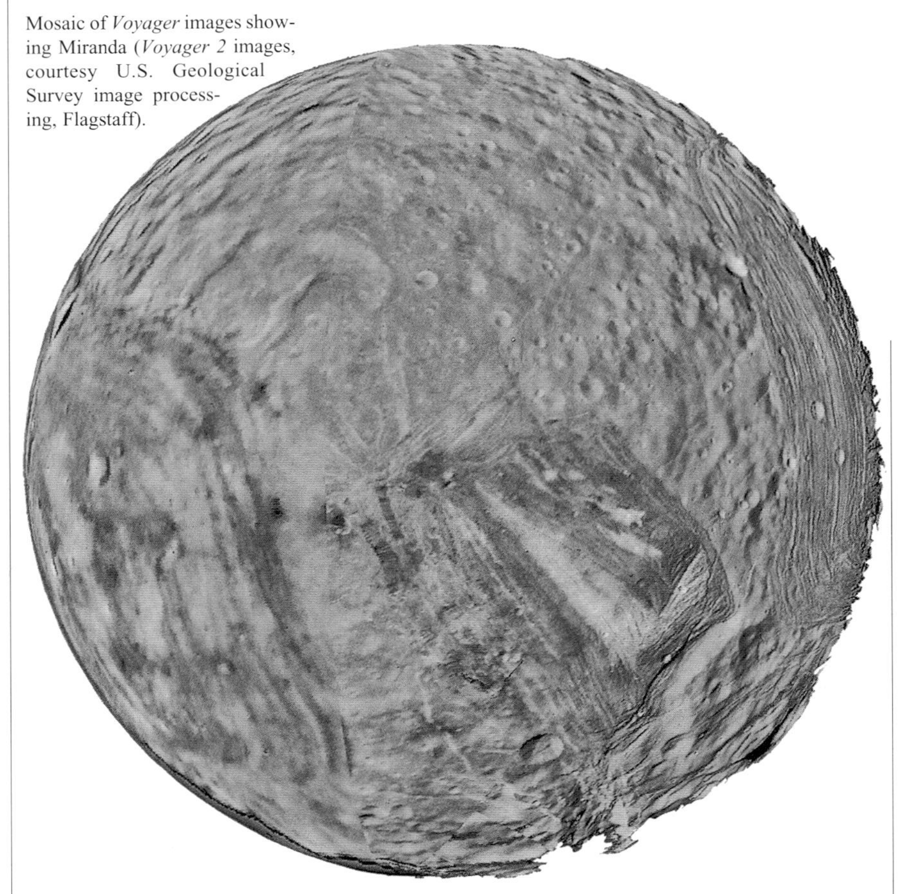

Mosaic of *Voyager* images showing Miranda (*Voyager 2* images, courtesy U.S. Geological Survey image processing, Flagstaff).

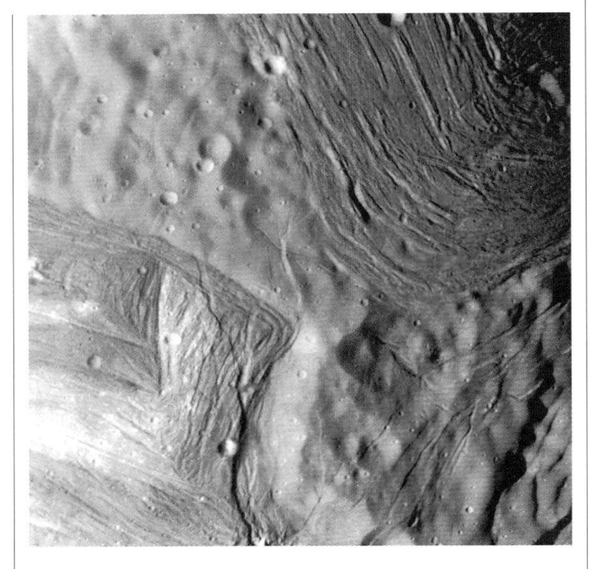

View of Miranda, showing Inverness Corona (the so-called "chevron" on left side) and part of Elsinore Corona; these terrains are considered to be parts of the crust disrupted by tectonic processes and possibly modified by the extrusion of slushy ice from the interior. Area shown is about 200 kilometers across (JPL photograph P-29515).

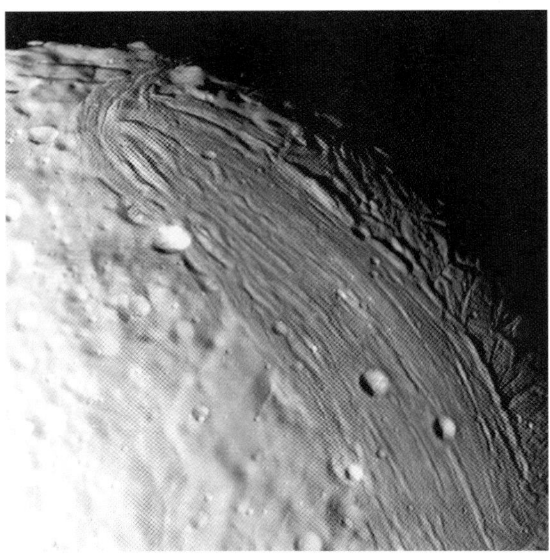

High-resolution image of Miranda showing ridge and grooved terrain of the Elsinore Corona region (right side) and older, more heavily cratered terrain. Area shown is about 150 kilometers across (*Voyager 2* photograph 26846.20).

Ariel; although some rare large craters are also present. This population of craters represents a later stage of impact events. Miranda appears to have both populations, with its old, heavily cratered terrain showing more craters that are 50 kilometers across than any surface on the other four satellites. All of the satellites, however, must have been exposed to essentially the same impacting objects. Consequently, the differences in the cratering record among the satellites must be related to differences in tectonic deformation and episodes of resurfacing. Both Oberon and Umbriel appear to have been modified only by limited faulting following the time of postaccretional heavy bombardment (Population I craters), whereas the crusts of Titania and Ariel have been extensively deformed and resurfaced by extrusive materials.

Despite their small sizes, Miranda and Ariel clearly have been subjected to substantial modification by internally generated processes. Extensive tectonism and volcanism suggest internal heat sources that have caused a reassessment of the mechanisms responsible for heat generation. As noted in the chapter on the Saturn system, prior to the *Voyager* discoveries, existing models suggested that because small objects would have a low amount of radioactive elements in the interior, and hence a low heat supply, they would exhibit few surface features indicative of volcanism and tectonism. However, the results from the *Voyager* mission have changed this perspective and shown that factors such as tidal stressing – and perhaps other yet unknown mechanics – can lead to internal activity.

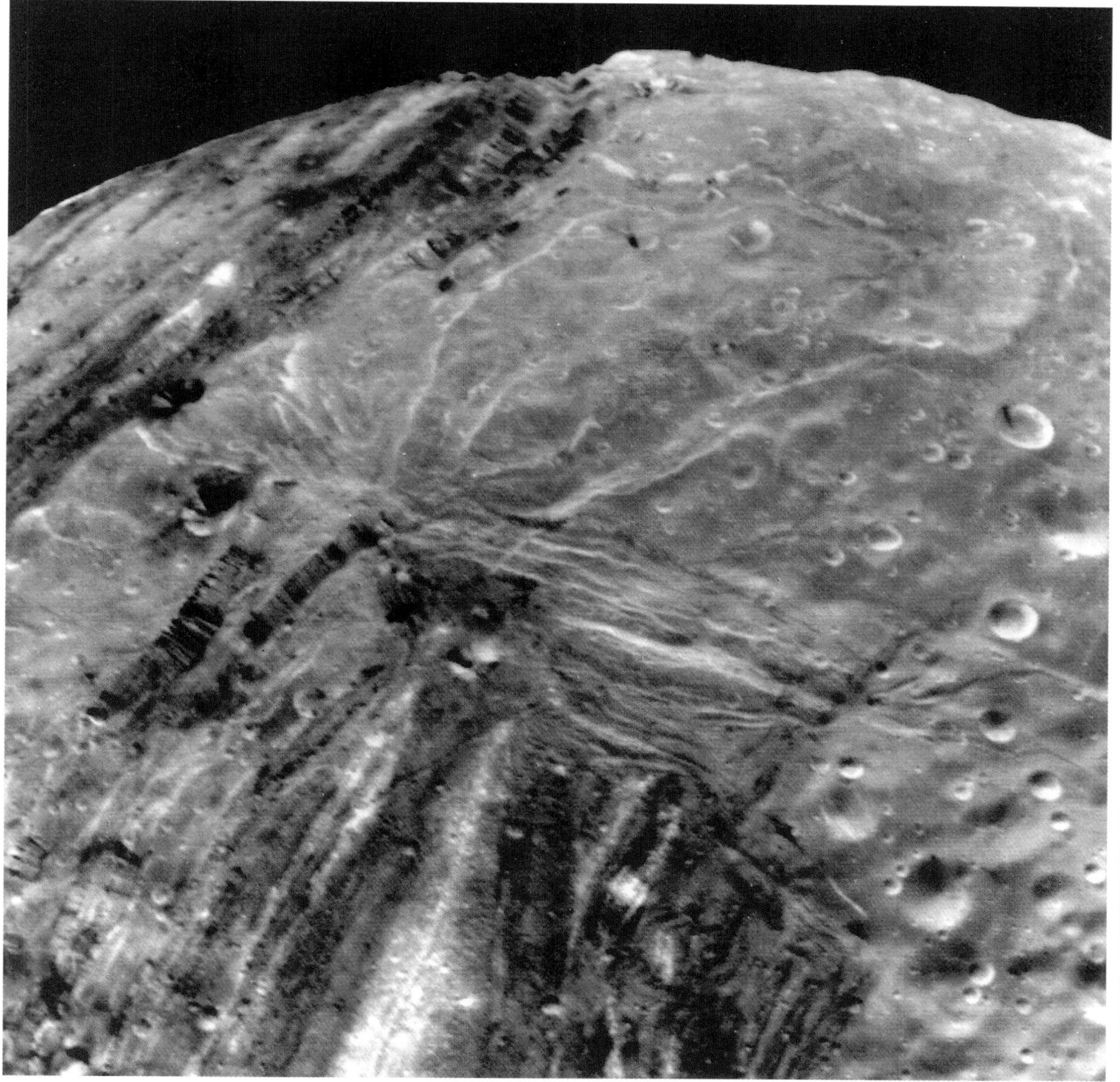

Computer-generated mosaic showing oblique views of heavily faulted terrain on Miranda. Fault scarps several kilometers high are visible on the horizon (*Voyager 2* JPL photographP-29513).

Montage of *Voyager 2* images showing Uranus, its rings, Miranda, and part of Miranda's canyon system (JPL photograph P-29549).

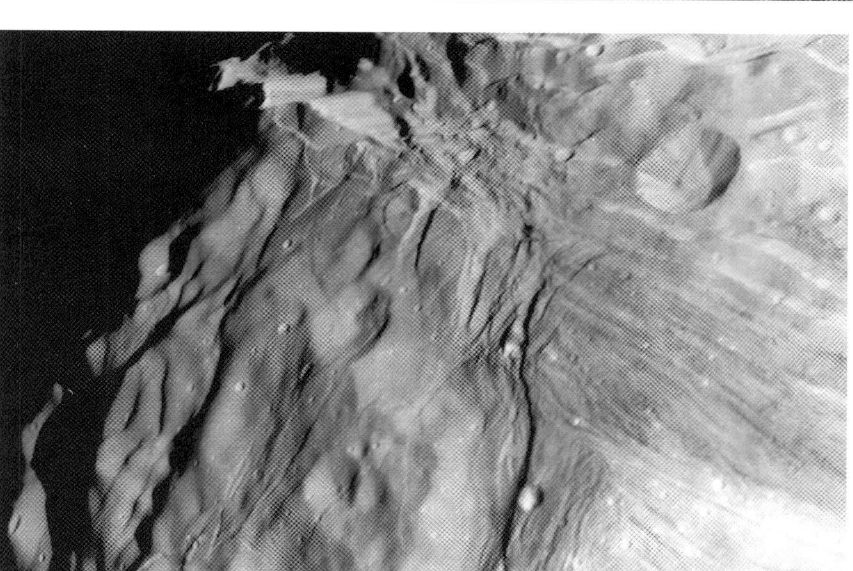

Part of Miranda, innermost of Uranus's large satellites, photographed at high resolution by *Voyager 2*, showing fault-dominated ridge and trough terrain. Prominant scarp in upper left is named Verona Rupes. The impact crater on the right is about 25 kilometers across (JPL photograph P-29512).

Miranda
Southern Hemisphere*
Geologic Map

Center lat. –90°
Center lon. 0°

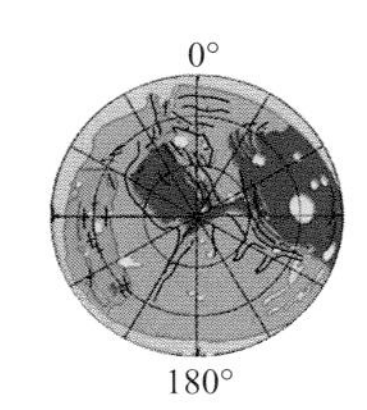

Miranda
Southern Hemisphere*
Reference Map

Center lat. –90°
Center lon. 0°

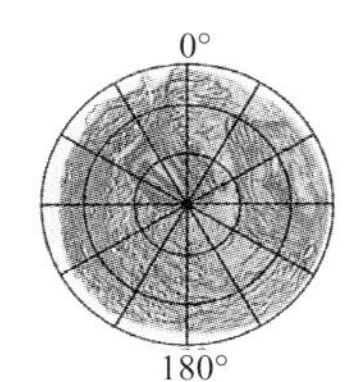

Miranda
Correlation of Map Units

Youngest at top

Impact materials | Cratered terrrain | Corona materials | Canyon materials

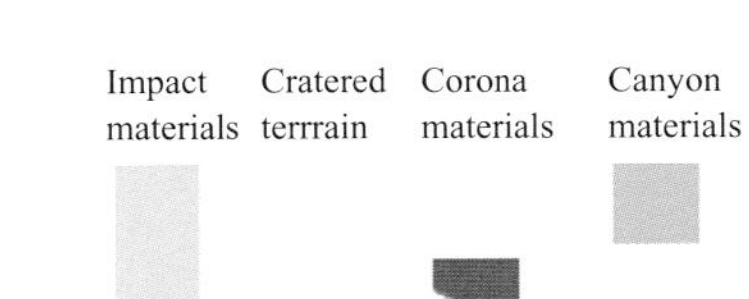

Miranda
Description of Map Units

Impact Materials

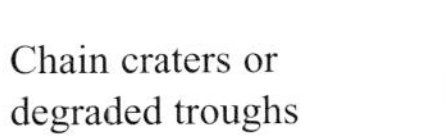

Fresh

Chain craters or degraded troughs

Cratered Terrain

Mantled

Corona Materials

Flow material

Ridged plains in Elsinore Corona

Arden and Inverness Coronae materials

Canyon Materials

Slope materials

Rough floor

Unmapped

Data resolution insufficient

Structure Symbols

Fault scarp

Trough

Ridge

Ariel
Southern Hemisphere*
Geologic Map

Center lat. –90°
Center lon. 0°

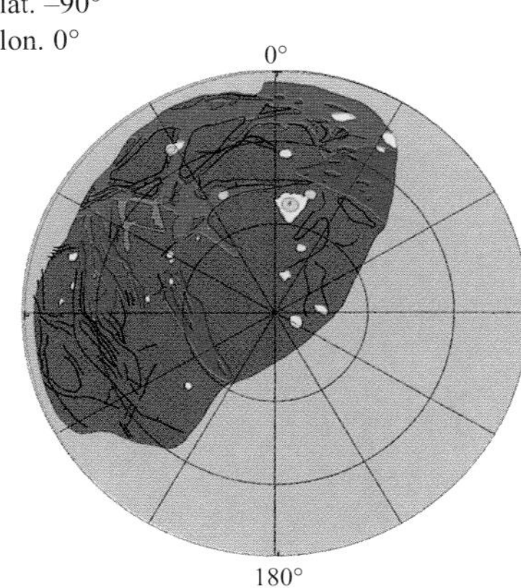

Ariel
Southern Hemisphere*
Reference Map

Center lat. –90°
Center lon. 0°

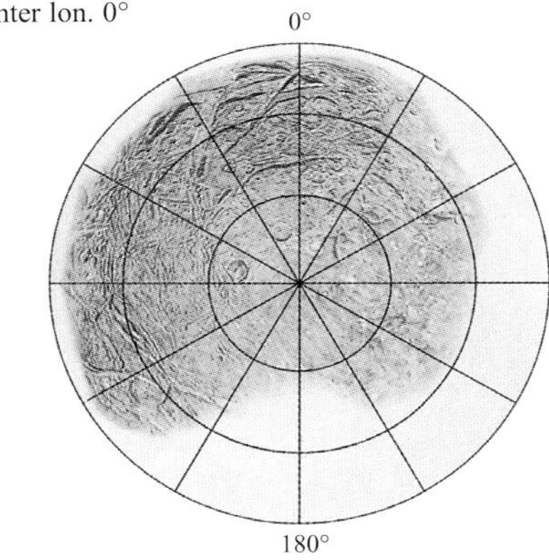

Ariel
Description of Map Units

Volcanic Materials

Younger age

Intermediate age

Older age

Sedimentary Materials

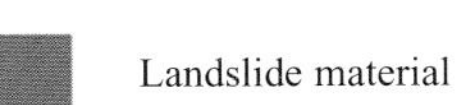

Landslide material

Impact Materials

Bright crater ejecta

Deep fresh craters

Plains Materials

Cratered plains – Pre-volcanic flow materials

Unmapped

Data resolution insufficient

Structure Symbols

Fault

Groove/trough

Ariel
Correlation of Map Units

Youngest at top

Volcanic materials | Sedimentary materials | Impact materials | Plains materials

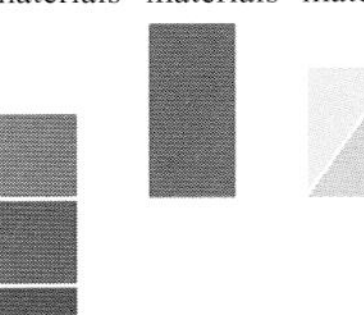

Umbriel
Correlation of Map Units

Youngest at top

Impact materials | Plains materials | Cratered terrain materials | Canyon materials

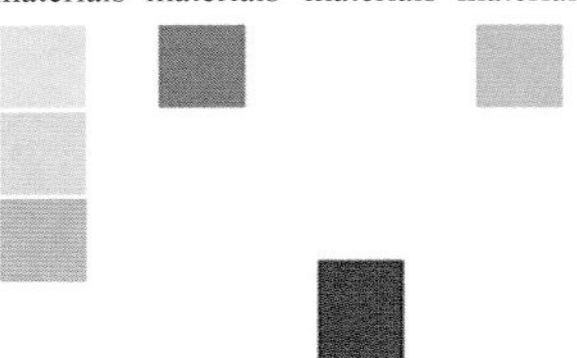

Umbriel
Southern Hemisphere*
Geologic Map

Center lat. –90°
Center lon. 0°

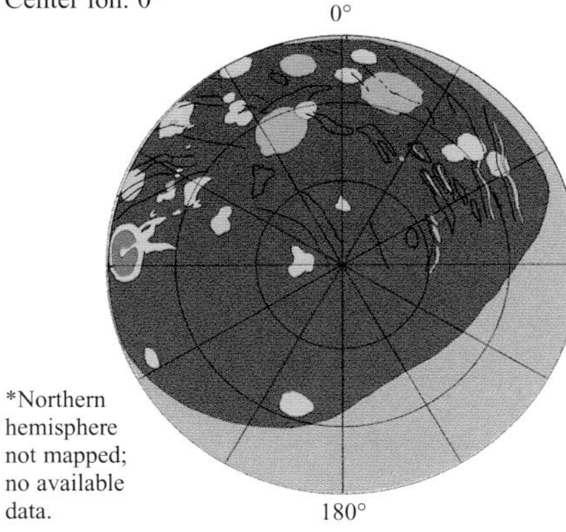

*Northern hemisphere not mapped; no available data.

Umbriel
Southern Hemisphere*
Reference Map

Center lat. –90°
Center lon. 0°

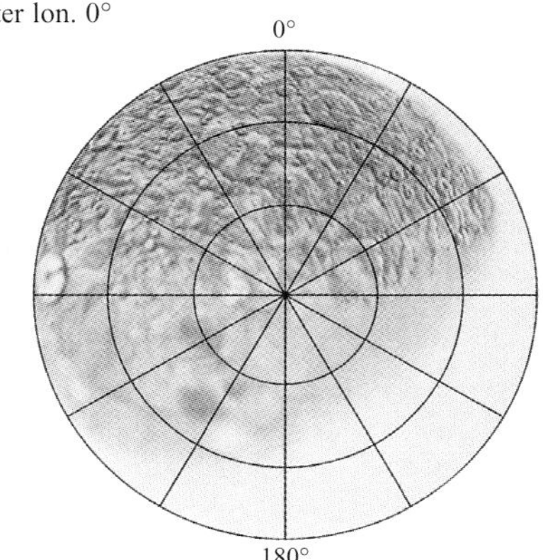

Umbriel
Description of Map Units

Impact Materials

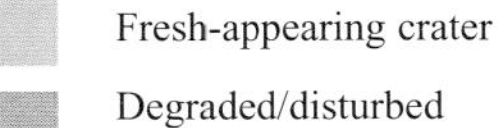

Rayed or bright deposit crater

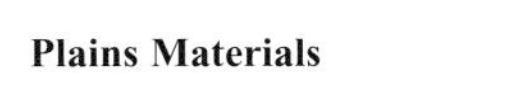

Fresh-appearing crater

Degraded/disturbed crater

Plains Materials

Bright smooth deposit

Cratered Terrain Materials

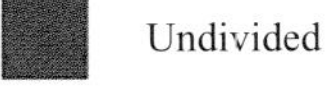

Undivided

Canyon Materials

Bright slope material

Unmapped

Data resolution insufficient

Structure Symbols

Fault

Graben

Ridge

Titania Southern Hemisphere* Geologic Map

Center lat. –90°
Center lon. 0°

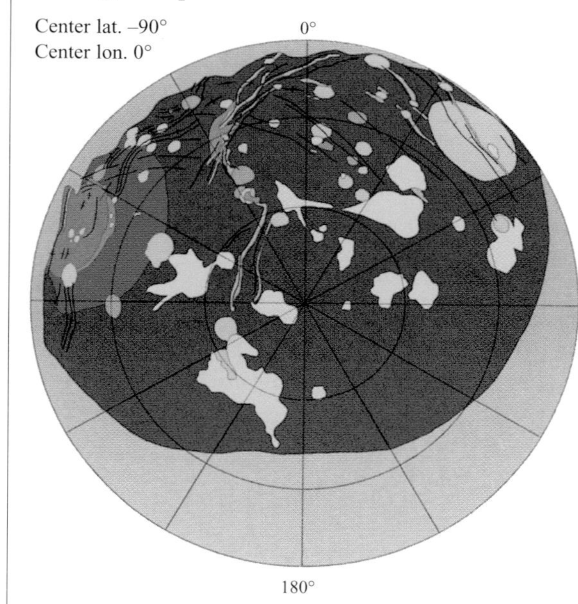

Titania Southern Hemisphere* Reference Map

Center lat. –90°
Center lon. 0°

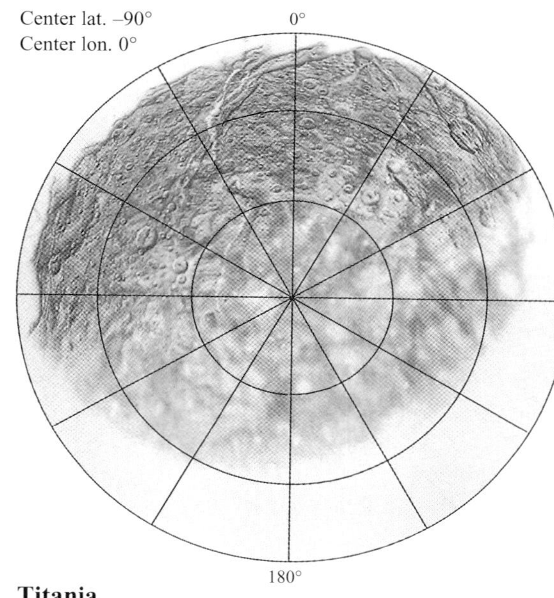

Titania Correlation of Map Units

Youngest at top

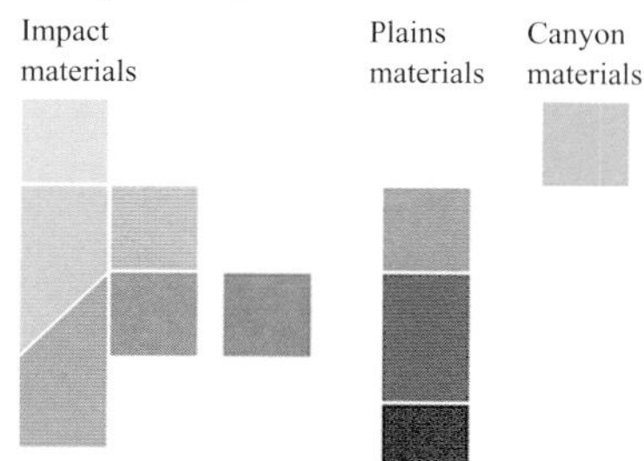

Titania Description of Map Units

Impact Materials

- Rayed or bright deposit crater
- Fresh-appearing crater, no rays
- Moderately degraded crater
- Elongate craters (secondaries?)
- Crater rim massif
- Crater floor material

Plains Materials

- Smooth
- Moderately cratered
- Heavily cratered

Canyon Materials

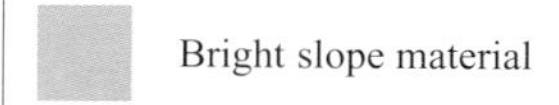
Bright slope material

Unmapped

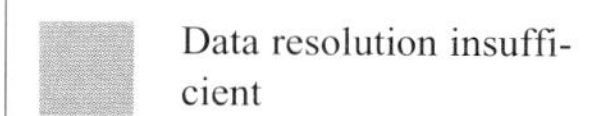
Data resolution insufficient

Structure Symbols

- Fault scarp
- Groove/trough
- Ridge

Oberon Southern Hemisphere* Geologic Map

Center lat. –90°
Center lon. 0°

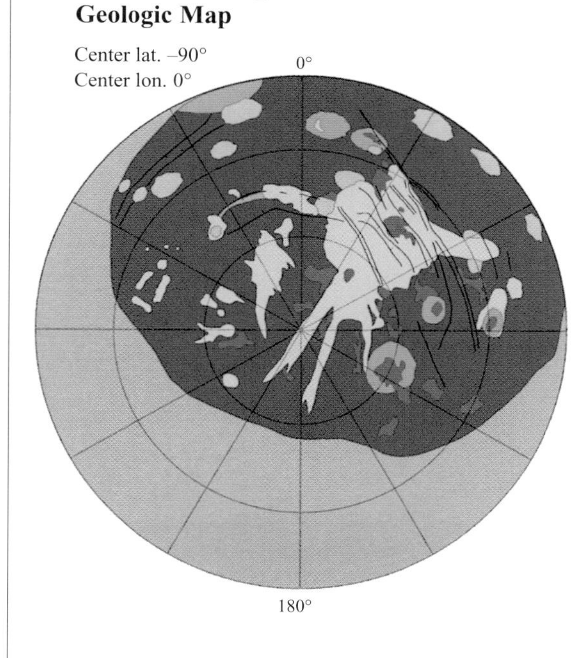

Oberon Southern Hemisphere* Reference Map

Center lat. –90°
Center lon. 0°

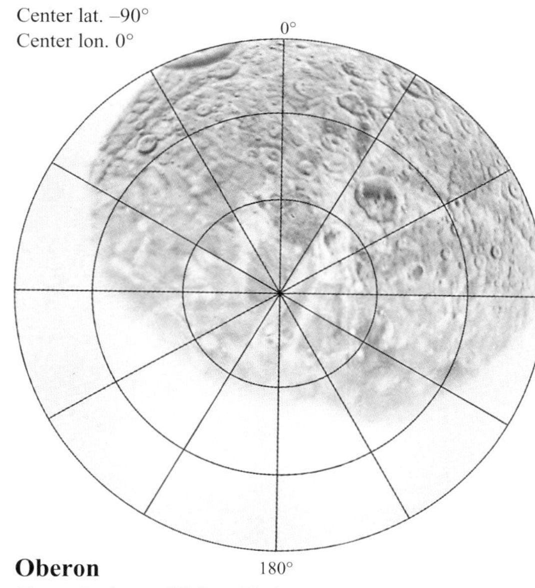

Oberon Correlation of Map Units

Youngest at top

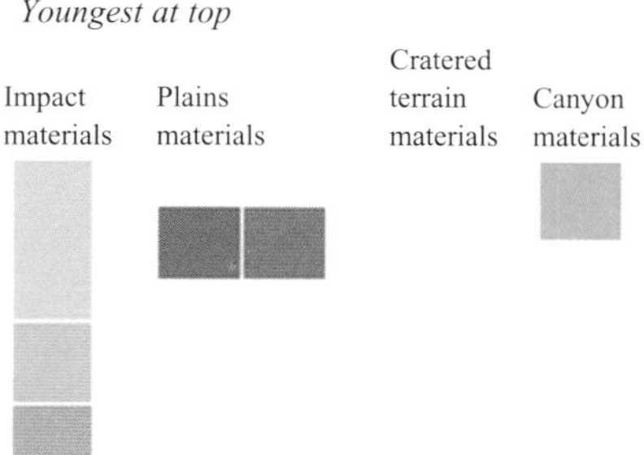

Oberon Description of Map Units

Impact Materials

- Fresh-appearing crater with rays
- Fresh-appearing crater, no rays
- Degraded crater

Plains Materials

- Dark
- Very dark

Cratered Terrain Materials

- Undivided

Canyon Materials

Bright slope material on canyon walls

Unmapped

Data resolution insufficient

Structure Symbols

- Fault scarp
- Deep trough or graben

*Northern hemisphere not mapped; no available data.

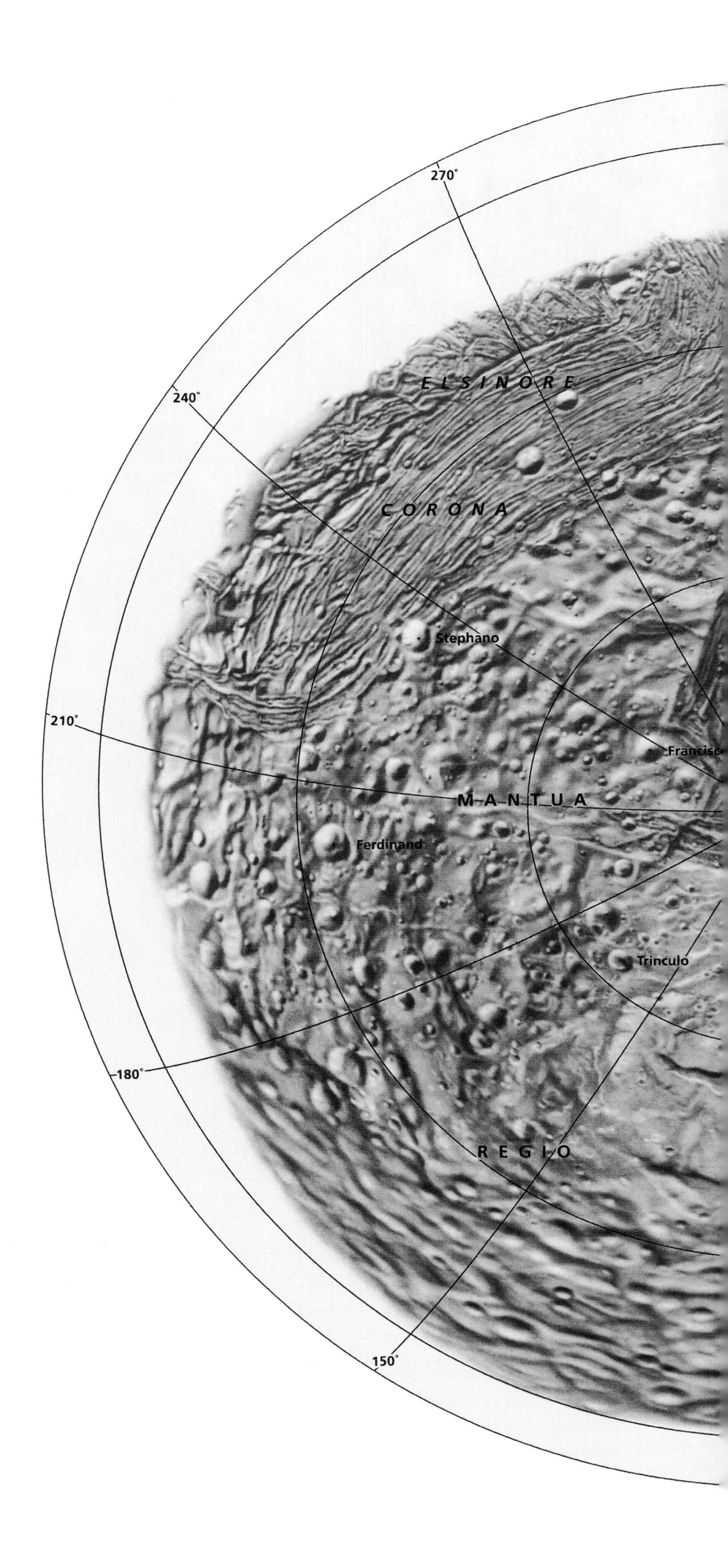

270°
240°
210°
180°
150°
ELSINORE
CORONA
Stephano
MANTUA
Ferdinand
Trinculo
REGIO

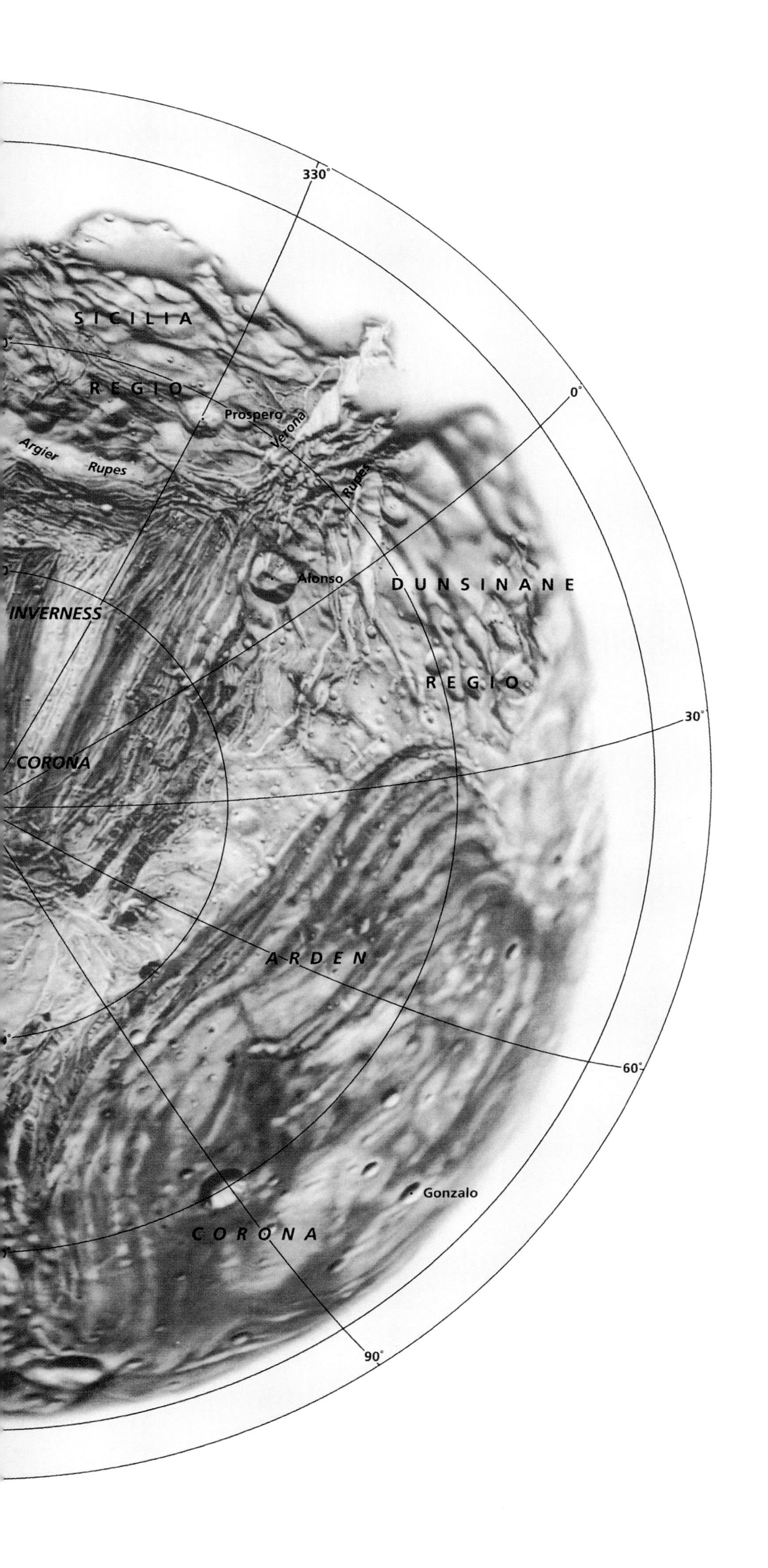

Miranda Airbrush Map

Center lat. –80°
Center lon. 300°

1 cm = 26 km
1 cm^2 = 676 km^2

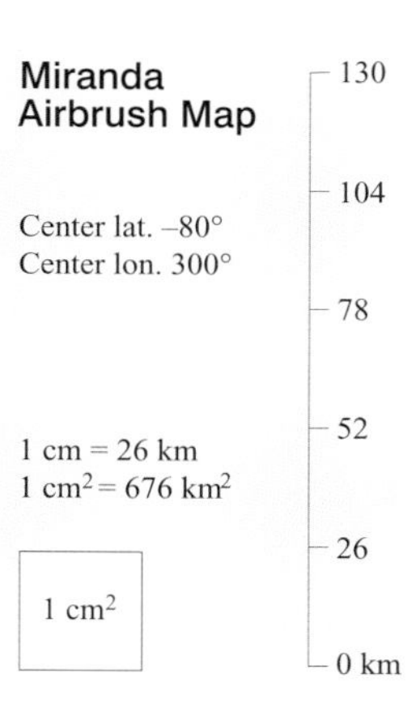

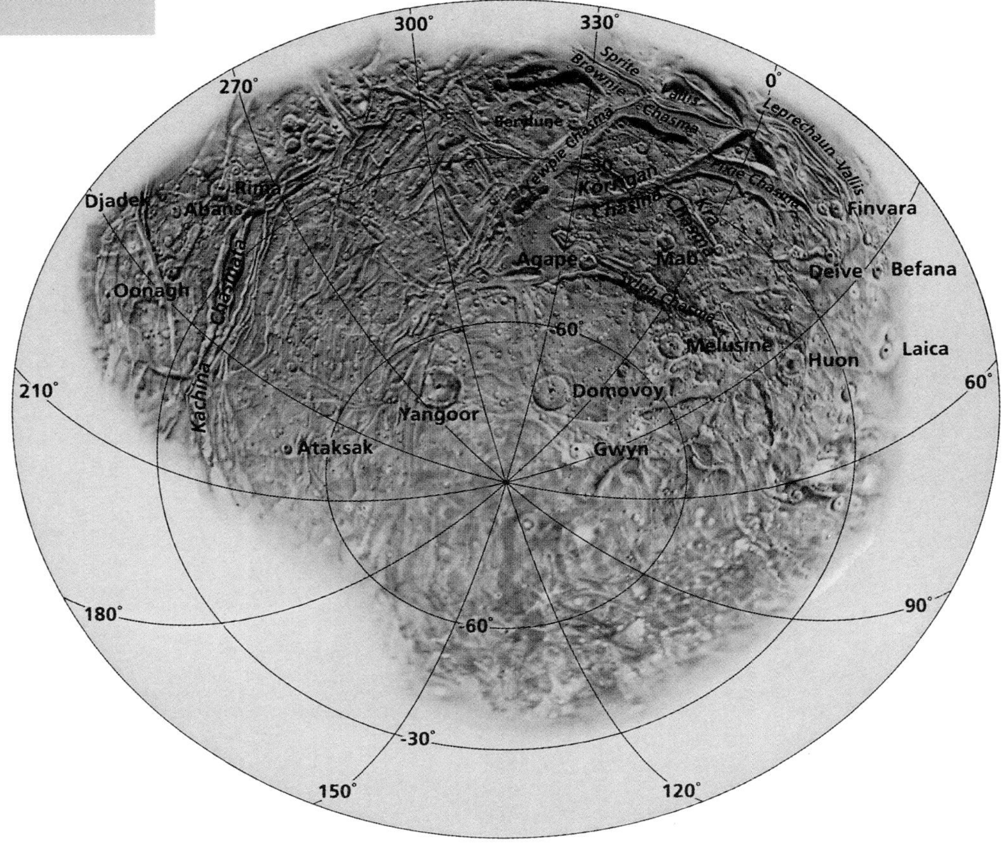

Ariel
Airbrush Map

Center lat. –50°
Center lon. 315°

1 cm = 132 km
1 cm^2 = 17,424 km^2

1 cm^2

660
528
396
264
132
0 km

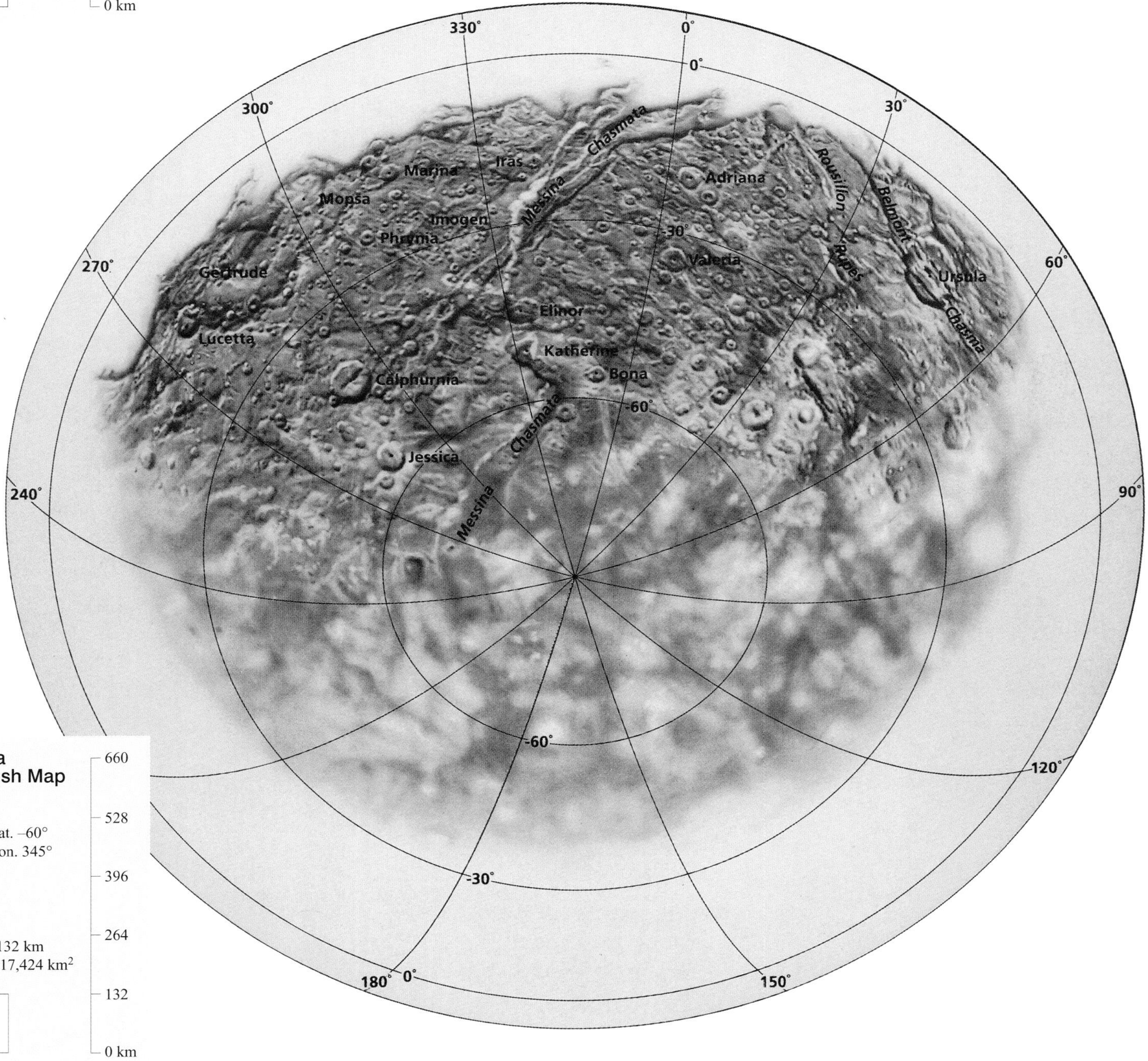

Titania
Airbrush Map

Center lat. –60°
Center lon. 345°

1 cm = 132 km
1 cm^2 = 17,424 km^2

1 cm^2

660
528
396
264
132
0 km

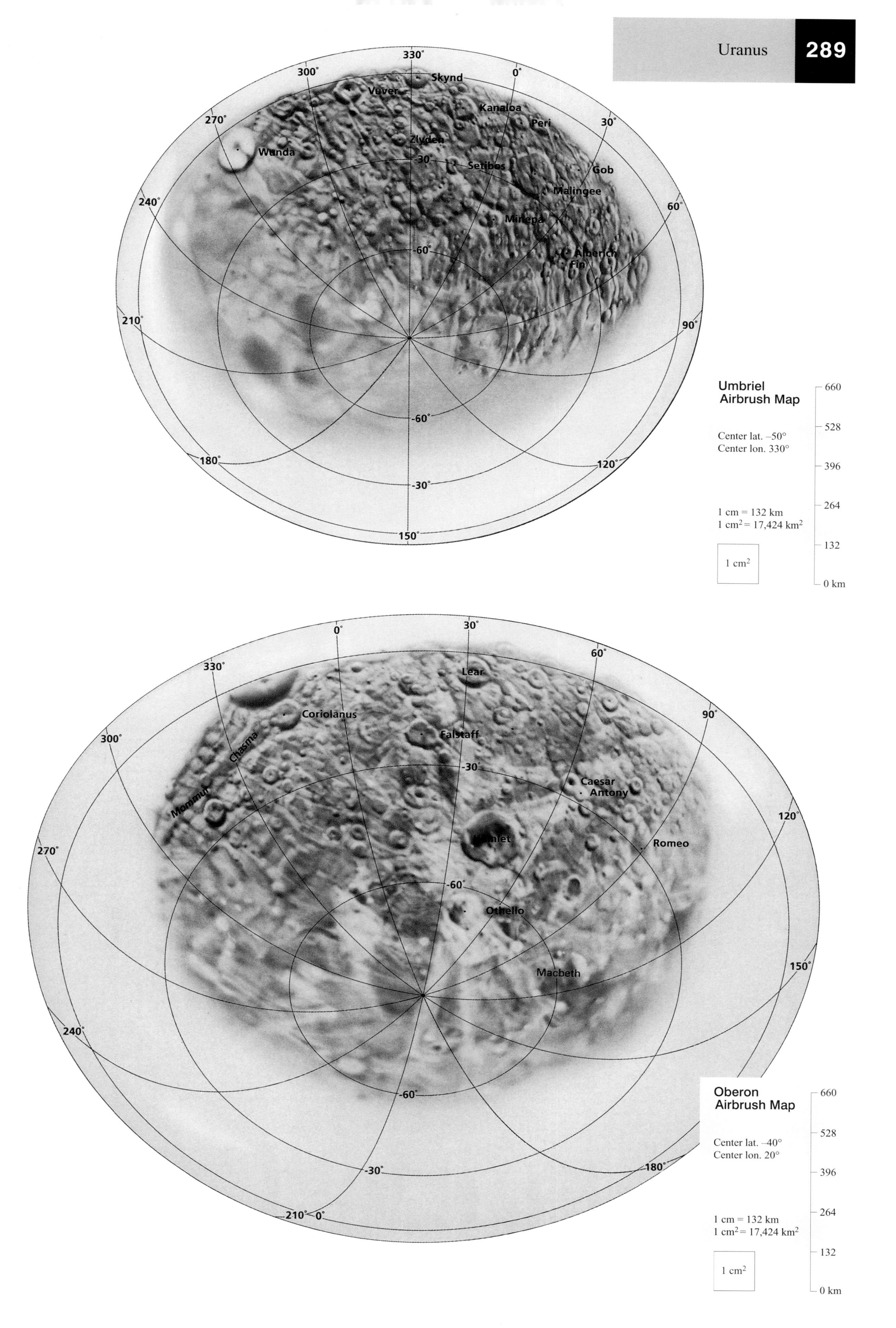
330°
300°
0°
Skynd
Vuver
Kanaloa
270°
Peri
30°
Zlyden
Wunda
-30°
Setibos
Gob
Malingee
240°
60°
Minepa
-60°
Alberich
Fin
210°
90°
-60°
180°
120°
-30°
150°
Umbriel
Airbrush Map
Center lat. –50°
Center lon. 330°
1 cm = 132 km
1 cm² = 17,424 km²
1 cm²
660
528
396
264
132
0 km
0°
30°
Lear
60°
330°
Coriolanus
90°
300°
Falstaff
Chasma
-30°
Caesar
Antony
Mommur
120°
Hamlet
Romeo
270°
-60°
Othello
150°
Macbeth
240°
-60°
Oberon
Airbrush Map
Center lat. –40°
Center lon. 20°
-30°
180°
210°
0°
1 cm = 132 km
1 cm² = 17,424 km²
1 cm²
660
528
396
264
132
0 km

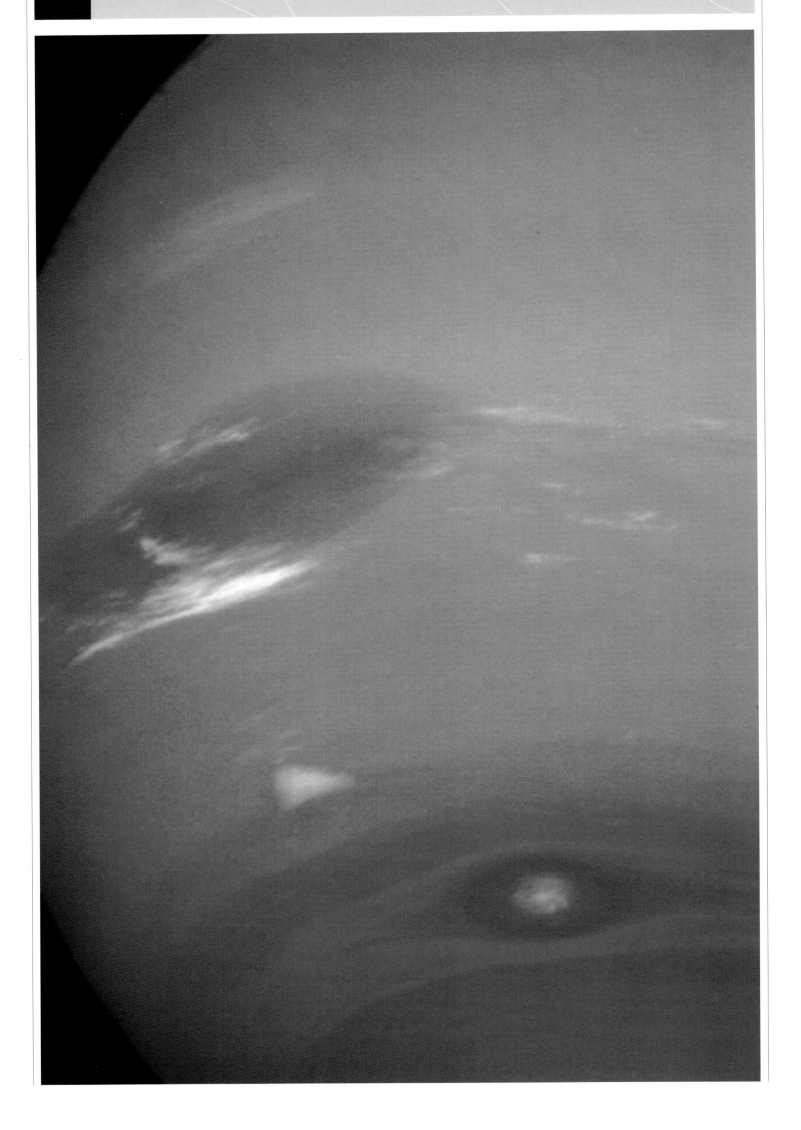

Neptune System

Neptune, the outermost giant gaseous planet, was proposed to exist before it was actually discovered. Telescopic observations of Uranus in the early 1800s showed notable discrepancies between its predicted position in orbit and its actual position. Independently of each other, two young mathematicians, John Couch Adams of Cambridge, England, and the Frenchman Urbain John Joseph LeVerrier, analyzed the problem and suggested both the position and the mass of an as-then unknown planet in the Solar System to account for the abnormal motions of Uranus.

Adams was essentially ignored by the English scientific establishment, while LeVerrier – although gaining the attention of the French Academy – was unable to convince French astronomers to search for the proposed new planet. LeVerrier did, however, stimulate the interest of a young German astronomer, Johann Gottfried Galle, who eagerly turned the prime telescope of the Berlin Observatory to the task of looking for the predicted planet. On the very first night of Galle's search, September 23, 1846, he found the eighth planet within 1° of its predicted position.

Both Adams and LeVerrier share the credit for the discovery of Neptune. Their story is a lesson to young scientists in perseverance and to the "establishment" in not hastily dismissing new ideas.

Only 17 days after the discovery of Neptune, its largest moon, Triton, was found by the Englishman, William Lassell. But it would be more than a century before another satellite of Neptune, Nereid, would be found by Gerard Kuiper in 1949. Neptune's ring-arcs were not discovered until the mid-1980s.

In its final phase of Solar System exploration, *Voyager 2* began observing the Neptune system in June 1989. During the few days of its encounter, this intrepid spacecraft returned an immense amount of information. Knowledge was gained on the magnetosphere and atmosphere of Neptune, its ring-arcs were defined, six new satellites were discovered, and, in a spectacular closing to what can only be described as a fantastic journey of discovery, cameras on board the *Voyager* spacecraft returned images of actively erupting geysers on Triton.

The Neptune system compared to the Earth/Moon system.

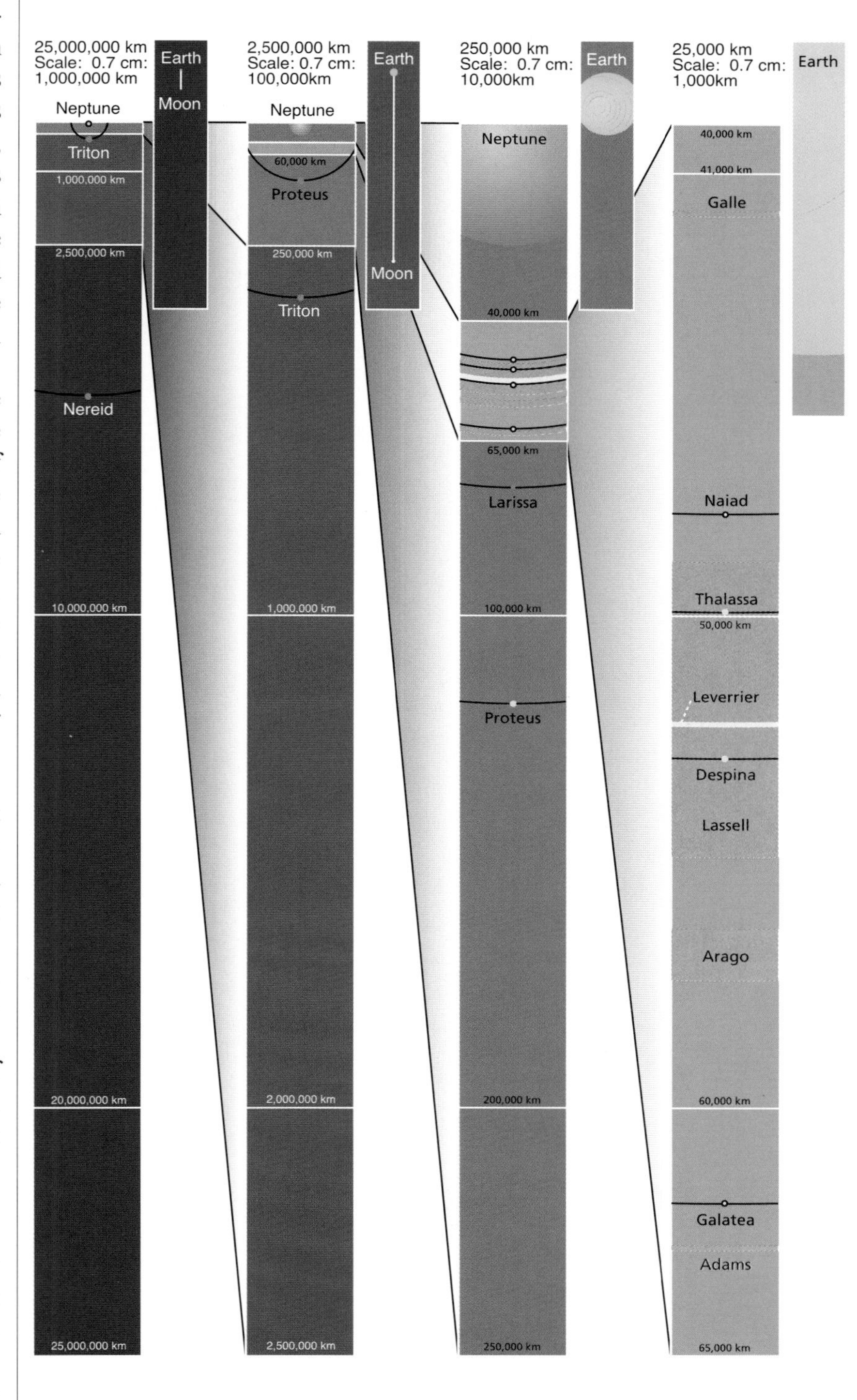

Facing page: color image of Neptune reconstructed from two images taken by the *Voyager 2* spacecraft. The Great Dark Spot is the large prominent feature accompanied by bright white clouds. South of the Great Dark Spot is a smaller white cloud and a second dark spot (*Voyager 2* JPL photograph P-34648).

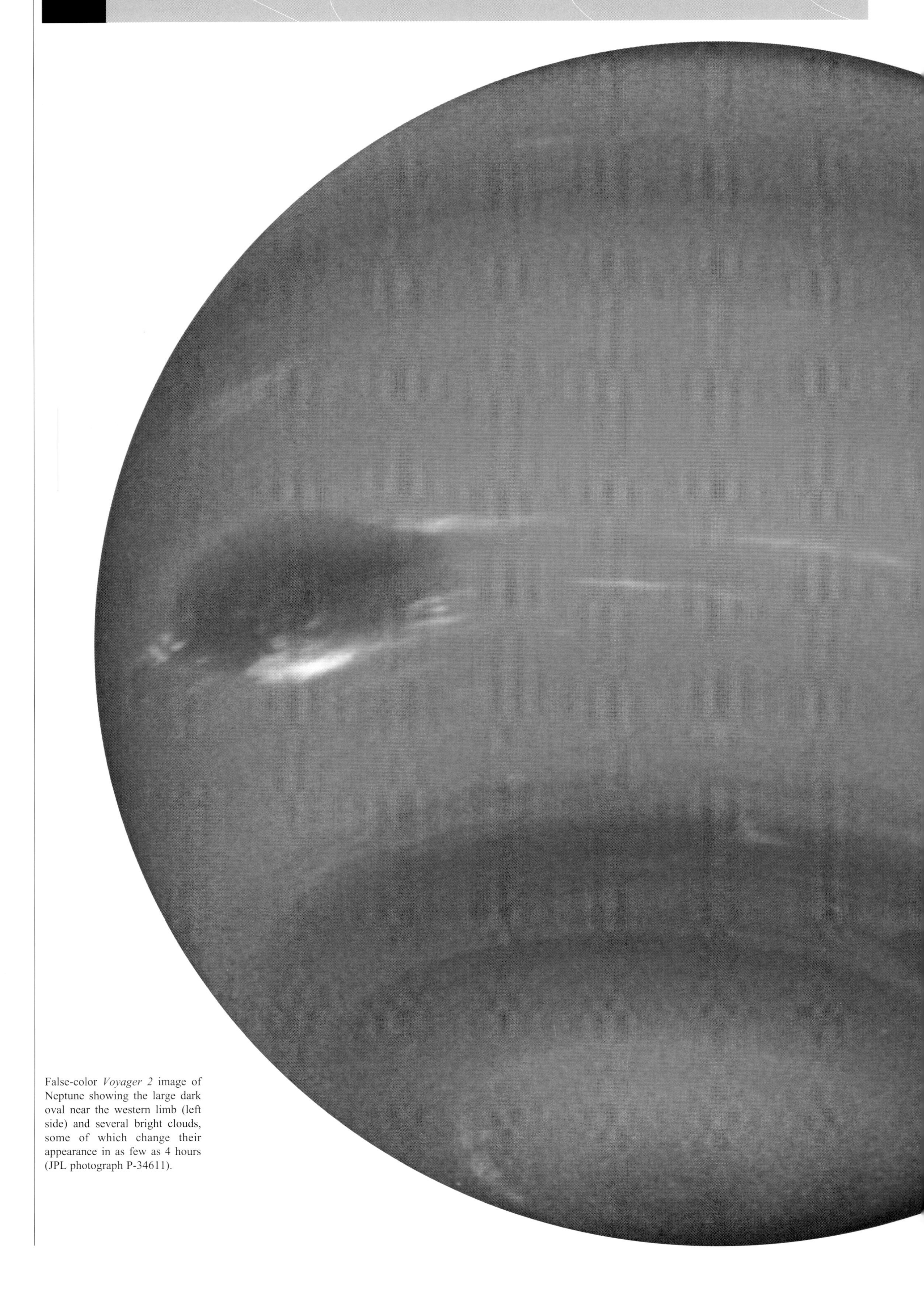

False-color *Voyager 2* image of Neptune showing the large dark oval near the western limb (left side) and several bright clouds, some of which change their appearance in as few as 4 hours (JPL photograph P-34611).

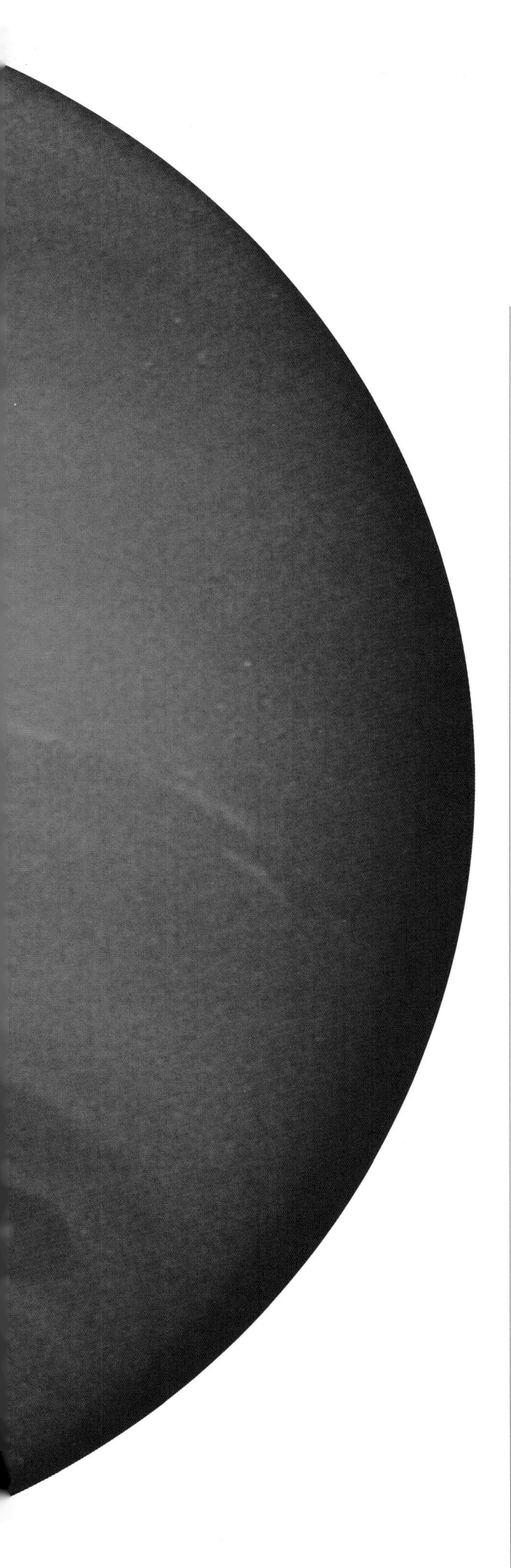

Neptune

After its discovery, later telescopic observations showed Neptune to have a nearly circular orbit. Because of the enormous distances in the outer Solar System, 165 years are required for Neptune to complete one trip around the Sun, and Neptune has yet to make one complete orbit since its discovery. The planet is about one-third the size of Jupiter and probably has a composition similar to that of Uranus, consisting mostly of hydrogen.

Neptune is the most dense of the Jovian planets, suggesting the presence of a relatively large core of heavy elements. Until the flyby of the *Voyager 2* spacecraft, many planetary scientists thought Neptune had a rocky core surrounded by a liquid mantle. Recent results, however, suggest that Neptune may have a low-density, liquid core. Laboratory experiments, using a mixture of water, ammonia, and alcohol to simulate the interiors of both Neptune and Uranus, show that under the high pressures and temperatures considered to exist in the outer planets, a fluid core can be maintained that matches the postulated densities of these giant planets.

In contrast to the bland appearance of Uranus's atmosphere, *Voyager* photographs of Neptune show a wealth of cloud structures and motions. Although the temperature in the upper atmosphere is a chilling 70 K, energy and heat released from the interior of the planet are apparently sufficient to drive the cloud motions, in contrast to conditions on Uranus. The overall appearance of Neptune is blue, similar to Uranus, and also a consequence of abundant methane in the atmosphere.

Neptune's atmosphere travels with the greatest speed of any in the Solar System. While the planet rotates from west to east (like the Earth), winds at the equator travel in the opposite direction. Careful tracking of the clouds seen on sequential photographs shows that they travel in excess of 2,000 kilometers per hour at the equator.

Several types of cloud features were discovered by *Voyager*. The most pronounced, dubbed the Great Dark Spot, was found in the southern hemisphere just below the equator at 20° S. This elongate feature is some 12,000

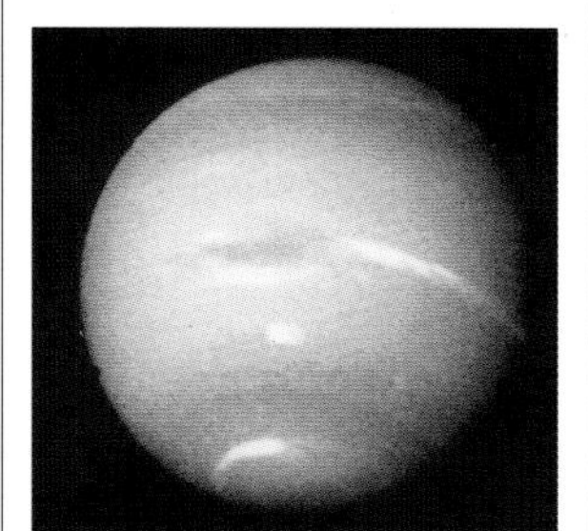

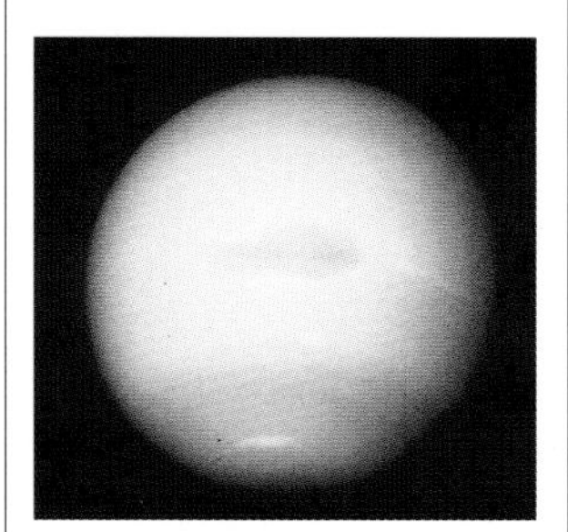

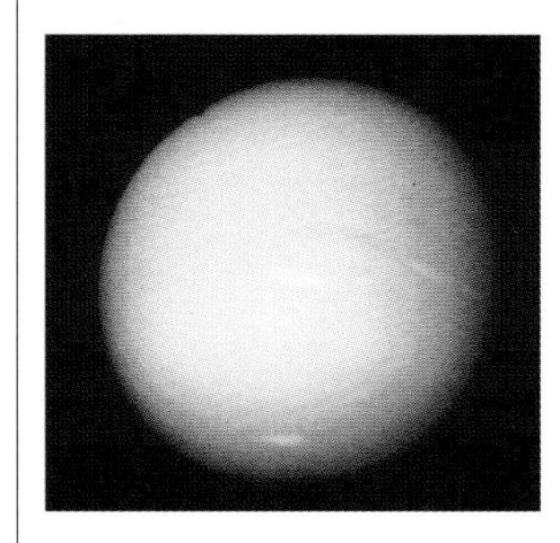

Images of Neptune taken through orange (top view), violet (middle view), and ultraviolet (bottom view) filters by the *Voyager 2* camera. Features that become less visible when imaged through the ultraviolet filter are thought to reside deeper in the atmosphere (JPL photograph P-34616).

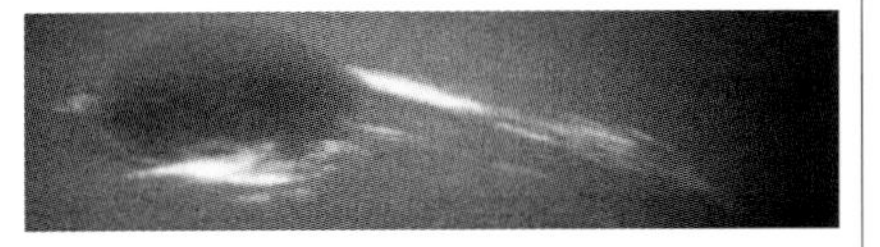

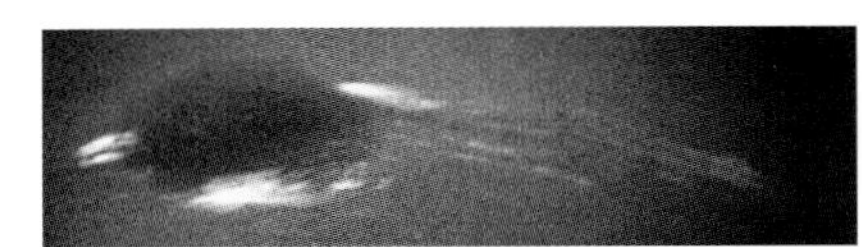

Sequential photographs of clouds near the Great Dark Spot, covering a period of about 36 hours, or two rotations of Neptune. The cirruslike clouds are composed of frozen methane (*Voyager 2* JPL photograph P-34622).

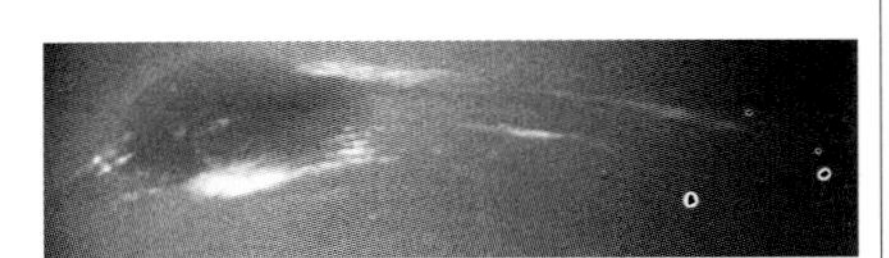

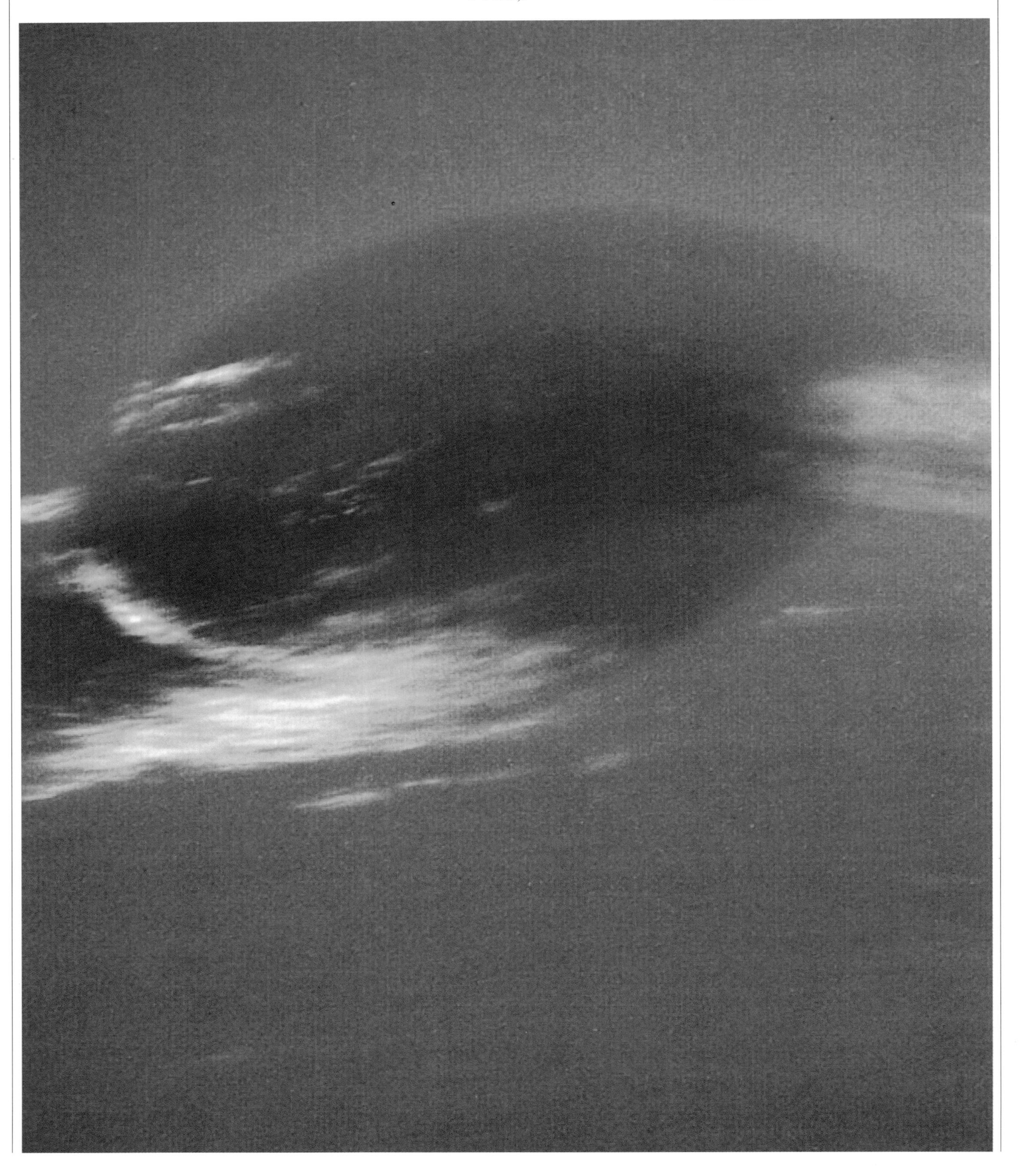

Close-up view of the Great Dark Spot. The spiral structure of the dark boundary and white cirrus clouds suggest a storm system rotating in a counter-clockwise direction (*Voyager 2* JPL photograph P-34672).

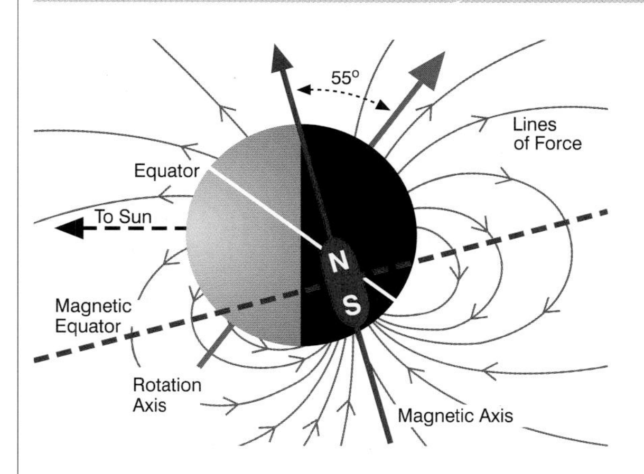

Diagram showing the orientation of the magnetosphere with respect to the spin axis of Neptune.

Following spread: *Voyager 2* view of the south pole of Neptune; clouds near the bright limb rotate eastward into Neptune's night side. The bright cloud visible to the left appears to be an organized storm system near the pole (JPL photograph P-34715).

kilometers wide (the size of the Earth) and exhibits counter-clockwise rotation with a period of 16 days. Another dark spot, also found in the southern hemisphere, appears to be a smaller version of the Great Dark Spot. Both dark spots are found at a lower altitude within the clouds and hazes of the upper atmosphere.

Bright cirruslike clouds are also visible on Neptune. They appear to be dense, upward extensions of the methane clouds. Some are found in association with the Great Dark Spot. Others occur as narrow belts in zones of wind shears in both the northern and southern hemispheres of the planet.

Neptune has a strong, complex magnetosphere and associated magnetotail. Measurements taken by the *Voyager 2* spacecraft show that the magnetic field is offset from the center of Neptune and is tilted about 47° from the rotational axis of the planet. In some ways, it is similar to the magnetic field observed at Uranus and is just as puzzling. Although Neptune may also be undergoing a reversal of magnetic poles (as described for Uranus), for both planets to experience this reversal at the same time seems unlikely. Explanations are currently being sought by planetary scientists to explain the offset of magnetic fields with regard to the planetary motions.

Rings

Rings (or more properly, segments of rings or ring-arcs) around Neptune were discovered independently by two teams of scientists in the mid-1980s. Andre Brahic and his colleagues and William Hubbard and Faith Vilas noted occultations of stars while observing Neptune. Such occultations, or blocking of light, were attributed to material in orbit around Neptune, similar to the rings around Uranus. Subsequent pictures taken by *Voyager* showed several rings. Informally called N63, N53, and N42 (referring to the distances in thousands of kilometers from Neptune), these rings are all in prograde orbits confined to the equatorial plane of Neptune.

Both N63 and N53 are narrow rings and appear to be composed of fine, dusty material, as evidenced by the way that they reflect sunlight. N42 is a broad, diffuse ring that also seems to be composed of dust and is similar to

False-color *Voyager 2* high-resolution image of the southern hemisphere of Neptune showing a large cloud system at the left (JPL photograph P-34649).

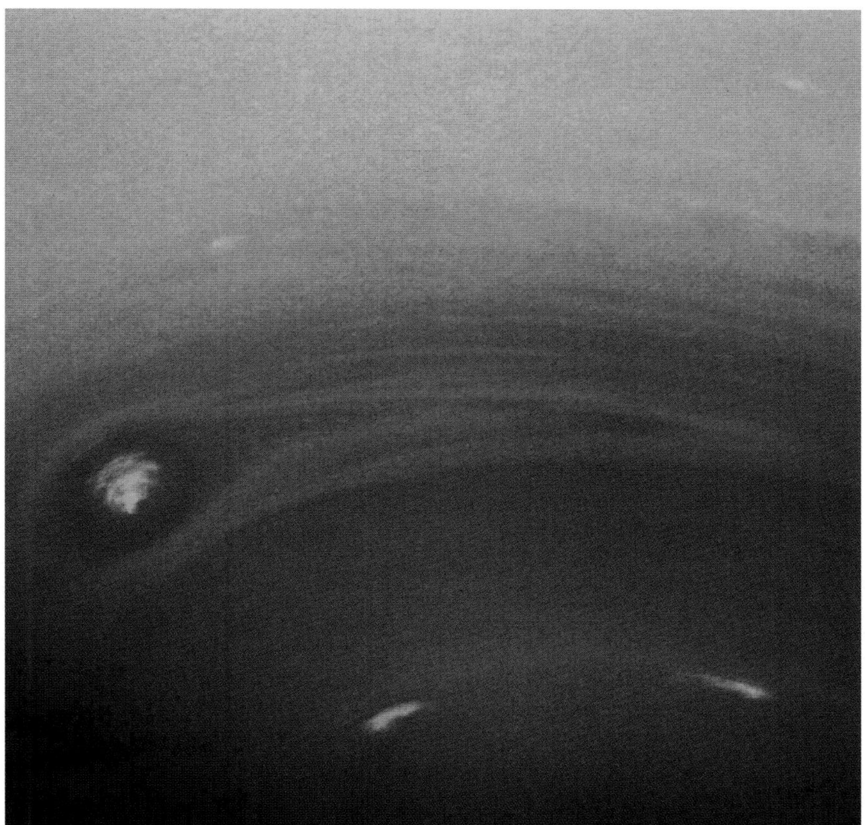

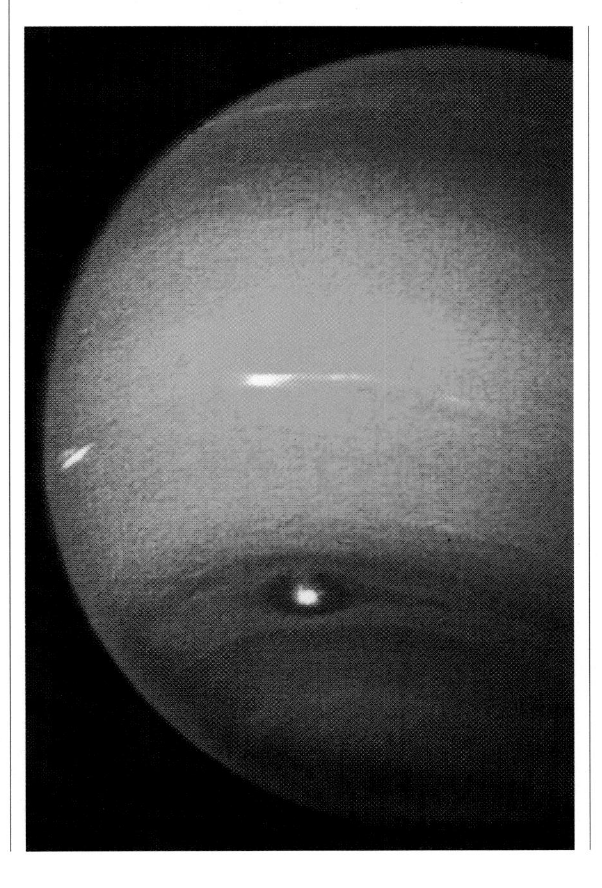

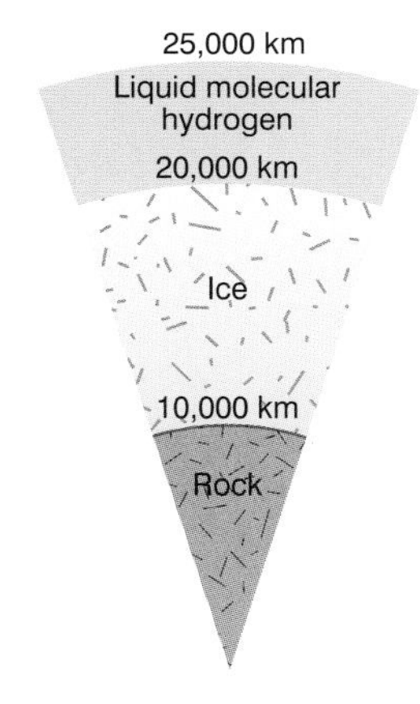

Diagram showing the interior configuration of Neptune.

False-color photograph of Neptune showing various cloud patterns; *Voyager* images have been processed in this view to reveal the ubiquitous haze that covers the planet (JPL photograph P-34705).

Voyager 2 image showing the two prominent rings, N53 and N63 (outer). Three "ring-arcs" (concentrations of dust particles) are seen in ring N63. The direction of motion is clockwise (*Voyager 2*, JPL photograph P-34712).

Image of the two main rings, N63 and N53, and the inner (faint) ring N42. The fourth ring forms a faint band that extends from between the N63 and N53 rings inward toward Neptune. The bright spots are stars in the background (*Voyager 2* JPL photograph P- 34726).

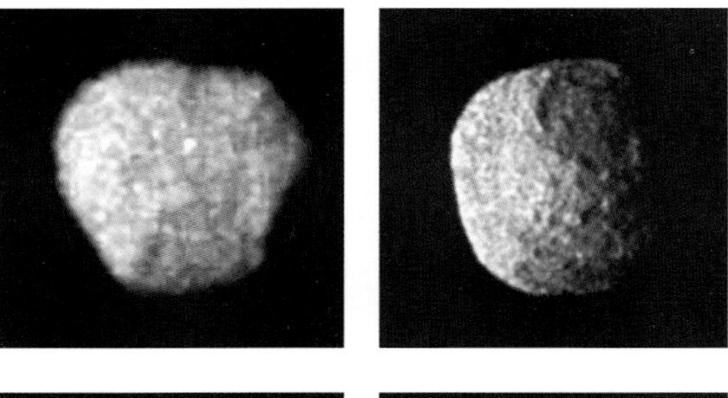

Neptune satellite, Proteus, photographed by *Voyager 2*; this object is about 416 kilometers in diameter and is pock-marked with impact craters (JPL photographs P-34681 and P-34727).

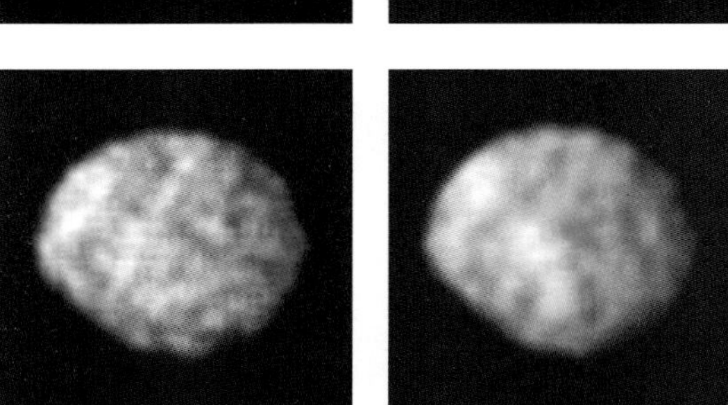

Voyager 2 images of the Neptune satellite, Larissa; this moon is about 210 by 190 kilometers across and shows numerous dark depressions 30 to 50 kilometers wide that are thought to be impact craters (JPL photograph P-34698 for both images).

Jupiter's ring and the G Ring of Saturn. The two outer rings include clumps of dust that form rings, arcs, or segments. However, analysis of *Voyager* pictures shows that the rings are complete with faintly visible material connecting the arcs.

A fourth ring mass forms a broad, diffuse sheet that extends throughout the inner Neptune system. Traveling from Neptune outward, one would encounter first ring material, then small satellites and ring material, then the larger satellites. The entire mass of ring material around Neptune is some 10,000 times less than that of the Uranus ring system and even less than Saturn's rings.

Satellites

Neptune has eight known satellites, six of which were discovered from *Voyager 2* data. The largest, Triton, is in the size class of Earth's Moon. The remainder are much smaller, ranging from 50 to 400 kilometers in diameter.

The four smallest satellites are in orbits within the ring system of Neptune but do not appear to be ring shepherds. All are dark objects that travel in prograde orbits around Neptune. Information for these satellites is very limited, but from the manner in which they reflect light, the satellites probably have irregular shapes.

The largest of the newly discovered satellites, Proteus, orbits Neptune very rapidly just outside the ring system, taking only 1.1 Earth days to make one complete orbit. Images show it to be irregular in shape and to have craters on the surface, some as large as 100 kilometers across. In addition, part of a large, nearly circular depression can be seen on the surface. This feature could be a degraded impact crater, a structure formed by tectonic processes, or deformation of the crust resulting from a large impact.

Nereid, the outermost of Neptune's satellites, is about 340 kilometers in diameter. It travels in a highly inclined, eccentric orbit and takes 359 days to complete one trip around Neptune. The *Voyager* spacecraft was unable to take high-resolution images of Nereid because it passed at too great a distance from the satellite, nor was any information obtained to allow estimates of Nereid's density. There are, however, some variations in brightness seen on the satellite; these could be attributed either to an irregular shape or to variations in the composition of its surface.

Triton

Triton is a fascinating moon that revealed many surprises to the cameras of the *Voyager* spacecraft. Not only does its surface show features unseen in the rest of the Solar System, but vigorously erupting geysers were discovered. Moreover, its orbital geometry is unique and poses some intriguing puzzles that have not yet been solved.

Triton travels around Neptune in a circular, synchronous orbit but, unlike any other large satellite of the giant planets, it is in retrograde motion (that is, it travels in the opposite direction to the rotation of Neptune). In addition, careful analysis of Triton's motions shows that the satellite is in a decaying orbit and is slowly being pulled toward Neptune.

The orbital geometry of Triton suggests to many planetary scientists that the satellite was formed elsewhere in the outer Solar System and was then gravitationally captured by Neptune. It is interesting to note that both Triton and Pluto share many characteristics, including size, mass, and probable compositions.

If Triton is a captured object, then to achieve its present orbit it must have experienced extreme tidal stresses exerted by Neptune following its capture. Such stresses undoubtedly would have generated large amounts of heat in the interior of the moon, perhaps leading to melting of the interior and compositional differentiation. Triton, at 2.1 grams per square centimeter, has the highest density of any of the outer planet satellites. This suggests that Triton is composed of about one-third ice and two-thirds rocky materials. If Triton did experience heating and differentiation, then we would expect it to have a rocky core, surrounded by a mantle that could still be liquid, and a crust of ice.

Spectral characteristics indicate that the surface of Triton is composed mostly of methane, nitrogen ices, and frost. Some of the areas are

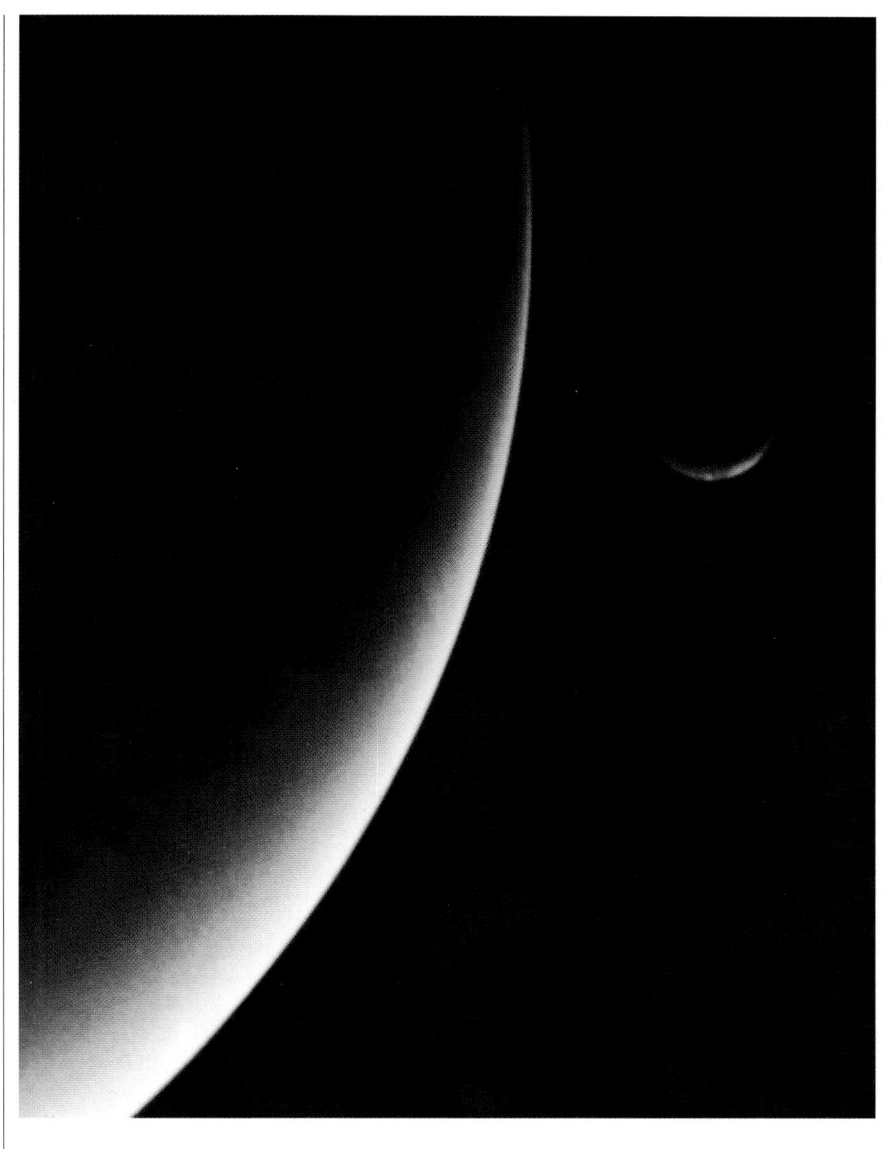

View of Neptune and its satellite Triton taken by *Voyager 2* about 3.3 days after closest approach and as the spacecraft traveled southward at an angle of 48° to the plane of the ecliptic (JPL photograph P-34761).

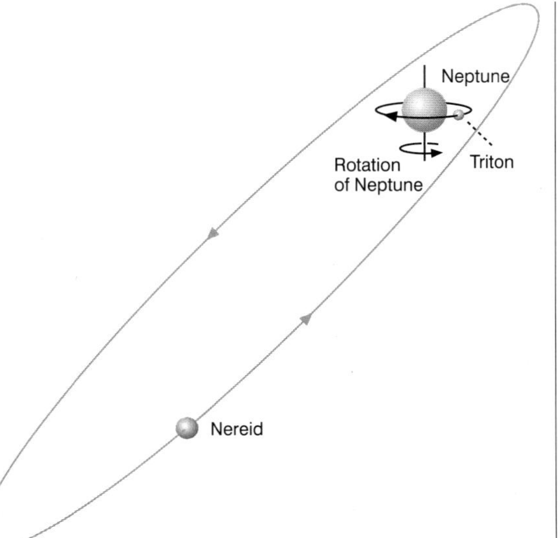

Diagram showing the orbits of Nereid and Triton around Neptune.

yellowish to peach colored, possibly representing methane and nitrogen ices that have been converted to complex organic compounds from bombardment by cosmic rays and ultraviolet radiation from the Sun. Two primary terrains are seen in *Voyager* images of Triton – an older rugged unit called cantaloupe terrain, and a younger plains unit probably of volcanic origin. Cantaloupe terrain is so named because of its resemblance to the skin of a melon and includes irregular pits, bumps, and linear veins. The pits and bumps may represent surface deformation and extrusion of viscous ice onto the surface from the interior. Linear features include grooves and troughs, some of which are partly filled with ridges that also seem to be the result of the extrusion of ice along fracture systems. Impact craters pockmark parts of Triton's surface and provide some indication of the age of the surface. Most of the craters are found on the leading hemisphere of Triton. The size and number of craters are about the same as for the dark areas on the Moon – the lunar maria. Because of the lack of large craters and the absence of heavily cratered terrain, the surface of Triton is geologically young.

A few patches of smooth plains cut the cantaloupe terrain. This relationship and the paucity of impact craters on some of the smooth plains shows that they are geologically younger than the cantaloupe terrain. The smooth plains are characterized by irregular-shaped depressions, some of which are 100 to 200 kilometers wide. In some cases, they may have been flooded with vast lakes of liquid water that erupted onto the surface and quickly froze. The volcanic plains and irregular depressions found in some areas may have been formed by the eruption of liquid water, methane, and nitrogen on the surface. Such a mixture would have a relatively low melting point and would have been easily produced in the interior of Triton.

Triton has a prominent southern ice cap, composed mostly of nitrogen ice, which extends from the pole nearly to the equator. Because the spin axis of Triton is inclined some 21° with respect to its orbital plane around Neptune, like Earth and Mars, Triton also has seasons. With the changing seasons, the ice cap presumably shifts from one pole to the other.

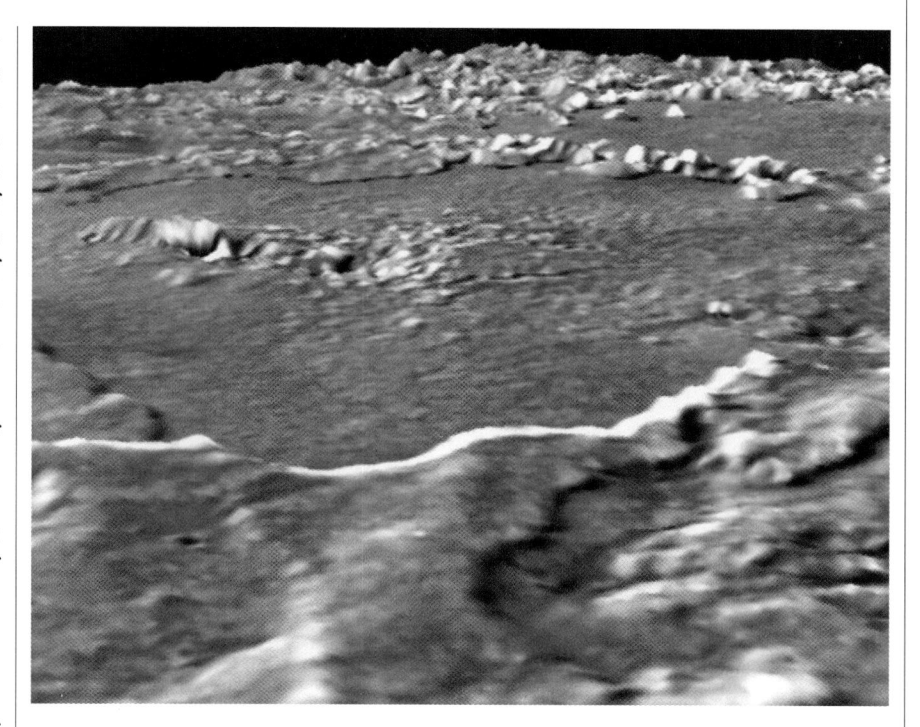

Computer-generated oblique view across one of the prominent caldera-like features on Triton. This feature is more than 200 kilometers across and is considered to have erupted slushy ice onto the surface. Terraces and benches around the margin of the depression may be remnants of earlier flooding (JPL photograph P-34751).

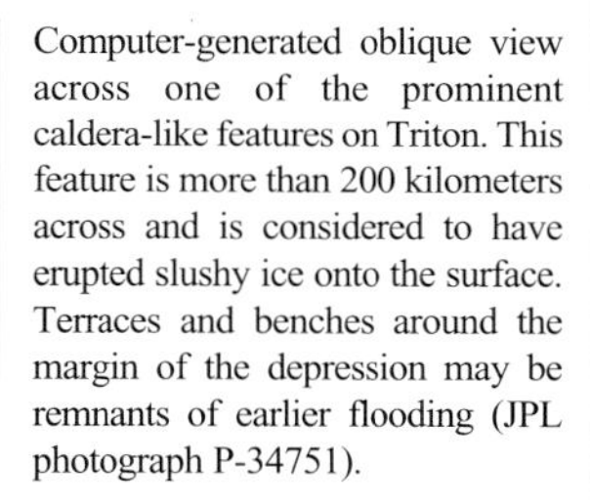

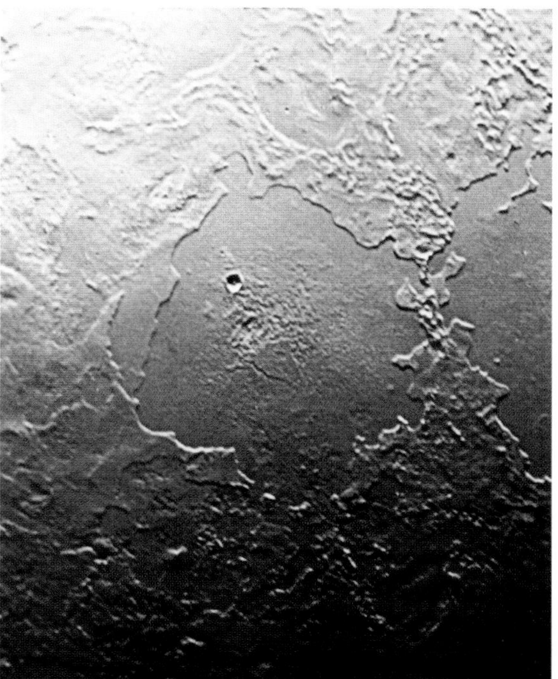

Smooth volcanic plains, named Rauch Planum, filling irregular calderalike depressions on Triton. Area shown is about 290 kilometers across; south is toward the top (U.S. Geological Survey photograph).

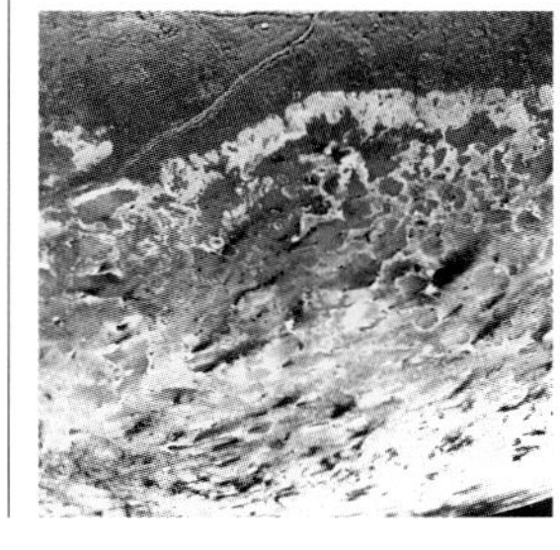

Voyager 2 image of the south polar region of Triton, showing numerous dark plumes, or "wind streaks," on the ice-rich surface. Some plumes are more than 60 kilometers long and are thought to be deposits of dark material erupted from geysers (JPL photograph P-34714).

Facing page: color composite of *Voyager* photographs showing the two principal terrains on Triton. Lightly cratered plains are shown to the north (top) of this view, while the southern hemisphere includes deposits of nitrogen ice. Also visible are dark plume-shaped deposits that are the result of eruptive geysers (photomosaic from U.S. Geological Survey).

Detail of the "cantaloupe" terrain of Triton, showing linear grooves and ridges and irregular depressions. The linear feature is about 35 kilometers wide and is probably a graben or fault block (*Voyager 2* JPL photograph P-34689).

Alternatively, Triton may have only one cap. At the time of the Voyager 2 flyby, it was summer in the southern hemisphere and the south polar cap appears to have been sublimating gases into the atmosphere. The boundary of the ice cap is very irregular and follows local terrain. Isolated patches of frost are found in depressions near the boundary.

Among the many spectacular discoveries made through the Voyager spacecraft, the active eruptions on Io and Triton are the highlights of the mission. During early analysis of *Voyager* data, two eruptive plumes were discovered on Triton. Later analysis of the data revealed two more active eruptions. The eruptions consist of dark geysers as wide as 1 kilometer, which rise vertically above the surface of Triton to an altitude of about 8 kilometers. At this height, wind shear in the upper, thin atmosphere apparently catches the plumes and carries them downwind hundreds of kilometers.

Many ideas have been proposed to explain how the eruptions on Triton occur. Their location in the southern hemisphere in association with the ice cap may provide some clues to the process. First, we know that nitrogen in its pure form is very transparent. It has been suggested that nitrogen ice serves as a kind of "solid-state greenhouse," in which solar energy is trapped beneath the ice and raises the temperature at the base of the ice layer. A rise in temperature of only 4 K is adequate to vaporize the nitrogen ice. As more gas is released, the expanding vapors exert very high pressures beneath the ice cap. Eventually the gas would rupture the ice and be explosively vented into the thin atmosphere above the surface. Dark material, perhaps silicates or methane ice particles darkened by ultraviolet radiation, is carried along with the expanding vapor into the atmosphere.

The presence of numerous dark streaks on the surface suggests that plume-producing processes are common on Triton. Mapping the orientation of these and other features related to the atmosphere shows that most streaks are oriented toward the northeast and east. Analysis of these features and cloud patterns suggests that winds 1 to 3 kilometers above the surface are predominantly eastward, whereas those at about 8 kilometers blow toward the west.

Triton
Sub-Neptunian Hemisphere*
Geologic Map

Center lat. 0°
Center lon. 0°

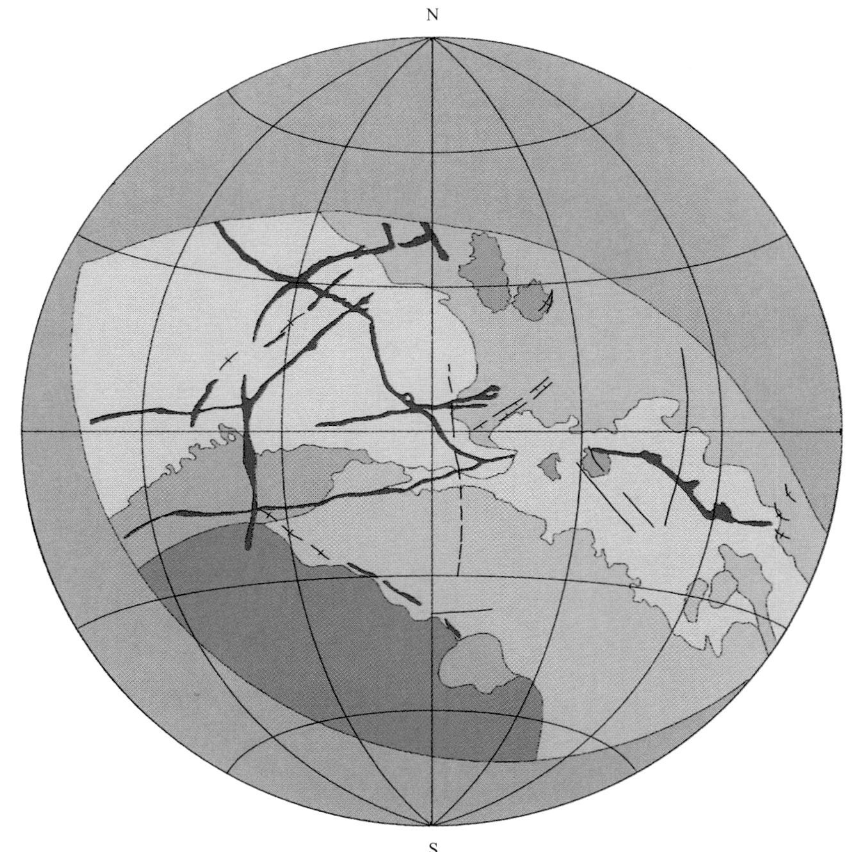

Triton
Sub-Neptunian Hemisphere*
Reference Map

Center lat. 0°
Center lon. 0°

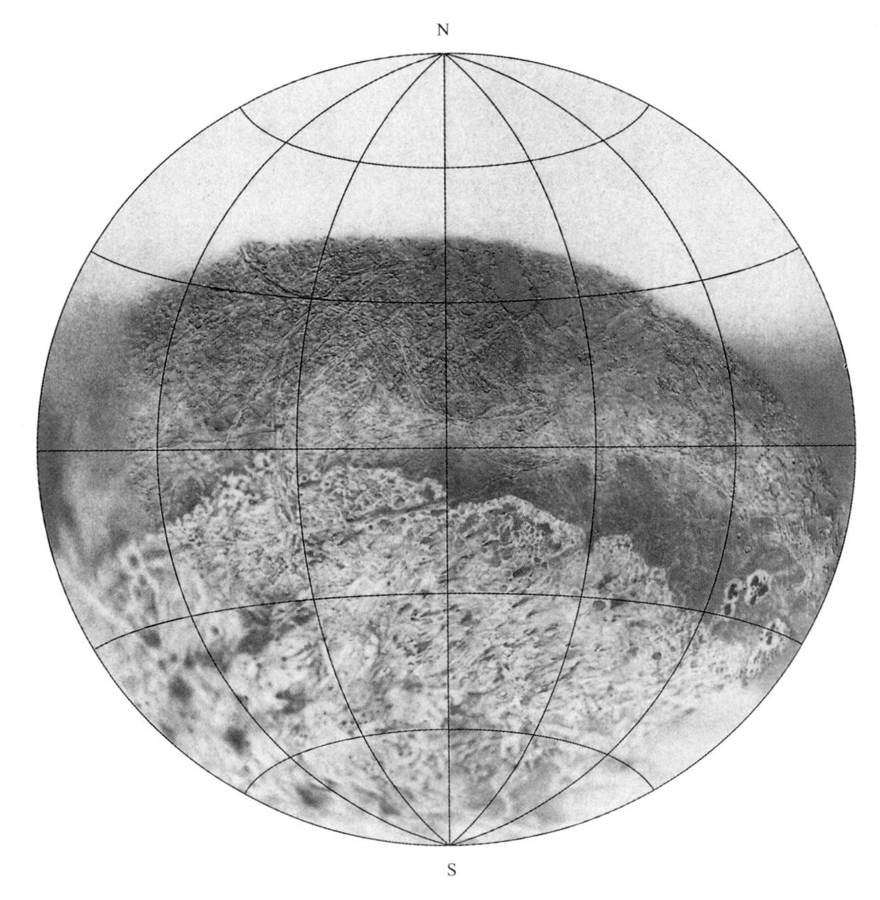

*Anti-Neptunian hemisphere not mapped; data resolution insufficient.

Correlation of Map Units

Youngest at top

Bright materials | Plains materials | Miscellaneous materials

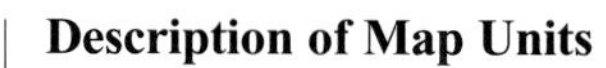

Description of Map Units

Bright Materials

 Spotted

 Rugged

 Streaked

Plains Materials

 Valley

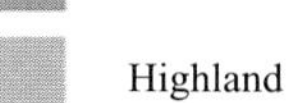 Highland

Miscellaneous Units

 Hummocky terrain

 Linear ridge materials

 "Cantaloupe" terrain

Unmapped

 Data resolution insufficient

Structure Symbols

Fault or scarp

 Graben

Ridge

270°
300°
330°
Slidr Sulci
Sulci
M
Dagon Cavus
BUBEMBE
Tano
Hekt Cavus
Kormet
Sulci
Apep Cavus
Sulci
Bheki Cavus
Hirugo Cavus
Ormet
Rem Maculae
Sulci
REGIO
Lo Sulci
Sulci
Yasu
Sulci
Ho
Ob
Mangwe Cavus
Sulci
Boynne
UHLANGA
330°
300°
270°
240°
210°

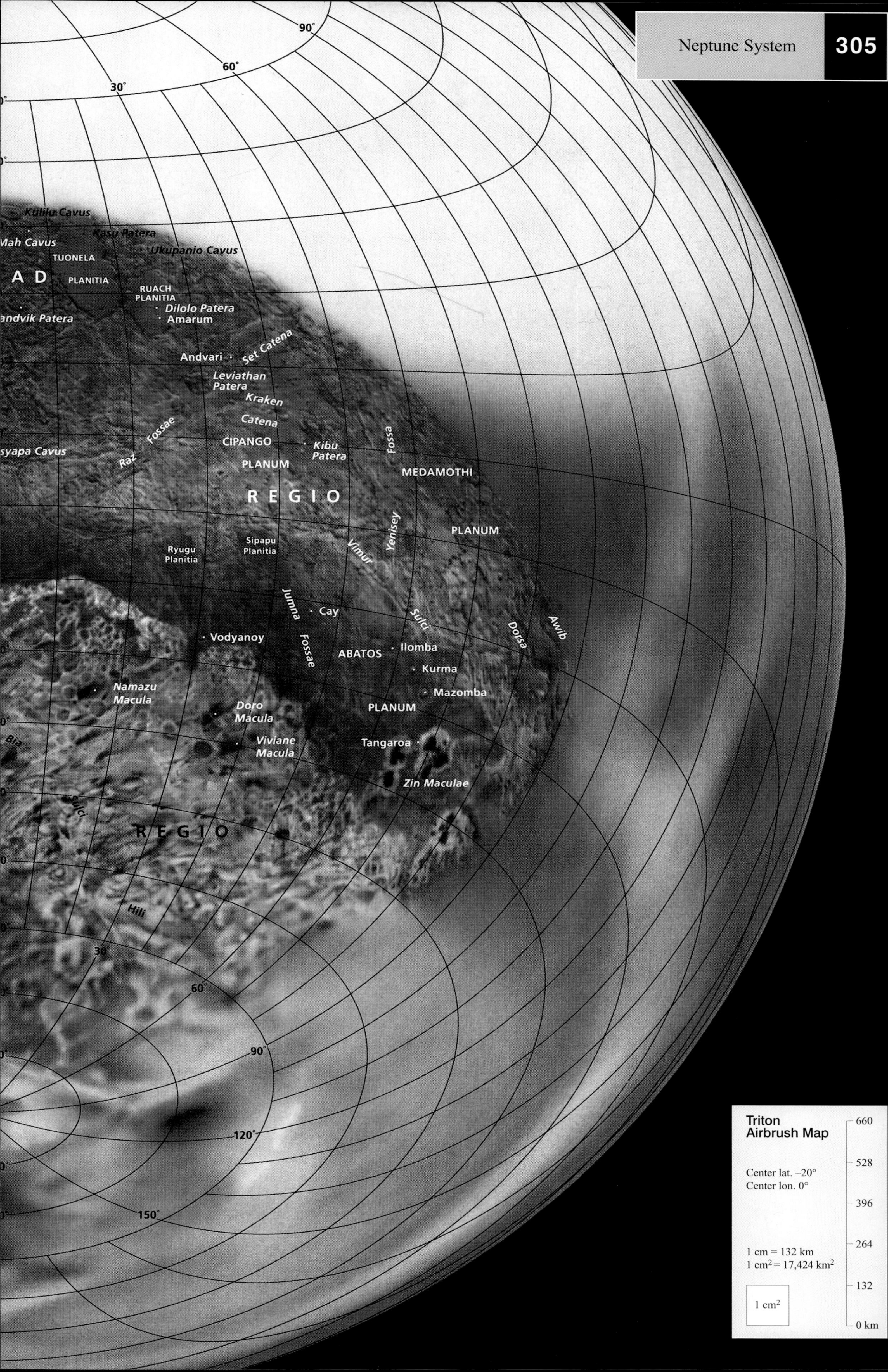

90°
60°
30°
Kulilu Cavus
Kasu Patera
Mah Cavus
Ukupanio Cavus
TUONELA
PLANITIA
A D
RUACH
PLANITIA
Dilolo Patera
Amarum
andvik Patera
Andvari
Set Catena
Leviathan
Patera
Kraken
Catena
Raz Fossae
syapa Cavus
CIPANGO
PLANUM
Kibu
Patera
Fossa
MEDAMOTHI
R E G I O
PLANUM
Yenisey
Vimur
Sipapu
Planitia
Ryugu
Planitia
Jumna
Fossae
Cay
Sulci
Dorsa
Awib
Vodyanoy
ABATOS
Ilomba
Kurma
Mazomba
PLANUM
Namazu
Macula
Doro
Macula
Viviane
Macula
Tangaroa
Zin Maculae
Bia
Sulci
R E G I O
Hili
30°
60°
90°
120°
150°
Triton
Airbrush Map
Center lat. –20°
Center lon. 0°
1 cm = 132 km
1 cm² = 17,424 km²
1 cm²
660
528
396
264
132
0 km

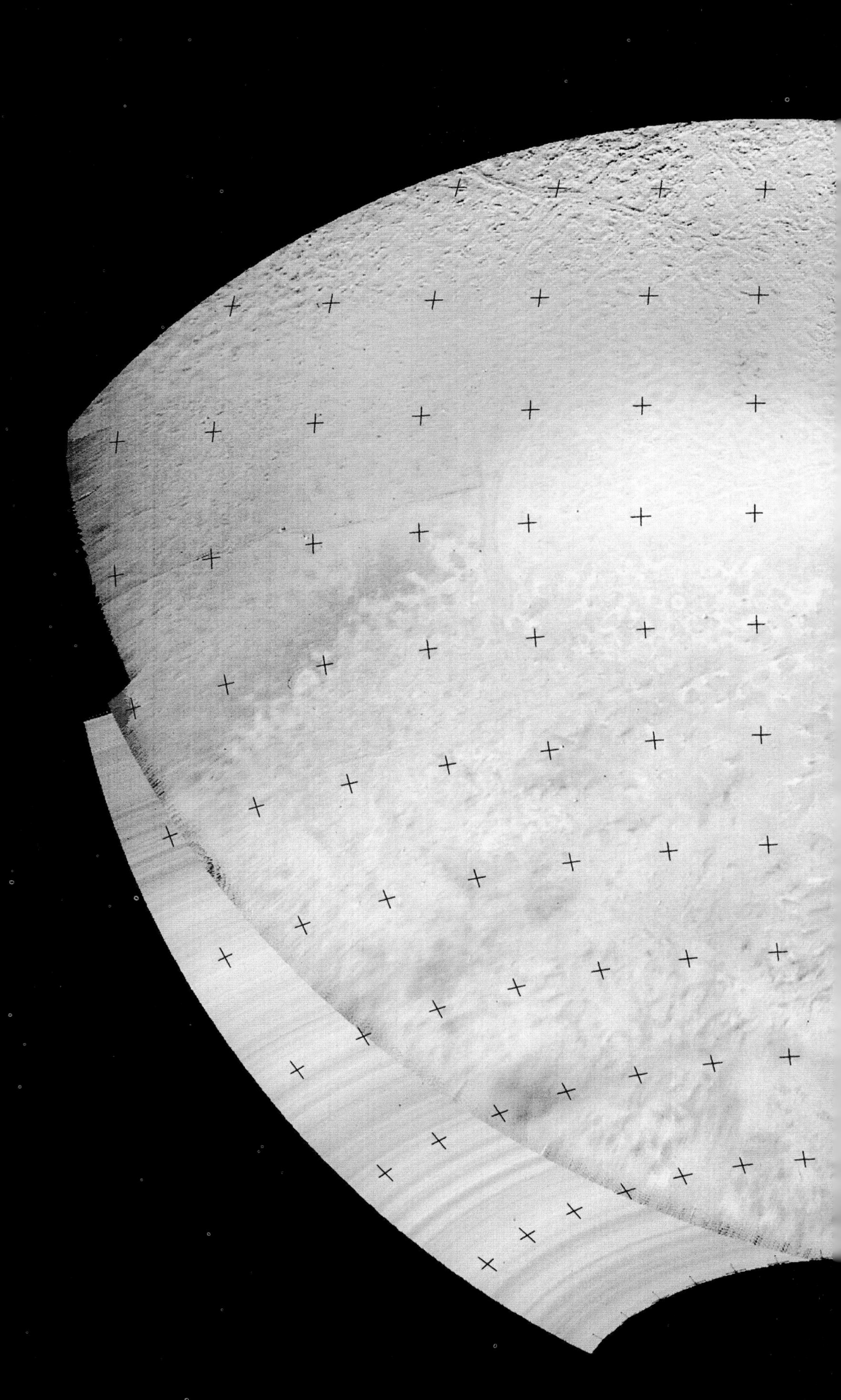

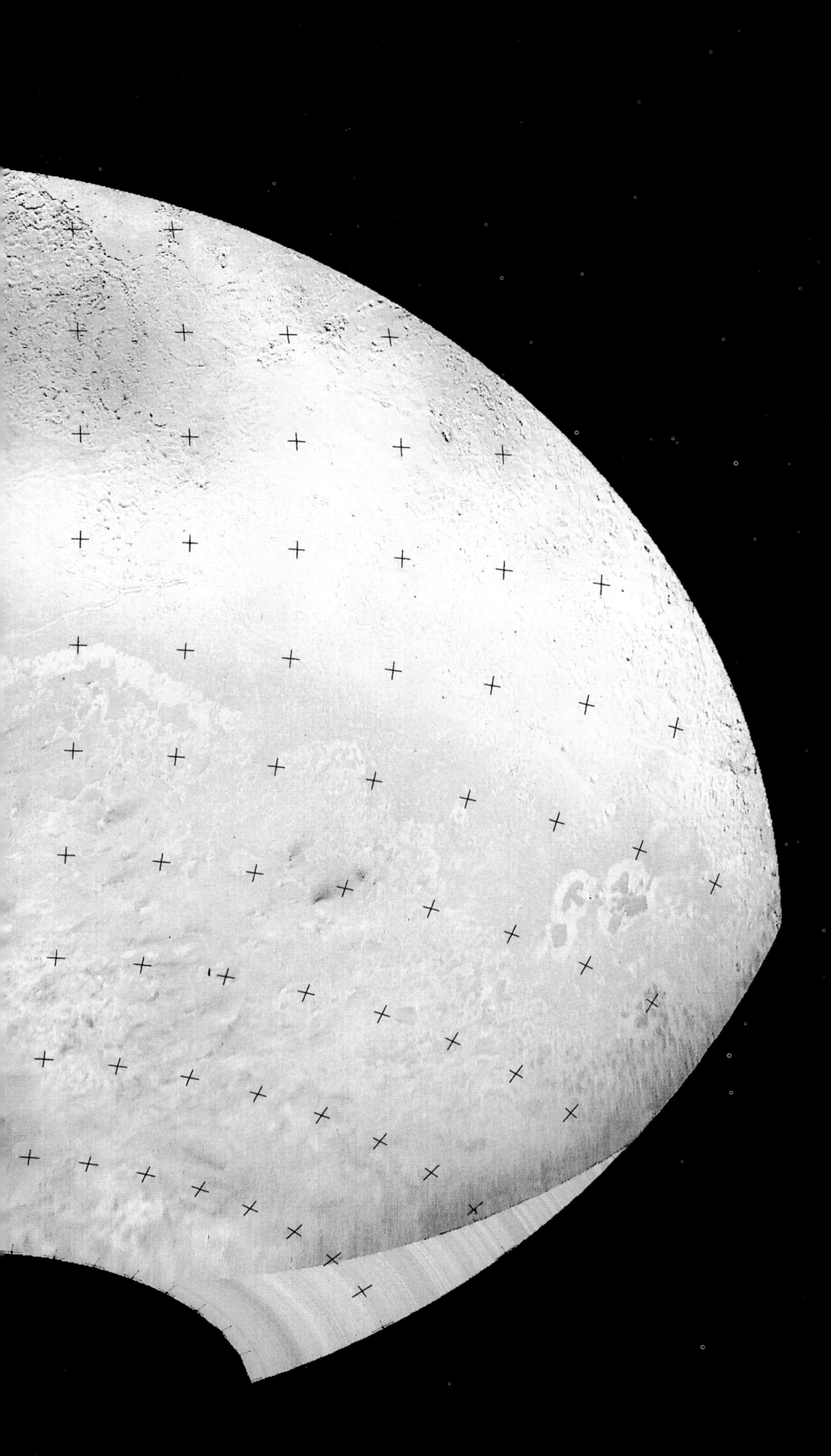

Triton Image Mosaic

Center lat. –10°
Center lon. 0°

1 cm = 132 km
1 cm^2 = 17,424 km^2

1 cm^2

660
528
396
264
132
0 km

Pluto, Asteroids, and Comets

Each of the major planets and planetary systems previously reviewed has been mapped using spacecraft data. This chapter presents data on Pluto – the only planet not photographed at high resolution – asteroids, and comets.

Pluto

To some extent, Pluto followed the same path of discovery as Neptune – its existence was suggested before it was found. Two Bostonians, William Pickering and Percival Lowell analyzed perturbations in the orbits of Uranus and Neptune and predicted that a planet some six times more massive than Earth should exist to account for the perturbations. The search for the proposed planet began in 1905 at the Mt. Wilson Observatory in California and at Lowell's observatory in Flagstaff, Arizona, but nearly a quarter of a century passed before the planet was located.

Although the predictive calculations were wrong – Pluto and its moon, Charon, are far too small to account for the perturbations seen in the orbits of Uranus and Neptune – the predictions prompted a rigorous, systematic search. Following Percival Lowell's death in 1916, his brother contributed a photographic telescope, now called the Pluto Telescope, to Lowell Observatory. In 1929, a young amateur astronomer at Lowell Observatory, Clyde W. Tombaugh, began searching for the planet by examining hundreds of photographic plates and separating candidates from more than 90 million star images. The search was aided by use of a blink comparator, a device that rapidly alternates two different photographs of the same star field. In this device, moving objects will shift position against the fixed background of stars and allow their detection. Although this device facilitated the search, the arduous task involved examination of every square millimeter over a total of more than 75 square meters. This is like looking at each single grain in a sand bed spread over the floor of a small home.

On February 18, 1930, Tombaugh found the planet, but not where it had been predicted to exist. After the discovery of the newly found planet was confirmed, its existence was announced on March 13, 1930 – Percival Lowell's birthday. The planet was

The Pluto system compared to the Earth/Moon system.

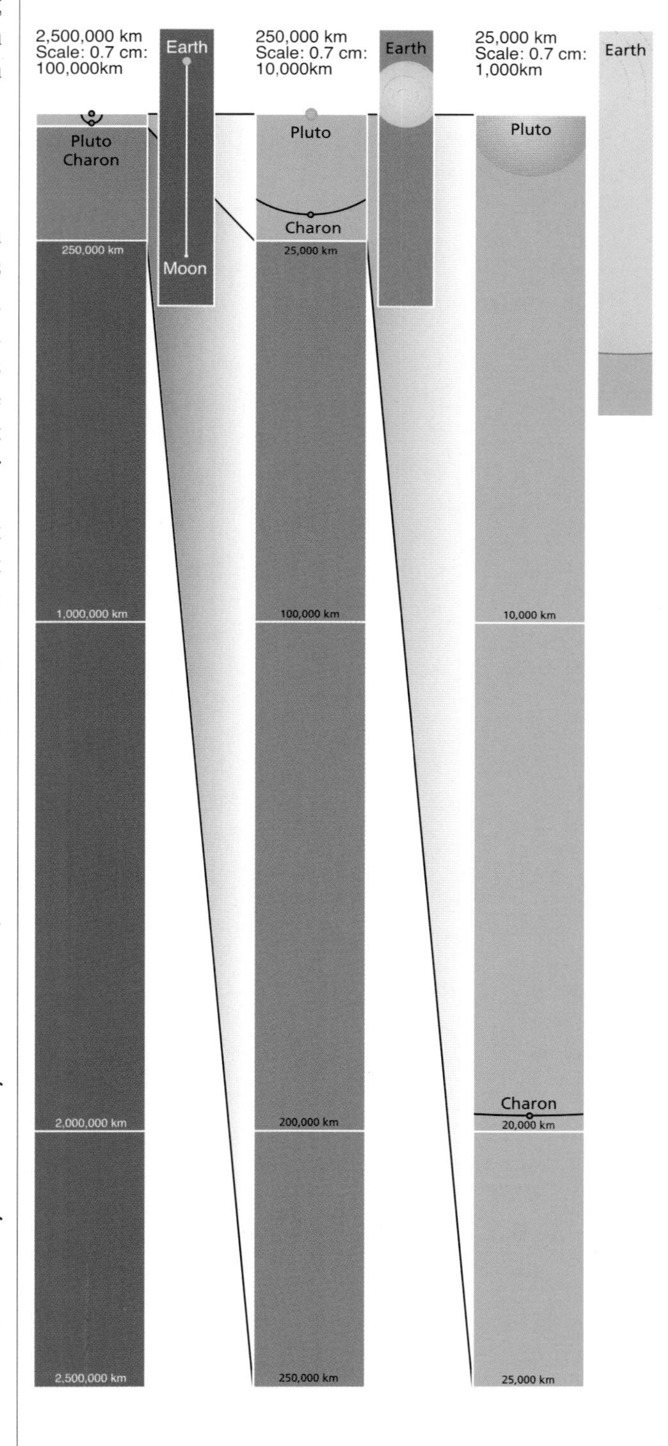

Facing page: a spacecraft may visit Pluto and its moon Charon early in the 21st century (JPL photograph P-35945).

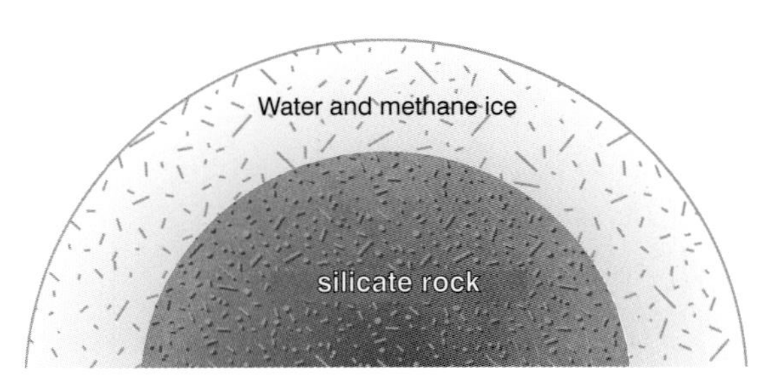

The interior of Pluto is considered to consist of a silicate rock core surrounded by water and methane ice.

later named Pluto for the Greek god of the underworld.

Although Pluto is the outermost of the Sun's known planets, its elliptical orbit occasionally brings Pluto inside the orbit of Neptune. For example, between the years 1989 and 1999, Neptune was farther from the Sun than Pluto and is the most distant planet. More than 248 Earth years are required for Pluto to complete one orbit and, since its discovery in 1930, Pluto has traveled less than one-third of an orbit around the Sun.

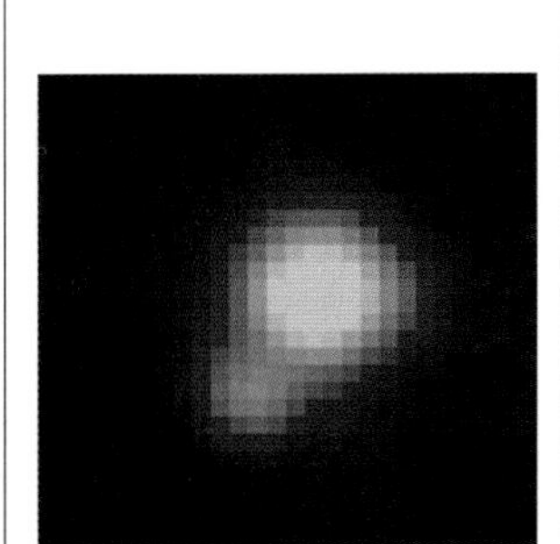

Images of Pluto and Charon, showing (upper) the best ground-based view and (lower) view of the clear separation of the planet and satellite taken by the Faint Object Camera of the Hubble Space Telescope (NASA and the European Space Agency photograph; ground-based photograph from the Canada-France--Hawaii Telescope).

In 1978, James Christy of the U.S. Naval Observatory discovered Pluto's satellite using ground-based observations made at Flagstaff, Arizona. Named Charon after the ferryman who crosses the River Styx in Greek mythology to reach the underworld, Pluto's moon is about 1,190 kilometers in diameter, or about half the size of its parent planet. Thus, even more than Earth and its satellite, the Pluto system is considered by many scientists to be a double planet. Moreover, Pluto and Charon are dynamically locked, always showing the same face to each other.

Data from ground-based observatories and the *Hubble Space Telescope* show that Pluto's surface contains nitrogen ice and some methane, while Charon is mostly water ice. Pluto also has a thin atmosphere of methane and may contain carbon dioxide or molecular nitrogen. Both Pluto and Charon are dense objects compared to outer planet satellites. At more than 2 grams per cubic centimeter, they appear more like Europa and are thought to consist of a rocky core (making up about 70 to 80 percent of the body), surrounded by a water-ice mantle.

The peculiar orbits and high densities of Pluto and Charon have led to speculation about their origin. In many respects, they have affinities with Neptune's moon, Triton. One idea suggests that Pluto and Charon were satellites of Neptune that escaped, perhaps by some disruption of their orbits. Alternatively, Pluto and Charon may represent a large satellite of Neptune that was struck by a collisional impact, broken apart, and knocked out of Neptune's orbit.

Pluto and Charon remain intriguing objects that in many respects are unique in the family of the Solar System objects. Although their great distance from Earth makes spacecraft exploration difficult, plans have already been formulated for such a mission in the future. Early in the twenty-first century expeditions allowing close observations may be undertaken.

Asteroids

From their discovery in 1801 until a few decades ago, asteroids were considered curiosities in the Solar System. Asteroids are currently recognized as important parts of the Solar System, providing insight into the nature of the formation and evolution of the inner planets.

In the mid-1700s, the astronomer Daniel Titius analyzed the systematic spacing of the known planets and noted an apparent "gap" between Mars and Jupiter. This assessment was refined and popularized by Johann Bode, and the sequence of planet spacing is known as the Titius-Bode "law." The apparent order of the spacing is without physical basis; however, in the late 1700s, there was interest in finding the "missing" planet between Mars and Jupiter, and efforts were organized to conduct a systematic search. On New Year's Day, 1801 – and before the formal search began – a Sicilian astronomer, Giuseppe Piazzi, located a bright object not identified on any of his star charts. At first, Piazzi thought he had seen another comet, and he communicated his findings to colleagues in Germany. Initially, some scientists thought that he had found the missing planet. But over the next few years, several more objects were seen between the orbits of Mars and Jupiter, and it was recognized that a new type of planetary object existed in the Solar System. William Herschel, discoverer of Uranus, proposed the term asteroid, meaning "starlike" in Greek, for these objects.

By convention, asteroids and similar small objects are given sequential numbers in order of their discovery and names that are assigned by their discoverer. Thus, the asteroid found by Piazzi is designated 1 Ceres, indicating that it was the first asteroid discovered and named Ceres after the principal Greek goddess of Sicily, Piazzi's home.

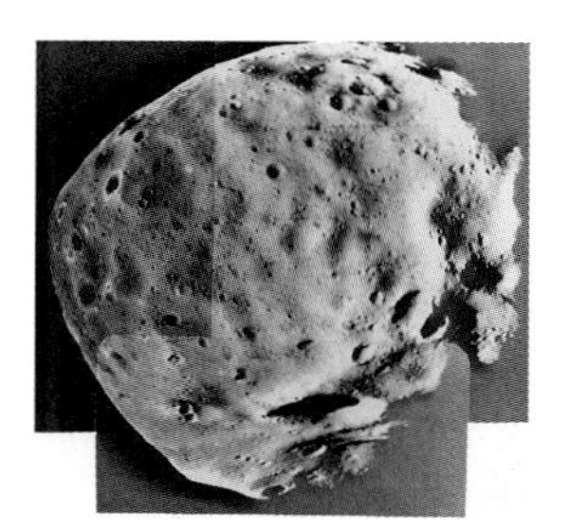

Phobos (left), a moon of Mars, was photographed by the *Viking Orbiter* spacecraft. This object is about 19 by 21 kilometers and may be a captured asteroid (JPL photograph P-18612).

Mathilde (right), a C-type asteroid, is about 50 by 50 by 70 km (NASA image from the *NEAR-Shoemaker* spacecraft); PIA 02477).

In all, more than 18,000 asteroids have been identified, of which the orbits of some 5,000 have been determined. Asteroids are classified by type – related to their inferred composition – and location in the Solar System. The characteristics of asteroids are difficult to determine because they are so small and, until recently, observations from spacecraft had not been made. Important clues to the characteristics of asteroids are derived from several sources. Spectral reflectance measurements made through telescopes show differences among groups of asteroids that can be related to their composition. Because many meteorites seem to be derived from asteroids, telescopic measurements can be compared with compositional signatures of meteorites obtained in the laboratory. On this basis, a simple classification scheme includes three groups: C-type for carbonaceous asteroids, rich in carbon and complex organic compounds, plus some chemically bound water; S-type for siliceous asteroids, composed of silicate materials lacking the carbon-rich compounds; and M-type for metallic asteroids. In addition, a unique asteroid, Vesta, appears to be composed of volcanic basaltic rocks, similar to the eucrite meteorites.

C-type objects are considered to be the most primitive, with unaltered material similar in composition to the inner planets. S-type materials may be slightly "processed," with carbon and other volatile materials released. Both the M-type asteroids and Vesta are considered to be chemically differentiated as the result of extreme heating. The metallic asteroids may have been derived from the core of a small, differentiated planetesimal, while Vesta could represent the crust of such an object. Eucrite meteorites, possibly associated with Vesta, consist of rocks that are essentially the same as basaltic lava flows found on Earth and the Moon, but that show ages of 4.5 billion years. Thus, eucrites were derived from some planetary object that was large enough to have experienced volcanism shortly after the formation of the Solar System 4.6 billion years ago.

Most asteroids are between the orbits of Mars and Jupiter in a main asteroid belt. It is

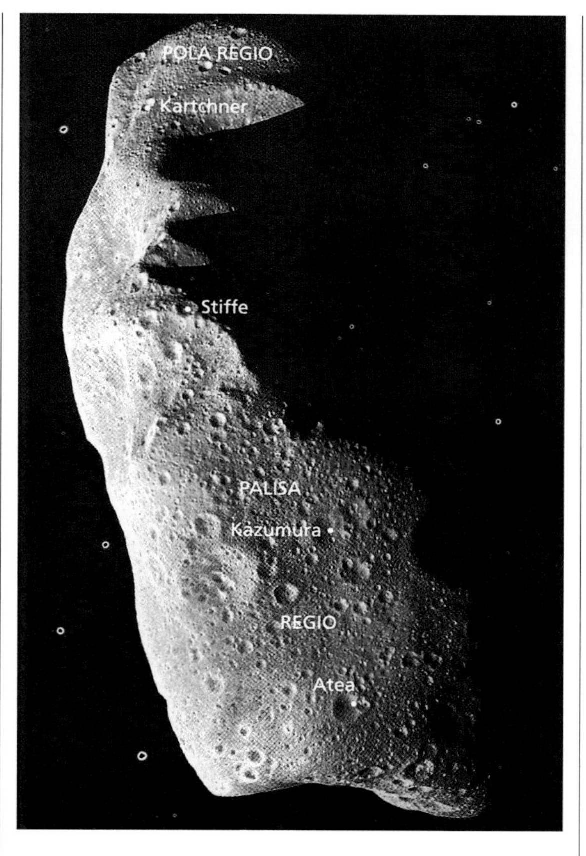

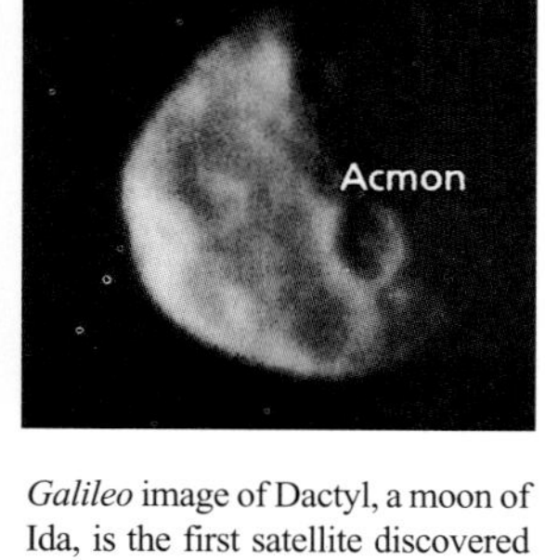

Galileo image of Dactyl, a moon of Ida, is the first satellite discovered orbiting an asteroid. Dactyl is approximately 1.4 kilometers across. (USGS, Flagstaff image 31338).

Mosaic of the asteroid Ida from *Galileo* images. Ida measures some 56 by 24 by 21 kilometers across. (USGS, Flagstaff image 31340).

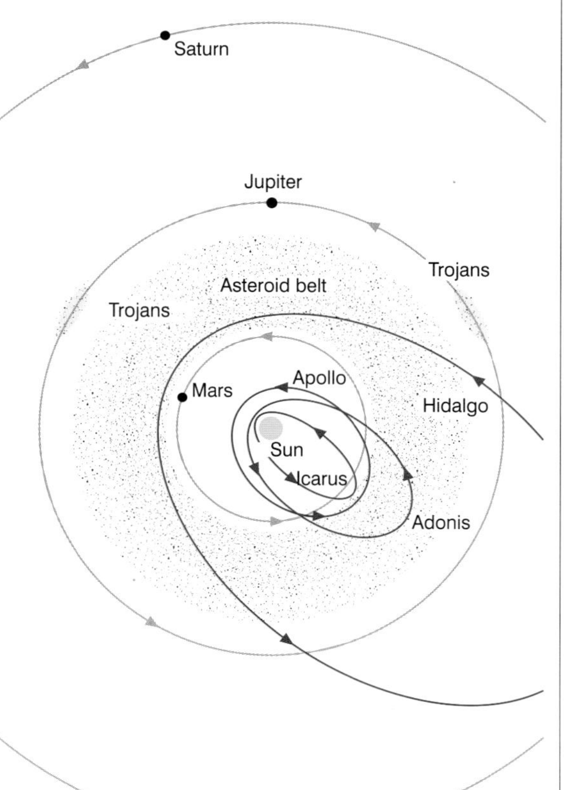

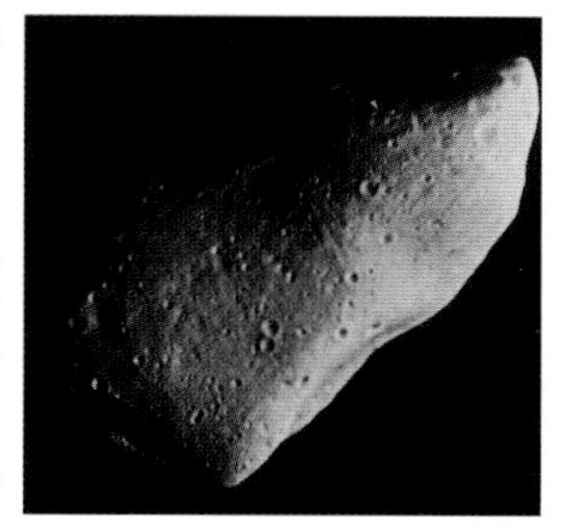

Galileo image of the asteroid, Gaspra. Gaspra measures some 19 by 12 by 11 kilometers across. Color variations are partly attributed to differences in composition and the surface texture related to impact crater deposits (NASA image 92-HC-389).

Asteroids' primary distribution in the main asteroid belt between the orbits of Mars and Jupiter, and the paths of the "Earth-crossing" Apollo asteroids. The Trojans are in the same orbit as Jupiter.

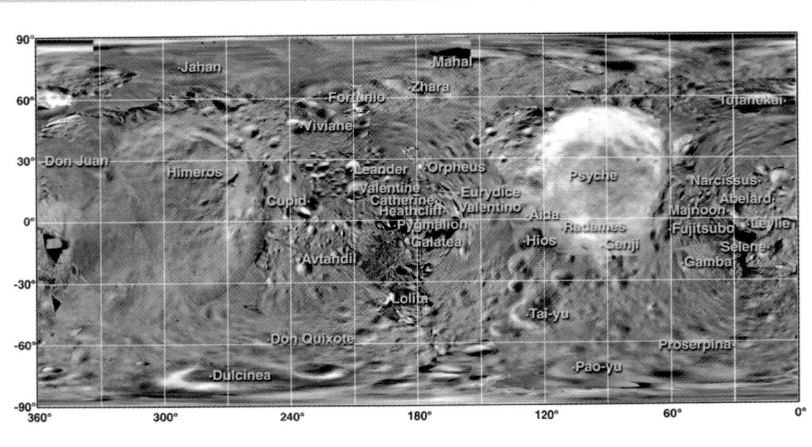

Cylindrical projection of image mosaic for Eros, showing proposed names of surface features.

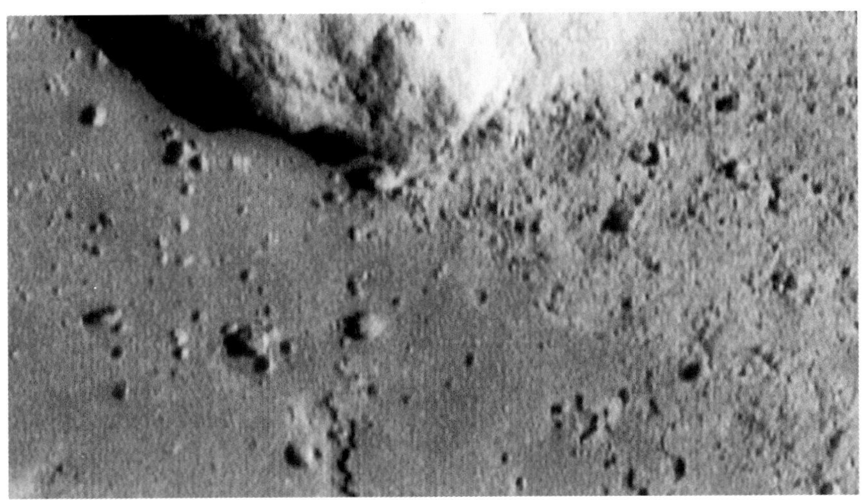

Last picture taken by *NEAR-Shoemaker* before landing on Eros, showing an area about 6 meters across.

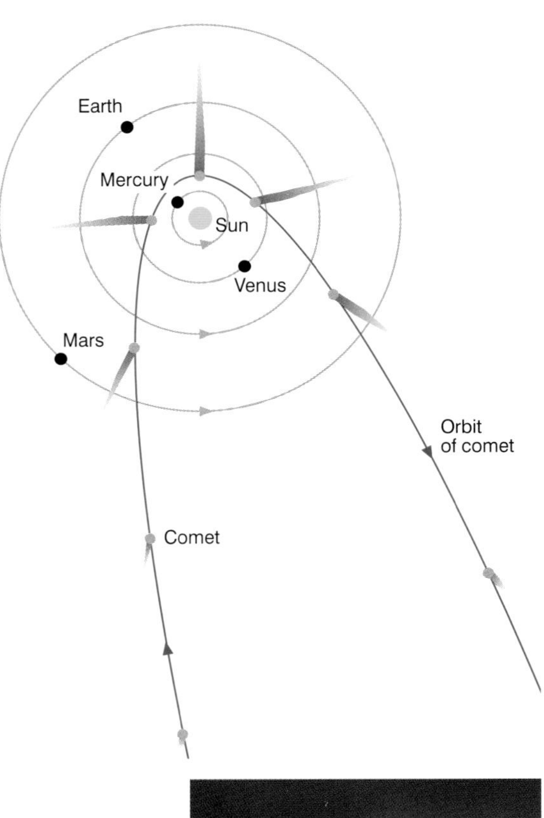

The path of a typical comet as it enters the inner Solar System. Note that the tail of the comet is oriented away from the Sun, independent of the direction of travel by the comet.

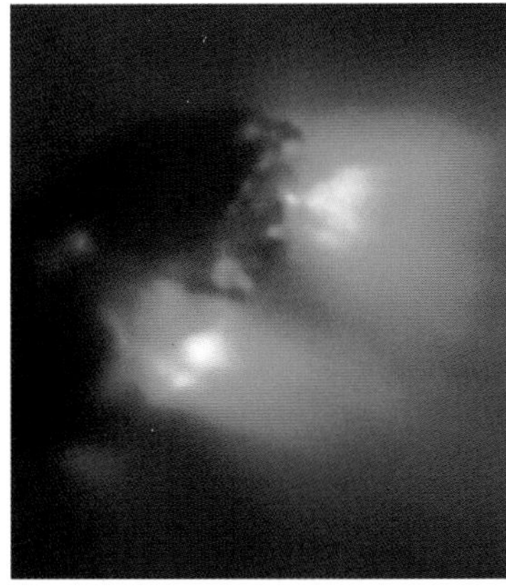

False-color image of Comet Halley. This view, taken in 1986 by the *Giotto* spacecraft, shows the dark nucleus of the comet and jets of gases on the side facing the Sun (right) (Max Planck Institut für Aeronomie and Ball Aerospace Corporation, courtesy of Harold Reitsema).

estimated that there are at least 1,000 objects larger than 30 kilometers, of which some 200 are larger than 100 kilometers. Probably a million asteroids in the main asteroid belt are larger than 1 kilometer across.

In addition to the main belt, asteroids are found in other groupings. Two sets of asteroids are in resonance with Jupiter. Termed Trojan asteroids, one set leads Jupiter in orbit by 60°, while the other set follows Jupiter in orbit by 60°, both in the stable Lagrangian points. Three populations of asteroids reside in the inner Solar System, all referred to as near-Earth asteroids. Two populations, the Atens and the Apollos, cross Earth's orbit and occasionally collide with the Earth and the Moon, while the third population, the Amors, are found between Earth and Mars. The lifetime of Earth-crossing asteroids is only about 10 to 100 million years, which is relatively short on Solar System time scales. Consequently, there is probably some mechanism to resupply this group of asteroids. Although several ideas have been suggested, two current thoughts are that asteroids may be comets that have been exhausted of their gases, or that Earth-crossing asteroids are resupplied from the main belt by some mechanism.

As discussed in the chapter on the Mars system, most scientists consider the moons of Mars – Phobos and Deimos – to be captured asteroids. This idea is supported by considerations of their spectral characteristics, which are similar to C-type asteroids. Consequently, images of Phobos and Deimos, taken by the *Viking Orbiter* spacecraft, may allow assessment of the surface characteristics of at least some asteroids.

The first close-up pictures of asteroids in the asteroid belt were obtained by the *Galileo* spacecraft in 1991 and 1993. The first asteroid observed, Gaspra, is about 12 by 20 by 11 kilometers and is an S-type object. Images reveal a cratered surface on an irregular-shaped body that may have been chipped from a larger object. Color images show distinctive colors that suggest differences in rock types. In particular, some crater rims show differences that may indicate exposed bedrock.

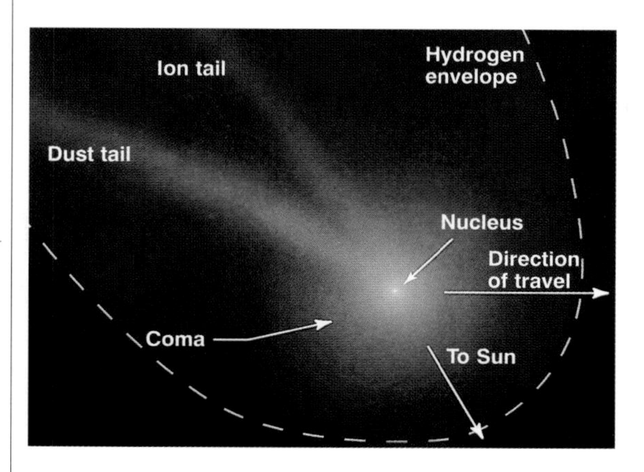

Diagram showing the principal components of a comet.

Ida was the second asteroid to be viewed up-close by a spacecraft. Based on the success of *Galileo* in observing Gaspra, the spacecraft was programmed to fly even closer past Ida, enabling higher resolution pictures to be taken. In 1993, not only was Ida seen in greater detail than Gaspra, but the pictures enabled the discovery of a tiny moon in orbit around the asteroid. Subsequently named Dactyl, the existence of such moons had long been speculated but was not proven until the *Galileo* data were returned to Earth.

Ida orbits in the outer part of the main asteroid belt and, like Gaspra, it is an S-type asteroid. Ida is a croissant-shaped, cratered object approximately 56 kilometers long by 24 kilometers by 21 kilometers. The size frequency distributions of craters suggests that the surface of Ida is about one billion years old, or twice that estimated for Gaspra.

NEAR-Shoemaker encountered asteroid Eros in February 2000 to begin a year-long orbit of this S-type object. The mission culminated with a soft touch-down on the surface in February 2001, marking another first in Solar System exploration. Like Ida, Eros has a density similar to solid rock. However, Eros' surface might be younger than that of Ida, suggested by its relative lack of small craters.

Ida's moon, Dactyl, is only about 1.4 kilometers in diameter. Despite its small size, like nearly all other solid-surface objects seen in the Solar System, its surface is also cratered by impacts.

In 1996 the *Near Earth Asteroid Rendezvous* (*NEAR*) spacecraft began its journey to explore asteroids in detail. Later named the *NEAR-Shoemaker* spacecraft to honor deceased planetary geologist Eugene Shoemaker, this mission provided the first close-up data for a C-type asteroid (Mathilde) in 1997. Mathilde was found to have a very low density, suggesting that it is something like a rubble pile.

Once referred to as the "vermin of the skies," asteroids lately have gained status. They probably have diverse origins, reflected in different compositions and orbits, and play a role in the evolution of planetary and satellite surfaces through collisional impacts. In at least some cases, as with Vesta, asteroids may represent the very early stages of formation of the terrestrial planets.

Comets

Records of comets date back 23 centuries to Chinese chronicles. The striking appearance of some comets and their seemingly irregular appearance in the sky probably struck awe and apprehension in the minds of most people, as references to comets abound in mythology. Even as recently as the appearance of Comet Halley in 1910, there was concern about comet-born diseases. In fact, comet gas masks were sold in the United States and England at the time. Rather than being objects to fear, comets are an important part of the family of Solar System objects, and, like asteroids, comets hold important records of the early Solar System.

The Danish astronomer, Tycho Brahe, is credited with the first systematic studies of a comet. In 1577, a particularly bright object was visible in the sky. At the time, these types of objects were thought to be a phenomenon in Earth's atmosphere. However, from his careful analysis, Brahe declared that the object was not near Earth's atmosphere but, in fact, appeared to orbit the Sun.

In the late 1600s, the British astronomer, Edmund Halley, drew upon the earlier work of Johannes Kepler and Isaac Newton to define the paths of comets. Halley noted that, like other planets, comets do orbit the Sun, but in highly eccentric paths. From his analysis, Halley was able to predict the reappearance of a comet in 1758. This object was subsequently named Comet Halley, one of the best known in the world. In later years, it was realized that comets travel in orbits far beyond Pluto and mark the extreme boundaries of the Solar System.

Drawing on results from decades of telescopic observations and considerations of the origin and evolution of the Solar System, in 1950 the American astronomer Fred Whipple established the foundation for modern concepts of comets. He proposed that comets are mostly water ice with bits of dust (comets are often called "dirty snowballs") and are made of materials remaining after the formation of the Solar System. These

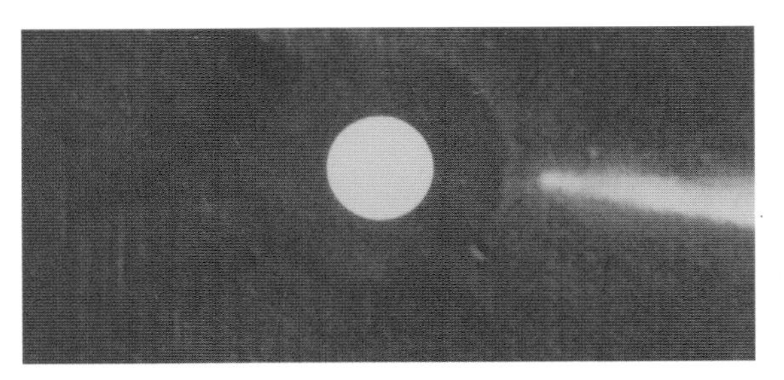

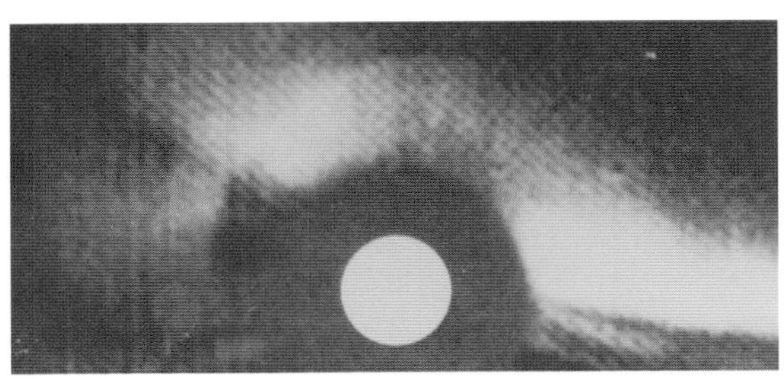

Photographs obtained by a U.S. Air Force satellite in August 1979. The sequence shows Comet Howard-Koomen-Michels (1979 XI) as it approaches the Sun. After being absorbed by the Sun, only the comet tail remained visible (right image) (photograph courtesy of D. J. Michels, U.S. Naval Research Laboratory).

concepts have been supported and essentially confirmed with new observations and information gained from spacecraft.

The main component of a comet is the nucleus, a mass of ice and dust – the dirty snowball – that typically is 1 to 10 kilometers across. As comets travel in their orbit toward the Sun, at about the distance of the orbit of Mars they begin to release gases in response to solar energy. The released gases form a glowing mass, known as the coma, surrounding the nucleus, 100,000 kilometers across. Streaming away from the coma is the cometary tail. It includes two components, the plasma tail composed of carbon dioxide, nitrogen, and water ions and a broad, slightly curved tail composed of dust grains. Comet tails may stream 1 to 10 million kilometers from the coma and are always oriented away from the Sun, regardless of the direction of travel by the comet. The first spacecraft observations of a comet occured in 1985, when the U.S. spacecraft, *International Sun-Earth Explorer*, was placed into a new trajectory to pass through the tail of Comet Giacobini-Zinner. After the maneuver, the spacecraft was renamed the *International Comet Explorer* (ICE) and was the first spacecraft to make measurements of a comet. Although not specifically instrumented for the study of comets, ICE data revealed a complex interaction between the comet tail and the solar wind and revealed that the amount of dust in the tail was far less than expected.

At the return of Comet Halley to the inner Solar System in 1986, an international armada of spacecraft met Comet Halley to make critical measurements. Because of budget constraints, the United States used the *ICE* spacecraft for long-range observations. Both the *ICE* and the Japanese *Sakagaki* spacecraft, the first of two sent by Japan, observed Comet Halley from several million kilometers. A second Japanese spacecraft, named *Suisei*, came within one million kilometers of the coma.

The primary observations of Comet Halley were made through the Soviet and European spacecraft. In late 1984, the Soviets launched two spacecraft, *Vega 1* and *2*, both of which carried an array of instruments and, after passing by Venus, continued on a path to encounter Comet Halley in March 1986. Both spacecraft measured the inner atmosphere and dust clouds of the comet to within 8,000 kilometers of the nucleus but were damaged from the impact of dust grains. Nonetheless, the spacecraft made key observations and paved the way for the European spacecraft, *Giotto*, named for the Italian artist who included a comet in his fresco, *The Adoration of the Magi*. The *Giotto* spacecraft carried instruments that measured the nucleus and coma and obtained high-resolution images.

Observations of Comet Halley through binoculars and small telescopes were disappointing. Expectations of a dazzling object clearly visible to the naked eye were not met. These disappointments, however, were more than offset by the first spacecraft analysis.

Photographs from the *Giotto* spacecraft show the nucleus of Comet Halley to be a peanut-shaped object 8 by 15 kilometers, rather than a spherical body as suggested previously. The nucleus rotates rather slowly, making one revolution in about 52 hours. Most of its surface is charcoal colored with gas jets feeding the coma originating as geyserlike eruptions from specific places on the surface. Only about 15 percent of the surface was active at any one time during the spacecraft observations. Temperature measurements showed the Sun-facing side to be about 350 K. Collectively, these observations suggest that the nucleus has an outer layer of carbonaceous material composed of fragments too large to be carried away by the gas jets. With each passage of Comet Halley bythe Sun, more material is probably concentrated on the surface.

Although water was long suspected to be a principal component of comets, *Giotto* instruments confirmed the presence of water in the gas jets. The gases are composed of about 80 percent water with smaller amounts of carbon dioxide and other gases.

What is the life history of comets? In 1950, the Dutch astronomer Jan Oort deduced that the gravitational influence of individual stars extends to about one-fifth of the distance between them and the next star. In our Solar System, the Sun gravitationally holds objects to a distance of 50,000 astronomical units (AU), or more than 7.5 trillion kilometers. This is about the same perihelion distance calculated

Telescopic photograph of Comet Mrkos taken in August 1957 showing the well-defined ion tail and more diffuse, slightly curved dust tail (Palomar Observatory photograph, California Institute of Technology).

for the orbits of comets. This outermost zone of the Solar System is named the Oort cloud, after Jan Oort, and is considered to be the source for comets. Unlike nearly all other Solar System objects, the comet orbits in the Oort cloud do not define a plane but are randomly positioned to form a spherical cloud around the Sun. The number of comets at this distance probably exceeds one trillion. Closer to the orbit of Pluto, a disklike zone of comets comprises the Kuiper belt, named after the astronomer, Gerard Kuiper, who suspected its existence. This zone of comets was confirmed in the early 1990s.

What brings comets into the inner Solar System where they may become visible? One idea is that perturbations from passing stars may reset the orbits of some comets residing in the Oort cloud. Once gyrating on an elliptical path through the inner Solar System, comets may not survive very long in terms of Solar System history. Many comets are thought to collide with planets, satellites, the Sun, or other objects. For example the collision of Comet Shoemaker-Levy 9 into Jupiter, described earlier, amply demonstrates the destruction of comets when they impact planets.

With each passage of the Sun, a comet loses material into deep space by gas jetting at a rate of a million tons per day, similar to that seen in Comet Halley. At this rate, comets would shrink by several meters with each orbit and would have lifetimes of only thousands of years.

Erosion of comets can be directly observed in shooting or falling stars. Typically, these represent the consumption of peasize remnants of comets passing through our atmosphere. Spectacular "fireballs" result from only slightly larger objects that are about the size of marbles.

Still other comets may become asteroids. Because the spectral properties of C-type asteroids and the nucleus of Comet Halley – as well as some other comets – are similar, some asteroids may represent comets that have lost their volatile components, such as ice, leaving behind rocky, carbonaceous chunks. These may continue to orbit the sun for millions of years.

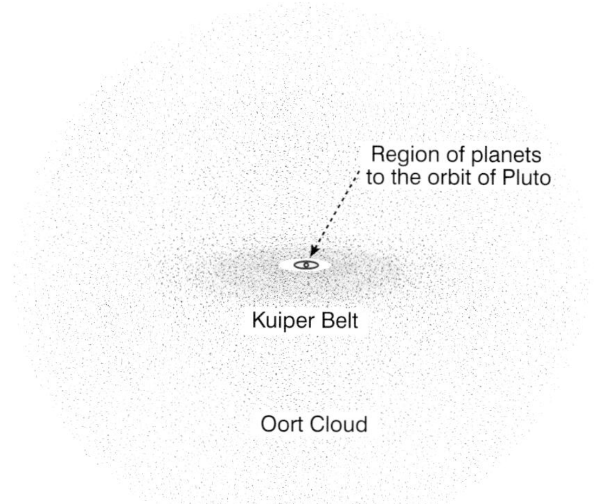

The distribution of comets in the Oort cloud in relation to the rest of the Solar System, out to the orbit of Pluto.

Understanding comets is integral to understanding the Solar System. Comets are representative of the basic building blocks from which the giant, outer planets and their satellites were formed, as well as possible sources for some asteroids. As the most abundant object in the Solar System and a frequent impacting agent through the evolution of the Solar System, comets could have implanted the volatiles on Earth, including the oceans and atmosphere. Because comets include organic compounds, life itself could have been transported by comets not only to Earth but throughout the Solar System.

Glossary

absolute zero
The theoretical temperature at which molecular motion vanishes and a body does not have heat energy, the zero point of the Kelvin (K) temperature scale.

achondrite
A stony meteorite that lacks chondrules; most achondrites appear to be the products of igneous differentiation.

accretion
The agglomeration of matter to form larger bodies such as stars, planets, and moons.

albedo
The ratio of the radiation reflected by a body to the amount incident upon it, often expressed as a percentage; as, the albedo of the Earth is 34 percent.

amors
Asteroids that cross the orbit of Mars and approach the orbit of Earth, 10 percent of which evolve into Earth-crossing orbits over short time scales – hundreds or thousands of years.

anorthosite
A granular plutonic igneous rock composed almost wholly of the mineral plagioclase.

anticline
Folded rocks generally convex upward, in which the core contains geologically older rocks; compare with syncline.

anticyclonic
Having a rotation about the local vertical opposite to the rotation of the planets; for example, anticyclonic motion would be clockwise in the northern hemisphere of Earth, the opposite of cyclonic.

antipode
Anything exactly opposite to something else, that point on a planet 180° from a given place.

aperture
An opening, particularly that opening in the front of a camera through which light rays pass when a picture is taken.

aphelion
That point in a solar orbit that is most distant from the Sun.

apoapsis
The orbital point farthest from the center of gravitational attraction.

apogee
That point in a geocentric orbit that is most distant from Earth, the opposite of perigee.

apollos
Asteroids that cross Earth's orbit. This family is composed of some of the closest planetary objects to Earth. A few known examples come even closer than the Moon.

asteroid
One of the many small bodies revolving around the Sun, with orbits generally between Mars and Jupiter.

asteroid belt
A region of space between Mars and Jupiter where most asteroids are found.

asthenosphere
A weak spherical shell below the lithosphere, in which isostatic adjustments take place, magmas may be generated, and seismic waves are strongly attenuated.

astronomical unit (abbr AU)
A unit of length defined as the distance from the Earth to the Sun, 149,599,000 kilometers.

astronomy
The science that treats the location, magnitude, motion, and constitution of celestial bodies and structures.

astrophysics
A branch of astronomy that treats the physical properties of celestial bodies, such as size, mass, density, temperature, and chemical composition.

atens
Asteroids with orbits that lie mostly inside Earth's orbit; at their farthest point from the Sun, they may cross Earth's orbit.

atmosphere
The body of gases surrounding or comprising any planet or other celestial body.

atmospheric pressure
The pressure at any point in an atmosphere due to the weight of the atmospheric gases above the point concerned.

axis
(1) A straight line about which a body rotates, or along which its center of gravity moves (axis of translation); (2) A straight line around which a plane figure may rotate to produce a solid, a line of symmetry; (3) One of a set of reference lines for a coordinate system.

bar
A unit of atmospheric pressure; 1 bar = 10^6 dyne cm^{-2} = 0.987 atmosphere.

basin (impact)
Term applied to impact structures larger than about 100 kilometers across.

basalt
A general term for dark-colored mafic igneous rocks, commonly volcanic but locally intrusive (such as dikes), composed chiefly of calcic plagioclase and clinopyroxene minerals.

bit
An abbreviation of binary digit.

blackbody
An idealized body that absorbs all radiation of all wavelengths incident on it; its radiation is a function of temperature only.

bolide
An exploding or exploded meteor or meteorite; a detonating fireball.

boundary layer
The layer of fluid in the immediate vicinity of a bounding surface; in fluid mechanics, the layer affected by viscosity of the fluid.

breccia
Rock composed of broken rock fragments surrounded by finer-grained material.

carbonaceous chondrite
A class of chemically primitive meteorites characterized by the presence of hydrated minerals and organic (carbon) compounds.

central peak
A central, high area produced in an impact crater by inward and upward movement of underlying material.

chondrite
A stony meteorite usually characterized by the presence of chondrules; carbonaceous chondrites are characterized by the presence of carbon compounds, while Type 1 or C1 chondrites contain no chondrules.

chondrules
Small spherical grains, usually composed of iron, aluminum, or magnesium silicates.

clast
A rock fragment produced by weathering of a larger rock, which has been incorporated in another rock.

coma
The gaseous envelope that surrounds the nucleus of a comet.

comet
A luminous member of the solar system composed of a head, or coma, and often with a spectacular gaseous trail extending a great distance from the head.

comminution
The reduction of a rock to progressively smaller particles by weathering, impacts, and/or erosion.

condensation
The physical process by which a vapor becomes a liquid or solid, the opposite of evaporation.

convection
Mass motions within a fluid resulting in transport and mixing of the properties of that fluid.

core (of a planet)
The central zone, or nucleus of a planet or satellite, consisting of higher density material.

Coriolis effect
The acceleration that a body in motion experiences when observed in a rotating frame. This apparent force acts at right angles to the direction of the angular velocity.

cosmic dust
Finely divided solid matter with particle sizes smaller than a micrometeorite, thus with diameters much smaller than a millimeter, moving in interplanetary space.

cosmic rays
Extremely high-energy subatomic particles, mostly protons, hydrogen nuclei, and heavier nuclei, which travel in the Solar System in all directions; rays originating outside the Solar System are galactic cosmic rays, those originating in the Sun are called solar cosmic rays.

creep
The slow but continuous deformation of a material under constant load or prolonged stress.

crust
The chemically-distinct outermost solid layer of a planet or satellite.

datum
A reference or a base for measurement of other quantities; for example, sea level is a common reference for elevations on Earth.

day
The duration of one rotation of Earth, or another celestial body, on its axis.

detector
(1) Sensor; (2) An instrument employing a sensor to detect the presence of something in the surrounding environment.

diapirism
The process of piercing or rupturing of domed or uplifted rocks by mobile material; by tectonic stresses, as in anticlinal folds; by the effect of geostatic load in sedimentary strata, as in salt domes and shale diapirs; or by igneous intrusion, forming diapiric structures such as plugs.

differentiation
(1) Processes by which planets and satellites develop concentric layers or zones of different chemical and mineralogical composition; (2) The process of developing more than one rock type from a common reservoir of molten rock.

diffusion
In an atmosphere or in any gaseous system, the exchange of fluid parcels between regions in apparently random motions of a scale too small to be treated by the equations of motion.

dike
The intrusion of magma into a crack that cuts across existing rocks.

diurnal
Having a period of, or related to, a day.

Doppler effect
The change in frequency with which energy reaches a receiver when the receiver and the energy source are in motion relative to each other; also called Doppler shift.

eccentricity
A number that defines the shape of an ellipse, the ratio of the distance from center to focus to the semimajor axis.

eclipse
(1) The reduction in visibility or disappearance of a body by passing into the shadow cast by another body; (2) The apparent reduction of light from a luminous body caused by a dark body coming between it and the observer.

eclogite
A granular rock composed essentially of garnet and pyroxene.

ejecta
The deposit surrounding an impact crater composed of material thrown from the crater during its formation.

ephemeris
A periodical publication tabulating the predicted positions of celestial bodies at regular intervals, such as daily, and containing other data of interest to astronomers.

equator
The primary great circle of a sphere or spheroid, such as the Earth, perpendicular to the polar axis; also a line resembling or approximating such a circle.

escape velocity
The radial speed that an object must attain to escape from the gravitational field of a planetary body or star.

facies
The aspect, appearance, and characteristics of a rock unit, usually reflecting the conditions of its origin.

fault
A fracture or zone of fractures along which the sides are displaced relative to one another.

fault, normal
Fault in which the rocks have been shifted vertically by extensional forces.

fault, reverse
Fault in which the rocks have been shifted vertically by compressional forces.

feldspar
Common aluminous silicate mineral in rocks and some meteorites; includes two principal groups, plagioclase and orthoclase.

fiducial mark
An identification mark on a film or an image; two or more are used to determine the geometric center of the film.

find
A meteorite that cannot be associated with an observed fall; also weathered meteorites recovered months or years after an observed fall.

fireball
A meteor that equals or exceeds the brightness of the brightest planets.

first quarter
The phase of the Moon where the western half is visible to an observer on Earth.

fission
The splitting of an atomic nucleus into two approximately equal fragments.

fissure eruption
An eruption that takes place from an elongate fracture in the lithosphere.

flightpath
The path made or followed by an aircraft, rocket, or spacecraft.

flux
The rate of flow of some quantity, often used in reference to the flow of some form of energy.

flyby
An interplanetary mission in which the vehicle passes close to the target but does not impact it or go into orbit around it.

formation
Geologic term used for a homogeneous rock unit formed typically by a single process, for a specific interval of geologic time, in a specific locality.

forsterite
A whitish or yellowish mineral of the olivine group: Mg_2SiO_4.

full Moon
The Moon at opposition, with a phase angle of 0°, when it appears as a round disk to an observer on Earth.

fumarole
A volcanic vent from which gases are emitted.

fusion
The combining of atoms and consequent release of energy.

G
The constant of proportionality in Newton's law of gravitation; $G=6.672 \times 10^{11}$ newton . m^2Kg^{-2}.

gabbro
A plutonic rock consisting of calcic plagioclase (commonly labradorite) and clinopyroxene, with or without orthopyroxene and olivine; apatite and magnetite or ilmenite are common accessories.

galaxy
A vast assemblage of stars and nebulae, composing an island universe separated from other such assemblages by great distances.

Galilean satellites
The four largest satellites of Jupiter: Io, Europa, Ganymede, and Callisto.

geocentric
Relative to the Earth as a center; measured from the center of the Earth.

geodesy
The science which deals mathematically with the size and shape of planets.

geographic coordinates
Coordinates defining a point on the surface of a planet, usually latitudeand longitude

geophysics
The physics of planetary bodies and their environment (earth, air, and, by extension, space).

giant planets
The planets Jupiter, Saturn, Uranus, and Neptune, also Jovian planets.

graben
An elongate crustal depression bounded by normal faults on its long sides.

gradation
Geologic process involving the weathering, erosion, transportation, and deposition of planetary materials by agents such as wind and water.

gravity
(1) Viewed from a frame of reference fixed in the Earth, force imparted by the Earth to a mass that is at rest relative to the Earth; (2) Acceleration of gravity; (3) By extension, the attraction of any heavenly body for any mass; as martian gravity.

greenhouse effect
The heating effect exerted by the atmosphere upon a planet by virtue of the fact that the atmosphere absorbs and emits infrared radiation.

grid
(1) A series of lines, usually straight and parallel, superimposed on a chart or plotting sheet to serve as a directional reference for navigation; (2) Two sets of mutually perpendicular lines dividing a map or chart into squares or rectangles to permit location of any point by a system of rectangular coordinates.

grooves
Curvilinear depressions or troughs found on some icy satellites.

heliocentric
Relative to the Sun as a center, as a heliocentric orbit.

highlands (lunar)
Rugged, heavily cratered, bright terrain on the Moon.

horst
An elongate block of uplift crust bounded by normal faults on its long sides.

igneous
Materials or processes related to magma.

impact
In planetology, the collision of objects ranging in size from tiny micrometeoroids to planetesimals.

inert gas
Any one of six gases: helium, neon, argon, krypton, xenon, and radon, all of whose shells of planetary electrons contain stable numbers of electrons so that the atoms are almost completely chemically inactive; also rare gas.

inferior conjunction
The conjunction of an inferior planet and the Sun when the planet is between the Earth and the Sun.

inner planets
The four planets nearest the Sun: Mercury, Venus, Earth, and Mars.

ion
A charged atom or molecularly bound group of atoms, sometimes also a free electron or other charged subatomic particle.

ionosphere
An atmospheric shell whose base and height are indefinite, and which is characterized by a high ion density.

isostasy
A supposed equality existing in vertical sections of the Earth, whereby the weight of any column from the surface of the Earth to a constant depth is approximately the same as that of any other column of equal area; the equilibrium being maintained by plastic flow of material from one part of the Earth to another.

isostatic
A substance subject to equal pressure from every side, in hydrostatic equilibrium.

isotope
One of several nuclides having the same number of protons in their nuclei, and hence belonging to the same element, but differing in the number of neutrons and therefore in mass number.

jet stream
A strong band of wind or winds in the upper troposphere or in the stratosphere, moving in a general direction from west to east on Earth and often reaching velocities of hundreds of kilometers an hour.

joint
Fracture or parting in rock, but without vertical or horizontal displacement of the rocks.

Jovian
Referring to Jupiter, Jupiter-like (derived from Jove, another name for the Roman god Jupiter).

Jovian planet
Any one of the giant planets: Jupiter, Saturn, Uranus, or Neptune. Usually in plural, Jovian planets.

KREEP
Lunar rock rich in radioactive elements (K for potassium, REE for rare Earth elements, P for phosphorus).

Lagrangian points
The five equilibrium points in the restricted three-body problem: two of the Lagrange points (L_4 and L_5) are located at the vertices of equilateral triangles formed by the two primaries (Sun and Saturn, or Saturn and satellite) and are stable; the other three are unstable and lie on the line connecting the two primaries.

lava
Magma that reaches the surface of a planet or satellite.

leading hemisphere
The side of a planet or satellite facing the direction of travel in the orbit around its primary object.

liberation
A small oscillation around an equilibrium configuration, such as the angular change in the face that a synchronously rotating satellite presents toward the focus of its orbit.

light year
A unit of length used in expressing distances equal to the distance electromagnetic radiation, light travels in 1 year (1 light year = 9.460×10^{12} kilometers = 63,239 astronomical units).

limb
The edge of the apparent disk of a celestial body, such as the Moon or Mars.

linea
Elongate markings.

lithosphere
The stiff upper layer of a planetary body, including (on Earth) the crust and part of the upper mantle, lying above the weaker asthenosphere; the solid part of a planet.

macula
A dark spot.

magma
Mobile or fluid rock material; generalized to refer to any material that behaves like silicate magma in the Earth.

magnetic north
The direction north at any point as determined by the Earth's magnetic lines of force, the reference direction for measurement of magnetic directions.

magnetosphere
The region of space surrounding a planet in which the planet's magnetic field dominates that of the solar wind.

magnetotail
The far downstream part of a magnetosphere within a flowing medium; the magnetotail contains magnetic fields oriented mainly toward or away from the central body.

major axis
The longest diameter of an ellipse or ellipsoid.

mantle
The interior zone of a planet or satellite below the crust and above the core; behaves plastically.

mare (pl. maria)
An area on the Moon that appears darker and smoother than its surroundings; composed primarily of basaltic lava flows.

mascons
The dozen or so large-scale gravity anomalies on the Moon, primarily gravity enhancements associated with maria.

meridian
A north-south reference line, particularly a great circle through the geographic poles of planetary body.

metamorphism
Solid-state recrystallization, and replacement of less stable with more stable mineral phases by heat and/or pressure.

meteor
A "shooting star" – the streak of light in the sky produced by the transit of a meteoroid through the Earth's atmosphere, also the glowing meteoroid itself; a "fireball" is a very bright meteor.

meteorite
Extraterrestrial material that survives falling to a planetary surface as a recoverable object.

meteoroid
A small particle orbiting the Sun.

Milky Way
The galaxy to which the Sun belongs.

mineral
A naturally occurring substance of a fairly specific chemical composition and crystal structure formed by inorganic processes.

minor axis
The shortest diameter of an ellipse or ellipsoid.

nadir
That point on the celestial sphere vertically below the observer, or 180° from the zenith.

north pole
In astronomy, that end of the axis of rotation of a celestial body at which, when viewed from above, the body appears to rotate in a clockwise direction.

nutation
The motion of the true axis of rotation of a planet or satellite about its mean position.

obliquity
The angle between an object's axis of rotation and the pole of its orbit.

occultation
The extinction of light or radiation from a body or spacecraft as it passes behind another body.

olivine
(1) Common rock-forming silicate mineral in some igneous rocks and meteorites; (2) The most abundant mineral in chondritic meteorites, $(Mg,Fe)_2SiO_4$. The ratio of metal oxides (MgO and FeO) to SiO_2 is 2:1; iron and magnesium ions freely substitute for one another.

Oort cloud
A region extending to more than 50,000 AU from the Sun, barely gravitationally bound; postulated as the "birth-place" of comets.

opposition
The situation of two celestial bodies having either celestial longitudes or sidereal hour angles differing by 180°; usually used in relation to the position of a planet or the Moon from the Sun.

optical depth
Optical thickness.

optical thickness
The mass of a given absorbing or emitting material lying in a vertical column of unit cross-sectional area and extending between two specific levels; also optical depth.

orbit
The path of a body or particle under the influence of a gravitational or other force; the path relative to another body around which it revolves.

orbital period
The interval between successive passages of a satellite through the same point in its orbit; also period.

orbital velocity
The average velocity at which an Earth satellite or other orbiting body travels around its primary.

outer planets
The planets with orbits larger than that of Mars: Jupiter, Saturn, Uranus, Neptune, and Pluto.

palimpsest
A roughly circular albedo spot lacking topographic structure, on icy satellites, that is presumed to mark the site of a former crater and its rim deposit.

parallel of latitude
A circle (or approximation of a circle) on the surface of an object, parallel to the equator, and connecting points of equal latitude; also parallel.

patera
A crater with irregular or scalloped edges; inferred to be volcanic.

path
(1) Satellite: the projection of the orbital plane on the surface of its primary; the locus of the satellite subpoint; (2) Meteor: the projection of the trajectory on the celestial sphere, as seen by the observer.

periapse
The orbital point nearest the focus of attraction.

perigee
The orbital point nearest the Earth when the Earth is the center of attraction.

perihelion
That point in a solar orbit that is nearest the Sun.

period
(1) The time interval needed to complete a cycle; (2) The orbital period; (3) The time interval between passages at a fixed point of a given phase of a simple harmonic wave, the reciprocal of frequency; (4) The time interval during which power level (flux) of a reactor changes by a factor of e (2.718, the base of natural logarithms).

phase angle
The angle between an observer, an object, and the Sun.

photogrammetry
The art or science of obtaining reliable measurements by means of photography.

photometry
The study of the measurement of the intensity of light, luminous intensity (intensity of light in the wavelength to which the eye is sensitive).

pit crater
An impact crater containing a central depression rather than a central peak.

plagioclase
A silicate mineral group, formula $(Na,Ca)Al(Si,Al)Si_2O_8$, a solid solution series from $NaAlSi_3O_8$ (albite) to $CaAl_2Si_2O_8$ (anorthite); common rock-forming minerals.

planet
Any of the nine large bodies in orbit around a star.

planetesimal
A hypothetical early body of intermediate (perhaps 1 to 100 meters) size out of which all Solar System members are presumed to have accumulated; bodies of intermediate (perhaps 1 meter to 100 meters) size, most of which finally accreted to larger bodies.

planetocentric
(1) Of or pertaining to a planet's center of mass; (2) Of or pertaining to the planet as the center of a system.

planum
A high plateau.

plasma
The completely ionized gas, the so-called fourth state of matter; in which the temperature is too high for atoms to exist and which consists of free electrons and free atomic nuclei.

pole
(1) The origin of a system of polar coordinates; (2) For any circle on the surface of a sphere, the point of intersection of the surface of the sphere and the normal line through the center of the circle.

protoplanet
The original material from which a planet condensed.

pyroxene
A group of common, rock-forming silicates that have ratios of metal oxides (MgO, FeO or CaO) to SiO_2 of 1:1. Pure members of this group are $MgSiO_3$ (enstatite), and $FeSiO_3$ (ferrosilite).

radar
(1) A method, system, or technique of using beamed, reflected, and timed radio waves for detecting, locating, or tracking objects (such as rockets)

andfor measuring altitude, in activities such as air traffic control or guidance; (2) The electronic equipment or apparatus used to generate, transmit, receive, and usually to display radio scanning or locating waves; a radar set.

radar reflectivity
The measure of the efficiency of a target in intercepting and returning a radar signal; dependent on the size, shape, aspect, and dielectric properties at the surface of the target.

radar shadow
A condition in which radar signals do not reach a region because of an intervening obstruction.

radiation belt
An envelope of charged particles trapped in the magnetic field of a spatial body.

radionuclide
An atom that emits radiation.

reflectance
The fraction of the total radiant flux incident upon something (such as a planetary atmosphere or surface) that is reflected; varies according to the wavelength distribution of the incident radiation; usually measured for a narrow range of wavelengths through a filter or other dispersive device.

refractory
A material of high vaporization point, or the property of resisting heat.

regio
A large area of distinctive albedo markings.

regolith
A layer of fragmentary debris produced by meteoritic impact on the surface of an object.

resolution
The ability of a film, a lens, a combination of both, or an imaging system to render distinguishable, a standard pattern of black and white lines.

retrograde motion
Motion in an orbit opposite to the usual orbital direction of celestial bodies within a given system; a satellite: motion in a direction opposite the direction of rotation of the primary.

rille
Trench or cracklike valleys, up to several hundred kilometers long and 1 to 2 kilometers wide.

Roche limit
The minimum distance at which a fluid satellite influenced by its own gravitation and that of a central mass can be in mechanical equilibrium; for a satellite of zero tensile strength and the same mean density as its primary, in a circular orbit around its primary, this critical distance is 2.46 times the radius of the primary.

rock
A substance composed of one or more minerals (see mineral).

rotation
(1) Turning of a body about an axis within the body, as the daily rotation of Earth; (2) One turn of a body about an internal axis, as a rotation of Earth.

satellite
An attendant body that revolves about another body, the primary.

scarp
A cliff produced by tectonic, impact, or erosion processes.

secondary crater
Satellite crater formed by ejecta thrown from a "primary" crater.

sensor
The component of an instrument that converts an input signal into a quantity that is measured by another part of the instrument.

shield volcano
A volcanic mountain in the shape of a broad, flattened dome.

sidereal
Of or pertaining to the stars.

sidereal period
The time it takes for a planet or satellite to make one complete circuit of its orbit (360°) relative to the stars.

silicate
A compound whose crystal structure contains silicon (Si) and oxygen (O) in a ratio of 1:4 as a tetrahedron structure, commonly in combination with elements such as iron, aluminum, sodium, and potassium.

silicic
A term applied to igneous rock or magma containing about two-thirds silicon dioxide.

sill
A tabular mass of igneous rock that is parallel to the rocks within which it lies.

small bodies
General term for various objects smaller than a few hundred kilometers across, such as some asteroids and satellites.

soft landing
The act of landing on the surface of a planet without damage to the vehicle or payload except possibly the landing gear.

solar
(1) Of or pertaining to the Sun or caused by the Sun, as solar radiation, solar atmospheric tide; (2) Relative to the Sun as a datum or reference, as solar time.

solar nebula
The gas and dust from which the solar system formed; also, the nebula surrounding the protosun when the planets and smaller bodies were still accreting.

Solar System
The Sun and other celestial bodies within its gravitational influence, including planets, asteroids, satellites, comets, and meteors.

solar wind
A radial outflow of energetic charged particles from the solar corona, carrying mass and angular momentum away from the Sun.

spectral
(1) Of or pertaining to a spectrum; (2) Thermal radiation properties, for ratios such as emittance, reflectance, and transmittance, at a specified wavelength; for powers, such as emissive power, within a narrow wavelength band centered on a specified wavelength.

sputtering
The expulsion of atoms or atomic fragments from a solid, caused by impact of energetic particles.

structure
Geologic term applied to the configuration of rocks; normal faults, and anticlines are types of structures.

sulcus
A complex area of subparallel furrows and ridges.

synchronous rotation
The rotation of a satellite or planet that has equal orbital and rotational periods; 1:1 spin-orbit coupling.

syncline
Folded rocks generally concave upward, whose core contains geologically younger rocks (see anticline).

tectonic
Deformation of planetary materials, as in faulting of Earth's crust.

terminator
The line of sunrise or sunset on a planet or satellite.

terrestrial
Of or pertaining to Earth or Earth-like.

trailing hemisphere
The side of a planet opposite to the direction of travel in the orbit around its primary object.

trajectory
The path traced by any body moving as a result of an externally applied force.

trojan asteroids
Two groups of minor planets that librate in long orbits around the stable Lagrangian points of the Sun and Jupiter.

vapor
Any substance in the gaseous state, a gasified liquid or solid.

weathering
Chemical and physical alteration of materials exposed to the environment on or near the surface of a planetary object.

xenolith
An inclusion or fragment not genetically related to the rock in which it is found.

Summary of Planetary Spacecraft Missions

Spacecraft	Launch	Arrival	Remarks
Mercury			
Mariner 10	Nov. 3, 1973	Mar. 29, 1974 Sept. 21, 1974 Mar. 16, 1975	Trajectory allowed 3 flybys
Venus			
*Sputnik 7**	Feb. 4, 1961		Attempted impact; failed to leave Earth orbit
*Venera 1**	Feb. 12, 1961	May 19, 1961	Closest approach: 99,800 km; lost contact at 7 million km
Mariner 1	July 22, 1962		Attempted flyby; launch failure
*Sputnik 23**	Aug. 25, 1962		Attempted flyby; failed to leave Earth orbit
Mariner 2	Aug. 26, 1962	Dec. 14, 1962	Flyby; closest approach: 34,830 km; first successful planetary flight
*Sputnik 24**	Sept. 1, 1962		Attempted lander; failed to leave Earth orbit
*Sputnik 25**	Sept. 12, 1962		Attempted flyby; failed to leave Earth orbit
*Cosmos 21**	Nov. 11, 1963		Possible attempted flyby (Venera test flight?); failed to leave Earth orbit
*Venera 1964A**	Feb. 19, 1964		Attempted flyby; launch failure
*Venera 1964B**	Mar. 1, 1964		Attempted flyby; launch failure
*Cosmos 27**	Mar. 27, 1964		Attempted flyby; failed to leave Earth orbit
*Zond 1**	April 2, 1964	July 14, 1964	Flyby at 100,000 km; contact lost soon after May 14, 1964
*Venera 2**	Nov. 12, 1965	Feb. 27, 1966	Closest approach: 23,950 km; failed to transmit data to Earth
*Venera 3**	Nov. 16, 1965	Mar. 1, 1966	Closest approach: impact; crushed by atmosphere at 32,000-m altitude; failed to return data
*Cosmos 96**	Nov. 23, 1965		Attempted lander; failed to leave Earth orbit
*Venera 1965A**	Nov. 26, 1965		Attempted flyby; launch failure
*Venera 4**	June 12, 1967	Oct. 18, 1967	Atmosphere probe; closest approach: impact (presumed); transmitted for 94 minutes during entry
Mariner 5	June 14, 1967	Oct. 19, 1967	Flyby; closest approach: 3,990 km
*Cosmos 167**	June 17, 1967		Attempted lander; failed to leave Earth orbit
*Venera 5**	Jan. 5, 1969	May 16, 1969	Atmosphere probe; closest approach: impact (presumed); penetrated deeper than *Venera 4*
*Venera 6**	Jan. 10, 1969	May 17, 1969	Atmosphere probe; closest approach: impact (presumed); penetrated deeper than *Venera 4*
*Venera 7**	Aug. 17, 1970	Dec. 15, 1970	Lander; transmitted data 23 minutes from surface
*Cosmos 359**	Aug. 22, 1970		Possible attempted lander; failed to leave Earth orbit
*Cosmos 482**	Mar. 03, 1972		Possible attempted atmospheric probe; failed to leave Earth orbit
*Venera 8**	Mar. 27, 1972	July 22, 1972	Lander; transmitted data 50 minutes from surface
Mariner 10	Nov. 3, 1973	Feb. 5, 1974	Flyby; closest approach: 5,310 km (enroute to Mercury); first spacecraft photographs of Venus
*Venera 9**	June 8, 1975	Oct. 23, 1975	Orbiter and lander; first photographs of surface
*Venera 10**	June 14, 1975	Oct. 26, 1975	Orbiter and lander; photographs of surface
Pioneer/Venus: orbiter	May 20, 1978	Dec. 4, 1978	Radar images of surface, studies of atmosphere, ionosphere
multiprobe	Aug. 8, 1978	Dec. 9, 1978	Five atmosphere probes
*Venera 11**	Sept. 9, 1978	Dec. 21, 1978	Flyby and lander
*Venera 12**	Sept. 14, 1978	Dec. 25, 1978	Flyby and lander
*Venera 13**	Oct. 30, 1981	Mar. 1, 1982	Lander, surface data
*Venera 14**	Nov. 4, 1981	Mar. 5, 1982	Lander, surface data
*Venera 15**	June 2, 1983	Oct. 10, 1983	Orbiter, radar images of surface
*Venera 16**	June 7, 1983	Oct. 14, 1983	Orbiter, radar images of surface
[1]*Vega 1**	Dec. 15, 1984	June 11, 1985	Lander and balloon atmosphere probe
[1]*Vega 2**	Dec. 21, 1984	June 15, 1985	Lander and balloon atmosphere probe
Magellan	May 4, 1989	Aug. 10, 1990	Radar-imaging orbiter
Galileo	Oct. 18, 1989	Feb. 10, 1990	Flyby within 16,000 km
Cassini	Oct. 15, 1997	April 26, 1998	1st flyby; closest approach: 284 km
		June 24, 1999	2nd flyby; closest approach: 600 km
Moon			
Thor-Able 1	Aug. 17, 1958		Attempt to reach the Moon; failed
*Luna 1958A**	Sept. 23, 1958		Attempted impact; launch failure
Pioneer 1	Oct. 11, 1958		Failed; sent data for 43 hours
*Luna 1958B**	Oct. 12, 1958		Attempted impact; launch failure
Pioneer 2	Nov. 8, 1958		Failed
*Luna 1958C**	Dec. 4, 1958		Attempted impact; launch failure
Pioneer 3	Dec. 6, 1958		Failed, but provided radiation data
*Luna 1**	Jan. 2, 1959	Jan. 4, 1959	Flyby of Moon at 6,000 km
Pioneer 4	Mar. 3, 1959		Flyby of Moon at 60,000 km
*Luna 1959A**	June 18, 1959		Attempted impact; launch failure

* Soviet spacecraft; all others are/were from the United States, except where noted.
[1] The *Vega* main spacecraft both continued on to Comet Halley after dropping probes at Venus.

Spacecraft	*Launch*	*Arrival*	*Remarks*
*Luna 2**	Sept. 12, 1959	Sept. 14, 1959	First probe to Moon
*Luna 3**	Oct. 4, 1959	Oct. 6, 1959	Photographed the farside (photos taken on Oct. 7, 1959)
Atlas-Able 4	Nov. 26, 1959		Failed
*Luna 1960A**	April 15, 1960		Attempted flyby; failed to reach or leave Earth orbit.
*Luna 1960B**	April 16, 1960		Attempted flyby; launch failure
Atlas-Able 5	Sept. 25, 1960		Failed
Atlas-Able 5B	Dec. 15, 1960		Failed
Ranger 3	Jan. 26, 1962	Jan. 28, 1962	Missed Moon by 36,800 km
Ranger 4	April 23, 1962	April 26, 1962	Hit the Moon, but cameras failed
Ranger 5	Oct. 18, 1962	Oct. 21, 1962	Missed Moon by 725 km
*Sputnik 33**	Jan. 4, 1963		Attempted lander; failed to leave Earth orbit
*Luna 1963B**	Feb. 2, 1963		Attempted lander; launch failure
*Luna 4**	April 2, 1963	April 6, 1963	Unsuccessful soft-lander
Ranger 6	Jan. 30, 1964	Feb. 2, 1964	Hit the Moon, but failed
*Luna 1964A**	Mar. 21, 1964		Attempted lander; launch failure
*Luna 1964 B**	April 20, 1964		Attempted lander; launch failure
*Zond 1964A**	June 4, 1964		Attempted lunar flyby to test the *Zond* spacecraft for future Mars missions; launch failure
Ranger 7	July 28, 1964	July 31, 1964	Returned 4,308 photographs; crashed in Mare Nubium.
Ranger 8	Feb. 17, 1965	Feb. 20, 1965	Returned 7,137 photographs; crashed in Mare Tranquillitatis.
*Cosmos 60**	Mar. 12, 1965		Attempted lander; failed to leave Earth orbit
Ranger 9	Mar. 21, 1965	Mar. 24, 1965	Returned 5,814 photographs; crashed in Alphonsus.
*Luna 1965A**	April 10, 1965		Possible attempted lander; launch failure believed
*Luna 5**	May 9, 1965	May 12, 1965	Failed soft-lander
*Luna 6**	June 8, 1965	June 11, 1965	Missed Moon
*Zond 3**	July 18, 1965	July 20, 1965	Flyby of Moon at 9,219 km; returned 25 pictures of far side
*Luna 7**	Oct. 4, 1965	Oct. 8, 1965	Crashed in Oceanus Procellarum; failed
*Luna 8**	Dec. 3, 1965	Dec. 7, 1965	Crashed in Oceanus Procellarum; failed
*Luna 9**	Jan. 31, 1966	Feb. 3, 1966	Successful soft-lander in Oceanus Procellarum; photographs returned
*Cosmos 111**	Mar. 1, 1966		Possible attempted orbiter; failed to leave Earth orbit
*Luna 10**	Mar. 31, 1966	April 3, 1966	Lunar satellite: minimum distance from Moon 350 km; contact maintained from 460 orbits in two months
*Luna 1966A**	April 30, 1966		Possible attempted orbiter; launch failure
Surveyor 1	May 30, 1966	June 2, 1966	Landed near Flamsteed; returned 11,240 photographs
Explorer 33	July 1, 1966		Failed lunar orbiter
Lunar Orbiter 1	Aug. 10, 1966	Aug. 14, 1966	Lunar satellite; returned photographs
*Luna 11**	Aug. 24, 1966	Aug. 28, 1966	Minimum distance 159 km. Transmitted until Oct. 1, 1966
Surveyor 2	Sept. 20, 1966		Unsuccessful lander
*Luna 12**	Oct. 22, 1966	Oct. 25, 1966	Transmitted data until Jan. 19, 1967
Lunar Orbiter 2	Nov. 6, 1966	Nov. 9, 1966	Lunar satellite; 422 photographs
*Luna 13**	Dec. 21, 1966	Dec. 24, 1966	Soft-landed in Oceanus Procellarum; soil studies
Lunar Orbiter 3	Feb. 5, 1967	Feb. 8, 1967	Lunar satellite; returned 307 photographs
Surveyor 3	April 17, 1967	April 20, 1967	Landed in Oceanus Procellarum; returned 6,326 photographs; soil measurement
Lunar Orbiter 4	May 4, 1967	May 8, 1967	Lunar satellite; returned 326 photographs
Surveyor 4	July 14, 1967		Failed
Lunar Orbiter 5	Aug. 1, 1967	Aug. 5, 1967	Lunar satellite
Surveyor 5	Sept. 8, 1967	Sept. 11, 1967	Landed in Mare Tranquillitatis; returned 19,000 photographs
Zond 1967A	Sept. 28, 1967		Attempted lunar test flight; launch failure
Surveyor 6	Nov. 7, 1967	Nov. 10, 1967	Landed in Sinus Medii; returned 30,065 photographs
Zond 1967B	Nov. 22, 1967		Attempted lunar test flight; launch failure
Surveyor 7	Jan. 7, 1968	Jan. 10, 1968	Landed on north rim of Tycho; returned 21,274 photographs; soil analyses
*Luna 1968A**	Feb. 7, 1968		Attempted orbiter; launch failure
*Zond 4**	Mar. 2, 1968		Attempted lunar test flight; launched to 300,000 km from Earth in a direction away from the Moon, failed on Earth reentry
Apollo 6	April 4, 1968		Failed to reach the Moon
*Luna 14**	April 7, 1968	April 10, 1968	Lunar satellite; minimum distance from Moon 160 km
*Zond 1968A**	April 23, 1968		Possible attempted lunar test flight; launch failure
*Zond 5**	Sept. 15, 1968	Sept. 18, 1968	Went around Moon and landed in Indian Ocean, Sept. 21, 1968
*Zond 6**	Nov. 10, 1968	Nov. 14, 1968	Encircled Moon and returned to Earth on Nov. 17, 1968
Apollo 8	Dec. 21, 1968	Dec. 24, 1968	Manned orbiter; 10 orbits completed
*Zond 1969A**	Jan. 20, 1969		Attempted lunar flyby and return; launch failure
*Luna 1969A**	Feb. 19, 1969		Attempted lunar rover; launch failure
*Zond L1S-1**	Feb. 21, 1969		Attempted lunar orbiter; launch failure
*Luna 1969B**	April 15, 1969		Possible attempted lunar sample return; launch failure
Apollo 10	May 18, 1969	May 22, 1969	Manned orbiter; came within 14.9 km of the Moon
*Luna 1969C**	June 14, 1969		Attempted lunar sample return; launch failure
*Zond L1S2**	July 3, 1969		Attempted lunar orbiter; launch failure
*Luna 15**	July 13, 1969	July 21, 1969	52 orbits; crashed in Mare Crisium
Apollo 11	July 16, 1969	July 20, 1969	First humans on Moon, landed in Mare Tranquillitatis
*Zond 7**	Aug. 8, 1969	Aug. 11, 1969	Encircled Moon and returned to Earth
Apollo 12	Nov. 14, 1969	Nov. 19, 1969	Manned lander in Oceanus Procellarum
*Luna 1970A**	Feb. 6, 1970		Possible attempted lunar sample return; launch failure
*Luna 1970B**	Feb. 19, 1970		Possible attempted lunar orbiter; launch failure
Apollo 13	April 11, 1970	April 14, 1970	Unsuccessful manned lander
*Luna 16**	Sept. 12, 1970	Sept. 20, 1970	Landed in Mare Fecunditatis; returned 100 g of soil
*Zond 8**	Oct. 20, 1970	Oct. 24, 1970	Orbited Moon
*Luna 17**	Nov. 10, 1970	Nov. 17, 1970	Carried *Lunokhod 1* to Moon; Mare Imbrium;

Spacecraft	Launch	Arrival	Remarks
			returned over 20,000 photographs
Apollo 14	Jan. 31, 1971	Feb. 5, 1971	Manned lander; Fra Mauro
Apollo 15	July 26, 1971	July 30, 1971	Manned lander; Hadley-Apennines
*Luna 18**	Sept. 2, 1971	Sept. 11, 1971	Crashed in Mare Fecunditatis after 54 orbits
*Luna 19**	Sept. 28, 1971	Oct. 3, 1971	Contact maintained for over a year and 4,000 orbits
*Luna 20**	Feb. 14, 1972	Feb. 21, 1972	Landed in Mare Fecunditatis; returned samples
Apollo 16	April 16, 1972	April 21, 1972	Manned lander in Descartes area
*Soyuz L3**	Nov. 23, 1972		Attempted lunar orbit test and return; launch failure
Apollo 17	Dec. 7, 1972	Dec. 11, 1972	Manned lander in Taurus-Littrow
*Luna 21**	Jan. 8, 1973	Jan. 16, 1973	Carried Lunokhod 2 to Lemonnier area
Explorer 49	June 10, 1973		Radio astronomy from far side
*Luna 22**	May 29, 1974		Transmitted data
*Luna 23**	Oct. 28, 1974	Nov. 6, 1974	Landed in Mare Crisium; sampling unsuccessful
*Luna 1975A**	Oct. 16, 1975		Attempted lunar sample return; launch failure
*Luna 24**	Aug. 9, 1976	Aug. 18, 1976	Landed in Mare Crisium; returned samples
Galileo	Oct. 18, 1989	Dec. 8, 1990	Flyby; images of farside area
Galileo	Oct. 18, 1989	Dec. 8, 1992	Flyby; images of north polar area
Hiten	Jan. 24, 1990	Feb. 15, 1992	Satellite and subsatellite orbiters (Japan)
Clementine	Jan. 25, 1994	Feb. 19, 1994	Orbiter; global imaging and other remote sensing measurements
Cassini	Oct. 15, 1997	Aug. 17, 1999	Flyby
Lunar Prospector	Jan. 6, 1998	Jan. 11, 1998	Low polar orbit
		July 31, 1999	Controlled descent at South Pole
Mars			
*Marsnik 1**	Oct. 10, 1960		Flyby; failed to reach Earth orbit
*Marsnik 2**	Oct. 14, 1960		Flyby; failed to reach Earth orbit
*Sputnik 29**	Oct. 24, 1962		Possible flyby; failed to reach Earth orbit
*Mars 1**	Nov. 1, 1962		Flyby; communications failed
*Sputnik 31**	Nov. 4, 1962		Attempted lander; failed to reach Earth orbit
Mariner 3	Nov. 5, 1964		Flyby; stuck shroud prevented flyby
Mariner 4	Nov. 28, 1964	July 14, 1965	Flyby at 9,920 km
*Zond 2**	Nov. 30, 1964		Flyby; communications failed
*Mars 1969*A*	Mar. 27, 1969		Flyby; failed to reach Earth orbit
Mariner 6	Feb. 24, 1969	July 1, 1969	Flyby at 3,437 km
Mariner 7	Mar. 27, 1969	Aug. 5, 1969	Flyby at 3,551 km
*Mars 1969*B*	April 2, 1969		Flyby; failed to reach Earth orbit
Mariner 8	May 8, 1971		Flyby; failed at launch
*Cosmos 419**	May 10, 1971		Orbiter/lander; failed to leave Earth orbit
*Mars 2**	May 19, 1971	Nov. 27, 1971	Orbiter and lander; lander crashed
*Mars 3**	May 28, 1971	Dec. 2, 1971	Orbiter and lander; lander transmitted 20 sec.
Mariner 9	May 30, 1971	Nov. 13, 1971	Orbiter; ceased functioning on Oct. 27, 1972
*Mars 4**	July 21, 1973	Feb. 10, 1974	Orbiter; failed to enter Mars orbit

Spacecraft	Launch	Arrival	Remarks
*Mars 5**	July 25, 1973	Feb. 12, 1974	Orbiter and lander; lander failed within seconds of touchdown
*Mars 6**	Aug. 5, 1973	Mar. 12, 1974	Orbiter and lander; lander descent data only; lander crashed
*Mars 7**	Aug. 9, 1973	Mar. 9, 1974	Orbiter and lander; lander missed planet
Viking 1	Aug. 20, 1975	June 19, 1976	Orbiter; ceased functioning on Aug. 17, 1980
		July 20, 1976	Lander; ceased operating Nov. 1982
Viking 2	Sept. 9, 1975	Aug. 7, 1976	Orbiter; ceased functioning on July 24, 1978
		Sept. 3, 1976	Lander; ceased functioning on April 12, 1980
*Phobos 1**	July 7, 1988		Lander and orbiter; contact lost Aug. 29, 1988
*Phobos 2**	July 12, 1988	Jan. 1989	Lander and orbiter; contact lost March 27, 1989, after initial return of data
Mars Observer	Sept. 25, 1992		Attempted orbiter lost Aug. 21, 1993, three days before scheduled orbit insertion
Mars 96 Orbiter	Nov. 16, 1996		Failed to achieve cruise trajectory, fell to Earth on Nov. 17, 1996 (Russia)
Mars Pathfinder	Dec. 4, 1996	July 4, 1997	Lander and rover; operated until Sept. 27, 1997 returning images, soil analyses, and surface weather conditions
Mars Global Surveyor	Nov. 7, 1996	Sept. 12, 1997	Orbiter; high and low resolution images, laser altimeter, and other remote sensing measurements
Nozomi	July 3, 1998	Dec. 2003	Orbiter, aeronomy mission (Japan)
Mars Climate Orbiter	Dec. 12, 1998	Sept. 23, 1999	Attempted orbiter lost on Sept. 23, 1999; presumed destroyed by atmospheric friction due to navigational error
Deep Space 2	Jan. 3, 1999	Dec. 3, 1999	Microprobes were to penetrate martian surface and send back data on sub-surface properties. Conatct lost Dec. 3, 1999. Failure unknown
Mars Polar Lander	Jan. 3, 1999	Dec. 3, 1999	Lander; targeted for southern polar terrain to record meteorological conditions, analyze soil samples and image surroundings. Contact lost Dec. 3, 1999. Failure unknown
2001 Mars Odyssey	April 2001	Oct. 2001	Orbiter; planned mineralogical analysis and measuring of radiation. To act as communications relay for future Mars missions
Jupiter			
Pioneer 10	Mar. 3, 1972	Dec. 3, 1973	Flyby
Pioneer 11	Apr. 5, 1973	Dec. 2, 1974	Flyby
Voyager 1	Sept. 5, 1977	Mar. 5, 1979	Flyby
Voyager 2	Aug. 20, 1977	July 9, 1979	Flyby
Galileo	Oct. 18, 1989	Dec. 7, 1995	Orbiter and probe
Cassini	Oct. 15, 1997	Dec. 30, 2000	Flyby; 9.7 million km
Saturn			
Pioneer 11	Apr. 4, 1973	Sept. 1, 1979	Flyby
Voyager 1	Sept. 5, 1977	Nov. 12, 1980	Flyby

Spacecraft	Launch	Arrival	Remarks
Voyager 2	Aug. 20, 1977	Aug. 25, 1981	Flyby
Cassini/ Huygens	Oct. 15, 1997	July 1, 2004	Saturn orbiter and Titan probe
Uranus			
Voyager 2	Aug. 20, 1977	Jan. 24, 1986	Flyby
Neptune			
Voyager 2	Aug. 20, 1977	Aug. 24, 1989	Flyby
Comets and Asteroids			
International Comet Explorer	Aug. 12, 1978	Sept. 11, 1985	Comet Giacobini-Zinner flyby
*Vega 1**	Dec. 15, 1984	Mar. 6, 1986	Comet Halley flyby
*Vega 2**	Dec. 21, 1984	Mar. 9, 1986	Comet Halley flyby
Giotto	July 2, 1985	Mar. 14, 1986	Comet Halley flyby (European Space Agency)
Sakigake	Jan. 7, 1985	Mar. 11, 1986	Comet Halley flyby (Japan)
Suisei	Aug. 18, 1985	Mar. 8, 1986	Comet Halley flyby (Japan)
Galileo	Oct. 18, 1989	Oct. 29, 1991	Asteroid Gaspra flyby (United States)
Giotto	July 2, 1985	July 10, 1992	Comet Grigg-Skjellerup flyby (European Space Agency)
Galileo	Oct. 18, 1989	Aug. 28, 1993	Asteroid Ida flyby
Deep Space 1	Oct. 24, 1998	July 29, 1999	Asteroid Braille flyby; closest approach 26 km
NEAR/ Shoemaker	Feb. 17, 1996	June 27, 1997	Asteroid Mathilde flyby
		Dec. 23, 1998	Asteroid Eros flyby
		Feb. 14, 2000	Insertion into Eros orbit
		Feb. 12, 2001	Controlled descent to surface of Eros
Stardust	Feb. 7, 1999	Jan. 2004	Comet P/Wild 2; to collect samples from coma for return to Earth

Data for Planets and Satellites

Basic Data for Planets

Name	Mean Distance from Sun (10^6 km)	(AU)	Revolution Period (year)	Diameter (km)	Rotation* (days)	Mass (kg)	Density (g/cm^3)	Escape Velocity (km/s)	Surface	Atmosphere
Mercury	57.91	0.39	0.24	2,439.7	58.65	3.30×10^{23}	5.43	4	silicates	trace Na
Venus	108.21	0.72	0.62	6,051.8	243.02 (R)	4.87×10^{24}	5.24	10	basalt, granite?	90 bar: 97% CO_2
Earth	149.60	1.00	1.00	6,378.14	1.00	5.97×10^{24}	5.52	11	basalt, granite, water	1 bar: 78% N_2, 21% O_2
Mars	227.94	1.52	1.88	3,397	1.02	6.42×10^{23}	3.94	5	basalt, clays	0.07 bar: 95% CO_2
Jupiter	778.30	5.20	11.86	71,492	0.41	1.90×10^{27}	1.33	60	none	H_2, He, CH_4, NH_3, etc.
Saturn	1,429.39	9.55	29.42	60,268	0.44	5.68×10^{26}	0.70	36	none	H_2, He, CH_4, NH_3, etc.
Uranus	2,875.04	19.22	83.75	25,559	0.72 (R)	8.66×10^{25}	1.30	21	none	H_2, He, CH_4, NH_3, etc.
Neptune	4,504.45	30.11	163.72	24,764	0.67	1.03×10^{26}	1.76	23	none	H_2, He, CH_4, NH_3, etc.
Pluto	5,915.80	39.54	248.02	1,195	6.39 (R)	1.50×10^{22}	1.1	1	CH_4 ice	trace CH_4

*(R) = retrograde

Data for Satellites

Planet	Satellite Name	Discovery[4]	Semi-Major Axis[2] (km x 1000)	Period[2] (days)	Radius[2] (km)	Mass[3] (10^{20} kg)	Density[3] (g/cm^3)	Surface Material
Earth	Moon	—	384.40	27.32	1737.4	734.9	3.34	silicates
Mars	Phobos	Hall (1877)	9.38	0.32	13.4 × 11.2 × 9.2	1.08×10^{-4}	1.9	carbonaceous
	Deimos	Hall (1877)	23.46	1.26	7.5 × 6.1 × 5.2	1.80×10^{-5}	1.76	carbonaceous
Jupiter	Io	Galileo (1610)	422	1.77	1830 × 1819 × 1815	893.3	3.53	silicates, sulfur, SO_2
	Europa	Galileo (1610)	671	3.55	1565	479.7	2.99	ice
	Ganymede	Galileo (1610)	1070	7.15	2634	1482	1.94	dirty ice
	Callisto	Galileo (1610)	1883	16.69	2403	1076	1.851	dirty ice
	Amalthea	Barnard (1892)	181	0.50	131 × 73 × 67	—	—	rock, sulfur
	Himalia	Perrine (1904)	11480	250.57	85	—	—	carbonaceous
	Elara	Perrine (1905)	11737	259.65	40	—	—	carbonaceous
	Pasiphae	Melotte (1908)	23500	735 (R)	18	—	—	?
	Sinope	Nicholson (1914)	23700	758 (R)	14	—	—	?
	Lysithea	Nicholson (1938)	11720	259.22	12	—	—	carbonaceous
	Carme	Nicholson (1938)	22600	692 (R)	15	—	—	?
	Ananke	Nicholson (1951)	21200	631 (R)	10	—	—	?
	Leda	Kowal (1974)	11094	238.72	5	—	—	?
	Thebe	Synnott (1980)	222	0.67	55 × 45	—	—	rock?
	Adrastea	Jewitt, Danielson (1979)	129	0.30	13 × 10 × 8	—	—	rock?

Sources

1. *Explanatory Supplement to the Astronomical Almanac 1992; Nautical Almanac Office, U.S. Government Printing Office, Washington D.C.*
2. *The Astronomical Almanac 2001; Nautical Almanac Office, U.S. Government Printing Office, Washington D.C.*
3. *Global Earth Physics, A Handbook of Physical Constants 1995; American Geophysical Union*
4. *International Astronomical Union Circular 2000; Smithsonian Astrophysical Observatory, Cambridge, MA*
5. All data comes from the *International Astronomical Union Circular; Smithsonian Astrophysical Observatory, Cambridge, MA*

Planet	Satellite Name	Discovery[4]	Semi-Major Axis[2] (km × 1000)	Period[2] (days	Radius[2] (km)	Mass[3] (10^{20} kg)	Density[3] (g/cm^3)	Surface Material
	Metis	Synnott (1980)	128	0.29	20	—	—	rock?
	S/1999 J1[5]	Scotti et al. (1999)	24103	758.76				
	S/1975 J1#[5]	Shepard et al. (2000)	7507	130.02				
	S/2000 J2[5]	Shepard et al. (2000)	23746	750.81				
	S/2000 J3[5]	Shepard et al. (2000)	20210	585.17				
	S/2000 J4[5]	Shepard et al. (2000)	22972	713.52				
	S/2000 J5[5]	Shepard et al. (2000)	21336	633.68				
	S/2000 J6[5]	Shepard et al. (2000)	23074	718.71				
	S/2000 J7[5]	Shepard et al. (2000)	21010	619.19				
	S/2000 J8[5]	Shepard et al. (2000)	23618	743.13				
	S/2000 J9[5]	Shepard et al. (2000)	22304	682.68				
	S/2000 J10[5]	Shepard et al. (2000)	20174	587.62				
	S/2000 J11[5]	Shepard et al. (2000)	12557	286.95				
Saturn	Mimas	Herschel (1789)	185.52	0.94	209 × 196 × 191	0.375	1.14	ice
	Enceladus	Herschel (1789)	238.02	1.37	256 × 247 × 245	0.73	1.12	pure ice
	Tethys	Cassini (1684)	294.66	1.89	536 × 528 × 526	6.22	1.00	ice
	Dione	Cassini (1684)	377.40	2.74	560	10.52	1.44	ice
	Rhea	Cassini (1672)	527.04	4.52	764	23.1	1.24	ice
	Titan	Huygens (1655)	1221.83	15.94	2575	1345.5	1.881	cloudy atmosphere
	Hyperion	Bond, Lassell (1848)	1481.1	21.28	180 × 140 × 113	—	—	dirty ice
	Iapetus	Cassini (1671)	3561.3	79.33	718	15.9	1.02	ice/carbonaceous
	Phoebe	Pickering (1898)	12952	550.48 (R)	110	—	—	carbonaceous?
	Janus	Dollfus (1966, 1980)	151.47	0.69	97 × 95 × 77	0.0198	0.65	ice?
	Epimetheus	Fountain et al. (1980)	151.42	0.69	69 × 55 × 55	0.0055	0.63	ice?
	Helene	Laques, Lecacheux (1980)	377.40	2.74	18 × 16 × 15	—	—	ice?
	Telesto	Smith et al. (1980)	294.66	1.89	15 × 12.5 × 7.5	—	—	ice?
	Calypso	Pascu et al. (1980)	294.66	1.89	15 × 8 × 8	—	—	ice?
	Atlas	Terrile (1980)	137.67	0.60	18.5 × 17.2 × 13.5	—	—	ice?
	Prometheus	Collins (1980)	139.35	0.61	74 × 50 34	0.001	0.27	ice?
	Pandora	Collins (1980)	141.70	0.63	55 × 44 × 31	0.001	0.42	ice?
	Pan	Showalter (1990)	133.58	0.58	10	—	—	?
	S/2000 S1[5]	Gladman (2000)	23076	1310.60				
	S/2000 S2[5]	Gladman (2000)	15172	685.89				
	S/2000 S3[5]	Gladman, Kavelaars (2000)	17251	826.02				
	S/2000 S4[5]	Gladman, Kavelaars (2000)	18231	924.58				
	S/2000 S5[5]	Gladman (2000)	11339	447.77				
	S/2000 S6[5]	Gladman, Kavelaars (2000)	11465	453.05				
	S/2000 S7[5]	Gladman, Kavelaars (2000)	20144	1068.06				
	S/2000 S8[5]	Gladman, Kavelaars (2000)	15676	730.84				

Planet	Satellite Name	Discovery[4]	Semi-Major Axis[2] (km × 1000)	Period[2] (days	Radius[2] (km)	Mass[3] (10^{20} kg)	Density[3] (g/cm^3)	Surface Material
	S/2000 S9[5]	Gladman, Kavelaars (2000)	18486	939.90				
	S/2000 S10[5]	Gladman, Kavelaars (2000)	17452	860.03				
	S/2000 S11[5]	Holman (2000)	17874	888.54				
	S/2000 S12[5]	Gladman (2000)	19747	1038.11				
Uranus	Ariel	Lassell (1851)	191.02	2.52	581 × 578 × 578	13.53	1.67	dirty ice
	Umbriel	Lassell (1851)	266.30	4.14	585	11.72	1.4	dirty ice
	Titania	Herschel (1787)	435.91	8.70	789	35.27	1.71	dirty ice
	Oberon	Herschel (1787)	583.52	13.46	761	30.14	1.63	dirty ice
	Miranda	Kuiper (1948)	129.39	1.41	240 × 234 × 233	0.659	1.2	dirty ice
	Cordelia	Terrile (1986)	49.77	0.34	13	—	—	?
	Ophelia	Terrile (1986)	53.79	0.38	15	—	—	?
	Bianca	*Voyager 2* (1986)	59.17	0.43	21	—	—	?
	Cressida	Synnott (1986)	61.78	0.46	31	—	—	?
	Desdemona	Synnott (1986)	62.68	0.47	27	—	—	?
	Juliet	Synnott (1986)	64.35	0.49	42	—	—	?
	Portia	Synnott (1986)	66.09	0.51	54	—	—	?
	Rosalind	Synnott (1986)	69.94	0.56	27	—	—	?
	Belinda	Synnott (1986)	75.26	0.62	33	—	—	?
	Puck	Synnott (1985)	86.01	0.76	77	—	—	carbonaceous?
	Caliban	Gladman et al. (1997)	7169	579.47 (R)	30			
	Sycorax	Gladman et al. (1997)	12214	1283.27 (R)	60			
	Prospero[5]	Holman et al. (1999)	16665	2037.14				
	Setebos[5]	Kavelaars et al. (1999)	17879	2273.34				
	Stephano[5]	Gladman et al. (1999)	7979	673.56				
	S/1986 U10[5]	Karkoschka (1999)	76.4	0.64	40			
Neptune	Triton	Lassell (1846)	354.76	5.88 (R)	1353	214.7	2.054	methane ice
	Nereid	Kuiper (1949)	5513.4	360.14	170	—	—	dirty ice
	Naiad	Terrile (1989)	48.23	0.29	29	—	—	?
	Thalassa	Terrile (1989)	50.07	0.31	40	—	—	?
	Despina	Synnott (1989)	52.53	0.33	74	—	—	?
	Galatea	Synnott (1989)	61.95	0.43	79	—	—	?
	Larissa	Reitsema et al. (1989)	73.55	0.55	104 × 89	—	—	?
	Proteus	Synnott (1989)	117.65	1.12	218 × 208 × 201	—	—	?
Pluto	Charon	Christy (1978)	19.6	6.39	593	—	—	ice

Metric and English Units of Measure

Conversions from Metric to English Measures

Symbol	When you know	Multiply by	To find	Symbol
Length				
mm	millimeters	0.04	inches	in
cm	centimeters	0.4	inches	in
m	meters	3.3	feet	ft
m	meters	1.1	yards	yd
km	kilometers	0.6	miles	mi
Area				
cm^2	square centimeters	0.16	square inches	in^2
m^2	square meters	1.2	square yards	yd^2
km^2	square kilometers	0.4	square miles	mi^2
ha	hectares (10,000 m^2)	2.5	acres	
Mass (weight)				
g	grams	0.035	ounces	oz
kg	kilograms	2.2	pounds	lb
t	metric tons	1.1	short tons	
Volume				
ml	milliliters	0.03	fluid ounces	fl oz
l	liters	2.1	pints	pt
l	liters	1.06	quarts	qt
l	liters	0.26	gallons	gal
m^3	cubic meters	35	cubic feet	ft^3
m^3	cubic meters	1.3	cubic yards	yd^3
Temperature				
°C	Celsius temperature	9/5 (then add 32)	Fahrenheit temperature	°F
K	Kelvin* *convert to °C by subtracting 273, then convert to Fahrenheit subtracting 32	9/5		

Conversions from English to Metric Measures

Symbol	When you know	Multiply by	To find	Symbol
Length				
in	inches	2.54	centimeters	cm
ft	feet	30	centimeters	cm
yd	yards	0.9	meters	m
mi	miles	1.6	kilometers	km
Area				
in^2	square inches	6.5	square centimeters	cm^2
ft^2	square feet	0.09	square meters	m^2
yd^2	square yards	0.8	square meters	m^2
mi^2	square miles	2.6	square kilometers	km^2
	acres	0.4	hectares	ha
Mass (weight)				
oz	ounces	28	grams	g
lb	pounds	0.45	kilograms	kg
	short tons (2000 lb)	0.9	metric tons	t
Volume				
gal	gallons	3.8	liters	l
ft^3	cubic feet	0.03	cubic meters	m^3
yd^3	cubic yards	0.76	cubic meters	m^3
Temperature				
°F	Fahrenheit temperature	5/9 (after subtracting 32)	Celsius temperature	°C

Additional Readings

Introduction

Abell, G.O., D. Morrison, and S.C. Wolff. 1991. *Exploration of the Universe*. 6th ed. Philadelphia: Saunders College Publishing.

Batson, R. M. 1987. Digital cartography of the planets: New methods, its status, and its future. *Photogrammetric Engineering and Remote Sensing* 53 (9):1211-1218.

Batson, R. M., K. Edwards, T. C. Duxbury. 1992. Geodesy and cartography of the martian satellites. In *Mars*, eds. H. Kieffer, B. Jakosky, C. Snyder, and M. Matthews. Tucson: University of Arizona Press.

Greeley, R., and R. M. Batson, eds. 1990. *Planetary Mapping*. London: Cambridge University Press.

Kopal, Z., and R. W. Carder. 1974. *Mapping of the Moon*. Dordrecht-Holland: D. Reidel.

The Solar System

Beatty, J.K., C.C. Petersen, and A. Chaikin, eds. 1999. *The New Solar System*. 4th ed. Cambridge, MA.: Sky Publishing Corporation and Cambridge: Cambridge University Press.

Carr, M.H., ed. 1984. *The Geology of the Terrestrial Planets*. National Aeronautics and Space Administration, Special Paper 469.

Christiansen, E. H., and Hamblin, W.K. 1995. *Exploring the Planets*. Englewood Cliffs: Prentice Hall.

Greeley, R. 1987. *Planetary Landscapes*. 2d ed. London: Allen and Unwin.

Morrison, D. 1993. Exploring Planetary Worlds. New York: W. H. Freeman.

Murray, B., ed. 1983. *The Planets: Readings from Scientific American*. San Francisco: W.H. Freeman and Co.

Sonett, C.P., M.S. Giampapa, and M.S. Matthews, eds. 1991. *The Sun in Time*. Tucson: University of Arizona Press.

Mercury

Davies, M.E., S.E. Dwornik, D.E. Gault, and R.G. Strom. 1978. *Atlas of Mercury*. National Aeronautics and Space Administration, Special Publication 423.

Gault, D.E., J.A. Burns, P. Cassen, and R.G. Strom. 1977. Mercury. In *Ann. Rev. Astron. Astrophys*. 15: 97-126.

Strom, R.G. 1987. *Mercury: The Elusive Planet*. Washington, D.C.: Smithsonian Institution Press.

Vilas, F., C.R. Chapman, and M.S. Matthews, eds. 1988. *Mercury*. Tucson: University of Arizona Press.

Venus

Barsukov, V.L., A.T. Basilevsky, V.P. Volkov, and V.N. Zharkov, eds. 1992. *Venus Geology, Geochemistry, and Geophysics*. Tucson: University of Arizona Press.

Basilevsky, A.T., and J.W. Head III. 1988. The geology of Venus. In *Annual Review of Earth and Planetary Sciences* 16, eds. G.W. Wetherill, A.L. Albee, and F.G. Stehli, 295-317.

Batson, R.M., Kirk, R.L., Edwards, K.F., and Morgan, H.F., 1994. Venus cartography. *Journal of Geophysical Research* 99: 21, 173 - 21, 182.

Catermole, P. 1994. Venus, The Geological Story. Baltimore: Johns Hopkins.

Hunten, D.M., L. Colin, T.M. Donahue, and V.I. Moroz, eds. 1983. *Venus*. Tucson: University of Arizona Press.

Magellan at Venus. 1992. *Science* 252: 247-312.

Magellan special issue, 1992. *Journal of Geophysical Research* 97: E-8.

Earth-Moon System

Hartmann, W.K., R.J. Phillips, and G.J. Taylor, eds. 1986. *Origin of the Moon*. Houston: Lunar and Planetary Institute.

Press, F., and R. Sevier. 1985. *Earth*. 4th ed. New York: W.H. Freeman and Co.

Smith, D.G., ed. 1982. *The Cambridge Encyclopedia of Earth Sciences*. Cambridge: Cambridge University Press.

Taylor, S.R. 1975. *Lunar Science: A Post-Apollo View*. New York: Pergamon.

Wilhelms, D.E. 1987. *The Geologic History of the Moon*. U.S. Geological Survey Professional Paper 1348.

Mars

Baker, V.R. 1982. *The Channels of Mars*. Austin: University of Texas Press.

Carr, M.H. 1981. *The Surface of Mars*. New Haven: Yale University Press.

Catermole, P. 1992. Mars, The Story of the Red Planet. London: Chapman and Hall.

Kieffer, H., B. Jakosky, C. Snyder, and M. Matthews, eds. 1992. *Mars*. Tucson: University of Arizona Press.

Mutch, T.A., R.E. Arvidson, J.W. Head, III, K.L. Jones, and R.S. Saunders. 1976. *The Geology of Mars*. Princeton: Princeton University Press.

Thomas, P., J. Veverka, and S. Dermott. 1986. Small satellites. In *Satellites*, ed. J.A. Burns, and M.S. Matthews, 802-835. Tucson: University of Arizona Press.

Veverka, J., and P. Thomas. 1979. Phobos and Deimos: A preview of what asteroids are like? In *Asteroids*, ed. T. Gehrels, 628-651. Tucson: University of Arizona Press.

Jupiter System

Belton, M.J.S., R.A. West, and J. Rahe, eds. 1989. *Time-variable Phenomena in the Jovian System*. National Aeronautics and Space Administration Special Publication 464.

Burns, J.A., and M.S. Matthews, eds. 1986. *Satellites*. Tucson: University of Arizona Press.

Ingersoll, A.P. 1981. Jupiter and Saturn. *Scientific American* 245:90-108.

Malin, M.C., and D.C. Pieri. 1986. Europa. In *Satellites*, ed. J.A. Burns and M.S. Matthews, 689-717. Tucson: University of Arizona Press.

Morrison, D. 1980. Four New Worlds, the Voyager Exploration of Jupiter's satellites. *Mercury* 9(3): 53-64.

Morrison, D., ed. 1982. *Satellites of Jupiter*. Tucson: University of Arizona Press.

Morrison, D., and J. Samz. 1980. *Voyage to Jupiter*. National Aeronautics and Space Administration Special Publication 439.

Soderblom, L.A. 1980. The Galilean Moons of Jupiter. *Scientific American* 242: 88-100.

Saturn System

Burns, J.A., and M.S. Matthews, eds. 1986. *Satellites.* Tucson: University of Arizona Press.

Gehrels, T. and M.S. Matthews, eds. 1984. *Saturn*. Tucson: University of Arizona Press.

Greenberg, R., and A. Brahic, eds. 1984. *Planetary Rings.* Tucson: University of Arizona Press.

Morrison, D. 1982. *Voyages to Saturn*. National Aeronautics and Space Administration Special Publication 451.

Soderblom, L.A., and T.V. Johnson. 1982. The Moons of Saturn. *Scientific American* 246:100-116.

Uranus System

Bergstrahl, J. T., E. D. Miner, and M .S. Matthews, eds. 1991. *Uranus*. Tucson: University of Arizona Press.

Miner, E. D. 1991. *Uranus: The Planets, Rings, and Satellites*. Chichester: Ellis Horwood Ltd.

Neptune System

Littman, Mark. 1990. *Planets Beyond*. 2d ed. New York: John Wiley and Sons, Inc.

Triton special issue. 1990. *Science* 250: 404-443.

Voyager 2 – Neptune special issue. 1989. *Science* 246: 1417-1501.

Pluto, Asteroids, and Comets

Binzel, R.P., T. Gehrels, M.S. Matthews, eds. 1990. *Asteroids II*. Tucson: University of Arizona Press.

Hoyt, W.G. 1980. *Planets "X" and Pluto*. Tucson: University of Arizona Press.

Kerridge, J.F., and M.S. Matthews, eds. 1988. *Meteorites and the Early Solar System*. Tucson: University of Arizona Press.

Wilkening, L.L., ed. 1982. *Comets*. Tucson: University of Arizona Press.

Sources for Planetary Images

Sources for most of the images in the Atlas are given in the figure captions. Except for NASA illustrations, individuals or institutions are indicated and can be contacted directly for additional information. NASA illustrations and images are identified by spacecraft or other frame number. Requests for items originating in the United States should be directed to:

National Space Science Data Center
National Aeronautics and Space Administration
Goddard Space Flight Center
Greenbelt, Maryland 20771 USA

Requests originating from outside the United States for NASA materials should be directed to:

World Data Center, A: Rockets and Satellites
National Aeronautics and Space Administration
Goddard Space Flight Center
Greenbelt, Maryland 20771 USA

Collections of planetary images are available for inspection at the facilities listed below. These organizations are staffed by data managers who can assist in locating planetary images. However, the organizations do not provide photographs for distribution.

International Locations

Southern Europe Regional Planetary Image Facility
C.N.R. Area ricerca di Roma Tor Vergata
Isituto di Astrofisica Spaziale
Via del Fosso de Cavaliere, 100
00133 Roma, Italia

German Aerospace Center
Institute of Space Sensor Technology and Planetary Exploration
Regional Planetary Image Facility
Rutherfordstr. 2 12489 Berlin, Germany

Regional Planetary Image Facility
Institute of Space and Astronomical Sciences
Division of Planetary Science
3-1-1 Yoshinodai
Sagamihara-Shi, Kanagawa 229, Japan

Israeli Regional Planetary Image Facility
Department of Geography and Environmental Development
Ben-Gurion University of the Negev
P.O. Box 653
Beer-Sheva 84105, Israel

Regional Planetary Image Facility
Department of Geological Sciences
University College London
Gower Street
London WC1E 6BT, UK

Planetary and Space Science Centre
Department of Geology
University of New Brunswick
P.O. Box 4400
Fredericton NB, Canada, E3B 5A3

Nordic Regional Planetary Image Facility
Department of Geosciences and Astronomy
Astronomy Division
University of Oulu
FIN-90570 Oulu, Finland

Phototheque Planetaire
Department des Sciences de la Terre
Laboratoire de Geologie Dynamique de da Terre et des Planetes
Université de Paris-Sud
Batîment 509 F-91405
Orsay, France

United States Locations

Space Photography Laboratory
Department of Geological Sciences
Box 871404
Arizona State University
Tempe AZ 85287-1404

Northeast Regional Planetary Data Center
Department of Geological Sciences
Box 1846
Brown University
Providence RI 02912-1846

Spacecraft Planetary Imaging Facility
Center for Radiophysics and Space Research
317 Space Sciences Building
Cornell University
Ithaca NY 14853-6801

Regional Planetary Image Facility
Mail Stop 202-100
Jet Propulsion Laboratory
4800 Oak Grove Drive
Pasadena CA 91109

Center for Information and Research Services
Lunar and Planetary Institute
3600 Bay Area Boulevard
Houston TX 77058-1113

Regional Planetary Image Facility
Center for Earth and Planetary Studies
NASM MRC 315
National Air and Space Museum
Washington DC 20560

Regional Planetary Image Facility
Branch of Astrogeology
United States Geological Survey
2255 North Gemini Drive
Flagstaff AZ 86001

Space Imagery Center
Lunar and Planetary Laboratory
University of Arizona
1629 E. University Boulevard
Tucson AZ 85721-0092

Pacific Regional Planetary Data Center
Hawai'i Institute of Geophysics and Planetology
School of Ocean and Earth Science and Technology
2525 Correa Road
Honolulu HI 96822

Regional Planetary Image Facility
Departmnet of Earth and Planetary Sciences
Campus Box 1169
Washington University
One Brookings Drive
St. Louis MO 63130-4899

Gazetteer

The names of planetary features are frequently updated by the *International Astronomical Union*. For the most recent version, see planetarynames.wr.usgs.gov
Dashes appear in the columns if the information for the feature is unknown or is not applicable.
Readers noting errors or omissions are urged to communicate them to the U.S. Geological Survey, Branch of Astrogeology, 2255 N. Gemini Drive, Flagstaff, Arizona 86001, or by e-mail: jblue@usgs.gov

MERCURY

Feature name	Lat. (°)	Long. (°)	Size (km)	Page	Description
Abu Nuwas	17.4N	20.4W	116	49	Arab poet (ca. 1756-810).
Adventure Rupes	65.1S	65.5W	–	48	English; one of Cook's ships on 2nd voyage to the Pacific (1772-1775).
Africanus Horton	51.5S	41.2W	135	49	(James Beale); Sierra Leonean author (1835-1883).
Ahmad Baba	58.5N	126.8W	127	46	Abu-a;-Abbas Aj,Ad Obm Aj,Ad a;-Takruri Al-Massufi; Sudanese writer (1556-1627).
Al-Akhtal	59.2N	97.0W	102	47	Arab poet (ca. 640-710).
Alencar	63.5S	103.5W	120	51	Jose de; Brazilian novelist (1829-1877).
Al-Hamadhani	38.8N	89.7W	186	47	Arab writer (died 1007).
Al-Jāhiz	1.2N	21.5W	91	49	Arab author (died 869).
Amru Al-Qays	12.3N	175.6W	50	50	Arab poet (pre-Islamic).
Andal	47.7S	37.7W	108	49	(18th century).
Antoniadi Dorsum	25.1N	30.5W	–	47	Italian astronomer; defined classical nomenclature for albedo features on Mars (1870-1944).
Apollonia	45.0N	315.0W	–	–	Albedo name for unimaged H-5 region.
Arecibo Vallis	27.5S	28.4W	–	49	Radio telescope in Puerto Rico.
Aristoxenus	82.0N	11.4W	69	47	Greek philosopher and musical theorist (flourished 4th century B.C.).
Astrolabe Rupes	42.6S	70.7W	–	48	French; D'Urville's ship to explore Antarctica (1838-1840).
Aśvaghosa	10.4N	21.0W	90	49	Indian philosopher and poet (flourished 80-150).
Aurora	45.0N	90.0W	–	–	Albedo name for H-2, Victoria region.
Australia	72.5S	0.0W	–	–	Albedo name for H-15, Bach region.
Bach	68.5S	103.4W	214	51	J. S.; German composer (1685-1750).
Balagtas	22.6S	13.7W	98	49	F.; Philippino writer (1788-1862).
Balzac	10.3N	144.1W	80	51	Honoré de; French novelist (1799-1850).
Barma	41.3S	162.8W	128	50	Yakovlev; 16th Century Russian architect.
Bartók	29.6S	134.6W	112	51	Béla; Hungarian composer (1881-1945).
Bashō	32.7S	169.7W	80	50	Matsuo; Japanese poet (1644-1694).
Beethoven	20.8S	123.6W	643	51	Ludwig van; German composer of Flemish descent (1770-1827).
Belinskij	76.0S	103.4W	70	51	Vissarion Grigoryevich; Russian literary critic and journalist (1811-1848).
Bello	18.9S	120.0W	129	51	Andres; Venezuelan poet and scholar (1781-1865).
Bernini	79.2S	136.5W	146	51	Gian Lorenzo; Italian sculptor and architect (1598-1680).
Bjornson	73.1N	109.2W	–	46	Bjørnstjerne; Norwegian poet and dramatist (1832-1910).
Boccaccio	80.7S	29.8W	142	49	Giovanni; Italian poet and novelist (1313-1375).
Boethius	0.9S	73.3W	129	48	Anicius Manlius Severinus; Roman scholar (ca. 480-524).
Borea	75.0N	0.0W	–	–	Albedo name for H-1, Borealis region.
Borealis Planitia	73.4N	79.5W	–	47	"Northern plain"; from classical albedo name for quadrangle.
Botticelli	63.7N	109.6W	143	46	Sandro; Italian painter (1445-1510).
Brahms	58.5N	176.2W	96	46	Johannes; German composer and pianist (1833-1897).
Bramante	47.5S	61.8W	159	48	Donato; Italian architect (1444-1514).
Brontë	38.7N	125.9W	60	46	Charlotte (1816-1855) and Emily (1818-1848); English novelists.
Brueghel	49.8N	107.5W	75	46	Pieter; Flemish painter (1525-1569).
Brunelleschi	9.1S	22.2W	134	49	Filippo; Italian architect (1377-1446).
Budh Planitia	22.0N	150.9W	–	46	Hindu word for Mercury.
Burns	54.4N	115.7W	45	46	Robert; Scottish national poet (1759-1796).
Byron	8.5S	32.7W	105	–	George Gordon; English poet (1788-1824).
Caduceata	45.0N	135.0W	–	–	Albedo name for H-3, Shakespeare region.
Callicrates	66.3S	32.6W	70	49	Greek architect (5th century B.C.).
Caloris Montes	39.4N	187.2W	–	46	"Hot mountains"; surface temperature hottest near this position.
Caloris Planitia	30.5N	189.8W	–	46	"Hot plain"; surface temperature hottest near this position.
Camões	70.6S	69.6W	70	48	Luiz Vaz de; Portuguese poet (ca. 1524-1580).
Carducci	36.6S	89.9W	117	48	Giosuè; Italian poet (1835-1907).

Feature name	Lat. (°)	Long. (°)	Size (km)	Page	Description
Cervantes	74.6S	122.0W	181	51	Miguel de; Spanish novelist, playwright and poet (1547-1616).
Cézanne	8.5S	123.4W	75	51	Paul; French painter (1839-1906).
Chaikovskij	7.4N	50.4W	165	49	Pyotr Ilyich; Russian composer (1840-1893).
Chao Meng-Fu	87.3S	134.2W	167	51	Chinese painter and calligrapher (1254-1322).
Chekhov	36.2S	61.5W	199	48	Anton; Russian playwright and writer (1860-1904).
Chiang K'ui	13.8N	102.7W	35	51	Chinese composer (12th century).
Chŏng Ch'ŏl	46.4N	116.2W	162	46	Korean poet (1536-1593).
Chopin	65.1S	123.1W	129	51	Frédéric; Polish-born French composer and pianist (1810-1849).
Chu Ta	2.2N	105.1W	110	51	Chinese painter (ca. 1625-1705).
Coleridge	55.9S	66.7W	110	48	Samuel Taylor; English poet (1772-1834).
Copley	38.4S	85.2W	30	48	John Singleton; American painter (1738-1815).
Couperin	29.8N	151.4W	80	46	François; French composer (1668-1733).
Cyllene	41.0S	270.0W	–	–	Albedo name for unimaged H-14 region.
Darío	26.5S	10.0W	151	–	Ruben; Nicaraguan poet, journalist (1867-1916).
Degas	37.4N	126.4W	60	46	Edgar; French painter (1834-1917).
Delacroix	44.7S	129.0W	146	51	Eugene; French painter (1798-1863).
Derzhavin	44.9N	35.3W	159	47	Gavrila Romanovich; Russian poet (1743-1816).
Despréz	80.8N	90.7W	50	47	Josquin; French composer (ca. 1440-1521).
Dickens	72.9S	153.3W	78	50	Charles; English novelist (1812-1870).
Discovery Rupes	56.3S	38.3W	–	49	English; Cook's ship on last voyage to Pacific (1776-1780).
Donne	2.8N	13.8W	88	49	John; English poet (1572-1631).
Dostoevskij	45.1S	176.4W	411	50	Fyodor Mikhaylovich; Russian novelist (1821-1881).
Dowland	53.5S	179.5W	100	50	John; English composer (1562-1626).
Dürer	21.9N	119.0W	180	46	Albrecht; German painter (1471-1528).
Dvorák	9.6S	11.9W	82	49	Antonín; Bohemian composer (1841-1904).
Echegaray	42.7N	19.2W	75	47	José; Spanish dramatist; Nobel laureate (1832-1916).
Eitoku	22.1S	156.9W	100	50	Pseudonym of Kano Kuninobu; painter (1543-1590).
Endeavour Rupes	37.5N	31.3W	–	47	English; Cook's ship to explore Tahiti, New Zealand, Australia (1768-1771).
Equiano	40.2S	30.7W	99	49	O.; West Africa (Benin) slave, writer (ca. 1750-1797).
Fet	4.9S	179.9W	24	50	Afanasy Afanasyevich; Russian poet (1820-1892).
Flaubert	13.7S	72.2W	95	48	Gustave; French novelist (1821-1880).
Fram Rupes	56.9S	93.3W	–	48	Norwegian; ship used in Arctic by Nansen, 1892-96, and by Sverdrup and Admundsen in Antarctica, 1909.
Futabatei	16.2S	83.0W	66	48	S.; Japanese novelist (1864-1909).
Gainsborough	36.1S	183.3W	100	50	Thomas; English painter (1727-1788).
Gauguin	66.3N	96.3W	72	47	Paul; French painter (1848-1903).
Ghiberti	48.4S	80.2W	123	48	Lorenzo; Italian sculptor (1378-1455).
Giotto	12.0N	55.8W	150	48	Italian painter (1266/1276-1337).
Gjöa Rupes	66.7S	159.3W	–	50	Norwegian; Amundsen's ship through Northwest passage (1903-1906).
Gluck	37.3N	18.1W	105	47	Christoph Willibald; German composer (1714-1787).
Goethe	78.5N	44.5W	383	47	Johann Wolfgang von; German poet and dramatist (1749-1832).
Gogol	28.1S	146.4W	87	51	Nikolay; Russian dramatist and novelist (1809-1852).
Goldstone Vallis	15.8S	31.7W	–	49	Radio telescope facility in California.
Goya	7.2S	152.0W	135	50	Francisco José de Goya y Lucientes; Spanish painter (1746-1828).
Grieg	51.1N	14.0W	65	47	Edvard; Norwegian composer (1843-1907).
Guido d'Arezzo	38.7S	18.3W	66	49	Italian musical theorist (ca. 990-1050).
Hals	54.8S	115.0W	100	51	Frans; Dutch painter (ca. 1581-1666).
Han Kan	71.6S	143.8W	50	51	Chinese painter (720-780).
Handel	3.4N	33.8W	166	49	G. F.; German-born British composer (1685-1759).
Harunobu	15.0N	140.7W	110	51	Suzuki (1720/1724-1770).
Hauptmann	23.7S	179.9W	120	50	Gerhart; German novelist and dramatist (1862-1946).
Hawthorne	51.3S	115.1W	107	51	Nathaniel; American novelist (1804-1864).
Haydn	27.3S	71.6W	270	48	J.; Austrian composer (1732-1809).
Haystack Vallis	4.7N	46.2W	–	49	Radio telescope facility in Massachusetts.
Heemskerck Rupes	25.9N	125.3W	–	46	Dutch; one of Tasman's ships to explore Australia, New Zealand, (1642-1643).
Heine	32.6N	124.1W	75	46	Heinrich; German poet (1797-1856).
Heliocaminus	40.0N	170.0W	–	–	Albedo feature; Antoniadi map.
Hero Rupes	58.4S	171.4W	–	50	American; Palmer's ship to explore Antarctic coast (1820-1821).
Hesiod	58.5S	35.0W	107	49	Greek poet (ca. 800 B.C.).
Hesperis	45.0S	355.0W	–	–	Albedo feature; Antoniadi map.
Hiroshige	13.4S	26.7W	138	49	A.; Japanese artist (1797-1858).
Hitomaro	16.2S	15.8W	107	49	Kakinomoto No; Japanese poet (ca. 655-700/710).
Holbein	35.6N	28.9W	113	47	Hans (ca. 1465-1524), and Hans (ca. 1497-1543); German painters.
Holberg	67.0S	61.1W	61	48	Ludwig; Norwegian-born Danish writer (1684-1754).
Homer	1.2S	36.2W	314	49	Greek epic poet (8th or 9th century B.C.).
Horace	68.9S	52.0W	58	49	Roman poet (65-8 B.C.).

Feature name	Lat. (°)	Long. (°)	Size (km)	Page	Description
Hugo	38.9N	47.0W	198	47	Victor; French writer, dramatist and poet (1802-1885).
Hun Kal	1.6S	21.4W	13	49	Means "20" in Mayan language; 20th meridian passes through this crater.
Ibsen	24.1S	35.6W	159	49	Henrik J.; Norwegian poet and dramatist (1828-1906).
Ictinus	79.1S	165.2W	119	50	Greek architect (5th century B.C.).
Imhotep	18.1S	37.3W	159	49	Egyptian physician and sage (ca. 2686-2613 B.C.).
Ives	32.9S	111.4W	20	51	Charles; American composer (1874-1954).
Janáček	56.0N	153.8W	47	46	Leos; Czechoslovakian composer (1854-1928).
Jókai	72.4N	135.3W	106	46	Mór; Hungarian novelist (1825-1904).
Judah ha-Levi	10.9N	107.7W	80	51	Jewish poet and religious philosopher (ca. 1075-1141).
Kālidāsā	18.1S	179.2W	107	50	Indian poet and dramatist (5th century ?).
Keats	69.9S	154.5W	115	50	John; English poet (1795-1821).
Kenkō	21.5S	16.1W	99	49	Yashida Ca; Japanese writer (1283-1352).
Khansa	59.7S	51.9W	111	49	Al; great Arab woman poet.
Kōshō	60.1N	138.2W	65	46	Japanese sculptor (13th century).
Kuan Han-Ch'ing	29.4N	52.4W	151	47	Chinese dramatist (ca. 1241-1320).
Kuiper	11.3S	31.1W	62	49	Gerard P.; Dutch-born American astronomer, member of original Mariner Venus-Mercury Imaging Team (1905-1973).
Kurosawa	53.4S	21.8W	159	49	K. (18th century).
Leopardi	73.0S	180.1W	72	50	Giacomo; Italian poet, scholar, and philosopher (1798-1837).
Lermontov	15.2N	48.1W	152	49	Mikhail Yuryevich; Russian poet (1814-1841).
Lessing	28.7S	89.7W	100	48	Gotthold Ephraim; German critic and dramatist (1729-1781).
Li Ch'ing-Chao	77.1S	73.1W	61	48	Chinese poet (1081-ca. 1141).
Li Po	16.9N	35.0W	120	49	Chinese poet (701-762).
Liang K'ai	40.3S	182.8W	140	50	Chinese painter (ca. 1140-1210).
Liguria	45.0N	225.0W	–	–	Albedo name for unimaged H-4 region.
Liszt	16.1S	168.1W	85	50	Franz; Hungarian piano virtuoso and composer (1811-1886).
Lu Hsun	0.0N	23.4W	98	49	Chinese writer; real name: Chou Shu-Jen (1881-1936).
Lysippus	0.8N	132.5W	140	51	Greek sculptor (4th century B.C.).
Ma Chih-Yuan	60.4S	78.0W	179	48	Chinese dramatist (flourished 1251).
Machaut	1.9S	82.1W	106	48	Guillaume de; French poet and composer (ca. 1300-1377).
Mahler	20.0S	18.7W	103	49	Gustav; Austrian composer (1860-1911).
Mansart	73.2N	118.7W	95	46	Jules Hardouin; French architect (ca. 1646-1708).
Mansur	47.8N	162.6W	100	46	Ustad; Indian painter (17th century).
March	31.1N	175.5W	70	46	Ausias; Spanish (Catalan) poet (1397-1459).
Mark Twain	11.2S	137.9W	149	51	(Samuel Clemens); American novelist, satirist (1835-1910).
Martí	75.6S	164.6W	68	50	José Julian Martí y Perez; Cuban poet and essayist (1853-1895).
Martial	69.1N	177.1W	51	46	Marcus Valerius; Roman epigrammist (ca. 40-ca. 103).
Matisse	24.0S	89.8W	186	48	Henri; French painter and sculptor (1869-1954).
Melville	21.5N	10.1W	154	47	Herman; American novelist (1819-1891).
Mena	0.2S	124.4W	52	51	Juan de; Spanish poet (1411-1456).
Mendes Pinto	61.3S	17.8W	214	49	F.; Portuguese prose author (ca. 1510-1583).
Michelangelo	45.0S	109.1W	216	51	Michelangelo di Lodovico Buonarroti Simoni; Italian painter, sculptor and architect (1475-1564).
Mickiewicz	23.6N	103.1W	100	46	Adam Bernard; Polish poet (1798-1855).
Milton	26.2S	174.8W	186	50	John; English poet (1608-1674).
Mirni Rupes	37.3S	39.9W	–	49	Russian; Bellingshausen's ship for Antarctic exploration (1819-1821).
Mistral	4.5N	54.0W	110	49	Gabriela; Chilean poet (1889-1957).
Mofolo	37.7S	28.2W	114	49	Thomas; South African (Lesotho) novelist (1876/77-1948).
Molière	15.6N	16.9W	132	49	Jean-Baptiste Poquelin; French dramatist and satirist (1622-1673).
Monet	44.4N	10.3W	303	47	Claude; French painter (1840-1926).
Monteverdi	63.8N	77.3W	138	47	Claudio; Italian composer (1567-1643).
Mozart	8.0N	190.5W	270	50	Wolfgang Amadeus; Austrian composer (1756-1791).
Murasaki	12.6S	30.2W	130	49	Shikibu; Japanese novelist and poet (978-1014/1026).
Mussorgskij	32.8N	96.5W	125	47	Modest; Russian composer (1839-1881).
Myron	70.9N	79.3W	31	47	Greek sculptor (flourished ca. 480-440 B.C.).
Nampeyo	40.6S	50.1W	52	49	Hopi potter (ca. 1860-1942).
Nervo	43.0N	179.0W	63	46	Amado; Mexican poet (1870-1919).
Neumann	37.3S	34.5W	120	49	B.; German architect (1687-1753).
Nizāmi	71.5N	165.0W	76	46	Elyas Yusof Ganjavi; Persian epic poet (ca. 1141-1209).
Odin Planitia	23.3N	171.6W	–	46	Norse god.
Ōkyo	69.1S	75.8W	65	48	Maruyama; Japanese painter (1733-1795).
Ovid	69.5S	22.5W	44	49	Roman poet (43 B.C.-A.D. 17).
Pentas	5.0N	310.0W	–	–	Albedo feature; Antoniadi map.
Petrarch	30.6S	26.2W	171	49	Francesco; Italian poet (1304-1374).
Phaethontias	0.0N	167.0W	–	–	Albedo name for H-8, Tolstoj region.
Phidias	8.7N	149.3W	160	50	Greek sculptor (flourished ca. 490-430 B.C.).

Feature name	Lat. (°)	Long. (°)	Size (km)	Page	Description
Philoxenus	8.7S	111.5W	90	51	Greek lyric poet (436-380 B.C.).
Pieria	0.0N	270.0W	–	–	Albedo name for unimaged H-10 region.
Pigalle	38.5S	9.5W	154	–	Jean Baptiste; French sculptor (1714-1785).
Pleias Gallia	25.0N	130.0W	–	–	Albedo feature; Antoniadi map.
Po Chü-I	7.2S	165.1W	68	50	Chinese poet (772-846).
Polygnotus	0.3S	68.4W	133	48	Greek painter (ca. 500-ca. 400 B.C.).
Pourquoi-Pas Rupes	58.1S	156.0W	–	50	French; Charcot's ship to explore Antarctica (1908-1910).
Po Ya	46.2S	20.2W	103	49	(8th century B.C.).
Praxiteles	27.3N	59.2W	182	47	Greek sculptor (flourished 370-330 B.C.).
Proust	19.7N	46.7W	157	49	Marcel; French novelist (1871-1922).
Puccini	65.3S	46.8W	70	49	Giacomo; Italian composer (1858-1924).
Purcell	81.3N	146.8W	91	46	Henry; English composer (ca. 1659-1695).
Pushkin	66.3S	22.4W	231	49	Aleksandr Sergeyevich; Russian poet (1799-1837).
Rabelais	61.0S	62.4W	141	48	Francóis; French writer (ca. 1483-1553).
Rajnis	4.5N	95.8W	82	48	Ya; Latvian poet (1865-1925).
Rameau	54.9S	37.5W	51	49	Jean-Philippe; French composer (1683-1764).
Raphael	19.9S	75.9W	343	48	Raffaello Sanzio; Italian painter (1483-1520).
Ravel	12.0S	38.0W	75	49	Maurice; French composer (1875-1937).
Renoir	18.6S	51.5W	246	49	Pierre-Auguste; French painter (1841-1919).
Repin	19.2S	63.0W	107	48	Ilya Yefimovich; Russian painter (1844-1930).
Resolution Rupes	63.8S	51.7W	–	49	English; one of Cook's ships, second expedition to Pacific (1772-1775).
Riemenschneider	52.8S	99.6W	145	48	Tilman; German sculptor (ca. 1460-1531).
Rilke	45.2S	12.3W	86	49	Rainer Maria; German poet (1875-1926).
Rimbaud	62.0S	148.0W	85	51	Arthur; French poet (1854-1891).
Rodin	21.1N	18.2W	229	47	Auguste; French sculptor (1840-1917).
Rubens	59.8N	74.1W	175	47	Peter Paul; Flemish painter (1577-1640).
Rublev	15.1S	156.8W	132	50	Andrey; Russian painter (ca.1370-1430).
Rūdaki	4.0S	51.1W	120	49	Persian poet (ca. 859-940/941).
Rude	32.8S	79.6W	75	48	François; French sculptor (1784-1855).
Rūmī	24.1S	104.7W	75	51	Jalal Ad-Din, also Mawlana; Persian poet and Sufi mystic (1207-1273).
Sadī	78.6S	56.0W	68	49	Persian poet (ca. 1213-1291/1292).
Saikaku	72.9N	176.3W	88	46	Ihara; Japanese novelist and poet (1642-1693).
Santa María Rupes	5.5N	19.7W	–	49	Spanish; Columbus' flagship, first voyage to America, 1492.
Sarmiento	29.8S	187.7W	145	50	Domingo Faustino; Argentine writer (1811-1888).
Sayat-Nova	28.4S	122.1W	158	51	Aruthin Sayadian; Armenian/Georgian song writer (1712-1795).
Scarlatti	40.5N	100.0W	129	47	Alessandro (1660-1725) and Domenico (1685--1757); Italian composers.
Schiaparelli Dorsum	23.0N	164.1W	–	46	Giovanni V.; Italian astronomer (1835-1910).
Schoenberg	16.0S	135.7W	29	51	Arnold; Austrian-born American composer (1874-1951).
Schubert	43.4S	54.3W	185	49	Franz Peter; Austrian composer (1797-1828).
Scopas	81.1S	172.9W	105	50	Greek sculptor and architect (flourished 4th century B.C.).
Sei	64.3S	89.1W	113	48	Shonagun; Japanese diarist and poet (ca. 966-1013).
Shakespeare	49.7N	150.9W	370	46	William; English poet and dramatist (1564-1616).
Shelley	47.8S	127.8W	164	51	Percy Bysshe; English poet (1792-1822).
Shevchenko	53.8S	46.5W	137	49	Taras Hryhorovych; Ukrainian poet (1814-1861).
Sholem Aleichem	50.4N	87.7W	200	47	(Yakov Rabinowitz); Yiddish writer (1859-1916).
Sibelius	49.6S	144.7W	90	51	Jean; Finnish composer (1865-1957).
Simeiz Vallis	13.2S	64.3W	–	48	Radio telescope facility, Crimea, Ukraine.
Simonides	29.1S	45.0W	95	49	Greek lyric poet (556-468 B.C.).
Sinan	15.5N	29.8W	147	49	Joseph; Turkish architect (1489-1588).
Sinus Argiphontae	10.0S	335.0W	–	–	Albedo feature; Antoniadi map.
Smetana	48.5S	70.2W	190	48	Bedrich; Czechoslovakian composer (1824-1884).
Snorri	9.0S	82.9W	19	48	Sturluson; Icelandic saga writer and poet (1179-1241).
Sobkou Planitia	39.9N	129.9W	–	46	Messenger god.
Solitudo Admetei	55.0N	90.0W	–	–	Albedo feature; Antoniadi map.
Solitudo Alarum	15.0S	290.0W	–	–	Albedo feature; Antoniadi map.
Solitudo Aphrodites	25.0N	290.0W	–	–	Albedo feature; Antoniadi map.
Solitudo Atlantis	35.0S	210.0W	–	–	Albedo feature; Antoniadi map.
Solitudo Criophori	0.0N	230.0W	–	–	Albedo name for unimaged H-9 region.
Solitudo Helii	10.0S	180.0W	–	–	Albedo feature; Antoniadi map.
Solitudo Hermae Trismegisti	45.0S	45.0W	–	–	Albedo name for H-11, Discovery region.
Solitudo Horarum	25.0N	115.0W	–	–	Albedo feature; Antoniadi map.
Solitudo Iovis	0.0N	0.0W	–	–	Albedo feature; Antoniadi map.
Solitudo Lycaonis	0.0N	107.0W	–	–	Albedo name for H-7, Beethoven region.
Solitudo Maiae	15.0S	155.0W	–	–	Albedo feature; Antoniadi map.
Solitudo Martis	35.0S	100.0W	–	–	Albedo feature; Antoniadi map.
Solitudo Neptuni	30.0N	150.0W	–	–	Albedo feature; Antoniadi map.
Solitudo Persephones	41.0S	225.0W	–	–	Albedo name for unimaged H-13 region.
Solitudo Phoenicis	25.0N	225.0W	–	–	Albedo feature; Antoniadi map.

MERCURY

Feature name	Lat. (°)	Long. (°)	Size (km)	Page	Description
Solitudo Promethei	45.0S	142.5W	–	–	Albedo name for H-12, Michelangelo region.
Sophocles	7.0S	145.7W	150	50	Greek dramatist (ca. 496-406 B.C.).
Sor Juana	49.0N	23.9W	93	47	Ines de la Cruz; Mexican writer (1651-1695).
Sōseki	38.9N	37.7W	90	47	Natsume (Kinosaka); Japanese novelist (1867-1916).
Sōtatsu	49.1S	18.1W	165	49	Tawakaya; Japanese artist (1600-1643).
Spitteler	68.6S	61.8W	68	48	Carl; Swiss epic poet (1845-1924).
Stravinsky	50.5N	73.5W	190	47	Igor Fyodorovich; Russian-born American composer (1882-1971).
Strindberg	53.7N	135.3W	190	46	August; Swedish playwright, novelist, and short-story writer (1849-1912).
Suisei Planitia	59.2N	150.8W	–	46	Japanese Messenger god.
Sullivan	16.9S	86.3W	145	48	Louis; American architect (1856-1924).
Sūr Dās	47.1S	93.3W	132	48	Indian poet (1483-1563).
Surikov	37.1S	124.6W	120	51	Vassily; Russian painter (1848-1916).
Takanobu	30.8N	108.2W	80	46	Fujiwara; Japanese poet and portrait artist (1142-1205).
Takayoshi	37.5S	163.1W	139	50	Japanese painter (12th century).
Tansen	3.9N	70.9W	34	48	Indian (Mogul) composer from the court of Akbar.
Thākur	3.0S	63.5W	118	48	R.; Bengalese poet and novelist, Nobel Peace laureate (1861-1941).
Theophanes	4.9S	142.4W	45	51	Byzantine painter (ca. 1330-1405).
Thoreau	5.9N	132.3W	80	51	Henry David; American poet and philosopher (1817-1862).
Tintoretto	48.1S	22.9W	92	49	Italian painter (1518-1594).
Tir Planitia	0.8N	176.1W	–	50	Norse word for Mercury.
Titian	3.6S	42.1W	121	49	Tiziano Vecellio; Italian Renaissance painter (ca. 1488/90-1576).
Tolstoj	16.3S	163.5W	390	50	Lev N.; Russian novelist (1828-1910).
Tricrena	0.0N	36.0W	–	–	Albedo name for H-6, Kuiper region.
Ts'ai Wen-Chi	22.8N	22.2W	119	47	Han Dynasty composer (A.D. 2nd century).
Ts'ao Chan	13.4S	142.0W	110	51	Chinese writer (ca. 1715-1763).
Tsurayuki	63.0S	21.3W	87	49	Ki (Kino); noted Japanese man of letters (ca. 945).
Tung Yüan	73.6N	55.0W	64	47	Chinese painter (10th century).
Turgenev	65.7N	135.0W	116	46	Ivan Sergeyevich; Russian writer (1818-1883).
Tyagaraja	3.7N	148.4W	105	50	Indian composer (1767-1847).
Unkei	31.9S	62.7W	123	48	Japanese sculptor (ca. 1148-1223).
Ustad Isa	32.1S	165.3W	136	50	Turkish/Persian architect (17th century).
Vālmiki	23.5S	141.0W	221	51	Sanskrit poet, author of the Ramayala (1st century B.C.).
Van Dijck	76.7N	163.8W	105	46	Anthony; Flemish painter (1599-1641).
Van Eyck	43.2N	158.8W	282	46	Jan; Flemish painter (ca. 1395-ca. 1441).
van Gogh	76.5S	134.9W	104	51	Vincent Willem; Dutch painter (1853-1890).
Velázquez	37.5N	53.7W	129	47	Diego; Spanish painter (1599-1660).
Verdi	64.7N	168.6W	163	46	Giuseppe; Italian composer (1813-1901).
Victoria Rupes	50.9N	31.1W	–	47	Spanish; Magellan's and Del Cano's ship, first voyage around the world.
Vincente	56.8S	142.4W	98	51	Gil; Portuguese dramatist (ca. 1465-1537).
Vivaldi	13.7N	85.0W	213	48	Antonio; Italian composer (1678-1741).
Vlaminck	28.0N	12.7W	97	47	Maurice de; French painter (1876-1958).
Vostok Rupes	37.7S	19.5W	–	49	Russian; Bellingshausen's ship for Antarctic exploration (1819-1821).
Vyāsa	48.3N	81.1W	290	47	Indian poet (flourished 1500 B.C.).
Wagner	67.4S	114.0W	140	51	Richard; German composer (1813-1883).
Wang Meng	8.8N	103.8W	165	51	Chinese painter (1308-1385).
Wergeland	38.0S	56.5W	42	48	Henrik Arnold; Norwegian poet (1808-1845).
Whitman	41.4N	110.4W	70	46	Walter; American poet (1819-1892).
Wren	24.3N	35.2W	221	47	Christopher; English architect (1632-1723).
Yeats	9.2N	34.6W	100	49	William Butler; Irish poet and dramatist (1865-1939).
Yun Sŏn-Do	72.5S	109.4W	68	51	Korean poet (1587-1671).
Zarya Rupes	42.8S	20.5W	–	49	USSR; Motor-sail schooner to study Earth's magnetic field, 1953.
Zeami	3.1S	147.2W	120	50	Motokiyo; Japanese dramatist and playwright (ca. 1363-1443).
Zeehaen Rupes	51.0N	157.0W	–	–	Dutch; one of Tasman's ships to explore Australia, New Zealand (1642-1643).
Zola	50.1N	177.3W	80	46	Émile; French novelist (1840-1902).

VENUS

Feature name	Lat. (°)	Long. (°)	Size (km)	Page	Description
Abaka	52.6S	104.2E	14	–	Mari first name.
Abigail	52.2S	111.1E	19	–	First name from Hebrew.
Abington	47.8S	277.8E	23	–	Frances; English actress (1737-1815).
Abundia Corona	18.5N	125.0E	250	–	Norse goddess of giving.
Adaiah	47.3S	253.3E	19	–	Hebrew first name.
Adamson	14.8S	29.6E	20	–	Joy; Austrian author, animal expert (1910-1980).
Addams	56.1S	98.9E	90	–	Jane; American social reformer (1860-1935).
Adivar	8.9N	75.9E	30	–	Halide; Turkish educator, author (1883-1964).
Aegina Farrum	35.5N	20.9E	60	–	Greek river nymph.
Aeracura Corona	19.0S	238.5E	250	–	Celtic earth goddess.
Aethelflaed	18.2S	196.5E	17	–	English leader of the Mercians (ca. 884-918).
Aglaonice	26.5S	339.9E	66	68	Ancient Greek astronomer.
Agnesi	39.5S	37.8E	40	–	Maria; Italian mathemetician (1718-1799).
Agraulos Corona	27.7S	165.8E	170	–	Greek fertility goddess.
Agrippina	33.2S	65.2E	37	–	Roman empress (ca. 13 B.C.-A.D. 33).

Feature name	Lat. (°)	Long. (°)	Size (km)	Page	Description
Ahsonnutli Dorsa	47.9N	194.8E	1708	66	Navajo (N. America) spirit of light and sky.
Aimee	16.2N	127.1E	18	–	French first name.
Aino Planitia	40.5S	94.5E	4983	69	Finnish heroine who became water spirit.
Aita	8.9N	270.7E	14	–	Estonian first name.
Aitchison Patera	16.7S	349.4E	28	–	Alison; American geographer.
Akeley	8.0N	244.4E	25	–	Delia; American explorer (1875-1970).
Akhmatova	61.1N	307.4E	42	–	Anna; Russian poet (1889-1966).
Akiko	30.7N	187.1E	24	–	Yosano; Japanese Tanka poetess (1878-1942).
Akkruva Colles	46.1N	115.5E	1059	66	Saami-Lapp fishing goddess.
Akna Montes	68.9N	318.2E	830	66	Mayan goddess of birth.
Aksentyeva	42.0S	271.9E	43	–	Zinaida; Soviet geophysicist, astronomer (1900-1969).
Alcott	59.5S	354.5E	71	67	Louisa M.; American author (1832-1888).
Ale Tholus	68.2N	247.0E	87	–	Igbo (Nigeria) goddess who created Earth and vegetation.
Alima	45.9S	229.2E	13	–	Tatar first name.
Alimat	29.6S	205.9E	14	–	Osset first name.
Alison	4.0S	165.6E	16	–	Irish first name.
Allat Dorsa	63.3N	71.3E	302	–	Arab sky goddess.
Allatu Corona	15.5N	114.0E	125	–	Akkadian earth goddess.
Alma	2.4S	228.7E	17	–	Kazakh first name.
Almeida	46.0N	123.0E	18	–	Portuguese first name.
Alpha Regio	25.5S	1.3E	1897	69	First letter in Greek alphabet.
Al-Taymuriyya	32.9N	336.2E	22	–	Ayesha; Egyptian author, feminist (1840-1902).
Al-Uzza Undae	67.7N	90.5E	150	–	Arabian desert goddess.
Amalasthuna	11.5S	342.4E	18	–	Ostrogoth queen (ca. 498-535).
Amaya	11.3N	89.1E	32	–	Carmen; Spanish Gypsy dancer (1913-1963).
Amenardes	15.0N	54.1E	25	–	Egyptian princess (718-655 B.C.).
Ament Corona	67.2S	217.9E	115	–	Egyptian earth goddess.
Anahit Corona	77.1N	277.3E	324	66	Armenian goddess of fertility.
Anala Corona	11.0N	14.0E	240	–	Hindu fertility goddess.
Ananke Tessera	53.3N	133.3E	1060	66	Greek goddess of necessity.
Anaxandra	44.2N	162.1E	21	–	Greek artist (flourished ca. 228 B.C.).
Andami	17.5S	26.3E	28	–	Iranian doctor.
Andreianova	3.0S	68.7E	70	–	Elena; Russian ballerina (ca. 1821-ca. 1855).
Anicia	26.4S	31.1E	30	–	Greek physician, poet (flourished ca. 300 B.C.).
Annapurna Corona	35.5S	152.0E	300	–	Indian goddess of wealth.
Annia Faustina	22.1N	4.6E	20	–	Roman empress, wife of Marcus Aurelius (125-175).
Anning Paterae	66.5N	57.8E	–	–	Mary; English paleontologist (1799-1847).
Anqet Farrum	33.6N	311.5E	125	–	Egyptian goddess of fertile waters.
Anthony Patera	48.0N	33.0E	–	–	Susan B.; American suffrage leader (1820-1906).
Antiope Linea	40.0S	350.0E	–	–	Amazon awarded to Theseus.
Antonina	27.6N	107.1E	18	–	Russian first name.
Anuket Vallis	66.7N	8.0E	350	–	Egyptian river goddess.
Anush	14.9N	86.5E	13	–	Armenian first name.
Anya	39.5N	298.2E	21	–	Russian first name.
Apgar Patera	43.5N	84.0E	–	–	Virginia; American doctor (1909-1974).
Aphrodite Terra	5.8S	104.8E	9999	69	Greek goddess of love.
Api Mons	38.9N	54.7E	190	–	Scythian goddess of earth.
Aramaiti Corona	26.3S	82.0E	350	–	Persian fertility goddess.
Aranyani Chasma	69.3N	74.4E	718	–	Indian forest goddess.
Ariadne	43.8N	180.0E	37	–	Greek first name; crater defines longitude.
Arianrod Fossae	37.0N	239.9E	715	–	Celtic warrior queen.
Artemis Chasma	41.2S	138.5E	3087	70	Greek goddess of hunt/moon.
Artemis Corona	35.0S	135.0E	2600	70	Named from associated chasma.
Aruru Corona	9.0N	262.0E	450	–	Sumerian earth goddess.
Ashnan Corona	50.2N	357.0E	300	–	Sumerian harvest goddess.
Ashtart Tholus	48.7N	247.0E	138	–	Phoenician goddess of love, fertility and war; personification of planet Venus.
Asmik	3.9N	166.4E	18	–	Armenian first name.
Aspasia Patera	56.4N	189.1E	150	–	One of most outstanding women of ancient Greece (ca. 470-410 B.C.).
Asteria Regio	21.6N	267.5E	1131	71	Greek Titaness.
Astrid	21.4S	335.5E	12	–	Scandanavian first name.
Atalanta Planitia	45.6N	165.8E	2048	66	Greek; huntress associated with golden apples.
Atargatis Corona	8.0S	8.6E	360	–	Hittite fertility goddess.
Atete Corona	16.0S	243.5E	600	71	Oromo (Ethiopia) fertility goddess.
Atira Mons	52.2N	267.6E	152	66	Pawnee (N. America) wife of Great Spirit Tirawa.
Atla Regio	9.2N	200.1E	3200	71	Norse giantess, mother of Heimdall.
Atropos Tessera	71.5N	304.0E	469	–	Greek; one of three Fates.
Atse Estsan Corona	8.5N	92.0E	150	–	Navajo fertility goddess.
Audhumla Corona	45.5N	12.0E	225	–	Norse primordial nourisher.
Audra Planitia	61.5N	71.5E	–	66	Lithuanian sea mistress.
Audrey	23.8N	348.1E	15	–	English first name.
Aurelia	20.3N	331.8E	31	–	Mother of Julius Caesar.
Auska Dorsum	60.5N	179.0E	–	–	Lithuanian goddess of sun rays.
Aušrā Dorsa	49.4N	25.3E	859	66	Lithuanian dawn goddess.
Austen	25.0S	168.3E	46	–	Jane; English novelist (1775-1817).
Avfruvva Vallis	2.0N	70.0E	70	–	Saami (Lapp) river goddess.
Avviyar	18.0S	353.6E	21	–	Tamil poet (ca. 100 B.C.).
Ayana	29.1S	175.5E	15	–	Altai first name.
Ayrton Patera	6.0N	228.3E	85	–	Hertha M.; English physicist (1854-1923).
Baba-Jaga Chasma	53.2N	49.5E	580	–	Slavic forest witch.
Bachue Corona	73.3N	261.4E	463	–	Chibcha (Columbia) goddess of fertility.
Badarzewska	22.6S	137.0E	28	–	Tekla; Polish composer (1834-1861).
Ba'het Patera	48.6N	0.6E	–	66	Egyptian who defeated Portugese.
Baker	62.6N	40.5E	105	–	Josephine; American expatriate dancer, singer (1906-1975).
Balch	29.9N	282.9E	37	–	Emily; American economist, Nobel laureate (1867-1961).
Baltis Vallis	37.3N	161.4E	6000	–	Syrian word for planet Venus.
Ban Zhao	17.2N	146.9E	38	–	Chinese historian (ca. 35-100).
Baranamtarra	17.9N	267.8E	25	–	Mesopotamian queen (ca. 2500 B.C.).
Barrera	16.6N	109.3E	25	–	Olivia; Spanish medical writer (born 1562).
Barrymore	52.3S	195.6E	50	–	Ethel; American actress (1879-1959).
Barsova	61.5N	223.3E	95	–	Valeria; Soviet singer (1892-1967).

Feature name	Lat. (°)	Long. (°)	Size (km)	Page	Description
Barto	45.2N	146.4E	54	–	Agniya; Soviet poet (1906-1981).
Barton	27.4N	337.5E	50	–	Clara; American Red Cross founder (1821-1912).
Bascom	10.3S	302.2E	36	–	Florence; American geologist (1862-1945).
Bashkirtseff	14.7N	194.0E	38	–	Marie; Russian painter, diarist (ca. 1859-1884).
Bassi	19.0S	64.6E	35	–	Laura; Italian physicist, mathematician (1711-1778).
Bast Tholus	57.8N	130.3E	83	–	Egyptian goddess of joy.
Bathsheba	15.1S	49.3E	36	–	Hebrew queen (ca. 1030 B.C.).
Bau Corona	53.0N	258.0E	–	66	Sumerian fertility goddess.
Bayara Vallis	45.6N	16.5E	500	–	Dogon (Mali) word for planet Venus.
Bécuma Mons	34.0N	22.0E	–	–	Irish goddess.
Beecher	13.1N	253.5E	35	–	Catherine; American educator, author (1800-1878).
Behn	32.5S	141.8E	25	–	Aphra; English novelist, poet, playwright (1640-1689).
Beiwe Corona	52.6N	306.5E	600	66	Saami (Lapp) fertility goddess.
Belet-Ili Corona	6.0N	20.0E	300	–	Mesopotamian nature/ fertility goddess.
Belisama Vallis	50.0N	22.5E	220	–	English Celtic river goddess.
Bell Regio	32.8N	51.4E	1778	69	English giantess.
Bellona Fossae	38.0N	222.1E	855	–	Roman war goddess, wife of Mars.
Ben Dorsa	71.2N	284.1E	628	–	Vietnamese sky goddess.
Bennu Vallis	1.3N	341.2E	710	–	Egyptian word for planet Venus.
Benten Corona	16.0N	340.0E	310	–	Japanese love/fertility goddess.
Bereghinya Planitia	28.6N	23.6E	3902	69	Slavic water spirit.
Berggolts	63.4S	53.0E	31	–	Olga; Russian poet (1910-1975).
Bernadette	46.7S	285.6E	15	–	French first name.
Bernhardt	31.4N	84.3E	45	–	Sarah; French actress (1844-1923).
Berta	62.0N	322.0E	20	–	Finnish first name.
Beruth Corona	19.0S	233.5E	350	–	Phoenician earth goddess.
Beta Regio	25.3N	282.8E	2869	68	Second letter in Greek alphabet.
Bethune Patera	47.0N	321.5E	–	66	Mary; American educator (1875-1955).
Bette	24.6S	347.9E	6	–	German first name (form of Elizabeth).
Beyla Corona	25.0N	15.5E	400	69	Norse earth goddess.
Bezlea Dorsa	30.4N	36.5E	807	–	Lithuanian evening light goddess.
Bhumidevi Corona	17.2S	343.6E	150	–	Hindu earth goddess.
Bhumiya Corona	15.0N	118.0E	100	–	Hindu earth goddess.
Bickerdyke	82.0S	170.8E	39	–	Mary; American Civil War nurse (1817-1901).
Birute	36.0N	33.5E	–	–	Lithuanian first name.
Blackburne	11.0N	183.8E	33	–	Anna; English biologist (1726-1793).
Blai Corona	0.4S	134.5E	125	–	Celtic fertility goddess.
Blanche	8.5S	157.8E	18	–	French first name.
Blathnat Corona	35.0N	293.8E	300	–	Celtic fertility goddess.
Blixen	59.9S	145.6E	22	–	Karen; Danish author (1885-1962).
Bly	37.7N	305.5E	20	–	Nellie; American journalist (1867-1892).
Boadicea Paterae	56.0N	96.0E	–	–	(Boudicca); queen and heroine of Iceni (English Celtic tribe) (died A.D. 62).
Boann Corona	27.0N	136.5E	300	–	Irish fertility goddess.
Boivin	4.3N	299.5E	18	–	Marie; French medical researcher (1773-1847).
Boleyn	24.4N	219.9E	69	–	Anne; English queen (1507-1536).
Bona Corona	24.0S	157.5E	275	–	Roman virgin/fertility goddess.
Bonnevie	36.1S	126.8E	85	70	Norwegian biologist.
Bonnin	6.2S	117.6E	40	–	Gertrude (Zitkala-sa); Dakota reformer, writer (1875-1938).
Boulanger	26.5S	99.3E	57	–	Nadia; French pianist, composer (1881-1979).
Bourke-White	21.2N	147.8E	31	–	Margaret; American photo- -journalist (1905-1971).
Boyd	39.3S	221.3E	25	–	Louise; American explorer (1887-1972).
Boye	9.6S	292.3E	30	–	Karen; Swedish poet, novelist (1900-1941).
Bradstreet	16.5N	47.6E	39	–	Anne; American poet (ca. 1612-1672).
Breksta Dorsa	35.9N	304.0E	700	–	Lithuanian night darkness goddess.
Bremer Patera	67.0N	64.0E	–	–	Frederika; Swedish writer, reformer, feminist (1801-1865).
Bridgit	45.3S	348.9E	11	–	Irish first name.
Brigit Tholus	49.0N	246.0E	–	–	Celtic goddess of wisdom, doctoring, smithing.
Britomartis Chasma	33.0S	130.0E	0	–	Greek/Cretan goddess of the hunt.
Brooke	48.0N	296.0E	26	–	Frances; Canadian novelist (1724-1789).
Browning	28.0N	5.0E	24	–	Elizabeth; British poet (1806-1861).
Bryce	62.6S	197.1E	25	–	Lucy; Australian medical pioneer (1897-1968).
Buck	5.7S	349.6E	22	–	Pearl S.; American writer (1892-1973).
Budevska	0.5N	143.0E	20	–	Adriana; Bulgarian actress (1878-1955).
Bugoslavskaya	23.0S	300.4E	30	–	Yevgenia; Soviet astronomer (1899-1960).
Bunzi Mons	46.0N	355.0E	–	–	Woyo (Zaire) rainbow goddess.
Caccini	17.4N	170.4E	38	–	Francesca; Italian poet, composer (ca. 1581-ca. 1640).
Cailleach Corona	48.0S	88.3E	125	–	Scottish Celtic fertility goddess.
Caitlin	65.3S	12.1E	14	–	Welsh first name.
Caiwenji	12.4S	287.5E	22	–	Chinese painter, calligrapher (907-960).
Calakomana Corona	6.5N	43.5E	575	69	Pueblo Indian corn goddess.
Caldwell	23.6N	112.1E	44	–	Taylor; American author (1900-1985).
Callas	2.4N	26.9E	30	–	Maria; American opera singer (1923-1977).
Callirhoe	21.3N	140.6E	32	–	Greek sculptor (ca. 600 B.C.).
Carmenta Farra	12.4N	8.0E	180	–	Roman goddess of springs.
Caroline	6.8N	306.3E	17	–	First name from French.
Carpo Corona	37.5S	3.0E	215	–	Greek fertility goddess.
Carr	24.0S	295.7E	30	–	Emily; Canadian artist (1871-1945).
Carreno	3.9S	16.1E	57	69	Teresa; Venezuelan pianist, composer (1853-1917).
Carriera Patera	48.5N	48.5E	–	–	Rosalba; Italian portrait painter (1675-1757).
Carson	24.2S	344.2E	41	–	Rachel; American biologist, author (1907-1964).
Carter	5.3N	67.2E	18	–	Maybelle; American singer, songwriter (1909-1978).
Cassatt Patera	65.5N	207.5E	–	–	Mary; American Impressionist painter (1844-1926).

Feature name	Lat. (°)	Long. (°)	Size (km)	Page	Description
Castro	3.3N	233.9E	23	–	Rosalie; Spanish poet, novelist (1837-1885).
Cather	47.1N	106.7E	30	–	Willa; American novelist (1876-1947).
Cauteovan Corona	31.5N	144.0E	–	–	Kataba (Columbia) fertility goddess.
Cavell Patera	38.0N	19.0E	–	–	Edith; British nurse, heroine (1865-1915).
Centlivre	19.1N	290.4E	26	–	Susannah; English actress, playwright (ca. 1667-1723).
Ceres Corona	16.0S	151.5E	675	–	Roman harvest goddess.
Chapelle	6.4N	103.8E	23	–	Georgette; Am. photo-journalist, killed in Viet Nam (1919-1965).
Chih Nu Dorsum	73.0S	195.0E	625	–	Chinese sky goddess.
Chiun Corona	18.3N	340.5E	150	–	Hebrew fertility goddess.
Chiyojo	47.8S	95.2E	35	–	Japanese poetess.
Chloe	7.4S	98.6E	19	–	First name from Greek.
Christie	28.1N	72.5E	34	–	Agatha; British novelist (1891-1976).
Citlalpul Valles	57.4S	185.0E	2350	–	Aztec name for planet Venus.
Ciuacoatl Corona	53.0N	150.9E	100	–	Aztec earth goddess.
Cleopatra	66.0N	8.0E	104	66	Egyptian queen, notable for love affairs with Julius Caesar and Mark Anthony (69-30 B.C.).
Cline	21.8S	317.0E	40	–	Patsy; American singer (1932-1963).
Clotho Tessera	56.4N	334.9E	289	66	Greek; one of three Fates.
Coatlicue Corona	63.2N	273.0E	199	66	Aztec earth goddess.
Cochran	52.0N	142.6E	124	66	Jacqueline; American aviator (ca. 1906-1980).
Cohn	33.2S	208.1E	21	–	Carola; Australian artist (1892-1964).
Colette Patera	66.5N	322.8E	149	66	Claudine; French novelist (1873-1954).
Colijnsplaat Corona	32.0S	151.0E	350	–	Teutonic fertility goddess.
Colleen	60.8S	162.2E	14	–	Irish first name.
Colonna	64.7N	216.8E	28	–	Vittoria; Italian poet (ca. 1490-1547).
Comnena	1.2N	343.7E	20	–	Anna; Byzantine princess, physician, writer (1083-1148).
Conway	48.3N	39.1E	50	–	Anne Finch; English natural scientist (1631-1679).
Copia Corona	42.5S	75.5E	500	69	Roman goddess of plenty.
Corday Patera	62.5N	40.0E	–	–	Charlotte; French patriot (1768-1798).
Cori	25.4N	72.7E	50	69	Gerty; Czech biochemist, Nobel laureate (1896-1957).
Corinna	22.8N	40.5E	21	–	Greek poet (flourished ca. 490 B.C.).
Corpman	0.3N	151.8E	52	–	Elizabeth; Polish astronomer, wife of Hevelius (17th century).
Cortese	11.4S	218.3E	28	–	Isabella; Italian physician, medical writer (died 1561).
Cotis Mons	44.1N	233.1E	62	–	Thracian goddess, mother of gods, similar to Cybele.
Cotton	71.0N	300.0E	40	–	Egenni; French physicist (1881-1967).
Cunitz	14.5N	350.9E	48	68	Maria; Polish astronomer--mathematician (1610-1664).
Cybele Corona	7.5S	20.7E	500	–	Phrygian fertility goddess.
Cynthia	16.7S	347.5E	19	–	First name from Greek.
Dali Chasma	17.6S	167.0E	2077	70	Georgian; goddess of hunt.
Danilova	26.4S	337.3E	50	68	Maria; Russian ballet dancer (born 1793).
Danu Montes	58.5N	334.0E	808	66	Celtic mother of gods.
Danute	63.5S	56.5E	14	–	Lithuanian first name.
Daphne	41.3N	280.4E	14	–	First name from Greek.
Darago Fluctus	11.5S	313.5E	775	–	Philippine volcano goddess.
Darline	19.3S	232.6E	13	–	Anglo-Saxon first name.
Dashkova	77.9N	305.7E	42	–	Yekaterina; Russian philologist (1743-1810).
Datsolalee	38.3N	171.6E	19	–	Washo Indian artist, basketmaker (1835-1925).
Daura Chasma	72.4N	53.8E	729	–	Hausa (W. Sudan) great huntress.
Davies Patera	47.0N	269.0E	–	–	Emily Sarah; British educator; college founder (1830-1921).
de Ayala	12.3N	31.9E	20	–	Josefa; Spanish painter (1630-1684).
de Beausoleil	5.0S	102.9E	30	–	Martine; French earth science researcher (17th century).
de Beauvoir	2.0N	96.0E	40	–	Simone; French writer (1908-1986).
de Lalande	20.3N	354.9E	20	–	Marie-Jeanne; French astronomer (1768-1832).
de Staël	37.4N	324.2E	25	–	Anne; French historian, novelist (1766-1817).
De Witt	6.5S	275.6E	21	–	Lydia; American pathologist (1859-1928).
Deken	47.0N	288.5E	58	–	Agatha; Dutch novelist (1741-1804).
Dekla Tessera	57.4N	71.8E	1363	66	Latvian goddess of fate.
Deledda	76.0N	127.5E	32	–	Grazia; Italian novelist (1871-1936).
Delilah	57.9S	250.5E	18	–	First name from Hebrew.
Deloria	32.0S	97.0E	38	–	Ella; Dakota (Sioux) anthropologist (1888-1971).
Demeter Corona	53.9N	294.8E	560	66	Greek goddess of fertility.
Dennitsa Dorsa	85.6N	205.9E	872	66	Slavic goddess of day, light.
Derceto Corona	46.8S	20.2E	200	–	Phillistine fertility goddess.
d'este	34.2S	238.7E	21	–	Isabella; Italian archaeologist, business-woman (1474-1539).
Devana Chasma	9.6N	284.4E	1616	68	Czechoslovakian goddess of hunt.
Devorah	22.5S	343.4E	6	–	Hebrew first name.
Devorguilla	15.3N	3.8E	22	–	Irish heroine (died 1193).
Dhisana Corona	14.5N	111.7E	100	–	Vedic goddess of plenty.
Diana Chasma	14.8S	154.8E	938	70	Roman goddess of hunt/moon.
Dickinson	74.7N	177.2E	54	–	Emily; American poet (1830-1886).
Dinah	62.8S	37.0E	19	–	Hebrew first name.
Dione Regio	31.5S	328.0E	2300	68	Greek Titaness; 1st wife of Zeus.
Dix	36.9S	329.1E	68	–	Dorothea; American nurse, reformer (1802-1887).
Dodola Dorsa	46.8N	272.6E	607	–	South Slavic rain goddess.
Dolores	51.5N	200.5E	16	–	Spanish first name.
Doris	2.3N	89.9E	16	–	First name from Greek.
Drena	20.6S	338.6E	2	–	Lithuanian first name.
du Chatelet	21.5N	165.0E	19	–	Emilie; French mathematician, physicist (1706-1749).
Duncan	67.9N	291.7E	38	–	Isadora; American dancer (1878-1927).
Durant	62.3S	227.5E	23	–	Ariel; American writer (1898-1981).
Duse	82.5S	358.0E	27	–	Eleonora; Italian actress (1858-1924).
Dyan-Mu Dorsa	78.2N	31.9E	687	–	Chinese lightning goddess.
Earhart Corona	70.1N	136.2E	414	66	Amelia; American aviatrix (1897-1937).
Edgeworth	32.0N	22.7E	35	–	Maria; British novelist (1767-1849).
Edinger	68.8S	208.3E	34	–	Tilly; American geologist (1897-1967).

Feature name	Lat. (°)	Long. (°)	Size (km)	Page	Description
Efimova	81.0N	224.0E	28	–	(Simonovich-Efimova) Nina; Soviet painter and puppet-theatre designer (1877-1948).
Egeria Farrum	43.6N	7.5E	40	–	Roman water nymph.
Eigin Corona	5.0S	175.0E	200	–	Celtic fertility goddess.
Eileen	22.8S	232.6E	15	–	Irish first name.
Eistla Regio	10.5N	21.5E	8015	69	Norse giantess.
Eithinoha Corona	57.0S	7.5E	500	67	Iroquois earth goddess.
Elena	18.3S	73.3E	18	–	Italian first name.
Eliot Patera	39.0N	79.0E	–	–	George (Mary Ann Evans); English writer (1819-1880).
Elza	34.4S	275.8E	17	–	Latvian first name.
Enyo Fossae	61.0S	344.0E	900	–	Greek war goddess.
Eostre Mons	45.0N	329.5E	–	–	Teutonic goddess of spring.
Epona Corona	28.0S	208.5E	225	–	Celtic horse/fertility goddess.
Ereshkigal Corona	21.0N	84.5E	320	–	Mesopotamian nature/ fertility goddess.
Erika	72.0N	176.0E	12	–	Hungarian, German first name.
Erin	47.0S	184.8E	14	–	Irish first name.
Erinna	78.0S	309.0E	33	–	Greek poet (ca. 500 B.C.).
Eriu Fluctus	35.0S	358.0E	1200	–	Irish earth mother.
Erkir Corona	16.3S	233.7E	275	–	Armenian earth goddess.
Ermolova	60.0N	154.0E	60	–	Mariya; Russian actress (1853-1928).
Erxleben	50.9S	39.3E	28	–	Dorothea; first woman Ph.D. in Germany (1715-1762).
Escoda	18.2N	149.4E	20	–	Philipino organizer of Girl Scouts of the Philippines (1898-1945).
Estelle	1.1N	93.7E	20	–	First name from Latin.
Esther	19.4N	21.8E	17	–	First name from Persian.
Eudocia	59.5S	201.8E	28	–	Byzantine empress (ca. 401-460).
Eurynome Corona	26.5N	94.5E	200	–	Greek mother earth goddess.
Evangeline	69.7N	221.8E	15	–	First name from Greek.
Eve Corona	32.0S	359.8E	330	68	Hebrew first name; name changed from Eve (crater).
Evika	5.1S	31.4E	16	–	Tatar first name.
Faiga	4.9N	170.9E	10	–	Anglo-Saxon first name.
Fakahotu Corona	59.3N	106.4E	290	66	Tuamotu earth mother.
Farida	4.8N	38.9E	20	–	Azerbaijan first name.
Fatima	17.8S	31.9E	15	–	Arabic first name.
Fatua Corona	16.3S	17.7E	400	–	Roman goddess of fertility.
Fedorets	59.8N	65.7E	45	–	Velentina; Soviet astronomer (1923-1976).
Fedosova	45.0N	171.8E	24	–	Irina; Russian folk poet (1831-1899).
Felesta Fossae	35.0N	46.5E	–	–	Amazon queen in Scythian epic tales.
Felicia	19.8S	226.4E	12	–	First name from Latin.
Ferber	26.4N	13.0E	23	–	Edna; American author (1887-1968).
Fernandez	76.3N	16.4E	26	–	M. A.; Spanish actress (18th century).
Feronia Corona	68.0N	281.7E	360	–	Ancient Italian goddess of spring and flowers.
Ferrier	15.8N	111.1E	30	–	Kathleen; English opera singer (1912-1953).
Festa	11.5N	27.2E	25	–	Italian painter.
Flagstad	54.3S	18.9E	48	–	Kirsten; Norwegian opera singer (1895-1962).
Flosshilde Farra	10.5N	279.4E	75	–	German water nymph.
Foquet	15.1S	203.5E	50	–	Marie; French medical writer, charity worker (17th century).
Fornax Rupes	30.3N	201.1E	729	71	Roman goddess of hearth and baking of bread.
Fortuna Tessera	69.9N	45.1E	2801	66	Roman goddess of chance.

Feature name	Lat. (°)	Long. (°)	Size (km)	Page	Description
Fossey	2.0N	188.7E	30	–	Dian; American zoologist, conservationist (1932-1985).
Fotla Corona	58.5S	163.5E	150	–	Celtic fertility goddess.
Francesca	28.0S	57.7E	18	–	Italian first name.
Frank	13.2S	12.9E	20	–	Anne; Dutch heroine, diarist (1929-1945).
Fredegonde	50.7S	92.9E	26	–	Frankish queen (died A.D. 597).
Freyja Montes	74.1N	333.8E	579	66	Norse, mother of Odin.
Friagabi Fossae	50.2N	109.5E	141	–	Old English goddess, connected with Mars.
Frida	68.0N	56.0E	24	–	Swedish first name.
Frigg Dorsa	51.2N	148.9E	896	66	Norse, wife of supreme god Odin.
Fukiko	23.2S	105.7E	15	–	Japanese first name.
Furki Mons	35.9N	236.4E	79	–	Chechen and Ingush (Caucasus) goddess, wife of thunder god Sela.
Gabie Rupes	67.5N	109.9E	350	–	Lithuanian goddess of fire and hearth.
Gabriela	17.9S	240.4E	19	–	First name from Hebrew.
Gaia	6.0N	21.5E	400	–	Greek earth/fertility goddess.
Galina	47.4N	307.0E	20	–	Bulgarian first name.
Galindo	23.3S	258.8E	24	–	Beatrix; Italian physician, educator (1473-1535).
Ganiki Planitia	25.9N	189.7E	5158	–	Orochian (Siberia) water spirit, mermaid.
Ganis Chasma	16.3N	196.4E	615	71	Western Lapp forest maiden.
Gautier	26.5N	42.8E	60	–	Judith; French novelist (1845-1917).
Gaze	17.9N	240.2E	30	–	Vera; Soviet astronomer (1899-1954).
Gefjun Corona	33.5S	98.5E	300	–	Norse fertility goddess.
Gentileschi	45.2N	260.5E	20	–	Artemisia; Italian painter (1593-ca. 1652).
Goppert-Mayer	59.8N	26.5E	35	–	Maria; Polish physicist, Nobel laureate (1907-1972).
Gerda	45.9N	91.0E	30	–	Danish, German first name.
Germain	38.0S	63.5E	33	–	Sophie; French mathematician (1776-1831).
Gertjon Corona	30.0S	276.0E	250	68	Teutonic goddess of fertility.
Giliani	72.9S	142.0E	27	–	Alessandra; Italian anatomist (1307-1326).
Gillian	15.2S	49.9E	16	–	First name from Latin.
Gilmore	6.6S	132.8E	23	–	Mary; Australian poet (1865-1962).
Gina	78.0N	76.0E	24	–	Italian first name.
Glaspell	58.4S	269.6E	26	–	Susan; American playwright, novelist (ca. 1876-1948).
Gloria	68.5N	94.5E	14	–	Portuguese first name.
Godiva	56.1S	251.5E	32	–	(Godgifu); Mercian (England) noblewoman (ca. 1040-1085).
Golubkina	60.0N	286.5E	27	–	Anna; Soviet sculptor (1864-1927).
Goncharova	63.0S	97.7E	30	–	Natalya; Russian artist (1881-1962).
Grace	13.9S	268.9E	19	–	First name from Greek.
Graham	6.0S	6.0E	75	–	Martha; American dancer, choreographer (1894-1991).
Grazina	72.5N	337.5E	16	–	Lithuanian first name.
Greenaway	22.9N	145.1E	85	70	Catherine (Kate); English author, illustrator (1846-1901).
Gregory	7.2N	95.8E	21	–	Isabella; Irish playwright (1852-1932).
Gretchen	59.6S	213.2E	20	–	German first name.
Grey	52.4S	329.2E	50	–	Jane; English noblewoman (1537-1555).
Grimke	17.3N	215.3E	37	–	Sarah; American abolishionist (1792-1873).

Feature name	Lat. (°)	Long. (°)	Size (km)	Page	Description
Guan Daosheng	61.1S	181.8E	46	–	Chinese painter, calligrapher (1262-1319).
Gudrun	10.6N	326.3E	15	–	First name from Norse.
Guilbert	57.9S	13.3E	30	–	Yvette; French singer (1865-1944).
Guinevere Planitia	21.9N	325.0E	7519	68	British, wife of Arthur.
Gula Mons	21.9N	359.1E	276	68	Babylonian earth mother, creative force.
Guor Linea	17.0N	2.6E	–	69	Northern European Valkyrie.
Gwynn	9.7N	37.2E	32	–	Nell; English actress, courtesan (1650-1687).
Habonde Corona	3.0N	81.8E	125	–	Danish goddess of abundance.
Halle	19.8S	145.4E	23	–	Wilhelmina; Austrian violinist (1839-1911).
Hallgerda Mons	55.0N	198.0E	–	–	Icelandic goddess of vanity.
Hannah	17.9N	102.6E	19	–	First name from Hebrew.
Hansberry	22.7S	324.1E	28	–	Lorraine; American playwright (1930-1965).
Hariasa Linea	19.0N	15.0E	–	–	German war goddess.
Hathor Mons	38.7S	324.7E	333	68	Egyptian sky goddess.
Hatshepsut Patera	28.1N	64.5E	118	69	Egyptian pharoah (1479 B.C.).
Haumea Corona	54.0N	21.8E	375	–	Polynesian fertility goddess.
Hayasi	53.7N	244.1E	38	–	Fumiko; Japanese writer (1903-1951).
Heather	6.7S	334.1E	12	–	English first name.
Hecate Chasma	18.2N	254.3E	3145	71	Greek moon goddess.
Heidi	23.6N	350.1E	14	–	First name; form of Hester.
Helen Planitia	51.7S	263.9E	4362	67	Greek; "the face that launched 1000 ships."
Hellman	4.8N	356.2E	24	–	Lillian; American playwright, author (1905-1984).
Heloise	40.0N	51.9E	40	–	French physician, hospital founder (ca. 1098-1164).
Hemera Dorsa	51.0N	243.4E	587	–	Greek goddess, personification of day.
Heng-o Chasma	6.6N	355.5E	734	68	Chinese moon goddess.
Heng-o Corona	2.0N	355.0E	1060	68	Named for associated chasma.
Henie	52.0S	145.8E	70	67	Sonja; Norwegian skater (1912-1969).
Henwen Fluctus	20.5S	179.9E	485	–	British Celtic sow-goddess.
Hepat Corona	2.0S	145.5E	150	–	Hittite mother goddess.
Hepworth	5.2N	94.7E	54	–	Barbara; English sculptor (1903-1975).
Hera Dorsa	36.4N	29.5E	813	–	Greek sky goddess, wife of Zeus.
Hervor Corona	25.5S	269.0E	250	71	Norse fertility goddess.
Hestia Rupes	6.0N	71.1E	588	69	Greek hearth goddess.
Hiei Chu Patera	48.3N	97.4E	139	–	Chinese, converted silk worm product into thread and material (2698 B.C.).
Higgins	7.6N	241.4E	40	–	Marguerite; American journalist (1920-1966).
Hildr Fossa	45.4N	159.4E	677	–	Norse mythological warrior.
Himiko	19.0N	124.2E	35	–	Japanese queen (4th century A.D.).
Hina Chasma	64.5N	20.0E	–	–	Hawaiian moon goddess.
Hippolyta Linea	42.0S	345.0E	–	–	Amazon queen.
Holde Corona	53.5N	155.8E	200	–	German fertility goddess.
Holiday	46.7S	12.7E	24	–	Billie; American singer (1915-1959).
Horner	23.4N	97.5E	28	–	Mary; 19th century English naturalist, geologist.
Howe	45.7S	174.6E	39	–	Julia; American biographer, poet (1819-1910).
Hroswitha Patera	35.8N	34.8E	163	–	German writer (ca. 935-975).
Hsueh T'ao	52.9S	13.7E	20	–	Chinese poet, artist (ca. A.D. 760).
Hua Mulan	86.8N	337.7E	23	–	Chinese warrior (ca. A.D. 590).
Huang Daopo	54.2S	165.1E	27	–	Chinese engineer.
Hull	59.4N	263.3E	48	–	Peggy; American war correspondent (1889-1967).
H'uraru Corona	9.0N	68.0E	150	–	Pawnee earth mother.
Hurston	77.7S	94.5E	65	–	Zora; American anthropologist, writer (ca. 1901-1960).
Hwangcini	6.3N	141.7E	30	–	Korean poet (16th century A.D.).
Hyndla Regio	22.5N	294.5E	2300	–	Norse wood giantess.
Ichikawa	61.6S	156.4E	36	–	Fusaye; Japanese feminist (1893-1981).
Idem-Kuva Corona	25.0N	358.0E	230	–	Finno-Ugraic harvest spirit.
Idunn Mons	46.5S	213.5E	250	–	Norse goddess.
Ilbis Fossae	71.9N	254.6E	512	–	Yakutian (Siberia) goddess of bloodshed.
Ilga	12.4S	307.4E	11	–	Latvian first name.
Ilithyia Mons	13.5S	315.5E	90	–	Greek goddess of childbirth.
Imdr Regio	43.0S	212.0E	1611	67	Norse giantess.
Inanna Corona	37.0S	35.9E	350	–	Semitic fertility goddess.
Inari Corona	18.0S	120.3E	300	–	Japanese rice goddess.
Indira	64.0N	289.5E	14	–	Hindu first name.
Indrani Corona	37.5S	70.5E	200	–	Hindu fertility goddess.
Ingrid	12.4S	308.8E	15	–	Scandanavian first name.
Inira	43.1S	239.2E	17	–	Eskimo first name.
Innini Mons	34.6S	328.5E	339	68	Babylonian earth mother worshipped at Kish.
Irene	49.8N	134.0E	13	–	First name from Greek.
Irina	34.8N	91.4E	22	–	Russian first name.
Iris Dorsa	52.7N	221.3E	2050	66	Greek goddess of rainbow.
Isabella	29.7S	204.1E	175	71	Of Castile; Spanish queen (1451-1504).
Isako	9.0S	277.9E	13	–	Japanese first name.
Ishtar Terra	70.4N	27.5E	5609	66	Babylonian goddess of love.
Isong Corona	12.0N	49.2E	540	69	Ibibio (Nigeria) fertility goddess.
Itzpapalotl Tessera	75.7N	317.6E	380	–	Aztec goddess of fate.
Ivka	68.0N	304.0E	16	–	Serbocroatian first name.
Iweridd Corona	21.0S	310.0E	500	–	Brythonic (English Celtic) earth goddess.
Ix Chel Chasma	10.0S	73.4E	503	69	Aztec wife of the sun god; probably moon goddess.
Iyele Dorsa	50.0N	278.7E	595	–	Moldavian witch who directed the winds.
Izumi Patera	50.3N	193.6E	74	–	Sikibu; Japanese writer (974-1036).
Jacqueline	70.0S	123.8E	17	–	First name from French.
Jadwiga	68.5N	91.0E	14	–	Polish first name.
Jael Mons	52.0N	121.0E	–	–	Hebrew goddess of dawn.
Javine Corona	5.5S	251.2E	450	–	Lithuanian harvest goddess.
Jeanne	39.9N	331.5E	27	–	French first name.
Jennifer	4.6S	99.8E	9	–	First name from Greek.
Jerusha	22.0S	342.7E	17	–	Hebrew first name.
Jex-Blake	65.5N	169.0E	–	–	Sophia; British pioneer woman physician (1840-1912).
Jhirad	16.8S	105.6E	50	–	Jerusha; Indian physician.
Jocelyn	33.2S	276.4E	14	–	German first name.
Johanna	19.5N	247.2E	18	–	Hebrew first name.
Johnson	51.9N	254.5E	25	–	Amy; Australian aviator (1903-1941).
Joliot-Curie	1.6S	62.1E	80	69	Irene; French physicist, Nobel laureate (1897-1956).
Jord Corona	58.5S	349.5E	130	–	Norse earth goddess.
Joshee	5.5N	288.8E	34	–	Anandibai; Indian pioneer physician (1865-1887).
Juanita	62.9S	89.9E	19	–	Spanish first name.
Judith	29.1S	104.5E	20	–	Hebrew first name.
Julie	51.0N	242.5E	16	–	Czech, German first name.
Juno Chasma	30.5S	111.1E	915	70	Roman sky goddess; sister and consort of Jupiter.

Feature name	Lat. (°)	Long. (°)	Size (km)	Page	Description
Juno Dorsum	31.0S	95.6E	1652	70	As above.
Junkgowa Corona	37.0N	257.0E	400	–	Yulengor (Australia) fertility goddess.
Jurate Colles	56.8N	153.5E	418	66	Lithuanian sea goddess.
Kahlo	59.9S	178.8E	36	–	Frida; Mexican artist (1910-1954).
Kaikilani	32.7S	163.1E	19	–	First female ruler of Hawaii (ca. 1555).
Kaiwan Fluctus	48.0S	1.5E	1200	–	Ethiopian earth mother.
Kala	1.5N	314.2E	17	–	Kamchatka first name.
Kalaipahoa Linea	60.5S	338.0E	2400	67	Hawaiian war goddess.
Kallistos Vallis	51.1S	21.5E	900	–	Ancient Greek name for planet Venus.
Kamadhenu Corona	21.0N	136.5E	400	–	Hindu goddess of plenty.
Kamari Dorsa	59.2N	55.8E	589	–	Georgian sky maiden, daughter of weather god.
Kamui-Huci Corona	63.5S	322.5E	300	–	Ainu (Japan) earth goddess.
Kanik	32.6S	249.8E	16	–	Sakhalin first name.
Kara Linea	44.0S	306.0E	–	–	Icelandic valkryie.
Kartini	57.8N	333.0E	24	–	Raden Adjeng; Javanese educator (1879-1904).
Kauffman	49.5N	27.0E	24	–	Angelica; German painter (1741-1807).
Kawelu Planitia	32.8N	246.5E	3910	71	Hawaiian mythological heroine, died and brought back to life.
Kayanu-Hime Corona	33.5N	57.0E	150	–	Shinto grain goddess.
Kaygus Chasmata	49.6N	52.1E	503	66	Ketian (Siberia) ruler of forest animals.
Kelea	8.9N	25.6E	25	–	Chieftess of Maui (ca. 1450).
Keller Patera	45.0N	273.5E	–	–	Helen; blind and deaf American lecturer (1880-1968).
Kelly	4.8S	359.2E	11	–	Gaelic first name.
Kemble	48.3N	14.4E	29	–	Frances (Fanny) Anne; English actress (1809-1893).
Kenny	44.3S	271.1E	55	–	Elizabeth; Australian nurse, therapist (1880-1952).
Khatun	40.3N	86.9E	37	–	Mihri; Turkish poet (1456-1514).
Khelifa	1.5S	129.8E	13	–	Arabic first name.
Khotun Corona	46.5S	81.5E	200	–	Yakut goddess of plenty.
Kingsley	22.6S	306.3E	24	–	Mary; English explorer, writer (1862-1900).
Kiris	20.9N	98.8E	13	–	Latvian first name.
Kitna	28.9S	277.3E	16	–	Kamchatka first name.
Klafsky	20.8S	188.0E	28	–	Katherina; Hungarian opera singer (1855-1896).
Klenova	78.1N	104.2E	125	–	Mariya; Soviet marine geologist (ca. 1910-1978).
Koidula	64.3N	139.1E	47	–	Lydia; Estonian poet (1843-1886).
Kollwitz	25.2N	133.6E	30	–	Käthe; German artist (1867-1945).
Konopnicka	14.5N	166.7E	20	–	Maria; Polish author (1842-1910).
Kottauer Patera	36.7N	39.6E	136	–	Helena; Austrian historical writer (1410-1471).
Kottravey Chasma	30.5N	76.8E	744	–	Dravidian (India) hunting goddess.
Kozhla-Ava Chasma	56.2N	50.6E	581	–	Marian (Volga Finn) forest goddess.
Krumine Corona	5.0S	261.5E	300	–	Lithuanian food goddess.
Kuanja Chasma	12.0S	99.5E	890	70	Mbundu goddess of the spirit of the hunt.
Kuan-Yin Corona	4.3S	10.0E	310	–	Chinese fertility goddess.
Kubebe Corona	15.5N	132.5E	125	–	Hittite mother earth goddess.
Kunapipi Corona	33.9S	86.0E	220	–	Australian mother earth goddess.
Kunhild Corona	19.3N	80.1E	200	–	German fertility maiden.
Kurukulla Mons	48.5N	103.0E	–	–	Etan goddess of wealth.

Feature name	Lat. (°)	Long. (°)	Size (km)	Page	Description
Kutue Tessera	39.5N	108.8E	653	70	Ulchian (Siberia) folklore toad that brings happiness.
La Fayette	70.2N	107.9E	68	–	Marie; French novelist (1634-1693).
Labé Patera	52.0N	273.0E	–	–	Etan (Tibet) goddess of wealth.
Lachappelle	26.7N	336.5E	35	–	Marie; French medical researcher (1769-1821).
Lachesis Tessera	44.4N	300.1E	664	68	Greek, one of three Fates.
Lada Terra	60.0S	20.0E	8614	67	Slavic goddess of love.
Lagerlöf	81.0N	285.3E	53	–	Selma; Swedish novelist (1858-1940).
Laima Tessera	55.0N	48.5E	971	–	Latvian and Lithuanian goddess of fate.
Lakshmi Planum	68.6N	339.3E	2343	66	Indian goddess of love and war.
Lampedo Linea	57.0N	295.0E	–	–	Scythian Amazon queen.
Landowska	84.5N	83.0E	45	–	Wanda; Polish pianist (1877-1959).
Langtry	17.0S	155.0E	52	–	Lillie; English actress (1853-1929).
Lasdona Chasma	69.3N	34.4E	697	–	Lithuanian main forest goddess.
Laulani	68.2S	121.3E	15	–	Hawaiian first name.
Laūma Dorsa	64.8N	190.4E	1517	66	Latvian witch who flies in the sky.
Laura	49.0N	141.0E	16	–	Spanish, Italian first name.
Laurencin	15.4S	46.4E	30	–	Marie; French painter (1885-1956).
Lavinia Planitia	47.3S	347.5E	2820	67	Roman; wife of Aeneas.
Lazarus	52.8S	127.2E	26	–	Emma; American poet (1849-1887).
Leah	34.2S	187.7E	15	–	Hebrew first name.
Lebedeva	45.2N	49.6E	42	–	Sarah; Russian sculptor (1881-1968).
Leda Planitia	44.0N	65.1E	2890	69	Mother of Helen, Castor.
Ledoux Patera	9.2S	224.8E	75	–	Jeanne; French artist (1767-1840).
Lehmann	44.1S	38.7E	20	–	Inge; Danish geophysicist (1888-?).
Leida	23.3S	266.5E	22	–	Estonian first name.
Leila	44.3S	86.6E	19	–	First name from Arabic.
Lena	39.2N	22.8E	25	–	Russian first name.
Lenore	38.7N	292.3E	16	–	Greek first name (form of Helen).
Leonard	73.8S	185.0E	37	–	Wrexie; American assistant to P. Lowell (1867-1937).
Letitia	34.6N	288.6E	16	–	First name from Latin.
Leyster	1.0N	259.9E	45	–	Judith; Dutch painter (1609-1660).
Li Qingzhao	23.7N	94.3E	21	–	Chinese essayist, scholar (1085-1151).
Liban Farra	23.9S	353.5E	100	–	Irish water goddess.
Libera Corona	12.5N	24.0E	350	–	Roman fertility goddess.
Lida	29.1S	94.5E	14	–	First name from Greek.
Lida	36.5N	274.0E	–	–	Russian name.
Lilian	25.6N	336.0E	14	–	First name from Hebrew.
Lilinau Corona	34.0N	22.0E	200	–	Native American fertility maiden.
Liliya	30.0N	31.1E	18	–	Russian first name.
Lilwani Corona	29.5S	271.5E	500	68	Hittite earth goddess.
Lind	50.2N	354.9E	44	–	Jenny; Swedish singer (1820-1887).
Lineta	5.0S	354.1E	15	–	Latvian first name.
Lo Shen Valles	12.8S	89.6E	225	–	Chinese river goddess.
Lockwood	32.8S	51.5E	23	–	Belva; American lawyer, feminist (1830-1917).
Lois	17.9S	214.7E	15	–	First name from Greek.
Lonsdale	55.6N	222.1E	45	–	Kathleen; English physicist, crystallographer (1903-1971).
Loretta	19.7S	202.5E	13	–	First name from Latin.
Lotta	51.0N	336.0E	12	–	Swedish first name.

Feature name	Lat. (°)	Long. (°)	Size (km)	Page	Description
Louhi Planitia	80.5N	120.5E	2441	66	Karelo-Finn mother of the North.
Lucia	62.1S	67.8E	17	–	First name from Latin.
Lukelong Dorsa	73.3N	178.8E	1566	66	Polynesian goddess, creator of heavens.
Lullin	23.1N	81.0E	24	–	Maria; Swiss entomologist (1750-1831).
Lydia	10.7N	340.8E	15	–	First name from Greek.
Lyon	66.5S	270.5E	14	–	Mary; American educator, college president (1797-1849).
Lyudmila	62.0N	330.0E	16	–	Russian first name.
Ma Shouzhen	35.7S	92.4E	20	–	Chinese poet, painter (1592-1628).
Maa-Ling	14.7S	359.5E	6	–	Chinese first name.
Maan-Eno Corona	40.8N	102.5E	300	–	Estonian harvest goddess.
Maat Mons	0.5N	194.6E	395	71	Ancient Egyptian goddess of truth and justice.
MacDonald	30.0N	120.7E	19	–	Flora; Scottish heroine (1722-1790).
Madeleine	4.7S	293.2E	18	–	French first name.
Magda	67.0N	329.5E	12	–	Danish first name.
Magnani	58.0N	337.0E	30	–	Anna; Italian actress (1908-1973).
Mahina	2.0S	182.2E	16	–	Hawaiian first name.
Mahuea Tholus	37.5S	164.7E	110	–	Maori fire goddess.
Makh Corona	48.7S	85.0E	200	–	Assyro-Babylonian goddess of fecundity.
Makola	3.8S	106.7E	18	–	Hawaiian first name.
Malintzin Patera	57.0N	82.0E	–	–	(Malina); Aztec Indian guide, interpreter (1501-1550).
Maltby	23.3S	119.8E	40	–	Margaret; American physicist (1860-1944).
Mama-Allpa Corona	27.0S	31.0E	300	–	Peruvian harvest goddess.
Manto Fossae	64.5N	60.0E	–	–	Greek war goddess.
Manton	9.3N	26.8E	18	–	Sidnie; English zoologist (1902-1980).
Manzan-Gurme Tesserae	39.5N	179.5E	–	–	Ancestress who possesses the book of fate in Mongol, Tibetan, Buriat mythologies.
Manzolini	25.7N	91.1E	42	–	Anna; Italian anatomist, teacher (1716-1774).
Maram Corona	7.5S	221.5E	600	71	Oromo (Ethiopia) fertility goddess.
Maranda	4.9N	169.8E	14	–	Latvian first name.
Mardezh-Ava Dorsa	32.4N	68.6E	906	69	Marian (Volga Finn) wind goddess.
Margarita	12.8N	9.2E	13	–	Greek first name.
Margit	60.0N	273.0E	14	–	Hungarian first name.
Mari Corona	54.0N	151.0E	200	–	Cretan goddess of plenty.
Marie	21.7S	232.4E	15	–	French first name.
Maria Celeste	23.5N	140.5E	90	70	Daughter of Galileo (died 1634).
Markham	4.1S	155.6E	69	–	Beryl; English aviator (1902-1986).
Marsh	63.7S	46.7E	35	–	Ngaio; New Zealand playwright, novelist (1899-1982).
Martinez	11.7S	174.7E	25	–	Maria; Pueblo artist, potter (1886-1980).
Marzhan	58.9S	248.5E	20	–	Karakal first name.
Masako	30.2S	53.1E	26	–	Hozyo; Japanese ruler (1157-1225).
Maslenitsa Corona	77.0N	202.5E	–	–	Slavonic personification of fertility.
Mawu Corona	31.5N	241.0E	–	–	Fon (Benin) goddess of fertility.
Maxwell Montes	65.2N	3.3E	797	66	James C.; British physicist (1831-1879).
Maya Corona	23.0N	98.0E	225	–	Hindu mother earth goddess.
Mayaeul Corona	27.5S	154.0E	200	–	Mexican goddess of plenty.
Mbokomu Mons	15.1S	215.2E	460	–	Ngombe (Zaire) ancestor/goddess.
Mead	12.5N	57.4E	280	69	Margaret; American anthropologist (1901-1978).
Medeina Chasma	46.2N	89.3E	606	–	Lithuanian forest goddess.
Medhavi	19.5S	40.6E	30	–	Ramabai; East Indian author, humanitarian (1858-1922).
Megan	61.7S	130.6E	18	–	Welsh first name.
Meitner	55.9S	321.8E	98	67	Lise; Austrian physicist (1878-1968).
Melanie	62.8S	144.3E	16	–	First name from Greek.
Melba	4.7N	193.4E	2	–	Nellie; Australian opera singer (1861-1931).
Melia Mons	62.8N	119.3E	311	–	Greek nymph.
Mena Colles	52.5S	160.0E	850	–	Roman goddess of menses.
Menat Undae	24.8S	339.4E	100	–	Arabian desert goddess.
Meni Tessera	48.1N	77.9E	454	–	Semitic goddess of fate.
Mentha Mons	43.0N	237.4E	79	–	Roman goddess, personification of the human mind.
Merian	34.5N	76.2E	20	–	Maria; Dutch entomologist (1647-1717).
Merit Ptah	11.3N	115.7E	19	–	Egyptian queen, physician (ca. 2700 B.C.).
Mesca Corona	27.0N	342.6E	190	–	Irish fertility goddess.
Meskhent Tessera	65.8N	103.1E	1056	66	Egyptian goddess of fortune.
Metis Regio	72.0N	256.0E	729	66	Greek Titaness.
Metra Corona	26.0N	98.0E	–	–	Persian fertility/moon goddess.
Mežas-Mate Chasma	51.0N	50.7E	506	–	Latvian forest goddess.
Michelle	19.5S	40.4E	14	–	First name from French.
Milda Mons	52.5N	159.5E	–	–	Lithuanian goddess of love.
Millay	24.4N	110.9E	45	–	Edna St. Vincent; American poet (1892-1950).
Minerva Fossae	64.5N	252.5E	–	–	Roman goddess of war.
Mirabeau	1.1N	284.3E	22	–	Sibylle; French writer (died 1932).
Miralaidji Corona	14.0S	163.8E	300	–	Aborigine fertility goddess.
Miriam	36.5N	48.2E	15	–	First name from Hebrew.
Misne Chasma	77.1N	316.5E	610	–	Mansi (Siberia) forest maiden.
Mist Fossae	39.5N	247.3E	244	–	Norse Valkyrie.
Mnemosyne Regio	65.8N	277.9E	–	–	Greek Titaness.
Moira Tessera	58.7N	310.5E	361	–	Greek fate goddess.
Mokosha Mons	57.7N	255.0E	270	66	East Slavic main goddess.
Molpadia Linea	48.0S	359.0E	–	–	Amazon.
Molpe Colles	76.5N	195.0E	–	–	Greek; mother of Sirens.
Mona Lisa	25.6N	25.3E	80	69	Leonardo da Vinci's model, real name Lisa Giacondo (born ca. 1474).
Monika	72.5N	122.0E	24	–	German first name.
Montagu	36.9N	177.5E	20	–	Mary; English medical pioneer, poet, writer (1689-1762).
Montessori	59.1N	280.1E	43	–	Maria; Italian educator (1870-1952).
Montez	17.9N	266.6E	20	–	Lola; Irish dancer (1818-1861).
Moore	30.3S	248.3E	21	–	Marianne; American poet, editor (1887-1972).
Morana Chasma	68.9N	24.0E	317	–	Czech moon goddess.
Morisot	61.2S	211.4E	55	–	Berthe; French artist (1841-1895).
Morrigan Linea	54.5S	311.0E	3200	–	Celtic war goddess.
Moses	34.3N	120.1E	35	–	A. "Grandma"; American painter (20th century).
Mots Chasma	51.9N	56.1E	464	–	Avarian (Caucasus) moon goddess.
Mowatt	14.7S	292.2E	40	–	Anna; American actress, playwright, author (1819-1870).
Mu Guiying	41.2N	80.7E	25	–	Chinese warrior.

Feature name	Lat. (°)	Long. (°)	Size (km)	Page	Description
Mukhina	29.5N	0.5E	24	–	Vera; Soviet sculptor (1889-1953).
Mumtaz-Mahal	30.3N	228.3E	39	–	Mogul empress for whom Taj Mahal was built (1592-1631).
Munter	15.3S	39.3E	36	–	Gabriele; German painter (1877-1962).
Muriel	41.7S	12.3E	19	–	First name from Greek.
Muta Mons	56.0N	359.0E	–	–	Roman goddess of silence.
Mylitta Fluctus	56.0S	353.5E	1250	–	Semitic mother goddess.
Nabuzana Corona	8.5S	47.0E	69	–	Ganda (Uganda) crop goddess.
Nadine	7.8N	359.1E	19	–	First name from French.
Nadira	44.0N	201.7E	36	–	Uzbek poet (1791-1842).
Nagavonyi Corona	18.5S	259.0E	190	–	Ganda (Uganda) crop goddess.
Nalkowska	28.2N	290.0E	26	–	Zofia; Polish novelist, playwright (1884-1954).
Nambi Dorsum	72.5S	213.0E	1125	–	Ugandan sky goddess.
Nana	50.0N	75.0E	10	–	Serbocroatian first name.
Naomi	6.0N	70.1E	18	–	First name from Hebrew.
Natalia	67.0N	273.0E	10	–	Romanian first name.
Navka Planitia	8.1S	317.6E	2100	68	Arab mother-goddess Allat as goddess of good fortune.
Neago Fluctūs	48.9N	349.5E	–	–	Senaca (U.S.A.) goddess of silence.
Nefertiti Corona	35.8N	48.0E	–	–	Beautiful Egyptian queen (ca. 1390-ca. 1354 B.C.).
Nehalennia Corona	14.0N	10.0E	345	–	Teutonic fertility goddess.
Němcová	5.9N	125.0E	24	–	Božena; Czech novelist, poet (1820-1882).
Nemesis Tessera	45.9N	192.6E	355	–	Greek goddess of fate.
Nephele Dorsa	39.7N	139.0E	1937	70	Greek cloud goddess.
Nepret Corona	53.0N	7.0E	–	–	Egyptian grain goddess.
Nepthys Mons	33.0S	317.5E	350	68	Egyptian goddess of barren lands.
Nertus Tholus	61.2N	247.9E	66	–	German/Norse vegetation goddess.
Nevelson	35.3S	307.8E	75	–	Louise; Russian-born American artist (1899-1988).
Neyterkob Corona	49.5N	204.5E	–	66	Masai earth/fertility goddess.
Nightingale Corona	63.6N	129.5E	471	66	Florence; English nurse (1820-1910).
Nijinskaya	25.9N	122.3E	30	–	Bronislava; Russian dancer (1891-1972).
Nike Fossae	62.0S	347.0E	850	–	Greek goddess of victory.
Nilsson	76.0S	277.7E	26	–	Christine; Swedish opera singer, violinist (1843-1921).
Nin	3.9S	266.4E	27	–	Anaïs; French-born American novelist (1903-1977).
Nina	55.5S	238.7E	23	–	First name from Russian.
Ningal Undae	9.0N	60.7E	225	–	Sumerian desert goddess.
Ningyo Fluctus	5.5S	206.0E	970	–	Japanese fish goddess.
Ninhursag Corona	38.0S	23.5E	125	–	Babylonian earth goddess.
Nintu Corona	19.2N	123.5E	75	–	Akkadian earth goddess.
Niobe Planitia	21.0N	112.3E	5008	70	Greek; her 12 children were killed by Artemis and Apollo.
Nishtigri Corona	24.5S	72.0E	275	–	Hindu earth mother.
Nissaba Corona	25.5N	355.5E	300	–	Mesopotamian wisdom/ fertility goddess.
Nofret	58.7S	252.0E	22	–	Egyptian queen (ca. 1900 B.C.).
Nokomis Montes	18.9N	189.9E	486	71	Algonquin (N. America) earth mother.
Noreen	33.5N	22.7E	20	–	Irish first name.
Noriko	5.3S	358.3E	7	–	Japanese first name.
Nsomeka Planitia	55.0S	170.0E	4500	67	Bantu culture heroine.
Nzingha Patera	69.0N	206.0E	–	–	(Ann Zingha) queen, head of Amazon band (1582-1663).
Oakley	29.3S	310.5E	22	–	Annie; American sharpshooter, entertainer (1860-1926).
Oanuava Corona	32.5S	255.5E	375	–	Gaulish Celtic earth goddess.
Obukhova	71.0N	289.0E	44	–	Nadezhda; Soviet singer (1886-1961).
O'Connor	26.0S	143.8E	27	–	Flannery; American novelist (1925-1964).
Odilia	81.5N	200.5E	20	–	Portuguese first name.
Oduduva Corona	11.0S	211.5E	150	–	Yoruba (Nigeria) fertility goddess.
Ohogetsu Corona	27.0S	85.7E	175	–	Japanese food goddess.
O'Keeffe	24.5N	228.7E	76	–	Georgia; American artist (1887-1986).
Okipeta Dorsa	66.0N	238.5E	1200	66	Greek goddess of whirlwind.
Olesnicka	18.3N	210.8E	33	–	Zofia; Polish poet (flourished ca. 1550).
Olga	26.2N	283.8E	17	–	Russian first name.
Olwen Corona	37.5N	67.5E	175	–	Brythonic goddess of spring growth.
Olya	51.0N	292.0E	14	–	Russian first name.
Omeciuatl Corona	16.5N	119.0E	175	–	Aztec generative power.
Onatah Corona	49.0N	5.0E	–	66	Iroquois corn spirit.
Ops Corona	68.5N	89.0E	–	–	Greek fertility goddess.
Orczy	3.6N	52.2E	29	–	Emmuska; Hungarian novelist, playwright (1865-1947).
Orlova	56.5N	235.0E	28	–	Lyubov; Soviet actress (1902-1975).
Oshun Farra	4.2N	19.3E	80	–	Yoruba (Nigeria) fresh water goddess.
Osipenko	71.0N	321.0E	30	–	Polina; Soviet aviator (1907-1939).
Otau Corona	67.8N	298.7E	172	–	Bini (S. Nigeria) goddess of fertility.
Otygen Corona	57.0S	30.5E	400	67	Mongolian earth mother.
Ovda Fluctus	6.1S	95.5E	310	–	Named from regio where feature is located.
Ovda Regio	2.8S	85.6E	5280	69	Marijian (Russian) forest giantees.
Ozza Mons	4.5N	201.0E	507	70	Persian goddess honored by the Koreishies.
Pamela	11.1N	238.5E	13	–	English first name.
Pandrosos Dorsa	58.2N	206.2E	1254	66	Greek dew goddess.
Pani Corona	19.9N	231.5E	320	–	Maori fertility goddess.
Parga Chasma	24.5S	271.5E	–	–	Samoyed forest spirit.
Parra	20.5N	78.1E	50	–	Chilean writer.
Patti	34.8N	301.6E	40	–	Adelina; Italian singer (1843-1919).
Pavlova Corona	14.3N	38.9E	37	69	Anna; Russian ballerina (1881-1931).
Peck	29.0S	294.2E	30	–	Annie; American mountaineer, educator (1850-1935).
Peggy	20.4S	357.2E	12	–	English first name (form of Margaret).
Peña	23.6S	190.6E	32	–	Tonita (Quah Ah); Pueblo artist (1895-1949).
Penardun Linea	54.0S	344.0E	975	67	Celtic sky goddess.
Perchta Corona	17.0N	234.5E	500	71	German fertility goddess.
Phaedra	35.9N	252.7E	15	–	First name from Greek.
Phoebe Regio	6.0S	282.8E	2852	68	Greek Titaness.
Phra Naret Corona	66.6S	209.6E	150	–	Thai fertility goddess.
Phryne	46.2S	314.8E	40	–	Greek model, courtesan (fourth century B.C.).
Phyllis	12.3N	132.4E	13	–	First name from Greek.
Piaf	0.8N	5.1E	30	–	Edith; French singer, songwriter (1915-1963).

Feature name	Lat. (°)	Long. (°)	Size (km)	Page	Description
Piret	38.0N	42.0E	–	–	Estonian first name.
Piscopia	1.5N	190.9E	26	–	Elena; Italian mathematician, educator (1646-1684).
Pocahontas Patera	65.0N	49.5E	–	–	Daughter of Powhatan Indian peacemaker (1595-1617).
Polina	42.2N	148.2E	24	–	Russian first name.
Pölöznitsa Corona	0.5N	302.0E	675	68	Finno-Ugric grain goddess.
Pomona Corona	79.3N	299.4E	315	66	Roman goddess of fruits.
Ponselle	63.0S	289.0E	53	–	Rosa; American opera singer (1897-1981).
Potanina	31.6N	53.1E	82	–	Aleksandra; Russian explorer (1843-1893).
Potter	7.2N	309.4E	52	–	Beatrix; English children's author (1866-1943).
Prichard	43.8N	11.1E	30	–	Catharina; Australian writer (1884-1969).
Purandhi Corona	26.1N	343.5E	170	–	Hindu goddess of plenty.
Qetesh Corona	20.5S	343.5E	80	–	Egyptian fertility goddess.
Quetzalpetlatl Corona	64.0S	354.5E	400	67	Aztec fertility goddess.
Quilla Chasma	23.7S	127.3E	973	70	Inca moon goddess.
Rachel	48.7S	13.5E	12	–	First name from Hebrew.
Radka	76.0N	95.0E	12	–	Bulgarian first name.
Raisa	27.5N	280.3E	13	–	Russian first name.
Rananeida Corona	62.6N	263.5E	448	66	Saami-Lapp goddess of spring and fertility.
Rand	63.8S	59.5E	27	–	Ayn; Russian-born American writer (1905-1982).
Rangrid Fossae	62.7N	356.4E	243	–	Norse Valkyrie.
Rani	64.5N	160.0E	12	–	Hindu first name.
Raskova Paterae	51.0S	222.8E	80	–	Marina M.; Russian aviator (1912-1943).
Rauni Corona	40.8N	271.9E	271	68	Finnish goddess of harvest, earth.
Razia Patera	46.2N	197.8E	157	–	Queen of Delhi Sultanate (India) (1236-1240).
Recamier	12.5S	57.9E	25	–	Jeann-François-Julie--Adélaïde; French patriot; defied Napoleon (ca. 1777-ca. 1849).
Regina	29.8N	147.4E	35	–	First name from Latin.
Renenti Corona	32.7N	326.2E	200	–	Egyptian goddess of abundance.
Renpet Mons	76.0N	236.4E	138	66	Egyptian goddess of springtime and youth.
Rhea Mons	32.4N	282.2E	217	68	Greek Titaness.
Rhoda	11.5N	347.7E	13	–	First name from Greek.
Rhys	8.6N	298.8E	45	–	Jean; Welsh writer (1894-1979).
Richards	2.5N	196.0E	29	–	Ellen; founder of science of ecology (1842-1911).
Rigatona Corona	33.5S	278.5E	300	68	Celtic fertility goddess.
Riley	14.0N	72.2E	25	–	Margaretta; English botanist (1804-1899).
Rita	71.0N	335.0E	10	–	Italian first name.
Romanskaya	23.2N	178.4E	31	–	Sofia; Soviet astronomer (1886-1969).
Rosa Bonheur	9.8N	288.7E	105	68	French painter (1822-1899).
Rose	35.2S	248.2E	15	–	German first name.
Rosmerta Corona	0.0N	124.5E	300	–	Celtic fertility/luck goddess.
Rossetti	57.0N	7.0E	25	–	Christina; English poet (1830-1894).
Rowena	10.4N	171.3E	18	–	Celtic first name.
Roxanna	26.5N	334.6E	9	–	First name from Persian.
Rudneva	78.0N	176.0E	30	–	Varvara; Russian medical doctor (1844-1899).
Rusalka Planitia	9.8N	170.1E	3655	70	Russian mermaid.
Ruslanova	84.0N	16.0E	19	–	Lidiya; Soviet singer (1900-1973).
Ruth	43.2N	19.8E	18	–	Hebrew first name.
Sabin	38.5S	274.6E	36	–	Florence; American medical researcher (1871-1953).
Sabira	5.8S	239.9E	14	–	Tatar first name.
Sacajawea Patera	64.3N	335.4E	233	–	Blackfoot Indian woman who guided Lewis & Clark expedition to the Pacific Northwest (1786-1812).
Sachs Patera	49.0N	324.0E	–	–	Nelly; German-born Swedish playwright, poet (1891-1970).
Saga Vallis	76.1N	340.6E	450	–	Norse goddess in the form of a waterfall.
Salika	5.0S	97.7E	14	–	Mari first name.
Salme Dorsa	58.0N	28.0E	–	–	Estonian sky maiden.
Samantha	45.5N	281.4E	16	–	First name from Aramaic.
Samintang	39.0S	80.6E	24	–	16th century Korean poet.
Samundra Vallis	24.1S	347.1E	110	–	Indian river goddess.
Sand Patera	42.0N	15.5E	–	–	George (Aurore Dupin); French novelist (1804-1876).
Sandel	45.7S	211.6E	20	–	Cora; Norwegian author (1880-1974).
Sanger	33.8N	288.5E	84	–	Margaret; American medical researcher (1883-1966).
Sanija	33.1N	250.0E	18	–	Tatar first name.
Sapas Mons	8.5N	188.3E	217	71	Phoenician goddess.
Sappho Patera	14.1N	16.5E	92	69	Lyric poet; Lesbos, Asia Minor (flourished ca. 610-ca. 580 B.C.).
Sarah	42.4S	1.7E	19	–	Hebrew first name.
Sarpanitum Corona	52.3S	14.6E	170	–	Babylonian fertility goddess.
Sartika	63.4S	67.1E	28	–	Ibu Dewi; Indonesian educator (1884-1942).
Saskia	28.6S	337.2E	40	–	Artist's model, wife of Rembrandt.
Sati Vallis	3.2N	334.4E	225	–	Egyptian river goddess.
Saule Dorsa	58.0S	206.0E	1375	–	Lithuanian sun goddess.
Sayers	67.5S	230.0E	90	–	Dorothy L.; English novelist, playwright (1893-1957).
Scarpellini	23.4S	34.4E	25	–	Caterina; Italian astronomer (19th century).
Schumann-Heink Patera	74.0N	215.0E	–	–	Ernestine; German singer (1861-1936).
Sedna Planitia	42.7N	340.7E	3572	68	Eskimo; her fingers became seals and whales.
Seia Corona	3.0S	153.0E	225	–	Roman grain goddess.
Sekmet Mons	44.2N	240.8E	338	71	Ancient Egyptian goddess of war and battle.
Sel-Anya Dorsa	79.4N	81.3E	975	66	Hungarian wind goddess.
Selma	68.5N	156.0E	12	–	First name from Celtic.
Selu Corona	42.5S	6.0E	300	–	Cherokee corn goddess.
Semele Tholi	64.3N	202.9E	194	–	Frygian (Phoenician) earth goddess.
Semuni Dorsa	75.9N	8.0E	514	–	Ulchian (Siberia) sky goddess.
Seoritsu Farra	30.0S	11.0E	230	–	Japanese stream goddess.
Seshat Mons	26.5N	33.0E	–	–	Egyptian goddess of writing.
Sévigné	52.5N	326.5E	30	–	Marie; French writer (1626-1696).
Seymour	18.2N	326.5E	65	–	Jane; English queen (ca. 1509-1537).
Shakira	3.0N	213.7E	19	–	Bashkir first name.
Shih Mai-Yu	18.4N	318.9E	25	–	Chinese physician (1873-1954).
Shimti Tessera	31.9N	97.7E	1275	70	Babylonian; Ishtar as the goddess of Fate.
Shiwanokia Corona	42.0S	279.8E	500	68	Zuni fertility goddess.

Feature name	Lat. (°)	Long. (°)	Size (km)	Page	Description
Sicasica Fluctus	52.0S	180.4E	175	–	Aymara (Bolivia) mountain goddess.
Siddons	61.6N	340.6E	47	–	Sarah; English actress (1755-1831).
Sidney	13.4N	199.6E	21	–	Mary; Elizabethan dramatist (1561-1621).
Sif Mons	22.0N	352.4E	200	68	Teutonic goddess, Thor's wife.
Sige Dorsa	32.0N	106.5E	–	–	Babylonian sky goddess.
Sigrid	63.5N	314.5E	20	–	Scandanavian first name.
Sigrun Fossae	52.3N	19.9E	971	66	Norse Valkyrie.
Simone	59.5N	82.0E	14	–	French first name.
Simonenko	26.9S	97.3E	35	–	Soviet astronomer.
Sinann Vallis	49.0S	270.0E	425	–	Irish river goddess.
Sirani	31.5S	230.4E	28	–	Elisabetta; Italian painter, etcher, printmaker (1638-1665).
Sith Corona	10.2S	176.5E	350	–	Norse harvest goddess.
Sitwell	16.7N	190.3E	35	–	Edith; English poet, critic (1887-1964).
Skadi Mons	64.0N	4.0E	40	–	Norse mountain goddess.
Snegurochka Planitia	86.6N	328.0E	2773	66	Snow maiden in Russian folktales, melted in spring.
Somagalags Corona	9.3N	348.5E	105	–	Bella Coola earth mother.
Sophia	28.7S	18.7E	17	–	First name from Greek.
Stanton	23.4S	199.9E	110	71	Elizabeth C.; American suffragist (1815-1902).
Stefania	51.0N	333.0E	12	–	Romanian first name.
Stein	30.0S	345.5E	24	–	Gertrude; American writer (1874-1946).
Steinbach	41.4S	256.8E	21	–	Sabina; German sculptor (ca. 1250).
Stina	37.2N	22.8E	38	–	Swedish first name.
Stopes Patera	42.5N	47.0E	–	–	Marie; English paleontologist (1880-1959).
Storni	9.8S	245.6E	25	–	Alfonsina; Argentine poet (1892-1938).
Stowe	43.2S	233.0E	82	–	Harriet B.; American novelist (1811-1896).
Stuart	30.8S	20.2E	67	69	Mary; Queen of Scots (1542-1587).
Suliko	9.6N	214.6E	19	–	Georgian first name.
Sullivan	1.3S	110.8E	22	–	Anne; American teacher of Helen Keller (1866-1936).
Sunrta Corona	8.3N	11.7E	170	–	Hindu fertility goddess.
Surija	5.3N	178.3E	15	–	Azerbaijani first name.
Susanna	6.0N	93.3E	16	–	First name from Hebrew.
Sveta	82.5N	271.0E	–	–	Russian first name.
Tacoma Corona	37.0S	288.0E	500	–	Earth goddess of Salish, Puyallup & Yakima Indians.
Taglioni	41.5N	122.8E	23	–	Maria; Italian ballet dancer (1804-1884).
Tai Shan Corona	32.5S	95.0E	175	–	Chinese fertility goddess.
Taira	1.5S	296.8E	19	–	Osset first name.
Takus Mana Corona	19.6S	345.3E	125	–	Hopi (USA) fertility goddess.
Talakin Mons	11.0S	355.4E	175	–	Navajo (USA) goddess.
Tamara	61.5N	317.5E	10	–	Georgian first name.
Tamfana Corona	36.3S	6.0E	400	–	Norse fertility goddess.
Tamiyo Corona	36.0S	297.5E	400	–	Japanese goddess of abundance.
Tanya	19.3S	282.7E	14	–	Russian first name.
Taranga Corona	16.5N	251.5E	525	–	Polynesian fertility goddess.
Tarbell Patera	58.2S	351.5E	80	–	Ida; American author, editor (1857-1944).
Tatyana	85.5N	217.0E	16	–	Russian first name.
Ta'urua Vallis	80.2S	247.5E	525	–	Tahitian word for the planet Venus.
Taussig	9.2S	228.9E	26	–	Helen; American pediatrician, heart researcher (1898-1986).
Teasdale Patera	67.6S	189.1E	75	–	Sara; American poet (1884-1933).

Feature name	Lat. (°)	Long. (°)	Size (km)	Page	Description
Tefnut Mons	38.6S	304.0E	182	68	Ancient Egyptian goddess of dew or rain.
Tellus Tessera	42.6N	76.8E	2329	69	Greek Titaness.
Tepev Mons	29.0N	44.3E	301	69	Quiche Mayan creator goddess.
Teresa	42.5S	9.9E	17	–	First name from Greek.
Teteoinnan Corona	38.5S	149.5E	125	–	Aztec fertility goddess.
Tethus Regio	66.0N	120.0E	–	–	Roman earth goddess.
Tey Patera	17.8S	349.1E	20	–	Josephine; Scottish author (1897-1952).
Tezan Dorsa	81.4N	47.1E	1079	–	Etruscan dawn goddess.
Thallo Mons	76.0N	233.5E	216	–	Greek goddess of flowering vegetation (Spring Hora).
Theia Mons	22.7N	281.0E	226	68	Greek Titaness.
Themis Regio	37.4S	284.2E	1811	68	Greek Titaness.
Thermuthis Corona	8.0S	33.0E	330	–	Egyptian fertility/harvest goddess.
Thetis Regio	11.4S	129.9E	2801	70	Greek Titaness.
Thomas	13.0S	272.6E	25	–	Martha; American college president (1857-1935).
Thouris Corona	6.5S	12.9E	190	–	Egyptian fertility goddess.
T'ien Hu Colles	30.8N	16.0E	–	–	Chinese sea goddess.
Tinatin Planitia	15.0S	15.0E	–	69	Georgian epic heroine.
Tipporah Patera	38.9N	43.0E	99	–	Hebrew medical scholar (1500 B.C.).
Tituba Patera	42.5N	214.0E	–	–	Nurse who started Salem witch hunt (ca. 1692).
Toklas	0.7N	273.2E	21	–	Alice; American writer, art patron (1877-1967).
Tomem Dorsa	31.2N	7.2E	970	–	Ketian (Siberia) Mother of the hot; lives in the sky, near the Sun.
Toyo-uke Corona	62.5S	41.5E	300	–	Shinto fertility goddess.
Trollope	54.8S	246.4E	26	–	Frances; English novelist (1780-1863).
Trotula Patera	41.3N	18.9E	146	69	Italian physician (A.D. 1097).
Truth	28.7N	287.7E	47	–	Sojourner; American abolishionist (1797-1883).
Tseraskaya	26.9N	78.8E	36	–	Lidiya; Soviet astronomer (1855-1931).
Tsiala	2.9N	100.0E	17	–	Georgian first name.
Tsvetayeva	64.0N	147.0E	40	–	Marina; Soviet poet (1892-1941).
Tubman	23.6N	204.5E	45	–	Harriet; American abolishionist (1820-1913).
Tumas Corona	16.3S	351.2E	200	–	Hopi (USA) fertility goddess.
Tünde	76.0N	197.0E	16	–	Hungarian first name.
Tusholi Corona	69.5N	101.2E	350	–	Chechen and Ingush (Caucasus) goddess of fertility.
Tussaud	21.8N	220.9E	24	–	Marie; Swiss wax artist (1760-1850).
Tuulikki Mons	10.3N	274.7E	520	68	Finnish wood goddess.
Udaltsova	20.3S	275.3E	28	–	Nadezhda; Russian artist (1885-1961).
Ukemochi Corona	39.0S	296.1E	300	–	Japanese fertility goddess.
Ulfrun Regio	20.5N	223.0E	3954	71	Norse giantess.
Ulrique	76.0N	55.5E	22	–	French first name.
Undset	52.0N	59.5E	28	–	Sigrid; Norwegian author (1882-1949).
Uni Dorsa	33.7N	114.3E	800	70	Etruscan goddess, same as Hera or Juno.
Uorsar Rupes	76.8N	341.2E	820	66	Adygan (Caucasus) goddess of hearth.
Upunusa Tholus	66.2N	252.4E	223	–	Earth goddess of Leti and Babar (southwestern islands, eastern Indonesia).
Ushas Mons	24.3S	324.6E	413	–	Indian goddess of dawn.
Ut Rupes	55.3N	321.9E	676	66	Siberian; Turco-Tatar goddess of the hearth fire.

Feature name	Lat. (°)	Long. (°)	Size (km)	Page	Description
Uvaysi	2.3N	198.2E	40	–	Uzbek poet (ca. 1780-ca. 1850).
Văcărescu	63.0S	199.6E	30	–	Helene; Rumanian poet, novelist (1866-1947).
Vacuna Corona	60.4N	96.0E	448	–	Sabinian (ancient Italy) goddess of harvest.
Vaiva Dorsum	53.2S	204.0E	520	–	Lithuanian rainbow goddess.
Vakarine Vallis	5.0N	336.4E	625	–	Lithuanian word for planet Venus.
Valadon	49.1S	167.5E	29	–	Suzanne; French painter (1865-1840).
Valborg	75.3N	272.0E	26	–	Danish first name.
Valentina	46.7N	143.2E	30	–	Latin first name.
Valkyrie Fossae	58.8N	7.5E	–	–	Norse battle maidens.
Vallija	26.4N	120.0E	16	–	Latvian first name.
Varma-Ava Dorsa	62.3N	267.7E	767	–	Mordvinian (Volga Finn) wind goddess.
Varz Chasma	71.3N	27.0E	346	–	Lezghin (Caucasus) moon goddess.
Vashti	6.8S	43.7E	18	–	Persian first name.
Vasudhara Corona	43.2N	2.7E	160	–	Buddhist female Bodhisattva of abundance.
Vedma Dorsa	49.8N	170.5E	3345	66	East Slav witch.
Vellamo Planitia	45.4N	149.1E	2154	70	Karelo-Finn mermaid.
Venilia Mons	32.7N	238.8E	320	–	Ancient Italian sea goddess.
Verdandi Corona	5.5S	65.2E	180	–	Norse bestower of blessings.
Veronica	38.1S	124.6E	16	–	First name from Latin.
Vesna	60.3S	220.4E	17	–	Slavic first name.
Vesta Rupes	58.3N	323.9E	788	66	Roman hearth goddess.
Vigier Lebrun	17.3N	141.3E	53	–	Marie; French painter (1755-1842).
Vihansa Linea	54.0N	20.0E	–	–	German war goddess.
Vinmara Planitia	53.8N	207.6E	1634	66	Swan maiden whom sea god Qat kept on Earth by hiding her wings (New Hebrides).
Vir-ava Chasma	14.7S	124.1E	416	70	Mordvinian forest mother.
Vires-Akka Chasma	75.6N	341.6E	742	–	Saami-Lapp forest goddess.
Virginia	52.9S	185.9E	18	–	First name from Latin.
Virilis Tesserae	56.1N	239.7E	782	66	One of the names of Fortuna, Roman goddess of chance.
Virve	5.1S	346.8E	19	–	Estonian first name.
Volkova	75.1N	242.1E	52	–	Anna; Russian chemist (1800-1876).
von Paradis	32.2S	314.8E	36	–	Maria; Austrian pianist (1759-1834).
von Schuurman	5.0S	190.9E	29	–	Anna; Dutch linguist, writer, artist (1607-1678).
von Siebold	52.0S	36.7E	36	–	Regina; German physician, educator (1771-1849).
von Suttner	10.7S	234.9E	23	–	Bertha; Austrian journalist, pacifist (1843-1914).
Voynich	35.2N	56.0E	36	–	Lilian; English writer (1864-1960).
Wanda	71.5N	323.0E	16	–	Polish first name.
Wang Zhenyi	13.2N	217.8E	25	–	Chinese astronomer, geophysicist (18th century).
Warren	11.8S	176.5E	53	–	Mercy; American colonial poet, playwright, historian (1728-1814).
Weil	19.4N	283.1E	25	–	Simone; French author (1909-1943).
Wen Shu	5.0S	303.7E	33	–	Chinese painter (1595-1634).
West	26.1N	303.0E	29	–	Rebecca; Irish novelist, critic, actress (1892-1983).
Wharton	56.0N	62.0E	78	–	Edith; American writer (1862-1937).
Wheatley	16.6N	268.0E	75	–	Phillis; first black writer of note in America (1753-1784).
Whiting	6.0S	128.0E	36	–	Sarah; American physicist, astronomer (1847-1927).
Whitney	30.1S	151.3E	45	–	Mary; American astronomer (1847-1921).
Wieck	74.2S	244.9E	21	–	Clara; German pianist, composer (1819-1896).
Wilder	17.4N	122.4E	35	–	Laura Ingalls; American author (1867-1957).
Willard	24.7S	296.1E	47	–	Emma; American educator (1787-1870).
Winema	3.1N	168.6E	22	–	Modoc Indian heroine, peacemaker (ca. 1848-1932).
Winnemucca	15.4S	121.1E	30	–	Sarah; Piute interpreter, activist (ca. 1844-1891).
Wollstonecraft	39.2S	260.7E	44	–	Mary; English author (1759-1797).
Woodhull Patera	37.5N	306.0E	–	–	Victoria; American-English lecturer (1838-1927).
Woolf	37.7S	27.1E	25	–	Virginia; British writer (1882-1941).
Workman	12.9S	299.9E	19	–	Fanny; American mountaineer, author (1859-1925).
Wu Hou	25.4S	317.4E	30	–	Chinese empress (ca. 624-705).
Wurunsemu Tholus	40.6N	209.9E	83	71	One of the main figures in Hatti (proto-Hittite) mythology, sun goddess and mother of gods.
Xantippe	10.8S	11.7E	41	–	Wife of Socrates (5th century B.C.).
Xiao Hong	43.6S	101.5E	37	–	Chinese novelist (1911-1942).
Xochiquetzal Mons	3.5N	270.0E	80	–	Aztec goddess of flowers.
Yablochkina	48.6N	195.5E	63	66	Aleksandra; Soviet actress (1866-1964).
Yale	13.4S	271.2E	20	–	Caroline; American educator of the deaf (1848-1933).
Yaroslavna Patera	38.8N	21.2E	112	69	Russian, wife of Price Igor; patiently waited for his return from captivity (12th century).
Ymoja Vallis	71.6S	204.8E	390	–	Yoruba (Nigeria) river goddess.
Yonge	14.0S	115.1E	26	–	Charlotte; English writer (1823-1901).
Yoshioka	32.4S	58.8E	20	–	Yayoi; Japanese physician, college founder (ca. 1871-1959).
Yumyn-Udyr Dorsa	78.0N	130.0E	–	–	Marian (Volga Finn) daughter of main god.
Yvonne	56.0S	298.3E	15	–	French first name.
Zamudio	9.6N	189.2E	19	–	Adela; Bolivian poet (1854-1928).
Zdravka	65.0N	299.0E	12	–	Bulgarian first name.
Zemina Corona	11.7S	186.0E	530	–	Lithuanian fertility goddess.
Zenobia	29.3S	28.5E	39	–	Queen of Palmyra (Syria) (third century A.D.).
Zhilova	66.3N	125.4E	45	–	Maria; Russian astronomer (1870-1934).
Zhu Shuzhen	26.5S	356.6E	32	–	Chinese poet (1126-1200).
Zija	3.5S	265.0E	18	–	Azerbaijani first name.
Zina	41.9N	319.9E	15	–	Romanian first name.
Zisa Corona	12.0N	221.0E	850	71	German harvest goddess.
Zlata	64.5N	334.0E	8	–	Serbocroatian first name.
Zorile Dorsa	39.9N	338.4E	1041	68	Moldavian dawn goddess.
Zorya Tholus	9.4S	335.3E	22	–	Slavic dawn goddess.
Zoya	68.0N	237.0E	22	–	Russian first name.
Zvereva	45.2N	282.9E	44	–	Lidiya; Russian aviator (1890-1916).

EARTH

Feature name	Lat. (°)	Long. (°)	Size (km)	Page	Description
Africa	5.0N	25.0E	–	100	
Agulhas Basin	45.0S	25.0E	–	100	
Agulhas Plateau	40.0S	27.0E	–	100	
Alaska Abyssal Plain	58.0N	145.0W	–	99	
Alaska Range	63.0N	150.0W	–	99	
Aleutian Basin	58.0N	165.0W	–	99	
Aleutian Trench	52.0N	175.0W	–	99	
Alps	44.0N	8.0E	–	99	
Amazon Basin	3.0S	60.0W	–	102	
Amsterdam Fracture Zone	28.0S	75.0E	–	100	
Andes	17.5S	68.0W	–	103	
Angola Abyssal Plain	10.0S	8.0E	–	102	
Antarctic Peninsula	70.0S	65.0W	–	104	
Antarctica	90.0S	–	–	104	
Apenines	42.0N	12.0E	–	100	
Appalachian Mountains	37.0N	82.0W	–	103	
Arabian Basin	18.0N	65.0E	–	100	
Arabian Peninsula	25.0N	45.0E	–	100	
Argentine Basin	45.0S	43.0W	–	103	
Asia	45.0N	90.0E	–	99	
Atlantic Ocean	10.0N	30.0W	–	103	
Atlantic-Indian Ridge	53.0S	18.0E	–	104	
Atlas Mountains	34.0N	0.0E	–	103	
Australia	25.0S	133.0E	–	101	
Baffin Bay	74.0N	65.0W	–	99	
Baffin Island	68.0N	74.0W	–	99	
Balkan Peninsula	42.0N	25.0E	–	100	
Barents Abyssal Plain	85.0N	60.0E	–	99	
Bay of Bengal	17.5N	90.0E	–	100	
Bering Abyssal Plain	55.0N	170.0E	–	99	
Bermuda Islands	32.0N	65.0W	–	103	
Black Sea	43.0N	35.0E	–	100	
Borneo	0.0N	114.0E	–	101	
Brazilian Highlands	18.0S	45.0W	–	103	
British Isles	54.0N	2.0W	–	99	
Brooks Range	68.5N	152.0W	–	99	
Campbell Plateau	49.0S	173.0E	–	104	
Canada Abyssal Plain	80.0N	135.0W	–	99	
Canadian Shield	50.0N	85.0W	–	99	
Canary Basin	25.0N	20.0W	–	103	
Cape Basin	35.0S	5.0E	–	103	
Carlsberg Ridge	5.0N	63.0E	–	100	
Caspian Sea	42.0N	51.0E	–	100	
Caucasus Mountains	42.0N	45.0E	–	100	
Chagos-Laccadive Plateau	7.0N	73.0E	–	100	
Chinook Trough	43.0N	175.0W	–	—	
Chukchi Range	68.0N	178.0E	–	99	
Clarion Fracture Zone	15.0N	135.0W	–	102	
Clipperton Fracture Zone	10.0N	132.0W	–	102	
Cocos Basin	4.0N	82.0W	–	—	
Cocos Ridge	4.0N	87.0W	–	102	
Columbia Seamount	22.0S	32.0W	–	103	
Congo Basin	0.0N	25.0E	–	100	
Crozet Basin	40.0S	62.0E	–	100	
Diamantina Trench	36.0S	110.0E	–	101	
Drakensberg	32.0S	27.0E	–	100	
East Pacific Basin	10.0S	130.0W	–	102	
East Pacific Rise	20.0S	110.0W	–	102	
Easter Fracture Zone	25.0S	100.0W	–	102	
Eltanian Fracture Zone	55.0S	135.0W	–	104	
Emperor Seamounts	40.0N	170.0E	–	101	
Enderby Abyssal Plain	60.0S	40.0E	–	104	
Ethiopian Highlands	8.0N	40.0E	–	100	
Europe	48.0N	15.0E	–	99	
Falkland Plateau	52.0S	50.0W	–	104	
Fiji Plateau	18.0S	178.0E	–	101	
Gaussberg Abyssal Plain	62.0S	70.0E	–	104	
Gobi Desert	42.0N	105.0E	–	101	
Great Australian Bight	34.0S	130.0E	–	101	
Great Dividing Range	29.0S	152.0E	–	101	
Great Sandy Desert	20.0S	125.0E	–	101	
Great Victoria Desert	28.0S	130.0E	–	101	
Greenland	70.0N	45.0W	–	99	
Guiana Highlands	5.0N	61.0W	–	103	
Guinea Basin	2.0N	2.0W	–	103	
Gulf of Mexico	25.0N	90.0W	–	102	
Hainan	19.0N	109.0E	–	101	
Hatteras Abyssal Plain	35.0N	74.0W	–	103	
Hawaiian Islands	21.0N	158.0W	–	102	
Himalayas	28.0N	85.0E	–	100	
Hudson Bay	60.0N	85.0W	–	99	
Iberian Peninsula	40.0N	5.0W	–	103	
Iceland	66.0N	18.0W	–	99	
India	23.0N	80.0E	–	100	
Indian Desert	28.0N	73.0E	–	100	
Indian Ocean	5.0S	80.0E	–	100	
Indochina Peninsula	15.0N	105.0E	–	101	
Izu Trench	30.0N	140.0E	–	101	
Japan Trench	35.0N	141.0E	–	101	
Java Trench	10.0S	112.0E	–	101	
Kalahari Desert	24.0S	22.0E	–	100	
Kamchatka Peninsula	55.0N	160.0E	–	99	
Kerguelen Plateau	55.0S	70.0E	–	104	
Kermadec Trench	30.0S	177.0W	–	101	
Kola Peninsula	67.0N	35.0E	–	99	
Kolyma Range	63.0N	160.0E	–	99	
Korean Peninsula	37.0N	127.0E	–	101	
Kuril Trench	47.0N	155.0E	–	99	
Kyushu-Palau Ridge	16.0N	130.0E	–	101	
Labrador Basin	59.0N	50.0W	–	99	
Libyan Desert	27.0N	25.0E	–	100	
Lord Howe Rise	37.0S	165.0E	–	101	
Louisville Ridge	35.0S	168.0W	–	102	
Macquarie Ridge	52.0S	162.0E	–	104	
Madagascar	18.0S	46.0E	–	100	
Manchurian Plain	44.0N	125.0E	–	99	
Manihiki Plateau	17.0S	162.0W	–	102	
Mariana Trench	16.0N	147.0E	–	100	
Marie Byrd Land	78.0S	125.0W	–	104	
Marquesas Fracture Zone	8.0S	125.0W	–	102	
Mediterranean Sea	35.0N	20.0E	–	100	
Menard Fracture Zone	60.0S	122.0W	–	104	
Mendocino Fracture Zone	40.0N	144.0W	–	102	
Mid-Atlantic Ridge	20.0N	40.0W	–	103	

Feature name	Lat. (°)	Long. (°)	Size (km)	Page	Description
Middle America Trench	15.0N	95.0W	–	102	
Molokai Fracture Zone	26.0N	140.0W	–	102	
Mornington Abyssal Plain	53.0S	80.0W	–	104	
Murray Fracture Zone	33.0N	134.0W	–	102	
Namib Desert	22.0S	14.0E	–	100	
Nazca Ridge	20.0S	80.0W	–	103	
New Guinea	6.0S	143.0E	–	101	
New Hebrides Trench	21.0S	172.0E	–	101	
Ninety East Ridge	8.0S	87.0E	–	100	
North America	45.0N	95.0W	–	102	
Novaya Zemlya	73.0N	57.0E	–	99	
Pacific Ocean	5.0N	175.0W	–	102	
Pacific-Antarctic Ridge	60.0S	155.0W	–	104	
Pampas	35.0S	63.0W	–	103	
Peru Basin	18.0S	92.0W	–	102	
Peru-Chile Trench	18.0S	72.0W	–	103	
Philippine Trench	10.0N	126.0E	–	101	
Polar Abyssal Plain	88.0N	15.0E	–	99	
Puerto Rico Trench	19.0N	66.0W	–	103	
Pyrenees	43.0N	1.0E	–	103	
Queen Maud Land	75.0S	20.0E	–	104	
Queensland Plateau	15.0S	148.0E	–	101	
Red Sea	21.0N	38.0E	–	100	
Reykjanes Ridge	60.0N	21.0W	–	99	
Rio Grande Rise	31.0S	36.0W	–	103	
Rocky Mountains	41.0N	107.0W	–	102	
Ross Ice Shelf	80.0S	178.0W	–	104	
Ryukyu Trench	27.0N	129.0E	–	101	
Sahara	22.0N	12.0E	–	100	
Scandanavia	65.0N	15.0E	–	99	
Shatskiy Rise	35.0N	160.0E	–	101	
Siberia	65.0N	110.0E	–	99	
Sierra Nevada	38.0N	120.0W	–	102	
Sikhote Alin Range	47.0N	138.0E	–	99	
Sohm Abyssal Plain	42.0N	55.0E	–	102	
Somali Basin	3.0S	45.0E	–	100	
Somali Peninsula	10.0N	50.0E	–	100	
South America	20.0S	60.0W	–	103	
South Australian Basin	35.0S	130.0E	–	101	
South Sandwich Trench	60.0S	26.0W	–	104	
Southeast Indian Ridge	46.0S	105.0E	–	—	
Southwest Pacific Basin	40.0S	155.0W	–	102	
Sumatra	0.0N	102.0E	–	101	
Syrian Desert	35.0N	40.0E	–	100	
Taklimakan Desert	38.0N	83.0E	–	100	
Tasman Abyssal Plain	35.0S	152.0E	–	101	
Tasman Basin	40.0S	160.0E	–	101	
Taymyr Peninsula	76.0N	105.0E	–	99	
Tian Shan	43.0N	82.0E	–	100	
Tibesti	21.0N	18.0E	–	100	
Tibetan Plateau	32.0N	87.0E	–	100	
Tonga Trench	21.0S	174.0W	–	101	
Trans-Antarctic Range	88.0S	140.0E	–	104	
Ural Mountains	58.0N	59.0E	–	99	
Victoria Island	72.0N	110.0W	–	99	
Walvis Ridge	25.0S	7.0E	–	103	
Weddell Abyssal Plain	70.0S	40.0W	–	104	
West Indies	19.0N	71.0W	–	103	
Wilkes Abyssal Plain	55.0S	110.0E	–	104	
Wilkes Land	70.0S	120.0E	–	104	
Zagros Mountains	32.0N	50.0E	–	100	

MOON

Feature name	Lat. (°)	Long. (°)	Size (km)	Page	Description
Abbe	57.3S	175.2E	66	–	Ernst K.; German optician, physician, astronomer (1840-1905).
Abbot	5.6N	54.8E	10	–	Charles Greeley; American astrophysicist (1872-1973).
Abel	34.5S	87.3E	122	111	Niels H.; Norwegian mathematician (1802-1829).
Abenezra	21.0S	11.9E	42	–	Abraham Bar Rabbi Ben Ezra; Spanish Jewish mathematician, astronomer (1092-1167).
Abetti	19.9N	27.7E	65	–	Antonio; Italian astronomer (1846-1928); Georgio; Italian astronomer (1882-1982).
Abul Wáfa	1.0N	116.6E	55	–	Persian mathematician, astronomer (940-998).
Abulfeda	13.8S	13.9E	65	–	Abu'L-fida, Ismail; Syrian geographer (1273-1331).
Catena Abulfeda	16.9S	17.2E	219	–	Named from nearby crater.
Acosta	5.6S	60.1E	13	–	Cristobal; Portuguese doctor, natural historian (1515-1580).
Adams	31.9S	68.2E	–	111	Charles H.; American astronomer (1868-1951); John Couch; British astronomer (1819-1892); Walter S.; American astronomer (1876-1956).
Lacus Aestatis	15.0S	69.0W	90	–	"Lake of Summer."
Sinus Aestuum	10.9N	8.8W	290	110	"Seething Bay."
Promontorium Agarum	14.0N	66.0E	70	–	Named from cape in Sea of Azov.
Promontorium Agassiz	42.0N	1.8E	20	–	Jean Louis Rodolphe; Swiss zoologist, geologist (1807-1873).
Agatharchides	19.8S	30.9W	48	–	Greek geographer (unknown-ca. 150 B.C.).
Rima Agatharchides	20.0S	28.0W	50	–	Named from nearby crater.
Mons Agnes	18.6N	5.3E	1	–	Greek female name.
Montes Agricola	29.1N	54.2W	141	–	Georgius; German earth scientist (1494-1555).
Rima Agricola	29.0N	53.0W	110	110	Named from nearby Montes.
Agrippa	4.1N	10.5E	44	–	Greek astronomer (unknown-flourished A.D. 92).
Airy	18.1S	5.7E	36	–	George Biddell; British astronomer (1801-1892).
Aitken	16.8S	173.4E	135	112	Robert G.; American astronomer (1864-1951).
Akis	20.0N	31.8W	2	–	Greek female name.
Alan	10.9S	6.1W	2	–	Irish male name.
Al-Bakri	14.3N	20.2E	12	–	A. A.; Spanish-Arabic geographer (1010-1094).
Albategnius	11.7S	4.3E	114	111	Al-Batani, Muhammed Ben Geber C.; Iraqi astronomer, mathematician (850-929).
Al-Biruni	17.9N	92.5E	77	112	Persian astronomer, mathematician, geographer (973-1048).

Feature name	Lat. (°)	Long. (°)	Size (km)	Page	Description
Alden	23.6S	110.8E	104	112	Harold L.; American astronomer (1890-1964).
Alder	48.6S	177.4W	77	–	Kurt; German organic chemist, Nobel laureate (1902-1958).
Aldrin	1.4N	22.1E	3	–	Edwin E., Jr.; American astronaut (1930-).
Dorsa Aldrovandi	24.0N	28.5E	136	–	Italian earth scientist (1522-1605).
Alekhin	68.2S	131.3W	70	–	Nikolaj P.; Soviet rocket designer, engineer (1913-1964).
Alexander	40.3N	13.5E	81	–	Alexander the Great, of Macedon; Greek geographer (356-323 B.C.).
Alfraganus	5.4S	19.0E	20	–	Al Fargani, Muhammed Ebn Ketir; Persian astronomer (unknown-ca. 840).
Alhazen	15.9N	71.8E	32	–	Abu Ali Al-Hasan Ibn Al Haitham; Iraqi mathematician (987-1038).
Aliacensis	30.6S	5.2E	79	–	D'Ailly, Pierre; French geographer (1350-1420).
Al-Khwarizmi	7.1N	106.4E	65	–	Iraqi mathematician (unknown-ca. 825).
Almanon	16.8S	15.2E	49	–	Abdalla Al Mamun; Persian astronomer (786-833).
Al-Marrakushi	10.4S	55.8E	8	–	Moroccan astronomer, mathematician (flourished ca. 1262).
Aloha	29.8N	53.9W	3	–	Hawaiian greeting.
Montes Alpes	46.4N	0.8W	281	108	Named from terrestrial Alps.
Vallis Alpes	48.5N	3.2E	166	108	"Alpine Valley."
Alpetragius	16.0S	4.5W	39	–	Nur Ed-Din Al Betrugi; Moroccan astronomer (unknown-ca. 1100).
Alphonsus	13.7S	3.2W	108	110	Alfonso X; Spanish astronomer (1223-1284).
Rimae Alphonsus	14.0S	2.0W	80	–	Within crater of same name.
Rupes Altai	24.3S	22.6E	427	111	Named from terrestrial Altai Mountains.
Alter	18.7N	107.5W	64	–	Dinsmore; American astronomer, meteorologist (1888-1968).
Ameghino	3.3N	57.0E	9	–	Fiorino (or Florentino); Italian natural historian (ca. 1854-1911).
Amici	9.9S	172.1W	54	–	Giovanni B.; Italian astronomer, botanist (1786-1863).
Ammonius	8.5S	0.8W	8	–	Greek philosopher (unknown-ca. 517).
Amontons	5.3S	46.8E	2	–	Guillaume; French physicist (1663-1705).
Sinus Amoris	18.1N	39.1E	130	111	"Bay of Love."
Mons Ampère	19.0N	4.0W	30	–	André Marie; French physicist (1775-1836).
Amundsen	84.3S	85.6E	101	109	Roald E.; Norwegian explorer (1872-1928).
Anaxagoras	73.4N	10.1W	50	–	Greek astronomer (500-428 B.C.).
Anaximander	66.9N	51.3W	67	–	Greek astronomer (ca. 611-547 B.C.).
Anaximenes	72.5N	44.5W	80	–	Greek astronomer (585-528 B.C.).
Anděl	10.4S	12.4E	35	–	Karel; Czechoslovakian astronomer (1884-1947).
Anders	41.3S	142.9W	40	–	William A.; American astronaut (1933-).
Anderson	15.8N	171.1E	109	112	John A.; American astronaut (1876-1959).
Andersson	49.7S	95.3W	13	–	Leif Erland; American astronomer (1943-1979).
Mons André	5.2N	120.6E	10	–	French male name.
Andronov	22.7S	146.1E	16	–	Aleksandr Aleksandrovich; Soviet physicist (1901-1952).
Dorsa Andrusov	1.0S	57.0E	160	–	Nicolai I.; Soviet geologist (1861-1924).
Ango	20.5N	32.3W	1	–	African male name.
Angström	29.9N	41.6W	9	–	Anders Jonas; Swedish physicist (1814-1874).
Mare Anguis	22.6N	67.7E	150	111	"Serpent Sea."
Ann	25.1N	0.1W	3	–	Hebrew female name.
Annegrit	29.4N	25.6W	1	–	German female name.
Ansgarius	12.7S	79.7E	94	111	St. Ansgar; German theologian (801-864).
Antoniadi	69.7S	172.0W	143	109	Eugēne M.; French astronomer (1870-1944).
Anuchin	49.0S	101.3E	57	–	Dimitri N.; Russian geographer (1843-1923).
Anville	1.9N	49.5E	10	–	Jean-Baptiste Bourguignon; French cartographer (1697-1782).
Apennine Front	25.9N	3.7E	6	–	Astronaut-named feature, Apollo 15 site.
Montes Apenninus	18.9N	3.7W	401	110	Named from terrestrial Apennines.
Apianus	26.9S	7.9E	63	–	Bienewitz, Peter; German mathematician, astronomer (1495-1552).
Apollo	36.1S	151.8W	537	112	Named to honor Apollo missions.
Apollonius	4.5N	61.1E	53	–	Apollonius of Perga; Greek mathematician (3rd century B.C.).
Rimae Apollonius	5.0N	53.0E	230	–	Named from nearby crater.
Appleton	37.2N	158.3E	63	112	Sir Edward V.; British physicist; Nobel laureate (1892-1965).
Arago	6.2N	21.4E	26	–	Dominique Franćois Jean; French astronomer (1786-1853).
Aratus	23.6N	4.5E	10	–	Greek astronomer (315-245 B.C. ?).
Promontorium Archerusia	16.7N	22.0E	10	–	Named from cape on the Black Sea.
Archimedes	29.7N	4.0W	82	110	Greek physicist, mathematician (ca. 287-212 B.C.).
Montes Archimedes	25.3N	4.6W	163	110	Named from nearby crater.
Rimae Archimedes	26.6N	4.1W	169	–	Named from nearby crater.
Archytas	58.7N	5.0E	31	–	Greek mathematician (428-347 B.C.).
Rima Archytas	53.0N	3.0E	90	–	Named from nearby crater.
Mons Ardeshir	5.0N	121.0E	8	–	Persian (Iranian) king's name.
Dorsum Arduino	24.9N	35.8W	107	–	Giovanni; Italian earth scientist (1713-1795).
Mons Argaeus	19.0N	29.0E	50	–	Named from peak in Asia Minor (now Erciyas Dagi).
Dorsa Argand	28.1N	40.6W	109	–	Emile; Swiss earth scientist (1879-1940).
Argelander	16.5S	5.8E	34	–	Friedrich Wilhelm August; German astronomer (1799-1875).
Ariadaeus	4.6N	17.3E	11	–	Arrhidaeus, Philipus; King of Babylon chronologer (unknown.-317 B.C.).
Rima Ariadaeus	6.4N	14.0E	250	111	Named from nearby crater.
Aristarchus	23.7N	47.4W	40	–	Greek astronomer (310-230 B.C.).
Rimae Aristarchus	26.9N	47.5W	121	–	Named from nearby crater.

Feature name	Lat. (°)	Long. (°)	Size (km)	Page	Description
Aristillus	33.9N	1.2E	55	111	Greek astronomer (flourished ca. 280 B.C.).
Aristoteles	50.2N	17.4E	87	108	Greek astronomer, philosopher (384-322 B.C.).
Arminski	16.4S	154.2E	26	–	Franciszek; Polish astronomer (1789-1848).
Armstrong	1.4N	25.0E	4	–	Neil A.; American astronaut (1930-).
Arnold	66.8N	35.9E	94	108	Christoph; German astronomer (1650-1695).
Arrhenius	55.6S	91.3W	40	–	Svante A.; Swedish chemist; Nobel laureate (1859-1927).
Artamonov	25.5N	103.5E	60	–	Nikolaj N.; Soviet rocket scientist (1906-1965).
Catena Artamonov	26.0N	105.9E	134	–	Named from nearby crater.
Artem'ev	10.8N	144.4W	67	–	Vladimir A.; Soviet rocket scientist (1885-1962).
Artemis	25.0N	25.4W	2	–	Greek moon goddess.
Artsimovich	27.6N	36.6W	8	–	Lev Andreevich; Soviet physicist (1909-1973).
Rima Artsimovich	27.0N	39.0W	70	–	Named from nearby crater.
Aryabhata	6.2N	35.1E	22	–	Indian astronomer, mathematician (476-ca. 550).
Arzachel	18.2S	1.9W	96	110	Al Zarkala; Spanish--Arabic astronomer (ca. 1028-1087).
Rimae Arzachel	18.0S	2.0W	50	–	Within crater.
Asada	7.3N	49.9E	12	–	Goryu; Japanese astronomer (1734-1799).
Asclepi	55.1S	25.4E	42	–	Giuseppe; Italian astronomer (1706-1776).
Ashbrook	81.4S	112.5W	156	–	Joseph; American astronomer (1918-1980).
Sinus Asperitatis	3.8S	27.4E	206	111	"Bay of Roughness."
Aston	32.9N	87.7W	43	–	Francis W.; British chemist, physicist, Nobel laureate (1877-1945).
Atlas	46.7N	44.4E	87	108	Mythological Greek Titan.
Rimae Atlas	47.5N	43.6E	60	–	Within crater.
Atwood	5.8S	57.7E	29	–	G.; British mathematician, physicist (1745-1807).
Mare Australe	38.9S	93.0E	603	112	"Southern Sea."
Autolycus	30.7N	1.5E	39	–	Greek astronomer (unknown-ca. 330 B.C.).
Lacus Autumni	9.9S	83.9W	183	110	"Lake of Autumn."
Auwers	15.1N	17.2E	20	–	Georg Friedrich Julius Arthur; German astronomer (1838-1915).
Auzout	10.3N	64.1E	32	–	Adrien; French astronomer, physicist (1622-1691).
Avery	1.4S	81.4E	9	–	Oswald Theodore; Canadian doctor (1877-1955).
Avicenna	39.7N	97.2W	74	–	Abu Ali Ibn Sīnā; Persian-born Arabian doctor (980-1037).
Avogadro	63.1N	164.9E	139	108	Amedeo (Conte di Quaregna y Ceretto); Italian physicist (1776-1856).
Dorsum Azara	26.7N	19.2E	105	–	Felix de; Spanish earth scientist (1746-1811).
Azophi	22.1S	12.7E	47	–	Al-Sufi, Abderrahman; Persian astronomer (903-986).
Baade	44.8S	81.8W	55	–	Walter; American astronomer (1893-1960).
Vallis Baade	45.9S	76.2W	203	–	Named from nearby crater.
Babakin	20.8S	123.3E	20	–	Georgij N.; Soviet space scientist (1914-1971).
Babbage	59.7N	57.1W	143	108	Charles; British mathematician (1792-1871).
Babcock	4.2N	93.9E	99	112	Harold D.; American astronomer, physicist (1882-1968).
Baby Ray	9.1S	15.4E	–	–	Astronaut-named feature, Apollo 16 site.
Back	1.1N	80.7E	35	–	Ernst E. A.; German physicist (1881-1959).
Backlund	16.0S	103.0E	75	–	Oscar A.; Russian astronomer (1846-1916).
Bacon	51.0S	19.1E	69	–	Roger; British philosopher and scientist, optician (ca. 1220-1292).
Baillaud	74.6N	37.5E	89	–	Benjamin; French astronomer (1848-1934).
Bailly	66.5S	69.1W	287	109	Jean Sylvain; French astronomer (1736-1793).
Baily	49.7N	30.4E	26	–	Francis; British astronomer (1774-1844).
Balandin	18.9S	152.6E	12	–	A. A.; Soviet chemist (1898-1967).
Balboa	19.1N	83.2W	69	–	Vasco N. de; Spanish explorer (1475-1519).
Baldet	53.3S	151.1W	55	–	François; French astronomer (1885-1964).
Ball	35.9S	8.4W	41	–	William; British astonomer (unknown-1690).
Balmer	20.3S	69.8E	138	111	Johann J.; Swiss mathematician, physician (1825-1898).
Banachiewicz	5.2N	80.1E	92	–	Tadeusz; Polish astronomer, mathematician (1882-1954).
Bancroft	28.0N	6.4W	13	–	W. D.; American chemist (1867-1953).
Banting	26.6N	16.4E	5	–	Sir Frederick Grant; Canadian doctor, Nobel laureate (1891-1941).
Barbier	23.8S	157.9E	66	–	Daniel; French astronomer (1907-1965).
Barkla	10.7S	67.2E	42	–	C. G.; British physicist, Nobel laureate (1877-1944).
Dorsa Barlow	15.0N	31.0E	120	–	William; British crystallographer (1845-1934).
Barnard	29.5S	85.6E	105	111	Edward E.; American astronomer (1857-1923).
Barocius	44.9S	16.8E	82	–	Francesco; Italian mathematician (unknown--flourished 1570).
Barringer	28.0S	149.7W	68	–	Daniel M.; American engineer, geologist (1860-1929).
Barrow	71.3N	7.7E	92	–	Isaac; British mathematician (1630-1677).
Bartels	24.5N	89.8W	55	–	Julius; German geophysicist (1899-1964).
Bawa	25.3S	102.6E	1	–	African male name.
Bayer	51.6S	35.0W	47	–	Johann; German astronomer (1572-1625).
Beals	37.3N	86.5E	48	–	Carlyle F.; Canadian astronomer (1899-1979).
Bear Mountain	20.0N	30.7E	–	–	Astronaut-named feature, Apollo 17 site.
Beaumont	18.0S	28.8E	53	–	Leonce Elie de; French geologist (1798-1874).
Becquerel	40.7N	129.7E	65	–	Antoine-Henri; French physicist; Nobel laureate (1852-1908).
Bečvář	1.9S	125.2E	67	–	Antonin; Czechoslovakian astronomer (1901-1965).
Beer	27.1N	9.1W	9	–	Wilhelm; German astronomer (1797-1850).

Feature name	Lat. (°)	Long. (°)	Size (km)	Page	Description
Behaim	16.5S	79.4E	55	–	Martin; German navigator, cartographer (1436-1506).
Beijerinck	13.5S	151.8E	70	–	Martinus W.; Dutch botanist (1851-1931).
Beketov	16.3N	29.2E	8	–	N. N.; Russian chemist (1827-1911).
Béla	24.7N	2.3E	11	–	Slavic female name.
Bel'kovich	61.1N	90.2E	214	108	Igor V.; Soviet astronomer (1904-1949).
Bell	21.8N	96.4W	86	–	Alexander G.; Scottish-born American inventor (1847-1922).
Bellinshauzen	60.6S	164.6W	63	–	Faddey F.; Russian explorer (1778-1852).
Bellot	12.4S	48.2E	17	–	Joseph René; French explorer (1826-1853).
Belopol'skiy	17.2S	128.1W	59	–	Aristarch A.; Russian astronomer (1854-1934).
Belyaev	23.3N	143.5E	54	–	Pavel I.; Soviet cosmonaut (1925-1970).
Bench	3.2S	23.4W	–	–	Astronaut-named feature, Apollo 12 site.
Benedict	4.4N	141.5E	14	–	F. G.; American chemist, physiologist (1870-1957).
Bergman	7.0N	137.5E	21	–	Torbern Olof; Swedish chemist, mineralogist, astronomer (1735-1784).
Bergstrand	18.8S	176.3E	43	–	Carl O. E.; Swedish astronomer (1873-1948).
Berkner	25.2N	105.2W	86	113	Lloyd V.; American geophysicist (1905-1967).
Berlage	63.2S	162.8W	92	–	Hendrik P.; Dutch geophysicist, meteorologist (1896-1968).
Bernouilli	35.0N	60.7E	47	–	Jacques; Swiss mathematician (1654-1705); Jean; Swiss mathematician (1667-1748).
Berosus	33.5N	69.9E	74	–	Berosus the Chaldean; Babylonian astronomer (unknown-ca. 250 B.C.).
Berzelius	36.6N	50.9E	50	–	Jöns Jakob; Swedish chemist (1779-1848).
Bessarion	14.9N	37.3W	10	–	Johannes; Greek scholar (ca. 1369-1472).
Bessel	21.8N	17.9E	15	–	Friedrich Wilhelm; German astronomer (1784-1846).
Bettinus	63.4S	44.8W	71	109	Mario; Italian mathematician, astronomer (1582-1657).
Bhabha	55.1S	164.5W	64	–	Homi J.; Indian physicist (1909-1966).
Bianchini	48.7N	34.3W	38	–	Francesco; Italian astronomer (1662-1729).
Biela	54.9S	51.3E	76	109	Wilhelm von; Austrian astronomer (1782-1856).
Bilharz	5.8S	56.3E	43	–	T.; German doctor (1825-1862).
Billy	13.8S	50.1W	45	–	Jacques de; French mathematician (1602-1679).
Rima Billy	15.0S	48.0W	70	–	Named from nearby crater.
Bingham	8.1N	115.1E	33	–	H.; American explorer (1875-1956).
Biot	22.6S	51.1E	12	–	Jean-Baptiste; French astronomer (1774-1862).
Birkeland	30.2S	173.9E	82	–	Olaf K.; Norwegian physicist (1867-1917).
Birkhoff	58.7N	146.1W	345	108	George D.; American mathematician (1884-1944).
Birmingham	65.1N	10.5W	92	–	John; Irish astronomer (1829-1884).
Birt	22.4S	8.5W	16	–	William R.; British selenographer (1804-1881).
Rima Birt	21.0S	9.0W	50	–	Named from nearby crater.
Bjerknes	38.4S	113.0E	48	–	Vilhelm F. K.; Norwegian physicist (1862-1951).
Black	9.2S	80.4E	18	–	Joseph; French chemist (1728-1799).
Blackett	37.5S	116.1W	141	113	Patrick Maynard Stuart; British physicist, Nobel laureate (1897-1974).
Blagg	1.3N	1.5E	5	–	Mary Adela; British astronomer (1858-1944).
Mont Blanc	45.0N	1.0E	25	–	Named for terrestrial mountain in Alps.
Blancanus	63.8S	21.4W	117	109	Biancani, Giuseppe; Italian mathematician, astronomer (1566-1624).
Blanchard	58.5S	94.4W	40	–	J.P.; French aeronaut (1753-1809).
Blanchinus	25.4S	2.5E	61	–	Bianchini, Giovanni; Italian astronomer (unknown-flourished 1458).
Blazhko	31.6N	148.0W	54	–	Sergej N.; Soviet astronomer (1870-1956).
Block	3.2S	23.4W	–	–	Astronaut-named feature, Apollo 12 site.
Bobillier	19.6N	15.5E	6	–	E.; French geometer (1798-1840).
Bobone	26.9N	131.8W	31	–	Jorge; Argentinian astronomer (1901-1958).
Bode	6.7N	2.4W	18	–	Johann Elert; German astronomer (1747-1826).
Rimae Bode	10.0N	4.0W	70	–	Named from nearby crater.
Boethius	5.6N	72.3E	10	–	Greek physicist (ca. 480-524).
Boguslawsky	72.9S	43.2E	97	–	Palm Heinrich Ludwig von; German astronomer (1789-1851).
Bohnenberger	16.2S	40.0E	33	–	Johann Gottlieb Friedrich von; German astronomer (1765-1831).
Bohr	12.4N	86.6W	71	–	Niels H. D.; Danish physicist; Nobel laureate (1885-1962).
Vallis Bohr	12.4N	86.6W	80	–	Named from nearby crater.
Bok	20.2S	171.6W	45	–	Priscilla Fairfield; American astronomer (1896-1975).
Boltzmann	74.9S	90.7W	76	–	Ludwig E.; Austrian physicist (1844-1906).
Bolyai	33.6S	125.9E	135	112	Janos; Hungarian mathematician (1802-1860).
Bombelli	5.3N	56.2E	10	–	R.; Italian mathematician (1526-1572).
G. Bond	32.4N	36.2E	20	–	George Philip; American astronomer (1826-1865).
Rima G. Bond	33.3N	35.5E	168	–	Named from nearby crater.
W. Bond	65.4N	4.5W	156	108	William Cranch; American astronomer (1789-1859).
Bondarenko	17.8S	136.3E	30	–	V. V.; Soviet student-cosmonaut (1937-1961).
Lacus Bonitatis	23.2N	43.7E	92	–	"Lake of Goodness."
Bonpland	8.3S	17.4W	60	–	Aime; French botanist (1773-1858).
Boole	63.7N	87.4W	63	–	George; British mathematician (1815-1864).
Borda	25.1S	46.6E	44	–	Jean Charles; French astronomer (1733-1799).
Borel	22.3N	26.4E	4	–	Felix Edouard Emile; French mathematician (1871-1956).
Boris	30.6N	33.5W	10	–	Russian male name.
Rupes Boris	30.5N	33.5W	4	–	Named from nearby crater.
Borman	38.8S	147.7W	50	–	Frank; American astronaut, engineer (1928-).

Feature name	Lat. (°)	Long. (°)	Size (km)	Page	Description
Born	6.0S	66.8E	14	–	Max; German physicist (1882-1970).
Boscovich	9.8N	11.1E	46	–	Ruggiero Giuseppe; Italian physicist (1711-1787).
Rimae Boscovich	9.8N	11.1E	40	–	Within crater.
Bose	53.5S	168.6W	91	109	Jagadis C.; Indian botanist, physicist (1858-1937).
Boss	45.8N	89.2E	47	–	Lewis; American astronomer (1846-1912).
Bouguer	52.3N	35.8W	22	–	Pierre; French hydrographer (1698-1758).
Boussingault	70.2S	54.6E	142	109	Jean Baptiste Dieudonne; French chemist (1802-1887).
Vallis Bouvard	38.3S	83.1W	284	110	Alexis; French astronomer, mathematician (1767-1843).
Bowditch	25.0S	103.1E	40	–	Nathaniel; American astronomer, mathematician (1773-1848).
Bowen	17.6N	9.1E	8	–	Ira Sprague; American astronomer (1898-1973).
Bowen-Apollo	20.3N	30.9E	–	–	Astronaut-named feature, Apollo 17 site.
Boyle	53.1S	178.1E	57	–	Robert; British physicist, chemist (1627-1691).
Brackett	17.9N	23.6E	8	–	Frederick Sumner; American physicist (1896-).
Mons Bradley	22.0N	1.0E	30	–	James; British astronomer (1692-1762).
Rima Bradley	23.8N	1.2W	161	110	Named from nearby Mons.
Bragg	42.5N	102.9W	84	–	William H.; English physicist; Nobel laureate (1862-1942).
Brashear	73.8S	170.7W	55	–	John A.; American astronomer (1840-1920).
Brayley	20.9N	36.9W	14	–	Edward William; British geographer (1801-1870).
Rima Brayley	21.4N	37.5W	311	–	Named from nearby crater.
Bredikhin	17.3N	158.2W	59	–	Fedor A.; Russian astronomer (1831-1904).
Breislak	48.2S	18.3E	49	–	Scipione; Italian chemist, geologist, mathematician (1748-1826).
Brenner	39.0S	39.3E	97	–	Leo; Austrian astronomer (1855-1928).
Brewster	23.3N	34.7E	10	–	David; Scottish optician (1781-1868).
Brianchon	75.0N	86.2W	134	108	Charles J.; French mathematician (1783-1864).
Bridge	26.0N	3.6E	1	–	Astronaut-named feature, Apollo 15 site.
Bridgman	43.5N	137.1E	80	–	Percy W.; American physicist, Nobel laureate (1882-1961).
Briggs	26.5N	69.1W	37	–	Henry; British mathematician (1556-1630).
Catena Brigitte	18.5N	27.5E	5	–	French female name.
Brisbane	49.1S	68.5E	44	–	Sir Thomas; Scottish astronomer (1770-1860).
Bronk	26.1N	134.5W	64	–	Detlev Wulf; American neurophysiologist (1897-1975).
Brontë	20.2N	30.7E	–	–	Astronaut-named feature, Apollo 17 site.
Brouwer	36.2S	126.0W	158	112	Dirk; American astronomer (1902-1966); Luitzen E.J.; Dutch mathematician (1881-1968).
Brown	46.4S	17.9W	34	–	Ernest William; British astronomer, mathematician (1866-1938).
Bruce	1.1N	0.4E	6	–	Catherine Wolfe; American philanthropist, astronomer (1816-1900).
Brunner	9.9S	90.9E	53	–	William O.; Swiss astronomer (1878-1958).
Buch	38.8S	17.7E	53	–	Christian Leopold von; German geologist (1774-1853).
Dorsum Bucher	31.0N	39.0W	90	–	W. H.; Swiss earth scientist (1889-1965).
Dorsum Buckland	20.4N	12.8E	380	111	William; British earth scientist (1784-1856).
Buffon	40.4S	133.4W	106	113	Georges-Louis Leclerc de; French natural historian (1707-1788).
Buisson	1.4S	112.5E	56	–	Henri; French physicist, astronomer (1873-1944).
Bullialdus	20.7S	22.2W	60	–	Boulliaud, Ismael; French astronomer (1605-1694).
Bunsen	41.4N	85.3W	52	–	Robert W.; German chemist (1811-1899).
Burckhardt	31.1N	56.5E	56	–	Johann Karl; German astronomer (1773-1825).
Bürg	45.0N	28.2E	39	–	Johann Tobias; Austrian astronomer (1766-1834).
Rimae Bürg	44.5N	23.8E	147	–	Named from nearby crater.
Dorsa Burnet	28.4N	57.0W	194	–	Thomas; British earth scientist (1635-1715).
Burnham	13.9S	7.3E	24	–	Sherburne Wesley; American astronomer (1838-1921).
Büsching	38.0S	20.0E	52	–	Anton Friedrich; German geographer (1724-1793).
Butlerov	12.5N	108.7W	40	–	Aleksandr M.; Russian chemist (1828-1886).
Buys-Ballot	20.8N	174.5E	55	–	C. H. D.; Dutch meteorologist (1817-1890).
Byrd	85.3N	9.8E	93	108	Richard E.; American explorer, aviator, navigator (1888-1957).
Byrgius	24.7S	65.3W	87	110	Burgi, Joost; Swiss horologist (1552-1632).
Cabannes	60.9S	169.6W	80	–	Jean; French physicist (1885-1959).
Cabeus	84.9S	35.5W	98	–	Cabeo, Niccolo; Italian astronomer (1586-1650).
Cajal	12.6N	31.1E	9	–	Santiago Ramon Y.; Spanish doctor; Nobel laureate (1852-1934).
Cajori	47.4S	168.8E	70	–	Florian; American mathematician (1859-1930).
Calippus	38.9N	10.7E	32	–	Calippus of Cyzicus; Greek astronomer (ca. 330 B.C.).
Rima Calippus	37.0N	13.0E	40	–	Named from nearby crater.
Camelot	20.2N	30.7E	1	–	Astronaut-named feature, Apollo 17 site.
Cameron	6.2N	45.9E	10	–	Robert Curry; American astronomer (1925-1972).
Campanus	28.0S	27.8W	48	–	Campano, Giovanni; Italian astronomer (ca. 1200-unknown).
Campbell	45.3N	151.4E	219	108	Leon; American astronomer (1881-1951); William W.; American astronomer (1862-1938).
Cannizzaro	55.6N	99.6W	56	–	Stanislao; Italian chemist (1826-1910).
Cannon	19.9N	81.4E	56	–	Annie J.; American astronomer (1863-1941).
Cantor	38.2N	118.6E	–	–	Georg; German mathematician (1845-1918); Moritz; German mathematician (1829-1920).
Capella	7.5S	35.0E	90	–	Martianus; Roman astronomer (ca. A.D. 400-unknown).

Feature name	Lat. (°)	Long. (°)	Size (km)	Page	Description
Vallis Capella	7.6S	34.9E	49	–	Named from nearby crater.
Capuanus	34.1S	26.7W	59	–	Francesco Capuano di Manfredonia; Italian astronomer (ca. 1400-unknown).
Cardanus	13.2N	72.4W	49	110	Cardano, Girolamo; Italian mathematician (1501-1576).
Rima Cardanus	11.4N	71.5W	175	–	Named from nearby crater.
Carlini	33.7N	24.1W	10	–	Francesco; Italian astronomer (1783-1862).
Carlos	24.9N	2.3E	4	–	Spanish male name.
Rima Carmen	19.8N	29.3E	10	–	Spanish female name.
Carmichael	19.6N	40.4E	20	–	Leonard; American psychologist (1898-1973).
Carnot	52.3N	143.5W	126	108	Nicolas L. S.; French physicist (1796-1832).
Carol	8.5N	122.3E	117	–	Latin female name.
Montes Carpatus	14.5N	24.4W	361	110	Named from terrestrial Carpathians.
Carpenter	69.4N	50.9W	59	–	James; British astronomer (1840-1899); Edwin F.; American astronomer (1898-1963).
Carrel	10.7N	26.7E	15	–	Alexis; French doctor, physiologist, Nobel laureate (1873-1944).
Carrillo	2.2S	80.9E	16	–	Flores Nabor; Mexican soil engineer (1911-1967).
Carrington	44.0N	62.1E	30	–	Richard Christopher; British astronomer (1826-1875).
Cartan	4.2N	59.3E	15	–	E. J.; French mathematician (1869-1951).
Carver	43.0S	126.9E	59	–	George W.; American botanist (ca. 1864-1943).
Casatus	72.8S	29.5W	108	109	Casati, Paolo; Italian mathematician (1617-1707).
Cassegrain	52.4S	113.5E	55	–	Giovanni D.; French astronomer, doctor (1625-1712).
Cassini	40.2N	4.6E	56	–	Giovanni Domenico; French astronomer (1625-1712); Jacques J.; French astronomer (1677-1756).
Catalán	45.7S	87.3W	25	–	Miguel A.; Spanish spectroscopist (1894-1957).
Catharina	18.1S	23.4E	104	111	St. Catherine of Alexandria; Greek theologian, philosopher (unknown-ca. 307).
Dorsa Cato	1.0N	47.0E	140	–	Marcus Porcius; Roman geological engineer (234-149 B.C.).
Montes Caucasus	38.4N	10.0E	445	111	Named from terrestrial Caucasus Mountains.
Cauchy	9.6N	38.6E	12	–	Augustin Louis; French mathematician (1789-1857).
Rima Cauchy	10.5N	38.0E	140	–	Named from nearby crater.
Rupes Cauchy	9.0N	37.0E	120	–	Named from nearby crater.
Cavalerius	5.1N	66.8W	57	–	Cavalieri, Buonaventura; Italian mathematician (1598-1647).
Cavendish	24.5S	53.7W	56	–	Henry; British chemist, physicist (1731-1810).
Caventou	29.8N	29.4W	3	–	Joseph Bienaimé; French chemist, pharmacologist (1795-1877).
Dorsum Cayeux	1.6N	51.2E	84	–	Lucien; French sedimentary petrographer (1864-1944).
Cayley	4.0N	15.1E	14	–	Arthur; British astronomer, mathematician (1821-1895).
Celsius	34.1S	20.1E	36	–	Anders; Swedish astronomer (1701-1744).
Censorinus	0.4S	32.7E	3	–	Roman astronomer (flourished 238-unknown).
Cepheus	40.8N	45.8E	39	–	Mythological astronomer, father of Andromeda.
Chacornac	29.8N	31.7E	51	–	Jean; French astronomer (1823-1873).
Rimae Chacornac	29.0N	32.0E	120	–	Named from nearby crater.
Chadwick	52.7S	101.3W	30	–	James; British physicist (1891-1974).
Chaffee	38.8S	153.9W	49	–	Roger B.; American aeronautic engineer, astronaut (1935-1967).
Challis	79.5N	9.2E	55	–	James; British astronomer, mathematician, physicist (1803-1862).
Chalonge	21.2S	117.3W	30	–	Daniel; French astronomer (1895-1977).
Chamberlin	58.9S	95.7E	58	–	Thomas C.; American geologist (1843-1928).
Champollion	37.4N	175.2E	58	–	Jean-François; French egyptologist (1790-1832).
Chandler	43.8N	171.5E	85	–	Seth C.; American astronomer (1846-1913).
Chang Heng	19.0N	112.2E	43	–	Chinese astronomer (78-139).
Chang-Ngo	12.7S	2.1W	3	–	Chinese female name.
Chant	40.0S	109.2W	33	–	Clarence A.; Canadian astronomer, physicist (1865-1956).
Chaplygin	6.2S	150.3E	137	112	Sergej A.; Soviet mathematician (1869-1942).
Chapman	50.4N	100.7W	71	–	Sydney; British geophysicist (1888-1970).
Chappe	61.2S	91.5W	59	–	d'Auteroche, Jean-Baptiste; French astronomer (1728-1769).
Chappell	54.7N	177.0W	80	–	James F.; American astronomer (1891-1964).
Charles	29.9N	26.4W	1	–	French male name.
Charlier	36.6N	131.5W	99	113	Carl W. L.; Swedish astronomer (1862-1934).
Chaucer	3.7N	140.0W	45	–	Geoffrey; British writer, astronomer (ca. 1340-1400).
Chauvenet	11.5S	137.0E	81	–	William; American astronomer, mathematician (1820-1870).
Chebyshev	33.7S	133.1W	178	113	Pafnutif L.; Russian mathematician (1821-1894).
Chernyshev	47.3N	174.2E	58	–	Nikolaj G.; Soviet rocket engineer (1906-1963).
Chevallier	44.9N	51.2E	52	–	Temple; British astronomer (1794-1873).
Ching-Te	20.0N	30.0E	4	–	Chinese male name.
Chladni	4.0N	1.1E	13	–	Ernst Florens Friedrich; German physicist (1756-1827).
Chrétien	45.9S	162.9E	88	–	Henri; French mathematician, astronomer (1870-1956).
Christel	24.5N	11.0E	2	–	German female name.
Cichus	33.3S	21.1W	40	–	Franceso degli Stabili (Cecco d'Ascoli); Italian astronomer (1257-1327).
Cinco	19.1S	15.5E	–	–	Astronaut-named feature, Apollo 16 site.
Clairaut	47.7S	13.9E	75	–	Alexis Claude; French mathematician (1713-1765).
Clark	38.4N	118.9E	49	–	Alvan; American astronomer, optician (1804-1887).

Feature name	Lat. (°)	Long. (°)	Size (km)	Page	Description
Clausius	36.9S	43.8W	24	–	Rudolf Julius Emmanuel; German physicist (1822-1888).
Clavius	58.8S	14.1W	245	109	Christopher Klau; German mathematician (1537-1612).
Cleomedes	27.7N	56.0E	125	111	Greek astronomer (unknown-ca. 50 B.C.).
Rima Cleomedes	27.0N	57.0E	80	–	Within crater.
Rima Cleopatra	30.0N	53.8W	14	–	Greek female name.
Cleostratus	60.4N	77.0W	62	–	Greek astronomer (unknown-ca. 500 B.C.).
Clerke	21.7N	29.8E	6	–	Agnes Mary; British astronomer (1842-1907).
Dorsum Cloos	1.0N	91.0E	100	–	Hans; German earth scientist (1885-1951).
Coblentz	37.9S	126.1E	33	–	William W.; American physicist, astronomer (1873-1962).
Cochise	20.2S	30.8E	1	–	Astronaut-named feature, Apollo 17 site.
Cockcroft	31.3N	162.6W	93	113	Sir John D.; British nuclear physicist; Nobel laureate (1897-1967).
Mare Cognitum	10.0S	23.1W	376	110	"Sea that has become known."
Collins	1.3N	23.7E	2	–	Michael; American astronaut (1930-).
Colombo	15.1S	45.8E	76	–	Columbus, Christopher; Spanish explorer (ca. 1446-1506).
Compton	55.3N	103.8E	182	108	Arthur H.; American physicist, Nobel laureate (1892-1962); Karl T.; American physicist (1887-1954).
Comrie	23.3N	112.7W	59	–	Leslie J.; British astronomer (1893-1950).
Comstock	21.8N	121.5W	72	–	George C.; American astronomer (1855-1934).
Sinus Concordiae	10.8N	43.2E	142	111	"Bay of Harmony."
Condon	1.9N	60.4E	34	–	Edward U.; American physicist (1902-1974).
Condorcet	12.1N	69.6E	74	–	Jean de; French mathematician (1743-1794).
Cone	3.7S	17.4W	–	–	Astronaut-named feature, Apollo 14 site.
Congreve	0.2S	167.3W	57	–	Sir William; British rocket engineer, inventor (1772-1828).
Conon	21.6N	2.0E	21	–	Conon of Samos; Greek astronomer (ca. 260 B.C.).
Rima Conon	18.6N	2.0E	30	–	Named from nearby crater.
Cook	17.5S	48.9E	46	–	James; British explorer (1728-1779).
Cooper	52.9N	175.6E	36	–	John C.; American humanitarian (1887-1967).
Copernicus	9.7N	20.1W	107	110	Nicolaus; Polish astronomer (1473-1543).
Montes Cordillera	17.5S	81.6W	574	110	Spanish for "mountain chain."
Cori	50.6S	151.9W	65	–	Gerty Theresa Radnitz; Czechoslovakian-born American physiologist; Nobel laureate (1896-1957).
Coriolis	0.1N	171.8E	78	–	Gaspard G. de; French physicist (1792-1843).
Couder	4.8S	92.4W	21	–	Andre; French astronomer (1897-1978).
Coulomb	54.7N	114.6W	89	–	Charles Augustin de; French physicist (1736-1806).
Courtney	25.1N	30.8W	1	–	English male name.
Cremona	67.5N	90.6W	85	–	Luigi; Italian mathematician (1830-1903).
Crescent	2.9S	23.4W	1	–	Astronaut-named feature, Apollo 12 site.
Crile	14.2N	46.0E	9	–	G.; American doctor (1864-1943).
Mare Crisium	17.0N	59.1E	418	111	"Sea of Crises."
Crocco	47.5S	150.2E	75	–	Gaetano A.; Italian aeronautical engineer (1877-1968).
Crommelin	68.1S	146.9W	94	–	Andrew Claude De La Cherois; British astronomer (1865-1939).
Crookes	10.3S	164.5W	49	–	Sir William; British physicist, chemist (1832-1919).
Crozier	13.5S	50.8E	22	–	Francis Rawdon Moira; British explorer (1796-1848).
Crüger	16.7S	66.8W	45	–	Peter; German mathematician (1580-1639).
Ctesibius	0.8N	118.7E	36	–	Egyptian physicist (unknown-ca. 100 B.C.).
Curie	22.9S	91.0E	151	112	Pierre; French physicist, chemist; Nobel laureate (1859-1906).
Curtis	14.6N	56.6E	2	–	Heber Doust; American astronomer (1872-1942).
Curtius	67.2S	4.4E	95	–	Curtz, Albert; German astronomer (1600-1671).
Cusanus	72.0N	70.8E	63	–	Nikolaus Krebs; German mathematician, philosopher (1401-1464).
Dorsum Cushman	1.0N	49.0E	80	–	J. A.; American micro-paleontologist (1881-1949).
Cuvier	50.3S	9.9E	75	–	Georges; French natural scientist, paleontologist (1769-1832).
Cyrano	20.5S	157.7E	80	–	Savinien de Cyrano de Bergerac ; French writer (1619-1655).
Cyrillus	13.2S	24.0E	98	–	Saint Cyril; Egyptian theologian, chronologist (unknown-A.D. 444).
Cysatus	66.2S	6.1W	48	–	Cysat, Jean-Baptiste; Swiss mathematician, astronomer (1588-1657).
da Vinci	9.1N	45.0E	37	–	Leonardo; Italian artist, inventor, mathematician (1452-1519).
Daedalus	5.9S	179.4E	93	112	Greek mythological character.
Dag	18.7N	5.3E	–	–	Scandinavian male name.
Daguerre	11.9S	33.6E	46	–	Louis; French artist, chemist, photographer (1789-1851).
Dale	9.6S	82.9E	22	–	Sir Henry Hallett; British physiologist; Nobel laureate (1875-1968).
D'Alembert	50.8N	163.9E	248	108	Jean-Le-Rond; French mathematician, physicist (1717-1783).
Dalton	17.1N	84.3W	60	–	John; British chemist, physicist (1766-1844).
Daly	5.7N	59.6E	17	–	Reginald Aldworth; Canadian geologist (1871-1957).
Damoiseau	4.8S	61.1W	36	–	Marie Charles Theodor de; French astronomer (1768-1846).
Dorsa Dana	3.0N	90.0E	70	–	James D.; American earth scientist (1813-1895).
Daniell	35.3N	31.1E	29	–	John Frederick; British physicist, chemist, meteorologist (1790-1845).
Rimae Daniell	37.0N	26.0E	200	–	Named from nearby crater.

Feature name	Lat. (°)	Long. (°)	Size (km)	Page	Description
Danjon	11.4S	124.0E	71	–	Andre; French astronomer (1890-1967).
Dante	25.5N	180.0W	54	–	Alighieri; Italian poet (1265-1321).
Darney	14.5S	23.5W	15	–	Maurice; French astronomer (1882-1958).
d'Arrest	2.3N	14.7E	30	–	Heinrich Ludwig; German astronomer (1822-1875).
d'Arsonval	10.3S	124.6E	28	–	Jacques Arsène; French physicist (1851-1940).
Darwin	20.2S	69.5W	120	110	Charles; British natural scientist (1809-1882).
Rimae Darwin	19.3S	69.5W	143	–	Named from nearby crater.
Das	26.6S	136.8W	38	–	Amil K.; Indian astronomer (1902-1961).
Daubrée	15.7N	14.7E	14	–	Gabriel-Auguste; French geologist (1814-1896).
Davisson	37.5S	174.6W	87	–	Clinton J.; American physicist; Nobel laureate (1881-1958).
Davy	11.8S	8.1W	34	–	Humphry; British chemist (1778-1829).
Catena Davy	11.0S	7.0W	50	–	Named from nearby crater.
Dawes	17.2N	26.4E	18	–	William Rutter; British astronomer (1799-1868).
Rima Dawes	17.5N	26.6E	15	–	Named from nearby crater.
Dawson	67.4S	134.7W	45	–	Bernhard H.; Argentinian astronomer (1890-1960).
Debes	29.5N	51.7E	30	–	Ernest; German cartographer (1840-1923).
Debye	49.6N	176.2W	142	108	Peter J. W.; Dutch physicist, chemist; Nobel laureate (1884-1966).
Dechen	46.1N	68.2W	12	–	Ernst Heinrich Karl von; German geologist, mineralogist (1800-1889).
De Forest	77.3S	162.1W	57	–	Lee; American inventor (1873-1961).
de Gasparis	25.9S	50.7W	30	–	Annibale; Italian astronomer (1819-1892).
Rimae de Gasparis	24.6S	51.1W	93	–	Named from nearby crater.
Delambre	1.9S	17.5E	51	–	Jean-Baptiste Joseph; French astronomer (1749-1822).
De La Rue	59.1N	52.3E	134	108	Warren; British astronomer (1815-1889).
Delaunay	22.2S	2.5E	46	–	Charles Eugene; French astronomer (1816-1872).
Delia	10.9S	6.1W	2	–	Greek female name.
Delisle	29.9N	34.6W	25	–	Joseph Nicolas; French astronomer (1688-1768).
Mons Delisle	29.5N	35.8W	30	–	Named from nearby crater.
Rima Delisle	31.0N	32.0W	60	–	Named from nearby crater.
Dellinger	6.8S	140.6E	81	–	John H.; American physicist (1886-1962).
Delmotte	27.1N	60.2E	32	–	Gabriel; French astronomer (1876-1950).
Delporte	16.0S	121.6E	45	–	Eugene J.; Belgian astronomer (1882-1955).
Deluc	55.0S	2.8W	46	–	Jean Andre; Swiss geologist, physicist (1727-1817).
Dembowski	2.9N	7.2E	26	–	Baron Ercole; Italian astronomer (1815-1881).
Democritus	62.3N	35.0E	39	–	Greek astronomer, philosopher (ca. 460-360 B.C.).
Demonax	77.9S	60.8E	128	109	Greek philosopher (unknown-ca. 100 B.C.).
De Moraes	49.5N	143.2E	53	–	A.; Brazilian astronomer (1916-1970).
De Morgan	3.3N	14.9E	10	–	Augustus; British mathematician (1806-1871).
Denning	16.4S	142.6E	44	–	William F.; British astronomer (1848-1931).
de Roy	55.3S	99.1W	43	–	Felix; Belgian astronomer (1883-1942).
Desargues	70.2N	73.3W	85	–	Gérard; French mathematician, engineer (1591-1662).
Descartes	11.7S	15.7E	48	–	René; French mathematician, philosopher (1596-1650).
Planitia Descensus	7.1N	64.4W	1	–	Luna 9 landing site ("plain of descent").
Deseilligny	21.1N	20.6E	6	–	Jules Alfred Pierrot; French selenographer (1868-1918).
de Sitter	80.1N	39.6E	64	–	Willem; Dutch astronomer (1872-1934).
Deslandres	33.1S	4.8W	256	110	Henri Alexandre; French astrophysicist (1853-1948).
Deutsch	24.1N	110.5E	66	–	Armin J.; American astronomer (1918-1969).
de Vico	19.7S	60.2W	20	–	Francesco; Italian astronomer (1805-1848).
Promontorium Deville	43.2N	1.0E	20	–	Sainte-Claire Charles; French geologist (1814-1876).
de Vries	19.9S	176.7W	59	–	Hugo M.; Dutch botanist (1848-1935).
Dewar	2.7S	165.5E	50	–	Sir James; Scottish chemist (1842-1923).
Diana	14.3N	35.7E	50	–	Latin female name.
Diderot	20.4S	121.5E	20	–	Denis; French philosopher (1713-1784).
Mons Dieter	5.0N	120.2E	20	–	German male name.
Mons Dilip	5.6N	120.8E	2	–	Indian male name.
Dionysius	2.8N	17.3E	18	–	St. Dionysius the Areopagite; Greek astronomer (9-120).
Diophantus	27.6N	34.3W	17	–	Greek mathematician (unknown-ca. A.D. 300).
Rima Diophantus	29.0N	33.0W	150	–	Named from nearby crater.
Dirichlet	11.1N	151.4W	47	–	Peter G. L.; German mathematician (1805-1859).
Dobrovol'skiy	12.8S	129.7E	38	–	Georgiy T.; Soviet aeronautical engineer (1928-1971).
Doerfel	69.1S	107.9W	68	–	Georg Samuel; German astronomer (1643-1688).
Dollond	10.4S	14.4E	11	–	John; British optician (1706-1761).
Lacus Doloris	17.1N	9.0E	110	–	"Lake of Sorrow."
Donati	20.7S	5.2E	36	–	Giovanni Battista; Italian astronomer (1826-1873).
Donna	7.2N	38.3E	2	–	Italian female name.
Donner	31.4S	98.0E	58	–	Anders; Finnish astronomer (1873-1949).
Doppelmayer	28.5S	41.4W	63	–	Johann Gabriel; German mathematician, astronomer (1671-1750).
Rimae Doppelmayer	25.9S	45.1W	162	–	Named from nearby crater.
Doppler	12.6S	159.6W	110	–	Christian Johann; Austrian physicist, mathematician, astronomer (1803-1853).
Doublet	3.7S	17.5W	–	–	Astronaut-named feature, Apollo 14 site.
Douglass	35.9N	122.4W	49	–	Andrew E.; American astronomer (1867-1962).
Dove	46.7S	31.5E	30	–	Heinrich Wilhelm; German physicist (1803-1879).
Draper	17.6N	21.7W	8	–	Henry; American astronomer (1837-1882).
Rima Draper	18.0N	25.0W	160	–	Named from nearby crater.

Feature name	Lat. (°)	Long. (°)	Size (km)	Page	Description
Drebbel	40.9S	49.0W	30	–	Cornelius; Dutch inventor (1572-1634).
Dreyer	10.0N	96.9E	61	–	John L. E.; British astronomer (1852-1926).
Drude	38.5S	91.8W	24	–	Paul K. L.; German physicist (1863-1906).
Dryden	33.0S	155.2W	51	–	Hugh L.; American physicist, engineer (1898-1965).
Drygalski	79.3S	84.9W	149	109	Erich D. von; German geographer, geophysicist (1865-1949).
Dubyago	4.4N	70.0E	51	–	Dmitrij I.; Russian astronomer (1850-1918); Alexander D.; Soviet astronomer (1903-1959).
Dufay	5.5N	169.5E	39	–	Jean C. B.; French astronomer (1896-1967).
Dugan	64.2N	103.3E	50	–	Raymond S.; American astronomer (1878-1940).
Dune	26.0N	3.7E	–	–	Astronaut-named feature, Apollo 15 site.
Dunér	44.8N	179.5E	62	–	Nils C.; Swedish astronomer (1839-1914).
Dunthorne	30.1S	31.6W	15	–	Richard; British astronomer (1711-1775).
Dyson	61.3N	121.2W	63	–	Sir Frank W.; British astronomer (1868-1939).
Dziewulski	21.2N	98.9E	63	–	Władysław; Polish astronomer (1878-1962).
Catena Dziewulski	19.0N	100.0E	80	–	Named from nearby crater.
Earthlight	26.1N	3.7E	–	–	Astronaut-named feature, Apollo 15 site.
Eckert	17.3N	58.3E	2	–	Wallace J.; American astronomer (1902-1971).
Eddington	21.3N	72.2W	118	110	Sir Arthur S.; British astrophysicist, mathematician (1882-1944).
Edison	25.0N	99.1E	62	–	Thomas A.; American inventor (1847-1931).
Edith	25.8S	102.3E	8	–	English female name.
Egede	48.7N	10.6E	37	–	Hans; Danish natural historian (1686-1758).
Ehrlich	40.9N	172.4W	30	–	Paul; German doctor, Nobel laureate (1854-1915).
Eichstadt	22.6S	78.3W	49	–	Lorentz; German mathematician, astronomer (1596-1660).
Eijkman	63.1S	143.0W	54	–	Christiaan H.; Dutch doctor; Nobel laureate (1858-1930).
Eimmart	24.0N	64.8E	46	–	Georg Christoph; German astronomer (1638-1705).
Einstein	16.3N	88.7W	198	110	Albert; German-born American physicist; Nobel laureate (1879-1955).
Einthoven	4.9S	109.6E	69	–	Willem; Dutch physiologist; Nobel laureate (1860-1927).
Elbow	26.0N	3.6E	–	–	Astronaut-named feature, Apollo 15 site.
Elger	35.3S	29.8W	21	–	Thomas Gwyn; British astronomer (1838-1897).
Ellerman	25.3S	120.1W	47	–	Ferdinand; American astronomer (1869-1940).
Ellison	55.1N	107.5W	36	–	Mervyn A.; British astronomer (1909-1963).
Elmer	10.1S	84.1E	16	–	Charles W.; American astronomer (1872-1954).
Elvey	8.8N	100.5W	74	–	Christian T.; American astronomer, geophysicist (1899-1970).
Emden	63.3N	177.3W	111	–	J. Robert; Swiss astrophysicist, meteorologist (1862-1940).
Emory	20.1N	30.8E	1	–	Astronaut-named feature, Apollo 17 site.
Encke	4.6N	36.6W	28	–	Johann Franz; German mathematician, astronomer (1791-1865).
End	8.9S	15.6E	–	–	Astronaut-named feature, Apollo 16 site.
Endymion	53.9N	57.0E	123	108	Greek mythological character.
Engel'gardt (Engelhardt)	5.7N	159.0W	43	–	Vasilij P.; Soviet astronomer (1828-1915).
Eötvös	35.5S	133.8E	99	112	Roland von; Hungarian physicist (1848-1919).
Palus Epidemiarum	32.0S	28.2W	286	110	"Marsh of Epidemics."
Epigenes	67.5N	4.6W	55	–	Greek astronomer (unknown-ca. 200 B.C.).
Epimenides	40.9S	30.2W	27	–	Greek philosopher, writer (unknown-flourished 596 B.C.).
Eppinger	9.4S	25.7W	6	–	H.; Czechoslovakian doctor (1879-1946).
Eratosthenes	14.5N	11.3W	58	110	Greek astronomer, geographer (ca. 276-194 B.C.).
Erro	5.7N	98.5E	61	–	Luis E.; Mexican astronomer (1897-1955).
Mons Esam	14.6N	35.7E	8	–	Arabic male name.
Esclangon	21.5N	42.1E	15	–	Ernest Benjamin; French astronomer (1876-1954).
Esnault-Pelterie	47.7N	141.4W	79	–	Robert A. C.; French rocket engineer (1881-1957).
Espin	28.1N	109.1E	75	–	Thomas H. E. C.; British astronomer (1858-1934).
Euclides	7.4S	29.5W	11	–	Euclid; Greek mathematician (flourished ca. 300 B.C.).
Euctemon	76.4N	31.3E	62	–	Greek astronomer (unknown-flourished 432 B.C.).
Eudoxus	44.3N	16.3E	67	–	Greek astronomer (ca. 408-355 B.C.).
Euler	23.3N	29.2W	27	–	Leonhard; Swiss mathematician (1707-1783).
Rima Euler	21.0N	31.0W	90	–	Named from nearby crater.
Evans	9.5S	133.5W	67	–	Sir Arthur; British archaeologist (1851-1941).
Evdokimov	34.8N	153.0W	50	–	Nikolaj N.; Soviet astronomer (1868-1940).
Evershed	35.7N	159.5W	66	–	John; British astronomer (1864-1956).
Ewen	7.7N	121.4E	3	–	Gaelic male name.
Dorsa Ewing	10.2S	39.4W	141	–	William Maurice; American geophysicist (1906-1974).
Lacus Excellentiae	35.4S	44.0W	184	–	"Lake of Excellence."
Fabbroni	18.7N	29.2E	10	–	Giovanni Valentino Mattia; Italian chemist (1752-1822).
Fabricius	42.9S	42.0E	78	–	Goldschmidt, David; Dutch astronomer (1564-1617).
Fabry	42.9N	100.7E	184	112	Charles; French physicist (1867-1945).
Fahrenheit	13.1N	61.7E	6	–	Gabriel Daniel; German physicist (1686-1736).
Fairouz	26.1S	102.9E	3	–	Arab female name.
Falcon	20.4N	30.3E	–	–	Astronaut-named feature, Apollo 17 site.
Family Mountain	20.4N	30.3E	7	–	Astronaut-named feature, Apollo 17 site.
Faraday	42.4S	8.7E	69	–	Michael; British chemist, physicist (1791-1867).
Faustini	87.3S	77.0E	39	–	Arnoldo; Italian polar geographer (1874-1944).

Feature name	Lat. (°)	Long. (°)	Size (km)	Page	Description
Fauth	6.3N	20.1W	12	–	Philipp Johann Heinrich; German selenographer (1867-1941).
Faye	21.4S	3.9E	36	–	Herve; French astronomer (1814-1902).
Fechner	59.0S	124.9E	63	–	Gustav T.; German physicist, psychologist (1801-1887).
Mare Fecunditatis	7.8S	51.3E	909	111	"Sea of Fecundity."
Fedorov	28.2N	37.0W	6	–	A. P.; Russian rocket scientist (1872-1920).
Lacus Felicitatis	19.0N	5.0E	90	–	"Lake of Happiness."
Felix	28.2N	37.0W	6	–	Latin male name.
Fényi	44.9S	105.1W	38	–	Gyula; Hungarian astronomer (1845-1927).
Feoktistov	30.9N	140.7E	23	–	Konstantin P.; Soviet cosmonaut (1926-).
Fermat	22.6S	19.8E	38	–	Pierre de; French mathematician (1601-1665).
Fermi	19.3S	122.6E	183	112	Enrico; Italian-born American physicist; Nobel laureate (1901-1954).
Fernelius	38.1S	4.9E	65	–	Jean; French doctor, astronomer (1497-1558).
Fersman	18.7N	126.0W	151	113	Aleksandr; Soviet geochemist (1883-1945).
Fesenkov	23.2S	135.1E	35	–	Vasiliy Grigor'evich; Soviet astrophysicist (1889-1972).
Feuillée	27.4N	9.4W	9	–	Louis; French natural scientist (1660-1732).
Sinus Fidei	18.0N	2.0E	70	–	"Bay of Trust."
Finsch	23.6N	21.3E	4	–	O. F. H.; German zoologist (1839-1917).
Finsen	42.0S	177.9W	72	–	Niels Ryberg; Danish physician; Nobel laureate (1860-1904).
Firmicus	7.3N	63.4E	56	–	Maternus; Italian astronomer (unknown-ca. 330).
Firsov	4.5N	112.2E	51	–	Georgij F.; Soviet rocketry engineer (1917-1960).
Fischer	8.0N	142.4E	30	–	Emil; German chemist (1852-1919); Hans; German organic chemist (1881-1945).
Fitzgerald	27.5N	171.7W	110	113	George F.; Irish physicist (1851-1901).
Fizeau	58.6S	133.9W	111	109	Armand H. L.; French physicist (1819-1896).
Flag	9.0S	15.5E	–	–	Astronaut-named feature, Apollo 16 site.
Flammarion	3.4S	3.7W	74	–	Camille; French astronomer (1842-1925).
Rima Flammarion	2.8S	5.6W	80	–	Named from nearby crater.
Flamsteed	4.5S	44.3W	20	–	John; British astronomer (1646-1720).
Flank	3.7S	17.4W	–	–	Astronaut-named feature, Apollo 14 site.
Fleming	15.0N	109.6E	106	112	Sir Alexander; British doctor; Nobel laureate (1881-1955); Williamina P.; American astronomer (1857-1911).
Florensky	25.3N	131.5E	–	–	Kirill P.; Soviet geologist (1915-1982).
Focas	33.7S	93.8W	22	–	Ionnas; French astronomer (1908-1969).
Rimae Focas	28.0S	98.0W	100	–	Named from nearby crater.
Fontana	16.1S	56.6W	31	–	Francesco; Italian astronomer (ca. 1585-1656).
Fontenelle	63.4N	18.9W	38	–	Bernard le Bovier de; French astronomer (1657-1757).
Foster	23.7N	141.5W	33	–	John S.; Canadian physicist (1890-1964).
Foucault	50.4N	39.7W	23	–	Jean-Bernard-Léon; French physicist (1819-1868).
Fourier	30.3S	53.0W	51	–	Jean-Baptiste Joseph; French mathematician (1768-1830).
Fowler	42.3N	145.0W	146	113	Alfred; British astronomer (1868-1940); Ralph H.; British mathematician, physicist (1889-1944).
Fox	0.5N	98.2E	24	–	Philip; American astronomer (1878-1944).
Fra Mauro	6.1S	17.0W	101	110	Italian geographer (unknown-1459).
Fracastorius	21.5S	33.2E	112	111	Fracastoro, Girolamo; Italian doctor, astronomer (1483-1553).
Franck	22.6N	35.5E	12	–	James; German-born American physicist; Nobel laureate (1882-1964).
Franklin	38.8N	47.7E	56	111	Benjamin; American inventor (1706-1790).
Franz	16.6N	40.2E	25	–	Julius Heinrich; German astronomer (1847-1913).
Fraunhofer	39.5S	59.1E	56	–	Joseph von; German astronomer, optician (1787-1826).
Fredholm	18.4N	46.5E	14	–	Erik Ivar; Swedish mathematician (1866-1927).
Promontorium Fresnel	29.0N	4.7E	20	–	Augustin Jean; French optician (1788-1827).
Rimae Fresnel	28.0N	4.0E	90	–	Named from nearby promontorium.
Freud	25.8N	52.3W	2	–	Sigmund; Austrian psychoanalyst (1856-1939).
Freundlich	25.0N	171.0E	85	–	Erwin (Finlay-); German-born British astronomer (1885-1964).
Fridman	12.6S	126.0W	102	113	Aleksandr; Soviet physicist (1888-1925).
Mare Frigoris	56.0N	1.4E	1596	108	"Sea of Cold."
Froelich	80.3N	109.7W	58	–	Jack E.; American rocket scientist (1921-1967).
Frost	37.7N	118.4W	75	–	Edwin B.; American astronomer (1866-1935).
Fryxell	21.3S	101.4W	18	–	R. H.; American geologist (1934-1974).
Furnerius	36.0S	60.6E	135	111	Furner, Georges; French mathematician (unknown-1643).
Rima Furnerius	35.0S	61.0E	50	–	Within crater.
Gadomski	36.4N	147.3W	65	–	Jan; Polish astronomer (1889-1966).
Gagarin	20.2S	149.2E	265	112	Yurij A.; Soviet cosmonaut (1934-1968).
Galen	21.9N	5.0E	10	–	Claudius; Greek doctor (ca. 129-200).
Galilaei	10.5N	62.7W	15	–	Galileo; Italian astronomer, physicist (1564-1642).
Rima Galilaei	11.9N	58.5W	89	–	Named from nearby crater.
Galle	55.9N	22.3E	21	–	Johann Gottfried; German astronomer (1812-1910).
Galois	14.2S	151.9W	222	113	Evariste; French mathematician (1811-1832).
Galvani	49.6N	84.6W	80	–	Luigi; Italian physicist (1737-1798).
Gambart	1.0N	15.2W	25	–	Jean Felix; French astronomer (1800-1836).
Gamow	65.3N	145.3E	129	108	George; American physicist (1904-1968).

Feature name	Lat. (°)	Long. (°)	Size (km)	Page	Description
Mons Ganau	4.8N	120.6E	14	–	African male name.
Ganskiy (Hansky)	9.7S	97.0E	43	–	Aleksey P.; Soviet astronomer (1870-1908).
Ganswindt	79.6S	110.3E	74	–	Hermann; German rocket engineer (1856-1934).
Garavito	47.5S	156.7E	74	–	J.; Colombian astronomer (1865-1920).
Gardner	17.7N	33.8E	18	–	Irvine Clifton; American physicist (1889-1972).
Gärtner	59.1N	34.6E	115	108	Christian; German mineralogist, geologist (ca. 1750-1813).
Rima Gärtner	59.0N	36.0E	30	–	Within crater.
Gassendi	17.6S	40.1W	101	110	Pierre; French astronomer, mathematician (1592-1655).
Rimae Gassendi	18.0S	40.0W	70	–	Within crater.
Dorsum Gast	24.0N	9.0E	60	–	Paul Werner; American geochemist, geologist (1930-1973).
Gaston	30.9N	34.0W	2	–	French male name.
Gator	9.0S	15.6E	1	–	Astronaut-named feature, Apollo 16 site.
Gaudibert	10.9S	37.8E	34	–	Casimir Marie; French astronomer (1823-1901).
Lacus Gaudii	16.2N	12.6E	113	–	"Lake of Joy."
Gauricus	33.8S	12.6W	79	–	Gaurico, Luca; Italian astronomer (1476-1558).
Gauss	35.7N	79.0E	177	111	Carl Friedrich; German mathematician (1777-1855).
Gavrilov	17.4N	130.9E	60	–	Aleksandr I.; Soviet rocket engineer (1884-1955); Igor B.; Soviet astronomer (1928-1982).
Gay-Lussac	13.9N	20.8W	26	–	Joseph-Louis; French physicist (1778-1850).
Rima Gay-Lussac	13.0N	22.0W	40	–	Named from nearby crater.
Geber	19.4S	13.9E	44	–	Gabir Ben Aflah; Spanish-born Arabian astronomer (unknown-ca. 1145).
Geiger	14.6S	158.5E	34	–	Johannes H. W.; German physicist (1882-1945).
Dorsa Geikie	4.6S	52.5E	228	111	Sir Archibald; Scottish geologist (1835-1924).
Geissler	2.6S	76.5E	16	–	Heinrich; German physicist (1814-1879).
Geminus	34.5N	56.7E	85	111	Greek astronomer (unknown-ca. 70 B.C.).
Gemma Frisius	34.2S	13.3E	87	–	Jemma, Reinier; Dutch doctor (1508-1555).
Gerard	44.5N	80.0W	90	–	Alexander; Scottish explorer (1792-1839).
Rimae Gerard	46.0N	84.0W	100	–	Named from nearby crater.
Gerasimovich	22.9S	122.6W	86	–	Boris P.; Soviet astronomer (1889-1937).
Gernsback	36.5S	99.7E	48	–	Hugo; American writer (1884-1967).
Gibbs	18.4S	84.3E	76	111	Josiah Willard; American mathematician and physicist (1839-1903).
Gilbert	3.2S	76.0E	112	111	Grove K.; American geologist (1843-1918).
Gill	63.9S	75.9E	66	–	Sir David; British astronomer (1843-1914).
Ginzel	14.3N	97.4E	55	–	Friedrich K.; Austrian astronomer (1850-1926).
Gioja	83.3N	2.0E	41	–	Flavio; Italian inventor (unknown-flourished 1302).
Giordano Bruno	35.9N	102.8E	22	–	Italian astronomer (1548-1600).
Glaisher	13.2N	49.5E	15	–	James; British meteorologist (1809-1903).
Glauber	11.5N	142.6E	15	–	Johann Rudolph; German chemist (ca. 1603-1670).

Feature name	Lat. (°)	Long. (°)	Size (km)	Page	Description
Glazenap	1.6S	137.6E	43	–	Sergej P.; Soviet astronomer (1848-1937).
Glushko	8.4N	77.6W	43	–	V.P.; Russian space scientist (1908-1989).
Goclenius	10.0S	45.0E	72	–	Gockel, Rudolf; German physicist, doctor, mathematician (1572-1621).
Rimae Goclenius	8.0S	43.0E	240	–	Named from nearby crater.
Goddard	14.8N	89.0E	89	–	Robert H.; American rocket scientist (1882-1945).
Godin	1.8N	10.2E	34	–	Louis; French astronomer, mathematician (1704-1760).
Goldschmidt	73.2N	3.8W	113	108	Hermann; German astronomer (1802-1866).
Golgi	27.8N	60.0W	5	–	Camillo; Italian doctor; Nobel laureate (ca. 1843-1926).
Golitsyn	25.1S	105.0W	36	–	Boris B.; Russian physicist (1862-1916).
Golovin	39.9N	161.1E	37	–	Nicholas E.; American rocket scientist (1912-1969).
Goodacre	32.7S	14.1E	46	–	Walter; British selenographer (1856-1938).
Gould	19.2S	17.2W	34	–	Benjamin Apthorp; American astronomer (1824-1896).
Dorsum Grabau	29.4N	15.9W	121	–	Amadeus W.; American earth scientist (1870-1946).
Grace	14.2N	35.9E	1	–	English female name.
Grachev	3.7S	108.2W	35	–	Andrej D.; Soviet rocket scientist (1900-1964).
Graff	42.4S	88.6W	36	–	Kasimir R.; Polish-born German astronomer (1878-1950).
Grave	17.1N	150.3E	40	–	Dmitriy Aleksandrovich; Soviet engineer (1863-1939); Ivan P.; Soviet engineer (1874-1960).
Greaves	13.2N	52.7E	13	–	William Michael Herbert; British astronomer (1897-1955).
Green	4.1N	132.9E	65	–	George; British mathematician (1793-1841).
Gregory	2.2N	127.2E	67	–	James; Scottish astronomer, mathematician (1638-1675).
Catena Gregory	0.6S	129.9E	152	112	Named from nearby crater.
Grigg	12.9N	129.4W	36	–	J.; New Zealand astronomer (1838-1920).
Grimaldi	5.5S	68.3W	172	110	Francesco Maria; Italian astronomer, physicist (1618-1663).
Rimae Grimaldi	9.0N	64.0W	230	–	Named from nearby crater.
Grissom	47.0S	147.4W	58	–	Virgil I.; American astronaut (1926-1967).
Grotrian	66.5S	128.3E	37	–	W.; German astronomer (1890-1954).
Grove	40.3N	32.9E	28	–	Sir William Robert; British physicist (1811-1896).
Gruemberger	66.9S	10.0W	93	–	Christoph; Austrian astronomer (1561-1636).
Gruithuisen	32.9N	39.7W	15	–	Franz von; German astronomer (1774-1852).
Mons Gruithuisen Delta	36.0N	39.5W	20	–	Named from nearby crater.
Mons Gruithuisen Gamma	36.6N	40.5W	20	–	Named from nearby crater.
Guericke	11.5S	14.1W	63	–	Otto von; German physicist, engineer, naturalist (1602-1686).
Dorsum Guettard	10.0S	18.0W	40	–	Jean-Etienne; French geologist, mineralogist (1715-1786).

Feature name	Lat. (°)	Long. (°)	Size (km)	Page	Description
Guillaume	45.4N	173.4W	57	–	Charles Edouard; French physicist; Nobel laureate (1861-1938).
Gullstrand	45.2N	129.3W	43	–	Allvar; Swedish ophthalmologist; Nobel laureate (1862-1930).
Gum	40.4S	88.6E	54	–	Colin; Australian astronomer (1924-1960).
Gutenberg	8.6S	41.2E	74	–	Johannes; German inventor (ca. 1390-1468).
Rimae Gutenberg	5.0S	38.0E	330	–	Named from nearby crater.
Guthnick	47.7S	93.9W	36	–	Paul; German astronomer (1879-1947).
Guyot	11.4N	117.5E	92	–	Arnold H.; Swiss-born American geographer, geologist (1807-1884).
Gyldén	5.3S	0.3E	47	–	Hugo; Swedish astronomer (1841-1896).
Mons Hadley	26.5N	4.7E	25	–	Hadley, John; British instrument maker (1682-1743).
Mons Hadley Delta	25.8N	3.8E	15	–	Named from nearby mountain.
Rima Hadley	25.0N	3.0E	80	–	Named from nearby Mons.
Montes Haemus	19.9N	9.2E	560	111	Named for range in the Balkans.
Hagecius	59.8S	46.6E	76	–	Hayek, Thaddaeus; Czechoslovakian astronomer, mathematician (1525-1600).
Hagen	48.3S	135.1E	55	–	Johann G.; Austrian astronomer (1847-1930).
Hahn	31.3N	73.6E	–	–	Graf Friedrich von; German astronomer (1741-1805); Otto; German chemist (1879-1968).
Haidinger	39.2S	25.0W	22	–	Wilhelm Karl von; Austrian geologist, physicist (1795-1871).
Hainzel	41.3S	33.5W	70	–	Paul; German astronomer (unknown-flourished 1570).
Haldane	1.7S	84.1E	37	–	John Burdon Sanderson; British doctor (1892-1964).
Hale	74.2S	90.8E	83	109	George Ellery; American astronomer (1868-1938); William; British rocket scientist (1797-1870).
Halfway	9.0S	15.5E	–	–	Astronaut-named feature, Apollo 16 site.
Hall	33.7N	37.0E	35	–	Asaph; American astronomer (1829-1907).
Halley	8.0S	5.7E	36	–	Edmond; British astronomer (1656-1742).
Halo	3.2S	23.4W	–	–	Astronaut-named feature, Apollo 12 site.
Hamilton	42.8S	84.7E	57	–	Sir William R.; Irish mathematician (1805-1865).
Hanno	56.3S	71.2E	56	–	Roman explorer (unknown-ca. 500 B.C.).
Hansen	14.0N	72.5E	39	–	Peter Andreas; Danish astronomer (1795-1874).
Hansteen	11.5S	52.0W	44	–	Christopher; Norwegian astronomer (1784-1873).
Mons Hansteen	12.1S	50.0W	30	–	Named from nearby crater.
Rima Hansteen	12.0S	53.0W	25	–	Named from nearby crater.
Montes Harbinger	27.0N	41.0W	90	–	Harbingers of dawn on crater Aristarchus.
Harden	5.5N	143.5E	15	–	Sir Arthur; British chemist; Nobel laureate (1865-1940).
Harding	43.5N	71.7W	22	–	Karl Ludwig; German astronomer (1765-1834).
Haret	58.4S	175.6W	29	–	Spiru; Rumanian astronomer (1851-1912).
Hargreaves	2.2S	64.0E	16	–	Frederick James; British astronomer, optician (1891-1970).
Dorsa Harker	14.5N	64.0E	197	–	Alfred; British petrologist (1859-1939).
Harkhebi	39.6N	98.3E	237	112	Egyptian astronomer (ca. 300 B.C.).
Harold	10.9S	6.0W	2	–	Scandinavian male name.
Harpalus	52.6N	43.4W	39	–	Greek astronomer (unknown-ca. 460 B.C.).
Harriot	33.1N	114.3E	56	–	Thomas; British mathematician, astronomer (1560-1621).
Hartmann	3.2N	135.3E	61	–	Johannes D.; German astronomer (1865-1936).
Hartwig	6.1S	80.5W	79	–	Karl E.; German astronomer (1851-1923).
Harvey	19.5N	146.5W	60	–	William; British doctor (1578-1657).
Hase	29.4S	62.5E	83	–	Johann Matthias; German mathematician (1684-1742).
Rimae Hase	29.4S	62.5E	83	–	Named from nearby crater.
Hatanaka	29.7N	121.5W	26	–	Takeo; Japanese astronomer (1914-1963).
Hausen	65.0S	88.1W	167	109	Christian August; German astronomer, mathematician, physicist (1693-1743).
Hayford	12.7N	176.4W	27	–	John F.; American civil engineer (1868-1925).
Hayn	64.7N	85.2E	87	108	Friedrich; German astronomer (1863-1928).
Head	3.0S	23.4W	–	–	Astronaut-named feature, Apollo 12 site.
Healy	32.8N	110.5W	38	–	Roy; American rocket scientist (1915-1968).
Heaviside	10.4S	167.1E	165	112	Oliver; British mathematician, physicist (1850-1925).
Hecataeus	21.8S	79.4E	167	111	Greek geographer (unknown-ca. 476 B.C.).
Hédervári	81.8S	84.0E	69	–	Peter; Hungarian geoscientist (1931-1984).
Hedin	2.0N	76.5W	150	–	Sven A.; Swedish explorer (1865-1952).
Dorsum Heim	32.0N	29.8W	148	–	Heim, Albert; Swiss earth scientist (1849-1937).
Heinrich	24.8N	15.3W	6	–	Wladimir W.; Czechoslovakian astronomer (1884-1965).
Heinsius	39.5S	17.7W	64	–	Gottfried; German astronomer (1709-1769).
Heis	32.4N	31.9W	14	–	Eduard; German astronomer (1806-1877).
Helberg	22.5N	102.2W	62	–	Robert J.; American aeronautical engineer (1906-1967).
Helicon	40.4N	23.1W	24	–	Greek astronomer, mathematician (unknown-ca. 400 B.C.).
Hell	32.4S	7.8W	33	–	Maximilian; Hungarian astronomer (1720-1792).
Helmert	7.6S	87.6E	26	–	Friedrich Robert; German astronomer, geodesist (1843-1917).
Helmholtz	68.1S	64.1E	94	109	Hermann von; German doctor (1821-1894).
Henderson	4.8N	152.1E	47	–	Thomas; Scottish astronomer (1798-1844).
Hendrix	46.6S	159.2W	18	–	Don O.; American optician (1905-1961).
Henry	24.0S	56.8W	41	–	Joseph; American physicist (1797-1878).

Feature name	Lat. (°)	Long. (°)	Size (km)	Page	Description
Henry Frères	23.5S	58.9W	42	–	Paul and Prosper; French astronomers (1848-1905), (1849-1903).
Henyey	13.5N	151.6W	63	–	Louis G.; American astronomer (1910-1970).
Promontorium Heraclides	40.3N	33.2W	50	–	Ponticus; Greek astronomer (ca. 388-310 B.C.).
Heraclitus	49.2S	6.2E	90	–	Greek philosopher (ca. 540-480 B.C.).
Hercules	46.7N	39.1E	69	–	Greek mythological hero.
Herigonius	13.3S	33.9W	15	–	Herigone, Pierre; French mathematician, astronomer (flourished 1644).
Rimae Herigonius	13.0S	37.0W	100	–	Named from nearby crater.
Hermann	0.9S	57.0W	15	–	Jacob; Swiss mathematician (1678-1733).
Hermite	86.0N	89.9W	104	108	Charles; French mathematician (1822-1901).
Herodotus	23.2N	49.7W	34	–	Of Halikarnassus; Greek historian (ca. 484-425 B.C.).
Mons Herodotus	27.5N	53.0W	5	–	Named from nearby crater.
Heron (Hero)	0.7N	119.8E	24	–	Egyptian inventor (unknown-ca. 100 B.C.).
Herschel	5.7S	2.1W	40	–	William; British astronomer (1738-1822).
C. Herschel	34.5N	31.2W	13	–	Caroline; British astronomer (1750-1848).
J. Herschel	62.0N	42.0W	165	108	John; British astronomer (1792-1871).
Hertz	13.4N	104.5E	90	–	Heinrich R.; German physicist (1857-1894).
Hertzsprung	2.6N	129.2W	591	113	Ejnar; Danish astronomer (1873-1967).
Hesiodus	29.4S	16.3W	42	–	Hesiod; Greek humanitarian (ca. 735 B.C.).
Rima Hesiodus	30.0S	20.0W	256	–	Named from nearby crater.
Hess	54.3S	174.6E	88	109	Harry H.; American geologist (1906-1969); Victor F.; Austrian physicist (1883-1964).
Hess-Apollo	20.1N	30.7E	1	–	Astronaut-named feature, Apollo 17 site.
Hevelius	2.2N	67.6W	115	110	Howelcke, Johann; Polish astronomer (1611-1687).
Rimae Hevelius	1.0N	68.0W	182	–	Named from nearby crater.
Heymans	75.3N	144.1W	50	–	Corneille-Jean-François; Belgian physiologist; Nobel laureate (1892-1968).
Heyrovsky	39.6S	95.3W	16	–	Jaroslav; Czechoslovakian chemist (1890-1967).
Lacus Hiemalis	15.0N	14.0E	50	–	"Wintry Lake."
Dorsum Higazy	28.0N	17.0W	60	–	Riad; Egyptian earth scientist (1919-1967).
Hilbert	17.9S	108.2E	151	112	David; German mathematician (1862-1943).
Hill	20.9N	40.8E	16	–	George William; American astronomer, mathematician (1838-1914).
Hind	7.9S	7.4E	29	–	John Russell; British astronomer (1823-1895).
Hippalus	24.8S	30.2W	57	–	Greek explorer (unknown-ca. 120).
Rimae Hippalus	25.5S	29.2W	191	–	Named from nearby crater.
Hipparchus	5.1S	5.2E	138	111	Greek astronomer (flourished 146-127 B.C.).
Hippocrates	70.7N	145.9W	60	–	Greek doctor (ca. 460-377 B.C.).
Hirayama	6.1S	93.5E	132	112	Kiyotsugu; Japanese astronomer (1874-1943); Shin; Japanese astronomer (1867-1945).
Hoffmeister	15.2N	136.9E	45	–	Cuno; German astronomer (1892-1968).
Hogg	33.6N	121.9E	38	–	Arthur R.; Australian astronomer (1903-1966); Frank S.; Canadian astronomer (1904-1951).
Hohmann	17.9S	94.1W	16	–	Walter; German space flight engineer (1880-1945).
Holden	19.1S	62.5E	47	–	Edward Singleton; American astronomer (1846-1914).
Holetschek	27.6S	150.9E	38	–	Johann; Austrian astronomer (1846-1923).
Hommel	54.7S	33.8E	126	109	Johann; Greek astronomer, mathematician (1518-1562).
Sinus Honoris	11.7N	18.1E	109	111	"Bay of Honor."
Hooke	41.2N	54.9E	36	–	Robert; British physicist, inventor (1635-1703).
Hopmann	50.8S	160.3E	88	–	Josef; Austrian astronomer (1890-1975).
Horatio	20.2N	30.7E	–	–	Astronaut-named feature, Apollo 17 site.
Hornsby	23.8N	12.5E	3	–	Thomas; British astronomer (1733-1810).
Horrebow	58.7N	40.8W	24	–	Peder; Danish astronomer (1679-1764).
Horrocks	4.0S	5.9E	30	–	Jeremiah; British astronomer (1619-1641).
Hortensius	6.5N	28.0W	14	–	Hove, Martin van den; Dutch astronomer (1605-1639).
Houtermans	9.4S	87.2E	29	–	Friedrich Georg; German physicist (1903-1966).
Houzeau	17.1S	123.5W	71	113	Jean C. (de Lehaie); Belgian astronomer (1820-1888).
Hubble	22.1N	86.9E	80	–	Edwin P.; American astronomer (1889-1953).
Huggins	41.1S	1.4W	65	–	Sir William; British astronomer (1824-1910).
Humason	30.7N	56.6W	4	–	Milton L.; American astronomer (1891-1972).
Humboldt	27.0S	80.9E	189	111	Wilhelm von; German philologist (1767-1835).
Catena Humboldt	21.5S	84.6E	165	111	Named from nearby crater.
Mare Humboldtianum	56.8N	81.5E	273	108	Humboldt, Alexander von; German naturalist (1769-1859).
Hume	4.7S	90.4E	23	–	David; Scottish philosopher (1711-1776).
Mare Humorum	24.4S	38.6W	389	110	"Sea of Moisture."
Hutton	37.3N	168.7E	50	–	James; Scottish geologist (1726-1797).
Huxley	20.2N	4.5W	4	–	Thomas Henry; British biologist (1825-1895).
Mons Huygens	20.0N	2.9W	40	–	Christian; Dutch astronomer, mathematician, physicist (1629-1695).
Hyginus	7.8N	6.3E	9	–	Caius Julius; Spanish astronomer (unknown-ca. A.D. 100).
Rima Hyginus	7.4N	7.8E	219	111	Named from nearby crater.
Hypatia	4.3S	22.6E	40	–	Egyptian mathematician (unknown-A.D. 415).
Rimae Hypatia	0.4S	22.4E	206	111	Named from nearby crater.
Ian	25.7N	0.4W	1	–	Scottish male name.
Ibn Battuta	6.9S	50.4E	11	–	Abu Abd Allah Mohammed Ibn Abd Allah; Moroccan geographer (1304-1377).
Ibn Firnas	6.8N	122.3E	89	–	Spanish-born Arabian humanitarian, technologist (unknown-ca. A.D. 887).

Feature name	Lat. (°)	Long. (°)	Size (km)	Page	Description
Ibn Yunus	14.1N	91.1E	58	–	C. Abu Muhammad Ibn-Rushd (Averrdes); Egyptian astronomer (950-1009).
Ibn-Rushd	11.7S	21.7E	32	–	Spanish-born Arabian philosopher, doctor (1126-1198).
Icarus	5.3S	173.2W	96	113	Greek mythical flyer.
Ideler	49.2S	22.3E	38	–	Christian Ludwig; German astronomer (1766-1846).
Idel'son	81.5S	110.9E	60	–	Naum I.; Soviet astronomer (1885-1951).
Il'in	17.8S	97.5W	13	–	N.Ja.; Soviet rocket scientist (1901-1937).
Mare Imbrium	32.8N	15.6W	1123	110	"Sea of Showers."
Ina	49.2S	22.3E	38	–	Latin female name.
Index	26.1N	3.7E	–	–	Astronaut-named feature, Apollo 15 site.
Ingalls	26.4N	153.1W	37	–	Albert L.; American optician (1888-1958).
Mare Ingenii	33.7S	163.5E	318	112	"Sea of Cleverness."
Inghirami	47.5S	68.8W	91	–	Giovanni; Italian astronomer (1779-1851).
Vallis Inghirami	43.8S	72.2W	148	110	Named from nearby crater.
Innes	27.8N	119.2E	42	–	Robert T. A.; Scottish astronomer (1861-1933).
Mare Insularum	7.5N	30.9W	513	110	"Sea of Islands."
Ioffe	14.4S	129.2W	86	–	Joffe, Abram F.; Soviet physicist (1880-1960).
Sinus Iridum	44.1N	31.5W	236	110	"Bay of Rainbows."
Isabel	28.2N	34.1W	1	–	Spanish female name.
Isaev	17.5S	147.5E	90	–	Aleksei M.; Soviet space scientist (1908-1971).
Isidorus	8.0S	33.5E	42	–	St. Isidore of Seville; Roman astronomer (ca. 570-636).
Isis	18.9N	27.5E	1	–	Egyptian goddess.
Ivan	26.9N	43.3W	4	–	Russian male name.
Izsak	23.3S	117.1E	30	–	Imre; Hungarian-born American astronomer (1929-1965).
Jackson	22.4N	163.1W	71	–	John; Scottish astronomer (1887-1958).
Jacobi	56.7S	11.4E	68	–	Karl Gustav Jacob; German mathematician (1804-1851).
Jansen	13.5N	28.7E	23	–	Janszoon, Zacharias; Dutch optician (1580-ca. 1638).
Rima Jansen	14.5N	29.0E	35	–	Named from nearby crater.
Jansky	8.5N	89.5E	72	–	Karl; American radio engineer (1905-1950).
Janssen	45.4S	40.3E	199	109	Pierre Jules; French astronomer (1824-1907).
Rimae Janssen	45.6S	40.0E	114	–	Within crater.
Jarvis	34.9S	148.9W	38	–	Gregory B.; member of the Challenger crew (1944-1986); previous designation Borman Z.
Jeans	55.8S	91.4E	79	–	Sir James H.; British physicist and astronomer (1877-1946).
Jehan	20.7N	31.9W	5	–	Turkish female name.
Jenkins	0.3N	78.1E	38	–	Louise F.; American astronomer (1888-1970).
Jenner	42.1S	95.9E	71	–	Edward; British doctor (1749-1823).
Jerik	18.5N	27.6E	1	–	Scandinavian male name.
Joliot	25.8N	93.1E	164	112	Frédéric Joliot-Curie; French physicist; Nobel laureate (1900-1958).
Jomo	24.4N	1.6W	7	–	African male name.
José	12.7S	1.6W	2	–	Spanish male name.
Joule	27.3N	144.2W	96	113	James P.; British physicist (1818-1889).

Feature name	Lat. (°)	Long. (°)	Size (km)	Page	Description
Joy	25.0N	6.6E	5	–	Alfred H.; American astronomer (1882-1973).
Jules Verne	35.0S	147.0E	143	112	French writer (1828-1905).
Julienne	26.0N	3.2E	2	–	French female name.
Julius Caesar	9.0N	15.4E	90	–	Roman emperor (ca. 102-44 B.C.).
Montes Jura	47.1N	34.0W	422	108	Named from terrestrial Jura Mountains.
Kaiser	36.5S	6.5E	52	–	Frederick; Dutch astronomer (1808-1872).
Kamerlingh Onnes	15.0N	115.8W	66	–	Heike; Dutch physicist; Nobel laureate (1853-1926).
Kane	63.1N	26.1E	54	–	Elisha Kent; American explorer (1820-1857).
Kant	10.6S	20.1E	33	–	Immanuel; German philosopher (1724-1804).
Kao	6.7S	87.6E	34	–	Ping-Tse; Taiwanese astronomer (1888-1970).
Kapteyn	10.8S	70.6E	49	–	Jacobus C.; Dutch astronomer (1851-1922).
Karima	25.9S	103.0E	3	–	Arabic female name.
Karpinskiy	73.3N	166.3E	92	–	Aleksey P.; Soviet geologist (1846-1936).
Karrer	52.1S	141.8W	51	–	Paul; Swiss biochemist,;Nobel laureate (1889-1971).
Kasper	8.3N	122.1E	12	–	Polish male name.
Kästner	6.8S	78.5E	108	111	Abraham Gotthelf; German mathematician, physicist (1719-1800).
Katchalsky	5.9N	116.1E	32	–	Katzir-Katchalsky, Aharon; Polish-born Israeli chemist (1914-1972).
Kathleen	5.9N	116.1E	32	–	Irish female name.
Kearons	11.4S	112.6W	23	–	William M.; American astronomer (1878-1948).
Keeler	10.2S	161.9E	160	112	James E.; American astronomer (1857-1900).
Kekulé	16.4N	138.1W	94	–	Friedrich A.; German chemist (1829-1896).
Keldysh	51.2N	43.6E	33	–	Mstislav V.; Soviet mathematician (1911-1978).
Promontorium Kelvin	27.0S	33.0W	50	–	William Thomson; British mathematician and physicist (1824-1907).
Rupes Kelvin	27.3S	33.1W	78	–	Named from nearby promontorium.
Kepinski	28.8N	126.6E	31	–	Felicjan; Polish astronomer (1885-1966).
Kepler	8.1N	38.0W	31	–	Johannes; German astronomer (1571-1630).
Khvol'son	13.8S	111.4E	54	–	Orest D.; Soviet physicist (1852-1934).
Kibal'chich	3.0N	146.5W	92	–	Nikolaj I.; Russian rocket scientist (1853-1881).
Kidinnu	35.9N	122.9E	56	–	Or Cidenas; Babylonian astronomer (unknown-ca. 343 B.C.).
Kies	26.3S	22.5W	45	–	Johann; German mathematician, astronomer (1713-1781).
Kiess	6.4S	84.0E	63	–	Carl Clarence; American astrophysicist (1887-1967).
Kimura	57.1S	118.4E	28	–	Hisashi; Japanese astronomer (1870-1943).
Kinau	60.8S	15.1E	41	–	C. A.; German botanist, selenographer (unknown-flourished 1850).
King	5.0N	120.5E	76	–	Arthur S.; American physicist (1876-1957); Edward S.; American astronomer (1861-1931).
Kira	17.6S	132.8E	3	–	Russian female name.

Feature name	Lat. (°)	Long. (°)	Size (km)	Page	Description
Kirch	39.2N	5.6W	11	–	Gottfried; German astronomer (1639-1710).
Kircher	67.1S	45.3W	72	–	Athanasius; German humanitarian (1601-1680).
Kirchhoff	30.3N	38.8E	24	–	Gustav Robert; German physicist (1824-1887).
Kirkwood	68.8N	156.1W	67	–	Daniel; American astronomer (1814-1895).
Kiva	8.6S	15.5E	1	–	Astronaut-named feature, Apollo 16 site.
Klaproth	69.8S	26.0W	119	109	Martin Heinrich; German chemist, mineralogist (1743-1817).
Klein	12.0S	2.6E	44	–	Hermann Joseph; German astronomer (1844-1914).
Kleymenov	32.4S	140.2W	55	–	Ivan T.; Soviet rocket scientist (1898-1938).
Klute	37.2N	141.3W	75	–	Daniel O.; American rocket scientist (1921-1964).
Knox-Shaw	5.3N	80.2E	12	–	Harold; British astronomer (1885-1970).
Koch	42.8S	150.1E	95	–	Robert; German doctor; Nobel laureate (1843-1910).
Kohlschütter	14.4N	154.0E	53	–	Arnold; German astronomer (1883-1969).
Kolhörster	11.2N	114.6W	97	–	Werner; German physicist (1887-1946).
Komarov	24.7N	152.5E	78	112	Vladimir M.; Soviet cosmonaut (1927-1967).
Kondratyuk	14.9S	115.5E	108	112	Yurij V.; Soviet rocket scientist (1897-1942).
König	24.1S	24.6W	23	–	Rudolf; Austrian mathematician, astronomer (1865-1927).
Konoplev	28.5S	125.5W	25	–	B. T.; Soviet radio engineer (1912-1960).
Konstantinov	19.8N	158.4E	66	–	Konstantin I.; Russian rocket scientist (1817-1871).
Kopff	17.4S	89.6W	41	–	August; German astronomer (1882-1960).
Rimae Kopff	17.4S	89.6W	41	–	Named from nearby crater.
Korolev	4.0S	157.4W	437	113	Sergej P.; Soviet rocketry scientist (1906-1966).
Kosberg	20.2S	149.6E	15	–	C. A.; Soviet aeronaut (1903-1965).
Kostinskiy	14.7N	118.8E	75	–	Sergej K.; Soviet astronomer (1867-1937).
Kovalevskaya	30.8N	129.6W	115	113	Sofia V.; Russian mathematician (1850-1891).
Koval'skij	21.9S	101.0E	49	–	Marian A.; Russsian astronomer (1821-1884).
Krafft	16.6N	72.6W	51	–	Wolfgang Ludwig; German astronomer, physicist (1743-1814).
Catena Krafft	15.0N	72.0W	60	–	Named from nearby crater.
Kramarov	2.3S	98.8W	20	–	G. M.; Soviet space scientist (1887-1970).
Kramers	53.6N	127.6W	61	–	Hendrik A.; Dutch physicist (1894-1952).
Krasnov	29.9S	79.6W	40	–	Aleksander V.; Russian astronomer (1866-1907).
Krasovskiy	3.9N	175.5W	59	–	Feodosiy N.; Soviet geodesist (1878-1948).
Kreiken	9.0S	84.6E	23	–	E. A.; Dutch astronomer (1896-1964).
Krieger	29.0N	45.6W	22	–	Johann Nepomuk; German selenographer (1865-1902).
Rima Krieger	29.0N	45.6W	22	–	Named from nearby crater.
Krishna	24.5N	11.3E	3	–	Indian male name.
Krogh	9.4N	65.7E	19	–	Schack August Steenberg; Danish zoologist, physiologist; Nobel laureate (1874-1949).
Krusenstern	26.2S	5.9E	47	–	Adam Johann, Baron von; Russian explorer (1770-1846).
Krylov	35.6N	165.8W	49	–	Aleksej N.; Soviet mathematician, mechanical engineer (1863-1945).
Kugler	53.8S	103.7E	65	–	F. X.; German chronologist (1862-1929).
Kuiper	9.8S	22.7W	6	–	Gerard Peter; Dutch-born American astronomer (1905-1973).
Kulik	42.4N	154.5W	58	–	Leonid A.; Soviet mineralogist (1883-1942).
Kundt	11.5S	11.5W	10	–	August; German physicist (1839-1894).
Kunowsky	3.2N	32.5W	18	–	Georg Karl Friedrich; German astronomer (1786-1846).
Kuo Shou Ching	8.4N	133.7W	34	–	Chinese astronomer (1231-1316).
Kurchatov	38.3N	142.1E	106	112	Igor' V.; Soviet nuclear physicist (1903-1960).
Catena Kurchatov	37.2N	136.3E	226	112	Named from nearby crater.
La Caille	23.8S	1.1E	67	–	Nicholas Louis de; French astronomer (1713-1762).
Lacchini	41.7N	107.5W	58	–	Giovanni; Italian astronomer (1884-1967).
La Condamine	53.4N	28.2W	37	–	Charles Marie de; French astronomer, physicist (1701-1774).
Lacroix	37.9S	59.0W	37	–	Sylvestre François de; French mathematician (1765-1843).
Lade	1.3S	10.1E	55	–	Heinrich Eduard von; German astronomer (1817-1904).
Lagalla	44.6S	22.5W	85	–	Giulio Cesare; Italian philosopher (1571-1624).
Lagrange	32.3S	72.8W	225	–	Joseph-Louis; French mathematician (1736-1813).
Mons La Hire	27.8N	25.5W	25	–	Philippe de; French mathematician, astronomer (1640-1718).
Lalande	4.4S	8.6W	24	–	Joseph Jerome Le François de; French astronomer (1732-1807).
Lallemand	14.3S	84.1W	18	–	Andre; French astronomer (1904-1978).
Lamarck	22.9S	69.8W	100	–	Jean-Baptiste de Monet de; French naturalist (1744-1829).
Lamb	42.9S	100.1E	106	112	Sir Horace; British mathematician, physicist (1849-1934).
Lambert	25.8N	21.0W	30	–	Johann Heinrich; German astronomer, mathematician, physicist (1728-1777).
Lamé	14.7S	64.5E	84	–	Gabriel; French mathematician (1795-1870).
Lamèch	42.7N	13.1E	13	–	Felix Chemla; French selenographer (1894-1962).
Lamont	4.4N	23.7E	106	–	John; Scottish-born German astronomer (1805-1879).
Lampland	31.0S	131.0E	65	–	Carl O.; American astronomer (1873-1951).
Landau	41.6N	118.1W	214	113	Lev D.; Soviet physicist; Nobel laureate (1908-1968).
Lander	15.3S	131.8E	40	–	Richard Lemon; British explorer (1804-1834).
Landsteiner	31.3N	14.8W	6	–	Karl; Austrian-born pathologist in America; Nobel laureate (1868-1943).

Feature name	Lat. (°)	Long. (°)	Size (km)	Page	Description
Lane	9.5S	132.0E	55	–	Jonathan H.; American physicist, astrophysicist (1819-1880).
Langemak	10.3S	118.7E	97	112	Georgij E.; Soviet rocket scientist (1898-1938).
Langevin	44.3N	162.7E	58	–	Paul; French physicist (1872-1946).
Langley	51.1N	86.3W	59	–	Samuel P.; American astronomer, physicist (1834-1906).
Langmuir	35.7S	128.4W	91	–	Irving; American physicist, chemist; Nobel laureate (1881-1957).
Langrenus	8.9S	61.1E	127	111	Langren, Michel Florent van; Belgian selenographer, engineer (ca. 1600-1675).
Lansberg	0.3S	26.6W	38	–	Philippe van; Belgian astronomer (1561-1632).
La Pérouse	10.7S	76.3E	77	–	Jean-François de Galoup, Comte de la Pérouse; French explorer (1741-1788).
Promontorium Laplace	46.0N	25.8W	50	–	Pierre-Simon; French mathematician, astronomer (1749-1827).
Lara	20.4N	30.5E	–	–	Astronaut-named feature, Apollo 17 site.
Larmor	32.1N	179.7W	97	113	Sir Joseph; British mathematician, physicist (1857-1942).
Lassell	15.5S	7.9W	23	–	William; British astronomer (1799-1880).
Last	26.1N	0.0W	–	–	Astronaut-named feature, Apollo 15 site.
Laue	28.0N	96.7W	87	–	Max T. F. von; German physicist; Nobel laureate (1879-1960).
Lauritsen	27.6S	96.1E	52	–	Charles C.; Danish-born American physicist (1892-1968).
Lavoisier	38.2N	81.2W	70	–	Antoine-Laurent; French chemist (1743-1794).
Lawrence	7.4N	43.2E	24	–	Ernest Orlando; American physicist; Nobel laureate (1901-1958).
Leakey	3.2S	37.4E	12	–	Louis Seymour Bazett; British paleontologist (1903-1972).
Leavitt	44.8S	139.3W	66	–	Henrietta S.; American astronomer (1868-1921).
Lebedev	47.3S	107.8E	102	109	Pëtr N.; Russian physicist (1866-1912).
Lebedinskiy	8.3N	164.3W	62	–	Aleksandr I.; Soviet astrophysicist (1913-1967).
Lebesgue	5.1S	89.0E	11	–	Henri Leon; French mathematician (1875-1941).
Lee	30.7S	40.7W	41	–	John; British astronomer, humanitarian (1783-1866).
Leeuwenhoek	29.3S	178.7W	125	113	Antonie van; Dutch microscopist (1632-1723).
Legendre	28.9S	70.2E	78	–	Adrien-Marie; French mathematician (1752-1833).
Le Gentil	74.6S	75.7W	128	–	Guillaume Hyazinthe; French astronomer (1725-1792).
Lehmann	40.0S	56.0W	53	–	Jakob Heinrich Wilhelm; German astronomer (1800-1863).
Leibnitz	38.3S	179.2E	245	112	Gottfried W.; German mathematician, physicist, philosopher (1646-1716).
Lemaître	61.2S	149.6W	91	109	Georges; Belgian astrophysicist (1894-1966).
Le Monnier	26.6N	30.6E	60	–	Pierre-Charles; French astronomer, physicist (1715-1799).
Lacus Lenitatis	14.0N	12.0E	80	–	"Lake of Softness."
Lents (Lenz)	2.8N	102.1W	21	–	H. F. Emil; Russian physicist (1804-1865).
Leonov	19.0N	148.2E	33	–	Aleksej A.; Soviet cosmonaut (1934-).
Lepaute	33.3S	33.6W	16	–	Nicole Reine de la Briere; French astronomer (1723-1788).
Letronne	10.8S	42.5W	116	110	Jean Antoine; French archaeologist (1787-1848).
Leucippus	29.1N	116.0W	56	–	Greek philosopher (unknown-flourished ca. 440 B.C.).
Leuschner	1.8N	108.8W	49	–	Armin O.; American astronomer (1868-1953).
Catena Leuschner (GDL)	4.7N	110.1W	364	113	Named from nearby crater; GDL=Gas Dynamics Laboratory.
Le Verrier	40.3N	20.6W	20	–	Urbain Jean; French astronomer, mathematician (1811-1877).
Levi-Civita	23.7S	143.4E	121	112	Tullio; Italian mathematician, physicist (1873-1941).
Lewis	18.5S	113.8W	42	–	Gilbert N.; American chemist (1875-1946).
Lexell	35.8S	4.2W	62	–	Anders Johann; Swedish mathematician, astronomer (1740-1784).
Ley	42.2N	154.9E	79	–	Willy; German-born American rocket scientist (1906-1969).
Licetus	47.1S	6.7E	74	–	Liceti, Fortunio; Italian physicist, philosopher, doctor (1577-1657).
Lichtenberg	31.8N	67.7W	20	–	Georg Christoph; German physicist (1742-1799).
Lick	12.4N	52.7E	31	–	James; American benefactor (1796-1876).
Liebig	24.3S	48.2W	37	–	Justus, Baron von Liebig; German chemist (1803-1873).
Rimae Liebig	20.0S	45.0W	140	–	Named from nearby crater.
Rupes Liebig	24.4S	48.5W	37	–	Named from nearby crater.
Light Mantle	20.2N	30.8E	4	–	Astronaut-named feature, Apollo 17 site.
Lilius	54.5S	6.2E	61	–	Luigi Giglio; Italian doctor, philosopher, chronologist (unknown-1576).
Linda	30.7N	33.4W	1	–	Spanish female name.
Lindbergh	5.4S	52.9E	12	–	Charles Augustus; American aviator (1902-1974).
Lindblad	70.4N	98.8W	66	–	Bertil; Swedish astronomer (1895-1965).
Lindenau	32.3S	24.9E	53	–	Bernhard von; German astronomer (1780-1854).
Lindsay	7.0S	13.0E	32	–	Eric M.; Irish astronomer (1907-1974).
Linné	27.7N	11.8E	2	–	Carl von; Swedish botanist (1707-1778).
Liouville	2.6N	73.5E	16	–	Joseph; French mathematician (1809-1882).
Lippershey	25.9S	10.3W	6	–	Hans (Jan Lapprey); Dutch optician (unknown-1619).
Lippmann	56.0S	114.9W	160	109	Gabriel; French physicist, Nobel laureate (1845-1921).
Lipskiy	2.2S	179.5W	80	–	Yuriy N.; Soviet selenographer (1909-1978).
Dorsa Lister	20.3N	23.8E	203	–	Martin; British stratigrapher, zoologist (1638-1712).

Feature name	Lat. (°)	Long. (°)	Size (km)	Page	Description
Litke (Lütke)	16.8S	123.1E	39	–	Fedor P.; Russian geographer (1797-1882).
Littrow	21.5N	31.4E	30	–	Johann Josef von; Czechoslovakian astronomer (1781-1840).
Catena Littrow	22.2N	29.5E	10	–	Named from nearby crater.
Rimae Littrow	22.1N	29.9E	115	–	Named from nearby crater.
Lobachevskiy	9.9N	112.6E	84	–	Nikolay I.; Russian mathematician (1792-1856).
Lockyer	46.2S	36.7E	34	–	Joseph Norman; British astrophysicist 1836-1920
Lodygin	17.7S	146.8W	62	–	Aleksandr N.; Russian inventor (1847-1923).
Loewy	22.7S	32.8W	24	–	Moritz; French astronomer (1833-1907).
Lohrmann	0.5S	67.2W	30	–	Wilhelm Gotthelf; German selenographer (1796-1840).
Lohse	13.7S	60.2E	41	–	Oswald; German astronomer (1845-1915).
Lomonosov	27.3N	98.0E	92	–	Mikhail V.; Russian cartographer (1711-1765).
Longomontanus	49.6S	21.8W	157	109	Christian Sorensen; Danish astronomer, mathematician (1562-1647).
Lorentz	32.6N	95.3W	312	113	Hendrik A.; Dutch physicist; Nobel laureate (1853-1928).
Louise	28.5N	34.2W	–	–	French female name.
Louville	44.0N	46.0W	36	–	Jacques D'Allonville, Chevalier de Louville; French astronomer, mathematician (1671-1732).
Love	6.3S	129.0E	84	–	Augustus E. H.; British mathematician, geophysicist (1863-1940).
Lovelace	82.3N	106.4W	54	–	William R.; American doctor, space scientist (1907-1965).
Lovell	36.8S	141.9W	34	–	James A., Jr.; American astronaut (1928-).
Lowell	12.9S	103.1W	66	112	Percival; American astronomer (1855-1916).
Lubbock	3.9S	41.8E	13	–	Sir John William; British astronomer, mathematician (1803-1865).
Lubiniezky	17.8S	23.8W	43	–	Stanislaus; Polish astronomer (1623-1675).
Lucian	14.3N	36.7E	7	–	Of Samosata; Greek writer (125-190).
Lucretius	8.2S	120.8W	63	–	Titus Lucretius Carus; Roman scientific philosopher (ca. 95-55 B.C.).
Catena Lucretius (RNII)	3.4S	126.1W	271	113	Named from nearby crater; RNII=Rocket Research Institute.
Ludwig	7.7S	97.4E	23	–	Carl Friedrich Wilhelm; German physiologist (1816-1895).
Lundmark	39.7S	152.5E	106	112	Knut E.; Swedish astronomer (1889-1958).
Sinus Lunicus	31.8N	1.4W	126	110	"Lunik Bay" landing area of Luna (Lunik) 2.
Luther	33.2N	24.1E	9	–	Robert; German astronomer (1822-1900).
Lacus Luxuriae	19.0N	176.0E	50	–	"Lake of Luxury."
Lyapunov	26.3N	89.3E	66	–	Aleksandr M.; Russian mathematician, engineer (1857-1918).
Lyell	13.6N	40.6E	32	–	Sir Charles; British geologist (1797-1875).
Lyman	64.8S	163.6E	84	109	Theodore; American physicist (1874-1954).
Lyot	49.8S	84.5E	132	109	Bernard F.; French astronomer (1897-1952).

Feature name	Lat. (°)	Long. (°)	Size (km)	Page	Description
Mach	18.5N	149.3W	180	113	Ernst; Austrian physicist, philosopher (1838-1916).
Mackin-Apollo (Mackin)	20.1N	30.7E	–	–	Astronaut-named feature, Apollo 17 site.
Maclaurin	1.9S	68.0E	50	–	Colin; Scottish mathematician (1698-1746).
MacLear	10.5N	20.1E	20	–	Sir Thomas; Irish astronomer (1794-1879).
Rimae MacLear	13.0N	20.0E	110	–	Named from nearby crater.
MacMillan	24.2N	7.8W	7	–	William Duncan; American mathematician, astronomer (1871-1948).
Macrobius	21.3N	46.0E	64	111	Ambrosius Aurelius Theodosius; Roman writer (unknown-flourished ca. 410).
Mädler	11.0S	29.8E	27	–	Johann Heinrich; German astronomer (1794-1874).
Maestlin	4.9N	40.6W	7	–	Michael; German mathematician (1550-1631).
Rimae Maestlin	2.0N	40.0W	80	–	Named from nearby crater.
Magelhaens	11.9S	44.1E	40	–	Fernão de (Ferdinand Magellan); Portuguese explorer (1480-1521).
Maginus	50.5S	6.3W	194	109	Magini, Giovanni Antonio; Italian astronomer, mathematician (1555-1617).
Main	80.8N	10.1E	46	–	Robert; British astronomer (1808-1878).
Mairan	41.6N	43.4W	40	–	Jean Jacques d'Ortous de; French geophysicist (1678-1771).
Rima Mairan	38.0N	47.0W	90	–	Named from nearby crater.
Maksutov	40.5S	168.7W	83	–	Dmitrij D.; Soviet optician (1896-1964).
Malapert	84.9S	12.9E	69	–	Charles; Belgian astronomer, mathematician, philosopher (1581-1630).
Mallet	45.4S	54.2E	58	–	Robert; Irish seismologist, engineer (1810-1881).
Malyy	21.9N	105.3E	41	–	Aleksandr L.; Soviet rocket scientist (1907-1961).
Mandel'shtam	5.4N	162.4E	197	112	Leonid I.; Soviet physicist (1879-1944).
Manilius	14.5N	9.1E	38	–	Marcus; Roman writer (unknown-ca. 50 B.C.).
Manners	4.6N	20.0E	15	–	Russell Henry; British astronomer (1800-1870).
Manuel	24.5N	11.3E	–	–	Spanish male name.
Manzinus	67.7S	26.8E	98	109	Manzini, Carlo Antonio; Italian astronomer (1599-1677).
Maraldi	19.4N	34.9E	39	–	Giovanni Domenico; Italian astronomer, geodesist (1709-1788); Jacques Philippe; French astronomer (1665-1729).
Mons Maraldi	20.3N	35.3E	15	–	Named from nearby crater.
Rima Marcello	18.6N	27.7E	2	–	Italian male name.
Marci	22.6N	167.0W	25	–	Jan Marek Marci von Kronland; Czechoslovakian physicist (1595-1667).
Marco Polo	15.4N	2.0W	28	–	Italian explorer (1254-1324).
Marconi	9.6S	145.1E	73	–	Guglielmo; Italian physicist, inventor; Nobel laureate (1874-1937).
Mare Marginis	13.3N	86.1E	420	111	"Sea of the Edge."
Marinus	39.4S	76.5E	58	–	Of Tyre; Greek geographer (unknown-ca. 100).
Mariotte	28.5S	139.1W	65	–	Edme; French physicist (1620-1684).
Marius	11.9N	50.8W	41	–	Mayer, Simon; German astronomer (1570-1624).

Feature name	Lat. (°)	Long. (°)	Size (km)	Page	Description
Rima Marius	16.5N	48.9W	121	–	Named from nearby crater.
Markov	53.4N	62.7W	40	–	Aleksandr V.; Soviet astrophysicist (1897-1968); Andrei A.; Russian mathematician (1856-1922).
Marth	31.1S	29.3W	6	–	Albert; German astronomer (1828-1897).
Mary	18.9N	27.4E	1	–	English form of Hebrew female name.
Maskelyne	2.2N	30.1E	23	–	Nevil; British astronomer (1732-1811).
Mason	42.6N	30.5E	33	–	Charles; British astronomer (1728-1786).
Maunder	14.6S	93.8W	55	–	Annie S. D. R.; British astronomer (1868-1947); Edward Walter; British astronomer (1851-1928).
Maupertuis	49.6N	27.3W	45	–	Pierre Louis de; French mathematician (1698-1759).
Rimae Maupertuis	52.0N	23.0W	60	–	Named from nearby crater.
Maurolycus	42.0S	14.0E	114	111	Maurolico, Francesco; Italian mathematician (1494-1575).
Maury	37.1N	39.6E	17	–	Matthew Fontaine; American oceanographer (1806-1873); Antonia C.; American astronomer (1866-1952).
Mavis	29.8N	26.4W	1	–	Scottish female name.
Dorsa Mawson	4.6S	55.7E	132	–	Douglas; Australian Antarctic explorer (1882-1958).
Maxwell	30.2N	98.9E	107	112	James Clerk; Scottish physicist (1831-1879).
C. Mayer	63.2N	17.3E	38	–	Christian; German astronomer, mathematician, physicist (1719-1783).
T. Mayer	15.6N	29.1W	33	–	Tobias; German astronomer (1723-1762).
Rima T. Mayer	13.0N	31.0W	50	–	Named from nearby crater.
McAdie	2.1N	92.1E	45	–	Alexander George; American meteorologist (1863-1943).
McAuliffe	33.0S	148.9W	19	–	Sharon Christa; civilian school teacher member of the Challenger crew (1948-1986); previous designation Borman Y.
McClure	15.3S	50.3E	23	–	Robert Le Mesurier; British explorer (1807-1873).
McDonald	30.4N	20.9W	7	–	William Johnson; American benefactor (1844-1926); Thomas Logie; Scottish selenographer (unknown-1973).
McKellar	15.7S	170.8W	51	–	Andrew; Canadian astronomer (1910-1960).
McLaughlin	47.1N	92.9W	79	–	Dean B.; American astronomer (1901-1965).
McMath	17.3N	165.6W	86	113	Francis C.; American engineer, astronomer (1867-1938); Robert R.; American astronomer (1891-1962).
McNair	35.7S	147.3W	29	–	Ronald Erwin; member of the Challenger crew (1950-1986); previous designation Borman A.
McNally	22.6N	127.2W	47	–	Paul A.; American astronomer (1890-1955).
Mechnikov	11.0S	149.0W	60	–	Il'ya I.; Russian-French bacteriologist; Nobel laureate (1845-1916).
Sinus Medii	2.4N	1.7E	335	111	"Bay of the Center."
Mee	43.7S	35.3W	126	110	Arthur Butler Phillips; Scottish astronomer (1860-1926).
Mees	13.6N	96.1W	50	–	C. E. Kenneth; American photographer (1882-1960).
Meggers	24.3N	123.0E	52	–	William F.; American physicist (1888-1968).
Meitner	10.5S	112.7E	87	–	Lise; German physicist (1878-1968).
Melissa	8.1N	121.8E	18	–	Greek female name.
Mendel	48.8S	109.4W	138	109	Gregor J.; Austrian biologist (1822-1884).
Mendeleev	5.7N	140.9E	313	112	Dmitry I.; Russian chemist (1834-1907).
Catena Mendeleev	6.3N	139.4E	188	–	Named from nearby crater.
Menelaus	16.3N	16.0E	26	–	Of Alexandria; Greek geometer, astronomer (ca. 98 A.D.).
Rimae Menelaus	17.2N	17.9E	131	–	Named from nearby crater.
Menzel	3.4N	36.9E	3	–	Donald Howard; American astrophysicist, Smithsonian researcher (1901-1976).
Mercator	29.3S	26.1W	46	–	Gerhard Kremer (Gerardus Mercator); Flemish cartographer, geographer, mathematician (1512-1594).
Rupes Mercator	31.0S	22.3W	93	–	Named from nearby crater.
Mercurius	46.6N	66.2E	67	–	Mercury; Roman mythical messenger.
Merrill	75.2N	116.3W	57	–	Paul W.; American astronomer (1887-1961).
Mersenius	21.5S	49.2W	84	110	Mersenne, Marin; French mathematician, physicist (1588-1648).
Rimae Mersenius	21.5S	49.2W	84	–	Named for nearby crater.
Meshcherskiy	12.2N	125.5E	65	–	Ivan V.; Russian mathematician (1859-1935).
Messala	39.2N	60.5E	125	111	(Ma-Sa-Allah); Jewish astronomer (unknown- ca. 815).
Messier	1.9S	47.6E	11	–	Charles; French astronomer (1730-1817).
Rima Messier	1.0S	45.0E	100	–	Named from nearby crater.
Metius	40.3S	43.3E	87	–	Adriaan Adriaanszoon; Dutch astronomer (1571-1635).
Meton	73.6N	18.8E	130	108	Greek astronomer (unknown- flourished 432 B.C.).
Mezentsev	72.1N	128.7W	89	–	Yurij B.; Soviet rocket scientist (1929-1965).
Michael	25.1N	0.2E	4	–	English male name.
Michelson	7.2N	120.7W	123	113	Albert A.; German-born American physicist; Nobel laureate (1852-1931).
Catena Michelson (GIRD)	1.4N	113.4W	456	113	Named from nearby crater; GIRD=Group for the Study of Reaction Motion.
Middle Crescent	3.2S	23.4W	–	–	Astronaut-named feature, Apollo 12 site.
Milankovič	77.2N	168.8E	101	108	M.; Yugoslavian astronomer (1879-1958).
Milichius	10.0N	30.2W	12	–	Milich, Jakob; German doctor, mathematician, astronomer (1501-1559).
Rima Milichius	8.0N	33.0W	100	–	Named from nearby crater.
Miller	39.3S	0.8E	61	–	William Allen; British chemist (1817-1870).
Millikan	46.8N	121.5E	98	108	Robert A.; American physicist; Nobel laureate (1868-1953).

Feature name	Lat. (°)	Long. (°)	Size (km)	Page	Description
Mills	8.6N	156.0E	32	–	Mark M.; American physicist (1917-1958).
Milne	31.4S	112.2E	272	112	E. Arthur; British mathematician, astrophysicist (1896-1950).
Mineur	25.0N	161.3W	73	–	Henri; French mathematician, astronomer (1899-1954).
Minkowski	56.5S	146.0W	113	109	Hermann; German mathematician (1864-1909); Rudolph L.B.; American astronomer (1895-1976).
Minnaert	67.8S	179.6E	125	109	Marcel G.; Dutch astronomer, astrophysicist (1893-1970).
Mitchell	49.7N	20.2E	30	–	Maria; American astronomer (1818-1889).
Mitra	18.0N	154.7W	92	–	S. K.; Indian physicist (1890-1963).
Möbius	15.8N	101.2E	50	–	August F.; German mathematician, astronomer (1790-1868).
Mohoróvičic	19.0S	165.0W	51	–	Andrija; Croatian geophysicist (1857-1936).
Moigno	66.4N	28.9E	36	–	François Napoleon Marie; French mathematician, physicist (1804-1884).
Moiseev	9.5N	103.3E	59	–	Nikolaj D.; Soviet astronomer (1902-1955).
Moissan	4.8N	137.4E	21	–	Ferdinand Frederic Henri; French chemist; Nobel laureate (1852-1907).
Moltke	0.6S	24.2E	6	–	Helmuth Karl, Graf von; German benefactor (1800-1891).
Monge	19.2S	47.6E	36	–	Gaspard; French mathematician (1746-1818).
Monira	12.6S	1.7W	2	–	Arabic female name.
Montanari	45.8S	20.6W	76	–	Geminiano; Italian astronomer, mathematician (1633-1687).
Montgolfier	47.3N	159.8W	88	–	Jacques E.; French inventor (1745-1799); Joseph M.; French inventor (1740-1810).
Moore	37.4N	177.5W	54	–	Joseph H.; American astronomer (1878-1949).
Moretus	70.6S	5.8W	111	109	Moret, Theodore; Belgian mathematician (1602-1667).
Morley	2.8S	64.6E	14	–	Edward Williams; American chemist (1838-1923).
Mons Moro	12.0S	19.7W	10	–	Antonio Lazzaro; Italian earth scientist (1687-1764).
Morozov	5.0N	127.4E	42	–	Nikolaj A.; Soviet natural scientist (1854-1945).
Morse	22.1N	175.1W	77	–	Samuel F. B.; American inventor (1791-1872).
Lacus Mortis	45.0N	27.2E	151	108	"Lake of Death."
Mare Moscoviense	27.3N	147.9E	277	112	"Sea of Muscovy."
Moseley	20.9N	90.1W	90	–	Henry G. J.; British physicist (1887-1915).
Mösting	0.7S	5.9W	24	–	Johan Sigismund von; Danish benefactor (1759-1843).
Mouchez	78.3N	26.6W	81	–	Ernest Amedee Barthelemy; French astronomer (1821-1892).
Moulton	61.1S	97.2E	49	–	Forest R.; American astronomer (1872-1952).
Müller	7.6S	2.1E	22	–	Karl; Czechoslovakian astronomer (1866-1942).
Murakami	23.3S	140.5W	45	–	H.; Japanese physicist, astronomer (1872-1947).
Murchison	5.1N	0.1W	57	–	Sir Roderick Impey; Scottish geologist (1792-1871).
Mutus	63.6S	30.1E	77	–	Vincente Mut, or Muth; Spanish astronomer (unknown-1673).
Nagaoka	19.4N	154.0E	46	–	Hantaro; Japanese physicist (1865-1940).
Nansen	80.9N	95.3E	104	108	Fridtjof; Norwegian explorer (1861-1930).
Nansen-Apollo	20.1N	30.5E	1	–	Astronaut-named feature, Apollo 17 site.
Naonobu	4.6S	57.8E	34	–	Ajima; Japanese mathematician (ca. 1732-1798).
Nasireddin	41.0S	0.2E	52	–	Nasir-Al-Din (Mohammed Ibn Hassan); Persian astronomer (1201-1274).
Nasmyth	50.5S	56.2W	76	–	James; Scottish engineer, astronomer (1808-1890).
Nassau	24.9S	177.4E	76	–	Jason J.; American astronomer (1892-1965).
Natasha	20.0N	31.3W	12	–	Russian female name.
Naumann	35.4N	62.0W	9	–	Karl Friedrich; German geologist (1797-1873).
Neander	31.3S	39.9E	50	–	Neumann, Michael; German mathematician (1529-1581).
Nearch	58.5S	39.1E	75	–	Greek explorer (unknown-flourished 325 B.C.).
Necho	5.0S	123.1E	30	–	Egyptian geographer (610-593 B.C.).
Mare Nectaris	15.2S	35.5E	333	111	"Sea of Nectar."
Neison	68.3N	25.1E	53	–	(Neville), Edmund; British astronomer, selenographer (1849-1940).
Neper	8.5N	84.6E	137	111	John; Scottish mathematician (1550-1617).
Nernst	35.3N	94.8W	116	113	Walther H.; German physical chemist; Nobel laureate (1864-1941).
Neujmin	27.0S	125.0E	101	112	Grigorij N.; Soviet astronomer (1885-1946).
Neumayer	71.1S	70.7E	76	–	Georg Balthasar von; German meteorologist, hydrographer (1826-1909).
Newcomb	29.9N	43.8E	41	–	Simon; Canadian-born American astronomer (1835-1909).
Newton	76.7S	16.9W	78	–	Isaac; British mathematician, physicist, astronomer (1642-1727).
Nicholson	26.2S	85.1W	38	–	Seth B.; American astronomer (1891-1963).
Dorsum Nicol	18.0N	23.0E	50	–	William; Scottish physicist (1768-1851).
Nicolai	42.4S	25.9E	42	–	Friedrich Bernhard Gottfried; German astronomer (1793-1846).
Nicollet	21.9S	12.5W	15	–	Jean Nicholas; French astronomer (1788-1843).
Nielsen	31.8N	51.8W	9	–	Harald Herborg; American physicist (1903-1973); Axel V.; Danish astronomer (1902-1970).
Niépce	72.7N	119.1W	57	–	Joseph N.; French physicist, photographer (1765-1833).
Dorsum Niggli	29.0N	52.0W	50	–	Paul; Swiss earth scientist (1888-1953).
Nijland	33.0N	134.1E	35	–	Albertus A.; Dutch astronomer (1868-1936).

Feature name	Lat. (°)	Long. (°)	Size (km)	Page	Description
Nikolaev	35.2N	151.3E	41	–	Andriyan G.; Soviet cosmonaut (1929-).
Nishina	44.6S	170.4W	65	–	Yoshio; Japanese physicist (1890-1951).
Nobel	15.0N	101.3W	48	–	Alfred B.; Swedish inventor (1833-1896).
Nobile	85.2S	53.5E	73	–	Umberto; Italian artic explorer (1885-1978).
Nobili	0.2N	75.9E	42	–	Leopoldo; Italian physicist (1784-1835).
Nöggerath	48.8S	45.7W	30	–	Johann Jakob; German geologist, mineralogist, seismologist (1788-1877).
Nonius	34.8S	3.8E	69	–	Nuñez, Pedro; Portuguese mathematician (ca. 1492-1578).
Norman	11.8S	30.4W	10	–	Robert; British natural scientist (unknown-flourished ca. 1590).
North Complex	26.2N	3.6E	2	–	Astronaut-named feature, Apollo 15 site.
North Massif	20.4N	30.8E	14	–	Astronaut-named feature, Apollo 17 site.
North Ray	8.8S	15.5E	1	–	Astronaut-named feature, Apollo 16 site.
Nöther (Noether)	66.6N	113.5W	67	–	Emmy; German mathematician (1882-1935).
Mare Nubium	21.3S	16.6W	715	110	"Sea of Clouds."
Numerov	70.7S	160.7W	113	109	Boris V.; Soviet astronomer (1891-1941).
Nunn	4.6N	91.1E	19	–	Joseph; American engineer (1905-1968).
Nüsl	32.3N	167.6E	61	–	Frantisek; Czechoslovakian astronomer (1867-1925).
Lacus Oblivionis	21.0S	168.0W	50	–	"Lake of Forgetfulness."
Obruchev	38.9S	162.1E	71	–	Vladimir A.; Soviet geologist (1863-1956).
O'Day	30.6S	157.5E	71	–	Marcus; American physicist (1897-1961).
Lacus Odii	19.0N	7.0E	70	112	"Lake of Hatred."
Oenopides	57.0N	64.1W	67	–	Of Chios; Greek astronomer, geometrician (ca. 500-430 B.C.).
Oersted	43.1N	47.2E	42	–	Hans Christian; Danish physicist, chemist (1777-1851).
Ohm	18.4N	113.5W	64	–	Georg Simon; German physicist (1787-1854).
Oken	43.7S	75.9E	71	–	(Okenfuss), Lorenz; German biologist, physiologist (1779-1851).
Olbers	7.4N	75.9W	74	110	Heinrich Wilhelm Malthaus; German astronomer, doctor (1758-1840).
Olcott	20.6N	117.8E	81	112	William T.; American astronomer (1873-1936).
Old Nameless	3.7S	17.5W	–	–	Astronaut-named feature, Apollo 14 site.
Olivier	59.1N	138.5E	69	–	Charles Pollard; American astronomer (1884-1975).
Omar Khayyam	58.0N	102.1W	70	–	Al-Khayyami; Persian mathematician, astronomer, poet (ca. 1050-1123).
Onizuka	36.2S	148.9W	29	–	Ellison Shoji; member of the Challenger crew (1946-1986).
Opelt	16.3S	17.5W	48	–	Friedrich Wilhelm; German astronomer (1794-1863).
Rimae Opelt	13.0S	18.0W	70	–	Named from nearby crater.
Dorsum Oppel	18.7N	52.6E	268	111	Albert; German paleontologist (1831-1865).
Oppenheimer	35.2S	166.3W	208	113	J. Robert; American physicist (1904-1967).
Oppolzer	1.5S	0.5W	40	–	Theodor Egon von; Czechoslovakian astronomer (1841-1886).
Rima Oppolzer	1.7S	1.0E	94	–	Named from nearby crater.
Oresme	42.4S	169.2E	76	–	Nicole; French mathematician (1323(?)-1382).
Mare Orientale	19.4S	92.8W	327	113	"Eastern sea."
Orlov	25.7S	175.0W	81	–	Aleksandr V.; Soviet astronomer (1880-1954); Sergei V.; Soviet astronomer (1880-1958).
Orontius	40.6S	4.6W	105	–	Finnaeus (Oronce Fine); French mathematician, cartographer (1494-1555).
Osama	18.6N	5.2E	–	–	Japanese male name.
Osiris	18.6N	27.6E	1	–	Egyptian god of the dead.
Osman	11.0S	6.2W	2	–	Turkish male name.
Ostwald	10.4N	121.9E	104	112	Wilhelm; German chemist; Nobel laureate (1853-1932).
Dorsum Owen	25.0N	11.0E	50	–	George; British earth scientist (1552-1613).
Palisa	9.4S	7.2W	33	–	Johann; Czechoslovakian-born Austrian astronomer (1848-1925).
Palitzsch	28.0S	64.5E	41	–	Johann Georg; German astronomer (1723-1788).
Vallis Palitzsch	26.4S	64.3E	132	111	Named from nearby crater.
Pallas	5.5N	1.6W	46	–	Peter Simon; German geologist, natural historian (1741-1811).
Palmetto	8.9S	15.5E	–	–	Astronaut-named feature, Apollo 16 site.
Palmieri	28.6S	47.7W	40	–	Luigi; Italian physicist, mathematician (1807-1896).
Rimae Palmieri	28.0S	47.0W	150	–	Named from nearby crater.
Paneth	63.0N	94.8W	65	–	Friedrich Adolf; German chemist (1887-1958).
Pannekoek	4.2S	140.5E	71	–	Antonie; Dutch astronomer (1873-1960).
Papaleksi	10.2N	164.0E	97	–	Nikolaj D.; Soviet physicist (1880-1947).
Paracelsus	23.0S	163.1E	83	–	Theophrastus B. von Hohenheim; Swiss-born American doctor, chemist (1493-1541).
Paraskevopoulos	50.4N	149.9W	94	–	John S.; Greek-born American astronomer (1889-1951).
Parenago	25.9N	108.5W	93	–	Pavel P.; Soviet astronomer (1906-1960).
Parkhurst	33.4S	103.6E	96	–	John A.; American astronomer (1861-1925).
Parrot	14.5S	3.3E	70	–	Johann Jacob F. W.; Russian doctor, physicist (1792-1840).
Parry	7.9S	15.8W	47	–	Sir William Edward; British explorer (1790-1855).
Rimae Parry	6.1S	16.8W	82	–	Named from nearby crater.
Parsons	37.3N	171.2W	40	–	John W.; American rocket scientist (1913-1952).
Pascal	74.6N	70.3W	115	108	Blaise; French mathematician (1623-1662).
Paschen	13.5S	139.8W	124	113	Friedrich; German physicist (1865-1940).
Pasteur	11.9S	104.6E	224	112	Louis; French chemist, microbiologist (1822-1895).
Patricia	25.0N	0.3E	5	–	English female name.
Patsaev	16.7S	133.4E	55	–	Viktor I.; Soviet engineer (1933-1971).

Feature name	Lat. (°)	Long. (°)	Size (km)	Page	Description
Pauli	44.5S	136.4E	84	112	Wolfgang; Austrian-born American physicist; Nobel laureate (1900-1958).
Pavlov	28.8S	142.5E	148	112	Ivan P.; Soviet physiologist; Nobel laureate (1849-1936).
Pawsey	44.5N	145.0E	60	–	Joseph L.; Australian radio astronomer (1908-1962).
Peary	88.6N	33.0E	73	–	Robert E.; American explorer (1856-1920).
Pease	12.5N	106.1W	38	–	Francis G.; American astronomer (1881-1938).
Peek	2.6N	86.9E	12	–	Bertrand Meigh; British astronomer (1891-1965).
Peirce	18.3N	53.5E	18	–	Benjamin; American mathematician, astronomer (1809-1880).
Peirescius	46.5S	67.6E	61	–	Peiresc, Nicolas Claude Fabri de; French astronomer, archaeologist (1580-1637).
Mons Penck	10.0S	21.6E	30	–	Albrecht; German geographer (1858-1945).
Pentland	64.6S	11.5E	56	–	Joseph Barclay; Irish geographer (1797-1873).
Perel'man	24.0S	106.0E	46	–	Yakov I.; Soviet rocket scientist (1882-1942).
Perepelkin	10.0S	129.0E	97	–	Evgenij J.; Soviet astrophysicist (1906-1940).
Perkin	47.2N	175.9W	62	–	Richard S.; American telescope manufacturer (1906-1969).
Perrine	42.5N	127.8W	86	–	Charles D.; American astronomer (1867-1951).
Lacus Perseverantiae	8.0N	62.0E	70	–	"Lake of Perseverance."
Petavius	5.1S	60.4E	188	–	Petau, Denis; French chronologist, astronomer (1583-1652).
Rimae Petavius	25.9S	58.9E	80	–	Within crater.
Petermann	74.2N	66.3E	73	–	August Heinrich; German geographer (1822-1878).
Peters	68.1N	29.5E	15	–	Christian August Friedrich; German astronomer (1806-1880).
Petit	2.3N	63.5E	5	–	Alexis Therese; French physicist (1771-1820).
Petrie	45.3N	108.4E	33	–	Robert M.; Canadian astronomer (1906-1966).
Petropavlovskiy	37.2N	114.8W	63	–	Boris S.; Soviet rocket engineer (1898-1933).
Petrov	61.4S	88.0E	49	–	Evgenij S.; Soviet rocket scientist (1900-1942).
Pettit	27.5S	86.6W	35	–	Edison; American astronomer (1889-1962).
Rimae Pettit	23.0S	92.0W	450	–	Named from nearby crater.
Petzval	62.7S	110.4W	90	109	Joseph von; Austrian optician (1807-1891).
Phillips	26.6S	75.3E	122	111	John; British geologist, astronomer (1800-1874).
Philolaus	72.1N	32.4W	70	–	Of Croton; Greek mathematician, astronomer, philosopher (unknown-flourished 400 B.C.).
Phocylides	52.7S	57.0W	121	109	Johannes Phocylides Holwarda (Jan Fokker); Dutch astronomer (1618-1651).
Piazzi	36.6S	67.9W	134	–	Giuseppe; Italian astronomer (1746-1826).
Piazzi Smyth	41.9N	3.2W	13	–	Charles; Scottish astronomer (1819-1900).
Picard	14.6N	54.7E	22	–	Jean; French astronomer (1620-1682).
Piccolomini	29.7S	32.2E	87	111	Alessandro; Italian astronomer (1508-1578).
Pickering	2.9S	7.0E	15	–	Edward Charles; American astronomer (1846-1919).
Mons Pico	45.7N	8.9W	25	–	Spanish for "peak."
Pictet	43.6S	7.4W	62	–	Pictet-Turretin, Marc-Auguste; Swiss physicist (1752-1825).
Catena Pierre	19.8N	31.8W	9	–	French male name.
Pikel'ner	47.9S	123.3E	47	–	Solomon Borisovich; Soviet astronomer, cosmologist (1921-1975).
Pilâtre	60.2S	86.9W	50	–	Rozier, F. de; French aeronaut (1753-1785).
Pingré	58.7S	73.7W	88	109	Alexandre Guy; French astronomer (1711-1796).
Pirquet	20.3S	139.6E	65	–	Baron Guido von; Austrian (spacecraft trajectories) astronaut (1867-1936).
Pitatus	29.9S	13.5W	106	110	Pitati, Pietro; Italian astronomer, mathematician (unknown.-flourished ca. 1500).
Rimae Pitatus	28.5S	13.8W	94	–	Within crater.
Pitiscus	50.4S	30.9E	82	–	Bartholemaeus; German mathematician (1561-1613).
Mons Piton	40.6N	1.1W	25	–	Named from Mt. Piton in Tenerife.
Pizzetti	34.9S	118.8E	44	–	P.; Italian geodesist (1860-1918).
Plain	26.2N	3.6E	2	–	Astronaut-named feature, Apollo 15 site.
Plana	42.2N	28.2E	44	–	Baron Giovanni Antonio Amedeo; Italian astronomer, geometrician (1781-1864).
Planck	57.9S	136.8E	314	109	Max Karl Ernst; German physicist; Nobel laureate (1858-1947).
Vallis Planck	58.4S	126.1E	451	109	Named from nearby crater.
Planté	10.2S	163.3E	37	–	Gaston; French physicist (1834-1889).
Plaskett	82.1N	174.3E	109	108	John S.; Canadian astronomer (1865-1941).
Plato	51.6N	9.4W	109	108	Greek philosopher (ca.428-ca.347 B.C.).
Rimae Plato	52.9N	3.2W	87	–	Named from nearby crater.
Playfair	23.5S	8.4E	47	–	John; Scottish mathematician, geologist (1748-1819).
Plinius	15.4N	23.7E	43	–	Gaius Plinius Secundus (The Elder); Roman natural scientist (23-ca. 79).
Rimae Plinius	17.9N	23.6E	124	–	Named from nearby crater.
Plum	9.0S	15.5E	–	–	Astronaut-named feature, Apollo 16 site.
Plummer	25.0S	155.0W	73	–	Henry C.; British astronomer (1875-1946).
Plutarch	24.1N	79.0E	68	–	Greek biographer (ca. 46-ca. 120).
Poczobutt	57.1N	98.8W	195	108	Marcin Odlanicki; Polish astronomer (1728-1810).
Pogson	42.2S	110.5E	50	–	Norman Robert; British astronomer (1829-1891).
Poincaré	56.7S	163.6E	319	109	Jules-Henri; French mathematician, physicist (1854-1912).
Poinsot	79.5N	145.7W	68	–	Louis; French mathematician (1777-1859).
Poisson	30.4S	10.6E	42	–	Simeon Denis; French mathematician (1781-1840).
Polybius	22.4S	25.6E	41	–	Greek historian (ca. 204-ca. 122 B.C.).
Polzunov	25.3N	114.6E	67	–	Ivan I.; Russian heat engineer (1728-1766).

Feature name	Lat. (°)	Long. (°)	Size (km)	Page	Description
Pomortsev	0.7N	66.9E	23	–	Mikhail Mikhailovich; Russian rocket scientist (1851-1916).
Poncelet	75.8N	54.1W	69	–	Jean V.; French mathematician, engineer (1788-1867).
Pons	25.3S	21.5E	41	–	Jean Louis; French astronomer (1761-1831).
Pontanus	28.4S	14.4E	57	–	Pontano, Giovanni Gioviani; Italian astronomer (1427-1503).
Pontécoulant	58.7S	66.0E	91	109	Philippe Gustave Doulcet, Comte de Pontécoulant; French mathematician (1795-1874).
Popov	17.2N	99.7E	65	–	Aleksandr S.; Russian physicist, engineer (1859-1905); C.; Bulgarian astronomer (1880-1966).
Porter	56.1S	10.1W	51	–	Russell W.; American telescope designer (1871-1949).
Posidonius	31.8N	29.9E	95	111	Of Apamea; Greek geographer (ca. 135-ca. 51 B.C.).
Rimae Posidonius	32.0N	28.7E	70	–	Within crater.
Powell	20.2N	30.8E	1	–	Astronaut-named feature, Apollo 17 site.
Poynting	18.1N	133.4W	128	113	John H.; British physicist (1852-1914).
Prager	3.9S	130.5E	60	–	Richard A.; German-born American astronomer (1884-1945).
Prandtl	60.1S	141.8E	91	–	Ludwig; German physicist (1875-1953).
Priestley	57.3S	108.4E	52	–	Joseph; British chemist (1733-1804).
Prinz	25.5N	44.1W	46	–	Wilhelm; German-born Belgian astronomer (1857-1910).
Rimae Prinz	27.0N	43.0W	80	–	Named from nearby crater.
Priscilla	–	–	–	–	Latin female name.
Oceanus Procellarum	18.4N	57.4W	2568	110	"Ocean of Storms."
Proclus	16.1N	46.8E	28	–	Diadochos (The Successor); Greek mathematician, astronomer, philosopher (410-485).
Proctor	46.4S	5.1W	52	–	Mary; American astronomer (1862-1957).
Protagoras	56.0N	7.3E	21	–	Greek philosopher (ca. 481-ca. 411 B.C.).
Ptolemaeus	9.3S	1.9W	164	110	Ptolemy, Claudius; Greek astronomer, mathematician, geographer (ca. 87-150).
Puiseux	27.8S	39.0W	24	–	Pierre; French astronomer (1855-1928).
Pupin	23.8N	11.0W	2	–	Michael Idvorsky; Yugoslavian-born American physicist (1858-1935).
Purbach	25.5S	2.3W	115	110	Georg von; Austrian mathematician, astronomer (1423-1461).
Purkyně	1.6S	94.9E	48	–	Jan Evangelista; Czechoslovakian doctor, physiologist (1787-1869).
Palus Putredinis	26.5N	0.4E	161	–	"Marsh of Decay."
Montes Pyrenaeus	15.6S	41.2E	164	111	Named from terrestrial Pyrenees.
Pythagoras	63.5N	63.0W	142	108	Of Samos; Greek philosopher, mathematician (unknown-flourished ca. 532 B.C.).
Pytheas	20.5N	20.6W	20	–	Of Marseilles; Greek navigator, geographer (born ca. 308 B.C.).
Quetelet	43.1N	134.9W	55	–	Lambert A. J.; Belgian statistician, astronomer (1796-1874).
Rabbi Levi	34.7S	23.6E	81	–	Gershon, Levi Ben; Spanish-born Jewish philosopher, mathematician, astronomer (1288-1344).
Racah	13.8S	179.8W	63	–	Giulio; Italian-born Israeli physicist (1909-1965).
Raimond	14.6N	159.3W	70	–	J. J., Jr.; Dutch astronomer (1903-1961).
Raman	27.0N	55.1W	10	–	Chandrasekhara V.; Indian physicist; Nobel laureate (1888-1970).
Ramsay	40.2S	144.5E	81	–	Sir William; British chemist; Nobel laureate (1852-1916).
Ramsden	32.8S	31.8W	24	–	Jesse; British instrument maker (1735-1800).
Rimae Ramsden	33.9S	31.4W	108	–	Named from nearby crater.
Rankine	3.9S	71.5E	8	–	William John M.; Scottish physicist, engineer (1820-1872).
Raspletin	22.5S	151.8E	48	–	Aleksandr Andreyevich; Soviet radio engineer (1908-1967).
Ravi	22.5S	151.8E	48	–	Indian male name.
Ravine	8.9S	15.6E	1	–	Astronaut-named feature, Apollo 16 site.
Rayet	44.7N	114.5E	27	–	George A. P.; French astronomer (1839-1906).
Rayleigh	29.3N	89.6E	114	111	John W. Strutt, Lord Rayleigh; British physicist; Nobel laureate (1842-1919).
Razumov	39.1N	114.3W	70	–	Vladimir V.; Soviet rocket builder (1890-1967).
Recht	9.8N	124.0E	20	–	Albert W.; American astronomer, mathematician (1892-1962).
Rupes Recta	22.1S	7.8W	134	110	Latin for "straight cliff" (the straight wall).
Montes Recti	48.0N	20.0W	90	–	Latin for "straight range."
Regiomontanus	28.3S	1.0W	108	110	Muller, Johann; German astronomer, mathematician (1436-1476).
Regnault	54.1N	88.0W	46	–	Henri Victor; French chemist, physicist (1810-1878).
Reichenbach	30.3S	48.0E	71	–	Georg von; German optician (1772-1826).
Rima Reiko	18.6N	27.7E	2	–	Japanese male name.
Reimarus	47.7S	60.3E	48	–	Baer, Nicolai Reymers; German mathematician (ca. 1550-ca.1600).
Reiner	7.0N	54.9W	29	–	Reinieri, Vincentio; Italian astronomer, mathematician (unknown-1648).
Reiner Gamma	7.5N	59.0W	70	–	Bright marking; named from nearby crater.
Reinhold	3.3N	22.8W	42	–	Erasmus; German astronomer, mathematician (1511-1553).
Repsold	51.3N	78.6W	109	108	Johann Georg; German inventor (1770-1830).
Rimae Repsold	50.6N	81.7W	166	–	Named from nearby crater.
Resnik	38.8S	150.1W	20	–	Judith Arlene; member of the Challenger crew (1949-1986); previous designation Borman X.
Respighi	2.8N	71.9E	18	–	Lorenzo; Italian astronomer (1824-1890).

Feature name	Lat. (°)	Long. (°)	Size (km)	Page	Description
Réumur	2.4S	0.7E	52	–	Rene Antoine Ferchault de; French physicist (1683-1757).
Rima Réumur	3.0S	3.0E	30	–	Named from nearby crater.
Rhaeticus	0.0N	4.9E	45	–	Georg Joachim von Lauchen of Rhaetia; Hungarian astronomer, mathematician (1514-1576).
Rheita	37.1S	47.2E	70	–	Anton Maria Schyrle of Rhaetia; Czechoslovakian astronomer, optician (ca. 1597-1660).
Vallis Rheita	42.5S	51.5E	445	111	Named from nearby crater.
Rhysling	26.1N	3.7E	–	–	Astronaut-named feature, Apollo 15 site.
Riccioli	3.3S	74.6W	139	110	Giovanni Battista; Italian astronomer (1598-1671).
Rimae Riccioli	2.0N	74.0W	400	–	Named from nearby crater.
Riccius	36.9S	26.5E	71	–	Ricci, Matteo; Italian mathematician, geographer (1552-1610).
Ricco	75.6N	176.3E	65	–	Annibale; Italian astronomer (1844-1911).
Richards	7.7N	140.1E	16	–	Theodore W.; American chemist; Nobel laureate (1868-1928).
Richardson	31.1N	100.5E	141	112	Sir Owen Willans; British quantum physicist; Nobel laureate (1879-1959).
Riedel	48.9S	139.6W	–	–	Klaus; German rocket scientist (1907-1944); Walter; German rocket scientist (1902-1968).
Riemann	38.9N	86.8E	163	111	Georg F. B.; German mathematician (1826-1866).
Montes Riphaeus	7.7S	28.1W	189	110	Named from range in Asia (now Ural Mountains).
Ritchey	11.1S	8.5E	24	–	George Willis; American astronomer, optician (1864-1945).
Rittenhouse	74.5S	106.5E	26	–	David; American inventor, astronomer, mathematician (1732-1796).
Ritter	2.0N	19.2E	29	–	Karl; German geographer (1779-1859); August; German astrophysicist (flourished 1890).
Rimae Ritter	3.0N	18.0E	100	–	Named from nearby crater.
Ritz	15.1S	92.2E	51	–	Walter; Swiss physicist (1878-1909).
Robert	19.0N	27.4E	1	–	English male name.
Roberts	71.1N	174.5W	89	–	Alexander W.; South African astronomer (1857-1938); Isaac; British astronomer (1829-1904).
Robertson	21.8N	105.2W	88	–	Howard P.; American physicist, mathematician (1903-1961).
Robinson	59.0N	45.9W	24	–	John Thomas Romney; Irish astronomer, physicist, meteorologist (1792-1882).
Rocca	12.7S	72.8W	89	–	Giovanni Antonio; Italian mathematician (1607-1656).
Rocco	28.9N	45.0W	4	–	Italian male name.
Roche	42.3S	136.5E	160	112	Edouard A.; French astronomer (1820-1883).
Romeo	7.5N	122.6E	8	–	Italian male name.
Römer	25.4N	36.4E	39	–	Ole; Danish astronomer (1644-1710).
Rimae Römer	27.0N	35.0E	110	–	Named from nearby crater.
Röntgen	33.0N	91.4W	126	113	Wilhelm C.; German physicist; Nobel laureate (1845-1923).
Montes Rook	20.6S	82.5W	791	110	Lawrence; British astronomer (1622-1666).
Sinus Roris	54.0N	56.6W	202	108	"Bay of Dew."
Rosa	20.3N	32.3W	1	–	Spanish female name.
Rosenberger	55.4S	43.1E	95	–	Otto August; German astronomer, mathematician (1800-1890).
Ross	11.7N	21.7E	24	–	Sir James Clark; British explorer (1800-1862); Frank E.; American astronomer, optician (1874-1966).
Rosse	17.9S	35.0E	11	–	William Parsons, Earl of Rosse; Irish astronomer (1800-1867).
Rosseland	41.0S	131.0E	75	–	Svein; Norwegian astrophysicist (1894-1985).
Rost	56.4S	33.7W	48	–	Leonhard; German astronomer (1688-1727).
Rothmann	30.8S	27.7E	42	–	Christopher; German astronomer (unknown-1600).
Rowland	57.4N	162.5W	171	108	Henry A.; American physicist (1848-1901).
Rozhdestvenskiy	85.2N	155.4W	177	108	Dmitriy S.; Soviet physicist (1876-1940).
Dorsa Rubey	10.0S	42.0W	100	–	William Walden; American geologist (1898-1974).
Rima Rudolf	19.6N	29.6E	8	–	German male name.
Rumford	28.8S	169.8W	61	–	Benjamin Thompson, Count Rumford; British physicist (1753-1814).
Mons Rümker	40.8N	58.1W	70	–	Karl Ludwig Christian; German astronomer (1788-1862).
Runge	2.5S	86.7E	38	–	Carl David Tolme; German mathematician (1856-1927).
Russell	26.5N	75.4W	103	110	Henry Norris; American astronomer (1877-1957); John; British selenographer (1745-1806).
Ruth	28.7N	45.1W	3	–	Hebrew female name.
Rutherford	10.7N	137.0E	13	–	Sir Ernest; British physicist; Nobel laureate (1871-1937).
Rutherfurd	60.9S	12.1W	48	–	Lewis Morris; American astronomer (1816-1892).
Rydberg	46.5S	96.3W	49	–	Johannes R.; Swedish physicist (1854-1919).
Rynin	47.0N	103.5W	75	–	Nikolaj A.; Soviet rocket scientist (1877-1942).
Sabatier	13.2N	79.0E	10	–	Paul; French chemist; Nobel laureate (1854-1941).
Sabine	1.4N	20.1E	30	–	Sir Edward; Irish physicist, astronomer (1788-1883).
Sacrobosco	23.7S	16.7E	98	–	John of Holywood, Johannes Sacrobuschus; British astronomer, mathematician (ca. 1200-1256).
Saenger	4.3N	102.4E	75	–	Eugen; German rocket scientist (1905-1964).
Săfărík	10.6N	176.9E	27	–	Vojtech; Czechoslovakian astronomer (1829-1902).
Saha	1.6S	102.7E	99	112	Meghnad; Indian astrophysicist (1893-1956).
Samir	28.5N	34.3W	2	–	Arabic male name.
Sampson	29.7N	16.5W	1	–	Ralph Allen; British astronomer, mathematician (1866-1939).
Sanford	32.6N	138.9W	55	–	Roscoe F.; American astronomer (1883-1958).

Feature name	Lat. (°)	Long. (°)	Size (km)	Page	Description
Santbech	20.9S	44.0E	64	–	Daniel Santbech Noviomagus; Dutch mathematician, astronomer (unknown-flourished 1561).
Santos-Dumont	27.7N	4.8E	8	–	Alberto; Brazilian-born French aeronautical engineer (1873-1932).
Sarabhai	24.7N	21.0E	7	–	Vikram Ambalal; Indian astrophysicist (1919-1971).
Sarton	49.3N	121.1W	69	–	George (A. L.); Belgian-born American historian of science (1844-1956).
Sasserides	39.1S	9.3W	90	–	Sasceride, Gellio; Danish astronomer, doctor (1562-1612).
Saunder	4.2S	8.8E	44	–	Samuel Arthur; British mathematician, selenographer (1852-1912).
Saussure	43.4S	3.8W	54	–	Horace Benedict de; Swiss geologist (1740-1799).
Scaliger	27.1S	108.9E	84	–	Joseph J.; French chronologist (1540-1609).
Scarp	20.3N	30.6E	8	–	Astronaut-named feature, Apollo 17 site.
Schaeberle	26.2S	117.2E	62	–	John M.; American astronomer (1853-1924).
Scheele	9.4S	37.8W	4	–	Carl Wilhelm; Swedish chemist (1742-1786).
Scheiner	60.5S	27.5W	110	109	Christopher; German astronomer (1575-1650).
Schiaparelli	23.4N	58.8W	24	–	Giovanni Virginio; Italian astronomer (1835-1910).
Schickard	44.3S	55.3W	206	110	Wilhelm; German astronomer, mathematician (1592-1635).
Schiller	51.9S	39.0W	180	109	Julius; German astronomer (unknown-flourished 1627).
Schjellerup	69.7N	157.1E	62	–	H. C.; Danish astronomer (1827-1887).
Schlesinger	47.4N	138.6W	97	–	Frank; American astronomer (1871-1943).
Schliemann	2.1S	155.2E	80	–	Heinrich; German archaeologist (1822-1890).
Schlüter	5.9S	83.3W	89	–	Heinrich; German astronomer (1815-1844).
Schmidt	1.0N	18.8E	11	–	Johann Friedrich Julius; German astronomer (1825-1884); Bernhard; German optician (1879-1935); Otto Y.; Soviet astronomer (1891-1956).
Schneller	41.8N	163.6W	54	–	Herbert; German astronomer (1901-1967).
Schomberger	76.7S	24.9E	85	–	Georg; Austrian astronomer, mathematician (1597-1645).
Schönfeld	44.8N	98.1W	25	–	Eduard; German astronomer (1828-1891).
Schorr	19.5S	89.7E	53	–	Richard; German astronomer (1867-1951).
Schrödinger	75.0S	132.4E	312	109	Erwin; Austrian physicist; Nobel laureate (1887-1961).
Vallis Schrödinger	67.0S	105.0E	310	109	Named from nearby crater.
Schröter	2.6N	7.0W	35	–	Johann Hieronymus; German astronomer (1745-1816).
Rima Schröter	24.0N	0.0W	200	–	Erroneous name for Vallis Schröteri on LTO 38B3.
Rima Schröter	6.0N	12.0W	40	–	Named from nearby crater.
Vallis Schröteri	26.2N	50.8W	168	110	Schröter's Valley.
Schubert	2.8N	81.0E	54	–	Theodor Friedrich von; Russian cartographer (1789-1865).
Schumacher	42.4N	60.7E	60	–	Heinrich Christian; German astronomer (1780-1850).
Schuster	4.2N	146.5E	108	112	Sir Arthur; British mathematician, physicist (1851-1934).
Schwabe	65.1N	45.6E	25	–	Heinrich; German astronomer (1789-1875).
Schwarzschild	70.1N	121.2E	212	108	Karl; German astronomer (1873-1916).
Dorsum Scilla	32.8N	60.4W	108	110	Agostino; Italian geologist (1639-1700).
Scobee	31.1S	148.9W	40	–	Francis Richard; member of the Challenger crew (1939-1986); previous designation Barringer L.
Scoresby	77.7N	14.1E	55	–	William; British explorer (1789-1857).
Scott	82.1S	48.5E	103	109	Robert F.; British explorer (1868-1912).
Sculptured Hills	20.3N	31.0E	8	–	Astronaut-named feature, Apollo 17 site.
Seares	73.5N	145.8E	110	108	Frederick H.; American astronomer (1873-1964).
Secchi	2.4N	43.5E	22	–	Pietro Angelo; Italian astronomer, astrophysicist (1818-1878).
Montes Secchi	3.0N	43.0E	50	–	Named from nearby crater.
Rimae Secchi	1.0N	44.0E	35	–	Named from nearby crater.
Sechenov	7.1S	142.6W	62	–	Ivan M.; Russian physiologist (1829-1905).
Seeliger	2.2S	3.0E	8	–	Hugo von; German astronomer (1849-1924).
Segers	47.1N	127.7E	17	–	Carlos; Argentinian astronomer (1900-1967).
Segner	58.9S	48.3W	67	–	Johann Andreas von; German physicist, mathematician (1704-1777).
Seidel	32.8S	152.2E	62	–	Ludwig P. von; German astronomer, mathematician (1821-1896).
Seleucus	21.0N	66.6W	43	–	Babylonian astronomer (unknown-flourished ca. 150 B.C.).
Seneca	26.6N	80.2E	46	–	Lucius Annaeus; Roman philosopher, natural scientist (4 B.C.- A.D. 65).
Mare Serenitatis	28.0N	17.5E	707	111	"Sea of Serenity."
Seyfert	29.1N	114.6E	110	112	Carl K.; American astronomer (1911-1960).
Shackleton	89.9S	0.0W	19	–	Earnest H.; English Antarctic explorer (1874-1922).
Shahinaz	7.5N	122.4E	15	–	Turkish female name.
Shakespeare	20.2N	30.8E	1	–	Astronaut-named feature, Apollo 17 site.
Shaler	32.9S	85.2W	48	–	Nathaniel S.; American geologist, paleontologist (1841-1906).
Shapley	9.4N	56.9E	23	–	Harlow; American astronomer (1885-1972).
Sharonov	12.4N	173.3E	74	–	Vsevolod V.; Soviet astronomer (1901-1964).
Sharp	45.7N	40.2W	39	–	Abraham; British astronomer, mathematician (1651-1742).
Rima Sharp	46.7N	50.5W	107	–	Named from nearby crater.
Sharp-Apollo	3.2S	23.4W	–	–	Astronaut-named feature, Apollo 12 site.
Shatalov	24.3N	141.5E	21	–	Vladimir A.; Soviet cosmonaut (1927-).
Shayn	32.6N	172.5E	93	112	Grigoriy A.; Soviet astrophysicist (1892-1956).

Feature name	Lat. (°)	Long. (°)	Size (km)	Page	Description
Sheepshanks	59.2N	16.9E	25	–	Anne; British benefactor (1789-1876).
Rima Sheepshanks	58.0N	24.0E	200	–	Named from nearby crater.
Sherlock	20.2N	30.8E	–	–	Astronaut-named feature, Apollo 17 site.
Sherrington	11.1S	118.0E	18	–	Sir Charles Scott; British neurophysiologist; Nobel laureate (1857-1952).
Shi Shen	76.0N	104.1E	43	–	Shi(H) Shen; Chinese astronomer (unknown-ca. 300 B.C.).
Shirakatsi	12.1S	128.6E	51	–	Anania; Armenian geographer (ca. 620-ca. 685).
Short	74.6S	7.3W	70	–	James; Scottish mathematician, optician (1710-1768).
Shorty	20.2N	30.6E	–	–	Astronaut-named feature, Apollo 17 site.
Shternberg (Sternberg)	19.5N	116.3W	70	–	Pavel K.; Russian astronomer (1865-1920).
Shuckburgh	42.6N	52.8E	38	–	Sir George; British geographer, benefactor (1751-1804).
Shulejkin	27.1S	92.5W	15	–	M. V.; Soviet radio engineer (1884-1939).
Siedentopf	22.0N	135.5E	61	–	H.; German astronomer (1906-1963).
Rima Siegfried	25.9S	103.0E	14	–	German male name.
Sierpinski	27.2S	154.5E	69	–	Waclaw; Polish mathematician (1882-1969).
Sikorsky	66.1S	103.2E	98	–	Igor Ivan; Russian-born American aeronautical engineer (1889-1972).
Silberschlag	6.2N	12.5E	13	–	Johann Essaias; German astronomer (1721-1791).
Simpelius	73.0S	15.2E	70	–	Sempill, Hugh; Scottish mathematician (1596-1654).
Sinas	8.8N	31.6E	11	–	Simon; Greek benefactor (1810-1876).
Sirsalis	12.5S	60.4W	42	–	Sersale, Gerolamo; Italian astronomer (1584-1654).
Rimae Sirsalis	15.7S	61.7W	426	110	Named from nearby crater.
Sisakyan	41.2N	109.0E	34	–	Norajr M.; Soviet doctor (1907-1966).
Sita	4.6N	120.8E	2	–	Indian female name.
Sklodowska	18.2S	95.5E	127	112	Maria (Madame Curie); Polish-born French physicist, chemist; Nobel laureate (1867-1934).
Slipher	49.5N	160.1E	69	–	Earl C.; American astronomer (1883-1964); Vesto M.; American astronomer (1875-1969).
Slocum	3.0S	89.0E	13	–	Frederick; American astronomer (1873-1944).
Dorsa Smirnov	27.3N	25.3E	156	111	Sergei S.; Soviet earth scientist (1895-1947).
Smith	31.6S	150.2W	34	–	Michael John; member of the Challenger crew (1945-1986); previous designation Barringer M.
Smithson	2.4N	53.6E	5	–	James; British chemist, mineralogist (1765-1829).
Smoky Mountains	8.8S	15.6E	3	–	Astronaut-named feature, Apollo 16 site.
Smoluchowski	60.3N	96.8W	83	–	Marian; Polish physicist (1872-1917).
Mare Smythii	1.3N	87.5E	373	111	Smyth, William Henry; British astronomer (1788-1865).
Snellius	29.3S	55.7E	82	–	Snell, Willebrod van Roijen; Dutch mathematician, astronomer, optician (1591-1626).
Vallis Snellius	31.1S	56.0E	592	111	Named from nearby crater.
Sniadecki	22.5S	168.9W	43	–	Jan; Polish astronomer, mathematician (1756-1830).
Snowman	3.2S	23.4W	1	–	Astronaut-named feature, Apollo 12 site.
Soddy	0.4N	121.8E	42	–	Frederick; British chemist; Nobel laureate (1877-1956).
Lacus Solitudinis	27.8S	104.3E	139	112	"Lake of Solitude."
Somerville	8.3S	64.9E	15	–	Mary Fairfax; Scottish physicist, mathematician (1780-1872).
Sommerfeld	65.2N	162.4W	169	108	Arnold J. W.; German physicist (1868-1951).
Sömmering	0.1N	7.5W	28	–	Samuel Thomas; German doctor (1755-1830).
Palus Somni	14.1N	45.0E	143	111	"Marsh of Sleep."
Lacus Somniorum	38.0N	29.2E	384	111	"Lake of Dreams."
Soraya	12.9S	1.6W	2	–	Persian female name.
Dorsa Sorby	19.0N	14.0E	80	–	Henry C.; British earth scientist (1826-1908).
Sosigenes	8.7N	17.6E	17	–	Greek astronomer, chronologist (unknown-flourished 46 B.C.).
Rimae Sosigenes	8.6N	18.7E	190	–	Named from nearby crater.
South	58.0N	50.8W	104	108	James; British astronomer (1785-1867).
South Cluster	26.0N	3.7E	2	–	Astronaut-named feature, Apollo 15 site.
South Massif	20.0N	30.4E	16	–	Astronaut-named feature, Apollo 17 site.
South Ray	9.2S	15.4E	1	–	Astronaut-named feature, Apollo 16 site.
Spallanzani	46.3S	24.7E	32	–	Lazzaro; Italian natural scientist, biologist (1729-1799).
Lacus Spei	43.0N	65.0E	80	–	"Lake of Hope."
Spencer Jones	13.3N	165.6E	85	–	Sir Harold; British astronomer (1890-1960).
Montes Spitzbergen	35.0N	5.0W	60	–	German for "sharp peaks," and named for resemblance to the terrestrial island group.
Spook	9.0S	15.5E	–	–	Astronaut-named feature, Apollo 16 site.
Spörer	4.3S	1.8W	27	–	Friederich Wilhelm Gustav; German astronomer (1822-1895).
Spot	9.0S	15.5E	–	–	Astronaut-named feature, Apollo 16 site.
Mare Spumans	1.1N	65.1E	139	111	"Foaming Sea."
Spur	27.9N	1.2W	13	–	Astronaut-named feature, Apollo 15 site.
Spurr	27.9N	1.2W	11	–	Josiah Edward; American geologist (1870-1950).
St. George	26.0N	3.5E	2	–	Astronaut-named feature, Apollo 15 site.
St. John	10.2N	150.2E	68	–	Charles E.; American solar physicist, astronomer (1857-1935).
Stadius	10.5N	13.7W	69	–	Stade, Jan; Belgian astronomer, mathematician (1527-1579).
Stark	25.5S	134.6E	49	–	Johannes; German physicist; Nobel laureate (1874-1957).
Statio Tranquillitatis	0.8N	23.5E	–	–	"Tranquility Base," Apollo 11 landing site.
Stearns	34.8N	162.6E	36	–	Carl Leo; American astronomer (1892-1972).
Stebbins	64.8N	141.8W	131	108	Joel; American astronomer (1878-1966).

Feature name	Lat. (°)	Long. (°)	Size (km)	Page	Description
Stefan	46.0N	108.3W	125	108	Josef; Austrian physicist (1835-1893).
Stein	7.2N	179.0E	33	–	J. W.; Dutch astronomer (1871-1951).
Steinheil	48.6S	46.5E	67	–	Karl August von; German astronomer, physicist (1801-1870).
Steklov	36.7S	104.9W	36	–	Vladimir A.; Soviet mathematician (1864-1926).
Stella	19.9N	29.8E	36	–	Latin female name.
Steno	32.8N	161.8E	31	–	Nicolaus; Danish doctor (1638-1686).
Steno-Apollo	20.1N	30.8E	1	–	Astronaut-named feature, Apollo 17 site.
Sternfeld	19.6S	141.2W	100	–	A. A.; Soviet space scientist (1905-1980).
Stetson	39.6S	118.3W	64	–	Harlan T.; American astronomer, geophysicist (1885-1964).
Stevinus	32.5S	54.2E	74	111	Stevin, Simon; Belgian mathematician, physicist (1548-1620).
Stewart	2.2N	67.0E	13	–	John Quincy; American astrophysicist (1894-1972).
Stiborius	34.4S	32.0E	43	–	Stoberl, Andreas; German astronomer, mathematician (1465-1515).
Dorsa Stille	27.0N	19.0W	80	–	Hans; German earth scientist (1876-1966).
Stöfler	41.1S	6.0E	126	111	Johann; German astronomer, mathematician (1452-1531).
Stokes	52.5N	88.1W	51	–	Sir George G.; British mathematician, physicist (1819-1903).
Stoletov	45.1N	155.2W	42	–	Aleksandr G.; Russian physicist (1839-1896).
Stone Mountain	9.1S	15.6E	5	–	Astronaut-named feature, Apollo 16 site.
Stoney	55.3S	156.1W	45	–	George J.; Irish physicist (1826-1911).
Störmer	57.3N	146.3E	69	–	F. Carl M.; Norwegian mathematician and astronomer, aurora research (1874-1957).
Strabo	61.9N	54.3E	55	–	Greek geographer (54 B.C.- A.D. 24).
Stratton	5.8S	164.6E	70	–	Frederick J. M.; British astronomer, astrophysicist (1881-1960).
Street	46.5S	10.5W	57	–	Thomas; British astronomer (1621-1689).
Strömgren	21.7S	132.4W	61	–	Elis; Danish astronomer (1870-1947).
Struve	22.4N	77.1W	164	110	Otto von; Russian astronomer (1819-1905); Otto; American astronomer (1897-1963); Friedrich G.W. von; German astronomer (1873-1964).
Stubby	9.1S	15.5E	1	–	Astronaut-named feature, Apollo 16 site.
Subbotin	29.2S	135.3E	67	–	Mikhail F.; Soviet astronomer (1893-1966).
Sinus Successus	0.9N	59.0E	132	111	"Bay of Success."
Suess	4.4N	47.6W	8	–	Eduard; Austrian geologist (1831-1914).
Rima Suess	6.7N	48.2W	165	–	Named from nearby crater.
Sulpicius Gallus	19.6N	11.6E	12	–	Gaius; Roman astronomer (unknown-flourished ca. B.C. 166).
Rimae Sulpicius Gallus	21.0N	10.0E	90	–	Named from nearby crater.
Sumner	37.5N	108.7E	50	–	Thomas H.; American crater.crater.geographer (1807-1876).
Catena Sumner	37.3N	112.3E	247	–	Named from nearby crater.
Sundman	10.8N	91.6W	40	–	K. F.; Finnish astronomer (1873-1949).
Sung-Mei	24.6N	11.3E	5	–	Chinese female name.
Rima Sung-Mei	24.6N	11.3E	40	–	Chinese female name; part of Lorca.
Surveyor	3.2S	23.4W	–	–	Astronaut-named feature, Apollo 12 site.
Susan	11.0S	6.3W	1	–	English female name.
Swann	52.0N	112.7E	42	–	William F. G.; British-born American physicist (1884-1962).
Swasey	5.5S	89.7E	23	–	Ambrose; American inventor (1846-1937).
Swift	19.3N	53.4E	10	–	Lewis; American astronomer (1820-1913).
Sylvester	82.7N	79.6W	58	–	James J.; British mathematician (1814-1897).
Catena Sylvester	81.4N	86.2W	173	–	Named from nearby crater.
Szilard	34.0N	105.7E	122	112	Leo; Hungarian-born American physicist (1898-1964).
Tacchini	4.9N	85.8E	40	–	Pietro; Italian astronomer (1838-1905).
Tacitus	16.2S	19.0E	39	–	Cornelius; Roman historian (ca. 55-ca. 120).
Tacquet	16.6N	19.2E	7	–	Andre; Belgian mathematician (1612-1660).
Promontorium Taenarium	19.0S	8.0W	70	–	Named from cape in Greece; now Matapan or Tainaron.
Taizo	16.6N	19.2E	6	–	Japanese male name.
Talbot	2.5S	85.3E	11	–	William Henry Fox; British photographer, physicist, archaeologist (1800-1877).
Tamm	4.4S	146.4E	38	–	Igor; Soviet physicist; Nobel laureate (1895-1971).
Tannerus	56.4S	22.0E	28	–	Tanner, Adam; Austrian mathematician (1572-1632).
Taruntius	5.6N	46.5E	56	–	Firmanus, Lucius; Roman philosopher (unknown-flourished 86 B.C.).
Catena Taruntius	3.0N	48.0E	100	–	Named from nearby crater.
Rimae Taruntius	5.5N	46.5E	25	–	Within crater.
Montes Taurus	28.4N	41.1E	172	111	Named from terrestrial Taurus Mountains.
Taurus-Littrow Valley	20.0N	31.0E	30	–	Astronaut-named feature, Apollo 17 site.
Taylor	5.3S	16.7E	42	–	Brook; British mathematician (1685-1731).
Tebbutt	9.6N	53.6E	31	–	John; Australian astronomer (1834-1916).
Teisserenc	32.2N	135.9W	62	–	de Bort, Leon-Philippe; French meteorologist (1855-1913).
Tempel	3.9N	11.9E	45	–	Ernst Wilhelm Leberecht; German astronomer (1821-1889).
Lacus Temporis	45.9N	58.4E	117	108	"Lake of Time."
Ten Bruggencate	9.5S	134.4E	59	–	P.; German astronomer (1901-1961).
Montes Teneriffe	47.1N	11.8W	182	–	Named from terrestrial island.
Tereshkova	28.4N	144.3E	31	–	Valentina Vladimirovna; Soviet cosmonaut (1937-).
Dorsum Termier	11.0N	58.0E	90	–	Pierre-Marie; French geologist (1859-1930).
Terrace	26.1N	3.7E	–	–	Astronaut-named feature, Apollo 15 site.

Feature name	Lat. (°)	Long. (°)	Size (km)	Page	Description
Tesla	38.5N	124.7E	43	–	Nikola; Croatian-born American inventor (1856-1943).
Dorsa Tetyaev	19.9N	64.2E	176	111	Mikhail Mikhailovich; Soviet geologist (1882-1956).
Thales	61.8N	50.3E	31	–	Of Miletos; Greek mathematician, astronomer, philosopher (ca. 636-546 B.C.).
Theaetetus	37.0N	6.0E	24	–	Greek geometrician (unknown-ca. 380 B.C.).
Rimae Theaetetus	33.0N	6.0E	50	–	Named from nearby crater.
Thebit	22.0S	4.0W	56	–	Ben Korra; Iraqi astronomer (826-901).
Theiler	13.4N	83.3E	7	–	Max; South African-born American bacteriologist; Nobel laureate (1899-1972).
Theon Junior	2.3S	15.8E	17	–	Of Alexandria; Greek astronomer (unknown-ca. 380).
Theon Senior	0.8S	15.4E	18	–	Of Smyrna; Greek mathematician (unknown-ca. 100).
Theophilus	11.4S	26.4E	110	111	Greek astronomer (died A.D. 412).
Theophrastus	17.5N	39.0E	9	–	Greek philosopher and naturalist (ca. 372-287 B.C.).
Dorsum Thera	24.4N	31.4W	7	–	Greek female name.
Thiel	40.7N	134.5W	32	–	Walter; German rocket builder (1910-1943).
Thiessen	75.4N	169.0W	66	–	E.; German astronomer (1914-1961).
Thomson	32.7S	166.2E	117	112	Sir Joseph John; British physicist; Nobel laureate (1856-1940).
Tikhomirov	25.2N	162.0E	65	–	Nikolaj I.; Soviet chemical engineer (1860-1930).
Tikhov	62.3N	171.7E	83	–	Gavrill A.; Soviet astronomer (1875-1960).
Tiling	53.1S	132.6W	38	–	Reinhold; German rocket scientist (1890-1933).
Timaeus	62.8N	0.5W	32	–	Greek astronomer (unknown-ca. 400 B.C.).
Timiryazev	5.5S	147.0W	53	–	Kliment A.; Russian botanist, physiologist (1843-1920).
Timocharis	26.7N	13.1W	33	–	Greek astronomer (unknown-flourished ca. 280 B.C.).
Catena Timocharis	29.0N	13.0W	50	–	Named from nearby crater.
Lacus Timoris	38.8S	27.3W	117	–	"Lake of Fear."
Tiselius	7.0N	176.5E	53	–	Arne Wilhelm Kaurin; Swedish biochemist, Nobel laureate (1902-1971).
Tisserand	21.4N	48.2E	36	–	Francois Felix; French astronomer (1845-1896).
Titius	26.8S	100.7E	73	–	Johann D.; German astronomer (1729-1796).
Titov	28.6N	150.5E	31	–	German S.; Soviet cosmonaut (1935-).
Tolansky	9.5S	16.0W	13	–	Samuel; British physicist (1907-1973).
Torricelli	4.6S	28.5E	22	–	Evangelista; Italian physicist (1608-1647).
Tortilla Flat	20.2N	30.7E	1	–	Astronaut-named feature, Apollo 17 site.
Toscanelli	27.4N	47.5W	7	–	Paolo del Pozza; Italian doctor, cartographer (1397-1482).
Rupes Toscanelli	27.4N	47.5W	70	–	Named from nearby crater.
Townley	3.4N	63.3E	18	–	Sidney Dean; American astronomer (1867-1946).
Tralles	28.4N	52.8E	43	–	Johann Georg; German physicist (1763-1822).
Mare Tranquillitatis	8.5N	31.4E	873	111	"Sea of Tranquility."
Trap	9.1S	15.4E	1	–	Astronaut-named feature, Apollo 16 site.
Trident	20.2N	30.8E	–	–	Astronaut-named feature, Apollo 17 site.
Triesnecker	4.2N	3.6E	26	–	Francis A. Paula; Austrian astronomer (1745-1817).
Rimae Triesnecker	4.3N	4.6E	215	–	Named from nearby crater.
Triplet	3.7S	17.5W	–	–	Astronaut-named feature, Apollo 14 site.
Trouvelot	49.3N	5.8E	9	–	Etiénne Leopold; French astronomer (1827-1895).
Trumpler	29.3N	167.1E	77	–	Robert J.; American astronomer (1866-1956).
Tsander	6.2N	149.3W	181	113	Friedrich A.; Soviet rocket scientist (1887-1933).
Tseraskiy (Cerasky)	49.0S	141.6E	56	–	Vitol'd K.; Polish-born Russian astronomer (1849-1925).
Tsinger	56.7N	175.6E	44	–	Nikolaj Ya.; Russian astronomer (1842-1918).
Tsiolkovskiy	21.2S	128.9E	185	112	Konstantin E.; Soviet physicist (1857-1935).
Tsu Chung-Chi	17.3N	145.1E	28	–	Chinese mathematician (430-501).
Tucker	5.6S	88.2E	7	–	Richard Hawley; American astronomer (1859-1952).
Turner	1.4S	13.2W	11	–	Herbert Hall; British astronomer (1861-1930).
Tycho	43.4S	11.1W	102	110	Tycho Brahe; Danish astronomer (1546-1601).
Tyndall	34.9S	117.0E	18	–	John; British physicist (1820-1893).
Ukert	7.8N	1.4E	23	–	Friedrich August; German historian, humanitarian (1780-1851).
Ulugh Beigh	32.7N	81.9W	54	–	Ulugh-Beg; Mongolian astronomer, mathematician (1394-1449).
Mare Undarum	6.8N	68.4E	243	111	"Sea of Waves."
Urey	27.9N	87.4E	38	–	H.; American chemist; Nobel laureate (1893-1981).
Mons Usov	12.0N	63.0E	15	–	Mikhail A.; Soviet geologist (1883-1933).
Väisälä	25.9N	47.8W	8	–	Yrjo; Finnish astronomer (1891-1971).
Valier	6.8N	174.5E	67	–	Max; German rocket engineer (1895-1930).
van Albada	9.4N	64.3E	21	–	Gale Bruno; Dutch astronomer (1912-1972).
Van Biesbroeck	28.7N	45.6W	9	–	George A.; Belgian-born American astronomer (1880-1974).
Van de Graaff	27.4S	172.2E	233	112	Robert J.; American physicist (1901-1967).
van den Bergh	31.3N	159.1W	42	–	G.; Dutch astronomer (1890-1966).
van den Bos	5.3S	146.0E	22	–	Willem Hendrik; South African astronomer (1896-1974).
Van der Waals	43.9S	119.9E	104	112	Johannes D.; Dutch physicist; Nobel laureate (1837-1923).
van Gent	15.4N	160.4E	43	–	H.; Dutch astronomer (1900-1947).
van Maanen	35.7N	128.0E	60	–	Adriaan; Dutch-born American astronomer (1884-1946).
van Rhijn	52.6N	146.4E	46	–	Pieter J.; Dutch astronomer (1886-1960).
Van Serg	20.2N	30.8E	–	–	Astronaut-named feature, Apollo 17 site.

Feature name	Lat. (°)	Long. (°)	Size (km)	Page	Description
Van't Hoff	62.1N	131.8W	92	–	Jacobus H.; Dutch chemist; Nobel laureate (1852-1911).
Van Vleck	1.9S	78.3E	31	–	John Monroe; American astronomer, mathematician (1833-1912).
Van Wijk	62.8S	118.8E	32	–	Uco; Dutch-born American astronomer (1924-1966).
Mare Vaporum	13.3N	3.6E	245	111	"Sea of Vapors."
Vasco da Gama	13.6N	83.9W	83	110	Portuguese navigator, explorer (1469-1524).
Rimae Vasco da Gama	10.0N	82.0W	60	–	Named from nearby crater.
Vashakidze	43.6N	93.3E	44	–	Mikhail A.; Soviet astronomer (1909-1956).
Vavilov	0.8S	137.9W	98	–	Nikolaj; Soviet botanist (1887-1943); Sergei I.; Soviet physiological optician (1891-1951).
Vega	45.4S	63.4E	75	–	Georg Freiherr von; German mathematician (1756-1802).
Vendelinus	16.4S	61.6E	131	111	Wendelin, Godefroid; Belgian astronomer (1580-1667).
Vening Meinesz	0.3S	162.6E	87	–	Felix A.; Dutch geophysicist, geodesist (1887-1966).
Ventris	4.9S	158.0E	95	–	Michael G. F.; British decipherer of Linear B Cretan script (1922-1956).
Vera	26.3N	43.7W	2	–	Latin female name.
Lacus Veris	16.5S	86.1W	396	110	"Lake of Spring."
Vernadskiy	23.2N	130.5E	91	112	Vladimir Ivanovich; Soviet mineralogist (1863-1945).
Verne	24.9N	25.3W	2	–	Latin male name.
Vertregt	19.8S	171.1E	187	112	M.; Dutch chemist (1897-1973).
Very	25.6N	25.3E	5	–	Frank Washington; American astronomer (1852-1927).
Vesalius	3.1S	114.5E	61	–	Andreas; Belgian doctor (1514-1564).
Vestine	33.9N	93.9E	96	–	Ernest H.; American physicist (1906-1968).
Vetchinkin	10.2N	131.3E	98	–	Vladimir P.; Soviet physicist, engineer (1888-1950).
Victory	20.2N	30.7E	1	–	Astronaut-named feature, Apollo 17 site.
Vieta	29.2S	56.3W	87	110	François; French mathematician (1540-1603).
Vil'ev	6.1S	144.4E	45	–	Mikhail; Russian chemist (1893-1919).
Mons Vinogradov	22.4N	32.4W	25	–	Aleksandr Pavlovich; Soviet geochemist and cosmochemist (1895-1975); formerly called Mons Euler.
Virchow	9.8N	83.7E	16	–	Rudolph Ludwig Karl; German doctor, pathologist (1821-1902).
Virtanen	15.5N	176.7E	44	–	Artturi Ilmari; Finnish agricultural biochemist; Nobel laureate (1895-1973).
Vitello	30.4S	37.5W	42	–	Witelo, Erazmus Ciokek; Polish physicist, mathematician (1210-1285).
Vitruvius	17.6N	31.3E	29	–	Vitruvius Pollio, Marcus; Roman engineer, architect (unknown-flourished ca. 25 B.C.).
Mons Vitruvius	19.4N	30.8E	15	–	Named from nearby crater.
Viviani	5.2N	117.1E	26	–	Vincenzo; Italian physicist, mathematician (1622-1703).
Vlacq	53.3S	38.8E	89	–	Adriaan; Dutch mathematician (ca. 1600-1667).
Rima Vladimir	25.2N	0.8W	10	–	Russian male name.
Vogel	15.1S	5.9E	26	–	Hermann Karl; German astronomer (1841-1907).
Volkov	13.6S	131.7E	40	–	Vladislav N.; Soviet engineer (1935-1971).
Volta	53.9N	84.4W	123	108	Conte Alessandro G. A. A.; Italian physicist (1745-1827).
Volterra	56.8N	132.2E	52	–	Vito; Italian mathematician, physicist (1860-1940).
von Behring	7.8S	71.8E	38	–	Emil Adolf; German bacteriologist; Nobel laureate (1854-1917).
von Békésy	51.9N	126.8E	108	–	Georg; Hungarian-born American otological physicist, Nobel laureate (1899-1972).
von Braun	41.1N	78.0W	60	–	Werner; German-born American rocket pioneer (1912-1977).
Dorsum Von Cotta	23.2N	11.9E	199	111	Carl Bernard; German earth scientist (1808-1879).
von der Pahlen	24.8S	132.7W	56	–	Emanuel; German astronomer (1882-1952).
Von Kármán	44.8S	175.9E	180	112	Theodore; American aeronautical scientist (1881-1963).
Von Neumann	40.4N	153.2E	78	112	John; American mathematician (1903-1957).
von Zeipel	42.6N	141.6W	83	–	E. H.; Swedish astronomer (1873-1959).
Voskresenskiy	28.0N	88.1W	49	–	Leonid A.; Soviet rocket scientist (1913-1965).
Walker	26.0S	162.2W	32	–	Joseph A.; American test pilot (1921-1966).
Wallace	20.3N	8.7W	26	–	Alfred Russel; British naturalist (1823-1913).
Wallach	4.9N	32.3E	6	–	Otto; German chemist; Nobel laureate (1847-1931).
Walter	28.0N	33.8W	1	–	German male name.
Walter (Walther)	33.1S	1.0E	128	111	Bernard; German astronomer (1430-1504).
Wan-Hoo	9.8S	138.8W	52	–	Chinese inventor (unknown-ca. 1500).
Rima Wan-Yu	20.0N	31.5W	12	–	Chinese female name.
Wargentin	49.6S	60.2W	84	–	Pehr Vilhelm; Swedish astronomer (1717-1783).
Warner	4.0S	87.3E	35	–	Worcester Reed; American inventor (1846-1929).
Waterman	25.9S	128.0E	76	–	Alan T.; American physicist (1892-1967).
Watson	62.6S	124.5W	62	–	James C.; American astronomer (1838-1880).
Watt	49.5S	48.6E	66	–	James; Scottish inventor (1736-1819).
Watts	8.9N	46.3E	15	–	Chester Burleigh; American astronomer (1889-1971).
Webb	0.9S	60.0E	21	–	Thomas William; British astronomer (1806-1885).
Weber	50.4N	123.4W	42	–	Wilhelm E.; German physicist (1804-1891).
Wegener	45.2N	113.3W	88	–	Alfred L.; German geophysicist, meteorologist (1880-1930).
Weierstrass	1.3S	77.2E	33	–	Karl; German mathematician (1815-1897).
Weigel	58.2S	38.8W	35	–	Erhard; German mathematician (1625-1699).

Feature name	Lat. (°)	Long. (°)	Size (km)	Page	Description
Weinek	27.5S	37.0E	32	–	Ladislaus; Czechoslovakian astronomer (1848-1913).
Weird	3.7S	17.5W	–	–	Astronaut-named feature, Apollo 14 site.
Weiss	31.8S	19.5W	66	–	Edmund; German astronomer, mathematician, physicist (1837-1917).
H. G. Wells	40.7N	122.8E	114	112	Herbert George; British scientific writer (1866-1946).
Werner	28.0S	3.3E	70	–	Johann; German mathematician (1468-1528).
Wessex Cleft	20.3N	30.9E	4	–	Astronaut-named feature, Apollo 17 site.
West	0.8N	23.5E	–	–	Astronaut-named feature, Apollo 11 site.
Wexler	69.1S	90.2E	51	–	Harry; American meteorologist (1911-1962).
Weyl	17.5N	120.2W	108	113	Hermann; German-born American mathematician (1885-1955).
Whewell	4.2N	13.7E	13	–	William; British philosopher (1794-1866).
Dorsa Whiston	29.4N	56.4W	85	–	William; British mathematician, astronomer (1667-1752).
White	44.6S	158.3W	39	–	Edward H. II; American astronaut (1930-1967).
Wichmann	7.5S	38.1W	10	–	Moritz Ludwig Georg; German astronomer (1821-1859).
Widmanstätten	6.1S	85.5E	46	–	Aloys B.; German physicist (1753-1849).
Wiechert	84.5S	165.0E	41	–	E.; German geophysicist (1861-1928).
Wiener	40.8N	146.6E	120	112	Norbert; American mathematician (1894-1964).
Wildt	9.0N	75.8E	11	–	Rupert; German-born American astronomer (1905-1976).
Wilhelm	43.4S	20.4W	106	110	Wilhelm IV, Landgraf of Hesse; German astronomer (1532-1592).
Wilkins	29.4S	19.6E	57	–	Hugh Percy; British selenographer (1896-1960).
Williams	42.0N	37.2E	36	–	Arthur Stanley; British astronomer (1861-1938).
Wilsing	21.5S	155.2W	73	–	J.; German astronomer (1856-1943).
Wilson	69.2S	42.4W	69	–	Alexander; Scottish astronomer (1714-1786); Charles T.R.; Scottish physicist (1869-1959); Ralph E.; American astronomer (1886-1960).
Winkler	42.2N	179.0W	22	–	Johannes; German rocket scientist (1897-1947).
Winlock	35.6N	105.6W	64	–	Joseph; American astronomer (1826-1875).
Winthrop	10.7S	44.4W	17	–	John; American astronomer (1714-1779).
Wöhler	38.2S	31.4E	27	–	Friedrich; German chemist (1800-1882).
Wolf	22.7S	16.6W	25	–	Maxmilian Franz Joseph Cornelius; German astronomer (1863-1932).
Mons Wolff	17.0N	6.8W	35	–	Christian, Baron von; German philosopher (1679-1754).
Wollaston	30.6N	46.9W	10	–	William Hyde; British chemist, physicist (1766-1828).
Woltjer	45.2N	159.6W	46	–	Jan; Dutch astronomer (1891-1946).
Wood	43.0N	120.8W	78	–	Robert W.; American physicist (1868-1955).
Wreck	9.1S	15.5E	1	–	Astronaut-named feature, Apollo 16 site.
Wright	31.6S	86.6W	39	–	Frederick E.; American astronomer (1878-1953); Thomas; British philosopher (1711-1786); William H.; American astronomer (1871-1959).
Wróblewski	24.0S	152.8E	21	–	Zygmunt; Polish physicist (1845-1888).
Wrottesley	23.9S	56.8E	57	–	John, Baron Wrottesley; British astronomer (1798-1867).
Wurzelbauer	33.9S	15.9W	88	–	Johann Philipp von; German astronomer (1651-1725).
Wyld	1.4S	98.1E	93	–	James H.; American rocket scientist (1913-1953).
Xenophanes	57.5N	82.0W	125	108	Of Colophon; Greek philosopher (ca. 570-ca. 478 B.C.).
Xenophon	22.8S	122.1E	25	–	Greek natural philosopher, historian (ca. 430-354 B.C.).
Yablochkov	60.9N	128.3E	99	–	Pavel N.; Russian electrical engineer (1847-1894).
Yakovkin	54.5S	78.8W	37	–	A. A.; Soviet astronomer (1887-1974).
Yamamoto	58.1N	160.9E	76	–	I.; Japanese astronomer (1889-1959).
Yangel'	17.0N	4.7E	8	–	Mikhail Kuzmich; Soviet rocket scientist (1911-1971).
Rima Yangel'	16.7N	4.6E	30	–	Named from nearby crater.
Yerkes	14.6N	51.7E	36	–	Charles T.; American benefactor (1837-1905).
Yoshi	24.6N	11.0E	1	–	Japanese male name.
Young	41.5S	50.9E	71	–	Thomas; British doctor, physicist (1773-1829).
Catena Yuri	24.4N	30.4W	5	–	Russian male name.
Zach	60.9S	5.3E	70	–	Freiherr von; Hungarian astronomer (1754-1832).
Zagut	32.0S	22.1E	84	–	Abraham Ben Samuel; Spanish-Jewish astronomer (unknown-ca. 1450).
Rima Zahia	25.0N	29.5W	16	–	Arabic female name.
Zähringer	5.6N	40.2E	11	–	Josef; German physicist (1929-1970).
Zanstra	2.9N	124.7E	42	–	Herman; Dutch astronomer (1894-1972).
Zasyadko	3.9N	94.2E	11	–	Alexander Dmitrievich; Russian rocket scientist, inventor (1779-1837).
Zeeman	75.2S	133.6W	190	109	Pieter; Dutch physicist; Nobel laureate (1865-1943).
Zelinskiy	28.9S	166.8E	53	–	Nikolay Dimitrievich; Soviet chemist (1860-1953).
Zeno	45.2N	72.9E	65	–	Of Citium; Greek philosopher (ca. 335-263 B.C.).
Zernike	18.4N	168.2E	48	–	Frits; Dutch physicist; Nobel laureate (1888-1966).
Zhiritskiy	24.8S	120.3E	35	–	Georgiy S.; Soviet rocketry scientist (1893-1966).
Zhukovskiy	7.8N	167.0W	81	–	Nikolay E.; Russian physicist (1847-1921).
Zinner	26.6N	58.8W	4	–	Ernst; German astronomer (1886-1970).
Dorsum Zirkel	28.1N	23.5W	193	110	Ferdinand; German geologist, mineralogist (1838-1912).

Feature name	Lat. (°)	Long. (°)	Size (km)	Page	Description
Zöllner	8.0S	18.9E	47	–	Johann Karl Friedrich; German astrophysicist, astronomer (1834-1882).
Zsigmondy	59.7N	104.7W	65	–	Richard Adolf; German chemist; Nobel laureate (1865-1929).
Zucchius	61.4S	50.3W	64	–	Zucchi, Niccolo; Italian mathematician, astronomer (1586-1670).
Zupus	17.2S	52.3W	38	–	Zupi, Giovanni Battista; Italian astronomer (ca. 1590-1650).
Rimae Zupus	15.0S	53.0W	120	–	Named from nearby crater.
Zwicky	15.4S	168.1E	150	112	Fritz; Swiss astrophysicist (1898-1974).

MARS

Feature name	Lat. (°)	Long. (°)	Size (km)	Page	Description
Abalos Undae	81.0N	83.1W	218	–	Classical albedo feature at 72N, 70W.
Aban	16.2N	249.1W	–	–	Town in Russia.
Abus Vallis	6.3S	147.5W	129	141	Classical name for Humber River in England.
Achar	45.7N	236.8W	6	–	Town in Uruguay.
Acheron Catena	38.2N	100.7W	554	132	Classical albedo feature at 35N, 140W.
Acheron Fossae	38.7N	136.6W	1120	135	Classical albedo feature at 35N, 140W.
Acidalia Planitia	54.6N	19.9W	2791	133	Classical albedo feature at 44N, 21W.
Mare Acidalium	45.0N	30.0W	–	–	Name for Acidalian (Venusian) fountain in Boeotia where the graces bathed.
Adamas Labyrinthus	36.5N	255.0W	664	137	Classical albedo feature name; "A River of Diamonds"; modern River Sarbarnarekha in India.
Adams	31.3N	197.1W	100	134	Walter S.; American astronomer (1876-1956).
Aeolis	5.0S	215.0W	–	–	Floating island where winds were kept in a cave.
Aeolis Mensae	3.7S	218.5W	841	143	Classical albedo feature name.
Aeria	10.0N	310.0W	–	–	Greek name for Egypt; "far land of mist."
Aesacus Dorsum	38.5N	206.6W	–	–	From albedo feature at 45N, 205W.
Aetheria	40.0N	230.0W	–	–	Upper world; land of living.
Aethiopis	10.0N	230.0W	–	–	Countries of the Ethiopians on southern edge of the Earth.
Aganippe Fossa	8.7S	126.0W	329	141	Classical albedo feature name.
Agassiz	70.1S	88.4W	104	146	Jean L.; American naturalist (1807-1873).
Airy	5.2S	0.0W	56	144	George B.; British astronomer (1801-1892).
Aki	36.0S	60.5W	8	–	Town in Japan.
Alba Catena	35.2N	114.6W	148	132	Classical albedo name.
Alba Fossae	43.4N	113.0W	2077	132	Classical albedo name.
Alba Patera	40.5N	109.9W	464	132	Classical albedo name.
Albany	23.3N	49.0W	2	–	American colonial town (New York).
Albi	41.9S	34.7W	8	–	Town in France.
Albor Fossae	18.6N	209.4W	113	140	From albedo feature at 20N, 205W.
Albor Tholus	19.3N	209.8W	165	140	Classical albedo feature name.
Alga	24.6S	26.5W	19	–	Town in Kazakhstan.
Alitus	35.3S	38.1W	29	–	Town in Lithuania.
Alpheus Colles	38.5S	300.6W	514	150	From albedo feature at 45S, 292W.
Al-Qahira Vallis	17.5S	196.7W	546	140	Word for "Mars" in Arabic, Indonesian, Malay.
Amazonis	0.0N	140.0W	–	–	Land of the Amazons, on island Hesperia.
Amazonis Planitia	16.0N	158.4W	2816	141	Classical albedo feature name; home of the Amazons.
Amazonis Sulci	3.3S	145.1W	243	141	Albedo feature name; home of the Amazons.
Amenthes	5.0N	250.0W	–	–	Egyptian name for place where souls of the dead go.
Amenthes Fossae	10.2N	259.3W	1592	142	Classical albedo feature name.
Amenthes Rupes	1.8N	249.4W	441	143	Classical albedo feature name.
Amet	23.7N	57.4W	–	–	Town in India.
Amphitrites Patera	59.1S	299.0W	138	151	Mare Amphitrites, classical albedo feature name.
Amsterdam	23.3N	47.0W	2	–	Dutch port.
Angusta Patera	80.7S	79.5W	–	–	Classical albedo name.
Aniak	32.1S	69.6W	58	146	Town in Alaska, USA.
Anio Vallis	38.1N	304.3W	–	–	Classical river in Italy; modern Aniene and Teverone rivers.
Annapolis	23.4N	47.8W	1	–	American colonial town (Maryland).
Anseris Cavus	30.0S	264.5W	35	–	Named for albedo feature Anseris Fons.
Anseris Mons	30.1S	263.2W	–	–	From albedo feature Anseris Fons.
Antoniadi	21.7N	299.0W	381	142	Eugene M.; French astronomer (1870-1944).
Aonia Terra	58.2S	94.8W	3372	146	Classical albedo feature name.
Aonium Sinus	45.0S	105.0W	–	–	Named for Aonides, or Muses.
Apia	37.7S	271.0W	–	–	Town in American Samoa.
Apodis Catena	27.2S	256.8W	–	–	Classical albedo name.
Apollinaris Patera	8.3S	186.0W	198	140	Classical albedo feature name.
Apollinaris Sulci	11.3S	183.4W	300	–	Albedo feature at 5S, 187W.
Apsus Vallis	35.1N	225.0W	137	134	Classical river in ancient Macedonia, present-day Greece.
Apt	40.1N	9.5W	10	–	Town in France.
Arabia	20.0N	330.0W	–	–	Country bordering on Aeria (Egypt).
Arabia Terra	25.0N	330.0W	6000	145	Classical albedo feature name.
Arago	10.5N	330.2W	154	145	Dominique F.; French astronomer (1786-1853).
Aram Chaos	2.7N	21.0W	388	144	Classical albedo feature name.
Arandas	42.6N	15.1W	22	–	Town in Mexico.
Arcadia	45.0N	100.0W	–	–	Mountainous region in southern Greece.
Arcadia Planitia	46.4N	152.1W	3052	135	From classical albedo feature at 45N, 120W.
Arda Valles	21.0S	32.4W	172	139	Ancient European river (Bulgaria).
Arena Colles	24.3N	278.7W	578	142	Classical albedo feature at 13N, 294W.
Arena Dorsum	13.0N	291.5W	205	142	Classical albedo feature name.
Ares Vallis	9.7N	23.4W	1690	144	Word for "Mars" in Greek.
Dorsa Argentea	77.6S	33.3W	749	147	Classical albedo feature name.
Argyre	45.0S	25.0W	–	–	"Silver" Island at mouth of Ganges River; present – day Arakan, Burma.

Feature name	Lat. (°)	Long. (°)	Size (km)	Page	Description
Argyre Cavi	49.1S	40.2W	–	–	Albedo name.
Argyre Planitia	49.4S	42.8W	868	147	Classical albedo feature name.
Argyre Rupes	63.3S	66.7W	592	146	Classical albedo feature name.
Ariadnes Colles	34.8S	188.0W	200	148	Classical albedo feature name.
Arica	24.0S	249.7W	–	–	Town in Colombia.
Arimanes Rupes	10.0S	147.5W	203	141	Albedo feature; classical Persian deity of wickedness.
Arkhangelsky	41.3S	24.6W	119	147	A. D.; Russian geologist.
Arnon	48.0N	335.0W	–	–	Classical name for present El-Mojib River, Jordan.
Arnus Vallis	13.3N	290.0W	288	142	Classical and present-day Arno River in Tuscany, Italy (previously named Arena Rupes).
Aromatum Chaos	1.0S	43.0W	115	139	Classical albedo feature name.
Arrhenius	40.2S	237.0W	132	148	Svante; Swedish physical chemist (1859-1927).
Arsia Chasmata	7.9S	119.4W	–	–	Albedo name.
Arsia Mons	9.4S	120.5W	485	141	From Arsia Silva, classical albedo feature name.
Arsia Sulci	6.4S	129.7W	–	–	Albedo name.
Arsinoes Chaos	7.9S	28.1W	203	144	Daughter of Ptolomy Lagun and Bernice.
Artynia Catena	47.4N	119.6W	275	132	From classical albedo feature at 54N, 137W.
Ascraeus Chasmata	8.7N	105.5W	–	–	Classical albedo name.
Ascraeus Mensa	11.7N	107.8W	–	–	Classical albedo name.
Ascraeus Mons	11.3N	104.5W	462	138	From Ascraeus Lacus, classical albedo feature name.
Ascraeus Sulci	11.9N	108.7W	–	–	Classical albedo name.
Ascuris Planum	39.5N	84.0W	–	–	Classical albedo name.
Asopus Vallis	4.4S	149.6W	35	–	Classical river in Greece; modern Hagios River.
Aspen	21.6S	22.9W	16	–	Town in Colorado, USA..
Astapas Colles	33.7N	271.1W	500	137	From albedo feature at 35N, 269W.
Atlantis Chaos	34.8S	176.7W	141	149	From albedo feature at 30N, 173W.
Atrax Dorsum	38.3N	89.1W	–	–	Classical town in Greece.
Auqakuh Vallis	28.6N	299.3W	195	–	Word for “Mars” in Quechua (Inca).
Aureum Chaos	4.1S	27.1W	468	144	Classical albedo feature name.
Aurorae Chaos	80.6S	34.5W	–	–	Classical albedo name.
Aurorae Planum	11.1S	50.2W	590	139	Classical albedo feature name.
Aurorae Sinus	15.0S	50.0W	–	–	“Bay of Rosy Dawn.”
Ausonia	40.0S	250.0W	–	–	Country of the Aruncii (Ausones in Greek).
Ausonia Cavus	32.2S	262.5W	–	–	Albedo feature Ausonia.
Ausonia Mensa	30.2S	262.5W	–	–	Albedo feature name.
Ausonia Montes	28.5S	260.0W	–	–	Albedo feature name.
Chasma Australe	82.9S	273.8W	491	151	Classical albedo feature name.
Mare Australe	60.0S	10.0W	–	–	“South Sea.”
Planum Australe	80.3S	155.1W	1313	148	Classical albedo feature.
Australis Patera	80.2S	51.5W	–	–	Classical albedo name.
Auxo Dorsum	56.1S	41.9W	–	–	One of the Graces.
Aveiro	21.5N	80.0W	–	–	Town in Portugal.
Avernus Cavi	3.8S	187.4W	60	–	Named for albedo feature at 10S, 195W.
Avernus Colles	2.0S	188.5W	100	–	Named for albedo feature at 10S, 195W.
Avernus Dorsa	4.7S	189.1W	130	140	From albedo feature at 4S, 190W.
Avernus Rupes	8.9S	186.4W	200	–	From albedo feature Avernus at 10S, 195W.
Axius Valles	55.8S	290.9W	316	151	Ancient European river (Vardar River of Greece).
Ayacucho	38.5N	92.0W	–	–	Town in Bolivia.
Ayr	39.5S	268.2W	–	–	Town in Queensland, Australia.
Azul	42.5S	42.3W	20	–	Town in Argentina.
Azusa	5.5S	40.5W	41	–	Town in California, USA.
Babakin	36.4S	71.4W	78	146	Soviet builder of unmanned space stations (1914-1970).
Bacht	18.9N	257.5W	7	–	Town in Uzbekistan.
Baetis Chasma	4.3S	64.8W	95	139	From classical albedo feature at 5S, 60W. Changed from Iamunae Chasma.
Baetis Mensa	5.4S	72.4W	181	139	From albedo feature at 7S, 63W.
Bahn	3.5S	43.5W	10	–	Town in Liberia.
Bahram Vallis	21.4N	58.7W	403	139	Word for “Mars” in Persian.
Bakhuysen	23.1S	344.3W	162	144	Henricus G.; Dutch astronomer (1838-1923).
Balboa	3.5S	34.0W	20	–	Town in the Panama Canal zone.
Baldet	23.0N	294.5W	195	142	Ferdnand; French astronomer (1885-1964).
Balta	24.1S	26.4W	15	–	Town in Ukraine.
Baltia	60.0N	50.0W	–	–	Name of large island in northern Europe where amber was found.
Baltisk	42.6S	54.5W	48	–	Town in Russia.
Bamba	3.5S	41.7W	21	–	Town in Zaire.
Bamberg	40.0N	3.0W	57	133	Town in Germany.
Banff	17.4N	30.5W	3	–	Town in Alberta, Canada.
Banh	19.4N	55.2W	13	–	Town in Burkina Faso (formerly Upper Volta).
Baphyras Catena	38.8N	84.0W	–	–	Classical river in Greece.
Bar	25.6S	19.3W	2	–	Town in Ukraine.
Barabashov	47.6N	68.5W	126	132	Nilolay P.; Russian astronomer (1894-1971).
Barnard	61.3S	298.4W	128	151	Edward E.; American astronomer (1857-1923).
Baro	24.9S	249.3W	–	–	Town in Iceland.
Basin	18.1N	253.2W	16	–	Town in Wyoming, USA.
Batoka	7.6S	36.8W	14	–	Town in Zambia.
Batoş	21.7N	29.5W	16	–	Town in Romania.
Baykonyr	46.7N	227.2W	4	–	Soviet launch site.
Bazas	28.0S	266.5W	–	–	Town in France.
Becquerel	22.3N	7.9W	167	144	Antoine-Henri; French physicist (1852-1908).
Beer	14.6S	8.2W	80	144	Wilhelm; German astronomer (1797-1850).
Belz	22.0N	43.1W	10	–	Town in Ukraine.
Bend	22.6S	27.5W	2	–	Town in Oregon, USA.
Bentham	56.0S	40.3W	–	–	Town in England.
Bentong	22.6S	18.9W	10	–	Town in Malaysia.
Bernard	23.8S	154.2W	129	141	P.; French atmospheric scientist.
Berseba	4.4S	37.7W	36	–	Town in Namibia.
Bhor	42.0N	225.5W	6	–	Town in India.
Bianchini	64.2S	95.1W	77	146	Francesco; Italian astronomer (1662-1729).
Biblis Patera	2.3N	123.8W	117	141	Classical albedo feature name.
Bigbee	25.0S	34.6W	18	–	Town in Mississippi, USA.
Bise	20.2N	56.5W	9	–	Town in Okinawa.
Bison	26.6S	29.0W	15	–	Town in Kansas, USA.
Bjerknes	43.4S	188.7W	89	148	Vilhelm F.; Norwegian physicist (1862-1951).
Bled	21.8N	31.4W	7	–	Town in Yugoslavia.
Blitta	26.3S	20.8W	11	–	Town in Togo.
Blois	24.6N	55.6W	11	–	Town in France.
Bluff	23.7N	250.1W	6	–	Town in New Zealand.
Boeddicker	14.8S	197.6W	107	140	Otto; German astronomer (1853-1937).
Bogra	24.4S	28.6W	18	–	Town in Bangladesh.
Bok	20.8N	31.6W	7	–	Town in New Guinea.
Bole	25.4N	53.7W	8	–	Town in Ghana.

Feature name	Lat. (°)	Long. (°)	Size (km)	Page	Description
Bombala	30.3S	253.9W	–	–	Town in New S. Wales, Australia.
Bond	33.3S	35.7W	104	147	George P.; American astronomer (1825-1865).
Bor	18.1N	33.8W	4	–	Town in Russia.
Bordeaux	23.4N	48.9W	2	–	French port.
Chasma Boreale	83.2N	45.0W	318	133	Classical albedo feature name.
Boreosyrtis	55.0N	290.0W	–	–	Northern continuation of Nilosyrtis, "Syrtis of the north."
Mare Boreum	60.0N	180.0W	–	–	"North Sea."
Planum Boreum	85.0N	180.0W	1066	132	Classical albedo feature name.
Boru	24.6S	27.7W	11	–	Town in Russia.
Bosporos Planum	33.4S	64.0W	330	146	Classical albedo feature name.
Bosporos Rupes	42.8S	57.2W	500	147	Classical albedo feature name.
Bouguer	18.6S	332.8W	106	145	Pierre; French physicist-hydrographer (1698-1758).
Boulia	23.1S	248.6W	–	–	Town in Queensland, Australia.
Bozkir	44.4S	32.0W	89	147	Town in Turkey.
Brashear	54.1S	119.2W	126	146	John A.; American physicist (1840-1920).
Brazos Valles	6.3S	341.7W	494	145	River in Texas, USA.
Bremerhaven	23.5N	48.6W	2	–	German port.
Dorsa Brevia	70.8S	300.3W	748	150	Classical albedo feature name.
Briault	10.1S	270.2W	100	142	P.; French astronomer (died 1922).
Bridgetown	22.2N	47.1W	2	–	Port of Barbados.
Bristol	22.4N	47.0W	3	–	English port.
Broach	23.5N	56.6W	11	–	Town in India.
Brush	21.9N	248.8W	5	–	Town in Colorado, USA.
Bulhar	50.6N	225.6W	19	–	Town in Somalia.
Bunge	34.0S	49.0W	78	147	Andrey Aleksandrovich; Russian zoologist, permafrost investigator (late 19th century).
Burroughs	72.5S	243.1W	104	151	Edgar R.; American novelist (1875-1950).
Burton	14.5S	156.3W	137	141	Charles E.; British astronomer (1846-1882).
Buta	23.5S	32.2W	12	–	Town in Zaire.
Butte	5.1S	39.1W	12	–	Town in Montana, USA.
Buvinda Vallis	33.3N	208.2W	–	–	Classical river in Hibernia; present Boyne River, Ireland.
Byrd	65.6S	231.9W	122	148	Richard E.; American aviator-explorer (1888-1975).
Byske	4.9S	31.1W	5	148	Town in Sweden.
Cádiz	23.4N	49.1W	1	–	Spanish port.
Cairns	24.0N	47.3W	10	–	Town in Australia.
Calbe	25.5S	28.7W	14	–	Town in Germany.
Calydon Fossa	7.5S	88.8W	265	–	The son of Ares and Astynome.
Camiling	0.8S	38.1W	21	–	Town in the Philippines.
Camiri	45.1S	41.9W	20	–	Town in Bolivia.
Campbell	54.0S	195.0W	123	148	John W.; Canadian physicist (1889-1955); William W.; American astronomer (1862-1938).
Campos	22.0S	27.6W	7	–	Town in Brazil.
Can	48.4N	14.6W	6	–	Town in Turkey.
Canas	31.4S	270.1W	–	–	Town in Puerto Rico.
Canaveral	47.1N	224.0W	3	–	American launch site.
Canberra	47.5N	227.3W	3	–	Australian tracking site.
Candor	3.0N	75.0W	–	–	Means "blaze" or "white" in Latin.
Candor Chaos	7.2S	72.7W	45	–	From albedo feature at 5N, 75W.
Candor Chasma	6.5S	71.0W	816	139	Classical albedo feature name.
Candor Labes	4.9S	76.0W	165	–	From classical albedo feature at 5N, 75W.
Candor Mensa	6.0S	73.0W	50	–	Classical albedo feature.
Cangwu	42.1N	89.5W	–	–	Town in China.
Canso	21.5N	60.6W	–	–	Town in Nova Scotia, Canada.
Capri Chasma	8.7S	42.6W	1498	139	Classical albedo feature name.
Cartago	23.6S	17.8W	33	–	Town in Costa Rica.
Casius	40.0N	260.0W	–	–	Epithet of Zeus; for his two sanctuaries in Egypt/Arabia and Syria.
Cassini	23.8N	327.9W	415	145	Giovanni; Italian astronomer (1625-1712).
Cave	21.9N	35.6W	8	–	Town in New Zealand.
Cavi Angusti	75.9S	71.9W	294	146	Classical albedo feature name.
Cavi Frigorēs	80.1S	67.9W	199	146	From classical albedo feature Polus Frigoris at 84S, 30W.
Caxias	29.5S	100.6W	–	–	Town in Brazil.
Cebrenia	50.0N	210.0W	–	–	Main country of the Trojan Plain.
Cecropia	60.0N	320.0W	–	–	Old name for Acropolis; previously meant Athens.
Centauri Montes	38.5S	263.1W	–	–	Albedo feature Centauri Lacus.
Ceraunius	20.0N	93.0W	–	–	"Thunderclap," named for Ceraunii Mountains on coast of Epirus, Greece.
Ceraunius Catena	37.4N	108.1W	51	132	From classical albedo feature at 35N, 96W.
Ceraunius Fossae	24.8N	110.5W	711	138	Classical albedo feature name.
Ceraunius Tholus	24.2N	97.2W	108	138	Classical albedo feature name.
Cerberus	15.0N	205.0W	–	–	Hound who had three heads; guarded gates of hell.
Cerberus Dorsa	13.8S	254.4W	407	143	Classical albedo feature name.
Cerberus Rupēs	8.4N	195.4W	1254	140	From albedo feature at 10N, 212W.
Cerulli	32.6N	337.9W	120	136	Vicenzo; Italian astronomer (1859-1927).
Ceti Chasma	5.2S	68.3W	45	–	From albedo feature at 10S, 74W.
Ceti Mensa	5.9S	76.3W	55	–	Albedo feature Ceti Lacus.
Chalce	50.0S	0.0W	–	–	Old name of island of Khalki, west of Rhodes.
Chalce Fossa	51.9S	40.0W	–	–	Albedo name.
Chalce Montes	53.8S	37.0W	–	–	Albedo name.
Chalcoporos Rupēs	54.9S	338.6W	380	150	From albedo feature at 50S, 6W.
Chamberlin	66.1S	124.3W	125	149	Thomas C.; American geologist (1843-1928).
Changsŏng	23.5N	57.1W	32	138	Town in the Republic of Korea.
Chapais	22.6S	20.4W	33	–	Town in Quebec, Canada.
Charis Dorsum	55.8S	41.2W	–	–	One of the Graces.
Charitum Montes	58.3S	44.2W	1412	147	Classical albedo feature name.
Charleston	22.9N	47.8W	2	–	American colonial town (South Carolina).
Charlier	68.6S	168.4W	100	149	Carl V.; Swedish astronomer (1862-1934).
Charlieu	38.3N	84.1W	–	–	Town in France.
Charybdis Scopulus	24.9S	339.9W	513	145	From albedo feature at 19S, 320W.
Chatturat	35.7N	94.8W	–	–	Town in Thailand.
Chauk	23.3N	55.6W	9	–	Town in Burma.
Cheb	24.5S	19.3W	8	–	Town in Czechoslovakia.

Feature name	Lat. (°)	Long. (°)	Size (km)	Page	Description
Chefu	23.1N	247.7W	–	–	Town in Mozambique.
Chekalin	24.7S	26.6W	87	144	Town in Turkmenistan.
Chersonesus	50.0S	260.0W	–	–	Gallipoli Peninsula.
Chia	1.6N	59.5W	94	–	Town in Spain.
Chimbote	1.5S	39.8W	65	139	Town in Peru.
Chincoteague	41.5N	236.0W	35	134	Town in Virginia, USA.
Chinju	4.7S	42.3W	69	139	Town in the Republic of Korea.
Chinook	22.5N	55.2W	17	–	Town in Alberta, Canada.
Chive	21.6N	55.7W	8	–	Town in Bolivia.
Choctaw	41.5S	37.0W	20	–	Town in Ohio, USA.
Chom	38.7N	2.2W	6	–	Town in China (Tibet).
Mare Chronium	58.0S	210.0W	–	–	"Cronian Sea," northern part of World Sea where eternal dead calm, dangerous to ships, prevailed.
Planum Chronium	62.0S	212.6W	1402	148	From albedo feature at 58S, 190W.
Chryse	10.0N	30.0W	–	–	Island rich in gold; region of Thailand/Malacca.
Chryse Planitia	27.0N	36.0W	1500	139	Classical albedo feature name.
Chrysokeras	50.0S	110.0W	–	–	"Golden Horn"; Byzantine (Turkey) Peninsula, or its inlet.
Chur	16.7N	29.2W	4	–	Town in Russia.
Terra Cimmeria	34.0S	215.0W	2285	148	Classical albedo feature name.
Mare Cimmerium	20.0S	220.0W	–	–	Cimmerians were ancient Thracian seafarers; "far western sea" (Homer).
Circle	22.4S	25.4W	11	–	Town in Montana, USA.
Clanis Valles	34.0N	301.5W	–	–	Classical river in Etruria; present Chiana River, Italy.
Claritas	35.0S	110.0W	–	–	Latin, meaning 'bright.'
Claritas Fossae	34.8S	99.1W	2033	146	Classical albedo feature name.
Claritas Rupes	26.0S	105.5W	–	–	From albedo feature at 25S, 110W.
Clark	55.7S	133.2W	93	149	Alvan; American optician-astronomer (1804-1887).
Clasia Vallis	33.8N	302.5W	240	136	Classical river in Umbria, Italy.
Cleia Dorsum	55.3S	46.3W	–	–	One of the Graces.
Clogh	20.8N	47.5W	10	–	Town in Ireland.
Clota Vallis	26.0S	20.8W	101	144	Ancient European river; present Clyde River, Great Britain.
Cluny	24.1S	27.1W	9	–	Town in France.
Cobalt	26.1S	26.8W	10	–	Town in Connecticut, USA.
Coblentz	55.3S	90.2W	111	146	William W.; American physicist (1873-1962).
Cobres	12.1S	153.7W	89	141	Village in Argentina.
Colles Nili	39.6N	295.3W	551	137	From classical albedo feature, Portus Nili at 38N, 295W.
Coloe Fossae	37.1N	303.9W	930	136	Classical albedo feature name.
Colón	23.4N	49.1W	2	–	Port of Panama.
Columbus	29.7S	165.8W	114	140	Christopher; Italian explorer (1451-1506).
Comas Sola	20.1S	158.4W	132	141	José; Spanish astronomer (1868-1937).
Conches	4.2S	34.3W	18	–	Town in France.
Concord	16.6N	34.1W	20	–	Town in Massachusetts, USA.
Cooma	24.0S	108.5W	–	–	Town in New South Wales, Australia.
Copais Palus	55.0N	280.0W	–	–	Named for marsh north of Mt. Helicon in Boeotia, Greece.
Copernicus	50.0S	168.6W	292	149	Nicolaus; Polish astronomer (1473-1543).
Coprates	15.0S	65.0W	–	–	Old name for Persian River Ab-I-Diz.
Coprates Catena	15.1S	59.4W	243	139	Classical albedo feature name.
Coprates Chasma	13.6S	60.7W	962	139	Classical albedo feature name.
Coprates Labes	13.5S	67.3W	115	–	From albedo feature at 14S, 65W.
Coracis Fossae	34.8S	78.6W	747	146	From albedo feature at 46S, 87W.
Corby	43.1N	222.4W	7	–	Town in England.
Coronae Mons	34.5S	271.5W	–	–	Albedo feature name.
Coronae Scopulus	33.9S	294.1W	366	151	From albedo feature at 26S, 276W.
Cost	15.3N	256.1W	10	–	Town in Texas, USA.
Cray	44.4N	16.2W	5	–	Town in England.
Creel	6.1S	39.0W	8	–	Town in Mexico.
Crewe	25.2S	19.4W	3	–	Town in England.
Crommelin	5.3N	10.2W	111	144	Andrew C.; British astronomer (1865-1939).
Cruls	43.2S	196.9W	83	148	Luiz; Brazilian astronomer (1848-1908).
Cruz	38.6N	1.7W	6	–	Town in Venezuela.
Cue	36.1S	266.8W	–	–	Town in Western Australia.
Curie	29.2N	4.9W	98	144	Pierre; French physicist-chemist (1859-1906).
Cusus Valles	13.9N	309.0W	234	–	Classical river; modern Hron River in Czechoslavakia.
Cyane Catena	37.0N	118.5W	143	132	Classical albedo feature Cyane Fons.
Cyane Fossae	36.0N	121.4W	292	135	Classical albedo feature name.
Cyane Sulci	25.5N	128.4W	286	141	Classical albedo feature name.
Cyclopia	5.0S	230.0W	–	–	Land where cyclops dwell.
Cydnus Rupes	59.6N	257.5W	430	137	From albedo feature at 70N, 248W.
Cydonia	40.0N	0.0W	–	–	Poetic term for Crete.
Cydonia Mensae	37.0N	12.8W	854	133	From albedo feature at 50N, 355W.
Cypress	47.6S	47.0W	11	–	Town in Illinois, USA.
Da Vinci	1.5N	39.1W	98	139	Leonardo; Italian artist-scientist (1452-1519).
Daan	40.9S	267.6W	–	–	Town in China.
Daedalia Planum	13.9S	138.0W	2477	141	Classical albedo feature name.
Daet	7.4S	41.9W	12	–	Town in the Philippines.
Daly	66.4S	23.0W	99	147	Reginald A.; Canadian geologist (1871-1975).
Dana	72.6S	33.1W	95	147	James D.; American geologist (1813-1895).
Dank	22.2N	253.2W	7	–	Town in Oman.
Dao Vallis	36.8S	269.4W	667	151	Word for "star" in Thai.
Darwin	57.2S	19.2W	166	147	George H.; British astronomer (1845-1912).
Dawes	9.3S	322.3W	191	145	William R.; British astronomer (1799-1868).
Deba	24.3S	17.1W	7	–	Town in Nigeria.
Dein	38.5N	2.4W	24	–	Town in New Guinea.
Dejnev	25.6S	164.5W	156	141	Semen Ivanovich; Russian geographer, explorer, and navigator (1605-1673).
Delta	46.3S	39.0W	5	–	Town in Louisiana, USA.
Deltoton Sinus	4.0S	305.0W	–	–	"Bay of the triangle"; makes triangle with Iapygia and Oenotria.
Denning	17.5S	326.6W	165	145	William F.; British astronomer (1848-1931).
Dese	45.8S	30.3W	12	–	Town in Ethiopa.
Dessau	43.1S	53.0W	9	–	Town in Germany.
Deucalionis Regio	15.0S	340.0W	–	–	"Deucalion's region"; Deucalion was King of

Feature name	Lat. (°)	Long. (°)	Size (km)	Page	Description
					Thessaly who saved himself from flood.
Deuteronilus	35.0N	0.0W	–	–	Designation of second part of old feature "Nilus."
Deuteronilus Colles	42.2N	338.3W	79	136	Classical albedo feature name.
Deuteronilus Mensae	45.7N	337.9W	627	136	From albedo feature at 35N, 355W.
Deva Valles	7.9S	156.9W	–	–	Expanded coordinates and pluralization of approved feature.
Deva Vallis	8.1S	156.7W	26	–	Classical name for Dee River in Scotland; new position and coordinates.
Dia-Cau	0.3S	42.8W	28	–	Town in the Socialist Republic of Vietnam.
Diacria	50.0N	180.0W	–	–	Highland area in northern Attica, Greece.
Diacria Patera	34.8N	132.7W	75	135	From albedo feature at 48N, 170W.
Dingo	24.0S	17.3W	13	–	Town in Australia.
Dinorwic	30.5S	101.4W	–	–	Town in Ontario, Canada.
Dioscuria	50.0N	320.0W	–	–	"Home of Dioscuri," Polydeuces and Pollux; symbolic name for Sparta.
Dison	25.4S	16.3W	20	–	Town in Belgium.
Dittaino Valles	0.5N	66.4W	–	–	Modern river in Italy.
Dixie	19.9N	55.9W	–	–	Town in Georgia, USA.
Doanus Vallis	63.2S	25.9W	128	147	Classical river shown in Ptolemy's map; may be modern Mekong River of Burma.
Dokuchaev	60.8S	127.1W	73	149	Vasily Vasil'evich; Russian soil scientist; founded modern genetical soil science (1840-1903).
Douglass	51.7S	70.4W	97	146	Andrew E.; American astronomer (1867-1962).
Drava Valles	48.5S	193.5W	162	–	Modern river in Yugoslavia.
Drilon Vallis	7.2N	52.4W	49	–	Classical river in Macedonia; present Drin River, Albania.
Dromore	20.2N	49.3W	14	–	Town in North Ireland.
Du Martheray	5.8N	266.4W	94	142	Maurice; Swiss astronomer (1892-1955).
Du Toit	71.8S	49.7W	75	147	Alexander L.; South African geologist (1878-1948).
Dubis Vallis	5.3S	148.1W	45	–	Classical river in France; modern Doubs River.
Dubki	37.0S	55.0W	–	–	Town in Russia.
Dunhuang	80.9S	48.7W	–	–	Town in China.
Dzeng	80.6S	70.5W	–	–	Town in Cameroon.
Dzigai Vallis	59.7S	31.3W	–	–	Word for "valley" in Navajo language.
E. Mareotis Tholis	36.1N	85.0W	–	–	Classical albedo name.
Eads	28.9S	29.8W	2	–	Town in Colorado, USA.
Eagle	44.0N	8.2W	12	–	Town in Idaho, USA.
Echt	22.2S	28.0W	2	–	Town in Scotland.
Echus Chaos	9.9N	74.6W	371	138	From albedo feature Echus Lacus at 1N, 90W.
Echus Chasma	3.7N	79.1W	613	138	Classical albedo feature name.
Echus Fossae	2.2N	77.1W	236	138	Classical albedo feature name.
Echus Montes	6.8N	78.2W	203	138	Albedo feature name.
Edam	26.6S	19.9W	20	–	Town in the Netherlands.
Eddie	12.5N	217.8W	90	143	Lindsay A.; South African astronomer (1845-1913).
Edom	0.0N	345.0W	–	–	Biblical country of Edomites; south of Judea.
Eger	48.7S	51.8W	12	–	Town in Hungary.
Eil	42.0N	9.7W	4	–	Town in Somalia.
Ejriksson	19.6S	173.7W	56	140	Leif; Norse explorer (ca. 1000).
Elath	46.1N	13.7W	13	–	Town in Israel.
Electris	45.0S	190.0W	–	–	Electra's island near River Eridanus; famous for amber formed from tears shed by Phaethon's sisters.
Ellsley	36.5N	83.0W	–	–	Town in England.
Ely	23.9S	27.1W	10	–	Town in Nevada, USA.
Elysium	25.0N	210.0W	–	–	Home of the blessed on western edge of world.
Elysium Catena	18.0N	210.4W	66	143	Albedo feature name.
Elysium Chasma	22.2N	218.0W	92	143	Albedo feature name.
Elysium Fossae	27.5N	219.5W	1114	143	Classical albedo feature name.
Elysium Mons	25.0N	213.0W	432	143	Classical albedo feature name.
Elysium Planitia	14.3N	241.1W	3899	143	Classical albedo feature name.
Elysium Rupes	25.4N	211.4W	180	143	Albedo feature name.
Enipeus Vallis	37.5N	93.1W	–	–	Classical river.
Eos Chaos	17.5S	46.5W	335	–	Greek name of Aurora; albedo feature.
Eos Chasma	12.6S	45.1W	963	139	Classical albedo feature name.
Erebus Montes	39.9N	170.5W	594	135	From albedo feature at 26N, 182W.
Eridania	45.0S	220.0W	–	–	Region on the Po River, Italy.
Eridania Scopulus	53.4S	217.9W	510	148	Classical albedo feature name.
Erythraea Fossa	27.8S	30.5W	248	139	Classical albedo feature name.
Mare Erythraeum	25.0S	40.0W	–	–	Indian Ocean.
Escalante	0.3N	244.8W	83	143	F.; Mexican astronomer (ca. 1930).
Escorial	77.0N	54.3W	–	–	Town in Spain.
Esk	45.4N	7.0W	4	–	Town in Australia.
Espino	19.9S	254.7W	–	–	Town in Venezuela.
Eudoxus	45.0S	147.2W	92	149	Greek astronomer (ca. 408-355 B.C.).
Eumenides Dorsum	5.7N	156.3W	599	141	Classical albedo feature name.
Eunostos	22.0N	120.0W	–	–	"Lucky journey or lucky return"; i.e., infernal regions beyond Elysium.
Euphrates	20.0N	335.0W	–	–	Biblical fourth river of Paradise.
Evpatoriya	47.3N	225.5W	6	–	Soviet tracking site.
Evros Vallis	12.6S	345.9W	335	144	River in Greece.
Faith	43.2N	11.9W	5	–	Town in North Dakota, USA.
Falun	24.2S	24.4W	10	–	Town in Sweden.
Faqu	24.7S	253.6W	–	–	Town in Jordan.
Fastov	25.4S	20.2W	12	–	Town in Ukraine.
Felis Dorsa	22.3S	65.5W	486	139	Classical albedo feature name.
Fenagh	34.6N	215.7W	–	–	Town in Ireland.
Fesenkov	21.9N	86.4W	86	138	Vasiliy G.; Russian astronomer (1889-1972).
Flammarion	25.7N	311.7W	160	145	Camille; French astronomer (1842-1925).
Flat	25.8S	19.4W	3	–	Town in Alabama, USA.
Flaugergues	17.0S	340.9W	235	145	Honore; French astronomer (1755-1835).
Flora	45.0S	51.2W	15	–	Town in Mississippi, USA.
Focas	33.9N	347.2W	82	136	Jean H.; Greco-French astronomer (1909-1969).
Fontana	63.2S	71.9W	78	146	Francesco; Italian astronomer (1585-1685).
Foros	34.0S	28.0W	23	–	Town in Ukraine.
Fortuna Fossae	5.1N	92.5W	309	–	Classical albedo feature name.

Feature name	Lat. (°)	Long. (°)	Size (km)	Page	Description
Fournier	4.2S	287.5W	112	142	Georges; French astronomer (1881-1954).
Freedom	43.6N	9.0W	12	–	Town in Oklahoma, USA.
Frento Vallis	50.5S	14.9W	203	147	Classical name for river in Italy.
Funchal	23.2N	49.5W	2	–	Port of Madeira Islands.
Gaan	38.5N	3.2W	3	–	Town in Somalia.
Gagra	20.9S	21.9W	14	–	Town in the Georgian Republic.
Gah	45.1S	32.3W	2	–	Town in Indonesia.
Galaxias Chaos	34.0N	212.4W	–	–	Albedo feature name.
Galaxias Colles	40.7N	209.0W	291	134	Albedo feature name.
Galaxias Fluctūs	32.3N	217.6W	–	–	Albedo feature name.
Galaxias Fossae	38.3N	218.8W	200	134	Albedo feature name.
Galaxias Mensae	36.5N	212.7W	327	134	Albedo feature name.
Galaxius Mons	35.0N	217.7W	–	–	Classical albedo name.
Gale	5.3S	222.3W	172	143	Walter F.; Australian astronomer (1865-1945).
Gali	44.1S	36.9W	24	–	Town in the Republic of Georgia.
Galilaei	5.8N	26.9W	124	144	Galileo; Italian astronomer and physicist (1564-1642).
Galle	50.8S	30.7W	230	147	Johann G.; German astronomer (1812-1910).
Galu	22.3S	21.5W	9	–	Town in Zaire.
Gander	31.5S	265.7W	–	–	Town in Newfoundland, Canada.
Gandu	45.8S	47.0W	5	–	Town in Brazil.
Gandzani	34.5N	90.8W	–	–	Town in the Republic of Georgia.
Ganges Catena	2.6S	69.3W	221	139	Classical albedo feature name.
Gangis Chasma	8.4S	48.1W	541	139	Classical albedo feature name.
Gardo	27.0S	24.6W	11	–	Town in Somalia.
Gari	36.0S	71.0W	9	–	Town in Russia.
Garm	48.4N	9.2W	5	–	Town in Tadzhikistan.
Gastre	24.8N	247.6W	5	–	Town in Argentina.
Gatico	21.2S	20.9W	18	–	Town in Chile.
Gehon	15.0N	0.0W	–	–	Biblical second river of Paradise; bordering Eden.
Geryon Montes	8.0S	80.7W	274	138	Classical albedo feature.
Gigas Fossae	4.5N	129.7W	328	141	Albedo feature name.
Gigas Sulci	9.9N	127.6W	467	141	Classical albedo feature name.
Gilbert	68.2S	273.8W	115	151	Grove K.; American geologist (1843-1918).
Gill	15.8N	354.5W	81	144	David; British astronomer (1843-1914).
Glazov	20.8S	26.4W	20	–	Town in Russia.
Gledhill	53.5S	272.9W	72	151	Joseph; British astronomer (1836-1906).
Glide	8.2S	43.3W	10	–	Town in Oregon, USA.
Globe	24.0S	27.1W	45	–	Town in Arizona, USA.
Goba	23.5S	20.9W	11	–	Town in Ethiopia.
Goff	23.5N	255.2W	6	–	Town in Somalia.
Gol	47.4N	10.7W	7	–	Town in Norway.
Gold	20.2N	31.3W	9	–	Town in Pennsylvania, USA.
Golden	22.2S	33.3W	19	–	Town in Illinois, USA.
Goldstone	48.1N	225.3W	1	–	American tracking site.
Gonnus Mons	41.6N	90.8W	–	–	Classical town.
Gordii Dorsum	5.3N	144.5W	624	141	Classical albedo feature name.
Sulci Gordii	17.9N	125.6W	316	141	Classical albedo feature name.
Gorgonum Chaos	39.4S	170.8W	83	149	From albedo feature at 24S, 154W.
Gori	23.2S	28.6W	6	–	Town in the Republic of Georgia.
Graff	21.4S	206.0W	157	140	Kasimir; German astronomer (1878-1950).
Granicus Valles	28.7N	227.6W	445	143	Ancient name for river in Turkey.
Green	52.4S	8.3W	184	147	Nathan E.; British astronomer (1823-1899).
Grójec	21.6S	30.6W	37	–	Town in Poland.
Groves	4.2S	44.6W	10	–	Town in Texas, USA.
Guaymas	26.2N	44.8W	20	–	Town in Mexico.
Guir	21.8S	20.4W	12	–	Town in Mali.
Gulch	16.1N	251.1W	7	–	Town in Ethiopia.
Gusev	14.6S	184.6W	166	140	Matwei M.; Russian astronomer (1826-1866).
Gwash	38.9N	3.0W	5	–	Town in Pakistan.
Hadley	19.3S	203.0W	113	140	George; British meteorologist (1685-1768).
Hadriaca Patera	30.6S	267.2W	451	151	Classical albedo feature name.
Mare Hadriacum	40.0S	270.0W	–	–	Adriatic Sea.
Haldane	53.0S	230.5W	72	148	John B.; British physiologist, geneticist (1892-1964).
Hale	36.1S	36.3W	136	147	George E.; American astronomer (1868-1938).
Halex Fossae	27.3N	126.1W	224	141	From albedo feature at 40N, 110W.
Halley	48.6S	59.2W	81	147	Edmund; British astronomer (1656-1742).
Ham	45.0S	32.2W	1	–	Town in France.
Hamaguir	49.0N	227.4W	3	–	Algerian launch site.
Hamelin	20.4N	32.8W	10	–	Old German town referred to in the Pied Piper fairy tale.
Handlová	37.9N	88.4W	–	–	Town in Czechoslovakia.
Haraḑ	27.8S	27.8W	8	–	Town in Saudi Arabia.
Harmakhis Vallis	39.1S	267.2W	585	151	Ancient Egyptian word for "Mars."
Hartwig	38.7S	15.6W	104	147	Ernst; German astronomer (1851-1923).
Heaviside	70.8S	94.8W	103	146	Oliver; British physicist (1850-1925).
Hebes Chasma	1.1S	76.1W	285	138	Classical albedo feature name.
Hebes Mensa	1.0S	76.8W	–	–	Classical albedo feature; name of goddess of youth.
Hebrus Valles	20.0N	233.9W	299	143	Ancient river in Greece.
Hecates Tholus	32.7N	209.8W	183	134	Classcial albedo feature name.
Heculaneum	19.2N	58.9W	–	–	Town in Italy.
Hegemone Dorsum	55.3S	44.9W	–	–	One of the Graces.
Heinlein	64.6S	243.8W	83	–	Robert A.; American author (1907-1988).
Hellas	40.0S	290.0W	–	–	Greece.
Hellas Chaos	46.0S	293.0W	900	–	Named for albedo feature Hellas.
Hellas Montes	37.9S	260.9W	–	–	Albedo feature Hellas.
Hellas Planitia	44.3S	293.8W	2517	151	Classical albedo feature name.
Depressio Hellespontica	60.0S	340.0W	–	–	Depression southwest of Hellespontus.
Hellespontus	50.0S	325.0W	–	–	The Dardanelles.
Hellespontus Montes	45.5S	317.5W	681	150	Classical albedo feature name.
Helmholtz	45.6S	21.1W	107	147	Hermann von; German physicist (1821-1894).
Henry	11.0N	336.8W	165	145	Paul; French astronomer (1848-1905); Prosper; French astronomer (1849-1903).
Hephaestus Fossae	20.5N	237.5W	528	143	Classical albedo feature name.
Her Desher Vallis	25.4S	47.8W	120	139	Egyptian name for Mars.
Hermus Vallis	5.4S	147.5W	52	–	Classical river in ancient Lydia, modern Turkey.
Herschel	14.9S	230.1W	304	143	John F.; British astronomer (1792-1871); William H.; British astronomer (1738-1822).

Feature name	Lat. (°)	Long. (°)	Size (km)	Page	Description
Hesperia	20.0S	240.0W	–	–	"The Occident"; name for Italy (Greek) or Spain (Roman) or west area of Ethiopians where sun sank.
Hesperia Dorsa	23.0S	245.0W	–	–	Albedo feature name.
Hesperia Planum	18.3S	251.6W	1869	143	Classical albedo feature name.
Hibes Montes	3.9N	188.9W	177	140	From albedo feature at 17N, 186W.
Hiddekel	15.0N	345.0W	–	–	Tigris River, Babylonia; Biblical third river of Paradise.
Hilo	44.8S	35.5W	20	–	Town in Hawaii, USA.
Himera Valles	21.1S	22.4W	218	144	Ancient name for Italian river.
Hipparchus	45.0S	151.1W	104	149	Greek astronomer (ca. 160-125 B.C.).
Hīt	47.3N	221.5W	7	–	Town in Iraq.
Holden	26.5S	33.9W	141	139	Edward S.; American astronomer (1846-1914).
Holmes	75.0S	293.9W	109	151	Arthur; British geologist (1890-1965).
Honda	22.7S	16.2W	6	–	Town in Colombia.
Hooke	45.0S	44.4W	145	147	Robert; British physicist-astronomer (1635-1703).
Hope	45.1N	10.3W	6	–	Town in British Columbia, Canada.
Horarum Mons	51.3S	36.4W	–	–	Albedo name.
Houston	48.5N	223.9W	2	–	American mission control site.
Hrad Vallis	37.8N	221.6W	719	134	Word for "Mars" in Armenian.
Hsūanch'eng	47.0N	227.2W	2	–	Chinese launch site.
Huancayo	3.7S	39.8W	25	–	Town in Peru.
Huggins	49.3S	204.3W	82	148	William; British astronomer (1824-1910).
Huo Hsing Vallis	31.5N	293.9W	340	–	Word for "Mars" in Chinese.
Hussey	53.8S	126.5W	100	149	William J.; American astronomer (1862-1926).
Hutton	71.9S	255.5W	99	151	James; British geologist (1726-1797).
Huxley	62.9S	259.2W	108	151	Thomas H.; British biologist (1825-1895).
Huygens	14.0S	304.4W	456	145	Christian; Dutch physicist-astronomer (1629-1695).
Hyblaeus Catena	21.8N	219.5W	–	–	Albedo feature name.
Hyblaeus Chasma	22.1N	218.9W	71	143	Albedo feature name.
Hyblaeus Fossae	21.7N	223.2W	405	143	Albedo feature name.
Hydaspis Chaos	3.4N	27.7W	363	144	Classical albedo feature name.
Hydrae Chasma	6.9S	61.9W	49	–	Classical albedo feature name.
Hydraotes Chaos	0.9N	34.3W	290	139	Classical albedo feature name.
Hypanis Valles	10.2N	46.6W	270	–	Classical river in Scythia; present Kuban River in Russia.
Hyperboreae Undae	77.5N	46.0W	301	133	Classical albedo feature name.
Hyperborei Cavi	79.6N	52.5W	–	–	Classical albedo feature name.
Hyperboreus Labyrinthus	79.8N	55.5W	–	–	Classical albedo feature name.
Hyperboreus (Lacus)	75.0N	60.0W	–	–	Far northern lake.
Hypsas Vallis	34.0N	302.1W	–	–	Classical name for river in Sicily.
Iani Chaos	2.3S	16.6W	565	144	Classical albedo feature name.
Iapygia (Iapigia)	20.0S	295.0W	–	–	Classically, Iapygia was all of Apulia, or just the Salentine Peninsula.
Iberus Vallis	21.4N	208.1W	148	140	Classical river in northeast Spain; present Ebro River.
Ibragimov	25.9S	59.5W	89	139	Nadir Baba Ogly; Soviet astronomer (1932-1977).
Icaria	40.0S	130.0W	–	–	Land where Icarus lived (Crete).
Icaria Fossae	53.7S	135.0W	2153	149	From albedo feature at 44S, 130W.
Icaria Planum	42.7S	107.2W	840	146	Classical albedo feature name.
Igol	20.4S	254.0W	–	–	Town in Hungary.
Indus Vallis	19.2N	322.1W	253	145	Ancient and modern river in Pakistan.
Innsbruck	6.5S	40.0W	64	139	Town in Austria.
Inta	24.6S	24.9W	14	–	Town in Russia.
Inuvik	79.1N	39.7W	–	–	Town in Northwest Territories, Canada.
Irbit	24.6S	24.7W	13	–	Town in Russia.
Irharen	34.8N	219.2W	–	–	Town in Algeria.
Isara Valles	5.4S	146.4W	13	–	Classical name for river in western Europe; modern Oise River in France.
Isidis Planitia	14.1N	271.0W	1238	142	Classical albedo feature name.
Isidis Regio	20.0N	275.0W	–	–	Isis' Region; Isis was Egyptian goddess of heaven and fertility.
Isil	27.1S	272.0W	–	–	Town in Spain.
Ismeniae Fossae	38.9N	326.1W	858	136	From albedo feature at 40N, 333W.
Ismenius Lacus	40.0N	330.0W	–	–	"Ismenian Lake"; Ismenia is poetic term for Thebes.
Issedon Paterae	38.8N	89.9W	–	–	Classical albedo name.
Issedon Tholus	36.3N	94.6W	–	–	Classical albedo name.
Ister Chaos	12.5N	56.5W	57	–	Classical albedo feature at 10N, 56W.
Ituxi Vallis	25.5N	206.9W	123	140	River in Brazil.
Ius Chasma	7.2S	84.6W	1003	138	Classical albedo feature name.
Ius Labes	7.3S	79.0W	129	138	Classical albedo feature name.
Izendy	29.4S	101.3W	–	–	Town in Russia.
Jal	26.6S	28.6W	5	–	Town in New Mexico, USA.
Jampur	38.8N	81.0W	–	–	Town in Pakistan.
Jamuna	10.0N	40.0W	–	–	Present Jumna River, India.
Janssen	2.8N	322.4W	166	145	Pierre J.; French astronomer (1824-1907).
Jarry-Desloges	9.5S	276.1W	97	142	Rene; French astronomer (1868-1951).
Jeans	69.9S	205.5W	71	148	James H.; British physicist, astronomer (1877-1946).
Jeki	23.9N	52.3W	–	–	Town in Ethiopia.
Jen	40.1N	10.6W	8	–	Town in Nigeria.
Ježža	48.8S	37.7W	6	–	Town in Russia.
Jijiga	25.2N	53.7W	14	–	Town in Ethiopia.
Jodrell	47.8N	227.6W	3	–	United Kingdom tracking site.
Johannesburg	48.2N	226.7W	1	–	Republic of South Africa tracking site.
Joly	74.6S	42.5W	81	147	John; Irish geologist (1857-1933).
Jones	19.1S	19.8W	85	144	Harold S.; British astronomer (1890-1960).
Jovis Fossae	19.5N	116.1W	412	138	From albedo feature at 16N, 111W.
Jovis Tholus	18.4N	117.5W	61	138	Classical albedo feature name.
Juventae Chasma	1.9S	61.8W	495	139	Classical albedo feature name.
Juventae Dorsa	1.3N	72.6W	325	139	From albedo feature at 4S, 63W.
Juventae Fons	5.0S	63.0W	–	–	The "Fountain of Youth," a fountain in India.
Kagoshima	47.6N	224.1W	1	–	Japanese launch site.

Feature name	Lat. (°)	Long. (°)	Size (km)	Page	Description
Kagul	24.0S	18.9W	8	–	Town in Moldova.
Kāid	4.6S	44.8W	7	–	Town in Iraq.
Kaiser	46.6S	340.9W	201	150	Frederick; Dutch astronomer (1808-1872).
Kaj	27.4S	29.2W	2	–	Town in Russia.
Kakori	41.9S	29.6W	25	–	Town in India.
Kaliningrad	48.8N	224.9W	2	–	Soviet mission control site.
Kamativi	20.7S	260.0W	–	–	Town in Zimbabwe.
Kamloops	53.9S	32.1W	–	–	Town in Canada.
Kampot	42.1S	45.4W	9	–	Town in Democratic Kampuchea (Cambodia).
Kanab	27.6S	18.8W	14	–	Town in Utah, USA.
Kansk	20.8S	17.1W	34	–	Town in Russia.
Kantang	24.8S	17.5W	64	144	Town in Thailand.
Karpinsk	46.0S	31.8W	28	–	Town in Russia.
Karshi	23.6S	19.2W	22	–	Town in Uzbekhistan.
Kartabo	41.2S	52.3W	17	–	Town in Guyana
Kasabi	28.0S	270.9W	–	–	Town in Zambia.
Kasei Valles	22.8N	68.2W	2222	139	Word for "Mars" in Japanese.
Kashira	27.5S	18.3W	68	144	Town in Russia.
Kasimov	25.0S	22.8W	92	144	Town in Russia.
Kaup	22.5N	33.6W	3	–	Town in New Guinea.
Kaw	16.7N	255.9W	10	–	Town in French Guiana.
Keeler	60.7S	151.2W	92	149	James E.; American astronomer (1857-1900).
Kem'	45.3S	32.7W	2	–	Town in Russia.
Kepler	47.2S	218.7W	219	148	Johannes; German astronomer (1571-1630).
Keul'	46.3N	237.7W	6	–	Town in Russia.
Kholm	7.3S	42.1W	10	–	Town in Russia.
Khurli	21.2S	252.0W	–	–	Town in Pakistan.
Kifrī	46.0S	54.1W	12	–	Town in Iraq.
Kimry	20.5S	16.2W	17	–	Town in Russia.
Kin	20.4N	33.4W	7	–	Town in Japan.
Kinda	26.0S	105.0W	–	–	Town in Zaire.
Kingston	22.4N	47.0W	2	–	Jamaican port.
Kinkora	25.3S	247.0W	–	–	Town in Prince Edward Island, Canada.
Kipini	26.0N	31.5W	75	139	Town in Kenya.
Kirs	26.7S	19.3W	3	–	Town in Russia.
Kirsanov	22.4S	25.0W	15	–	Town in Russia.
Kisambo	34.3N	89.0W	–	–	Town in Zaire.
Kita	23.1S	17.0W	11	–	Town in Mali.
Knobel	6.6S	226.9W	127	143	Edward B.; British astronomer (1841-1930).
Koga	29.3S	103.6W	–	–	Town in Tanzania.
Kok	15.7N	28.1W	6	–	Town in Malaysia (Sarawak).
Kong	5.4S	38.7W	10	–	Town in Ivory Coast.
Korolev	72.9N	195.8W	84	134	Sergey P.; Russian engineer (1906-1966).
Kourou	47.0N	227.1W	2	–	French Guianan launch site.
Koval'skiy	30.0S	141.4W	299	141	M. A.; Russian astronomer (1821-1884).
Krasnoye	36.1N	216.2W	–	–	Town in Russia.
Kribi	43.4S	43.4W	8	–	Town in the United Republic of Cameroon.
Krishtofovich	48.6S	262.6W	111	151	Afrikan Nikolaevich; Soviet paleobotanist (1885-1953).
Kuba	25.6S	19.5W	25	–	Town in Azerbaijan.
Kufra	40.6N	239.7W	32	134	Town in Libya.
Kuiper	57.3S	157.1W	86	149	Gerard P.; American astronomer (1905-1973).
Kumak	36.0S	67.5W	15	–	Town in Russia.
Kumara	43.3N	231.4W	12	–	Town in New Zealand.
Kunes	25.5S	252.0W	–	–	Town in Norway.
Kunowsky	57.0N	9.0W	60	133	George K.; German astronomer (1786-1846).
Kushva	44.3S	35.4W	39	–	Town in Russia.
La Paz	21.3N	49.0W	1	–	Mexican port.
Labeatis Catenae	18.8N	95.1W	318	138	Classical albedo feature name.
Labeatis Fossae	20.9N	95.0W	616	138	Previously named feature at 30N, 75W; expanded coordinates.
Labeatis Mensa	25.6N	74.9W	120	139	Albedo feature name.
Labeatis Mons	37.8N	75.9W	28	–	Named for albedo feature Labeatis Lacus.
Labou Vallis	8.8S	154.3W	201	141	French word for Mars.
Labria	35.3S	48.0W	60	147	Town in Brazil.
Lachute	4.3S	39.9W	13	–	Town in Canada.
Ladon Valles	22.3S	28.4W	238	144	Ancient European river (Greece).
Laestrygon (Laestrigon)	0.0N	200.0W	–	–	Man-eating giants who lived in the west.
Laf	48.3N	5.9W	6	–	Town in the United Republic of Cameroon.
Lagarto	50.0N	8.4W	19	–	Town in Brazil.
Lamas	27.4S	20.5W	21	–	Town in Peru.
Lambert	20.1S	334.6W	87	145	Johann H.; German physicist (1728-1777).
Lamont	58.3S	113.3W	72	146	Johann von; Scottish-born German astronomer (1805-1879).
Lampland	36.0S	79.5W	71	146	Carl O.; American astronomer (1873-1951).
Land	48.4N	8.8W	5	–	Town in Alabama, USA.
Lar	26.1S	28.8W	6	–	Town in Iran.
Lassell	21.0S	62.4W	86	139	William; British astronomer (1799-1880).
Lasswitz	9.4S	221.6W	122	143	Kurd; German author (1848-1910).
Lau	74.4S	107.3W	109	–	Hans E.; Danish astronomer (1879-1918).
Lebu	20.6S	19.4W	20	–	Town in Chile.
Leleque	36.7N	221.9W	–	–	Town in Argentina.
Lemgo	42.9S	34.5W	11	–	Town in Germany.
Lemuria	70.0N	200.0W	–	–	Designation of purported drowned continent south of India; also relates to lemur monkeys found on Madagascar and Sunda island.
Lenya	27.1S	106.7W	–	–	Town in Burma.
Le Verrier	38.2S	342.9W	139	150	Urbain J.; French astronomer (1811-1877).
Lexington	22.0N	48.7W	5	–	American colonial town (Massachusetts).
Li Fan	47.4S	153.0W	103	149	Chinese astronomer (ca. A.D. 85).
Liais	75.4S	252.9W	128	151	Emmanuel; French astronomer (1826-1900).
Libertad	23.3N	29.4W	31	–	Town in Venezuela.
Libya	0.0N	270.0W	–	–	Area from W. Egypt to greater Syrtis.
Libya Montes	2.7N	271.2W	1229	142	Classical albedo feature name.
Licus Vallis	3.1S	234.2W	244	143	Ancient river in France; modern Lech River.
Linpu	18.3N	247.0W	17	–	Town in China (Chekiang).
Lins	15.9N	29.8W	6	–	Town in Brazil.
Liris Valles	9.6S	302.6W	616	145	Ancient river in Italy; modern name is Liri.
Lisboa	21.5N	47.6W	1	–	Portuguese port.
Liu Hsin	53.7S	171.4W	129	149	Chinese astronomer (died A.D. 22).
Livny	27.5S	28.9W	10	–	Town in Russia.
Llanesco	28.6S	101.2W	–	–	Town in Spain.
Lobo Vallis	27.3N	61.4W	–	–	Modern river in Ivory Coast.
Locana	3.5S	38.3W	7	–	Town in Italy.
Lockyer	28.2N	199.4W	74	140	Joseph N.; British astronomer (1836-1920).
Locras Valles	8.1N	311.6W	148	145	Ancient European river (Corsica, France).
Lod	21.2N	31.6W	7	–	Town in Israel.

Feature name	Lat. (°)	Long. (°)	Size (km)	Page	Description
Lodwar	55.4S	43.0W	–	–	Town in Kenya.
Lohse	43.7S	16.4W	156	147	Oswald; German astronomer (1845-1915).
Loire Valles	18.5S	16.4W	670	144	Modern river in France.
Loja	41.4N	223.8W	10	–	Town in Ecuador.
Lomela	81.7S	56.0W	–	–	Town in Zaire.
Lomonosov	64.8N	8.8W	151	133	Mikhail V.; Russian chemist (1711-1765).
Longa	20.9S	25.7W	10	–	Town in Angola.
Loon	19.0S	246.5W	–	–	Town in Ontario, Canada.
Lorica	20.1S	28.3W	67	144	Town in Colombia.
Los	35.5S	76.5W	–	–	Town in Sweden.
Lota	46.5N	11.9W	13	–	Town in Chile.
Loto	22.2S	22.3W	22	–	Town in Zaire.
Louros Valles	8.7S	81.9W	423	138	Modern river in Greece.
Lowbury	42.8N	93.0W	–	–	Town in New Zealand.
Lowell	52.3S	81.3W	201	146	Percival; American astronomer (1855-1916).
Luck	17.4N	36.9W	7	–	Town in Wisconsin, USA.
Luga	44.6S	47.2W	42	–	Town in Russia.
Luki	30.0S	37.0W	20	–	Town in Ukraine.
Lunae Mensa	24.1N	62.3W	106	139	Albedo feature name.
Lunae Palus	15.0N	65.0W	–	–	Roman moon goddess Luna; "Luna's Swamp," or African Lunae Mountains where Nile was thought to originate.
Lunae Planum	9.6N	66.6W	1845	139	Classical albedo feature name.
Lutsk	38.5N	2.9W	5	–	Town in Ukraine.
Luzin	27.3N	328.8W	86	145	N. N.; Russian mathematician (1883-1950).
Lycus Sulci	29.2N	139.8W	1639	141	Classical albedo feature name.
Lyell	70.0S	15.6W	134	147	Charles; British geologist (1797-1875).
Lyot	50.7N	330.7W	220	136	Bernard; French astronomer (1897-1952).
Ma'adim Vallis	21.0S	182.7W	861	140	Word for "Mars" in Hebrew.
Mädler	10.8S	357.3W	100	144	Johann H. von; German astronomer (1794-1874).
Madrid	48.7N	224.4W	3	–	Spanish tracking site.
Mad Vallis	58.0S	283.0W	460	–	Modern river, Vermont, USA.
Mafra	44.4S	53.0W	12	–	Town in Brazil.
Magadi	34.8S	46.0W	57	147	Town in Kenya.
Magelhaens	32.9S	174.5W	102	149	Fernão de; Portuguese navigator (1480-1521).
Maggini	28.0N	350.4W	146	144	Mentore; Italian astronomer (1890-1941).
Maidstone	41.9S	54.1W	9	–	Town in England.
Main	76.8S	310.9W	102	150	Robert; British astronomer (1808-1878).
Maja Valles	15.4N	56.7W	1311	139	Nepali word for "Mars."
Malea Planum	65.9S	297.4W	1068	151	From albedo feature at 60S, 290W.
Mamers Vallis	41.7N	344.5W	945	136	Word for "Mars" in Oscan.
Manah	4.7S	33.8W	9	–	Town in Oman.
Mangala Fossa	16.5S	148.8W	–	–	Named for nearby valles.
Mangala Valles	7.4S	150.3W	880	141	Word for "Mars" in Sanskrit.
Manti	3.7S	37.7W	–	–	Town in Utah, USA.
Manzī	22.3S	27.3W	7	–	Town in Burma.
Maraldi	62.2S	32.1W	119	147	Giacomo F.; French astronomer (1665-1729).
Marbach	17.9N	249.2W	20	–	Town in Switzerland.
Marca	10.4S	158.2W	83	141	Village in Peru.
Mareotis Fossae	45.0N	79.3W	795	132	From albedo feature at 32N, 96W.
Margaritifer Chaos	9.1S	21.4W	500	144	Classical albedo feature name.
Margaritifer Sinus	10.0S	25.0W	–	–	"Pearl Bay," after Pearl Coast, South India.
Margaritifer Terra	16.2S	21.3W	2049	144	Classical albedo feature name.
Mari	52.4S	45.7W	–	–	Ruined city in Syria.
Mariner	35.2S	164.3W	151	149	Named for Mariner IV spacecraft.
Marth	13.1N	3.6W	104	144	Albert; British astronomer (1828-1897).
Marti Vallis	11.0N	182.0W	1700	–	Spanish word for Mars.
Martz	35.2S	215.8W	91	148	Edwin P.; American physicist (1916-1967).
Matrona Vallis	7.9S	183.7W	50	–	Classical river in France; present Marne River.
Maumee Valles	18.6N	55.2W	116	139	North American river (Indiana, Ohio).
Maunder	50.0S	358.1W	93	150	Edward W.; British astronomer (1851-1928).
Mawrth Vallis	22.4N	16.1W	575	144	Welsh word for "Mars."
McLaughlin	22.1N	22.5W	90	144	Dean B.; American astronomer (1901-1965).
Medrissa	18.8N	56.6W	–	–	Town in Algeria.
Medusae Fossae	2.0S	162.1W	291	141	Classical albedo feature name.
Medusae Sulci	5.3S	159.8W	91	141	Albedo feature name.
Mega	1.5S	37.0W	13	–	Town in Ethiopia.
Melas Chasma	10.5S	72.9W	526	139	Classical albedo feature name.
Melas Dorsa	17.8S	70.9W	501	139	Classical albedo feature name.
Melas Fossae	24.8S	72.4W	313	139	Classical albedo feature name.
Melas Labes	9.1S	71.2W	128	139	From albedo feature at 10S, 74W.
Mellish	72.9S	24.0W	99	–	John E.; American amateur astronomer (1886-1970).
Memnonia	20.0S	150.0W	–	–	Land of Memnon, King of Ethiopia.
Memnonia Fossae	21.9S	154.4W	1370	141	Classical albedo feature name.
Memnonia Sulci	6.5S	175.6W	361	140	Albedo feature name.
Mena	32.5S	18.5W	31	–	Town in Russia.
Mendel	59.0S	198.5W	82	148	Gregor J.; Austrian biologist (1822-1884).
Mendota	36.1N	221.8W	–	–	Town in Illinois, USA.
Sinus Meridiani	5.0S	0.0W	–	–	"Middle Bay," from Flammarion map.
Terra Meridiani	7.2S	356.0W	1622	144	Classical albedo feature name.
Meroe	35.0N	285.0W	–	–	Ethiopian island on Nile; now Atbar.
Meroe Patera	7.2N	291.5W	60	142	Classical albedo feature name.
Mie	48.6N	220.4W	93	134	Gustav; German physicist (1868-1957).
Mila	27.5S	20.6W	10	–	Town in Algeria.
Milankovič	54.8N	146.6W	113	135	Milutin; Yugoslav geophysicist, astrophysicist (1879-1958).
Milford	52.6S	45.1W	–	–	Town in Utah, USA.
Millman	54.4S	149.6W	82	–	Peter; Canadian astronomer (1906-1990).
Millochau	21.5S	274.7W	102	142	Gaston; French astronomer (born 1866).
Minio Vallis	5.0S	151.1W	73	141	Classical river in Italy.
Misk	0.9S	35.5W	10	–	Town in Turkey.
Mistretta	25.0S	109.1W	–	–	Town in Sicily.
Mitchel	67.8S	284.0W	141	151	Ormsby M.; American astronomer (1809-1862).
Mliba	39.9S	272.1W	–	–	Town in Swaziland.
Moab	20.0N	350.0W	–	–	Biblical town bordering Edom.
Moeris Lacus	8.0N	270.0W	–	–	"Moeris Lake"; Moeris was Egyptian lake in Libyan Desert.

Feature name	Lat. (°)	Long. (°)	Size (km)	Page	Description
Mohawk	43.2N	5.4W	11	–	Town in New York, USA.
Molesworth	27.8S	210.6W	175	143	Percy B.; British astronomer (1867-1908).
Moreux	42.2N	315.5W	138	136	Theophile; French astronomer and meteorologist (1867-1954).
Morpheos Rupes	36.2S	234.1W	321	148	Classical albedo feature name.
Mosa Vallis	14.3S	338.5W	134	145	Modern river in Western Europe.
Moss	19.4N	250.7W	8	–	Town in Norway.
Müller	25.9S	232.0W	120	143	Karl H.; German astronomer (1851-1925).
Munda Vallis	5.4S	146.1W	22	–	Classical river in ancient Lusitania, modern Mondega River in Portugal.
Murgoo	24.0S	22.3W	24	–	Town in Australia.
Musmar	24.8N	55.7W	–	–	Small town in Sudan.
Mut	22.7N	35.7W	7	–	Town in Turkey.
Mutch	0.6N	55.1W	200	139	Thomas A.; American geologist, Viking Lander Imaging Team leader (1931-1980).
N. Mareotis Tholus	36.8N	86.0W	–	–	Classical albedo name.
Naar	23.2N	42.1W	12	–	Town in Egypt.
Naic	24.7N	252.8W	8	–	Town in the Philippines.
Nain	41.7N	233.2W	7	–	Town in Newfoundland, Canada.
Naju	45.3N	237.1W	8	–	Town in the Republic of Korea.
Naktong Vallis	5.2N	326.7W	807	145	Modern river in Korea.
Nan	27.1S	19.8W	2	–	Town in Thailand.
Nanedi Valles	5.5N	48.7W	470	139	Word for "planet" in Sesotho, national language of Lesotho, Africa.
Nansen	50.5S	140.3W	82	149	Fridtjof; Norwegian explorer (1861-1930).
Nardo	27.8S	32.7W	23	–	Town in Italy.
Naro Vallis	3.5S	299.6W	175	–	Ancient river in Yugoslavia, modern name is Neretva.
Naukan	21.4N	30.6W	6	–	Town in Russia.
Navan	26.2S	23.2W	25	–	Town in Ireland.
Nazca	31.9S	266.1W	–	–	Town in Peru.
Nectar	28.0S	72.0W	–	–	Named for the drink of the gods.
Nectaris Fossae	24.7S	57.4W	664	139	Classical albedo feature name.
Negele	36.0S	264.0W	–	–	Town in Ethiopia.
Neith Regio	38.0N	272.0W	–	–	Lower world in Egyptian mythology.
Nema	20.9N	52.1W	14	–	Town in Russia.
Nepa	25.3S	19.5W	15	–	Town in Russia.
Nepenthes	20.0N	260.0W	–	–	Egyptian drug of forgetfulness.
Nepenthes Mensae	9.3N	241.7W	1704	143	Classical albedo feature name.
Nereidum Fretum	45.0S	55.0W	–	–	"Straits of the Nereids" (sea nymphs).
Nereidum Montes	41.0S	43.5W	1677	147	Classical albedo feature name.
Nestus Valles	7.3S	158.3W	34	–	Classical river in Macedonia (Greece).
New Bern	21.8N	49.2W	2	–	American colonial town (North Carolina).
New Haven	22.3N	49.3W	2	–	American colonial town (Connecticut).
Newcomb	24.1S	359.0W	259	–	Simon; American astronomer (1835-1909).
Newport	22.5N	49.0W	2	–	American colonial town (Rhode Island).
Newton	40.8S	157.9W	287	149	Isaac; British physicist (1643-1727).
Nhill	28.9S	103.2W	–	–	Town in Victoria, Australia.
Nia Fossae	14.4S	72.4W	333	139	Classical name for River Gambia, West Africa.
Nia Vallis	54.3S	33.0W	–	–	Lowell canal name; also classical river name.
Nicer Vallis	7.3S	158.2W	44	–	Classical name for present Neckar River, Germany.
Nicholson	0.1N	164.5W	114	141	Seth B.; American astronomer (1891-1963).
Niesten	28.2S	302.1W	114	145	Louis; Belgian astronomer (1844-1920).
Nif	20.2N	56.3W	7	–	Town in the Caroline Islands (Yap).
Niger Vallis	35.2S	267.5W	230	–	River in Africa.
Nili Fossae	24.0N	283.0W	709	142	Classical albedo feature name.
Nili Patera	9.2N	293.0W	70	142	Classical albedo feature name.
Niliacus Lacus	30.0N	30.0W	–	–	"Lake of the Nile."
Nilokeras	30.0N	55.0W	–	–	"Horn of the Nile," part of "Nulus" Canal.
Nilokeras Fossa	25.4N	56.9W	148	139	Classical albedo feature name.
Nilokeras Mensae	32.9N	51.1W	327	133	Albedo feature name.
Nilokeras Scopulus	31.5N	57.2W	1064	133	From albedo feature at 30N, 55W.
Nilosyrtis	42.0N	290.0W	–	–	Syrtis of the Nile; part of Nilus (Nile) Canal.
Nilosyrtis Mensae	35.4N	293.6W	693	137	Classical albedo feature name.
Nilus Chaos	25.5N	77.6W	280	138	Named for albedo feature at 20N, 65W.
Nilus Dorsa	22.3N	79.4W	188	138	Named for albedo feature at 20N, 65W.
Nilus Mensae	22.1N	71.8W	196	–	Named for albedo feature at 20N, 65W.
Nipigon	34.0N	81.9W	–	–	Town in Ontario, Canada.
Nirgal Vallis	28.3S	41.9W	511	139	Word for "Mars" in Babylonian.
Nitro	21.5S	23.8W	28	–	Town in West Virginia, USA.
Nix Olympica	20.0N	130.0W	–	–	"Snows of Olympus"; Olympus was mountain home of gods in Greece.
Njesko	35.4S	274.8W	–	–	Town in Czechoslovakia.
Noachis	45.0S	330.0W	–	–	Biblical; "Noah's (region)."
Noachis Terra	45.0S	350.0W	3500	150	Classical albedo feature name.
Noctis Fossae	3.3S	99.0W	692	138	From classical albedo feature at 10S, 96W.
Noctis Labyrinthus	7.2S	101.3W	976	138	Classical albedo feature name.
Noma	25.7S	24.0W	38	–	Town in Namibia.
Nordenskiöld	53.0S	158.7W	87	149	Nils Adolf Erik; Swedish geologist, geographer, arctic researcher (1832-1901).
Nune	17.7N	38.6W	9	–	Town in Mozambique.
Nutak	17.6N	30.3W	11	–	Town in Newfoundland, Canada.
Ocampo	32.9N	221.7W	–	–	Town in Mexico.
Oceanidum Fossa	61.7S	29.6W	165	147	Classical albedo feature name.
Oceanidum Mons	55.2S	41.4W	36	–	Name change from Charitum Tholus.
Ochakov	42.5S	31.6W	30	–	Town in Ukraine.
Ochus Valles	7.0N	45.2W	123	–	Classical name for present Hari-Rud River in Turkmenistan.
Octantis Cavi	52.7S	45.6W	–	–	Albedo name.
Octantis Mons	55.5S	42.5W	–	–	Albedo name.
Oenotria Scopulus	11.2S	283.2W	1438	142	Classical albedo feature name.

Feature name	Lat. (°)	Long. (°)	Size (km)	Page	Description
Oglala	3.2S	38.3W	15	–	Town in South Dakota, USA.
Ogygis Regio	45.0S	65.0W	–	–	"Land of Ogygos"; Ogygos was ancient King of Thebes or Athens, Greece.
Ogygis Rupes	34.1S	54.9W	225	147	Classical albedo feature name.
Okhotsk	23.2N	47.4W	2	–	Russian port.
Oltis Valles	23.5S	21.6W	190	144	Ancient name forr modern Lot River, France.
Olympia	80.0N	200.0W	–	–	Ancient and modern Greek city.
Olympia Planitia	81.5N	180.0W	–	134	Classical albedo feature name.
Olympica Fossae	24.4N	115.3W	573	138	From albedo feature at 17N, 134W.
Olympus Mons	18.4N	133.1W	624	141	Classical albedo feature name.
Olympus Rupes	17.2N	133.9W	1819	141	Classical albedo feature name.
Ōmura	25.7S	25.0W	8	–	Town in Japan.
Oodnadatta	52.7S	34.8W	–	–	Town in Australia.
Ophir	10.0S	65.0W	–	–	Biblical land (probably India) to which King Solomon sent naval expedition.
Ophir Chasma	4.0S	72.5W	251	139	Classical albedo feature name.
Ophir Labes	11.1S	68.3W	–	–	From albedo feature at 10S, 65W.
Ophir Planum	9.6S	62.2W	1068	139	Classical albedo feature name.
Oraibi	17.4N	32.4W	31	–	Town in Arizona, USA.
Orcus Patera	14.4N	181.5W	381	140	Classical albedo feature name.
Ore	16.9N	33.9W	7	–	Town in Nigeria.
Orinda	45.6N	233.0W	9	–	Town in California, USA.
Ortygia	60.0N	0.0W	–	–	Floating island (present Delos) where Leto bore Apollo and Artemis.
Ostrov	26.9S	28.0W	67	144	Town in Russia.
Osuga Valles	16.0S	39.1W	155	139	River in Russia.
Oti Fossae	10.0S	117.7W	231	138	Classical albedo feature.
Ottumwa	24.9N	55.7W	55	139	Town in Iowa, USA.
Oudemans	10.0S	91.7W	121	138	Jean A.; Dutch astronomer (1827-1906).
Oxia Colles	20.2N	27.0W	433	144	From albedo feature at 25N, 24W.
Oxia Palus	8.0N	180.0W	–	–	Lake (swamp into which Oxus River flows; i.e., Sea of Aral).
Oxus	20.0N	12.0W	–	–	Present Amu-Darya River.
Pabo	27.3S	22.9W	9	–	Town in Uganda.
Padus Vallis	4.6S	150.0W	53	–	Classical name for modern Po River in Italy.
Paks	7.8S	42.2W	7	–	Town in Hungary.
Pallacopas Vallis	54.5S	21.2W	–	–	Lowell canal name; also name of classical river.
Panchaia	60.0N	200.0W	–	–	1) Situated near Heliopolis, Egypt; or 2) Island in Red Sea rich in frankincense, gold, silver; Utopia.
Pandorae Fretum	25.0S	316.0W	–	–	Woman who let loose evils of world when she opened a box.
Paraná Valles	21.6S	10.7W	219	144	Ancient and modern name for South American river (Brazil, Argentina).
Pāros	22.2N	98.1W	40	–	Famous in antiquity for its marble quarries.
Pasithea Dorsum	55.7S	41.8W	–	–	One of the Graces.
Pasteur	19.6N	335.5W	114	145	Louis; French chemist (1822-1895).
Patapsco Vallis	23.9N	208.0W	149	140	Modern river in Maryland, USA.
Pavonis Chasma	3.8N	111.1W	–	–	Albedo name.
Pavonis Fossae	4.2N	111.2W	–	–	Albedo name.
Pavonis Mons	0.3N	112.8W	375	138	Classical albedo feature name.
Pavonis Sulci	3.9N	117.6W	–	–	Albedo name.
Peixe	20.8N	47.6W	9	–	Town in Brazil.
Peneus Mons	31.2S	273.9W	–	–	Albedo feature name.
Peneus Patera	58.0S	307.4W	123	150	From albedo feature at 48S, 290W.
Peraea Cavus	29.9S	264.7W	–	–	Albedo feature name.
Peraea Mons	31.3S	273.9W	–	–	Albedo feature name.
Perepelkin	52.8N	64.6W	112	132	Evgenii J.; Russian astronomer (1906-1940).
Peridier	25.8N	276.0W	99	142	Julien; French astronomer (1882-1967).
Perrotin	3.0S	77.8W	95	138	Henri A.; French astronomer, studied dark lineations on Mars (1845-1904).
Pettit	12.2N	173.9W	104	140	Edison; American astronomer (1890-1962).
Phaenna Dorsum	54.3S	43.2W	–	–	One of the Graces.
Phaethontis	50.0S	155.0W	–	–	"Of Phaethon;" who recklessly drove Chariot of Sun.
Philadelphia	22.0N	48.0W	2	–	American colonial town (Pennsylvania).
Phillips	66.4S	45.1W	183	147	John; British geologist (1800-1874); Theodore E.; British astronomer (1868-1942).
Phison	20.0N	320.0W	–	–	Biblical first river of Paradise.
Phison Rupes	26.6N	309.4W	149	145	Classical albedo feature name.
Phlegethon Catena	40.5N	101.8W	875	132	From albedo feature at 38N, 125W.
Phlegra	31.3N	187.8W	134	–	"Burning plain"; in Chalcidian Peninsula of Greece where Zeus hurled thunderbolts at Titans to support Hercules.
Phlegra Montes	40.9N	197.4W	1310	134	Classical albedo feature name.
Phoenicis Lacus	12.0S	110.0W	–	–	"Lake of the Phoenix"; Arabia or India.
Phon	15.8N	257.3W	8	–	Town in Thailand.
Phrixi Regio	40.0S	70.0W	–	–	"Phrixus' Land"; Phrixus and sister Helle escaped sacrifice in Boeotia on back of ram with golden fleece.
Pickering	34.4S	132.8W	112	149	Edward C.; American astronomer (1846-1919); William H.; American astronomer (1858-1938).
Pindus Mons	39.7N	88.9W	–	–	Mountains near Vale of Tempe.
Pinglo	3.0S	36.9W	18	–	Town in China (Ningsia).
Pityusa Rupes	63.0S	328.0W	290	150	From albedo feature at 58S, 319W.
Piyi	23.0S	253.3W	–	–	Town in Cyprus.
Planum Angustum	80.0S	83.9W	208	146	Classical albedo feature name.
Playfair	78.0S	125.5W	68	149	John; British geologist and mathematician (1748-1819).
Plum	26.4S	18.9W	3	–	Town in Wisconsin, USA.
Podor	44.6S	43.0W	25	–	Town in Senegal.
Polotsk	20.1S	26.1W	23	–	Town in Beloruss.
Pompeii	19.0N	59.1W	–	–	Ancient town in Italy.
Poona	24.0N	52.3W	20	–	Town in India.

Feature name	Lat. (°)	Long. (°)	Size (km)	Page	Description
Port-Au-Prince	21.3N	48.2W	1	–	Port of Hispaniola Island, Haiti.
Porter	50.8S	113.8W	113	146	Russell W.; American astronomer (1871-1949).
Porth	21.5N	255.9W	10	–	Town in Wales.
Portsmouth	22.8N	49.1W	2	–	American colonial town (New Hampshire).
Porvoo	43.7S	40.6W	9	–	Town in Finland.
Poti	36.5S	273.4W	–	–	Town in Armenia.
Poynting	8.4N	112.8W	80	138	John Henry; English physicist (1852-1914).
Priestley	54.3S	229.3W	40	148	Joseph; British chemist (1733-1804).
Princeton	21.9N	49.1W	2	–	American colonial town (New Jersey).
Proctor	47.9S	330.4W	168	150	Richard A.; British astronomer (1837-1888).
Promethei Rupes	76.8S	286.8W	1491	151	Classical albedo feature name.
Promethei Sinus	65.0S	280.0W	–	–	Prometheus' Bay; Greek mythological character.
Promethei Terra	52.9S	262.2W	2761	151	Classical albedo feature name.
Propontis	45.0N	185.0W	–	–	The Sea of Marmora, Asia Minor.
Protei Regio	23.0S	50.0W	–	–	"Proteus' Land"; Proteus was a sea deity with the gift of prophecy.
Protonilus	42.0N	315.0W	–	–	Designation of first (eastern) part of "Nilus" or Nile Canal.
Protonilus Mensae	44.2N	309.4W	592	136	From albedo feature at 42N, 315W.
Protva Valles	30.1S	60.0W	213	147	River in Russia.
Ptolemaeus	46.4S	157.5W	184	149	Claudius; Greek-born astronomer (ca. 90-160).
Pulawy	36.6S	76.7W	51	146	Town in Poland.
Punsk	20.7N	41.2W	10	–	Town in Poland.
Pylos	16.9N	30.1W	29	–	Town in Greece.
Pyramus Fossae	52.6N	293.5W	165	137	From albedo feature at 65N, 300W.
Pyrrhae Chaos	10.5S	28.6W	174	144	Albedo feature name.
Pyrrhae Regio	15.0S	38.0W	–	–	"Pyrrha's Region"; wife of Deucalion.
Quenisset	34.7N	319.4W	127	136	Ferdinand J.; French astronomer (1872-1951).
Quibā	17.4N	103.0E	–	–	Town in Saudi Arabia.
Quick	18.5N	48.8W	8	–	Town in British Columbia, Canada.
Quines	42.1S	270.6W	–	–	Town in Argentina.
Quorn	5.6S	33.8W	5	–	Town in Australia.
Rabe	44.0S	325.2W	99	150	Wilhelm F.; German astronomer (1893-1958).
Radau	17.3N	4.7W	115	144	Rodolphe; French astronomer (1835-1911).
Rakke	4.7S	43.5W	18	–	Town in Estonia.
Rana	26.0S	21.6W	12	–	Town in Norway.
Raub	42.6N	224.9W	7	–	Town in Malaysia.
Rauch	21.6N	57.8W	8	–	Town in Argentina.
Ravi Vallis	0.4S	40.7W	169	139	Ancient Pakistani river.
Ravius Valles	47.0N	111.6W	243	132	Classical name for river in northwest Ireland.
Rayadurg	18.6S	257.6W	–	–	Town in India.
Rayleigh	75.7S	240.1W	153	151	Strutt, John W., third baron Rayleigh; British physicist (1842-1919).
Redi	60.6S	267.1W	62	151	Francesco; Italian physicist (1626-1697).
Renaudot	42.5N	297.4W	69	137	Gabrielle; French astronomer (1877-1962).
Rengo	43.9S	43.5W	12	–	Town in Chile.
Reull Vallis	41.6S	258.3W	972	151	Word for "planet" in Gaelic.
Reuyl	9.5S	193.1W	74	140	Dirk; American physicist (1906-1972).
Revda	24.6S	28.3W	26	–	Town in Russia.
Reykholt	40.8N	85.8W	–	–	Town in Iceland.
Reynolds	75.1S	157.6W	91	149	Osborne; British physicist (1842-1912).
Rhabon Valles	21.7N	90.3W	204	138	Classical river in Dacia (Romania).
Ribe	16.6N	29.2W	11	–	Town in Denmark.
Richardson	72.6S	180.3W	82	148	Lewis F.; British meteorologist, chemist (1881-1953).
Rimac	45.2N	223.8W	7	–	Town in Peru.
Rincon	8.1S	43.1W	13	–	Town in Netherlands Antilles (Bonaire Island).
Ritchey	28.9S	50.9W	82	139	George W.; American astronomer (1864-1945).
Roddenberry	49.9S	4.5W	140	–	Gene; American engineer, television producer (1921-1991).
Romny	25.7S	18.0W	5	–	Town in Russia.
Rong	22.7N	45.3W	9	–	Town in China (Tibet).
Rongxar	26.5N	55.4W	–	–	Small village in Tibet, near Mt. Everest.
Ross	57.6S	107.6W	88	146	Frank E.; American astronomer (1874-1966).
Rossby	47.8S	192.2W	82	148	Carl G.; Swedish-born American meteorologist (1898-1957).
Rubicon Valles	44.7N	116.8W	228	132	Ancient river in Italy.
Ruby	25.6S	16.9W	25	–	Town in South Carolina, USA.
Rudaux	38.6N	309.0W	52	136	Lucien; French astronomer (1874-1947).
Runa Vallis	28.6S	36.7W	36	–	Name proposed by Soviets.
Rupes Tenuis	81.9N	65.2W	–	–	Classical albedo feature name.
Russell	55.0S	347.4W	138	150	Henry N.; American astronomer (1877-1957).
Rutherford	19.2N	10.6W	116	144	Ernest; British physicist (1871-1937).
Ruza	34.3S	52.8W	20	–	Town in Russia.
Rynok	44.3N	238.2W	9	–	Town in Russia.
Rypin	1.3S	41.0W	18	–	Town in Poland.
Terra Sabaea	10.4S	330.6W	1367	145	Classical albedo feature name.
Sinus Sabaeus	8.0S	340.0W	–	–	Today's Red Sea; Saba was part of southern Arabian Peninsula.
Sabis Vallis	5.8S	152.7W	286	140	Classical name for present Sambre River in France and Belgium.
Sabrina Vallis	11.1N	48.9W	294	–	Classical name for present Severn River, England.
Sacra Dorsa	18.6N	67.0W	535	139	From albedo feature at 20N, 67W.
Sacra Fossae	16.5N	72.0W	461	139	Classical albedo feature name.
Sacra Mensa	25.0N	69.2W	602	139	Albedo feature name.
Sacra Sulci	25.0N	71.5W	–	–	From albedo feature at 20N, 67W.
Salaga	47.6S	51.0W	28	–	Town in Ghana.
Samara Valles	24.2S	18.9W	575	144	Ancient name for modern Somme River, France.
San Juan	23.1N	48.1W	1	–	Puerto Rican port.
Sandila	25.9S	30.2W	8	–	Town in India.
Sangar	27.9S	24.1W	28	–	Town in Russia.
Santa Cruz	21.5N	47.3W	2	–	Port of Canary Islands.
Santa Fe	19.5N	48.0W	20	–	Town in New Mexico, USA.
Santaca	41.6S	272.5W	–	–	Town in Mozambique.
Sarn	77.5S	54.5W	–	–	Town in Wales.
Sarno	44.7S	54.0W	20	–	Town in Italy.
Satka	43.0S	36.7W	14	–	Town in Russia.
Sauk	45.0S	32.3W	2	–	Town in Wisconsin, USA.

Feature name	Lat. (°)	Long. (°)	Size (km)	Page	Description
Savannah	22.3N	47.8W	1	–	American colonial town (Georgia).
Savich	27.8S	263.5W	–	–	A. M.; Russian astronomer.
Say	28.5S	29.5W	15	–	Town in Niger.
Scamander Vallis	16.1N	331.1W	272	145	Ancient name of river in Troy (modern Turkey).
Scandia	60.5N	147.9W	–	–	Southern Scandinavia.
Scandia Colles	66.0N	135.4W	1232	135	From albedo feature name.
Schaeberle	24.7S	309.8W	160	145	John M.; American astronomer (1853-1924).
Schiaparelli	2.5S	343.4W	461	145	Giovanni V.; Italian astronomer (1835-1910).
Schmidt	72.2S	77.5W	194	146	Johann F.; German astronomer (1825-1884); Otto Y.; Russian geophysicist (1891-1956).
Schöner	20.4N	309.5W	185	145	Johannes; German geographer (1477-1547).
Schroeter	1.8S	303.6W	337	145	Johann H.; German astronomer (1745-1816).
Scylla Scopulus	25.2S	342.0W	474	145	From albedo feature at 19S, 320W.
Sebec	40.0S	260.5W	–	–	Town in Maine, USA.
Secchi	57.9S	257.7W	218	151	Angelo; Italian astronomer (1818-1878).
Semeykin	41.8N	351.2W	71	136	Boris Evgen'evich; Soviet astronomer (1900-1937).
Seminole	24.5S	18.9W	21	–	Town in Florida, USA.
Senus Vallis	5.3S	146.9W	27	–	Classical river in Ireland.
Sepik Vallis	1.0S	65.8W	–	–	River in New Guinea.
Mare Serpentis	30.0S	320.0W	–	–	Named for Constellation Serpens.
Sevel	78.2N	39.5W	–	–	Town in Denmark.
Sfax	7.8S	43.5W	7	–	Town in Tunisia.
Shalbatana Vallis	5.6N	43.1W	687	139	Word for "Mars" in Akkadian.
Shambe	20.7S	30.5W	29	–	Town in Sudan.
Sharonov	27.3N	58.3W	95	139	Vsevolod V.; Russian astronomer (1901-1964).
Shatskiy	32.4S	14.7W	69	147	N. S.; Russian geologist.
Shawnee	22.7N	31.5W	16	–	Town in Ohio, USA.
Sibu	23.3S	19.6W	31	–	Town in Malaysia.
Sigli	20.5S	30.6W	31	–	Town in Indonesia.
Simud Vallis	11.5N	38.5W	1074	139	Word for "Mars" in Sumerian.
Sinai	20.0S	70.0W	–	–	Biblical; named for area on Mars next to Mare Erythraeum (Indian Ocean).
Sinai Dorsa	13.1S	81.4W	189	138	Classical albedo feature name.
Sinai Planum	12.5S	87.1W	1064	138	Classical albedo feature name.
Singa	22.8S	17.2W	11	–	Town in Sudan.
Sinop	23.5S	249.3W	–	–	Town in Turkey.
Sirenum Fossae	34.5S	158.2W	2712	149	Classical albedo feature name.
Mare Sirenum	30.0S	155.0W	–	–	"Sea of the Sirens."
Terra Sirenum	37.0S	160.0W	2165	149	Classical albedo feature name.
Sisyphi Cavi	71.9S	3.0W	362	147	Classical albedo feature name.
Sisyphi Montes	69.8S	346.9W	198	150	From albedo feature at 67S, 348W.
Sithonius Lacus	45.0N	245.0W	–	–	Region inhabited by Sithonii; synonym of Thrace.
Sitka	4.3S	39.3W	11	–	Town in Alaska, USA.
Sklodowska	33.8N	2.8W	116	133	Maria; Polish-born French chemist (Mme P. Curie) (1867-1934).
Slipher	47.7S	84.5W	129	146	Vesto M.; American astronomer (1875-1969).
Smith	66.1S	102.8W	71	146	William; British geologist, engineer (1769-1839).
Sögel	21.7N	55.1W	29	–	Town in Germany.
Sokol	42.8S	40.5W	20	–	Town in Russia.
Solano	27.0S	251.0W	–	–	Town in Phillipines.
Solis Dorsa	23.4S	78.9W	771	138	Classical albedo feature name.
Solis Lacus	28.0S	90.0W	–	–	"Lake of the Sun;" the so-called "Eye of Mars" connecting East with West.
Solis Planum	20.9S	94.6W	1464	138	Classical albedo feature name.
Soochow	16.8N	28.9W	30	–	Town in China (Kiangsu).
South	77.0S	338.0W	111	150	James; British astronomer (1785-1867).
Spallanzani	58.4S	273.5W	72	151	Lazzaro; Italian biologist (1729-1799).
Spry	3.8S	38.6W	6	–	Town in Utah, USA.
Spur	22.2N	52.3W	7	–	Town in Texas, USA.
Srīpur	31.1S	100.6W	–	–	Town in Bangladesh.
Stege	2.6N	58.4W	72	139	Town in Denmark.
Steno	68.0S	115.3W	105	146	Nicolaus; Danish geologist (1638-1686).
Stobs	5.0S	38.4W	10	–	Town in Scotland.
Stokes	56.0N	189.0W	70	134	George G.; British physicist (1819-1903).
Ston	47.2N	237.4W	7	–	Town in Yugoslavia.
Stoney	69.8S	138.4W	177	149	George J.; Irish physicist (1826-1911).
Stura Vallis	22.9N	217.6W	–	–	Classical river east of Rome, Italy.
Stygis Catena	23.6N	209.6W	99	140	From albedo feature at 30N, 200W.
Stygis Fossae	27.2N	210.3W	290	142	Albedo feature name.
Styx	30.0N	200.0W	–	–	Great river around Nether region which souls must cross on journey from Earth.
Styx Dorsum	31.4N	208.3W	–	–	Albedo feature name.
Suata	19.2S	253.3W	–	–	Town in Venezuela.
Subur Vallis	11.6N	53.1W	30	–	Classical river in Mauritania.
Sucre	24.0N	54.5W	9	–	Town in Colombia.
Suess	67.1S	178.4W	72	149	Eduard; Austrian geologist, engineer (1831-1914).
Sūf	16.6N	38.2W	9	–	Town in Jordan.
Sulak	18.3N	78.6W	–	–	Town in Russia.
Sumgin	37.0S	48.6W	83	147	M. I.; Russian cryopedologist.
Surinda Valles	29.3S	34.8W	101	139	Name proposed by Soviets; found on Mars-5 Map.
Surius Vallis	60.3S	51.0W	–	–	Lowell canal name.
Surt	17.0N	30.6W	9	–	Town in Libya.
Suzhi	27.6S	273.1W	–	–	Town in China.
Syria	20.0S	100.0W	–	–	Province in Near East including Phoenicia; or one of the Cyclades (Homer).
Syria Planum	12.0S	103.9W	757	138	Classical albedo feature name.
Syrtis Major	10.0N	290.0W	–	–	Libyan Gulf, now Gulf of Sirte.
Syrtis Major Planum	9.5N	289.6W	1356	142	Albedo feature name; changed from Planitia to Planum.
Sytinskaya	42.8N	52.8W	90	133	Nadezhda Nikolaevna; Soviet astronomer (1906-1974).
Tabor	36.0S	58.5W	20	–	Town in Czechoslovakia.
Tabou	45.5S	34.8W	8	–	Town in Ivory Coast.
Tader Valles	50.0S	152.5W	275	–	Ancient name for present Segura River, Spain.
Taejin	35.6S	274.3W	–	–	Town in Korea.
Tagus Valles	7.3S	246.0W	35	–	Ancient and modern river in Spain, Portugal.
Tak	26.4S	28.5W	5	–	Town in Thailand.

Feature name	Lat. (°)	Long. (°)	Size (km)	Page	Description
Tala	20.6S	252.2W	–	–	Town in Tunisia.
Talsi	41.9S	49.1W	9	–	Town in Latvia.
Tame	23.0S	107.8W	–	–	Town in Colombia.
Tanaica Montes	39.7N	90.8W	–	–	Classical albedo feature name.
Tanais	50.0N	70.0W	–	–	Present River Don, Russia.
Tanais Fossae	38.6N	85.3W	–	–	Classical albedo feature name.
Tantalus Fluctus	0.0N	0.0W	–	–	Albedo feature name.
Tantalus Fossae	44.5N	102.4W	1990	132	From albedo feature at 35N, 110W.
Tara	44.4S	52.7W	27	–	Town in Ireland.
Tarakan	41.6S	30.1W	37	–	Town in Indonesia (Borneo).
Tarata	3.8S	41.3W	12	–	Town in Bolivia.
Tarsus	23.5N	40.2W	19	–	Town in Turkey.
Tartarus Colles	17.2N	188.2W	593	140	From albedo feature at 2N, 183W.
Tartarus Montes	25.1N	188.7W	1011	140	Classical albedo feature name.
Tartarus Rupes	6.6S	184.4W	81	140	Named for classical albedo feature.
Tartarus Scopulus	6.4S	181.7W	127	–	Named for classical albedo feature.
Taus Vallis	4.9S	148.3W	–	–	Classical river in Caledonia (Scotland).
Taxco	20.8N	40.0W	17	–	Town in Mexico.
Taza	44.0S	45.1W	22	–	Town in Morocco.
Tecolote	24.9S	106.7W	–	–	Town in New Mexico, USA.
Teisserenc de Bort	0.6N	315.0W	118	145	Leon P.; French meteorologist (1855-1913).
Tem'	42.1N	9.5W	4	–	Town in Russia.
Tempe	40.0N	70.0W	–	–	Greek valley south of Mt. Olympus noted for its beauty.
Tempe Colles	33.9N	82.7W	–	–	Classical albedo feature name.
Tempe Fossae	40.2N	74.5W	1553	132	From albedo feature at 40N, 70W.
Tempe Mensa	28.1N	71.5W	84	139	From albedo feature at 40N, 70W.
Tempe Terra	41.3N	70.5W	2055	132	From albedo feature at 40N, 70W.
Terby	28.2S	286.0W	135	142	Franćois J.; Belgian astronomer (1846-1911).
Termes Vallis	11.3S	157.1W	–	–	Classical river in ancient Lusitania, present Tormes River, Spain.
Tharsis	0.0N	0.0W	–	–	Connecting link between East and West; ancient Spanish town Tartessus.
Tharsis Montes	2.8N	113.3W	2105	138	Classical albedo feature name.
Tharsis Tholus	13.4N	90.8W	153	138	Classical albedo feature name.
Thaumasia	35.0S	85.0W	–	–	Named for Thaumas, Arabian god of clouds.
Thaumasia Fossae	45.9S	97.3W	1118	146	Classical albedo feature name.
Thaumasia Planum	22.0S	65.0W	930	–	Albedo feature at 30S, 75W.
Thom	41.5S	267.5W	–	–	Town in Thailand.
Thoth	30.0N	255.0W	–	–	Egyptian messenger god.
Thule	23.6S	25.5W	14	–	Town in Greenland.
Thyle I, II	70.0S	180.0W	–	–	Named for Thule; may be middle Norway; used to indicate far northern lands.
Thyles Rupes	73.2S	205.6W	269	148	Name and feature changed 1984 from Ultimi Cavi and Thyles Chasma.
Thymiamata	10.0N	10.0W	–	–	"Land of sweet-scented perfumes;" South Yemen or India.
Tignish	31.1S	273.0W	–	–	Town in Prince Edward Island, Canada.
Tikhonravov	13.7N	324.1W	390	145	M. K.; Russian rocket scientist (1851-1916).
Tikhov	51.2S	254.1W	107	151	Gavril A.; Russian astronomer (1875-1960).
Tile	17.8N	28.6W	8	–	Town in Somalia.
Timaru	25.6S	22.2W	8	–	Town in New Zealand.
Timbuktu	5.7S	37.7W	63	139	Town in Mali.
Timoshenko	42.1N	63.9W	84	132	Ivan Fedorovich; Soviet astronomer (1918-1941).
Tinia Valles	4.7S	148.9W	20	–	Classical river in Italy.
Tinjar Valles	38.2N	235.7W	390	134	Modern river in Sarawak, Malaysia.
Tisia Valles	11.6S	313.9W	390	145	Ancient name for modern Tisza River, Ukraine.
Tithoniae Catena	5.5S	82.0W	380	–	Classical albedo feature name.
Tithoniae Catenae	5.1S	78.2W	101	138	Classical albedo feature name.
Tithoniae Fossae	1.7S	81.9W	549	138	Classical albedo feature name.
Tithoniae Fossae	6.4S	82.5W	492	–	Classical albedo feature name.
Tithonium Chasma	4.6S	86.5W	904	138	Classical albedo feature name.
Tithonius Lacus	5.0S	85.0W	–	–	"Tithonian Lake;" Tithonus, received from his wife Eos eternal life but not eternal youth.
Tiu Vallis	8.6N	34.8W	970	139	Word for "Mars" in old English (West Germanic).
Tiwi	27.9S	24.6W	17	–	Town in Oman.
Alexey Tolstoy	47.6S	234.6W	94	148	Soviet writer (1882-1945).
Tombe	42.8S	44.4W	12	–	Town in Sudan.
Tōno	45.2S	52.2W	10	–	Town in Japan.
Torsö	44.7S	51.0W	14	–	Town in Sweden.
Torup	28.1S	262.2W	–	–	Town in Sweden.
Tractus Albus	30.0N	80.0W	–	–	"White tract."
Tractus Catena	27.9N	103.2W	1234	138	Classical albedo feature name.
Tractus Fossae	27.1N	101.1W	535	138	Classical albedo feature name.
Trebia Valles	32.4N	209.9W	–	–	Classical name for modern Trebbia River, Italy.
Trinacria	25.0S	268.0W	–	–	Classical name for Sicily.
Trinidad	23.9S	251.0W	–	–	Town in Peru.
Trivium Charontis	20.0N	198.0W	–	–	"Crossroad of Charon;" meeting place of several nether world canals.
Troika	17.1N	255.0W	13	–	Town in Russia.
Trouvelot	16.3N	13.0W	168	144	Etienne L.; French-born American astronomer (1827-1895).
Troy	23.3N	52.4W	10	–	Town in Idaho, USA.
Trud	17.7N	30.8W	5	–	Town in Russia.
Trumpler	61.7S	150.6W	77	149	Robert J.; American astronomer (1886-1956).
Tsau	49.8N	238.9W	6	–	Town in Botswana.
Tsukuba	48.9N	225.9W	2	–	Japanese mission control site.
Tugaske	32.2S	101.0W	–	–	Town in Saskatchewan, Canada.
Tungla	41.1S	270.1W	–	–	Town in Nicaragua.
Tura	27.0S	21.8W	14	–	Town in Russia.
Turbi	40.9S	51.2W	25	–	Town in Kenya.
Tuskegee	2.9S	36.2W	69	139	Town in Alabama, USA.
Tycho Brahe	49.5S	213.8W	108	148	Danish astronomer (1546-1601).
Tyndall	40.1N	190.4W	79	134	John; British physicist (1820-1893).
Tyras Vallis	8.3N	50.1W	140	–	Classical name for present Dniester River, Ukraine.
Tyrrhena Dorsa	20.5S	243.7W	–	–	Albedo feature name.
Tyrrhena Fossae	22.1S	254.5W	–	–	Classical albedo name.
Tyrrhena Mons	24.5S	258.7W	–	–	Classical albedo name.

Feature name	Lat. (°)	Long. (°)	Size (km)	Page	Description
Tyrrhena Patera	21.9S	253.2W	597	143	Classical albedo feature name.
Tyrrhena Terra	14.3S	278.5W	2817	142	Classical albedo feature name.
Mare Tyrrhenum	20.0S	255.0W	–	–	"Tyrrhenian Sea;" area between Italy and Sicily.
Uchronia	70.0N	260.0W	–	–	"Land of Agelessness."
Ulya	18.0S	253.1W	–	–	Town in Russia.
Ulysses Fossae	11.4N	123.3W	720	141	From albedo feature name.
Ulysses Patera	2.9N	121.5W	112	141	Classical albedo feature name.
Ulyxis Rupes	68.2S	198.8W	303	148	Classical albedo feature name.
Umatac	42.7N	222.8W	12	–	Town in Guam, USA.
Umbra	50.0N	290.0W	–	–	Means "shadow" in Latin.
Uranius Dorsum	23.2N	76.3W	352	138	Named for albedo feature.
Uranius Fossae	25.8N	90.1W	438	138	Classical albedo feature name.
Uranius Patera	26.7N	92.6W	276	138	Classical albedo feature name.
Uranius Tholus	26.5N	97.8W	71	138	Classical albedo feature name.
Utopia	50.0N	250.0W	–	–	Greek, means "nowhere;" ideal state.
Utopia Planitia	47.6N	277.3W	3276	137	Classical albedo feature name.
Uzboi Vallis	30.9S	35.6W	340	139	Dry riverbed in Russia.
Vaals	4.0S	33.1W	8	–	Town in the Netherlands.
Valga	44.6S	36.3W	16	–	Town in Estonia.
Valles Marineris	11.6S	70.7W	4128	138	General name of the system of canyons honoring the scientific team of the Mariner 9 program.
Valverde	20.3N	55.8W	35	–	Town in the Dominican Republic.
Varus Valles	8.9S	155.9W	–	–	Classical name for present Var River, France.
Vastitas Borealis	67.5N	180.0W	9999	132	Classical albedo feature name.
Vätö	43.9S	53.2W	10	–	Town in Sweden.
Vaux	18.1N	42.8W	–	–	Town in France.
Vedra Valles	19.8N	55.7W	155	139	Ancient European river (Great Britain).
Verlaine	9.4S	295.9W	42	–	Town in France.
Very	49.8S	176.8W	127	149	Frank W.; American astronomer (1852-1927).
Viana	19.5N	255.3W	29	–	Town in Brazil.
Vik	36.0S	64.0W	25	–	Town in Iceland.
Vils	39.2N	11.7W	6	–	Town in Austria.
Vinogradov	20.3S	38.0W	191	139	Aleksander P.; Soviet geochemist (1895-1975).
Vinogradsky	56.3S	216.0W	61	148	Sergei N.; Russian microbiologist (1856-1953).
Virrat	31.1S	102.9W	–	–	Town in Sweden.
Vishniac	76.7S	276.1W	76	151	Wolf V.; American microbiologist (1922-1974).
Vistula Valles	14.2N	52.1W	190	–	Classical name for modern Wisła River, Poland.
Vivero	49.4N	241.3W	64	–	Town in Spain.
Voeykov	32.5S	76.1W	67	146	A. I.; Russian climatologist, geographer (1842-1916).
Vogel	37.0S	13.2W	124	147	Hermann; German astronomer (1841-1907).
Volgograd	48.4N	224.8W	2	–	Soviet launch site.
Vol'sk	23.2N	51.3W	8	–	Town in Russia.
Von Karman	64.3S	58.4W	100	147	Theodore; American aeronautical engineer (1881-1963).
Voo	27.3S	19.8W	2	–	Town in Kenya.
Vulcani Pelagus	35.0S	15.0W	–	–	Named for Vulcan, Roman god of fire.
W. Mareotis Tholus	35.8N	87.5W	–	–	Classical albedo name.
Wabash	21.5N	33.7W	42	–	Town in Indiana, USA.
Wahoo	23.5N	33.6W	66	139	Town in Nebraska, USA.
Wajir	27.2S	254.5W	–	–	Town in Kenya.
Wallace	52.8S	249.2W	159	151	Alfred R.; British biologist (1823-1913).
Wallops	46.9N	227.2W	2	–	American launch site.
Warra	20.8N	37.5W	11	–	Town in Australia.
Warrego Valles	43.0S	93.1W	164	146	Modern Australian river.
Waspam	20.7N	56.6W	40	–	Town in Nicaragua.
Wassamu	25.5N	53.0W	17	–	Town in Japan.
Wau	45.2S	42.4W	3	–	Town in New Guinea.
Weer	19.8N	51.4W	9	–	Town in the Netherlands.
Wegener	64.3S	4.0W	70	147	Alfred L.; German geophysicist (1880-1930).
Weinbaum	65.9S	245.5W	86	151	Stanley G.; American novelist (1902-1935).
Wells	60.1S	237.4W	94	148	Herbert G.; British novelist (1866-1946).
Wer	45.6N	6.2W	3	–	Town in India.
Wicklow	2.0S	40.7W	21	–	Town in Ireland.
Wien	10.5S	220.1W	105	143	Wilhelm; German physicist (1864-1928).
Williams	18.8S	164.1W	125	141	Arthur S.; British astronomer (1861-1938).
Wilmington	21.9N	47.5W	1	–	American colonial town (Delaware).
Windfall	2.1S	43.5W	20	–	Town in Alberta, Canada.
Wink	6.6S	41.5W	8	–	Town in Texas, USA.
Wirtz	48.7S	25.8W	128	147	Carl Wilhelm; German astronomer (1876-1939).
Wislicenus	18.3S	348.7W	138	144	Walter; German astronomer (1859-1905).
Woolgar	34.8N	85.5W	–	–	Town in Australia.
Woomera	48.4N	227.3W	3	–	Australian launch site.
Worcester	26.8N	50.3W	–	–	Town in New York, USA.
Wright	58.6S	150.8W	106	149	William H.; American astronomer (1871-1959).
Wukari	32.2S	102.8W	–	–	Town in Nigeria.
Xanthe	10.0N	50.0W	–	–	"Golden-Yellow Land."
Xanthe Dorsa	21.3N	45.5W	664	139	Classical albedo feature name.
Xanthe Terra	4.2N	46.3W	3074	139	Classical albedo feature name.
Yakima	43.3N	3.2W	12	–	Town in Washington, USA.
Yala	17.5N	38.5W	19	–	Town in Thailand.
Yaonis Regio	40.0S	320.0W	–	–	Named for Chinese Emperor Yao; flood occurred during his reign.
Yar	22.5N	39.1W	5	–	Town in Russia.
Yat	18.3N	29.1W	8	–	Town in Niger.
Yegros	22.5S	23.5W	14	–	Town in Paraguay.
Yorktown	23.1N	48.7W	8	–	American colonial town (Virginia).
Yoro	23.0N	28.0W	8	–	Town in Honduras.
Yungay	44.3S	44.6W	17	–	Town in Peru.
Yuty	22.4N	34.1W	18	–	Town in Paraguay.
Zea Dorsa	49.8S	280.5W	320	151	Classical albedo feature name.
Zephyria	0.0N	195.0W	–	–	Land of the west wind.
Zephyria Mensae	10.1S	188.1W	230	140	Albedo feature name.
Zephyrus Fossae	23.9N	214.4W	452	143	Albedo feature name.
Zhigou	29.5S	102.5W	–	–	Town in China.
Zilair	32.0S	33.0W	43	–	Town in Russia.
Zir	18.7N	36.6W	6	–	Town in Turkey.
Zongo	32.0S	42.0W	23	–	Town in Zaire.
Žulanka	2.3S	42.3W	47	–	Town in Russia.
Zuni	19.3N	29.6W	25	–	Town in New Mexico, USA.

PHOBOS

Feature name	Lat. (°)	Long. (°)	Size (km)	Page	Description
D'Arrest	35.0S	185.0W	–	162	Heinrich L.; German astronomer (1822-1875).
Hall	75.0S	225.0W	–	162	Asaph; American astronomer (1829-1907).

Feature name	Lat. (°)	Long. (°)	Size (km)	Page	Description
Kepler Dorsum	–	–	–	162	Johannes; German astronomer (1571-1630).
Roche	50.0N	185.0W	–	162	Edourd; French astronomer (1820-1883).
Sharpless	25.0S	155.0W	–	162	Bevan P.; American astronomer (1904-1950).
Stickney	5.0S	55.0W	–	162	Angeline; American wife of astronomer A. Hall (died 1938).

ASTEROID BELT

IDA

Feature name	Lat. (°)	Long. (°)	Size (km)	Page	Description
Afon	6.5S	0.0E	25	–	Cave in Russia.
Atea	5.7S	18.9E	2	311	Cave in the Muller Range of Papua New Guinea.
Azzurra	30.5N	217.2E	9	–	Flooded cave (known as the Blue Grotto) on the island of Capri in Southern Italy.
Bilemot	27.8S	29.2E	2	–	Lava tube in Korea.
Castellana	13.4S	335.2E	5	–	Cave in Puglia region, Italy.
Choukoutien	12.8N	23.6E	1	–	Site where Peking Man was discovered.
Fingal	13.2S	39.9E	1	–	Cave in the Hebrides.
Kartchner	7.0S	179.0E	1	311	Cave in Arizona.
Kazumura	32.0S	41.1E	2	311	Lava tube in Hawaii.
Lascaux	0.8N	161.2E	12	–	Cave noted for its prehistoric paintings.
Lechuguilla	7.9N	357.1E	1	–	Cave in Carlsbad National Park, New Mexico.
Mammoth	18.3S	180.3E	10	–	Longest limestone cavern known on Earth.
Manjang	28.3S	90.5E	1	–	Lava tube in Korea.
Orgnac	6.3S	202.7E	10	–	Cave in France, more than 1 million years old.
Padirac	4.3S	5.2E	2	–	Cave with underground river in France.
Palisa Regio	23.0S	34.0E	23	311	Johann; Austrian astronomer, discovered Ida (1848-1925).
Pola Regio	11.0S	184.0E	8	311	Place where Palisa (discoverer of Ida) observed.
Postojna	42.9S	359.9E	6	–	Large cave in Slovenia.
Sterkfontein	4.1S	54.1E	5	–	Cave in South Africa.
Stiffe	27.9S	126.5E	1	311	Karst cave in Sulmona, Italy.
Townsend Dorsum	25.0N	30.0E	40	–	Tim E.; Galileo imaging team participant (died 1989).
Undara	2.0N	113.8E	8	–	Lava tube from Undara Volcano, North Queensland, Australia.
Vienna Regio	8.0N	2.0E	13	–	Where Palisa discovered Ida.
Viento	12.2N	343.9E	1	–	Lava tube in Spain.

JUPITER

AMALTHEA

Feature name	Lat. (°)	Long. (°)	Size (km)	Page	Description
Gaea	80.0S	90.0W	–	–	Greek mother earth goddess who brought Zeus to Crete.
Ida Facula	20.0N	175.0W	–	–	Greek; mountain where Zeus played as a child.
Lyctos Facula	20.0S	120.0W	–	–	Greek; area in Crete where Zeus was raised.
Pan	55.0N	35.0W	–	–	Greek; goat-god, son of Amalthea and Hermes in some legends, also Zeus' foster brother.

Feature name	Lat. (°)	Long. (°)	Size (km)	Page	Description
Todd	5.0S	160.0W	–	162	David; American astronomer (1855-1939).
Wendall	0.0N	140.0W	–	162	Oliver C.; American astronomer (1845-1912).

DEIMOS

Feature name	Lat. (°)	Long. (°)	Size (km)	Page	Description
Swift	–	–	–	–	Jonathan; British writer (1667-1745).
Voltaire	–	–	–	–	Franćios M.; French writer (1694-1778).

DACTYL

Feature name	Lat. (°)	Long. (°)	Size (km)	Page	Description
Acmon	39.0S	138.0E	–	311	One of the original three dactyls.
Celmis	46.0S	220.0E	–	–	One of the original three dactyls.

GASPRA

Feature name	Lat. (°)	Long. (°)	Size (km)	Page	Description
Aix	47.9N	160.3W	6	–	Spa in France.
Alupka	65.0N	65.0W	–	–	Spa in Crimea, Ukraine.
Baden-Baden	46.0N	55.0W	–	–	Spa in Germany.
Badgastein	25.0N	3.0W	–	–	Spa in Austria.
Bagnoles	55.0N	122.0W	–	–	Spa in France.
Bath	13.4N	9.7W	10	311	Spa in England.
Beppu	3.9N	58.4W	5	–	Spa on Kyushu, Japan.
Brookton	27.7N	103.3W	6	–	Spa in New York, USA.
Calistoga	30.0N	2.0W	–	–	Resort in California, USA.
Carlsbad	29.7N	88.8W	5	–	Spa in Czech Republic.
Charax	8.6N	0.0W	11	–	
Dunne Regio	15.0N	15.0W	–	311	James; Galileo project planner (1934-1992).
Helwan	22.4N	118.9W	6	–	Spa in Egypt.
Ixtapan	11.9N	86.9W	5	–	Spa in Mexico.
Katsiveli	55.0N	65.0W	–	–	Spa in Crimea, Ukraine.
Krynica	49.0N	35.0W	–	–	Health resort in Poland.
Lisdoonvarna	16.5N	358.1W	10	–	Spa in Ireland.
Loutraki	42.0N	140.0W	–	–	Spa in Greece.
Mandal	23.5N	46.5W	5	–	Spa in Norway.
Manikaran	62.0N	155.0W	–	–	Spa in India.
Marienbad	35.4N	81.8W	5	–	Spa in Czech Republic.
Miskhor	15.0N	65.9W	5	–	Spa in Crimea, Ukraine.
Moree	15.1N	164.4W	6	–	Spa in Australia.
Neujmin Regio	2.0N	80.0W	–	–	Grigorij N.; Russian astronomer; discoverer of Gaspra (1885-1946).
Ramlösa	15.0N	4.9W	10	–	Spa in Australia.
Rio Hondo	31.7N	20.7W	7	–	Spa in Argentina.
Rotorua	18.8N	30.7W	6	–	Spa in New Zealand.
Saratoga	50.0N	270.0W	–	–	Spa in New York, USA.
Spa	51.5N	152.0W	6	311	Health resort in Belgium.
Tang-Shan	59.0N	256.0W	–	–	Spa in China.
Yalova	29.0N	10.0W	–	–	Health resort in Turkey.
Yalta	57.6N	261.3W	5	–	Spa in Crimea, Ukraine.
Yeates Regio	65.0N	75.0W	–	311	Clayne; Galileo project manager (1936-1991).
Zohar	23.0N	118.0W	–	–	Spa in Israel.

IO

Feature name	Lat. (°)	Long. (°)	Size (km)	Page	Description
Agni Patera	40.5S	334.2W	20	197	Hindu god of fire.
Amaterasu Patera	37.7N	306.6W	100	197	Japanese sun goddess.
Amirani	25.9N	114.5W	–	199	Georgian god of fire.
Angpetu Patera	21.2S	10.6W	45	196	Dakota name meaning the sun.
Apis Tholus	11.2S	348.8W	–	197	Greek; name for Epaphus, son of Io and Zeus.
Aramazd Patera	73.4S	337.7W	38	200	Armenian thunder god.
Argos Planum	47.0S	318.2W	140	200	Where Io was captured by Zeus.
Asha Patera	8.6S	225.8W	90	198	Persian spirit of fire.
Ātar Patera	30.2N	278.9W	125	197	Iranian personification of fire.

Feature name	Lat. (°)	Long. (°)	Size (km)	Page	Description
Aten Patera	47.9S	310.1W	40	200	Egyptian sun god.
Babbar Patera	39.5S	272.1W	95	197	Sumerian sun god.
Bactria Regio	45.8S	123.4W	–	200	Io passed through this area of ancient Iran in her wanderings.
Bochica Patera	61.0S	20.6W	50	200	Chibcha sky god.
Boösaule Montes	4.4S	270.1W	590	198	Cave where Io bore Epaphus.
Carancho Patera	1.5N	317.2W	30	197	Bolivian legendary hero who received fire from an owl.
Cataquil Patera	24.2S	18.7W	125	196	Inca god of thunder and lightning.
Chalybes Regio	45.5N	83.2W	–	195	Greek; Io passed through here in her wanderings.
Colchis Regio	5.3N	199.8W	–	198	Greek; Io passed through this part of Asia Minor in her wanderings.
Creidne Patera	52.4S	343.5W	125	200	Celtic smith god.
Crimea Mons	76.1S	244.8W	–	200	Io passed by here in her wanderings.
Culann Patera	19.9S	158.7W	100	199	Celtic smith god.
Daedalus Patera	19.1N	274.3W	40	197	Greek hero, smith; father of Icarus.
Danube Planum	20.9S	258.7W	150	198	Io passed by here in her wanderings.
Dazhbog Patera	54.0N	301.6W	–	195	Slavonic sun god.
Dingir Patera	4.0S	342.1W	40	197	Sumerian sun god; means "shining."
Dodona Planum	56.8S	352.9W	390	200	Greek; Io went there after the death of Argus.
Echo Mensa	79.6S	357.4W	–	200	Mother of Iynx.
Emakong Patera	3.2S	119.1W	80	199	Sulca (New Britain) man who brought fire.
Epaphus Mensa	53.5S	241.3W	–	200	"Child of touch;" son of Io and Zeus.
Ethiopia Planum	44.9S	27.0W	105	196	Io passed through here in her wanderings.
Euboea Fluctūs	45.1S	351.3W	–	–	Io passed through here in her wanderings.
Euboea Montes	46.3S	339.9W	–	200	Io passed through here in her wanderings.
Fuchi Patera	28.4N	327.9W	45	197	Ainu fire goddess.
Galai Patera	10.7S	288.3W	90	197	Mongol fire god.
Gibil Patera	14.9S	294.9W	95	197	Sumerian fire god.
Haemus Montes	68.9S	46.6W	–	200	Io passed by here in her wanderings.
Hatchawa Patera	58.2S	35.1W	75	200	Yaroro (Slavic) god who, in form of a boy, gave fire to mankind.
Heiseb Patera	29.7N	244.8W	60	198	Bushman devil who represents fire.
Heno Patera	56.5S	312.1W	65	200	Iroquois god of thunder.
Hephaestus Patera	2.0N	289.5W	50	197	Greek smith god.
Hiruko Patera	65.1S	329.3W	80	200	Japanese sun god.
Horus Patera	9.6S	338.6W	125	197	Egyptian falcon-headed solar god.
Huo Shen Patera	15.1S	329.3W	90	197	Chinese god of fire.
Hybristes Planum	54.0S	21.1W	150	200	Io passed by here in her wanderings.
Ilmarinen Patera	14.0S	2.8W	40	196	Finnish blacksmith with supernatural creative powers.
Inachus Tholus	15.9S	348.9W	–	197	Greek; river god, father of Io.
Inti Patera	68.1S	348.6W	75	200	Inca sun god.
Iopolis Planum	34.5S	333.5W	125	197	Town where Io was worshipped as moon goddess (present-day Antioch).
Iynx Mensa	61.1S	304.6W	–	200	Cast a spell on Zeus so he fell in love with Io.
Kane Patera	47.8S	13.4W	115	200	Hawaiian god of sunlight.
Kava Patera	16.5S	342.1W	75	197	Persian blacksmith.
Khalla Patera	6.0N	303.4W	80	197	Bushman sun in form of man often referred to as the hunter.
Kibero Patera	11.6S	305.5W	60	197	Yaroro toad who lives in underworld giving mankind fire.
Lerna Regio	64.0S	292.6W	–	200	Greek; meadows of Lyrcea.
Loki	17.9N	302.6W	–	197	Norse blacksmith, trickster god.
Loki Patera	12.6N	308.8W	250	197	Norse blacksmith, trickster god.
Lu Huo Patera	38.4S	354.1W	90	197	Stove fire associated with Chinese god of the hearth fire.
Lyrcea Planum	40.3S	269.3W	310	198	Plain where Io was born.
Maasaw Patera	40.0S	340.4W	30	197	Hopi (U.S.A.) god of fire and death.
Mafuike Patera	13.9S	260.0W	110	198	Hawaiian demigoddess whose fingers held fire.
Malik Patera	34.2S	128.5W	85	199	Babylonian, Caananite sun god.
Mama Patera	10.6S	356.5W	30	197	Chagaba (Chibcha, Colombia) word for sun.
Manua Patera	35.5N	322.0W	110	–	Hawaiian sun god.
Marduk	27.1S	207.5W	–	198	Sumerian-Akkadian fire god.
Masaya Patera	22.5S	348.1W	125	197	Nicaraguan smith god.
Masubi	43.6S	54.7W	–	–	Japanese fire god.
Maui	20.0N	122.0W	–	199	Hawaiian demigod who sought fire from Mafuike.
Maui Patera	16.5N	124.0W	45	199	Hawaiian demigod who sought fire from Mafuike.
Mazda Catena	8.6S	313.5W	–	197	Babylonian sun god.
Mbali Patera	31.5S	7.1W	60	196	Pygmy word representing fire itself.
Media Regio	4.6N	58.8W	–	196	Greek; Io passed through this part of Iran in her wanderings.
Menahka Patera	30.7S	346.2W	40	197	Mandan (U.S.A.) name for the sun.
Mihr Patera	16.4S	305.6W	60	197	Armenian fire god.
Mithra Patera	58.8S	267.9W	25	200	Persian god of light.
Mycenae Regio	37.3S	165.9W	–	199	Greek; in some legends, Io was transformed there.
Nemea Planum	73.3S	275.5W	500	200	Greek; where Io was turned into a cow by Zeus and given to Hera.
Nina Patera	38.3S	164.2W	425	199	Inca fire god.
Ninurta Patera	16.5S	315.3W	75	197	Babylonian god of the spring sun.
Nusku Patera	64.7S	4.6W	90	200	Assyrian fire god.
Nyambe Patera	0.6N	343.9W	50	197	Zambezi sun god.
Païve Patera	45.5S	0.0W	69	200	Saami-Lapp sun god.
Pan Mensa	49.5S	35.4W	–	200	Greek; father of Iynx.
Pautiwa Patera	32.8S	347.9W	40	197	Hopi (U.S.A.) name for the sun.
Pele	18.6S	257.8W	–	198	Hawaiian goddess of the volcano.
Podja Patera	18.2S	304.9W	60	197	Tungu spirit who keeps the fire.
Prometheus	1.6S	153.0W	–	199	Greek fire god.
Purgine Patera	2.6S	297.7W	20	197	Mordvinian (Russia) thunder god.
Pyerun Patera	56.0S	252.2W	40	200	Slavonic god of thunder.
Ra Patera	8.6S	325.3W	30	197	Egyptian sun god.
Reiden Patera	13.4S	235.7W	70	198	Japanese thunder god.
Reshet Catena	0.8N	305.6W	–	197	Aramaic sun god.
Ruwa Patera	0.4N	3.0W	50	196	African sun god associated with Mt. Kilimanjaro.
Sźd Patera	2.8S	303.8W	70	197	Phoenician chariot rider of the Sun.
Sengen Patera	32.8S	304.3W	55	197	Japanese; deity of Mt. Fugiyama.

Feature name	Lat. (°)	Long. (°)	Size (km)	Page	Description
Shakuru Patera	23.6N	266.4W	70	198	Pawnee (U.S.A.) sun god of the East; gives light and heat.
Shamash Patera	33.7S	152.1W	110	199	Assyro-Babylonian sun god.
Shoshu Patera	19.6S	324.1W	50	197	Caucasian patron of fire.
Silpium Mons	52.6S	272.9W	–	200	Greek; where Io dies of grief in some legends.
Siun Patera	49.8S	1.4W	50	200	Nanai (Siberia) sun god.
Sui Jen Patera	19.2S	4.3W	40	196	Chinese hero who discovered fire.
Surt	45.5N	337.9W	–	195	Icelandic volcano god.
Svarog Patera	48.3S	267.5W	70	200	Russian smith god.
Talos Patera	26.1S	356.5W	20	197	Nephew of Daedalus; also a blacksmith.
Taranis Patera	70.8S	28.6W	105	200	Celtic thunder god.
Tarsus Regio	43.7S	61.4W	–	196	Io passed through here in her wanderings.
Taw Patera	33.3S	0.0W	15	197	Monguor word for fire or hearth.
Tohil Patera	26.3S	156.5W	20	199	Central American god who gave fire to man.
Tol-Ava Patera	1.7N	322.0W	70	197	Mordvinian (Russia) goddess of fire.
Tung Yo Fluctus	16.4S	357.8W	–	197	Chinese fire god.
Tung Yo Patera	18.7S	2.5W	20	196	Chinese fire god.
Ülgen Patera	40.4S	288.0W	49	197	Siberian progenitor god who struck first fire.
Uta Fluctus	32.6S	19.2W	–	196	Sumerian sun god.
Uta Patera	35.3S	24.9W	30	196	Sumerian sun god.
Vahagn Patera	23.8S	351.7W	70	197	Armenian fire god.
Viracocha Patera	61.2S	281.7W	55	200	Qechua sun god.
Volund	25.0N	175.0WW	–	199	Germanic supreme smith of the gods.

EUROPA

Feature name	Lat. (°)	Long. (°)	Size (km)	Page	Description
Adonis Linea	51.8S	113.2W	758	209	Greek; son of Phoenix, nephew of Europa.
Agenor Linea	43.6S	208.2W	1326	208	Greek; Europa's father.
Alphesiboea Linea	28.0S	182.6W	1642	208	Son of Phoenix, nephew of Europa.
Argiope Linea	8.2S	202.2W	934	208	Greek; another name for Telephassa.
Asterius Linea	17.7N	265.6W	2753	208	Greek; Europa's husband after Zeus.
Astypalaea Linea	76.5S	220.3W	1030	–	Sister of Europa.
Belus Linea	11.8N	228.3W	2580	208	Greek; Agenor's twin brother.
Boeotia Macula	54.0S	166.0W	22	–	Place where Cadmus led cow before it stopped at site of Thebes.
Cadmus Linea	27.8N	173.1W	1212	209	Greek; brother of Europa.
Cilicia Flexus	47.6S	142.6W	639	209	Greek; land named for Cilix on his search for Europa.
Cilix	1.2N	181.9W	23	208	Brother of Europa; Rand control point crater.
Cyclades Macula	64.0S	192.0W	105	–	Islands where Rhadamanthys reigned.
Delphi Flexus	69.7S	172.3W	1125	209	Where the cow led Cadmus before it stopped at the site of Thebes.
Echion Linea	13.1S	184.3W	1217	208	Survivor of the men Cadmus sowed with the dragon's teeth; a founder of Thebes.
Gortyna Flexus	42.4S	144.6W	1261	209	Place on Crete where Zeus brought Europa.
Libya Linea	56.2S	183.3W	452	208	Greek; Agenor's mother.
Minos Linea	45.3N	195.7W	2134	208	Greek; son of Europa and Zeus.
Morvran	5.7S	152.2W	25	209	Celtic; ugly son of Tegid.
Pelorus Linea	17.1S	175.9W	1770	209	Greek; survivor of the men Cadmus sowed with the dragon's teeth; a founder of Thebes.
Phineus Linea	33.0S	269.2W	1984	208	Greek; brother of Europa.
Phocis Flexus	48.6S	197.2W	298	208	Where the cow lead Cadmus before it stopped at the site of Thebes.
Phoenix Linea	14.5N	184.7W	732	208	Brother of Europa.
Rhadamanthys Linea	18.5N	200.8W	1780	208	Son of Europa and Zeus.
Rhiannon	81.8S	199.7W	25	208	Celtic heroine.
Sarpedon Linea	42.2S	89.4W	940	209	Greek; son of Europa and Zeus.
Sidon Flexus	64.5S	170.4W	1216	209	Greek; another name for Tyre; where Europa was born.
Taliesin	23.2S	137.4W	48	209	Celtic; magician son of Bran.
Tectamus Linea	17.9N	181.9W	719	208	Father of Asterius.
Tegid	0.6S	164.0W	29	209	Celtic hero who lived in Bula Lake.
Telephassa Linea	2.8S	178.8W	800	209	Europa's mother.
Thasus Linea	68.7S	187.4W	1027	208	Greek; brother of Europa.
Thera Macula	47.7S	180.9W	78	208	Place where Cadmus stopped in his search for Europa.
Thrace Macula	46.6S	171.2W	173	209	Place in northern Greece where Cadmus stopped in his search for Europa.
Thynia Linea	57.9S	148.6W	398	209	Peninsula between Black and Marmara Seas, where Phineus sought Europa.
Tyre Macula	31.7N	147.0W	148	209	Greek; the seashore from which Zeus abducted Europa.

GANYMEDE

Feature name	Lat. (°)	Long. (°)	Size (km)	Page	Description
Abydos Facula	34.1N	154.0W	165	214	Egyptian town where Osiris was worshipped.
Achelous	60.3N	13.5W	51	212	Greek river god; father of Callirrhoe, Ganymede's mother.
Adad	55.9N	180.0W	59	–	Assyro-Babylonian god of thunder.
Adapa	71.3N	30.2W	54	212	Assyro-Babylonian; lost immortality when, at Ea's advice, he refused food of life.
Agreus	15.2N	235.4W	72	217	Hunter god in Tyre.
Agrotes	62.5N	199.5W	61	217	Tyre; greatest god of Gebal; farmer god.
Ammura	30.9N	344.2W	61	213	Phoenician; god of the west.
Amon	33.4N	223.3W	102	217	Theban king of gods.
Anat	3.1S	127.9W	28	220	Assyro-Babylonian goddess of dew; Rand control point crater.
Andjeti	52.4S	159.1W	65	220	Egyptian; first god of Busirus.
Anshar Sulcus	21.5N	202.9W	1181	217	Assyro-Babylonian; celestial home of Lakhmu and Lakhamu.
Antum	4.4N	220.0W	21	217	Babylonian; wife of Anu.
Anu	63.6N	346.3W	57	213	Sumerian-Akkadian god of power, of heavens.
Anubis	82.7S	118.5W	97	221	Egyptian jackal-headed god who opened the underworld to the dead.
Apsu Sulci	34.8S	235.5W	1281	223	Sumero-Akkadian; primordial ocean.
Aquarius Sulcus	50.0N	11.5W	1341	212	Greek; Zeus set Ganymede among the stars as the constellation Aquarius, the water carrier.

Feature name	Lat. (°)	Long. (°)	Size (km)	Page	Description
Arbela Sulcus	22.3S	353.6W	1896	219	Assyrian town where Ishtar was worshipped.
Ashîma	37.7S	122.4W	82	220	Semitic-Arab god of fate.
Asshur	53.0N	335.7W	23	213	Assyro-Babylonian warrior god.
Aya	66.1N	326.2W	42	213	Assyro-Babylonian; wife of Shamash.
Ba'al	24.0N	331.6W	52	213	Phoenician; Canaanite god.
Barnard Regio	0.8N	1.0W	2547	212	Edward E.; American astronomer (1857-1923).
Bau	24.1N	53.3W	81	212	Sumerian goddess who breathed into men the breath of life; daughter of Anu and patroness of Lagash.
Bes	25.8S	180.4W	220	223	Egyptian god of marriage.
Bubastis Sulci	79.8S	263.1W	2197	222	Town in Egypt where Bast was worshipped.
Busiris Facula	14.9N	216.1W	348	217	Town in lower Egypt where Osiris was first installed as local god.
Buto Facula	12.6N	204.3W	236	217	Swamp where Isis hid Osiris' body.
Coptos Facula	9.4N	209.8W	332	217	Early town from which caravans departed.
Danel	4.3S	25.2W	54	218	Phoenician; mythical hero versed in art of divination.
Dardanus Sulcus	39.3S	20.2W	2559	218	Greek; where Ganymede was abducted by Zeus disguised as an eagle.
Dendera Facula	0.0N	257.0W	114	222	Town where Hathor was chief goddess.
Diment	22.4N	353.6W	47	213	Egyptian goddess of the dwelling place of the dead.
Dukug Sulcus	81.3N	352.7W	467	213	Sumerian holy cosmic chamber of the gods.
Edfu Facula	26.8N	147.7W	187	214	Egyptian town where Horus was worshipped.
Elam Sulci	57.4N	205.5W	1866	217	Ancient Babylonian seat of sun worship, in present-day Iran.
Enkidu	27.9S	328.4W	121	219	Friend of Gilgamesh.
Enlil	53.9N	314.8W	43	213	Assyro-Babylonian; god of the air, hurricanes, and nature.
Erech Sulcus	5.4S	175.7W	998	220	Akkadian town that was built by Marduk.
Eshmun	17.8S	191.5W	99	223	Phoenician; divinity of Sidon.
Etana	72.5N	344.1W	49	213	Assyro-Babylonian; asked the eagle for an herb to give him an heir.
Gad	12.4S	137.5W	68	220	Semitic god of fate or good fortune.
Galileo Regio	35.7N	137.6W	3142	214	Galilaei; Italian astronomer (1564-1642).
Geb	58.3N	187.2W	62	217	Heliopolis earth god.
Geinos	18.0N	221.0W	45	217	Tyre; god of brick making.
Gilgamesh	61.7S	123.9W	145	220	Assyro-Babylonian; sought immortality after Enkidu died.
Gir	35.2N	146.6W	77	214	Sumerian god of summer heat.
Gula	62.7N	13.9W	38	212	Assyro-Babylonian; health god.
Halieus	35.2N	168.0W	90	214	Tyre; fisherman god.
Hapi	31.3S	212.4W	85	223	Egyptian god of the Nile.
Harpagia Sulcus	14.9S	319.1W	1398	219	Greek; where Ganymede was abducted by an eagle.
Hathor	70.4S	268.1W	59	222	Egyptian goddess of joy and love.
Hursag Sulcus	10.7S	234.5W	928	223	Sumerian mountain where winds dwell.
Ilah	22.8N	161.0W	71	214	First Sumerian sky god.
Ilus	11.5S	110.8W	41	221	Ganymede's brother.
Irkalla	31.1S	114.7W	116	221	Sumerian goddess of underworld, seen by Enkidu in a dream.
Ishkur	0.1N	11.5W	83	212	Sumerian god of rain.
Isimu	8.1N	2.5W	90	212	Sumerian god of vegetation.
Isis	67.9S	197.2W	68	223	Egyptian goddess; wife of Osiris.
Kadi	48.8N	181.0W	94	217	Babylonian goddess of justice.
Keret	16.0N	35.2W	36	212	Phoenician hero.
Khonsu	38.0S	189.7W	86	223	Egyptian moon god.
Khumbam	25.3S	338.7W	60	219	Assyro-Babylonian; Elamite creator god.
Kingu	35.7S	227.4W	91	223	Assyro-Babylonian; conquered leader of Tiamat's forces whose blood was used to create man.
Kishar	70.7N	352.5W	79	213	Assyro-Babylonian; terrestrial progenitor goddess.
Kishar Sulcus	8.3S	218.7W	1213	223	Assyro-Babylonian; terrestrial home of Lakhmu and Lakhamu.
Kittu	0.5S	336.6W	33	219	Assyro-Babylonian god of justice.
Kulla	34.8N	115.0W	82	215	Sumerian god of brick making.
Lagash Sulcus	10.4S	162.8W	1407	220	Early Babylonian town.
Lakhamu Fossa	12.5S	228.3W	392	223	Dragon monster, or divine natural force produced by Apsu and Tiamat.
Lakhmu Fossae	30.3N	142.3W	2871	214	Dragon monster, or divine natural force produced by Apsu and Tiamat.
Lumha	37.3N	155.2W	71	214	Title of Enki as patron of singers; also Babylonian priest.
Marius Regio	12.1N	199.3W	3572	217	Simon; German astronomer (1570-1624).
Mashu Sulcus	31.1N	209.2W	3030	217	Assyro-Babylonian; mountain with twin peaks where sun rose and set.
Mehit	29.7N	165.1W	50	214	Egyptian lion-headed goddess; Anhur's wife.
Melkart	10.0S	185.8W	111	223	Phoenician; divinity of Tyre.
Memphis Facula	15.4N	132.5W	344	214	Ancient capitol of Egyptian lower kingdom.
Min	28.7N	3.2W	35	212	Egyptian fertility god.
Mir	4.0S	231.4W	22	223	West Semitic god of wind.
Misharu	5.3S	338.3W	95	219	Assyro-Babylonian god of law.
Mor	29.5N	329.3W	40	213	Phoenician; spirit of the harvest.
Mush	13.5S	115.0W	97	221	Sumerian male deity; upper parts are human, lower parts a serpent.
Mysia Sulci	9.6S	28.6W	4221	218	Greek; where Ganymede was abducted by an eagle.
Nabu	47.3S	10.1W	44	218	Sumerian god of intellectual activity.
Namtar	62.4S	351.6W	58	219	Assyro-Babylonian plague demon.
Nanna	18.5S	244.4W	48	222	Sumerian moon god; god of wisdom.
Neheh	70.8N	57.3W	57	212	Egyptian god of eternity.
Neith	28.9N	9.0W	93	212	Egyptian warrior goddess; goddess of domestic arts.
Nicholson Regio	34.0S	356.7W	3719	219	Seth B.; American astronomer (1891-1963).
Nidaba	19.0N	123.8W	188	214	Sumerian grain goddess.

Feature name	Lat. (°)	Long. (°)	Size (km)	Page	Description
Nigirsu	61.2S	327.4W	70	219	Assyro-Babylonian; god of the fields, war god.
Ninkasi	56.8N	53.8W	75	212	Sumerian goddess of brewing.
Ninki	6.6S	120.9W	170	220	Consort to Ea, Babylonian god of water..
Ninlil	7.6N	118.7W	91	215	Chief Assyrian goddess; Asshar's consort.
Ninsum	13.3S	140.1W	91	220	Minor Babylonian goddess of wisdom; Gilgamesh's mother.
Nippur Sulcus	40.9N	191.5W	2158	217	Sumerian city.
Nun Sulci	49.3N	318.8W	1090	213	Egyptian; chaos, primordial ocean; held germ of all things.
Nut	60.1S	268.0W	93	222	Egyptian goddess of the sky.
Ombos Facula	3.8N	238.6W	90	217	Egyptian town where Sebek's triad was worshipped; present Kom Ombo.
Osiris	37.8S	165.2W	109	220	Egyptian god of the dead.
Perrine Regio	38.8N	30.0W	2145	212	Charles D.; American astronomer (1867-1951).
Philus Sulci	44.0N	212.0W	473	217	Greek; where Ganymede and Hebe were worshipped as rain-givers.
Phrygia Sulcus	12.4N	19.3W	3205	212	Greek; kingdom in Asia Minor where Ganymede was born.
Ptah	67.0S	214.3W	45	223	Egyptian; sovereign god of Memphis; patron of artisans.
Punt Facula	26.1S	242.2W	228	222	Land east of Egypt where Bes originated.
Ruti	11.9N	310.7W	35	213	Phoenician; Byblos god.
Sais Facula	37.9N	14.2W	137	212	Capital of Egypt in mid-7th century B.C.
Sapas	56.5N	37.9W	65	212	Assyro-Babylonian; torch of the gods.
Sati	30.5N	14.9W	98	212	Wife of Khnum, Egyptian god of the Cataracts.
Sebek	59.5N	178.9W	70	–	Egyptian crocodile god.
Seima	16.6N	217.4W	33	217	Mother goddess of the Arameans.
Seker	40.8S	351.0W	117	219	Egyptian god of the dead at Memphis.
Selket	16.7N	107.4W	140	215	Egyptian tutelary goddess who guarded intestines of the dead.
Shu	42.5N	357.9W	54	213	Egyptian god of air.
Sicyon Sulcus	36.5N	12.0W	1146	212	Greek; where Ganymede and Hebe were worshipped as rain-givers.
Sin	51.9N	359.2W	30	213	Babylonian moon god.
Sippar Sulcus	15.8S	191.0W	1539	223	Ancient Babylonian town.
Siwah Facula	7.5N	143.2W	220	214	Oasis oracle of Zeus at Ammon; visited by Alexander.
Tammuz	12.9N	232.7W	44	217	Akkadian youthful god of vegetation; Ishtar's son.
Tanit	56.7N	40.7W	45	212	Assyro-Babylonian; Carthaginian goddess.
Ta-Urt	26.5N	306.5W	85	213	Babylonian moon god.
Teshub	72.3S	281.0W	60	–	Elamite god of the tempest.
Tettu Facula	38.6N	160.9W	86	214	Egyptian town where Hatmenit and Osiris were worshipped.
Thebes Facula	6.0N	202.4W	475	217	Ancient capitol of upper kingdom.
Thoth	42.4S	146.0W	107	220	Egyptian moon god; invented all arts and sciences.
Tiamat Sulcus	3.2N	209.2W	1310	217	Assyro-Babylonian; tumultuous sea from which everything was generated.
Tros	11.0N	31.1W	109	212	Greek; father of Ganymede.
Ur Sulcus	56.5N	177.3W	1774	214	Ancient Sumerian seat of moon worship.
Uruk Sulcus	8.4N	169.0W	2456	214	Babylonian city ruled by Gilgamesh.
Zaqar	57.5N	41.3W	52	212	Assyro-Babylonian; Sin's messenger who brought dreams to men.
Zu Fossae	53.0N	129.4W	1386	214	Dragon of chaos slain by Marduk.

CALLISTO

Feature name	Lat. (°)	Long. (°)	Size (km)	Page	Description
Adal	75.4N	80.8W	40	229	Norse; son of Karl and Erna.
Adlinda	56.6S	23.1W	274	232	Eskimo; place in ocean depths where souls are imprisoned after death.
Ägröi	43.3N	11.0W	55	226	Finno-Ugric god of twins.
Ahti	41.8N	103.1W	52	229	Finnish god of water, sends fish to the fisherman.
Ajleke	22.4N	101.3W	46	229	Saami god of holidays.
Akycha	72.5N	318.6W	67	227	Alaskan name of the sun.
Alfr	9.7S	222.8W	60	237	Norse dwarf.
Áli	59.3N	56.2W	61	–	Norse; strongest of men.
Ánarr	44.1N	0.6W	47	226	Norse dwarf.
Aningan	50.5N	8.2W	287	226	Moon god of Greenland Eskimos.
Asgard	32.0N	139.8W	1347	228	Norse home of the gods.
Askr	51.7N	324.1W	64	227	Norse; first man, created from a log drifted ashore on a beach.
Aziren	35.4N	178.3W	64	228	Estonian spirit of death.
Balkr	29.1N	11.9W	64	226	Norse; Ottar's ancestor.
Bavörr	49.2N	20.3W	84	226	Norse dwarf.
Beli	62.6N	81.7W	50	229	Celtic; father of Caswallawn.
Bragi	75.7N	61.7W	65	229	Skaldic; god of poetry.
Brami	28.9N	19.2W	67	226	Norse; Ottar's ancestor.
Bran	24.3S	207.7W	89	237	Celtic; omnipotent god who watched over people.
Buga	22.2N	323.9W	54	227	Tungu heaven god.
Buri	38.7S	46.2W	98	232	Norse dwarf.
Burr	42.5N	135.5W	74	228	Norse giant; his sons raised up heaven's vault and shaped the Earth.
Dag	58.6N	74.2W	40	229	Norse; Ottar's ancestor.
Danr	62.5N	77.8W	48	229	Norse; king against whom Konr marched.
Dia	73.0N	50.4W	35	226	Greek; Callisto's sister.
Dryops	77.6N	21.3W	42	226	Greek; son of Dia by Apollo.
Durinn	67.0N	90.1W	49	229	Norse dwarf.
Egdir	33.9N	35.9W	58	226	Norse; shepherd for the giants.
Egres	42.5N	176.6W	38	228	Karelian deity of the harvest of beans.
Erlik	66.8N	1.3W	39	226	Russian first man who became a devil.
Fadir	56.4N	12.7W	81	226	Norse farmer.
Fili	64.3N	349.5W	42	227	Norse dwarf.
Finnr	15.5N	4.3W	65	226	Norse dwarf.
Freki	79.9N	352.0W	48	227	Norse; wolf's name meaning "insatiable."
Frodi	68.3N	139.1W	44	228	Norse; Hledis' father.
Fulla	73.5N	103.7W	45	229	Norse; maid to Frigg, queen of the gods.
Fulnir	60.3N	35.5W	46	226	Norse; son of Thrael and Thyr.
Geri	66.7N	354.2W	38	227	Norse; wolf's name meaning "greedy."
Gipul Catena	70.2N	48.2W	588	226	Norse river.
Gisl	57.2N	34.8W	39	226	Norse; steed ridden by Aesir.

JUPITER

Feature name	Lat. (°)	Long. (°)	Size (km)	Page	Description
Gloi	49.0N	245.7W	112	230	Norse dwarf.
Göll	57.5N	319.5W	55	227	Norse; servant to the gods.
Göndul	59.9N	115.5W	50	229	Norse; a Valkyrie.
Grimr	41.6N	215.2W	90	231	Norse; a name for Odinn.
Gunnr	64.7N	106.2W	58	229	Norse; a Valkyrie.
Gymir	63.8N	49.2W	40	226	Norse; another name for the sea-god, Legir.
Hábrok	74.8N	129.3W	40	228	Norse; a hawk.
Haki	24.9N	315.1W	69	227	Norse giant.
Hár	3.6S	357.9W	44	233	Norse; a name for Odin.
Hepti	64.5N	23.9W	41	226	Norse dwarf.
Hijsi	61.6N	169.3W	51	228	Karelian deity of hunting.
Hödr	69.0N	91.0W	76	229	Norse; Baldr's blind brother who shot Baldr unknowingly.
Hoenir	33.9S	261.2W	84	236	Norse; god who gave souls to first humans.
Högni	13.5S	4.5W	65	232	Norse; Ottar's ancestor.
Höldr	44.1N	109.2W	61	229	Son of Karl and Snor in Rigdismal.
Igaluk	5.6N	315.9W	105	227	Alaskan name of the Moon.
Ilma	30.0S	167.2W	53	235	A celestial divinity of air.
Ivarr	6.1S	321.5W	68	233	Norse; Ottar's ancestor.
Jumal	58.8N	119.5W	60	229	Estonian sky god.
Jumo	65.9N	12.3W	37	226	Finno-Ugric heaven god.
Kári	48.2N	117.9W	32	229	Ottar's ancestor.
Karl	56.4N	330.7W	42	227	Norse; Rigr's son with Amma.
Kul'	62.7N	123.5W	43	228	Komi wood spirit.
Lempo	25.6S	319.5W	46	233	Finno-Ugric evil spirit.
Ljekio	47.9N	161.4W	36	228	Finnish god of grass, roots of trees.
Lodurr	51.2S	270.8W	76	236	Norse; god who gave first humans goodly color.
Loni	3.6S	214.9W	86	237	Norse dwarf.
Losy	65.3N	323.1W	62	227	Mongolian; evil snake who tried to kill all living things.
Maderatcha	30.5N	95.9W	57	229	Saami sky god.
Mera	64.1N	75.8W	36	229	Greek; another nymph of Artemis seduced by Zeus.
Mimir	32.6N	53.2W	51	226	Norse giant.
Mitsina	57.5N	104.7W	43	229	Alaskan old man who perished while hunting on ice.
Modi	66.4N	120.9W	43	229	Norse; son of Thor and Sif.
Nama	57.2N	331.3W	61	227	Altaic hero who built ark to save his family from the flood.
Nár	1.7S	46.4W	63	232	Norse dwarf.
Nerivik	17.1S	57.1W	37	232	Alaskan name of Sedna.
Nerkes	29.8N	163.6W	48	228	Karelian patron of squirrel hunting.
Nidi	66.7N	96.5W	43	229	Norse dwarf.
Njord	16.5N	132.8W	34	228	Nordic gods called the Vanir; pacific, benevolent, guardians to man.
Nori	45.4N	343.5W	86	227	Norse dwarf.
Norov-Ava	54.6N	113.7W	47	229	Mordvinian mistress of the field.
Nuada	62.1N	273.2W	66	230	Irish chieftan god.
Omol'	42.2N	118.4W	60	229	Komi wood spirit.
Oski	57.2N	269.3W	57	230	Norse; a name for Odin.
Ottar	61.5N	104.8W	50	229	Innsteinn's son and Freyja's favorite.
Pekko	18.3N	5.4W	61	226	Finno-Ugric god of barley.
Reginn	39.7N	90.8W	51	229	Norse dwarf.
Rigr	70.9N	245.0W	54	230	Norse; another name for the god Heimdall.
Rongoteus	53.5N	106.8W	35	229	Karelian deity of the harvest of rye.
Rota	27.8N	109.8W	55	229	Deity of the underground world.
Saga	0.0N	326.0W	–	–	Scandinavian goddess, wife of Odin; Rand control point crater.
Sarakka	3.7S	53.7W	56	232	Finno-Ugric goddess of childbirth.
Seqinek	55.5N	25.5W	80	226	Eskimo; the sun.
Sholmo	53.9N	16.4W	58	226	Finno-Ugric heaven god.
Sigyn	35.8N	29.2W	44	226	Norse; Loki's wife.
Sköll	55.6N	315.3W	55	227	Norse wolf.
Skuld	10.1N	37.7W	81	226	Norse; maiden living near Yggdrasill who governed the fate of humans.
Sudri	55.4N	137.1W	69	228	Norse dwarf.
Sumbur	67.1N	324.8W	38	227	Russian (Buriat) world mountain.
Tapio	30.6N	109.7W	56	229	Finnish deity of the wood who sent game to the hunter.
Tindr	2.5S	355.5W	64	233	Norse; Ottar's ancestor.
Tontu	27.6N	100.3W	47	229	Finnish god of housekeeping.
Tornarsuk	28.7N	128.6W	104	228	Greenland legendary hero.
Tyll	43.3N	165.4W	65	–	Estonian epic hero; struggled with a giant.
Tyn	70.8N	233.6W	60	231	Great god of Germanic peoples.
Valfödr	1.2S	247.8W	81	236	Norse; a name for Odin, god of wisdom.
Valhalla	15.9N	56.6W	2748	226	Norse; Odin's hall.where he received the souls of slain warriors.
Vali	9.8N	325.2W	46	227	Norse; Ottar's ancestor.
Vanapagan	38.1N	158.0W	62	228	Estonian; a wicked giant.
Veralden	33.2N	96.1W	75	229	Saami god of fertility.
Vestri	45.3N	52.8W	75	226	Norse dwarf.
Vidarr	11.9N	193.6W	84	231	Norse god.
Vitr	22.4S	349.3W	76	233	Norse dwarf.
Vu-Murt	22.9N	170.9W	79	228	Estonian spirit of water.
Vutash	31.9N	102.9W	55	229	Estonian spirit of water.
Ymir	51.4N	101.3W	77	229	Norse; giant from whom Earth was created.

SATURN

EPIMETHEUS

Feature name	Lat. (°)	Long. (°)	Size (km)	Page	Description
Hilairea	–	–	–	–	Greek; sister of Phoibe, daughter of Leukippos.
Pollux	–	–	–	–	Latin name for Polydeukes, Castor's twin.

JANUS

Feature name	Lat. (°)	Long. (°)	Size (km)	Page	Description
Castor	–	–	–	–	One of the Dioscuri; famous as a tamer of horses.
Idas	–	–	–	–	Twins; cousins of Gemini.
Lynceus	–	–	–	–	One of twin cousins of Gemini.
Phoibe	–	–	–	–	Daughter of Leukippos.

MIMAS

Feature name	Lat. (°)	Long. (°)	Size (km)	Page	Description
Accolon	65.7S	184.7W	–	259	Companion of Arthur's; he is tricked into jousting with Arthur.
Arthur	33.2S	195.6W	–	258	King of the Round Table assemblage.
Avalon Chasma	39.6N	147.8W	–	–	Arthurian paradise.
Balin	17.1N	86.8W	–	256	Knight of "matchless courage and virtue."
Ban	39.2N	156.4W	–	256	King of Benwick; father of Sir Launcelot. Ally of Arthur in the battle of Bedgrayne.
Bedivere	9.5N	152.3W	–	259	Arthurian knight.

Feature name	Lat. (°)	Long. (°)	Size (km)	Page	Description
Bors	39.0N	166.0W	–	256	King of Gaul; father of Sir Ector de Marys, Sir Bors, Sir Lyonel.
Camelot Chasma	38.4S	27.6W	–	257	Home of the Round Table assemblage.
Dynas	3.8N	82.9W	–	256	A knight of the Round Table.
Elaine	44.5N	108.2W	–	256	Daughter of King Pelles, lover of Sir Launcelot and mother, by him, of Sir Galahad.
Gaheris	40.9S	298.4W	–	261	Older son of King Lot; killed by Sir Launcelot in his rescue of Gwynevere from burning.
Galahad	47.0S	135.0W	–	259	Bastard son of Launcelot and Elaine. He went on the quest to find the Holy Grail.
Gareth	40.9S	288.8W	–	261	Youngest son of King Lot; killed by Sir Launcelot in his rescue of Gwynevere from burning.
Gawain	58.4S	260.6W	–	258	Eldest son of King Lot; Arthur's favorite cousin.
Gwynevere	16.8S	324.0W	–	260	Queen; wife of Arthur; lover of Launcelot.
Herschel	2.9N	109.5W	–	256	William; German-British astronomer; discovered Mimas and Enceladus (1738-1822).
Igraine	42.1S	231.1W	–	258	Wife of Uther; mother of Arthur.
Iseult	46.4S	36.4W	–	257	Loved by Tristram.
Kay	47.0N	126.4W	–	259	Royal seneschal at Arthur's court.
Lamerok	61.3S	291.4W	–	261	Pellinore's son; sends testing horn to King Mark to expose adultery of Sir Tristram.
Launcelot	10.0S	328.2W	–	260	King Arthur's favorite; champion and lover of Queen Gwynevere.
Lot	31.1S	231.6W	–	258	Leader of the rebel kings of the north and west. Married Margawse and begat Sir Gawain, Sir Aggravayne, Sir Gaheris.
Mark	29.7S	300.1W	–	261	King of Cornwall.
Merlin	37.7S	219.5W	–	258	Magician and prophet; Arthur's mentor.
Mordred	5.0N	213.0W	–	258	Arthur's bastard son and mortal enemy; delivered fatal wound to Arthur but was killed by him.
Morgan	23.6N	242.6W	–	258	Arthur's half sister; enchantress; plots to destroy Arthur but failed.
Oeta Chasma	32.2N	117.5W	–	256	Shook by a Titan in the war between Titans and Olympians.
Ossa Chasma	20.6S	307.8W	–	261	Mt. Pelion piled on top of it in war between Titans and gods.
Palomides	4.1N	168.5W	–	259	Saracen enemy of Tristam.
Pangea Chasma	22.6S	348.8W	–	260	Picked up by a Titan in the war with the gods.
Pelion Chasma	24.0S	248.3W	–	258	Mountain piled up with Mt. Ossa in war with gods.
Pellinore	29.5N	139.5W	–	256	King whose duty was to pursue the questing beast and either run it to earth or lose his strength.
Percivale	1.0S	180.2W	–	259	Very pure knight; accomplished quest of Holy Grail.
Tintagil Chasma	49.7S	205.0W	–	258	Home of Igraine, Arthur's mother.
Tristram	52.2S	28.1W	–	257	Saved Iseult; fell in love with her.
Uther	34.9S	251.1W	–	258	Ruler of all Britain; Arthur's father.

ENCELADUS

Feature name	Lat. (°)	Long. (°)	Size (km)	Page	Description
Ahmad	57.4N	305.4W	17	262	Youngest son; brings father a magic apple; marries the Genie Peri Banu.
Aladdin	63.1N	16.9W	34	264	Hero of the tale; he has the magic lamp.
Ali Baba	57.2N	12.0W	35	264	Hero of tale who found a great treasure owned by 40 thieves.
Bassorah Fossa	45.4N	6.3W	131	264	Town from which Sindbad embarked on his 3rd voyage.
Dalilah	52.9N	246.4W	14	266	Crafty old crone who fools several men.
Daryabar Fossa	9.7N	359.1W	201	264	"Ocean region"; land from which Princess Daryabar came.
Diyar Planitia	0.5N	239.7W	311	266	Country where Khudadad's father ruled.
Duban	58.3N	277.5W	23	262	Sage who cured King Yunan of leprosy.
Dunyazad	42.6N	196.5W	30	262	Sister of Shahrazad.
Gharib	81.3N	251.7W	20	262	Hero of many tales.
Harran Sulci	26.7N	237.6W	276	262	City where Khudadad's father ruled.
Isbanir Fossa	12.6N	354.0W	132	264	Fakir Taj's home; may be ancient Ctesiphon.
Julnar	54.2N	342.0W	20	265	The seaborn; heroine of nights 738 to 756.
Musa	73.8N	6.5W	22	264	Goes to get the vessels that contain Jinni in "The City of Brass."
Peri-Banu	63.1N	317.9W	18	263	Genie who marries Ahmad and helps him fulfill the demands of his father.
Salih	6.5S	0.0W	4	264	Brother of Julnar.
Samad	62.3N	355.1W	16	264	Shaykh who guides Musa and Talib to the mountains in "The City of Brass."
Samarkand Sulci	30.5N	326.8W	383	265	Country ruled over by Zaman, brother of Shahryar.
Sarandib Planitia	4.4N	298.0W	200	265	Ceylon; the island visited by Sindbad on his 6th voyage.
Shahrazad	48.2N	195.1W	20	262	Heroine who tells King Shahryar "The Tales of a Thousand Nights."
Shahryar	59.7N	225.0W	21	262	King whom Shahrazad beguiles with the tales of a thousand nights and a night.
Sindbad	68.9N	211.4W	23	262	Voyager who had many marvelous adventures on seven voyages.

TETHYS

Feature name	Lat. (°)	Long. (°)	Size (km)	Page	Description
Ajax	29.1S	282.0W	–	268	Greek hero second only to Achilles.
Anticleia	52.3N	34.4W	–	268	Mother of Odysseus.
Antinous	62.0S	275.0W	–	–	Chief of the wooers; slain by Odysseus.
Arete	4.6S	299.7W	–	268	Wife of Alcinous, mother of Nausicaa.

Feature name	Lat. (°)	Long. (°)	Size (km)	Page	Description
Circe	12.1S	53.7W	–	268	Changed Odysseus' companions into swines.
Elpenor	54.8N	263.3W	–	268	Follower of Odysseus.
Eumaeus	22.8N	53.4W	–	268	Faithful swineherd who greets Odysseus, gives him warm cloak and guides him to palace.
Eurycleia	52.7N	245.9W	–	268	Faithful old nurse of Odysseus.
Ithaca Chasma	10.3S	3.0W	–	268	An Ionian island, home of Odysseus.
Laertes	47.6S	66.4W	–	269	Father of Odysseus.
Melanthius	62.0S	204.0W	–	–	Disloyal goatherd; insults Odysseus and is slain.
Mentor	1.3S	45.0W	–	268	Friend of Odysseus.
Nausicaa	82.3N	357.3W	–	268	Daughter of Alcinous who advised Odysseus.
Nestor	54.6S	61.7W	–	269	A wise old king.
Odysseus	30.0N	130.0W	–	269	Hero of Odyssey.
Penelope	11.5S	248.0W	–	268	Faithful wife of Odysseus.
Phemius	12.0N	285.8W	–	268	Minstrel to the wooers; spared by Odysseus.
Polyphemus	4.6S	282.8W	–	268	Cyclops battled by Odysseus.
Teiresias	59.6N	5.7W	–	268	Aged prophet; Odysseus consults him among the dead.
Telemachus	54.0N	338.7W	–	268	Son of Odysseus.

DIONE

Feature name	Lat. (°)	Long. (°)	Size (km)	Page	Description
Adrastus	61.7S	45.9W	31	270	King of Argos, one of the seven against Thebes, and the only one to return alive.
Aeneas	26.1N	46.3W	166	270	Hero of the Aeneid. The son of Anchises and Venus and a member of the royal family of Troy.
Amata	7.7N	285.3W	231	270	Mother of Lavinia, wife of Aeneas.
Anchises	33.7S	66.1W	42	270	Aeneas' father.
Antenor	6.5S	10.4W	82	270	Nephew of Priam. He escaped the fall of Troy and reached Italy before Aeneas. There he founded Padua.
Butes	64.2N	48.8W	26	270	A famous boxer who had been defeated by Dares.
Caieta	23.3S	80.5W	70	–	A nurse of Aeneas.
Carthage Linea	12.7N	321.9W	318	270	A Punic (Phoenician) city in North Africa.
Cassandra	39.5S	244.1W	36	271	Daughter of Priam; she could foretell the future.
Catillus	1.6S	273.0W	35	271	Brother of Tiburtus and twin brother of Coras.
Coras	0.6N	266.4W	37	271	Brother of Tiburtus and twin brother of Catillus. He was founder of Tibur and an ally of Turnus against Aeneas.
Creusa	48.0N	76.9W	35	270	Daughter of Priam; first wife of Aeneas.
Dido	23.7S	18.5W	118	270	Tyrian princess who founded Carthage.
Halys	59.1S	53.6W	29	270	A Trojan defending Aeneas' camp against the Rutulian attack. He was killed by Turnus.
Ilia	0.1N	346.0W	51	270	Also known as Rhea Silvia; mother by Mars of Romulus and Remus, the founders of Rome.
Italus	18.1S	77.5W	40	270	Ancient hero, eponymous ancestor of the Italians.
Larissa Chasma	30.2N	71.1W	315	270	A town in Thessaly, Achilles' native region.
Latagus	15.7N	26.4W	37	270	Soldier of Aeneas.
Latium Chasma	21.2N	69.5W	381	270	The Trojans' promised land in Italy.
Lausus	36.2N	23.2W	28	270	Son of Mezentius, killed by Aeneas.
Magus	19.3N	24.4W	41	270	A soldier Turnus, killed by Aeneas.
Massicus	34.8S	56.0W	43	270	An Etruscan ally of Aeneas.
Padua Linea	20.0S	210.7W	780	271	City in northern Italy founded by Antenor.
Palatine Chasma	75.6S	25.1W	394	–	One of the Seven Hills of Rome.
Palatine Linea	40.6S	305.4W	645	270	One of the Seven Hills of Rome.
Palinurus	4.0S	61.4W	33	270	Pilot of Aeneas' fleet.
Remus	13.2S	31.1W	69	270	Brother of Romulus, founder of Rome.
Ripheus	56.1S	35.5W	32	270	A Trojan. He fought at the side of Aeneas during Troy's last night.
Romulus	7.3S	26.5W	81	270	Mythical founder of Rome in 754 or 753 B.C., son of Mars by Ilia (Rhea Silvia).
Sabinus	47.8S	175.6W	79	271	Fabled ancestor of the Sabines.
Tibur Chasmata	57.2N	69.1W	156	270	Ancient town of Italy not far from Rome on the river Arno.
Turnus	16.2N	344.6W	97	270	Rutililan king; Aeneas' rival for hand of Lavinia.

RHEA

Feature name	Lat. (°)	Long. (°)	Size (km)	Page	Description
Aananin	34.9N	339.9W	–	272	Korean god of the Heavens.
Adjua	40.2N	118.9W	–	272	Mythical heroine and ancestor of the Ulci tribe.
Agunua	63.3N	66.2W	–	272	San Cristobal (Melanesia) god who made sea, land, people.
Ameta	53.3N	21.9W	–	272	Ceram (Indonesia) ancestor whose blood made Hainuwele.
Arunaka	15.3S	22.1W	–	272	Inca creator of all things.
Atum	47.1S	1.1W	–	272	Old creator god of Heliopolis; became son of Ptah.
Bulagat	38.2S	15.2W	–	272	Mythological ancestor of the Buriat tribe.
Bumba	63.1N	50.4W	–	272	Bushongo; dwelt in primordial waters; vomited up sun, moon, stars, animals, and men. Showed man how to make fire.
Burkhan	66.8N	310.6W	–	272	Buriat (Siberia) god who created world.
Con	25.8S	12.7W	–	272	Inca coastal creator god.
Djuli	31.2S	46.7W	–	272	Neghidahan (Ukrainian) first man who was ancestor of the people.
Ellyay	71.4N	91.8W	–	272	Yakutian ancestor of the people.
Faro	45.3N	114.0W	–	272	Mande; his sacrificial killing in heaven atoned for Pemba's sin; purified Earth.
Haik	36.6S	29.3W	–	272	Mythological ancestor of the Armenian people.
Haoso	8.3N	12.5W	–	272	Manchurian creator of all things.
Heller	10.1N	315.1W	–	272	Auracanin creator of men and bringer of civilization.
Iraca	39.4N	112.1W	–	272	Incan creator god who became the moon.
Izanagi	49.4S	310.3W	–	272	Japanese creator god, brother of Izanami.

Feature name	Lat. (°)	Long. (°)	Size (km)	Page	Description
Izanami	46.3S	313.4W	–	272	Sister and wife of Izanagi; creator goddess.
Jumo	52.8N	66.5W	–	272	Marijan sky god.
Karora	5.9N	20.1W	–	272	Aranda (Australia) ancestor who, in his dreams, gives birth to animals and male children.
Khado	41.6N	359.1W	–	272	Nanajan; mythological hero who built the world. The first shaman.
Kiho	11.1S	358.7W	–	272	Tuamotu (Society Islands) progenitor being; existed in void; made land, sea.
Kumpara	9.6N	327.1W	–	272	Jivaro (Ecuador) creator god.
Kun Lun Chasma	46.0N	307.5W	–	–	Chinese; mountain dwelling place of the immortals.
Leza	21.8S	309.2W	–	272	Tonga originator of the conditions of life.
Lowa	40.9N	16.6W	–	272	Marshall Islands (Melanesia) great creator god.
Malunga	65.1N	56.2W	–	272	Yao (Bantu); creator god; left Earth to live in sky when man was cruel to animals.
Manoid	29.5N	8.5W	–	272	Negrito (Malay Peninsula) female progenitress god; wife of Pedn.
Melo	53.2S	7.1W	–	272	Minyong (India); original male.
Mubai	55.8N	20.2W	–	272	Tibetan heavenly god.
Num	24.0N	92.7W	–	272	Samoyed god of heaven.
Ormazd	52.5N	58.5W	–	272	Persian progenitor god of light.
Pan Ku	65.7N	107.7W	–	272	Miao; creator of all things.
Pedn	46.0N	351.7W	–	272	Negrito (Malay Peninsula) god who created first men.
Pu Chou Chasma	26.1N	95.3W	–	272	Mountain attacked by Kung Chung.
Qat	23.8S	351.6W	–	272	New Hebrides (Melanesia); born from a stone; formed men out of trees.
Sholmo	12.0N	346.4W	–	272	Buriat (Siberia) devil who creates.
Taaroa	16.5N	95.5W	–	272	Tahitian god imminent in all creation; existed alone in the void.
Thunupa	45.6N	21.3W	–	272	Inca creator of all things.
Tika	25.1N	84.1W	–	272	Abkhazian (Georgian-eastern Black Sea region) supreme being.
Tirawa	34.2N	151.7W	–	272	Great spirit of Pawnee Tribe; created first men.
Tore	0.0N	335.6W	–	272	Pygmy lord of the world, creator of all things.
Torom	68.1S	343.5W	–	272	Ostyak (Western Siberia) sky god.
Uku	78.7N	95.5W	–	272	Estonian super god.
Whanin	66.9N	115.0W	–	272	Korean creator of all things.
Wuraka	25.1N	4.0W	–	272	Kakadu (Australia) ancestor of all people; a giant.
Xamba	2.1N	349.7W	–	272	Bushman supreme being, creator of all things.
Xu	55.0N	71.9W	–	272	Bushman creator.
Yu-Ti	50.1N	81.5W	–	272	"August Personage of Jade;" supreme primal Chinese god.

HYPERION

Feature name	Lat. (°)	Long. (°)	Size (km)	Page	Description
Bahloo	36.0N	196.0W	–	–	The moon; Aborigine maker of girl babies.
Bond-Lassell Dorsum	48.0N	143.5W	–	–	G.P. Bond (American) and William Lassell (British); discovered Hyperion on the same night in 1848.
Helios	71.0N	132.0W	–	–	Greek sun god; son of Hyperion.
Jarilo	61.0N	183.0W	–	–	East Slavic god of the sun, fertility, and love.
Meri	3.0N	171.0W	–	–	Bororo folk hero; the sun.

IAPETUS

Feature name	Lat. (°)	Long. (°)	Size (km)	Page	Description
Almeric	53.4N	276.6W	43	273	French, 1 of 12 peers, killed by Marsilion.
Baligant	16.4N	224.9W	66	273	Emir of Babylon; Marsilion enlisted his help against Charlemagne.
Basan	33.3N	194.7W	76	273	French baron; murdered while serving as Ambassador of Marsilon.
Berenger	62.1N	219.7W	84	273	1 of 12 peers; killed Estramarin; killed by Grandoyne.
Besgun	76.0N	309.8W	56	273	Chief cook for Charlemagne's army; he guarded Ganelon after Ganelon's treachery was discovered.
Cassini Regio	28.1S	92.6W	–	273	Italian-born French astronomer who discovered Iapetus in 1671, Rhea in 1672, Tethys and Dione in 1684 (1625-1712).
Charlemagne	55.0N	258.8W	95	273	Emperor of France and Germanic nations; his forces fought the Saracens in Spain.
Geboin	58.6N	173.4W	81	273	Guarded French dead; became leader of Charlemagne's second column.
Godefroy	71.9N	249.1W	63	273	Standard of Charlemagne; brother of Tierri, Charlemagne's defender against Pinabel.
Grandoyne	17.7N	214.5W	65	273	Son of Cappadocian King Capuel; killed Gerin, Gerier, Berenger, Guy St. Antoine, Duke Astorge; killed by Roland.
Hamon	10.6N	270.0W	96	273	Joint commander of Charlemagne's Eighth Division.
Lorant	65.2N	159.8W	44	273	French commander of one of first divisions against Baligant; killed by Baligant.
Marsilion	39.2N	176.1W	136	273	Saracen king of Spain; Roland wounded him and he died of wound.
Milon	67.9N	270.2W	119	273	Guarded French dead while Charlemagne pursued Saracen forces.
Ogier	42.5N	275.1W	100	273	Dane who led third column in Charlemagne's army against Baligant's forces.
Oliver	62.5N	200.8W	113	273	Roland's friend; mortally wounded by Marganice.
Othon	33.3N	347.8W	86	273	1 of 12 peers; guarded French dead while Charlemagne pursued Saracen forces; sixth column leader.
Roland	73.3N	25.2W	144	273	Charlemagne's nephew; led rear guard of French forces; hero in song of Roland.
Roncevaux Terra	37.0N	239.5W	1284	273	Pass where Roland and his forces were ambushed by the Saracens.
Turpin	47.7N	1.4W	87	273	Archbishop of Rheims in Song of Roland.

Feature name	Lat. (°)	Long. (°)	Size (km)	Page	Description
URANUS PUCK					
Bogle	–	–	–	–	Scottish mischievous spirits.
Butz	–	–	–	–	German roguish or evil spirits.
Lob	–	–	–	–	British mischievous spirits.
MIRANDA					
Alonso	44.0S	352.6E	25	287	King of Naples in "The Tempest."
Arden Corona	29.1S	73.7E	318	287	Forest, location of "As You Like It."
Argier Rupes	43.2S	322.8E	141	287	Location of early action in "The Tempest."
Dunsinane Regio	31.5S	11.9E	244	287	Location of castle where Macbeth was defeated.
Elsinore Corona	24.8S	257.1E	323	286	Location of Hamlet's castle.
Ferdinand	34.8S	202.1E	17	286	Son of King of Naples; loves Miranda in "The Tempest."
Francisco	73.2S	236.0E	14	286	A Lord of Naples in "The Tempest."
Gonzalo	11.4S	77.0E	11	287	Honest old counselor of Naples in "The Tempest."
Inverness Corona	66.9S	325.7E	234	287	Location of Macbeth's castle.
Mantua Regio	39.6S	180.2E	399	286	Location of part of "Two Gentlemen From Verona."
Prospero	32.9S	329.9E	21	287	Rightful Duke of Milan in "The Tempest."
Sicilia Regio	30.0S	317.2E	174	287	Location of "Winter's Tale."
Stephano	41.1S	234.1E	16	286	A drunken butler in "The Tempest."
Trinculo	63.7S	163.4E	11	286	A jester in "The Tempest."
Verona Rupes	18.3S	347.8E	116	287	Where Romeo and Juliet lived.
ARIEL					
Abans	15.5S	251.3E	20	288	Spirit of the iron mines.
Agape	46.9S	336.5E	34	288	Spirit in Spenser's "Fairy Queen."
Ataksak	53.1S	224.3E	22	288	Eskimo benevolent spirit.
Befana	17.0S	31.9E	21	288	Good spirit who fills Italian children's stockings with toys on Twelfth Night.
Berylune	22.5S	327.9E	29	288	Good spirit in Maeterlinck's "The Bluebird."
Brownie Chasma	16.0S	337.6E	343	288	German good spirits who live in woods.
Deive	22.3S	23.0E	20	288	Spirit of beautiful maiden.
Djadek	12.0S	251.1E	22	288	Czech ancestral benevolent spirit and household guardian.
Domovoy	71.5S	339.7E	71	288	Slavic spirit protector of home.
Finvara	15.8S	19.0E	31	288	Irish king of spirits; provided horses and wine to men.
Gwyn	77.5S	22.5E	34	288	Irish god of battle; leads mens' souls to Annwn.
Huon	37.8S	33.7E	40	288	Replaced Oberon as King of Spirits when Oberon died.
Kachina Chasmata	33.7S	246.0E	622	288	Pueblo (U.S.A.) good spirits who bring rain or other blessings.
Kewpie Chasma	28.3S	326.9E	467	288	British race of quaint spirit babies.
Korrigan Chasma	27.6S	347.5E	365	288	French wind spirits who cure diseases.
Kra Chasma	32.1S	354.2E	142	288	Vital spirits (Gold Coast).
Laica	21.3S	44.4E	30	288	Inca good spirit.
Leprechaun Vallis	10.4S	10.2E	328	288	Spirits or dwarfs.
Mab	38.8S	352.2E	34	288	Queen of Spirits who dethroned Titania.
Melusine	52.9S	8.9E	50	288	Spirit heroine of medieval French story.
Oonagh	21.9S	244.4E	39	288	Irish Queen of Fairies.
Pixie Chasma	20.4S	5.1E	278	288	British spirits that live in rocks.
Rima	18.3S	260.8E	41	288	Spirit in Hudson's "Green Mansions."
Sprite Vallis	14.9S	340.0E	305	288	Earth spirits.
Sylph Chasma	48.6S	353.0E	349	288	British air spirits who influence the temperaments of man.
Yangoor	68.7S	279.7E	78	288	Spirit that brings day.
UMBRIEL					
Alberich	33.6S	42.2E	52	289	Dwarf who guarded Niebelung gold, also had a mantle of invisibility.
Fin	37.4S	44.3E	43	289	Troll who helped build a church in Kallundburg, Zealand.
Gob	12.7S	27.8E	88	289	King of gnomes.
Kanaloa	10.8S	345.7E	86	289	Polynesian chief evil spirit.
Malingee	22.9S	13.9E	164	289	Aboriginal spirit who travels at night.
Minepa	42.7S	8.2E	58	289	Macouas and Banayis evil spirit.
Peri	9.2S	4.3E	61	289	Persian evil spirit who disguised malevolence by charm; disturbed natural elements and heavenly bodies.
Setibos	30.8S	346.3E	50	289	Chief devil.
Skynd	1.8S	331.7E	72	289	Troll who stole three wives of a man living in Englerup.
Vuver	4.7S	311.6E	98	289	Volga Finn evil spirit.
Wokolo	30.0S	1.8E	208	–	Baramba (West Africa) devil spirit.
Wunda	7.9S	273.6E	131	289	Australian dark spirit.
Zlyden	23.3S	326.2E	44	289	Slavic evil spirit.
TITANIA					
Adriana	20.1S	3.9E	50	288	Wife of Antipholus of Ephesus in "Comedy of Errors."
Belmont Chasma	8.5S	32.6E	258	288	Location in "As You Like It."
Bona	55.8S	351.2E	51	288	Sister of the French queen, "Henry VI," part 3.
Calphurnia	42.4S	291.4E	100	288	Wife of Julius Caesar.
Elinor	44.8S	333.6E	74	288	Mother of King John.
Gertrude	15.8S	287.1E	326	288	Mother of Hamlet.
Imogen	23.8S	321.2E	28	288	Cymbelline's daughter.
Iras	19.2S	338.8E	33	288	Attendant to Cleopatra in "Anthony and Cleopatra."
Jessica	55.3S	285.9E	64	288	Shylock's daughter in "The Merchant of Venice."
Katherine	51.2S	331.9E	75	288	Henry VIII's first queen.
Lucetta	14.7S	277.1E	58	288	Waiting women to Julia in "Two Gentlemen of Verona."
Marina	15.5S	316.0E	40	288	Daughter to Pericles in "Pericles, Prince of Tyre."
Messina Chasmata	33.3S	335.0E	1492	288	Location in "Much Ado About Nothing."
Mopsa	11.9S	302.2E	101	288	Shepardess in "The Winter's Tale."
Phrynia	24.3S	309.2E	35	288	Alcibiades' mistress in "Timon of Athens."

Feature name	Lat. (°)	Long. (°)	Size (km)	Page	Description
Rousillon Rupes	14.7S	26.5E	402	288	Location in "All's Well That Ends Well."
Ursula	12.4S	45.2E	135	288	Attendant to r Hero and Beatrice in "Much Ado about Nothing."
Valeria	34.5S	4.2E	59	288	Friend to Vergilia in "Coriolanus."

OBERON

Feature name	Lat. (°)	Long. (°)	Size (km)	Page	Description
Antony	27.5S	65.4E	47	289	Shakespearean hero in "Anthony and Cleopatra."
Caesar	26.6S	61.1E	76	289	Shakespearean hero in "Julius Caesar."
Coriolanus	11.4S	345.2E	120	289	Shakespearean hero.
Falstaff	22.1S	19.0E	124	289	Shakespearean character in "Merry Wives of Windsor."
Hamlet	46.1S	44.4E	206	289	Shakespearean hero.
Lear	5.4S	31.5E	126	289	Shakespearean hero in "King Lear."
Macbeth	58.4S	112.5E	203	289	Shakespearean hero.
Mommur Chasma	16.3S	323.5E	537	289	Spirit place, forest home of Oberon in "Midsummer Night's Dream."
Othello	66.0S	42.9E	114	289	Shakespearean hero.
Romeo	28.7S	89.4E	159	289	Shakespearean character in "Romeo and Juliet."

NEPTUNE

PROTEUS

Feature name	Lat. (°)	Long. (°)	Size (km)	Page	Description
Pharos	–	–	–	–	Island where Proteus reigned.

TRITON

Feature name	Lat. (°)	Long. (°)	Size (km)	Page	Description
Abatos Planum	21.5S	58.0E	–	305	Egyptian sacred island in Nile; "paradise."
Akupara Maculae	27.5S	63.0E	–	–	Tortoise upholding the world (India).
Amarum	26.0N	24.5E	–	305	Quecha (Ecuador) water boa.
Andvari	20.5N	34.0E	–	305	Norse fish-shaped dwarf.
Apep Cavus	20.0N	301.5E	–	304	Egyptian dragon of darkness.
Awib Dorsa	7.0S	80.0E	–	305	Nama Bushman word for rain.
Bheki Cavus	16.0N	308.0E	–	304	Frog symbolizing the sun on the horizon (India).
Bia Sulci	38.0S	3.0E	–	305	Yoruba; river named for obedient son of god.
Boynne Sulci	13.0S	350.0E	–	304	Celtic mythological river.
Bubembe Regio	18.0N	335.0E	–	304	Island location of temple of Mukasa (Uganda).
Cay	12.0S	44.0E	–	305	Mayan diety.
Cipango Planum	11.5N	34.0E	–	305	Legendary island described by Marco Polo.
Dagon Cavus	29.0N	345.0E	–	304	Babylonian fertility god, represented as a fish.
Dilolo Patera	26.0N	24.5E	–	305	Angolan sacred lake.
Doro Macula	27.5S	31.7E	–	305	Nanay mistress of fishing, Sea of Okhotsk.
Gandvik Patera	28.0N	5.5E	–	305	Tortuous sea; literally, "Serpent Bay" (Norse).
Hekt Cavus	26.0N	342.0E	–	304	Egyptian frog goddess.
Hili	57.0S	35.0E	–	305	Zulu water sprite.
Hirugo Cavus	14.5N	345.0E	–	304	Japanese diety born in shape of jellyfish.
Ho Sulci	2.0N	305.0E	–	304	Chinese sacred river.
Ilomba	14.5S	57.0E	–	305	Lozi (Zambia) water snake linked with destruction.
Jumna Fossae	13.5S	44.0E	–	305	Hindu river goddess.
Kasu Patera	39.0N	14.0E	–	305	Sacred lake of Zoroastrianism.
Kasyapa Cavus	7.5N	358.0E	–	304	The god Prajapati as a tortoise (India).
Kibu Patera	10.5N	43.0E	–	305	Mabuiag (Melanesia) island of the dead.
Kikimora Maculae	31.0S	78.0E	–	–	Slavic spirit of swamps, household spirit.
Kormet Sulci	23.0N	335.5E	–	304	Norse river through which dead must pass.
Kraken Catena	14.0N	35.5E	–	305	Giant Norse sea monster.
Kulilu Cavus	41.0N	4.0E	–	305	Babylonian destructive fish-man spirit.
Kurma	16.5S	61.0E	–	305	Vishnu in the form of a tortoise.
Leipter Sulci	7.0N	9.0E	–	–	Norse sacred river.
Leviathan Patera	17.0N	28.5E	–	305	Hebrew sea monster upholding earth.
Lo Sulci	3.8N	321.0E	–	304	Chinese sacred river.
Mah Cavus	38.0N	6.0E	–	305	Fish that holds up the universe (Persian).
Mahilani	50.5S	359.5E	–	304	Tonga sea spirit.
Mangwe Cavus	7.0S	343.0E	–	304	Ila (Zambia), "the flooder."
Mazomba	18.5S	63.5E	–	305	Chaga (Tanzania) mythical large fish.
Medamothi Planum	3.5N	69.0E	–	305	French fictional island, meaning "nowhere."
Monad Regio	20.0N	37.0E	–	305	Chinese symbol of duality in nature.
Namazu Macula	25.5S	14.0E	–	305	Japanese mythic fish, maker of earthquakes.
Ob Sulci	6.0S	328.0E	–	304	Mouth of this river is Ostiak entrance to underworld.
Ormet Sulci	17.0N	337.0E	–	304	Norse river through which dead pass.
Ravgga	3.0S	71.5E	–	–	Finnish fortune-telling fish god.
Raz Fossae	8.0N	21.5E	–	305	Breton bay of souls.
Rem Maculae	13.0N	349.5E	–	304	Egyptian fish who wept fertilizing tears.
Ruach Planitia	28.0N	24.0E	–	305	French isle of winds.
Ryugu Planitia	5.0S	27.0E	–	305	Japanese undersea dragon palace.
Set Catena	22.0N	33.5E	–	305	Egypian water monster, personification of evil.
Sipapu Planitia	4.0S	36.0E	–	305	Pueblo (U.S.A.) hole or lake of emergence from underworld.
Slidr Sulci	23.5N	350.0E	–	304	Norse river of daggers and spears.
Tangaroa	25.0S	65.5E	–	305	Maori fishing and sea god.
Tano Sulci	33.5N	337.0E	–	304	Yoruba; river named for willful son of god.
Tuonela Planitia	34.0N	14.5E	–	305	Underground realm across Black River (Finnish).
Uhlanga Regio	37.0S	357.0E	–	304	Zulu reed from which humanity sprang.
Ukupanio Cavus	35.0N	23.0E	–	305	Hawaiian shark god.
Vimur Sulci	11.0S	59.0E	–	305	Greatest of Elivagar rivers, a stream of ice (Norse).
Viviane Macula	31.0S	36.5E	–	305	Amour of Merlin (Welsh).
Vodyanoy	17.0S	28.5E	–	305	Slavic water spirit.
Yasu Sulci	2.0N	347.0E	–	304	Japanese heavenly river; literally, "peace."
Yenisey Fossa	3.0N	56.2E	–	305	Siberian mythical holy river.
Zin Maculae	24.5S	68.0E	–	305	Niger water spirits.

Index

(Italic page numbers refer to figures; place names shown on maps are given in the Gazetteer)